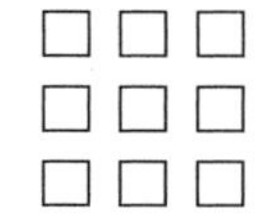

Nuclear Safeguards, Security, and Nonproliferation

Nuclear Safeguards, Security, and Nonproliferation

Achieving Security with Technology and Policy

Editor
James E. Doyle

AMSTERDAM • BOSTON • HEIDELBERG • LONDON •
NEW YORK • OXFORD PARIS • SAN DIEGO • SAN FRANCISCO •
SYDNEY • TOKYO
Butterworth-Heinemann is an imprint of Elsevier

Butterworth-Heinemann is an imprint of Elsevier
30 Corporate Drive, Suite 400, Burlington, MA 01803, USA
Linacre House, Jordan Hill, Oxford OX2 8DP, UK

Library of Congress Cataloging-in-Publication Data
Application submitted

British Library Cataloguing-in-Publication Data
A catalogue record for this book is available from the British Library.

ISBN: 978-0-7506-8673-0

For information on all Butterworth–Heinemann publications visit our Web site at www.books.elsevier.com

Typeset by Charon Tec Ltd (A Macmillan Company), Chennai, India
www.macmillansolutions.com

Printed in The United States of America

08 09 10 11 12 13 10 9 8 7 6 5 4 3 2 1

Dedication

For individuals everywhere working to secure the peaceful uses of nuclear energy and prevent its use for military purposes.

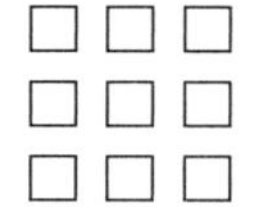

Contents

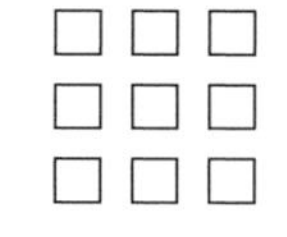

Foreword

With the development of thermonuclear weapons—the so-called hydrogen bombs—of unprecedented and almost unimaginable destructive power, mankind can for the first time in history threaten the survival of civilization as we know it. During the Cold War the threat of a nuclear holocaust was indeed a major concern. Although that danger has been greatly reduced since the collapse of the Soviet Union 17 years ago, the nuclear weapons that currently exist still present a mortal danger to the world.

We enter the 21st century witnessing the global spread of advanced nuclear-related technology and a growing danger of the proliferation of nuclear weapons, leading to an increasing likelihood that the world's most horrific weapons may be acquired by dangerous hands, including terrorist organizations. What good options can we pursue for reducing nuclear threats? Is the Cold War model of U.S.-Soviet mutual deterrence the best we can do, or even adequate, to meet the emerging dangers? There is no doubt about the difficulty of the challenge to do better, or of the need for perseverance, strong will, and new thinking on the many important scientific/technical and strategic/diplomatic dimensions of this challenge. This in turn will require maintaining strong cadres of scholars and scientists engaged in work on nuclear security.

Nuclear Safeguards, Security, and Nonproliferation: Achieving Security with Technology and Policy addresses this audience. It is encyclopedic in introducing the reader to the broad range of important concepts and techniques that constitute nuclear security policy and science. As such, this volume will be invaluable to members of the two communities of natural and social scientists who seek to enter into careers in these fields. It is essential for both the scientific and policy communities to understand the daunting challenges each faces so that they can work together effectively. More broadly this volume will be valuable for all who seek simply to understand one of the major problems facing the world. Consider some of the issues before us:

- Reducing relevance of nuclear weapons for maintaining strategic stability
- Reducing risks of unauthorized or accidental use of nuclear weapons
- Preventing the spread of nuclear weapons
- Developing best practices in securing nuclear materials
- Ensuring adequate means of verifying compliance with negotiated limits on nuclear arsenals and activities
- Controlling weapons material and countering nuclear terrorism in an era of growing interest in nuclear energy and other peaceful applications of nuclear technology

This book is the result of a massive and critical effort by James E. Doyle and his colleagues to present their collective expertise in essays that are both informative and accessible. Having read most of the manuscript in earlier drafts, I believe it will prove to be of immense value in introducing and helping a new generation of scholars, both natural and social scientists, to address the complex issues of nuclear security. This book is a pioneering effort to bridge the policy and technical worlds on the subject of nuclear security. More broadly it should be a source of valuable insights for an informed citizenry and governments. I hope it reaches the wide audience it merits.

—*Sidney D. Drell*

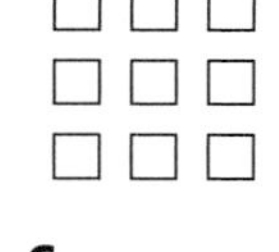

Preface

Since the end of the Cold War, terrorism and proliferation have been seen as very serious, growing threats to national and international security. The potentially catastrophic nexus between the proliferation of weapons of mass destruction (WMD) and terrorism is of particular concern. The active pursuit of WMD and the technologies and materials needed to produce them by so-called "rogue" states and terrorists has been abetted both by other suspect states and by nonstate actors such as the A. Q. Khan network. The risks to national and international security stemming from terrorism and proliferation are compounded as growth in the peaceful pursuits of nuclear energy — so dramatic it is deemed a "nuclear renaissance"— increases the global movement of nuclear materials and technology.

Clearly, in the post-Cold War, post-9/11 international security environment, the issues of nuclear energy, nuclear proliferation, and terrorism intersect. This complex and dynamic new environment demands that we deal responsibly with all weapons-usable materials, both in the United States and globally, while meeting our international security commitments and nonproliferation obligations. To do so is a daunting task involving, *inter alia,* the leveraged use of technologies for enhanced safeguards, proliferation resistance, and physical protection, many of which have their origins in the U.S. nuclear-weapon program. Of equal import is the need for crafting or strengthening nonproliferation and counterterrorism norms, institutions, and treaties. However, neither of these approaches will work without dedicated and knowledgeable people.

Those who undertake these critical tasks must be fully aware of their immense responsibility and prepared to act with empowering knowledge. In fact, the development of human capital is one of the most critical and pressing of future needs for the nuclear security enterprise. *Nuclear Safeguards, Security, and Nonproliferation: Achieving Security with Technology and Policy* provides an introduction for those who will toil on nuclear technology and policy issues in the future and provides a needed complement to traditional nuclear science and technology education as these fields grow in the future. It will also be a useful reference for all people working in this field.

—*Michael R. Anastasio*
Director, Los Alamos National Laboratory

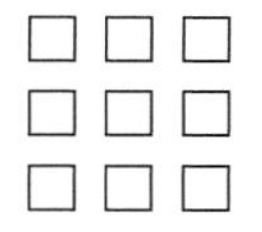

Acknowledgments

The completion of this project was possible due to the contributions of many people. I am grateful to them all, especially the contributing authors and to my colleagues, for their professional expertise, insight, guidance, and perseverance. I share credit with Bill Charlton at Texas A&M University for the original idea and concept for this book.

At the U.S. National Nuclear Security Administration I wish to thank Adam Scheinman, Cynthia Lersten, Monte Mallin, John McClelland-Kerr, Rich Goorevich, Dunbar Lockwood, Todd Perry, and Bret Palmer.

My colleagues at Los Alamos National Laboratory, Michael Anastasio, Paul White, Doug Beason, Joe Pilat, Sara Scott, John Szymanski, Nancy Jo Nicholas, Mike Weaver, Paige Harper, John Gustafson, Diana Hollis, and Jim Danneskiold—all contributed their support and involvement.

The book benefited tremendously from an informal advisory committee at the Stanford University Center for International Security and Cooperation (CISAC) that reviewed many of the early drafts. I wish to thank those who participated, including Scott Sagan, Sig Hecker, David Hafemeister, Lynn Eden, Michael May, Rodger Speed, David Holloway, Pavel Podvig, Chaim Braun, George Bunn, Lewis Franklin, Steve Steadman, and Belkis Cabrera–Palmer.

Finally, Pam Chester and Melinda Ritchie and their colleagues at Elsevier Inc. deserve credit for the skillful production of such a large and technical volume.

The book was almost three years in the making. I apologize for overlooking the names of many others who I know have supported this project in some fashion, and I am grateful to them as well. Special thanks go to my family and to my parents, Alfred and Laura Doyle, for their love and encouragement.

—James E. Doyle

Contributors

Mark E. Abhold has been a technical staff member at Los Alamos National Laboratory since 1995. Mark led the U.S.-Japanese safeguards program, designed many nuclear safeguards systems now in use by the International Atomic Energy Agency, and served as the Department of Homeland Security National thrust leader for radiation detection technology and is now the LANL Chief Scientist for the DOE Second Line of Defense program. Mark earned his Ph.D. in Nuclear Engineering at the University of Washington.

Galya I. Balatsky has been a staff member at Los Alamos National Laboratory (LANL) for 10 years and holds a Ph.D. in industrial psychology. She has worked on a variety of nonproliferation and threat reduction projects.

George T. Baldwin is a member of the technical staff in Global Security Programs at Sandia National Laboratories. His technical background is in nuclear and radiation measurements research and development; his current professional interests include international nuclear safeguards, the Additional Protocol, fissile material cutoff, disposition of excess materials, and related issues. He had worked previously for the Safeguards Department of the International Atomic Energy Agency in Vienna.

Wyn Bowen is Professor of Non-Proliferation & International Security in the Department of War Studies at King's College, London, and Director of the King's Centre for Science and Security Studies. In 1997–98 he served as a weapons inspector in Iraq with the U.N. Special Commission and has also worked as a consultant to the International Atomic Energy Agency.

Brian Boyer is a technical staff member at the Los Alamos National Laboratory, where he works in international safeguards and nonproliferation efforts. He has been in the international safeguards field since 1997 following a five-year term working as a safeguards inspector and analyst at the International Atomic Energy Agency. His areas of focus include IAEA gas centrifuge safeguards and the Additional Protocol.

Michael Browne joined the staff of the Safeguards Science and Technology group at Los Alamos National Laboratory in 2001. His focus has been on development of nuclear materials measurement technology, software, and safeguards systems. He holds a Ph.D. in Physics from North Carolina State University.

Dr. Kory Budlong Sylvester has been a staff member at Los Alamos National Laboratory since 1998, where he has contributed to variety of nuclear nonproliferation and national security projects. He has served as a senior technical advisor to the Office of International Regimes & Agreements at the U.S. National Nuclear Security Administration, was a fellow on the Senate Appropriations Committee, and spent a year on the staff of the Prevention of Nuclear and Biological Attack Subcommittee of the House Committee on Homeland Security. He received his Ph.D. in Nuclear Engineering from the Massachusetts Institute of Technology in 1997.

Tom Burr received a Ph.D. in statistics from Florida State University and joined Los Alamos National Laboratory in 1992. His research areas include statistical aspects of nuclear material assay and accounting and data fusion of various types, including the use of networked sensors.

Loren Byers is an analyst in the International and Applied Technology Division, Los Alamos National Laboratory. He served as an intern in the International Atomic Energy Agency Office of External Relations and holds an M.A. in international policy from the Monterey Institute of International Studies, with a Certificate in Nonproliferation Studies from the Center for Nonproliferation Studies (2001).

Bill Charlton is an Associate Professor in the Nuclear Engineering Department at Texas A&M University (TAMU) and also serves as the Director of the Nuclear Security Science and Policy Institute at TAMU. At TAMU, he coordinates research and teaches courses in the area of nuclear nonproliferation and nuclear material safeguards. He worked as a technical staff member in the Nonproliferation and International Security Division at Los Alamos National Laboratory from 1997–2000.

James E. Doyle has worked in the Nuclear Nonproliferation Division at Los Alamos National Laboratory since 1997. His professional focus is on systems analysis, strategic planning, and policy development. Dr. Doyle holds a Ph.D. in international security studies from the University of Virginia. He has managed cooperative projects with Russia's nuclear weapons institutes on the development of technologies and procedures for monitoring the dismantlement and storage of nuclear warheads and fissile materials. Previously, Dr. Doyle was a senior policy analyst at Science Applications International Corporation.

Sidney D. Drell is Professor of Physics Emeritus at Stanford University's Linear Accelerator Center and a senior fellow at its Hoover Institution. For many years he has been an adviser to the U.S. government on technical national security and arms control issues and currently serves on the Board of Governors for the operators of both Los Alamos and Lawrence Livermore national laboratories.

Stacey Eaton currently serves as the Deputy Group Leader for Safeguards and Security Systems at Los Alamos National Laboratory. Previously, she served as the project lead for Nuclear Diversions as part of the Sentry program. She assessed nuclear smuggling and illicit trafficking for seven years, including evaluating hundreds of cases, analyzing trends, and presenting her conclusions to numerous governmental audiences.

Norbert Ensslin was a member of the Los Alamos Safeguards Science and Technology Group, where he worked primarily in the area of neutron-based nondestructive analysis (NDA). He initiated the development of passive and active neutron multiplicity counting for measurement of bulk plutonium and uranium items. His research and development focus has been on NDA and materials control and accountability (MC&A).

Charles D. Ferguson is a Fellow for Science and Technology at the Council on Foreign Relations, an Adjunct Assistant Professor in the Security Studies Program at Georgetown University, and an Adjunct Lecturer in Homeland Security at Johns Hopkins University. Previously he was the Scientist-in-Residence in the Monterey Institute's Center for Nonproliferation Studies, where he co-authored (with William C. Potter) *The Four Faces of Nuclear Terrorism* (Routledge, 2005). After graduating with distinction from the United States Naval Academy, he served as a nuclear engineering officer on a ballistic-missile submarine. He holds a Ph.D. in physics from Boston University.

Mary Lynn Garcia has worked for the past 22 years at Sandia National Laboratories in international safeguards and physical security. Ms. Garcia has been a Certified Protection

Professional since November 1, 1997, and is the author of two textbooks on physical security; *The Design and Evaluation of Physical Protection Systems* (2001) and *Vulnerability Assessment of Physical Protection Systems* (2006). Currently she is developing risk management techniques and tools for use in government high-security applications.

Siegfried S. Hecker is Director Emeritus, Los Alamos National Laboratory, and Professor (Research) in the Department of Management and Engineering, Senior Fellow of the Freeman Spogli Institute for International Studies, and Co-Director of the Center for International Security and Cooperation at Stanford University.

Carol Kessler is the Director of the Pacific Northwest Center for Global Security at Pacific Northwest National Laboratory (PNNL) in Seattle. The Center conducts international security policy projects informed by the science and technology expertise of the Lab. Ms. Kessler was the Deputy Director General of the Nuclear Energy Agency at the Organization for Economic Cooperation and Development in Paris from 2001–2003. Previously Ms. Kessler led U.S. efforts in the G-7 to improve the safety of Soviet-designed nuclear plants and to close those which could not meet international standards.

Sara Z. Kutchesfahani is currently a Ph.D. candidate in political science and a Teaching Fellow in international security at University College, London. Prior to her doctoral studies, Sara worked in nuclear nonproliferation policy at Los Alamos National Laboratory and at the IISS in London. She will be a RAND Fellow in Washington, D.C., in the summer of 2008.

Leroy Leonard has been an advocate for the safe and secure management of radiological materials since 1989. Until his retirement from Los Alamos National Laboratory in 2006 he was the project leader for the U.S. effort to recover and secure excess and unwanted radiological sources. He currently lives in Vienna, where he serves as a consultant to the IAEA on disused radiological source issues.

Marcie Lombardi is currently a nuclear engineering Ph.D. student at the University of New Mexico. Her research is being performed with Los Alamos National Laboratory's Advanced Nuclear Technology group. Previous degrees include a B.S. in physics from Rutgers University and an M.S. in health physics with a nuclear nonproliferation focus from Georgetown University.

Christopher Lovejoy is a member of the International Threat Reduction Group at Los Alamos National Laboratory. His background includes degrees in chemistry and physics, and he is heavily involved in equipment design and testing. His recent focus has been on evaluation of detection algorithms for passive radiation detection systems.

Arvid Lundy did electronic nuclear instrumentation work for over 20 years before transferring to the Los Alamos International Technology Division in 1982, where he was a group leader, a program manager, and a technical expert on nuclear export controls. He served as the lead U.S. technical expert in the development of the Nuclear Supplier Groups Dual-Use List. In recent years, as a Los Alamos consultant, he has worked at the IAEA on the use of open-source technical information to assess foreign nuclear programs.

Carrie Mathews is the Nonproliferation Regimes and Agreements Program Manager at Pacific Northwest National Laboratory. Her current interests include the nonproliferation issues associated with a nuclear energy renaissance, enhancing international safeguards and increasing the global pool of safeguards experts, and providing technical support to the development of U.S. nonproliferation policy.

Frank Pabian is a senior nonproliferation infrastructure analyst at Los Alamos National Laboratory. He has over 36 years experience in the nuclear nonproliferation and imagery analysis fields, including having served as a chief inspector for the International Atomic Energy Agency during United Nations inspections in Iraq from 1996–1998. He is a Certified Mapping Scientist: Remote Sensing with the American Society of Photogrammetry and Remote Sensing.

Todd Perry joined government in 1999 as a policy analyst for DOE's Material Protection, Control, and Accounting (MPC&A) Program, and has directed the U.S. Department of Energy's International Nonproliferation Export Control Program (INECP) since 2002. Previously, he worked as an analyst for the Union of Concerned Scientists and the Institute for Science and International Security. Dr. Perry received his Ph.D. in government and politics from the University of Maryland in 2003. His doctoral thesis was on the origins and implementation of the 1992 Nuclear Suppliers Group Agreement.

Susan E. Pickett wrote her chapter during her work at Los Alamos National Laboratory as Project Leader for the International Nuclear Safeguards and Engagement Program. She has a dual M.S. in nuclear engineering and technology and policy from MIT and a Ph.D. from the University of Tokyo. Susan is grateful to many of her colleagues in N Division, Los Alamos National Laboratory, for their input and support.

Joseph F. Pilat is a Senior Advisor in the Director's Office of the Los Alamos National Laboratory. He served as Representative of the Secretary of Defense to the Fourth Review Conference of the Nuclear Nonproliferation Treaty (NPT) and as an adviser to the U.S. Delegation at the 1995 NPT Review and Extension Conference. Dr. Pilat has held positions in the Pentagon and the Congressional Research Service and has taught at Cornell University, Georgetown University, and the College of William and Mary. He is the editor of *Atoms for Peace: A Future After Fifty Years?* (2007).

Douglas Reilly has a B.A. from Amherst College (1964) and a Ph.D. from Case Western Reserve University (1969). He joined the Los Alamos safeguards program in 1969 and worked there until 2007. During that period he spent three years at the Joint Research Centre in Ispra, Italy, three years at DOE headquarters, and four years at the International Atomic Energy Agency in Vienna. His principal interests were gamma-ray spectroscopy and developing the safeguards training program.

Mark Schanfein is the Los Alamos National Laboratory's Program Manager for Nonproliferation and Security Technology, which includes both international and domestic safeguards and security. He has over eight years of experience working at the International Atomic Energy Agency, where he served four years as a safeguards inspector group leader and another four years as the unit head for Unattended Monitoring Systems in the Department of Safeguards.

William Severe is a Project Leader with the Global Threat Reduction Program Office at Los Alamos National Laboratory (LANL). Before joining LANL in 2005, he acquired more than 15 years of experience in nuclear-related matters with the U.S. Department of State, specializing in nuclear nonproliferation policy and Russian nuclear affairs, and 10 years of experience with the International Atomic Energy Agency (IAEA) in safeguards implementation.

Amy Seward is a staff member in the Global Security, Technology, and Policy Group at Pacific Northwest National Laboratory. She specializes in international security issues, focusing on nonproliferation and emerging threats to global security, especially environmental and energy security.

Elena Sokova is the Assistant Director of the James Martin Center for Nonproliferation Studies at the Monterey Institute of International Studies. Her research focuses on WMD nonproliferation issues in Russia and the NIS, nuclear trafficking, nuclear fuel cycle, and related nonproliferation policy issues.

Charles Streeper is a post-graduate researcher for the Radioactive Source Recovery Project at Los Alamos National Laboratory. His areas of nonproliferation policy expertise include Russia, India, and China and the history of nuclear weapons-free zones. He also improves global information resources on cases of radioactive material misuse.

James Tape has more than 30 years of experience in nuclear nonproliferation, safeguards and security, arms control verification technology development, and policy analysis. He retired from Los Alamos National Laboratory in 2005 and continues to work as an independent consultant. In 1994 and 1995 Jim was elected president of the Institute of Nuclear Materials Management (INMM), and the Institute elevated him to the rank of Fellow in 2000. Jim is currently the U.S. member of the IAEA Director General's Standing Advisory Group on Safeguards Implementation (SAGSI).

Carlton E. Thorne, since his retirement from the Department of State, has been a consultant to several government agencies on nuclear nonproliferation issues. He is known for his involvement with the Nuclear Suppliers Group and the Zangger Committee, where he was U.S. representative for many years and was the chief negotiator of many improvements in international nuclear export controls.

Dr. Gregory J. Van Tuyle manages the Los Alamos National Laboratory Program, focused on Nuclear, Radiological, and Explosives Threat Reduction for the Department of Homeland Security (DHS), including the Domestic Nuclear Detection Office (DNDO). His Ph.D. is in nuclear engineering from the University of Michigan (1978), and his research background includes identification and detection of threats, advanced nuclear reactors, and particle accelerator applications.

Richard Wallace is the Group Leader for Safeguards and Security Systems at the Los Alamos National Laboratory. He has worked in the nuclear weapons program at Los Alamos, has acted as a technical advisor to DOE, and was a project leader helping Russia secure and account for nuclear material during the 1990s. He spent three years as a Senior Analyst in the Safeguards Information Technology Department at the International Atomic Energy Agency in Vienna, specializing in open-source information collection and analysis, and still consults for the Agency.

Dr. Jooho Whang is a Professor in the Department of Nuclear Engineering at Kyung Hee University in South Korea. His special area of expertise is in radioactive waste management and radiation engineering. He is currently General Secretary of the Korean Nuclear Society and formerly served as the Nuclear Program Coordinator for the Korean Institute of Science and Technology. His Ph.D. (Nuclear Engineering) is from Georgia Tech.

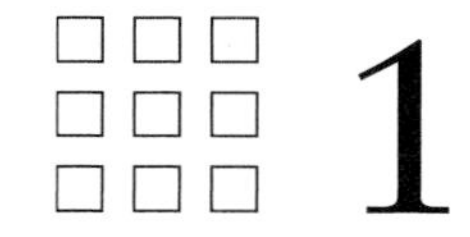 1

Introduction: Nuclear Security in the Twenty-First Century

James E. Doyle

Objectives for This Book

Nuclear security today is considerably more complex than it was during the Cold War. The superpower nuclear confrontation has been replaced by greater concerns about the proliferation of nuclear materials or weapons to states and nonstate groups. The specter of nuclear terrorism is of particular concern following the horrific events of September 11, 2001. The need for scientific understanding of the evolving nuclear threat is critical to informing policy decisions and diplomacy. The scientific underpinnings for such an understanding are remarkably broad, ranging from nuclear physics and engineering to chemistry, metallurgy and materials science, risk assessment, large-scale computational techniques, modeling and simulation, and detector development, among others. These areas constitute what we term *nuclear security science.*

The objective of this book is to present these subjects in a form that will be useful to academic studies in the area of nuclear security. These topics form a necessary foundation for students interested in nuclear weapons policies, nuclear proliferation, nuclear terrorism, nuclear energy and other peaceful applications of nuclear technologies. The scientific areas must be complemented by scholarly studies of public policy, with focus on areas such as political science, international relations, energy policies, economics, history, and regional studies. *Nuclear Safeguards, Security, and Nonproliferation: Achieving Security with Technology and Policy* has been written recognizing the importance of combining the social sciences with the physical sciences when addressing issues of nuclear security. It is our hope that this book will provide the necessary foundation in nuclear security science for undergraduate and graduate courses dealing with these complex issues.

Another objective of the book is to expose students and practitioners of nuclear security to some of the fundamental disciplines of their craft and provide an understanding of the unique challenges that arise when we apply these fundamentals to specific real-world problems. To this end, the major parts of the book progress from the introduction of concepts and techniques to case studies that provide a picture of real policy and technical approaches. For example, Part I describes the state of the art for modern, comprehensive nuclear safeguards systems that integrate physical protection and nuclear materials control and accounting. It also stresses how essential it is for global nuclear security that every state possessing nuclear material implement a comprehensive safeguards system that is open to some kind of international evaluation. We have done this by including technical chapters on nuclear materials measurements and the design of physical security systems and chapters on the historical development of the international safeguards system. Two case studies describing the application of safeguards at two very different nuclear facilities, a shut-down experimental reactor in Kazakhstan and a large plutonium reprocessing plant in Japan, illustrate how technology is used together with legal and administrative procedures to provide security and accountability.

To effectively deal with problems of nuclear security, it is important to balance the risks posed by nuclear technologies against their benefits. For example, whereas a significant expansion of nuclear power globally may help slow global climate change, it may also increase the risk of nuclear proliferation and terrorism. It will be important to conduct credible risk assessments to guide the public policy discussions of the expansion of nuclear power.

Nuclear Safeguards, Security, and Nonproliferation: Achieving Security with Technology and Policy was inspired in part by a previous study by Carter, Steinbruner, and Zraket that focused on the organizational requirements for safely maintaining a nuclear arsenal during the Cold War era.[1] Their 1987 book, *Managing Nuclear Operations*, is still an authoritative source on the challenges and requirements for the operational maintenance of a nuclear arsenal. The authors addressed what they perceived to be an imbalance in the study of nuclear security at the time. The imbalance they identified stemmed from a dominant focus on the doctrine of nuclear deterrence and the capabilities of nuclear weapons systems on one hand and a relative neglect of the process of managing the nuclear arsenals on the other. The authors believed this imbalance was both troubling and worthy of attention because they felt that overall security in the nuclear age of the time depended less on nuclear strategy and the capabilities of the weapons than on the effectiveness of human organizations to handle managerial problems that were more demanding and complex and with higher risks than any previously encountered. Their concern has been greatly magnified by the end of the Cold War and the challenges of dealing with the huge nuclear arsenal and nuclear complex of the former Soviet Union during the chaotic transition of governments in the 1990s.

This book takes a similar perspective, with a focus on the importance of effective policies, organizational systems, and procedures to provide nuclear security in the 21st century. It also stresses that these skills must be paired with innovative and reliable technologies to address a much broader range of nuclear security challenges than those prevailing during the Cold War period. Those challenges, which include preventing nuclear terrorism and expanding the use of nuclear energy while reducing the dangers of nuclear proliferation, demand the successful integration of effective policy and appropriate technology.

As *Nuclear Safeguards, Security, and Nonproliferation: Achieving Security with Technology and Policy* attempts to provide comprehensive coverage of the challenges of nuclear security today, it necessarily omits certain topics. Some topics, though vital to nuclear security in the traditional sense, did not fit with our focus on post-Cold War challenges. For example, we provide no detailed treatments of nuclear doctrine, deterrence, or force structure. Some other important topics were deemed too sensitive for treatment in the open literature. These include the specifics of nuclear weapons security, including unique physical protection methods or so-called "use control" features. These are aspects of a weapons design that prohibit any unauthorized party from gaining access to or detonating the weapons. Specific details of physical security measures taken during transportation of nuclear weapons or weapon-usable nuclear materials were also purposely omitted.

Finally, some topics were considered adequately covered in the existing literature and, hence, were described only briefly in the book. These include assessments of and recommendations for various government programs to improve nuclear materials security, such as the National Nuclear Security Administration's (NNSA) Nuclear Materials Protection, Control, and Accounting (MPC&A) program and the Second Line of Defense program.[2] Similarly,

---

[1]Ashton B. Carter, John D. Steinbruner, and Charles A. Zraket, *Managing Nuclear Operations*, Washington, D.C.: Brookings Institution Press, 1987, pp. 1–3.

[2]One of the most authoritative sources is the series *Securing the Bomb,* by Matthew Bunn and Anthony Wier. These studies were conducted by the Managing the Atom Project at the Belfer Center at Harvard University and are available in full on the Website of the Nuclear Threat Initiative (NTI), www.nti.org/e_research/cnwm/overview/cnwm_home.asp (Dec. 2006). Other sources are as follows: U.S. Department of Energy, *2006 Strategic Plan: Office of International Material Protection and Cooperation,* National Nuclear Security Administration (Washington, D.C.: DOE, 2006), and U.S. National Research Council, *Strengthening Long-Term Nuclear Security: Protecting Weapon-Usable Material in Russia* (Washington, D.C.: National Academy Press, 2005; available at http://fermat.nap.edu/catalog/11377 (April 2006).

historical treatments of traditional U.S.-Soviet and U.S.-Russian strategic nuclear arms control or speculation on what next steps, if any, might be taken along these lines are not included because several recent authoritative studies cover these subjects in depth.[3]

Nuclear Security in the Twenty-First Century

A New Nuclear Age

Nuclear security during the Cold War was dominated by superpower confrontation and reliance on the strategy of classic nuclear deterrence. There is a large body of excellent studies and reports on these topics, issued over the full span of the atomic age.[4] The focus of this book is on three interrelated objectives that we perceive as the major new nuclear security challenges in the post-Cold War and post-9/11 era:

1. *The states that possess nuclear weapons, materials, and knowledge must effectively control and protect them from theft or misuse.* Achieving this objective is imperative for all states that have developed nuclear weapons or nuclear energy. Loss of control of nuclear weapons or materials could cause international instability or conflict. All states are vulnerable to nuclear terrorism. The greatest challenge is that while huge quantities of nuclear weapons and materials have been produced across the globe for many years, the technical difficulties in securing them are not sufficiently appreciated and the resources and expertise for securing them have often been inadequate.
2. *The proliferation of nuclear weapons, both within states that possess them and to additional states, should be prevented.* There is a strong international consensus that the spread of nuclear weapons to additional states creates instability in the international system and increases the threat of nuclear war and nuclear terrorism. This is unequivocally the position of the United States and its major allies. Many other states, including a majority of those not possessing nuclear weapons, believe that the continued possession of nuclear weapons by those states that have them also represents a threat to international security. Therefore, stocks of nuclear weapons should be reduced to minimal levels. The objective of nuclear nonproliferation also includes the obligation of states that have pledged not to develop nuclear weapons to cooperate in verifying that all their nuclear activities are for nonweapons purposes.
3. *All states must make efforts to prevent nuclear terrorism.* Terrorists appear motivated to conduct ever more destructive attacks. The greatest obstacle to their capability to conduct an attack using a nuclear explosive is the acquisition of sufficient quantities of highly enriched uranium (HEU) or plutonium. These materials exist in great quantities and are used widely throughout the world. The ability to secure them from theft or smuggling is a long-term challenge critical to global nuclear security.

---

[3]See National Academy of Sciences, *A Comprehensive Nuclear Arms Reduction Regime: Interim Report* (Washington, D.C.: National Academy of Sciences, 2001), http://books.nap.edu/books/NI000347/html/1.html#pagetop (Jan. 2003), and Committee on International Security and Arms Control, *Monitoring Nuclear Weapons & Nuclear-Explosive Materials,* National Academy of Sciences, April 2005, http://books.nap.edu/catalog/11265.htm (June 2005). See also the White House, "Joint Statement on Parameters on Future Reductions in Nuclear Forces," Press Release, March 21, 1997.

[4]The literature on nuclear deterrence is vast. Five notable works in this area are Robert Jervis, *The Meaning of the Nuclear Revolution: Statecraft and the Prospect of Armageddon*, Ithaca: Cornell University Press, 1989; Scott D. Sagan and Kenneth N. Waltz, *The Spread of Nuclear Weapons: A Debate*, New York: W. W. Norton, 1995; Alexander L. George and Richard Smoke, *Deterrence in American Foreign Policy: Theory and Practice*, New York: Columbia University Press, 1974; Thomas Schelling, *Arms and Influence,* New Haven: Yale University Press, 1966; and Richard Betts, *Nuclear Blackmail and Nuclear Balance*, Washington, D.C.: Brookings Institution Press, 1987.

To describe some of the science, technologies, and practices used to pursue these objectives, this book is divided into three parts. Part I is devoted to technologies and processes for protecting, controlling, and accounting for nuclear material. Part II focuses on detecting nuclear proliferation and verifying the elimination of nuclear weapons programs. Part III concentrates on preventing nuclear terrorism and illicit nuclear trade. In each of these parts we include chapters that provide an outline of the basic security challenges, some background on the history and current status of these challenges, and a sample of the technical and political responses to them. Many of the chapters also include discussion of the likely future nuclear security challenges within their general subject area and how they might be managed.

Securing Nuclear Materials

Part I emphasizes the importance of maintaining the safety and security of nuclear materials. This challenge will continue to be shared by all states possessing nuclear materials for military or civil purposes for as long as they possess them. This task is critical because nuclear materials are the essential raw materials for nuclear explosive devices or radiological weapons and because they present public health and environmental hazards. A world with a high degree of "nuclear security" would be one in which all states possessing nuclear materials know to a high level of precision how much nuclear material they have, what form it is in, where it located, and whether it is adequately secured from theft or loss on a continuous, near-real-time basis. It would also be a world where all states had effective, enforceable laws criminalizing the unauthorized possession or trafficking of nuclear materials as well as possessing effective export and border controls to prevent illegal transfer of nuclear materials or the technologies and knowledge necessary for their production.

This is not the world we inhabit today. The legacy of the Cold War nuclear arms race between the United States and the Soviet Union has left behind vast stockpiles of directly weapons-usable nuclear materials, much of it surplus to current military needs. Although some of this material is being converted to forms that are less usable for weapons and can even be converted into fuel for nuclear reactors, hundreds of metric tons of plutonium and highly enriched uranium exist in these two countries alone.

It is therefore difficult to describe accurately the magnitude of the nuclear materials security challenge. The Institute of Science and International Security (ISIS) reports that as of late 2005, approximately 1.9 million kilograms of HEU and 1.83 million kilograms of plutonium existed worldwide in more than 50 countries, some with difficult social environments and very limited resources to devote to their security.[5] Approximately 1.4 million kilograms of this plutonium are found in highly radioactive spent fuel, which must be processed to be made suitable for a nuclear weapon. The technology for this process, though expensive and complex, is well known and well within the industrial capabilities of many states.

Highly enriched uranium (equal to or greater than 20% ^{235}U) and separated plutonium are the essential materials for nuclear explosives. Worldwide, plutonium and HEU are used in weapons, research, power reactors, and some industrial applications in forms that can be turned into weapons-usable materials via routine chemical processing. Such materials are processed, shaped, transported, stored, and used, and some inevitably wind up in waste streams.[6] For this reason it is difficult for countries to develop high confidence in a baseline inventory of these nuclear materials. In addition, much larger quantities of nonweapons-usable nuclear materials are in use and circulation worldwide, including low-enriched uranium and a broad range of radioactive source materials. In Part I of the book, technical descriptions of methods for measuring and accounting for nuclear materials are provided by Doug Reilly and Mark Abhold.

---

[5]David Albright, *Global stocks of nuclear explosive materials: Summary tables and charts*, Washington, D.C.: Institute for Science and International Security, July 12, 2005, updated Sept. 7.

[6]Sigfried Hecker, "Towards a Comprehensive International Safeguards System," *The Annals of the American Academy of Political and Social Science*, Sept. 2006, vol. 607, pp. 121–132.

State or "Domestic" Safeguards

Currently the security of nuclear materials is the responsibility of the states possessing them, and states have a variety of approaches to this task. There are no legally binding requirements for maintaining standard high levels of security, nor is there any multinational authority that inspects and evaluates the effectiveness of nuclear safeguards in each state. However, there is a Convention on the Physical Protection of Nuclear Material to which nearly all states with nuclear materials are party. These states agree to follow technical guidelines for adequate physical protection of nuclear materials during storage and transportation. The International Atomic Energy Agency (IAEA) provides these guidelines to all states through its document *Physical Protection of Nuclear Material and Nuclear Facilities* (INFCIRC/225/Rev.4) but does not require inspections or enforcement of the convention.[7] The result is that nuclear materials are protected with varying degrees of effectiveness by countries around the world, and very little is known about the specific security measures that are taken by some states. This is because the details of security and accounting for nuclear materials in many states are secret. This is universally the case for nuclear weapons and nuclear materials in most military programs.

Adequate nuclear security requires rigorous application of state or "domestic" safeguards. The U.S. domestic safeguard system is designed to protect nuclear materials against external threats such as terrorists and against insider threats. External threats are those posed by adversaries external to the nuclear facility, such as foreign commandos, criminal gangs, terrorists, or radical protesters. Insider threats are characterized by an individual or group of individuals who are either authorized to have access to the facility or have special knowledge of procedures and security measures that allow them to provide key aid to an adversarial plot to steal, divert, or sabotage nuclear material. The principal safeguard against external threats is physical protection. Part I includes an overview of the design and evaluation of physical protection systems by Mary Lynn Garcia.

Nuclear facilities that require physical protection include all research, development, production, and storage sites; nuclear reactors; fuel cycle facilities; and spent fuel storage and disposal facilities. Physical protection measures include a highly trained guard force, fences, and exclusion areas around facilities, in addition to perimeter and interior intrusion detection systems. Measures also include limited access and egress to facilities, buildings, and rooms. Finally, nuclear material and metal detectors at points of egress add an important element of defense.

The more insidious insider threat also requires additional rigorous internal controls and accounting. Modern safeguard systems combine physical protection with MPC&A. MPC&A systems are designed to deter and prevent loss or misuse of nuclear materials, provide timely and localized detection of unauthorized removal of materials, and ensure in near real time that all nuclear materials are accounted for. Proper materials control limits the handling of nuclear materials to only authorized areas and properly identified personnel and ensures that two people are present during nuclear materials transactions. It helps track nuclear materials from one site to another, from facility to facility, and from room to room. It ensures that there are a limited number of entries to and exits from the locations where nuclear materials are stored, and it alerts authorities to potential theft or diversion. It identifies nuclear materials for tracking purposes.

MPC&A programs are founded on a graded approach: The requirements for the facility's program vary depending on the types, attractiveness for weapons purposes, and amounts of nuclear materials used at the facility. In other words, the level of protection should be commensurate with the consequence of loss of the nuclear materials. The greater the risk to security and public safety that would result from the loss of certain materials at a facility, the more robust and effective should be the requirements for the MPC&A system and physical security system at that facility. On the other hand, safeguards and security systems may be minimal at some facilities because the materials are not considered significant from the point

---

[7]See www.iaea.org/worldatom/Programmes/Protection/inf225rev4/rev4_content.html (Oct. 2006).

of view of attractiveness for theft or diversion. Most states and the IAEA categorize nuclear materials according to the risk they present.[8]

Modern materials accounting also employs statistical and computer-based measures to maintain knowledge of quantities of nuclear materials present in each area of a facility. Some of these statistical methods are described by Tom Burr in Part I. The accounting system relies on inventories and material balances to verify the presence of material or to detect a loss. In the United States, the Nuclear Materials Management and Safeguards System (NMMSS) implemented in 1976 contains current and historical data on inventories and transactions involving source and special nuclear materials within the United States and on all exports and imports. It tracks all transactions, including domestic and foreign transfers, operating losses, inventory differences, and burn-up (transmutation and fission).[9] In U.S. facilities and those of several other states, operators must account for every gram of these materials in virtual real time. To declare any of it as an "inventory difference" or "waste" requires rigorous justification and verification. It is imperative that each state with nuclear facilities implement its own rigorous, comprehensive safeguards system to prevent theft or diversion of weapons-usable materials.

Preventing Nuclear Proliferation

International Safeguards

The International Atomic Energy Agency (IAEA) has the traditional responsibility for detecting the diversion of nuclear materials from civilian to military purposes in states that have joined the Treaty on the Nonproliferation of Nuclear Weapons (NPT) and pledged not to acquire nuclear weapons. To fulfill this mission the IAEA has created an organization of international safeguards inspectors who utilize dozens of specialized procedures and technologies to maintain confidence that member states are not diverting nuclear materials. The IAEA now has the additional task of determining the completeness of states' declarations of their nuclear activities and developing confidence that no undeclared nuclear activities exist within member states. James Tape and Joseph Pilat provide an informative history of the IAEA's evolving role in Part I.

For a good portion of its history, the IAEA carried out its traditional nuclear inspection role without much fanfare. A major change came after the discovery of Iraq's nuclear weapons development program after the country was defeated in the 1991 Gulf War. Iraq had foiled the international safeguards system by deceiving the IAEA and the international community. Instead of remaining a nonnuclear weapons state in good standing, as was its obligation under the NPT, Iraq had for years conducted secret work toward a nuclear weapons capability. This clandestine effort allowed Iraq to continue receiving aid for peaceful nuclear activities while making progress toward the bomb.

Iraq's illicit bomb program was the first time that a violation of IAEA safeguards was confirmed anywhere. More important, perhaps, the Iraqi program was largely based on clandestine, undeclared nuclear facilities that were not subject to IAEA inspection and could not have been detected under the existing international safeguards system. The discovery of the Iraqi deception resulted in sharp criticism of the international safeguards system and a long-term effort to strengthen that system.

In response to the weaknesses highlighted by Iraq's deception, the IAEA is adopting a fundamentally new approach to implementing safeguards. It is recognized that an effective,

---

[8]For example, such a system is implemented by the U.S. Nuclear Regulatory Commission. See www.nrc.gov/security/snm.html (Aug. 2007). Also see "The Physical Protection of Nuclear Material and Nuclear Facilities," IAEA INFCIRC/225/Rev.4, www.iaea.org/Publications/Documents/Infcircs/1999/infcirc225r4c/rev4_content.html (Aug. 2007).

[9]For a brief history of NMMSS, see www.nmmss.com/history.html (Aug. 2007).

strengthened international safeguards system, with a strong focus on searching for undeclared nuclear materials and activities, is essential to provide confidence that shared nuclear technologies and expertise, as well as nuclear materials themselves, are not being diverted to weapons programs. Verifying the "completeness" as well as the "correctness" of a state's declaration is now acknowledged as a critical objective of the IAEA safeguards system.

To achieve this new objective, the IAEA is encouraging member states to sign an Additional Protocol to their safeguards agreements, permitting the IAEA greater access to inspections and information regarding nuclear activities in that state. Although most states with significant nuclear activities have now signed the Additional Protocol, a large number of states have not yet ratified the protocol nor brought it into force on their territories.[10] Fundamental to the new approach to IAEA safeguards are information acquisition, evaluation, and analysis along with inspections. The new approach is designed to provide an evaluation of the nuclear program of a state as a whole and not just each of its declared nuclear facilities.

The goal of comprehensive international safeguards agreements is to detect the diversion of significant quantities of nuclear material from peaceful purposes within certain time periods. A significant quantity of nuclear material is defined by the IAEA as the approximate amount of nuclear material from which a nuclear explosive device could be manufactured. For highly enriched uranium, this quantity is 25 kilograms; for plutonium it is 8 kilograms.[11]

The IAEA has established safeguards criteria for each type of nuclear facility under safeguards. These criteria are used as templates for defining safeguards activities at specific facilities within a country, including the scope, the normal frequency, and the extent of the verification activities needed to achieve the inspection goals. The implementation of IAEA safeguards and the use of remote monitoring technologies are described by Mark Schanfein and Brian Boyer in Part I. For comprehensive safeguards agreements, the technical objectives of safeguards are the timely detection of the diversion of nuclear material from peaceful uses and the deterrence of such diversion by the risk of early detection. These objectives also include the detection of undeclared production or separation of direct-use material at reactors, reprocessing facilities, facilities with hot cells, and enrichment installations.

Nuclear materials accounting records of all nuclear materials on inventory and inventory changes are maintained by operators for each facility under safeguards. This information should be identical to that which exists in each state's "domestic" system of accounting and control. This inventory information and safeguards-relevant design information are transmitted through the state authorities to the IAEA. These state declarations on the nuclear materials present at facilities and the facility operations provide a baseline for the IAEA's verification activities. For comprehensive safeguards agreements *with* additional protocols, the overall objective is to provide credible assurance of both the nondiversion of nuclear material from declared activities and of the absence of undeclared nuclear material and activities in the state as a whole.

The concept of voluntary implementation of the Additional Protocol among IAEA member states includes the idea of incentives. Some states, after having implemented the Additional Protocol on all their nuclear facilities for several years, can eventually have a decreased safeguards burden in terms of IAEA presence within their facilities. This is because they will have reached a status where "integrated" safeguards, including facility inspections, complimentary access, and information analysis, are now believed to be in effect. This allows some inspections activities to be reduced within that country without a loss of confidence that its safeguards agreement is being fulfilled.

---

[10]For a summary of the status of Additional Protocol implementation, see www.iaea.org/OurWork/SV/Safeguards/sg_protocol.html (Aug. 2007).

[11]See "The Structure and Content of Agreements between the Agency and States Required in Connection with the Treaty on the Non-Proliferation of Nuclear Weapons," INFCIRC 153 (corrected), www.iaea.org/Publications/Documents/Infcircs/Others/infcirc153.pdf (Aug. 2007).

Limitations of International Safeguards System in the New Security Environment

The international safeguards system as implemented by the IAEA, although very important, is not intended to provide guarantees that no nuclear materials have been stolen by insiders or transferred to terrorists. As stated, the primary objective of the international safeguards systems is timely detection of significant quantities of nuclear materials and, in states implementing Additional Protocols, the detection of undeclared nuclear activities. The system offers limited detection capability against the threat of very small amounts of nuclear materials being diverted by insiders over time or the dedicated efforts of a nation to keep some nuclear activities secret.

Moreover, only a very small portion of the world's inventory of weapons-grade nuclear material is even subject to IAEA inspections. IAEA safeguards agreements with more than 130 states provide for inspections at some 900-plus nuclear facilities and locations. However, less than a third of the global inventory of roughly 3.73 million kilograms of fissile materials is subject to international safeguards. For example, nuclear materials in military programs are not subject to international safeguards. Because of the limitations of international safeguards to detect losses of nuclear materials to insiders or to prevent or detect clandestine nuclear activities, it is vital that states develop their own capabilities for these tasks and cooperate to reduce these threats.

Detecting Undeclared Nuclear Activities

Every state that has developed nuclear weapons has done so in secret, at least initially. Consistent with this trend there now exists a clear record of states that have accepted legal obligations not to develop nuclear weapons but nevertheless pursued nuclear weapons development in violation of those obligations. This list includes Iraq, Libya, North Korea, and most likely Iran. The secret development of nuclear weapons by additional states, especially by those that have formally rejected these weapons, can cause tension, regional instability, and distrust that increase proliferation incentives for neighboring or rival states. Because this negative pattern may continue, it is important to the security of the United States and the international community that tools are available for the detection and analysis of secret or undeclared nuclear activities. Such tools will also be important for monitoring the nuclear weapons programs of Israel, India, and Pakistan, states that have not signed the NPT.

Since the beginning of the atomic age, nations have devoted considerable effort to learning about the secret nuclear weapons development plans of their rivals and potential rivals. There has been mixed success in this effort. Certainly major aspects of the Israeli, South African, Iraqi, Indian, and Pakistani nuclear weapons efforts were not detected ahead of time.[12] Now that the IAEA has acknowledged the role of assessing the "completeness" of states' declarations in their safeguards agreements, it has been creating capabilities to detect undeclared nuclear activities and it receives periodic assistance in this effort from the national intelligence agencies of NPT member states.

The IAEA approach to assessing the presence or absence of undeclared nuclear activities derives from the fact that a state's nuclear program (past, present, and future) involves an interrelated set of nuclear and nuclear-related activities that require and/or are indicated by the presence of certain equipment, a specific infrastructure, observable traces of materials in the environment, and predictable use of nuclear material. The picture presented by these features provides the basis for an assessment of, first, the internal consistency of the state's

[12] An excellent reference assessing the efforts of U.S. intelligence agencies to detect nuclear programs is *Spying on the Bomb: American Nuclear Intelligence from Nazi Germany to Iran and North Korea*, by Jeffrey Richelson, W. W. Norton & Co., 2006.

declarations to the agency and, second, the consistency between the state's declarations and other information available to the agency.

It is in the assessment of other categories of information available to the agency that a large amount of innovation and progress has taken place in recent years. Some of the techniques for information analysis regarding proliferation activities are described in Part II by Rick Wallace, Arvid Lundy, and Frank Pabian. These cover open source analysis, including the use of commercially available satellite imagery and methodologies for assessing the technological capabilities of states that might be pursuing nuclear weapons. Another innovative approach to assessing a nation's commitment to its nonproliferation obligations is to look at its record of behavior with respect to the obligations it has accepted under various treaties and agreements and elements of its foreign policy. Such an approach is described in Part II by Carol Kessler, Carrie Mathews, and Amy Seward.

Confirming the Elimination of Nuclear Weapons Programs

Another important aspect of nuclear security in the 21st century will, hopefully, be the challenge of confirming that nations that have started nuclear weapons programs and then pledged to abandon them have actually done so. It is clear that the set of factors that lead a nation to acquire a nuclear arsenal can change, prompting the nation to decide that nuclear weapons are no longer in its interest. Several specific cases have demonstrated this situation since the early 1990s, and the IAEA has taken an active role in them. The latest such case is Libya, as described by Wyn Bowen in Part II.

The degree of progress a nation has made in developing the infrastructure to support a nuclear arsenal is one key factor in determining how difficult it will be to confirm its elimination. For example, if North Korea adheres to the six-party September 19, 2005, pledge for a denuclearized Korean Peninsula and rejoins the NPT, the IAEA will have to verify the elimination of North Korea's stocks of military plutonium, its plutonium production capability, its nuclear weapons assembly and testing facilities, and any uranium enrichment facilities that have a weapons-related function. This would likely be a very complex and time-consuming process requiring intrusive on-site inspection and significant decommissioning of several large contaminated nuclear facilities. In Part II George Baldwin of Sandia National Laboratory provides an in-depth discussion of the potential elimination of North Korea's plutonium production facilities. In a nation with an extensive nuclear weapons infrastructure, there is a risk that failure to verifiably eliminate that infrastructure or place it under safeguards will leave in place weapons capabilities that can be reconstituted.

The other major factor determining the difficulty of verifying the elimination of a nuclear weapons program is the degree of cooperation demonstrated by the government that has agreed to the elimination process. In the case of South Africa, the IAEA was able to verify, effectively and quickly, the elimination of a program that had produced six workable nuclear weapons and spanned more than 15 years. The South African government was cooperative and transparent regarding all information on its past program, including technical aspects of the facilities that produced the weapons and detailed accounting of the quantities of nuclear materials that were produced and then eliminated. It allowed IAEA inspectors sufficient access to facilities, documentation, and personnel. A full discussion of the South African case is provided by Sara Kutchesfahani and Marci Lombardi in Part II.

By contrast, Iraq's behavior following its defeat in the 1991 Gulf War and passage of several United Nations Security Council Resolutions mandating the elimination of its nuclear weapons programs was noncooperative. Information the Iraqis provided was often tardy, incomplete, confusing, and suspect. IAEA inspectors were taken to facilities and then denied entry or told to wait for hours. Iraqi cooperation with IAEA and U.N. inspection efforts was so poor that suspicions (later disproved) that Iraq was reconstituting its nuclear weapons program were among the reasons cited by the U.S. government for the invasion of Iraq in March 2003.

Preventing Nuclear Terrorism

The possibility that terrorists might acquire and use nuclear weapons is an urgent challenge to global security. Today, a terrorist nuclear attack is thought to be more likely than an exchange of nuclear weapons with another state.[13] Terrorist networks have proven that they are capable of sophisticated attacks involving dozens of heavily armed assailants.[14] There is a very strong consensus among terrorist experts that anti-Western Islamic extremism will persist for many decades, and many such experts predict that it will become more widespread and more violent and will concentrate on attacks with weapons of mass destruction, including nuclear weapons.[15]

The detonation of a 10-kiloton nuclear device in an urban area would kill hundreds of thousands of people instantly and overwhelm the medical and emergency response capabilities of even the most developed nations. A successful attack on Washington, D.C., could destroy elements of the U.S. national leadership and degrade the capabilities of some federal agencies. The toll from human losses and the psychological impact would be incalculable. Economic consequences would dwarf those resulting from the 9/11 attacks. Fundamental civil liberties, free social patterns, and open global commerce would be challenged and constrained in the aftermath of a devastating attack. In short, a successful nuclear terrorist attack could severely disrupt even the most powerful nation and degrade the quality of life of hundreds of millions of people.

It would be difficult for terrorists to obtain nuclear explosives, but it is not impossible. For this reason, Part I of this book highlights the need for nuclear materials security and the methods for maintaining it. Part III of the book explores the nuclear terrorist threat in more detail and provides information on some of the technologies that could help reduce the chances that a terrorist nuclear plot would succeed. As the chapter by William Potter and Charles Ferguson makes clear, the most difficult step for terrorists to complete in their plan to conduct a nuclear attack is likely to be the acquisition of a suitable quantity of fissile material. The nuclear materials security measures described in Part I are therefore the primary defense against this step, and their importance to nuclear security cannot be overstated.

A much less devastating but still extremely disruptive attack could be conducted by terrorists with radiological materials using a radiological dispersal device (RDD), or so-called "dirty bomb." An RDD causes no nuclear explosion but instead uses chemical explosive to spread harmful radioactivity. It could cause severe economic disruption and panic in an urban area, without causing large numbers of fatalities. Due to their common use in industry, agriculture, and medicine, these materials would be easier for terrorists to obtain. The IAEA reports that more than 100 countries may have inadequate controls to prevent or even detect the theft of radioactive materials needed for an RDD.[16] The threat posed by RDDs and steps to reduce it are described in detail by Greg van Tuyle and Lee Leonard.

Even if nuclear materials or a nuclear weapon escape the custody of those authorized to possess them, there are still some legal, administrative, and technical means that could prevent terrorists from acquiring the materials, building a weapon, and successfully delivering it to a target. One technical capability that can play a significant role in the defense against nuclear terrorism is the ability to detect nuclear materials during transit from one

---

[13]James Sterngold, "Kerry, Bush agree on peril of nuclear terrorism, but candidates have different policies on how to mitigate the global threat," *San Francisco Chronicle*, Oct. 28, 2004.

[14]Three examples are the attacks of September 11, 2001, in the United States, the October 2002 theater attack in Moscow, and the September 1, 2004, school assault in Beslan, Russia.

[15]Jessica Stern, "Terrorist Motivations and Unconventional Weapons," in Peter R. Lavoy, Scott D. Sagan, and James J. Wirtz (eds.), *Planning the Unthinkable* (Ithaca: Cornell University Press, 2000), p. 215.

[16]International Atomic Energy Agency, Division of Public Information, PR 2003/03 (March 13, 2003).

place to another. The science and technology used to build radiation detection and measurement equipment for the purpose of nuclear safeguards described in Part I has provided a foundation for developing nuclear materials detectors that might foil a plot to smuggle nuclear materials across borders or place them near a target.

In Part III, Mark Abhold and Chris Lovejoy describe the challenges of detecting and characterizing nuclear materials in the field and some technical options for accomplishing this task. One of the toughest challenges is that HEU, most likely the material of choice for a terrorist improvised nuclear device (IND), is easily shielded from current passive radiation detection equipment. Another challenge is that of making quick determinations between nuclear materials that present a threat and those that are transported or used daily across the world for beneficial commercial purposes. A third central challenge of radiation detection systems is how to respond. For example, if an automated system screening vehicles on a highway detects radiation, how do authorities isolate, track, and interdict the vehicle containing nuclear material? These are only a few of the hurdles involved in creating an effective system of radiation detection for defense against nuclear terrorism.[17]

Another tactic that can be employed, and one that would help create a more effective technical defense against nuclear terrorism, is to study and understand trends in illegal trafficking in nuclear materials. In Part III Galya Balatsky, Stacey Eaton, and William Severe provide a summary of nuclear smuggling incidents that have been reported in the open press. They offer observations on the trends of such activity over the past 15 years or so and the possible motivations of the traffickers. It is virtually impossible to stop all smuggling in nuclear materials, just as illegal trade in drugs, weapons, and human beings has been a historic problem throughout the world. However, the details of every case of nuclear smuggling, particularly in weapons-usable nuclear materials, are worth the attention of all institutions and law-enforcement agencies involved with nuclear security. Such information could be invaluable in identifying key individuals, facilities, and transshipment methods that were involved in nuclear smuggling and help direct resources to disrupt these crimes and improve weak security at particular locations.

Another great challenge for nuclear security in the 21st century will be improving the integration or coupling of technical and administrative systems that are responsible for securing nuclear materials at their authorized locations, prosecuting individuals or groups who seek such materials, and detecting their illegal shipment. The interrelationship of these problems is clear, but the global infrastructure to prevent them still operates like three or more disconnected organizations. For example, those responsible for physical security at a nuclear facility should know in detail whether or not nuclear smuggling rings or terrorists have been active in their location, but this information has traditionally resided with law enforcement agencies. Another example is that customs or border officials operating nuclear materials detection systems could do so much more effectively if they knew what type of nuclear materials were missing from what locations or how they were being transported. The laws that require nuclear facility managers to report missing materials within a certain time or to a certain degree of detail vary widely across the world. These and many other shortcomings need attention in the years ahead.

One of the most disturbing possibilities influencing the risk of a successful terrorist nuclear attack is whether or not a national government would aid or enable such an attack. Nuclear weapons and nuclear materials are regulated by national authorities. If any national authority transferred a nuclear weapon or the materials needed to build one or helped construct a weapon, the most difficult step for terrorists would be overcome. Of course, a government's decision to aid terrorists in this manner carries great risk. The country that became the victim of such a nuclear terror attack would have every right to respond as though the enabling government directly launched the attack. The motivation for a government to

[17]For an overview of these challenges, see James E. Doyle, "Needed: A Nuclear Dragnet for Homeland Security?" *The Nonproliferation Review*, Fall-Winter 2003 (vol. 10, no. 3).

consider such a risk could come from the hope that its actions remained secret, and it could plausibly deny involvement in the attack.

Governments must be deterred from believing that such activities would go undetected. The best way to accomplish this goal is to aggressively monitor and disrupt all cooperation between terrorist organizations and national governments. Also, governments should be made aware now that if any evidence suggests they have collaborated with nuclear terrorists, the governments would be held culpable. If a government proceeded to aid a nuclear terrorist attack despite the risks, it should know that there are means by which its involvement could be discovered. Forensic examination of trace nuclear material after an attack could help determine the material's origin and other details of the attack. This process, known as *nuclear attribution*, could be applied after a nuclear detonation with an IND or radiological attack with an RDD. In Part III William Charlton offers a model for the technical aspects of nuclear attribution.

Although proving the national origin of nuclear materials used in an attack could be a key piece of evidence in determining who was responsible, it would not be enough to justify retaliation against that nation. A much more complete understanding of the plot and its perpetrators would be needed, including clear evidence that the national government had aided the attack and had not simply been unable to prevent its nuclear materials from being stolen by terrorists. A premature tendency to hold a state responsible without complete confidence could itself be a destabilizing act. It could also lead to increased motivation for states who believed they could hide their actions. For example, Nation A could try a "false flag" operation by secretly stealing nuclear materials from rival Nation B and then providing them to terrorists for an attack on Nation C. The hope would be that after investigators assessed the origin of the nuclear materials used, Nation C would mistakenly retaliate against Nation B even though it was Nation A who facilitated the attack. Terrorists could use such a strategy to start a catalytic war between states.

A declared policy of retaliation against the country of origin of nuclear materials used in a terrorist attack could also create incentives for additional countries to acquire nuclear weapons. Such incentives could result if states believed that they might be attacked or coerced into making reparations for failing to prevent nuclear materials from being stolen from their facilities, even when they did nothing to aid terrorists' theft or hostile use of those materials. States might decide it is in their interest to acquire nuclear weapons in order to deter such action against them.

Despite these concerns, the ability to determine the origin of nuclear material after an attack could help deter states from ever helping terrorists with a nuclear attack. Perhaps even more relevant are the potential advantages that effective attribution could have on judging the credibility of future attacks and identifying vulnerable stocks of nuclear materials.

One capability that can make a contribution to efforts to both stop the proliferation of nuclear weapons and prevent nuclear terrorism is the implementation of more effective export controls. The improvement of multinational nuclear export controls has a long history, and a great deal of progress has been made over the years. This history includes changes in the definitions of items and categories of information to be controlled in order to more comprehensively regulate nuclear commerce and prevent weapons proliferation. This history is summarized by Carl Thorne in Part III.

Because so many of the materials and equipment used in a nuclear weapons program have other civilian uses, nonproliferation export control is a very information-intensive undertaking. Declared uses of dual-use equipment, the bona fides of end users, and the trustworthiness of manufacturers and shippers all must be verified, especially in the case of very sensitive technologies and materials. Many nations view the required level of intrusiveness as an unjustifiable interference with commerce or even as industrial espionage. Nevertheless, much of the history of nuclear export control is the history of a growing consensus among nations that export controls serve their security interests and that, to be effective, they must be enforced as consistently and uniformly as possible by all suppliers of nuclear materials, technology, and information. The consequences of not doing so, as described by Sara Kutchesfahani in Part III, were sharply revealed by the A. Q. Khan network.

Major nuclear suppliers have realized this for many years, and most have well-developed export control systems. As Todd Perry makes clear in Part III, one vital mission of states that have strong export control infrastructures is to assist in improving similar capabilities in other states. Customs and border services in many nations have not traditionally been the primary implementing agencies of nuclear export controls. Strengthening their participation through legal reform and education can have significant payoff in terms of effectiveness. Specific training in information analysis, interagency and international communications, and commodities identification are some of the activities that the United States and other leading nuclear suppliers can provide.

Summary

The management of nuclear security in the post-Cold War era is a complex and evolving challenge for the international community. Fortunately, there has so far been no catastrophic use of nuclear weapons or materials or the confirmed theft of a nuclear weapon or large quantity of weapons-usable nuclear materials. However, there have been many cases of states violating their legal nonproliferation obligations and of individuals conducting illegal trade in nuclear materials, technology, and knowledge. Even more troubling is that many nations have nuclear materials security and export control systems that require significant improvement to effectively prevent illegal loss or trafficking of nuclear materials, technology, and knowledge. The IAEA has limited resources and political authority to address these shortcomings. Meanwhile, large stocks of excess Cold War fissile materials have yet to be rendered nonweapons-usable, even as civilian stocks of plutonium continue to grow.

While there are encouraging international developments such as U.N. Security Council Resolution 1540, the IAEA's Additional Protocol, and the Proliferation Security Initiative, these efforts are in their early stages, and their full implementation is uncertain. In addition, the international security environment remains such that additional states will probably seek nuclear weapons or the capability to acquire them rapidly for reasons of security and influence. In short, there is clearly a need for continuous, vigorous efforts to improve the technological and human capital that managing nuclear security will require. It is our hope that *Nuclear Safeguards, Security, and Nonproliferation: Achieving Security with Technology and Policy* will contribute to this effort by helping to inform a new generation of nuclear security students and practitioners.

PART

I

Technologies and Processes for the Protection, Control, and Accounting of Nuclear Material

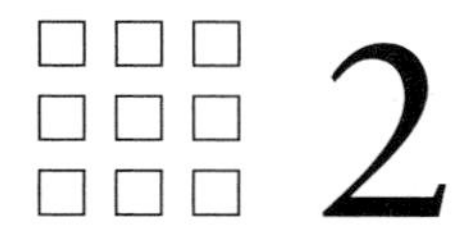

Nuclear Safeguards and the Security of Nuclear Materials

James Tape and Joseph Pilat

Introduction

Nuclear technology is Janus-headed; it is a dual-use technology with both peaceful and military applications. Concerns about the misuse of peaceful applications of nuclear energy were at first focused on states seeking nuclear weapons. The first concepts for restricting nuclear energy to peaceful purposes were proposed in the context of a broad international agreement under the auspices of the newly formed United Nations. The term *safeguards* in relation to peaceful uses referred to institutional, legal, and technical mechanisms to prevent the misuse of nuclear technologies and nuclear materials for military applications. Domestic security measures employed by states developing nuclear technologies were designed to counter commercial or military espionage or theft of materials by agents of other countries. The increase in concern about threats from nonstate groups and terrorists began to significantly impact the nuclear industry in the early 1970s as the specter of international terrorism grew, from the attacks at the 1972 Munich Olympics to the attacks on the World Trade Center and the Pentagon on September 11, 2001. This chapter deals with the history of measures to counter the proliferation of nuclear weapons by states and nonstate groups or terrorists.

International Control or Secrecy and Denial: From 1945 to Atoms for Peace

Even before the end of the Second World War, the scientists and political leaders who knew the secret of the U.S. Manhattan Project to build a nuclear weapon debated how to control the technology they had created and at the same time to realize its civilian benefits.[1] In a major political commitment, the President of the United States and the Prime Ministers of the United Kingdom and Canada issued an Agreed Declaration on November 15, 1945, which described three reasons to seek international control of nuclear activities: the massive destructive power of nuclear weapons; the likely futility of defense against such weapons; and the fact that no state could hope to have a monopoly on such weapons.

[1]For a discussion of the debates surrounding possible mechanisms of control, including the Acheson-Lilienthal Report, the Baruch Plan, and the Atoms for Peace proposal, see Richard Rhodes, *The Making of the Atomic Bomb* (New York: Simon and Schuster, 1986); Rhodes, *Dark Sun* (New York: Simon and Schuster, 1995); Kai Bird and Martin J. Sherwin, *American Prometheus: The Triumph and Tragedy of J. Robert Oppenheimer* (New York: Knopf, 2005); and Joseph F. Pilat, editor, *Atoms for Peace: A Future After Fifty Years?* (Baltimore: Johns Hopkins University Press/Woodrow Wilson Center Press, 2007).

Several months later, on January 7, 1946, the U.S. Secretary of State appointed a committee chaired by Dean Acheson with the following terms of reference:

> Anticipating favorable action by the United Nations Organization on the proposal for the establishment of a commission to consider the problems arising as to the control of atomic energy and other weapons of possible mass destruction, the Secretary of State has appointed a Committee of five members to study the subject of controls and safeguards necessary to protect this Government so that the persons hereafter selected to represent the United States on the United Nations Commission can have the benefit of the study.

The U.S. State Department committee subsequently appointed a Board of Consultants, including David Lilienthal and J. Robert Oppenheimer. The product of this committee, the so-called Acheson-Lilienthal Report, is remarkable for its vision and anticipation of problems that remain difficult and only partially solved today.[2]

The U.S. committee proceeded from the position that it was in the interests of the United States to seek international control of nuclear energy and weapons. They examined in some detail the possible treaty regimes and "safeguards" that would be necessary to enforce international control, including the role of inspections. Inspections to confirm the absence of nuclear weapons proliferation alone were seen as inadequate; additional legal and technical measures would be needed for effective international control of nuclear energy. The production of nuclear materials such as uranium and plutonium was noted as a technically difficult and strategically critical capability that should be a logical focus for international controls. The effective management and protection of nuclear materials was identified as a central objective and remains a primary mechanism of all nuclear safeguards efforts today.

The Acheson-Lilienthal Report recommended a distinction between "safe" and "dangerous" nuclear activities. Safe activities included use of tracer isotopes and small quantities of nuclear materials. Dangerous activities were uranium mining and refining; uranium enrichment; the operation of plutonium production reactors and associated reprocessing plants; and nuclear explosive research and development. Although a different list of safe and dangerous activities might be chosen today in light of the advances in nuclear technology and the wide availability of information, the process of determining proliferation risk associated with different elements of the nuclear fuel cycle remains central to current nonproliferation efforts.

The Acheson-Lilienthal Report recommended the creation of an international authority, an "international monopoly," to conduct all intrinsically dangerous operations in the nuclear field, with individual states and their citizens free to conduct, under license and a minimum of inspection, all nondangerous, or safe, operations. The proposed body would have authority to own and lease property and to carry on mining, manufacturing, research, licensing, inspecting, selling, or any other necessary operations. The analyses reflected in the Acheson Lilienthal report are an excellent example of the importance of combining technical and political expertise in dealing with nuclear nonproliferation.

The recommendations of the Acheson-Lilienthal Report were the basis for a presentation to the United Nations in 1946 by U.S. representative to the U.N. Atomic Energy Commission Bernard Baruch that became known as the Baruch Plan. The plan languished at the U.N. due to obstruction by the Soviet Union and its satellites. The Soviets had obtained the secrets of the Manhattan Project through espionage, and it is highly unlikely that any plan for international control would have been acceptable to them prior to their mastering nuclear weapons technology. The first Soviet atomic bomb was detonated in August 1949.

After the failure of the Baruch Plan, the United States followed a policy of maintaining secrecy around all nuclear matters and began slowly to expand its stockpile of nuclear weapons. However, many of the ideas considered in the Acheson-Lilienthal Report and the Baruch Plan

---

[2] *A Report on the International Control of Atomic Energy*, Prepared for the Secretary of State's Committee on Atomic Energy, U.S. Government Printing Office, Washington, D.C., March 16, 1946 (the Acheson-Lilienthal Report).

would reemerge as part of President Dwight D. Eisenhower's Atoms for Peace initiative and would provide a foundation for the development of the nuclear nonproliferation regime.

Another important decision related to the future of nuclear energy occurred in 1946 when the U.S. Congress established the United States Atomic Energy Commission (AEC) to foster and control the peacetime development of atomic science and technology. The U.S. Atomic Energy Act of August 1, 1946, transferred U.S. control of atomic energy from military to civilian hands. This action reflected the view that atomic energy should be employed not only for national defense, but also to promote world peace, improve the public welfare, and strengthen free competition in private enterprise. The signing was the culmination of long months of intensive debate among politicians, military planners, and atomic scientists over the fate of this new energy source. President Harry S. Truman appointed David Lilienthal as the first Chairman of the AEC.

International Collaboration and Technology Sharing: Atoms for Peace to the Late 1960s

By 1953 the Soviet Union and the United Kingdom had tested nuclear weapons and the U.S. was fully engaged in the development of thermonuclear warheads. An arms race between the United States and the Soviet Union had developed. It was clear that the U.S. policy of secrecy and denial was not having much effect on the control of nuclear weapons. On December 8, 1953, President Eisenhower delivered an address on peaceful uses of atomic energy to the U.N. General Assembly. The ideas outlined in this speech and its follow-on policies became known as the Atoms for Peace initiative.

The Atoms for Peace initiative included the following key proposals:

- States with nuclear materials should make joint contributions from their stockpiles of normal uranium and fissionable materials to an international Atomic Energy Agency that should be set up under the aegis of the United Nations.
- The Atomic Energy Agency could be made responsible for storing and protecting the contributed fissionable and other materials.
- The more important responsibility of this Atomic Energy Agency would be to devise methods whereby this fissionable material would be allocated to serve the peaceful pursuits of mankind, including agriculture, medicine, and the provision of electrical energy in the power-starved areas of the world.
- Also central to the Atoms for Peace initiative was the idea that states receiving assistance in peaceful uses of nuclear energy would allow inspections to ensure that the nuclear technology and materials were not used for military purposes.

The dramatic change in U.S. policy arising from the Atoms for Peace initiative had wide-ranging impacts on the development of both domestic and international nuclear safeguards. Atoms for Peace and the changes to the Atomic Energy Act of 1954 permitting international collaborations and private ownership of nuclear materials in the U.S. would require expanded thinking about how to manage nuclear materials outside secret government installations, as well as how to manage exports of technology and materials.

The shift in U.S. policy was motivated partly by the now confirmed view that nuclear technology could not be kept secret. It was also based on the idea that the United States and its allies could use their advanced nuclear capabilities to strike a bargain with the developing world. The basic structure of the bargain is the provision of nuclear materials and technology by the United Sates and others to the developing nuclear states in exchange for verifiable assurances that the recipient states would only use nuclear energy for civilian, not military, purposes.

International Nuclear Safeguards

Prior to 1953, the concept of international safeguards had not yet evolved in international discourse, and no international nuclear safeguards were applied at any facilities. International developments followed two paths: the negotiation of a statute to create the IAEA, and U.S.

bilateral agreements for cooperation involving sharing with certain countries carefully selected nuclear technologies. Early bilateral agreements with close allies required minimal or no safeguards, but later agreements included bilateral (really unilateral) inspection provisions. The specific inclusion of safeguards in the IAEA Statute and the use of inspections in bilateral agreements for cooperation laid a sound foundation for international safeguards.[3]

In 1957, the European Community established under Chapter VII of the Euratom Treaty a nuclear material control system. Euratom safeguards are designed to ensure that nuclear materials were not diverted from their intended use and to guarantee "that the Community complies with its international obligations concerning the supply and use of nuclear materials."[4] Supply agreements with Euratom employed Euratom safeguards in lieu of bilateral safeguards, in recognition of the multinational character of its safeguards system.[5] After the full development of IAEA safeguards, special arrangements and cooperative mechanisms between Euratom and IAEA inspections were worked out and continue to evolve. The safeguards and inspection arrangements that originated from bilateral nuclear agreements and the Euratom safeguards measures provided useful experience and a context for the development of safeguards agreements between the IAEA and individual member states. All these international safeguards activities had the common objective of providing assurances from the state in which the safeguards were applied to other states or an international organization that nuclear technology was not being misused for military purposes.

IAEA safeguards began very slowly, in part because the agreements for cooperation among states included bilateral nuclear safeguards provisions and partly because of strong Soviet opposition to the safeguards role of the IAEA. However, in the early 1960s, the Soviet position shifted from one of opposition to safeguards as an "imperialistic mechanism" to hold back nuclear have-not countries to one of cautious support.[6] This shift, coupled with U.S. encouragement to shift the implementation of safeguards under agreements for cooperation to the IAEA, provided strong support for the further development of the international safeguards system.[7]

Domestic Safeguards

As mentioned, changes to the U.S. Atomic Energy Act of 1954 permitting private ownership of nuclear materials in the United States also required new approaches for managing nuclear materials outside of secret government installations.[8] Such safeguards originally consisted of nuclear process operating records, with little or no independent verification of nuclear materials inventories.[9]

The original foundation for United States domestic nuclear safeguards within the emerging private civil sector licensed by the Atomic Energy Commission (AEC) was simply the health risks and intrinsic monetary value of these still rare materials. Physical security was little

---

[3]Myron Kratzer, "The Origin of International Safeguards," *Journal of Nuclear Materials Management*, special issue: "20 Years of Safeguards at Los Alamos National Laboratory," vol. XV, no. 4, July 1987, pp. 27–33.

[4]European Commission, Directorate-General for Energy and Transport, "Nuclear Safeguards—Europe remains vigilant," 2006; see http://europa.eu.int/comm/energy/nuclear/safeguards/doc/2006_brochure_nuclear_safeguards_en.pdf.

[5]Kratzer, JNMM, p. 31.

[6]Kratzer, JNMM, p. 32.

[7]Scheinman, *The International Atomic Energy Agency and World Nuclear Order*, pp. 36–37.

[8]See, for example, www.eh.doe.gov/oepa/laws/aea.html.

[9]Samuel C. T. McDowell, "U.S. Safeguards Before DOE," *Journal of Nuclear Materials Management*: special issue, "20 Years of Safeguards at Los Alamos National Laboratory," vol. XV, no. 4, July 1987, pp. 34–36.

more than standard industrial security designed to keep the public away from hazardous and valuable operations.[10] However, in the middle of the 1960s, concern began to grow about accounting and control for all nuclear materials in AEC contractor and licensee facilities. A major finding of nuclear material unaccounted for (MUF) at the NUMEC Apollo, Pennsylvania, plant that processed strategically and financially valuable materials provided major impetus for independent safeguards arrangements, oversight, and regulations as well as measurement capabilities to detect and account for special nuclear materials.[11] It was during this period that the AEC established formal domestic safeguards offices and a safeguards R&D program. Eventually, there evolved a comprehensive system of regulations and inspections of the safeguards and security measures at U.S. commercial nuclear facilities. This pattern has been followed in much of the world. There is also a similar, if not more stringent, system of safeguards in place at U.S. government-owned nuclear facilities.

Efforts to Stem Nuclear Proliferation: The 1970s Through the 1980s

Since the dawn of the nuclear age, most strategic thinkers concluded that these weapons were so destructive that their uncontrolled proliferation would create insecurity in the international system and be unacceptable. This view motivated early proposals for international nuclear controls such as the Baruch Plan. In the 1960s and 1970s, these concerns grew in the strategic community. National security strategists like Albert and Roberta Wohlstetter warned of "life in a nuclear-armed crowd."[12] The dangers of increased proliferation included increased chance of nuclear accidents, miscalculation, and regional arms races in addition to the heightened possibility of nuclear use in conflict or the loss of control of nuclear weapons. In 1963 President John F. Kennedy claimed that by 1975, 15 to 20 countries might possess nuclear weapons.[13] Kennedy urged all nations to act to slow the spread of nuclear weapons and sought to curb the arms race with the Soviet Union.

As a consensus regarding the dangers of nuclear proliferation was emerging, the U.N. General Assembly adopted a resolution proposed by Ireland in 1961 that called for the "prevention of the wider dissemination of nuclear weapons." The desire to address the issue of nuclear proliferation continued to evolve in the General Assembly and the Eighteen-Nation Committee on Disarmament (ENDC). By 1965, the General Assembly adopted Resolution 2028, setting out five principles on which a treaty to prevent the proliferation of nuclear weapons should be based:

- The treaty should be void of any loopholes which might permit nuclear or nonnuclear powers to proliferate, directly or indirectly, nuclear weapons in any form.
- The treaty should embody an acceptable balance of mutual responsibilities and obligations of the nuclear and nonnuclear powers.
- The treaty should be a step toward the achievement of general and complete disarmament and, more particularly, nuclear disarmament.

---

[10]W. C. Myre and J. M deMontmollin, "History of Physical Security R&D," *Journal of Nuclear Materials Management*: special issue, "20 Years of Safeguards at Los Alamos National Laboratory," vol. XV, no. 4, July 1987, pp. 61–63.

[11]McDowell, JNMM, p. 35.

[12]Albert Wohlstetter et al., *Moving Toward Life in a Nuclear Armed Crowd? Report to the U.S. Arms Control and Disarmament Agency* (Los Angeles: Pan Heuristics, 1976).

[13]Public Papers of the Presidents of the United States: John F. Kennedy, 1963 (Washington, D.C.: USGPO, 1964), p. 2890. Also see National Planning Association, *1970 Without Arms Control*, Planning Pamphlet 104 (Washington, D.C.: NPA, 1958), p. 42, and National Planning Association, *The Nth Country Problem and Arms Control*, Planning Pamphlet 108 (Washington, D.C.: NPA 1960), p. 27.

- There should be acceptable and workable provisions to ensure the effectiveness of the treaty.
- Nothing in the treaty should adversely affect the right of any group of states to conclude regional treaties in order to ensure the total absence of nuclear weapons in their respective territories.[14]

In early 1968, the ENDC submitted to the General Assembly a draft treaty incorporating these principles; the Assembly adopted a resolution commending the Treaty on the Non-Proliferation of Nuclear Weapons (NPT) and expressing the desire that it be joined by the greatest possible number of states. By the time the NPT was opened for signature in 1970, five states had exploded nuclear weapons: the United States, the Soviet Union, the United Kingdom, France, and China. But several key states, some with emerging nuclear weapons programs, did not sign the NPT at the time, including Israel, China, India, Pakistan, Brazil, and Argentina. In addition, growing commerce in nuclear technology and materials was making nuclear technology more available.

The successful negotiation and initial signing of the NPT marked a major milestone in the evolution of the nonproliferation regime and international nuclear safeguards. With its requirement that all nonnuclear weapon state parties place under IAEA safeguards all their peaceful nuclear activities, the treaty provided further support and challenges to the still embryonic international safeguards system.[15] When the NPT was negotiated, the IAEA safeguards system was conducted according to procedures described in an IAEA document known as Information Circular (INFCIRC)/66.[16] However, a number of states wanted to revisit the Agency safeguards system to be implemented under the NPT. The result of extensive negotiations was a new document, INFCIRC/153, which has become the cornerstone of international safeguards.[17]

---

[14]United Nations Committee on Disarmament, http://disarmament.un.org/wmd/npt/nptbi.html (Feb. 2007).

[15]NPT Article III.1: Each nonnuclear-weapon State Party to the Treaty undertakes to accept safeguards, as set forth in an agreement to be negotiated and concluded with the International Atomic Energy Agency in accordance with the Statute of the International Atomic Energy Agency and the Agency's safeguards system, for the exclusive purpose of verification of the fulfillment of its obligations assumed under this Treaty with a view to preventing diversion of nuclear energy from peaceful uses to nuclear weapons or other nuclear explosive devices. Procedures for the safeguards required by this Article shall be followed with respect to source or special fissionable material whether it is being produced, processed or used in any principal nuclear facility or is outside any such facility. The safeguards required by this Article shall be applied on all source or special fissionable material in all peaceful nuclear activities within the territory of such State, under its jurisdiction, or carried out under its control anywhere.

[16]The Agency's Safeguards System (1965, as Provisionally Extended in 1966 and 1968), as approved by the Board of Governors in 1965 and provisionally extended in 1966 and 1968. INFCIRC/66/rev 2. The development of the system from 1961 onward has been as follows:

- The Agency's Safeguards System (1961) INFCIRC/26
- The 1961 system as extended to cover large reactor facilities: The Agency's Safeguards System (1961, as Extended in 1964) INFCIRC/26 and Add.1
- The revised system: The Agency's Safeguards System INFCIRC/66 (1965)
- The revised system with additional provisions for reprocessing plants: The Agency's Safeguards System (1965 as Provisionally Extended in 1966) INFCIRC/66/Rev.1
- The revised system with further additional provisions for safeguarded nuclear material in conversion plants and fabrication plants: The Agency's Safeguards System (1965, as Provisionally Extended in 1966 and 1968) INFCIRC/66/Rev.2

[17]INFCIRC/153: The Structure and Content of Agreements Between the Agency and States Required in Connection with the Treaty on the Non-Proliferation of Nuclear Weapons.

This document became the basis for all Comprehensive Safeguards Agreements between NPT member states and the IAEA. These agreements have a number of important features worth highlighting. One is the requirement to place under safeguards *all* nuclear materials in peaceful uses in the state, which would later prove to have significance in determining the Agency's authority to search for undeclared nuclear materials and activities. A second feature is the requirement for states to establish so-called State's Systems of Accounting and Control (SSACs) to track domestic inventories of nuclear materials and provide reports to the IAEA. In many countries, these SSACs are also the national authorities regulating nuclear activities, including domestic safeguards and security. A third feature worthy of mention is that the agreement obligates the IAEA to apply safeguards with all states that have such agreements. This requirement has implications for IAEA budgets and the funding of safeguards. Part II of INFCIRC/153 outlines detailed procedures for the application of IAEA safeguards under the agreement.

The international safeguards system evolved considerably during the period from the 1970s to the early 1990s. Among the developments were technologies for the independent detection and assay of nuclear materials by Agency inspectors, nuclear materials containment and surveillance systems, and the development of systematic approaches to safeguards at the nuclear facility types that were being constructed and operated around the world. IAEA safeguards inspectors were becoming expert in their profession, with considerable technical and training support from key member states. In short, the NPT and the IAEA's international safeguards system became one of the primary international mechanisms designed to prevent the spread of nuclear weapons to additional states.

India's nuclear test explosion in 1974 sent shockwaves through the nuclear nonproliferation community. Most of India's civilian nuclear facilities were not under safeguards, and it had built several secret facilities with the help of technology obtained from commercial partners, including Canada and the United States. This event spurred greater interest on controlling nuclear trade, with the emergence of what was to become the Nuclear Suppliers Group, an association of states exporting nuclear technology that would agree to enforce similar rules and require similar commitments to nonproliferation from recipient states. In the U.S., the Nuclear Nonproliferation Act (NNPA) of 1978, which was a direct outgrowth of growing concerns about proliferation, driven in large part by India's nuclear test, placed new restrictions on international nuclear activities.[18]

Domestic safeguards and security were also evolving in the United States and elsewhere. At U.S. nuclear facilities licensed by the NRC and contractor facilities under the Department of Energy (DOE), requirements for increased quality and timeliness of nuclear materials accounting and control, in particular to deal with insider threats, as well as increased security against outsider threats, were promulgated and implemented. States which signed the NPT and brought into force comprehensive safeguards agreements with the IAEA had to develop national systems of accounting and control (SSACs) for their inventories of nuclear materials. The terrorist kidnapping and killing of Israeli athletes at the Munich Olympics in 1972, although having no nuclear dimension whatsoever, heightened concern about terrorism and greatly spurred interest in increasing physical security around nuclear installations of all kinds.[19] The NNPA required the U.S. DOE to conduct training in establishing SSACs and in physical protection for key individuals from developing nuclear states.

Safeguards and security R&D budgets were increasing from the 1970s through the 1980s, with new technologies applied to both domestic and international safeguards challenges. By the

---

[18]Public Law 95-242 (3/10/78) Nuclear Non-Proliferation Act: Declares it U.S. policy to: (1) pursue the establishment of international controls of nuclear equipment, material, and technology; (2) enhance the reliability of the United States as a supplier of nuclear reactors and fuels; (3) encourage ratification of the Treaty on the Non-Proliferation of Nuclear Weapons; and (4) aid other nations in identification and adaptation of appropriate energy production technology.

[19]Myre, JNMM.

end of the 1980s, advanced instruments operating in unattended conditions in nuclear facilities had been demonstrated, containment and surveillance systems were tracking material and people in real time, and the quantity of nuclear material being independently verified by domestic and international inspectors had grown dramatically.[20] Safeguards technology and implementation appeared to be keeping up with the growth of materials and facilities to be safeguarded. That perception, however, was about to change.

Cheaters, Rogue States, and Terrorists: The Early 1990s to 2006

The inspections in Iraq under the authority of U.N. Security Council Resolutions following the end of the Gulf War in 1991 provided a new shock to the nonproliferation regime in general and to IAEA safeguards in particular. The Iraqis had run an extensive clandestine nuclear weapons development program right under the noses of the IAEA inspectors who had been dutifully inspecting declared inventories of nuclear materials but were unaware of Iraq's clandestine activities. Later, the Democratic People's Republic of Korea (DPRK) would also present major challenges to the IAEA safeguards system by denying inspectors the ability to verify their declared nuclear activities and eventually withdrawing from the NPT. The discoveries in Iraq after the 1991 Gulf War and the actions by the DPRK shattered the assumption that the threats to the nuclear nonproliferation regime lay only in states that had refused to sign the NPT. It became clear that some states would try to deceive the international community by joining the treaty and conducting secret efforts to develop military nuclear capabilities.

These events resulted in an acknowledged need for a major strengthening of IAEA safeguards designed to detect states' efforts to conduct undeclared nuclear activities. Efforts to strengthen the international safeguards system focused on increased access to information about a state and its nuclear enterprise, increased access to locations in a state (not just those with declared nuclear materials), and increased access to the U.N. Security Council to follow up on evidence of safeguards violations. Strengthening measures that could be implemented under existing IAEA authorities included the use of environmental sampling to find evidence of undeclared nuclear activities at declared sites, earlier provision of nuclear facility design information to the IAEA, and the use of open source and third-party information in assessing safeguards compliance. Additional strengthening measures such as requiring states to provide additional information on nuclear R&D not involving nuclear materials and providing broader access to declared sites and other locations were deemed to require additional authorities beyond those contained in INFCIRC/153. In 1997, the IAEA Board of Governors approved the so-called Additional Protocol INFCIRC/540, which provides for these additional measures for states that sign and ratify it.[21]

When states bring into force an Additional Protocol (AP), the IAEA has available to it an expanded set of safeguards tools that have the potential to provide greater confidence in both the correctness and completeness of the state's declarations of its nuclear materials and activities. Verifying the correctness of a state's declaration requires confirming that its description of its nuclear activities and quantities of nuclear materials is accurate. Verifying the completeness of the declaration requires developing confidence that the state has faithfully informed the IAEA of all its nuclear activities and is not concealing any efforts to use nuclear technology for military purposes or for purposes unknown.

The IAEA safeguards system in general, and for states under the AP in particular, is evolving to one that looks at the "state as a whole." All relevant information available to the

---

[20]For a case study describing a modern unattended monitoring system, see *Remote and Unattended Monitoring Systems*, by Mark Schanfein.

[21]"Model Protocol Additional to the Agreement(s) Between State(s) and the International Atomic Energy Agency for the Application of Safeguards," INFCIRC/540 (Corrected), 1997.

Agency about a state is examined and evaluated to reach safeguards conclusions. Safeguards inspectors visiting nuclear facilities and conducting complementary access "visits" to additional locations under the AP are of central importance to the effectiveness of the safeguards system. Curious and observant humans inside the facilities and "under the roof" can provide information not available through other means. However, the role of open source information, including satellite imagery and the Internet, has also grown enormously in importance to safeguards and is now a key focus of the IAEA.[22] The three types of assessment—facility inspections, complementary access visits, and open source information analysis—are the major tools of the strengthened IAEA safeguards system.

A part of the IAEA's approach for assessing the completeness of a state's safeguards declaration is based on the concept of a state's nonproliferation bona fides. This approach involves looking at a broad range of information on a state's past behavior with respect to its nuclear declarations, its compliance with international treaties, its nuclear export behavior and the effectiveness of export control systems, enforcement of domestic safeguards, and counterterrorist activities. Although the IAEA cannot verify intent, if a positive assessment can be reached on all or most of these categories it logically increases confidence in a state's willingness to uphold its nonproliferation obligations and the completeness of its safeguards declaration. More important, such positive nonproliferation behavior provides the IAEA with essential information about all aspects of the state's nuclear enterprise. As a consequence, assessing states' nonproliferation bona fides can thus provide important supporting information that, combined with the results of technical inspections at facilities, could help reach safeguards conclusions. Conversely, a negative change in the assessment could raise concern that a greater safeguards effort or increased access may be needed to verify the correctness and completeness of declarations.

When states have had the AP in force for a number of years and the IAEA has been able to use the additional information and access available to it, as called for in the AP, the IAEA can, in principle, draw a positive conclusion about both the correctness and completeness of the declaration. In these situations, the state can come under so-called *integrated safeguards*, in which the IAEA makes use of the optimum combination of measures available to it, with the possibility of reducing some traditional safeguards efforts on some declared materials. States and the IAEA thus can benefit from integrated safeguards conclusions through reduced impacts on a state's nuclear facilities and cost savings for the IAEA on implementing safeguards.

Implementing the new measures in the Additional Protocol, as well integrating INFCIRC/153 and INFCIRC/540 safeguards, remains a work in progress. Although most states with significant nuclear activities have now brought the Additional Protocol into force, there remain a large number of states that have not yet ratified it. The Agency and IAEA member states are trying to remedy this situation and the problem of the universality of comprehensive safeguards agreements as well.

More Wake-Up Calls for Nuclear Security

The terrorist attacks in the United States on September 11, 2001, were a wake-up call to the international nuclear community. As noted previously, concerns about terrorism and its impact on the nuclear industry had been growing since the Munich Olympics in 1972, but the dedication, sophistication, and planning evident in the 9/11 attacks challenged many assumptions regarding the severity of the terrorist threat to nuclear facilities around the world as well as possible terrorist interests in using nuclear weapons or radiological materials. In the immediate aftermath of the September attacks, rapid assessments of nuclear security were conducted on many fronts. It was also recognized that although nuclear security is first a

[22]For a detailed description of both open source analysis and the use of satellite imagery, see Chapter 11, by Arvid Lundy and Rick Wallace, and Chapter 12, by Frank Pabian.

sovereign responsibility, the prospect of nuclear terrorist attacks threatens all states. This led many leading states' to realize that it was in their security interests to provide technical and financial assistance for improving nuclear materials security to states that have difficulty with this task.[23] The IAEA also has a significant program to provide information and training to help states improve nuclear materials security and to detect and respond to incidents of illicit nuclear trafficking.[24]

The late 2003 disclosure of a major clandestine nuclear trade network supplying Libya with nuclear materials, uranium enrichment technology, and nuclear weapon designs provided another wake-up call to the nonproliferation regime and to the international safeguards system. The same network, run by Pakistani nuclear scientist A. Q. Khan, is also suspected of supplying similar technology and information to Iran and North Korea.[25] Iran's admission to the IAEA that it had operated a secret enrichment R&D effort for more than 18 years, the DPRK's withdrawal from the NPT, and the revelations about Libya's weapons program were all factors in stimulating new measures to strengthen the nonproliferation regime and international safeguards. Furthermore, growing concerns about a "nexus" of proliferation and terrorism also began to blur the boundaries between domestic and international safeguards.

In response to these new challenges, in October 2003 the IAEA Director General, Mohamed ElBaradei, called for limiting the processing of weapons-usable material in civilian nuclear programs as well as new production of such materials by restricting these operations to facilities under multilateral control; deploying nuclear energy systems that avoid the use of materials that may be directly applied to making nuclear weapons; and consideration of multinational approaches to the management and disposal of spent fuel and radioactive waste.[26]

In a major nonproliferation policy address delivered in February 2004, President George W. Bush outlined a broad seven-point program to strengthen the nonproliferation regime and to counter nuclear terrorism. The President's proposals called for strengthening the Proliferation Security Initiative, calling on all states to strengthen laws and international controls that govern proliferation, expanding cooperative threat reduction activities, imposing restraints on enrichment and reprocessing coupled with nuclear fuel supply assurances, making the adoption of the AP a condition of nuclear supply to all states, creating a special committee of the IAEA Board of Governors to focus on safeguards and verification, and excluding states under investigation for safeguards violations from participating in IAEA Board decisions.[25]

In April 2004 the U.N. Security Council passed Resolution 1540, which, among other things, declares that all states shall:

- Refrain from providing any form of support to nonstate actors in acquiring or using nuclear, chemical, or biological weapons and their means of delivery.

---

[23]Charles B. Curtis, "Reducing the Nuclear Threat in the 21st Century," *Proceedings of the Symposium on International Safeguards: Verification and Nuclear Material Security*, International Atomic Energy Agency, Vienna, Austria, Oct. 29, 2001, IAEA-SM-367/1/04.

[24]This is the IAEA's Nuclear Security Plan and it is implemented by the Agency's Office of Nuclear Security. See http://www-ns.iaea.org/security/default.htm.

[25]The White House, Remarks by the President on Weapons of Mass Destruction Proliferation, National Defense University, Washington, D.C., Feb. 11, 2004, www.whitehouse.gov/news/releases/2004/02/print/20040211-4.html.

[26]Mohamed ElBaradei, "Towards a Safer World," op-ed piece published in *The Economist*, Oct. 16, 2003, www.iaea.org/NewsCenter/Statements/2003/ebTE20031016.html. The IAEA subsequently convened an international group to examine fuel cycle issues and published "Multilateral Approaches to the Nuclear Fuel Cycle: Expert Group Report Submitted to the Director General of the International Atomic Energy Agency," INFCIRC/640, Feb. 22, 2005.

- Adopt and enforce laws which prohibit any nonstate actor from acquiring or using such WMD and means of delivery.
- Take and enforce effective measures to establish domestic controls to prevent proliferation of WMD, including measures to account for and secure relevant materials, physical protection measures, border controls and law enforcement, and effective export controls.[28]

The first steps in implementing Resolution 1540 have been taken, with most U.N. member states submitting reports to a U.N. "1540 committee" on the status of their efforts to meet the resolution's objectives. States are also identifying areas in which they might need assistance or offering to provide needed assistance to other member states.

In February 2006, the Bush administration announced the Global Nuclear Energy Partnership (GNEP), which envisions major new nuclear energy technology developments closely coupled with nonproliferation measures. Prominent among the nonproliferation features are the concept of a small number of states that possess fuel cycle facilities employing advanced technologies to provide assured nuclear fuel cycle services, including fresh reactor fuel supply and spent fuel take-back services, to a much larger number of states that have foregone sensitive fuel cycle technologies and that use a range of tailored reactors to meet their energy demands.[29]

Although some of the nonproliferation and safeguards challenges of this most recent period are new, such as the discovery of an active international black market in sensitive nuclear technologies and the threat of sophisticated international terrorism, many were anticipated in the Acheson-Lilienthal report 60 years ago. The ElBaradei and Bush proposals to limit the spread of enrichment and reprocessing are simply more modern attempts to address some of the "dangerous" activities described in 1946 within the constraints of today's realities.

It remains uncertain, however, whether the majority of states will see advantages in some of these new proposals for preventing nuclear proliferation. The recommendations by both the IAEA Director General and President Bush are widely seen to impact the basic structure of rights and responsibilities under the NPT. Those rights have been interpreted by some as allowing any state that upholds its NPT obligations to develop the full nuclear fuel cycle, including the right to produce plutonium and highly enriched uranium, the materials that can be used to manufacture nuclear weapons. Any proposals that appear to compromise these rights by restricting the production of nuclear materials to international centers or existing "supplier states" is likely to be opposed by many NPT members, presenting challenges to the treaty. This is the way international efforts to stem Iran's nuclear weapon ambitions are being portrayed by Iran. The IAEA has suspicions regarding Iran's nuclear intentions and has evidence of safeguards violations that it has not yet fully resolved.[30] Iran for its part insists that it is using nuclear energy only for peaceful purposes and will move ahead with the completion of a large-scale uranium enrichment plant.

The Continuing Evolution of IAEA Safeguards

Because the international nuclear safeguards system has its legal foundation in the IAEA Statute and the NPT, it is a truly international approach and has developed widespread support in the international community over the years, even after the problems uncovered in the wake of the Gulf War and subsequently. The IAEA's safeguards system demonstrates to the world that relatively intrusive on-site inspections can be manageable, not only theoretically but by building confidence in on-site inspections through the experience of states with safeguards that are cost-effective, politically acceptable, and technically workable.

IAEA safeguards inspections have also been used to verify compliance with other treaties, such as the nuclear weapons-free zones agreed to in Latin America, the South Pacific,

---

[28]United Nations Security Council S/RES/1540 (2004).

[29]DOE Website for the Global Nuclear Energy Partnership, www.gnep.energy.gov.

[30]See Chapter 12, "Commercial Satellite Imagery: Another Tool in the Nonproliferation Verification and Monitoring Toolkit," by Frank Pabian.

Africa, and Southeast Asia. The Fissile Material Production Cutoff Treaty (FMCT) proposed by the United States does not have verification provisions, but many believe that if an FMCT with verification provisions were agreed, it would be logical to verify it through IAEA safeguards.

The IAEA's experience with respect to protocols and for inspections has been utilized in many areas, including allowing key breakthroughs in certain regions such as South America. In 1990, Argentina and Brazil signed an agreement to create a joint system for accounting and control of nuclear materials in the two countries. The agreement is administered by the Argentinean-Brazilian Agency for Accounting and Control (ABACC). In December 1990, the two countries signed a quadripartite agreement with ABACC and the IAEA for application of safeguards on existing nuclear material in Brazil and Argentina. Regimes in other arenas, including chemical weapons disarmament efforts under the Chemical Weapons Convention (CWC), closely follow the techniques and organizational structures that have proved effective through years of IAEA experience.

There has also been significant innovation and improvement in safeguards procedures and technologies allowing the IAEA to meet new challenges. These improvements to safeguards have been made on a continuous basis, and the IAEA has built up an unparalleled technical base in this area. For example, innovations in nondestructive assay equipment provided inspectors with rapid *in situ* determinations of the concentration, enrichment, isotopics, and masses of nuclear materials that would be expensive, time consuming, and in some cases impractical to obtain by other means. These instruments include neutron coincidence counters for quantitative measurements of unirradiated plutonium in a variety of forms as well as gamma spectroscopy instruments for determining isotopics of plutonium and uranium.[31] Advances in miniaturization of these instruments have provided inspectors with more portable measurement methods that are useful for both routine inspections in declared facilities and for in-field application.

Continuous unattended monitoring of activities in nuclear facilities has improved the efficiency of inspections by reducing the time inspectors spend at facilities. Examples of this technology are video surveillance devices that monitor spent fuel ponds at reactors, core discharge monitors that monitor fuel movements in on-load reactors, and electronic seals that record the time of application. All these devices have been important in providing assurances of material integrity during an inspector's absence by recording surveillance data for periodic review. Further gains in efficiency were provided by automated review stations. In addition, the Agency developed technology for secure remote transmission of these data that would further reduce the need for inspector presence in facilities.[32]

In addition to technology advances, the IAEA has made innovations in procedures that enhance effectiveness and efficiency. Examples include new ways of working with regional safeguards systems such as the New Partnership Agreement with Euratom, where the agency saved significant numbers of inspection days through coordinating activities and sharing equipment and duties with Euratom inspectors; application of randomized inspections to verify the material flows at low-enriched uranium fuel fabrication plants; expanded reporting requirements for states, especially in the area of imports and exports of nuclear technology; and earlier reporting requirements for design information relating to new facilities. The agency also looked to other international agreements with on-site inspections, such as the Chemical Weapons Convention, and sought to incorporate certain inspection features into its own safeguards where appropriate and acceptable.

The new IAEA system that is emerging is more flexible and should be better suited than the old to allocating scarce resources to where they are needed most in countering proliferation risk. To deal with the anticipated growth in nuclear energy use worldwide, it is essential

---

[31]For more detailed discussion of these measurement instruments, see Chapters 3B and 3C, by Mark E. Abhold and Doug Reilly, respectively.

[32]See Chapter 6, "International Atomic Energy Agency Unattended Monitoring Systems," by Mark Schanfein.

that the international safeguards system be both credible and effective. The IAEA, however, faces limits on safeguards inspections inherent to the agreements that authorize them. For example, even the Additional Protocol's complementary access authorities allow only limited access to locations in a state other than those declared as nuclear facilities. This is far short of "anytime, anyplace" inspections that have been called for in some cases. Limits to the effectiveness of safeguards in a given state can also stem from residual cultural issues, gaps in available technology such as wide-area environmental sampling, and cost issues. These limits are exacerbated by the fact that the Agency does not fully use all its authorities, especially the authority to conduct special inspections. And the IAEA has limited technological tools to address such issues as detection of undeclared facilities/activities, especially related to uranium enrichment and bulk-handling facilities.

The Future of Domestic and International Safeguards

Improving the responsiveness of both domestic and international safeguards to identified emerging threats and to future, unanticipated threats remains a critical challenge for global nuclear security. To achieve this goal, the IAEA must constantly seek ways that it can strengthen its management of the inspections process and utilize its authorities with NPT member states. This includes full implementation of the Additional Protocol in all member states. The IAEA and key member states with advanced fuel cycles should continue to make appropriate investments in new safeguards technologies and apply them efficiently. The international community needs to consider new political and legal mechanisms that can make nonproliferation safeguards more efficient and manageable as the global use of nuclear energy expands. By what authority will enrichment and reprocessing capabilities be limited to "supplier" or "fuel cycle" states? Is it possible to establish and enforce international standards for the physical protection of nuclear materials?

If advanced nuclear fuel cycle technologies are deployed as envisioned by the U.S. Global Nuclear Energy Partnership and the nuclear development plans of other states, several specific safeguards challenges either exist now or can be anticipated. For example, safeguards technologies for large, increasingly complex new facilities with high material throughputs will be needed and where current technology cannot meet IAEA detection goals. New techniques will be needed for difficult-to-measure nuclear materials such as those that will result from advanced fuel reprocessing, pyroprocessing, and electrorefining. These new technologies will have to operate reliably in harsh environments with high radiation dose rates and temperatures. They will need to be capable of measurements of both continuous flows of nuclear materials in various forms (solid, liquid, gas) and of nonnuclear process parameters such as temperature, density, and flow rate, which can help confirm safeguards declarations. New technologies and procedures will also be needed to detect possible nuclear materials diversions without physical changes to a plant through process controls, chemistry, and advanced surveillance techniques. The current state of the art for safeguards technologies and some advanced concepts are discussed in subsequent chapters of this volume.

Summary

The international nuclear safeguards system faces institutional, political, and technical challenges in its efforts to ensure that states are meeting their safeguards obligations. To a large extent, there is also a vital feedback loop that the IAEA must maintain with the domestic nuclear safeguards systems. For some less developed states with limited nuclear infrastructures, interaction with the IAEA through SSAC development and other technical assistance is the primary mechanism via which to improve their domestic nuclear security. In a related fashion, the advanced domestic nuclear safeguards capabilities in states with highly developed nuclear fuel cycles are often adopted by the IAEA for use in safeguarding other facilities. Thus the benefits flow both ways, with the IAEA serving a positive integrating function.

The IAEA is a vital institution for global efforts to maintain nuclear security that enjoys the support of the leading nuclear powers. It needs to remain innovative and flexible as the global nuclear energy sector expands and additional states potentially reconsider their commitment to foreswear nuclear weapons.

Further Readings

IAEA/SG/INF/2, IAEA Safeguards Guidelines for States' Systems of Accounting for and Control of Nuclear Materials, International Atomic Energy Agency, Vienna, 1980.

NUREG/BR-0137, Rev. 4, Nuclear Material Safety and Safeguards, USNRC, Office of Public Affairs, Washington, D.C., 20555-0001, Aug. 2001.

NUREG-1065, Rev. 2, Acceptable Standard Format and Content for the Fundamental Nuclear Material Control (FNMC) Plan Required for Low-Enriched Uranium Facilities, Dec. 1995.

D.R. Joy, Division of Fuel Cycle Safety and Safeguards, Office of Nuclear Material Safety and Safeguards, U.S. Nuclear Regulatory Commission, Washington, D.C.

MCA-101DC, Introduction to Materials Control and Accountability, U.S. Department of Energy National Training Center, Safeguards and Security Central Training Academy, April 2002.

DOE O 474.1A, Control and Accountability of Nuclear Materials, U.S. Department of Energy, Office of Security, Washington D.C., 2000 (11-20-00).

DOE M 474.1-1B, Manual for Control and Accountability of Nuclear Materials, U.S. Department of Energy, Office of Security, Washington D.C., 2003 (6-13-03).

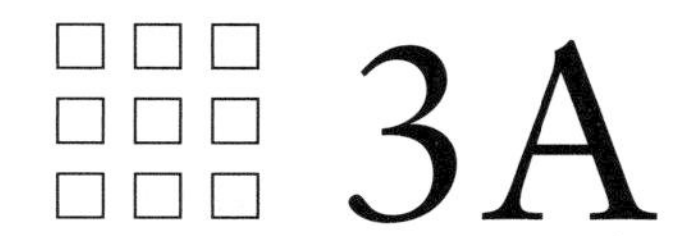

Nuclear Material Measurement Technologies

Douglas Reilly and Norbert Ensslin

Introduction

The accurate measurement of nuclear materials contributes to nuclear safety and security, nuclear safeguards and nonproliferation, and detection of illicit trafficking of nuclear materials and radioactive sources. Nuclear material control and accounting (MC&A) is quite similar to a banking system. They both have physical protection (guards, guns, and surveillance cameras) and various levels of accounting. One large difference involves the input data to the accounting systems; at least in principle, money can be counted exactly, whereas nuclear material often comes in forms that cannot be counted exactly (e.g., powders, liquids, metal pieces, etc.). Nuclear materials must often be measured by techniques that provide uncertain answers. Because measurements are the input to the accounting system, techniques must be developed that provide accurate results. This section covers the principal technologies involved in measurement of nuclear materials. These include both destructive analysis (DA) and nondestructive analysis or assay (NDA) with an emphasis on NDA because it is more often developed for portable, real-time applications. Destructive analysis refers to analytical chemistry and mass spectrometry, which are typically the most accurate measurement techniques available. However, they require a fixed laboratory to receive and analyze nuclear material samples. Sample transport and analysis usually involve a significant time before results are available.

NDA usually measures the entire item rather than a small part of it and provides immediate results. It is typically less accurate than DA. In a few cases, the performance of NDA approaches that of DA and NDA is usually quicker and less expensive. Nondestructive measurements are often used where it is impossible to sample an item for DA, such as waste and scrap and product materials such as nuclear fuel rods and assemblies. NDA techniques are applied for process control, criticality safety, waste and holdup assay, safeguards inspections, and customs inspections. This section will describe gamma-ray (γ-ray) spectroscopy, neutron counting, and calorimetry. It will also cover portable instruments and nuclide identification. Reference 1 presents a good overall discussion of many NDA methods.[1]

Destructive Analysis

Analytical chemistry usually provides the most accurate techniques to analyze pure metals and compounds. Most techniques require the destruction and analysis of a small sample of

[1]D. Reilly, N. Ensslin, and H. Smith, Jr., "Passive Nondestructive Assay of Nuclear Materials," Nuclear Regulatory Commission document NUREG/CR-5550, March 1991; also Los Alamos National Laboratory report LA-UR-90-732 (1990).

homogeneous, bulk material. Inhomogeneous materials cannot yield representative samples and, hence, cannot be analyzed. Many product materials, e.g., a finished fuel assembly, also cannot be sampled. We will briefly describe the measurement procedure and typical precision and accuracy for gravimetry, titrimetry, and mass spectrometry.[2]

Gravimetry, as the name implies, refers to the very accurate weighing of pure uranium or plutonium that has been burned to U_3O_8 or PuO_2. Nonvolatile impurities are determined spectrographically and the results corrected accordingly. The precision of either determination can have a relative standard deviation (RSD) of 0.05%.

There are a variety of reduction-oxidation (redox) titration procedures. In titrations, the U or Pu concentration in solution is determined by the slow addition of a very well-calibrated reagent (titrant) that reacts with the unknown ion until all of it has reacted with the reagent. The concentration of the U or Pu ion can then be calculated from the measured addition of titrant. The end point of the reaction is detected by observing a color change or by various electrical means. Generally, uranium or plutonium is first reduced to the U(IV) or Pu(III) oxidation states with a substance such as zinc amalgam. It is then oxidized to U(VI) or Pu(IV) with potassium dichromate or ceric sulfate. The most common procedure for analyzing uranium is the Davies-Gray method. In this method, U(VI) is reduced to U(IV) with Fe(II) in an H_3PO_4 solution, followed by oxidation of the excess Fe(II) with HNO_3 in the presence of a Mo(VI) catalyst and titration with $K_2Cr_2O_7$ to a colorimetric end point. Fully automatic titration systems for uranium analysis have been in use for almost 30 years; one such system is capable of completing 44 measurements in eight hours. The IAEA developed such an instrument in their Seibersdorf laboratory to measure uranium samples collected during safeguards inspections. In controlled-potential coulometry, U(IV) or Pu(III) is oxidized to U(VI) or Pu(IV) at a platinum electrode with a potential chosen to eliminate interfering electrode reactions. The current is integrated to the oxidation end-point to determine the uranium or plutonium concentration. The precision of these determinations can be 0.02% for uranium and 0.04% for plutonium.

Mass spectrometry is a very highly developed measurement procedure to determine the isotopic composition of uranium and plutonium samples. Commercial thermal ionization mass spectrometers (TIMS) are most commonly used for high-precision determination of uranium and plutonium isotopics. Carefully prepared uranium or plutonium samples are deposited on a special filament that is inserted in the spectrometer. The filament is slowly heated by an electrical current, "boiling off" ions of U or Pu. The ions are accelerated by an electric field and pass through a strong magnetic field at a right angle (orthogonal) to the trajectory of the ions. The trajectory of the ions through the magnetic sector curves with a radius that is a function of the mass of the ions.[3] Ions of differing mass leave the magnetic sector on slightly different trajectories. At some distance beyond the magnetic sector, small collection cups are positioned at locations corresponding to the paths of the ions of interest. The numbers of ions incident on the collection cups are used to determine the isotopic composition of U or Pu. These instruments can analyze ^{235}U enrichments to a precision of 0.014% and ^{239}Pu to 0.02%. Gas mass spectrometers, such as are used at uranium enrichment facilities, can achieve even higher precisions. By using the technique of Isotope Dilution Mass Spectrometry (IDMS), it is also possible to accurately determine total U or Pu concentration in addition to isotopic composition. A well-known ^{233}U or ^{242}Pu spike is added to the dissolved uranium or plutonium sample and this is then deposited on a filament. The ratio of the various U or Pu isotopes to the spike allows the analyst to calculate total U or Pu concentration.

---

[2] "Safeguards Techniques and Equipment — 2003 Edition," *International Nuclear Verification Series No. 1* (Revised), IAEA/NVS/1 (revised), August 2003.

[3] Strictly speaking, mass spectrometers separate ions by their mass (m) to charge (Z) ratio, m/Z. The charge mass ratio is often expressed as a multiple of the charge of an electron, m/z, where $Z = ze$. For a monoenergetic beam passing through an orthogonal uniform magnetic field B, $m/z = eR^2B^2/2V$, where R is the radius of curvature of the ion trajectory, and V is the ion accelerating voltage. Care is taken in the design of the instrument to ensure the beam consists of singly charged ions. Multiply charged ions are not collected and are "lost."

Table 3A.1 Analytical techniques used by SAL and NWAL.

Analytical Technique	Analysis	Material Type	Random Error (%)	Systematic Error (%)
Davies and Gray	U	U, MOX	0.05	0.05
MacDonald & Savage	Pu	Pu materials	0.1	0.1
Controlled potential coulometry	Pu	Pure Pu materials	0.1	0.1
Ignition gravimetry	U, Pu	Oxides	0.05	0.05
K x-ray fluorescence	Pu	Pu materials	0.2	0.2
Isotope dilution mass spectrometry	U, Pu	Pu, MOX, spent fuel	0.1	0.1
Pu(IV) spectrophotometry	Pu	Pu, MOX	0.2	0.2
Alpha spectrometry	Np, Am, Cm	Spent-fuel input	5.0	5.0
Thermal ionization mass spectrometry	U, Pu	Pure U, Pu	0.05	0.05

The International Atomic Energy Agency (IAEA), an affiliate of the United Nations, has the responsibility of verifying compliance with the Treaty on Nonproliferation of Nuclear Weapons (also known as the Non-Proliferation Treaty, or NPT). The IAEA's Department of Safeguards has a staff of approximately 650, about 240 of whom are safeguards inspectors who travel to nuclear facilities in all nations that have signed the NPT, to verify the accounting of nuclear materials. To perform this task they use both destructive and nondestructive analysis. Small nuclear material samples are collected at different points throughout the process and shipped to the IAEA's Safeguards Analytical Laboratory (SAL), which is located in Seibersdorf, about one hour from the headquarters in downtown Vienna, Austria. Some samples are analyzed at SAL and others are sent to laboratories in the Network of Analytical Laboratories (NWAL) in member states that have been certified by the IAEA. Table 3A.1 shows the principal DA techniques used by SAL and NWAL.[2]

Laser-induced breakdown spectroscopy (LIBS) is another technique that may have safeguards applications; the IAEA is now considering its possible uses.[4] A pulsed laser is focused on a small sample of the material to be analyzed. When fired the laser converts a very small quantity of material into a hot plasma and breaks the various chemical bonds. The resulting excited atoms and ions emit light at very precise wavelengths (energies) characteristic of the elements in the sample. The characteristic wavelengths span the near infrared through the visible and into the near ultra-violet (200–980 nm). The emitted light is transmitted through an optical fiber to a grating spectrometer and analyzed for the contained elements. The technique is simple enough to be contained in a very portable instrument that could be carried into the field by an inspector. A schematic drawing of LIBS is given in Figure 3A.1.

Nondestructive Assay (NDA)

Nondestructive assay or analysis (NDA) was developed after most destructive analysis techniques had fully evolved. Quite simply, NDA techniques measure nuclear materials without alteration or direct contact with the item under analysis. Most NDA techniques measure radiation, spontaneous or stimulated, from nuclear material items. Passive NDA techniques measure the radiation that is spontaneously emitted during nuclear decay. Active techniques measure radiation that is stimulated by neutron or γ-ray irradiation. The principal radiations,

[4] R. S. Harmon, F. C. De Lucia, et al., "Laser-induced breakdown spectroscopy (LIBS) — an emerging field-portable sensor technology for real-time, *in-situ* geochemical and environmental analysis," *Geochemistry: Exploration, Environment, Analysis*, vol. 5, 2005, pp. 21–28.

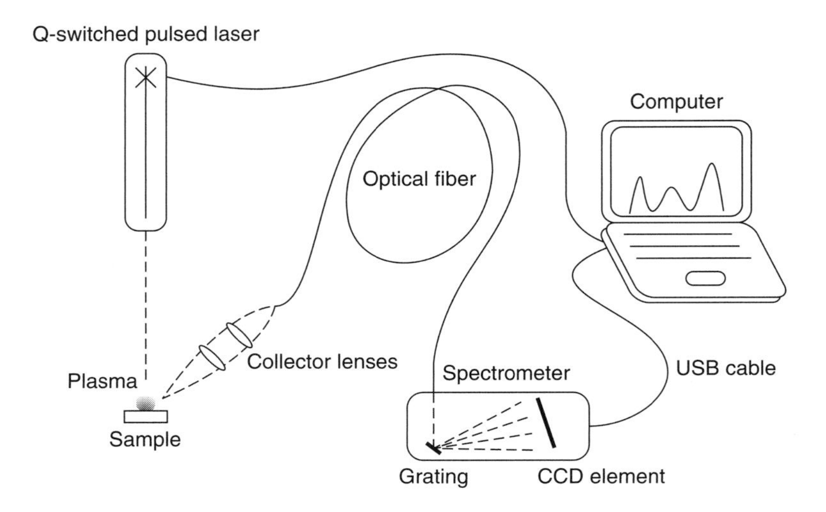

Figure 3A.1 Schematic of LIBS technique for elemental analysis.

spontaneous or stimulated, from nuclear materials are alpha (α) particles, beta (β) particles, γ rays, x rays, and neutrons. The first two, α and β, do not penetrate sufficiently in bulk material to be useful for assay. The other three radiations, all electrically neutral, do penetrate bulk material and all are used in NDA techniques.

The following sections describe the principal NDA techniques of γ-ray spectroscopy, neutron counting, and calorimetry. They describe basic techniques, typical instrumentation, and measurement procedures, but they do not attempt to list all existing NDA instruments.

Gamma-Ray Spectroscopy

Nuclear material nuclides usually decay by emitting α or β particles (Figure 3A.2). Most decays are accompanied by γ-ray and x-ray emission. Gamma rays and x-rays are high-energy photons of energies far above the visible light spectrum. Energy, in this case, is analogous to color in the visible spectrum. Visible light and γ rays are both electromagnetic radiations, as are radio waves, television signals, microwaves, radar, and infrared and ultraviolet light. X-rays are emitted during changes in the energy state of atomic electrons. Gamma rays are emitted when there is a change in the energy state of a nucleus. Thus, x-rays are useful for identifying elements, whereas γ rays are useful for identifying individual nuclides present in the materials. Both radiations are used in NDA techniques. In nuclear materials, x-rays are typically in the energy range 80–120 keV and γ rays are in the range 60–1000 keV. Thorium materials emit γ rays with energies as high as 2600 keV.

Gamma rays have very precise energies and intensities that are unique to each nuclide. Their energies provide a signature for the nuclides present[5] and their intensities, when properly interpreted, provide information regarding mass or concentration. A principal use of γ radiation is to determine the isotopic composition of uranium or plutonium samples. In certain situations, γ rays can be used to determine isotopic mass. Isotopic composition techniques are discussed first, followed by mass measurements.

[5]Consider the following analogy: One is presented with two gas discharge tubes, one hydrogen and one mercury, and asked to identify each. Because of their characteristic visible/ultraviolet emission spectra, one can quickly and correctly identify each tube. U and Pu nuclides are identified analogously by γ-ray spectroscopy.

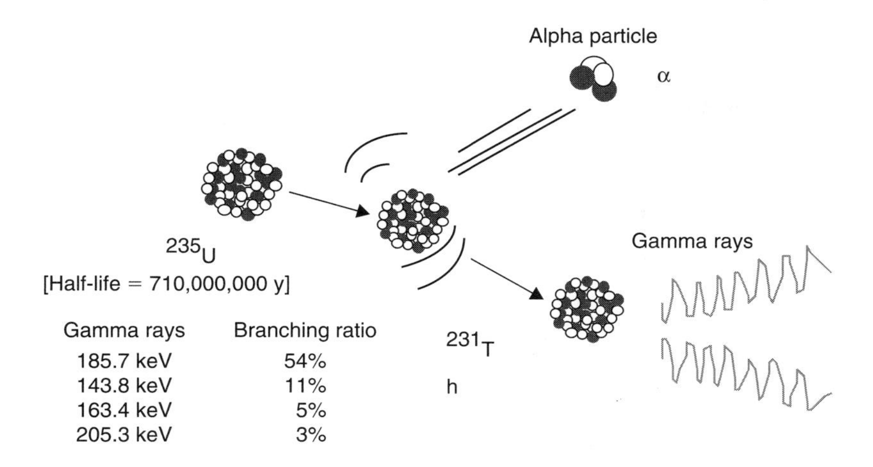

FIGURE 3A.2 Alpha decay of ^{235}U. Four of the most intense γ rays are listed on the left.

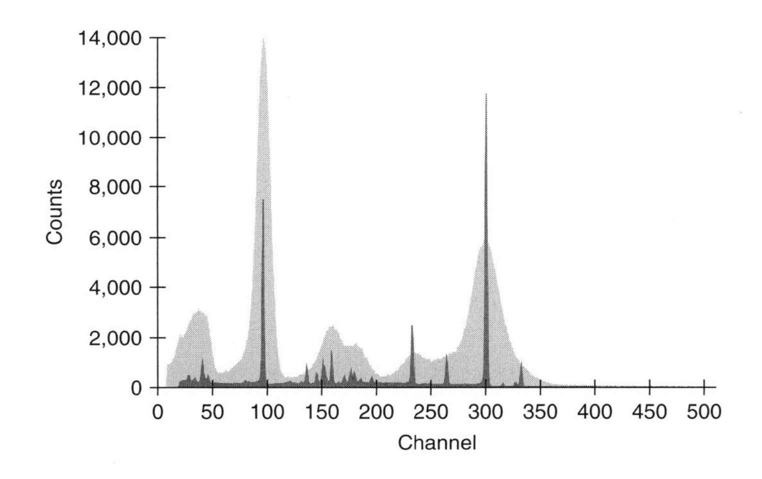

FIGURE 3A.3 γ-ray spectrum from highly enriched uranium and ^{241}Am. The light spectrum is from a NaI detector and the dark spectrum is from a Ge detector. The peak at channel 300 is the 185.7-keV γ ray of ^{235}U and that at channel 100 is the 59.6-keV γ ray from ^{241}Am.

Measurement of Uranium Enrichment

Uranium enrichment (% ^{235}U) can be measured several ways using different γ-ray detectors. The simplest method involves the measurement of the intensity of the 185.7-keV γ ray from ^{235}U (Figure 3A.3). Almost any γ-ray detector can be used for this method: NaI scintillators, CdZnTe semiconductors, or Ge detectors. Because of the high attenuation of this relatively low-energy γ ray by uranium, its intensity is directly proportional to the ^{235}U enrichment of most items. Well-characterized reference standards are required to calibrate a system using this method.

Resolution is a concept used in all types of spectroscopy from visual light to γ rays. It describes how well a given spectrometer can distinguish one color from another or one energy from another. Figure 3A.3 shows a spectrum of enriched uranium as measured by a NaI detector and Ge detector. The spikes, or "peaks," in the spectra correspond to the full energy of γ rays and x-rays that have interacted with the detector. Some features are visible in both spectra, but clearly the Ge detector is able to resolve many more γ rays. Work on a revolutionary

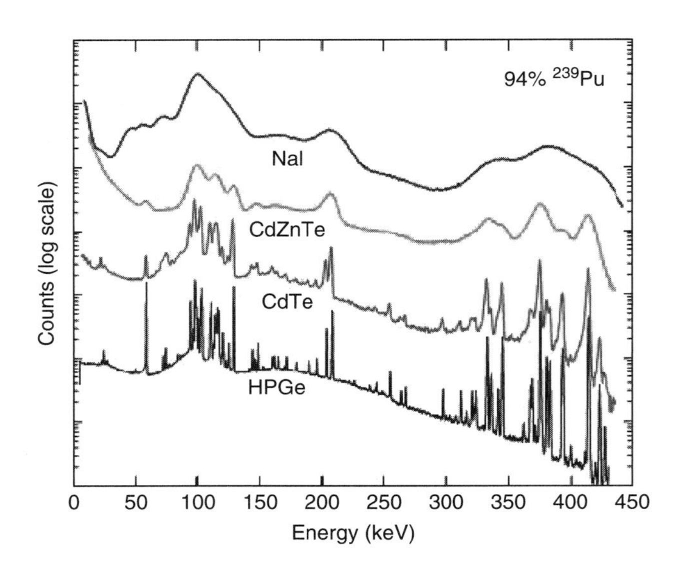

FIGURE 3A.4 Comparison of the γ-ray spectra from a sample containing 94% ^{239}Pu using four different detectors.

new γ-ray detector, based on cryogenic microcalorimeters, was recently published.[6] Although individual sensing element are very small ($\sim 1\,mm^3$) and inefficient, the reported resolution of these devices is an extraordinary 52 eV at 104 keV! (The resolution of a high-quality Ge detector at this same energy is ~500 eV.) Currently, these detectors are sensitive to energies of about 200 keV and below, though work on doubling this threshold to about 400 keV is ongoing. As research progresses on these devices, dramatic new possibilities may arise for γ-ray spectroscopy of nuclear materials. Using existing high-resolution Ge detectors, it is possible to determine complete isotopic compositions (e.g., ^{234}U, ^{235}U, ^{236}U and ^{238}U). Two computer programs that perform this analysis are called FRAM[7] and MGAU.[8] Both of these programs analyze spectral regions that contain γ rays from all of the uranium nuclides. Neither of these programs requires the use of reference standards; all the needed information is obtained from the γ-ray spectrum. A 300-s measurement will determine ^{235}U to better than 2%. In special situations, enrichment can be determined to better than 0.2%, which approaches the accuracy of destructive analysis.

Measurement of Plutonium Isotopic Composition

Measurement of plutonium isotopic composition can only be done using a detector with higher resolution than NaI. Figure 3A.4 shows the plutonium spectrum measured using NaI, CdZnTe, CdTe (Peltier cooled), and Ge. Obviously, the Ge detector has the best resolution and is the preferred detector. However, it must be cooled to liquid nitrogen temperatures (77 K) and this is not always practical. Also the Ge detector is often too heavy for portable

---

[6] J. N. Ullom, et al., "Development of Large Arrays of Microcalorimeters for Precision Gamma-Ray Spectroscopy," The Conference Record of the IEEE Nuclear Science Symposium, Puerto Rico, Oct. 23–29, 2005.

[7] T. E. Sampson, T. A. Kelley, and D. T. Vo, "Application Guide to Gamma-Ray Isotopic Analysis Using the FRAM Software," Los Alamos National Laboratory report LA-14018 (Sept. 2003).

[8] R. Gunnink, MGA: "A Gamma-Ray Spectrum Analysis Code for Determining Plutonium Isotopic Abundances, Volume I, Methods and Algorithms," Lawrence Livermore National Laboratory report UCRL-LR-103220, Vol. I (April 1990).

Table 3A.2 Intrinsic gamma-ray intensities of major plutonium gamma rays.

Region	^{238}Pu		^{239}Pu		^{240}Pu		^{241}Pu-^{237}U(*)		^{241}Am	
(keV)	(keV)	γ/s/g	(keV)	γ/s/g	(keV)	γ/s/g	(keV)	γ/s/g	(keV)	γ/s/g
40–60	43.5	2.5 e8	51.6	6.2 e5	45.2	3.8 e6			59.5	4.5e10
90–105	99.9	4.6 e7	98.8	2.8 e4	104.2	5.9 e5	103.7	3.9 e6	98.9	2.6 e7
									103.0	2.5 e7
120–450	152.7	6.1 e6	129.3	1.4 e5	160.3	3.4 e4	148.6	7.2 e6	125.3	5.2 e6
			375.0	3.6 e4			*208.0	2.0 e7	335.4	6.3 e5
			413.7	3.4 e4			*332.4	1.1 e6		
450–800	766.4	1.4 e5	646.0	3.4 e2	642.5	1.0 e3			662.4	4.6 e5
									722.0	2.5 e5

*The indicated γ rays come from ^{237}U which is in the weak α-decay branch of ^{241}Pu.

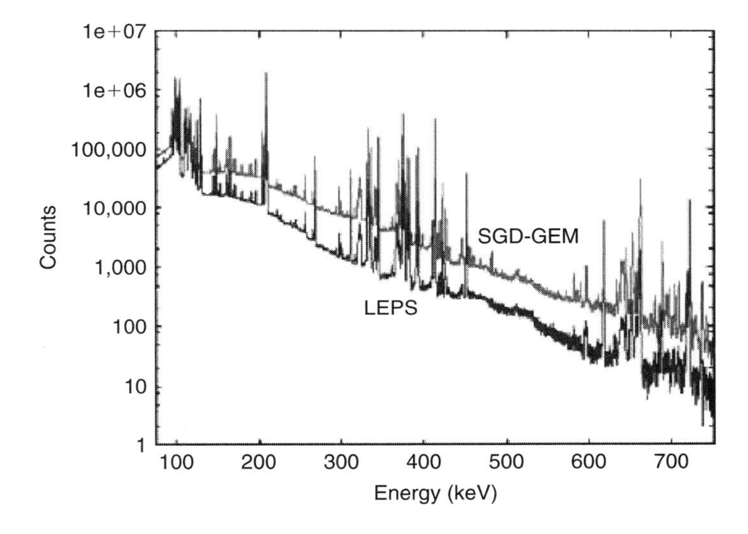

FIGURE 3A.5 Plutonium γ-ray spectrum. The low-energy photon spectrometer (LEPS) is a thin Ge detector optimized for resolution at low energies (<200 keV). SGD-GEM is a proprietary name of ORTEC for a detector that attempts to combine the properties of a LEPS detector and a large coaxial Ge.

applications. The CdTe detector, even with its cooler and power supply, is quite light and portable and has sufficient resolution to determine plutonium isotopic composition.

The principal γ rays used to determine plutonium isotopic composition are listed in Table 3A.2. A picture of the plutonium γ-ray spectrum is shown in Figure 3A.5. FRAM and MGA are the two most common computer programs available to determine plutonium isotopic composition. MGA was the first program developed for plutonium analysis; MGAU is a modification for uranium analysis. FRAM and MGA can analyze both pure plutonium and mixed oxide (MOX) samples. MGA originally analyzed only the γ rays and x-rays in the energy region 94–104 keV that are the most intense in the plutonium spectrum. Recent versions permit analysis up to ~850 keV to handle materials in highly attenuating containers. FRAM permits analysis in the 94–104 keV region and in other regions up to ~850 keV. Both programs use a procedure called *response-function fitting* to analyze the plutonium spectrum. A response function is a mathematical description of the spectrum expected from a pure single isotope. Figure 3A.6 shows measurement equipment that can be used for either FRAM or MGA.

MGA and MGAU analyze the γ rays between 90–105 keV; therefore, these radiations must be able to escape from the measured item. If the plutonium or uranium is in a thick-walled

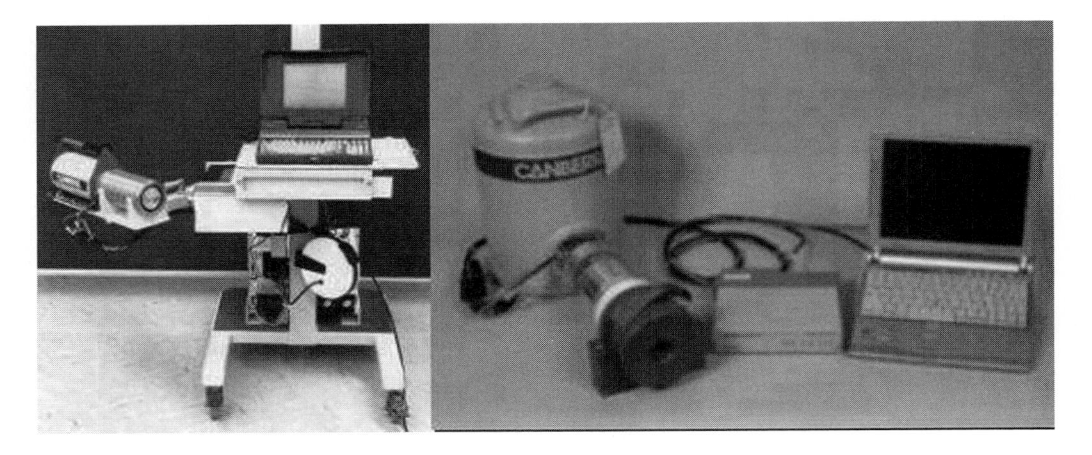

FIGURE 3A.6 Gamma-ray equipment for use with FRAM or MGA. On the left is a mobile system on a commercial thyroid scanner cart. On the right is a Ge spectroscopy system that can run either program.

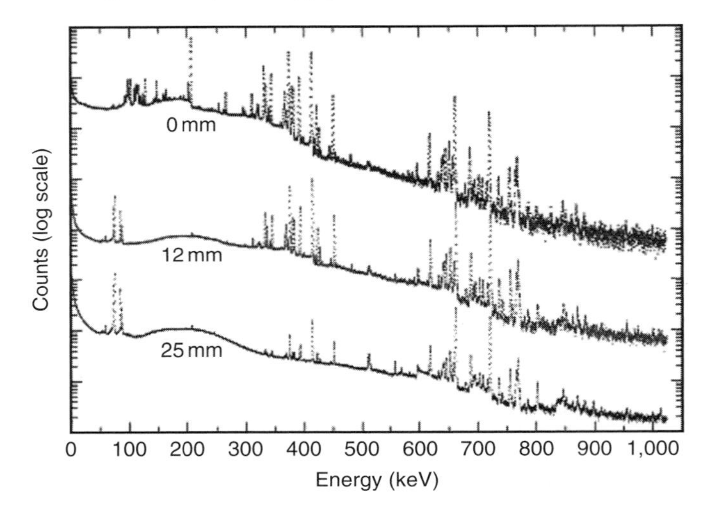

FIGURE 3A.7 Ge γ-ray spectra from a plutonium sample shielded with 0, 12, and 25 mm of lead. The peaks around 75 keV are Pb x rays from the lead shielding in front of the detector.

container or shielded by lead, the γ rays may not be able to penetrate to the detector. The limiting thickness is about 10 mm of steel or 1 mm of lead. Because FRAM and later versions of MGA are able to analyze higher energies, they are able to determine isotopic composition over a wider range of containers. Figure 3A.7 shows the effect of varying lead thickness on a plutonium spectrum.

Measurement of Nuclide Mass

Under certain conditions, it is possible to measure individual uranium and plutonium nuclide masses with γ rays. The principal limitation is that nuclear materials are high-Z and usually very dense; therefore, they readily scatter and absorb their own γ rays. A correction must be made for this absorption (attenuation). The most useful correction method involves measuring the transmission of an external γ-ray source through one or more regions of the measured item. A simple diagram of this procedure is given in Figure 3A.8, which shows the measurement of an item that is assumed to be uniform. The measured item might be a U- or Pu-bearing oxide, e.g., incinerator ash, in a produce can (plutonium is at least doubly contained). The detector

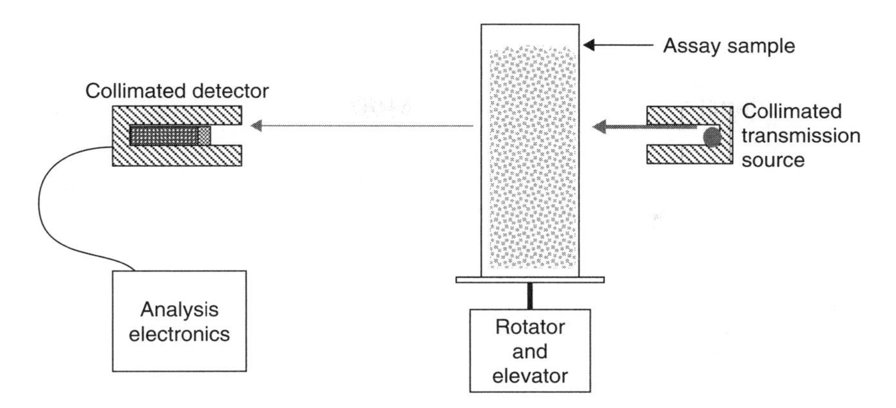

FIGURE 3A.8 Diagram of a simple far-field, transmission-corrected assay of a uniform U or Pu item.

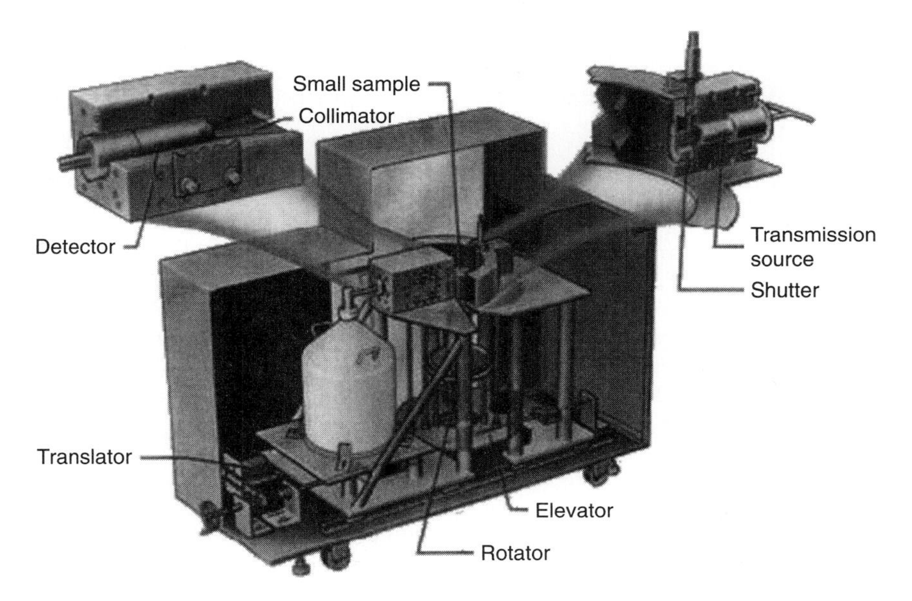

FIGURE 3A.9 Diagram of a segmented gamma scanner for uranium-bearing items up to 20 L.

views the entire sample and measures the 413.7-keV γ ray from ^{239}Pu or the 185.7-keV γ ray from ^{235}U. It also detects γ rays from the transmission source (e.g., ^{75}Se at 401 keV) that pass through the sample without being scattered or absorbed. The transmission measures the effective linear absorption coefficient of the item and can be used in a simple formula to determine its attenuation correction factor. This is called *far-field assay* because the detector-to-sample distance is much greater than any dimension of the detector or the sample.

Many nuclear material items are not uniform, e.g., scrap and waste materials, and several variations of the simple transmission-corrected assay have been developed. These procedures divide the measured item into a number of elements that are measured individually. The initial procedure, known as *segmented γ-ray scanning*, severely collimates the detector along the vertical axis, rotates the sample, and scans it vertically. In effect this treats a cylindrical sample as a series of disk-shaped slices, each of which is measured separately for Pu or U activity and attenuation. Figure 3A.9 shows a diagram of a segmented gamma-ray scanner (SGS) that was built to measure low-density waste for the Fuel Manufacturing Facility at Savannah River. SGSs are built to measure items up to 208 L; several hundred SGSs have been built commercially. In careful measurements of well-known standards, SGSs have

a.

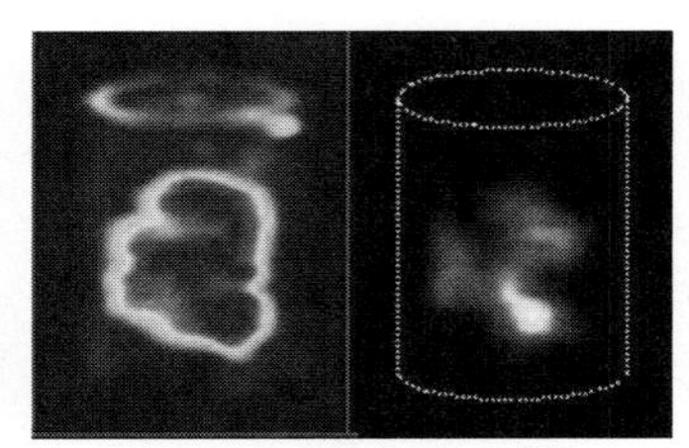

Rashing rings/23.2 g ^{235}U

b.

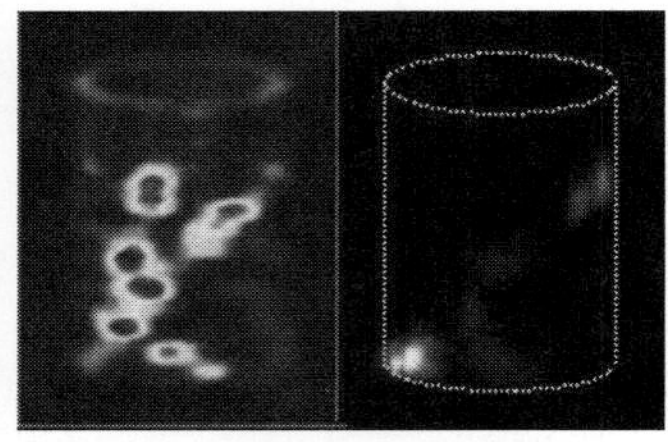

Light metals/30.7 g ^{235}U

c.

FIGURE 3A.10 Tomographic gamma scanner (TGS). Figures on the right show the reconstructed distributions of absorbing and emitting material in two 208-L waste drums.

demonstrated accuracies and precisions better than 1%. When measuring samples up to ~5 L, precisions of 1–5% are common. Measurements of 208-L waste drums can achieve accuracies in the range of 25%.

When modern computing power became available, the idea of the SGS could be carried even further using the principles of tomography, similar to the computerized axial tomography (CAT scan) used in today's hospitals. The tomographic gamma-ray scanner (TGS),[9] shown in Figure 3A.10, was designed so that Pu- or U-bearing samples up to 208 L could be rotated, translated vertically, translated horizontally, and assayed as a large number of small pieces (volume elements, or "voxels"). Separate scans are made to measure the transmission through the item using a ^{75}Se source and the emission of 414-keV γ-ray activity from ^{239}Pu. The TGS can measure scrap, waste, and residue drums or cans with densities higher than those possible with an SGS. The technique can be used to measure uranium also, provided that the material densities allow the 185.7-keV γ rays from ^{235}U to penetrate the container and its contents. The transmission source of choice for ^{235}U assay is ^{169}Yb (177.2 keV and 198.0 keV).

The accuracy and precision of the TGS was tested as part of the Performance Demonstration Program to certify Department of Energy (DOE) facilities and their NDA instruments to measure transuranic waste drums for storage in the Waste Isolation Pilot Project in Carlsbad, New Mexico. The TGS could assay containers of electro-refining salts to ±2.6%, combustibles in drums to ±3%, and heavy drums of sludge to ±18%.

Gamma-Ray Solution Assay

The γ-ray assay methods already discussed can be applied to containers of uranium and/or plutonium solutions to measure ^{235}U or ^{239}Pu abundances. In addition, there are two techniques used to measure total U and/or Pu concentration in solutions: x-ray fluorescence (XRF) and absorption-edge densitometry (or K-edge densitometry, KED). For XRF (see Figure 3A.11), a γ-ray source irradiates the solution sample ionizing K electrons in the U or Pu atoms that then emit K x-rays that are detected in a Ge detector. The excitation source,

---

[9]J. S. Hansen, "Application Guide to Tomographic Gamma Scanning of Uranium and Plutonium," Los Alamos National Laboratory report LA-UR-04-7014 (2004).

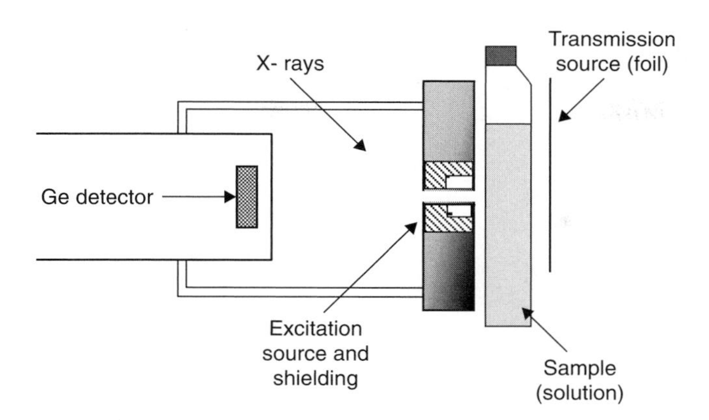

FIGURE 3A.11 Diagram of x-ray fluorescence measurement of U or Pu solution.

usually ^{57}Co (122.06 keV and 136.47 keV), is shielded from the detector, which measures K x-rays emitted in the back direction. The intensity of the x-rays is proportional to the concentration of U or Pu in the sample. The K x-rays of uranium and plutonium are sufficiently different in energy that they can be resolved by the Ge detector, permitting MOX solutions to be assayed. The solutions are contained in accurately fabricated vials, often vials for optical spectroscopy. Solution standards must be fabricated to calibrate the XRF measurement.

Examine Figure 3A.11 again and consider measuring the γ rays that are transmitted through the sample without interaction. The transmitted fraction is related exponentially to the product of the mass attenuation coefficient, the density, and the sample thickness. The mass attenuation coefficient is a smoothly varying function of energy except at the absorption edges of the elements in solution. The right-hand figure of Figure 3A.12 shows an x-ray spectrum from an x-ray generator with no solution absorber (reference spectrum) and with a 197 g/L uranium solution. The K edge of uranium is at 115.61 keV and its effect shows as a step in the lower spectrum. If the x-ray energy falls below the K-edge energy, the K electrons can no longer interact with the x-ray and the transmission rises dramatically. The height of this step is proportional to the uranium concentration in the solution. In addition, the attenuation of the solvent, usually nitric acid, is almost the same on either side of the absorption edge, so its effect is minimal. Absorption-edge densitometry can also be performed with discreet γ-ray sources by choosing two sources above and below the K edge. For example, the plutonium K edge is at 121.82 keV and two possible transmission sources are ^{57}Co (122.1 keV) and ^{75}Se (121.1 keV). Densitometry is also possible at the L absorption edge (17.17 keV for U and 18.05 keV for Pu) using an appropriate x-ray machine. In MOX solutions, both uranium and plutonium concentration can be measured with KED using an x-ray generator.

The left-hand figure in Figure 3A.12 shows a hybrid KED/XRF densitometer[10] that was designed to measure U, Pu, and MOX solution samples, including highly radioactive dissolver solution samples at a reprocessing facility. The demonstrated precision of this system ranges from 0.2% to 1% for samples containing as little as 2 g/L. Similar systems are in routine use in Japan and Europe.

Isotope dilution gamma-ray spectrometry (IDGS) is a method using γ-ray spectrometry to measure plutonium and uranium in solutions, especially dissolver solutions from the accountability tank of a reprocessing plant.[11] IDGS can determine Pu and U concentration

[10] S.-T. Hsue, "KED/KXRF Hybrid Densitometer," Los Alamos National Laboratory application note LALP-96-49 (May 1996).

[11] Duc T. Vo and Tien K. Li, "Generalization of the IDGS Technique," Los Alamos National Laboratory report LA-UR-04-4186, Proceedings of the 45th Annual Meeting of the Institute of Nuclear Materials Management, Orlando, Florida, July 2004.

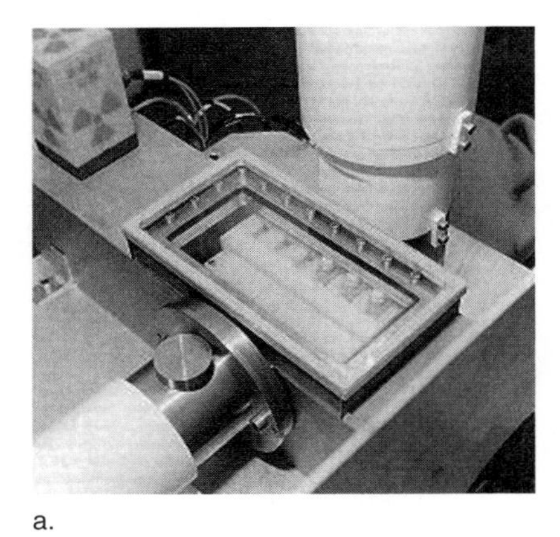

a.

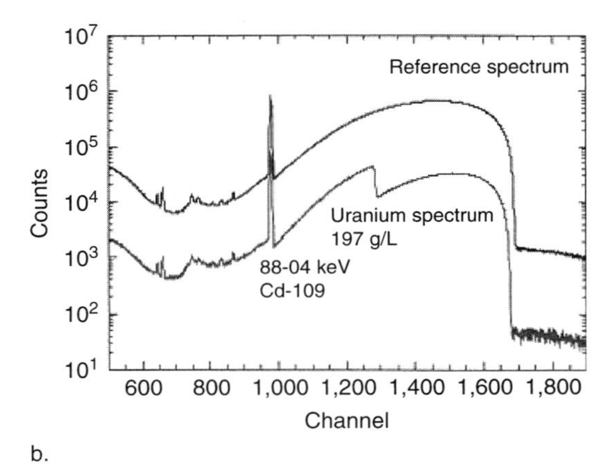

b.

FIGURE 3A.12 a. Hybrid KED/KXRF densitometer with x-ray generator. b. Spectra illustrate K-adsorption-edge densitometry to measure uranium solution.

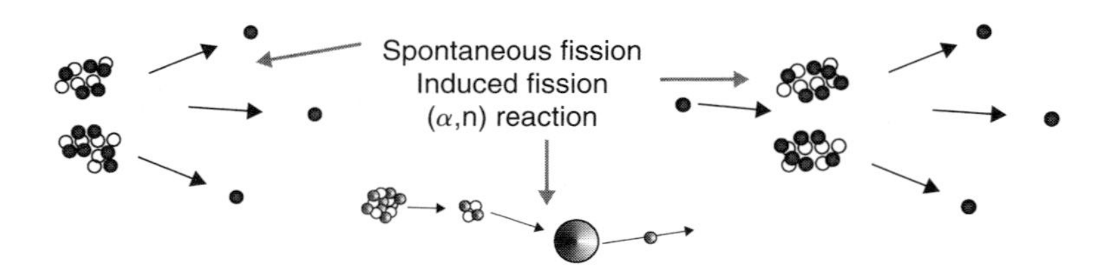

FIGURE 3A.13 Three ways to produce neutrons. Even-numbered, high-mass (A) nuclei can spontaneously split or fission. Odd-numbered, high-mass nuclei can absorb a neutron and fission (induced). Alpha particles from high-mass nuclear decay can interact with low mass nuclei and produce single neutrons.

along with isotopic composition. The unknown solution sample is spiked with plutonium of accurately known mass and isotopic composition and then measured with a high-resolution Ge spectrometer and analyzed with an isotopic composition program such as FRAM. The plutonium concentration in the unknown sample is determined by calculating the differences among the ^{239}Pu weight percent and isotopic ratios of the spike, the spiked solution, and the unknown solution.

Neutron Assay

Neutron Coincidence Counting

Neutrons are electrically neutral like γ rays, but they can penetrate high-Z, high-density materials better than γ rays. Neutrons have a rest mass almost identical to that of protons, whereas γ rays have zero rest mass. Neutrons are used to assay materials that are more dense, or stored in larger containers, than can be assayed with γ rays. Neutrons carry little or no discernable information about their origins, and are simply counted. Although the detectors used for nuclear material assay cannot distinguish one neutron from another, there are several useful signatures based on the neutron intensity or on time correlations between neutrons. Neutrons and γ rays are complementary because the interpretation of neutron measurements always requires information on isotopic composition that can only come from mass or γ-ray spectroscopy. Another useful property of neutrons is that there are few natural background sources, other than solar neutrons and neutrons from cosmic-ray interactions in the atmosphere.

Nuclear materials produce neutrons in three ways (see Figure 3A.13): (α,n) reactions on low-Z elements such as oxygen and fluorine, spontaneous fission, and induced fission.

Table 3A.3 Principal neutron production rates.

Spontaneous fission **Isotope**	**Neutrons/g-s**
^{238}U	0.011
^{238}Pu	2,500
^{240}Pu	1,020
^{242}Pu	1,700
^{244}Cm	11,000,000
(α,n) Neutrons **Material**	**Neutrons/g-s**
^{240}Pu oxide	170
^{240}Pu fluoride	16,000
^{235}U fluoride	580

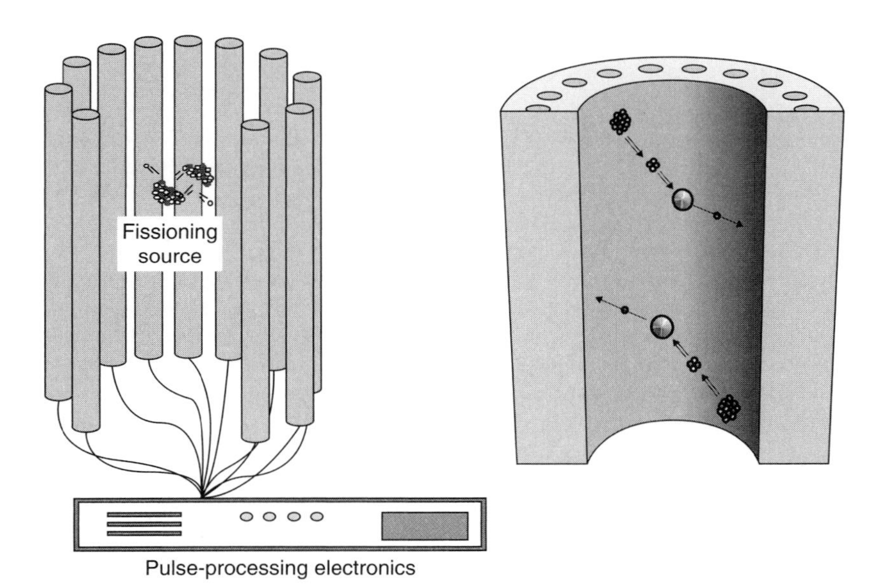

FIGURE 3A.14 Schematic drawing of a passive neutron coincidence counter showing ^{3}He tubes (left) and polyethylene moderator (right).

The even-numbered nuclides of plutonium undergo spontaneous fission and emit two to four neutrons per fission, on average. ^{238}U also decays by spontaneous emission, but with a very low intensity. The odd-numbered nuclides of plutonium and uranium can be induced to fission if they are irradiated with an external neutron source. The principal neutron production rates are shown in Table 3A.3.

While there are some applications where total (singles) neutrons are measured, most nuclear material measurements count coincident neutrons from fission. It is usually difficult to interpret the singles neutron rate, because the (α,n) reaction depends critically on the nature of the target nucleus and the physical coupling with the α-particle source. One interesting exception to this is UF_6, which has a significant neutron signal from the (α,n) reaction on fluorine. This is almost the only case where uranium provides a passive neutron signature.

Most neutron detectors for nuclear material assay use ^{3}He gas proportional counters embedded in polyethylene (Figure 3A.14). The ^{3}He has a high cross-section (probability) for

FIGURE 3A.15 Two neutron coincidence counters. On the left, a high-level neutron coincidence counter (HLNC-II). On the right, a Pu canister assay system (PCAS).

capturing thermal neutrons yielding a proton and a triton that share 765 keV and create an electronic pulse to be counted. The polyethylene serves as a moderator to reduce the high initial neutron energy (1–2 MeV) to thermal energy (0.025 eV).

An extensive effort has been devoted to develop neutron coincidence counters of all shapes and sizes to measure different forms of nuclear material.[1] Figure 3A.15 shows two neutron coincidence counters, one for produce-size cans containing kilogram quantities of PuO_2 and the other for measuring up to 14 kg of PuO_2 shipped in a special "canister" from a reprocessing plant to a MOX fuel fabrication plant. The Plutonium Fuel Production Facility (PFPF) in Tokai-Mura, Japan, is a prototype MOX fabrication plant that accepts plutonium from the Tokai Reprocessing Plant and the French reprocessing plant in Cap de l'Hague to produce PWR fuel assemblies for use in Japanese power plants. PFPF is a completely automated plant and requires automated measurement equipment for its MC&A system. More than 20 instruments have been installed in PFPF for joint use by the IAEA and the plant operator. PFPF is the prototype for a large MOX fabrication plant, JMOX, to be constructed in Rokkasho-Mura, Japan.

Counters such as shown in Figure 3A.15 can measure large samples of PuO_2 to a precision of 0.5–2% in a 300-s count. The average die-away time of a neutron in a coincidence counter is typically 25–50 μs. The neutron is quickly thermalized and undergoes many scatterings before it is captured by the moderator material, is captured by a ^{3}He tube, or escapes from the counter. Because of this process, neutrons that are born simultaneously in a fission reaction are detected at different times. The present coincidence circuit uses a shift register (integrated circuit) that serves as a short-term memory; typically it keeps a record of the neutrons that have been detected in the most recent 64 μs. From this record of neutron events, the shift register determines a net count rate that is proportional to the total fission rate in the item. With an appropriate calibration, this fission rate can be related to the effective ^{240}Pu mass ($^{240}Pu_{eff} = 2.52\ ^{238}Pu + ^{240}Pu + 1.68\ ^{242}Pu$).[12] The present-day shift-register

[12]The effective plutonium-240 mass is a weighted average of the mass of each of the plutonium isotopes. The weighting is equal to the spontaneous fission neutron yield of each isotope relative to that of Pu-240. Since only the even-numbered isotopes have significant spontaneous fission rates, the effective Pu-240 mass is given approximately by $^{240}Pu_{eff} = 2.52\ ^{238}Pu + ^{240}Pu + 1.68\ ^{242}Pu$. These coefficients are only known to about 5% accuracy.

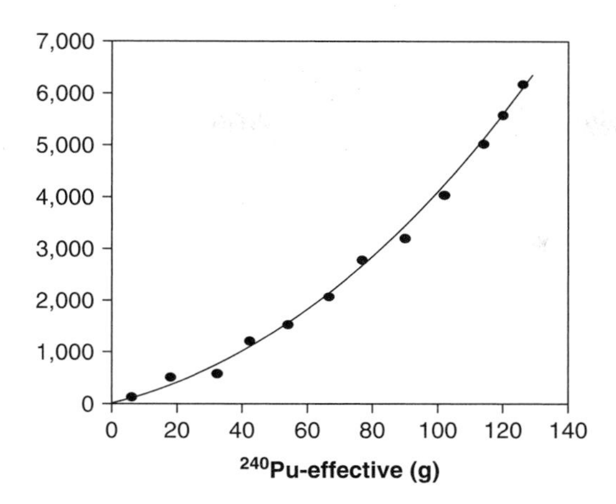

FIGURE 3A.16 Calibration curve for HLNC-II measuring pure PuO_2. (~16.5% ^{240}Pu). The total Pu mass at the upper end of the curve is ~1 kg.

circuit is controlled by a computer that uses the International Neutron Coincidence Code (INCC) program to do all the necessary data analysis and calibration procedures.[13]

All neutron coincidence counters that only measure two-fold coincidences require calibration using well-known reference standards that simulate the measured materials. The production of these standards is often difficult and costly. Figure 3A.16 shows a typical calibration curve for pure plutonium oxide; the measured responses are plotted along with the fit to a quadratic equation. The curve is concave upward because of neutron multiplication in the standards. There are several methods for treating multiplication that are described in more detailed references.[1, 14]

Multiplicity Counting

The neutron coincidence counters we've described require careful calibration with reference standards and can only handle neutron multiplication for relatively pure metal or oxide samples. The fundamental problem is that conventional coincidence counters only provide two measured quantities, singles and doubles. However, most impure plutonium items have at least three major variables that affect neutron assay: mass ($^{240}Pu_{eff}$), multiplication (M), and the ratio of (α,n) neutrons to spontaneous fission neutrons (called α). Therefore, it is usually not possible to obtain accurate assays of impure samples with conventional coincidence counting.

This problem led to the development of passive neutron multiplicity counting as an extension of neutron coincidence counting.[14,15,16] The basic principle of neutron multiplicity counting is the use of a third measured parameter, called *triple coincidences*, so that one can solve for the three unknown sample properties. The availability of a third measured parameter makes it possible to correctly assay many Pu-bearing materials without prior knowledge of the

---

[13]M. S. Krick, W. C. Harker, P. M. Rinard, T. R. Wenz, W. Lewis, P. Pham, and P. De Ridder, "The IAEA Neutron Coincidence Counting (INCC) and the DEMING Least-Square Fitting Programs," Los Alamos National Laboratory report LA-UR-98-2378, Proc. 39th Annual INMM Meeting, July 26–30, 1998, Naples, Florida.

[14]N. Ensslin, W. Harker, M. Krick, D. Langner, M. Pickrell, and J. Stewart, "Application Guide to Neutron Multiplicity Counting," Los Alamos National Laboratory report LA-13422-M, Nov. 1998.

[15]"Standard Test Method for Nondestructive Assay of Plutonium by Passive Neutron Multiplicity Counting," American Society for Testing and Manufacturing, ASTM International, C1500-02, Subcommittee C26.10 on NDA in the Nuclear Fuel Cycle, 2002 (www.astm.org).

[16]N. Ensslin, M. S. Krick, D. G. Langner, M. M. Pickrell, T. D. Reilly, and J. E. Stewart, "Passive Neutron Multiplicity Counting," Los Alamos National Laboratory report LA-UR-07-1402.

sample matrix, including moist or impure plutonium oxide, oxidized metal, and some categories of scrap and waste.

Passive multiplicity counting has applications in a number of different areas: improved materials accountability measurements, verification measurements, confirmatory measurements, and excess weapons materials inspections. Although the historical motivation for developing the technique was improved accountability measurements of impure plutonium in processing facilities, new applications have arisen in the areas of physical inventory verification and shipper/receiver confirmation because the technique does not require prior calibration with a complete set of representative physical standards; instead, the initial calibration can be performed with a known ^{252}Cf source. Measurement precision and accuracy is in the range of 1 to 3% (1σ) even for relatively impure plutonium materials. As a result, neutron multiplicity counters are now used in U.S., European, Japanese, Russian, and other international facilities for NDA of impure Pu metals, oxides, mixed oxides, residues, and wastes. In parallel with the development of passive multiplicity counting, an active multiplicity technique has been developed for uranium that may provide a multiplication correction for assay of bulk items.

The new multiplicity technique requires a new data analysis approach, new electronics, and new software.[13] The distribution of the number of neutrons emitted in spontaneous fission is called the *multiplicity distribution;* it can vary from zero to eight. Multiplicity counting utilizes a new shift register electronics package that sums up separately the number of 0, 1, 2, 3, 4, etc. neutrons within the coincidence resolving time, or *gate width*. This measures the multiplicity distribution of the neutrons that are emitted, detected, and counted. The data analysis is usually not based directly on the observed multiplicity distribution, but on its factorial moments. The first moment is the "singles" or "totals," the second factorial moment is the "double coincidences" or "reals," and the third factorial moment is the "triple coincidences." Neutron multiplicity analysis works with all three of these moments, whereas conventional coincidence counting only uses the singles and doubles.

The advent of multiplicity counting has led to the development of a new generation of thermal neutron multiplicity counters. Like conventional coincidence counters, multiplicity counters utilize polyethylene-moderated ^{3}He proportional counters. However, multiplicity counters are designed to maximize neutron counting efficiency and minimize neutron die-away time. Neutron multiplicity counting requires higher detection efficiency because the efficiency for an nth-order coincidence goes as the singles efficiency (ε) to the nth power (consider a detector with a singles efficiency of 20%; its triples efficiency is only $(20\%)^3 = 0.8\%$). Figure 3A.17 shows an early multiplicity counter, the Pyrochemical Neutron Multiplicity Counter, that has 126 ^{3}He tubes in four concentric rings. The singles efficiency is 55%, and the efficiency to detect a triple coincidence is ~17%. Multiplicity counters are more costly than conventional coincidence counters, in part because of the high number of ^{3}He tubes. Also, the measurement time for good precision on triples, typically 1000–2000 s, is longer than the 100–300 s counting time used for most conventional coincidence assays.

In the design of multiplicity counters, Monte Carlo (MCNP) codes are used to obtain high neutron detection efficiencies that are nearly independent of emitted neutron energy and sample matrix effects. The codes can be used to study design choices such as tube placement; number, size, and gas pressure of tubes; tube bank layout; placement of different neutron moderator or reflector materials; use of cadmium liners; and so on. A recent version of the Monte Carlo code, MCNPX, can directly simulate the singles, doubles, and triples count rates from a known neutron source.[17] This code can be used to simulate detector bias effects, calibration results, and actual item measurements to help improve assay performance.[18] Figure 3A.17 includes a schematic used in the Monte Carlo design of the pyrochemical neutron multiplicity counter.

---

[17] J. S. Hendricks et al., "Monte Carlo Neutron-Photon Extended Code," Los Alamos National Laboratory report LA-UR-04-0570 (2004).

[18] W. H. Geist, M. R. Mahmoud, and O. S. Seo, "IAEA Multiplicity Measurements at the KAMS Facility," Los Alamos National Laboratory report LA-UR-03-4727, Proc. 44th Annual INMM Meeting, Phoenix, AZ, July 13–17, 2003.

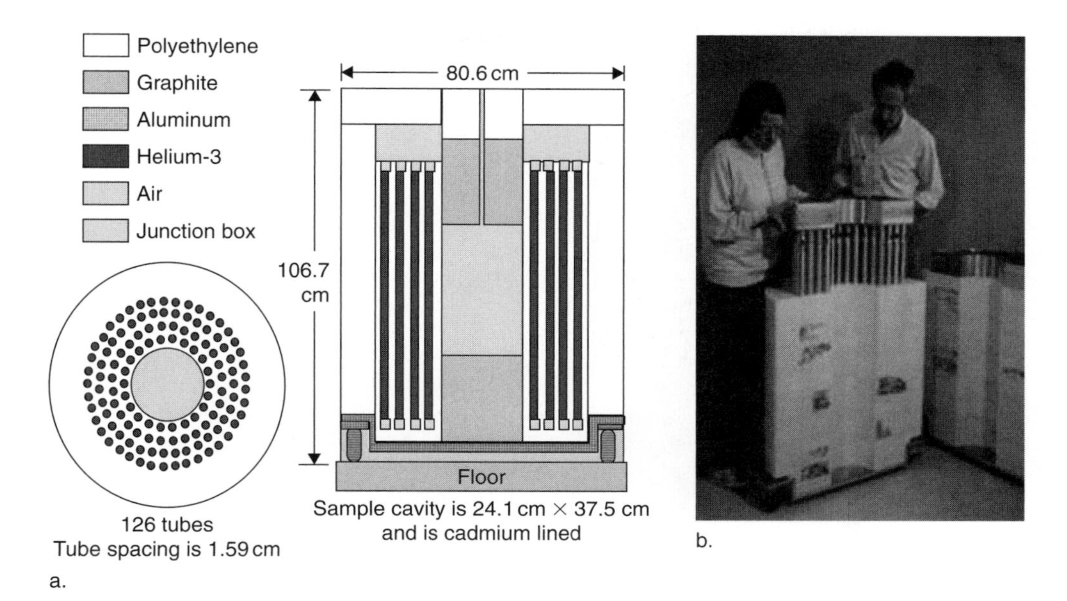

FIGURE 3A.17 Monte Carlo design schematic for the pyrochemical neutron multiplicity counter (left) and construction photo (right). This counter has 126 ^{3}He tubes and a singles detection efficiency of 55%.

The highest efficiency and shortest die-away time counter designed and built to date is the epithermal neutron multiplicity counter (ENMC) shown in Figure 3A.18. This counter uses 121 ^{3}He tubes filled to 10 atm (normal ^{3}He tubes are filled to 4 atm) and has a singles efficiency of 65% and a die-way time of 22 μs. The counter can be equipped with end plugs that contain AmLi sources for very fast active coincidence or active multiplicity measurements. There is an insert that can be added to the ENMC measurement cavity that increases the number of tubes to 142 and the efficiency to 80%. This enables very precise assays of small Pu samples. Figure 3A.18 (right) also shows the results of a performance test on the ENMC using 45 samples of different plutonium materials.[14, 19]

Active Neutron Assay Techniques

Most plutonium materials can be measured with passive coincidence counters such as those illustrated. Uranium, because of the longer half-lives of its principal isotopes, emits very few neutrons through either spontaneous fission or (α,n) reactions. However, when a ^{235}U nucleus captures a neutron, especially a thermal neutron, there is a high probability that it will fission and emit multiple neutrons simultaneously. Therefore, if one places a random neutron source in the cavity of a coincidence counter, source neutrons can cause fissions in the ^{235}U in the cavity and the coincidence counter can statistically separate the fission rate from the random source neutron rate. The random interrogating source creates a high singles background in the counter, but the shift register coincidence circuit is still able to measure the doubles rate from the ^{235}U fissions. The usual interrogation source is AmLi which provides a sufficiently intense neutron flux ($\sim 5 \times 10^{4}$ n/s). In addition, the mean neutron energy from AmLi is 0.5 MeV, which is below the 1 MeV fission threshold of ^{238}U.

Figure 3A.19 shows an active well coincidence counter (AWCC), one of the primary instruments for measuring highly enriched uranium (HEU). It is larger than the HLNC-II and has 42 ^{3}He tubes instead of 18; this creates a higher efficiency than the HLNC-II. It can

[19] J. E. Stewart, H. O. Menlove, D. R. Mayo, W. H Geist, and N. Ensslin, "Epithermal Neutron Multiplicity Counter (ENMC): Current Developments and Applications," Proc. 41st Annual INMM Meeting, New Orleans, July 2000.

a.

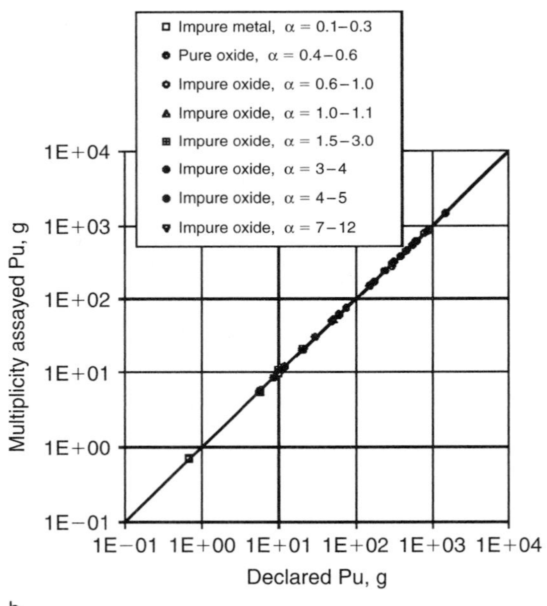

b.

FIGURE 3A.18 The epithermal neutron multiplicity counter (ENMC) is shown on the left in use in a training course for IAEA safeguards inspectors. The figure on the right shows ENMC measurements of a variety of impure plutonium materials.

be configured in several ways depending on the material to be assayed. The AWCC can be tipped on its side, the end plugs removed, and the AmLi sources placed in a special polyethylene cylinder that allows the measurement of materials testing reactor (MTR) fuel assemblies. The AWCC can, of course, also be used as a passive neutron coincidence counter for plutonium assay by removing the AmLi sources.

FIGURE 3A.19 An active well coincidence counter (AWCC) that uses two AmLi sources to induce fission in ^{235}U. It can assay kg quantities of HEU to a precision of 1–5% in 1,000 s. It can also measure LEU oxide or pellets to 5–10% in 1,000 s. The polyethylene cylinder on the left is an insert for measuring MTR fuel assemblies.

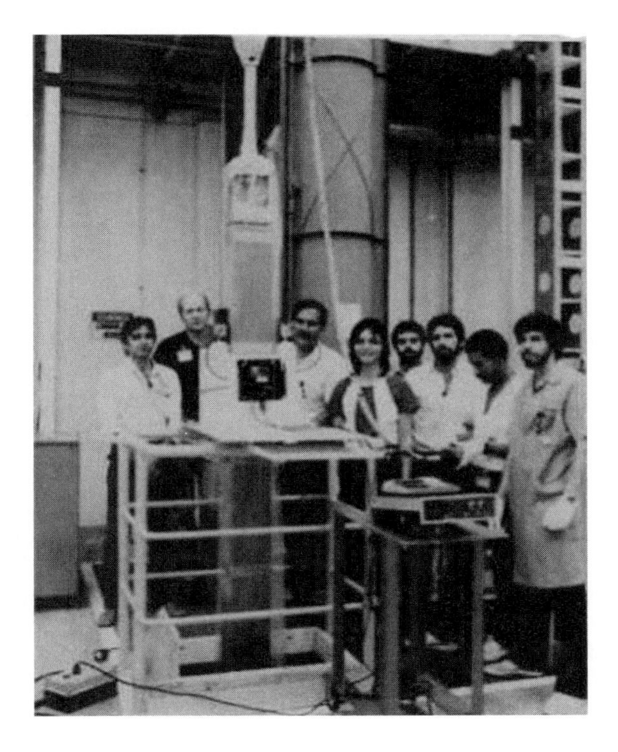

FIGURE 3A.20 The uranium neutron coincidence collar in use at Resende, Brazil, to measure PWR fuel assemblies.

Figure 3A.20 shows another instrument that operates on the same physical principles as the AWCC that measures power reactor fuel assemblies (BWR, PWR, WWER, etc.). It is called the uranium neutron coincidence collar (UNCL). The figure shows the collar measuring a PWR fuel assembly at a fuel fabrication plant in Resende, Brazil. The collar has four polyethylene sides, three of which contain six ^{3}He tubes each, and the fourth contains the AmLi source that irradiates the fuel assembly. The shift register electronics for all these

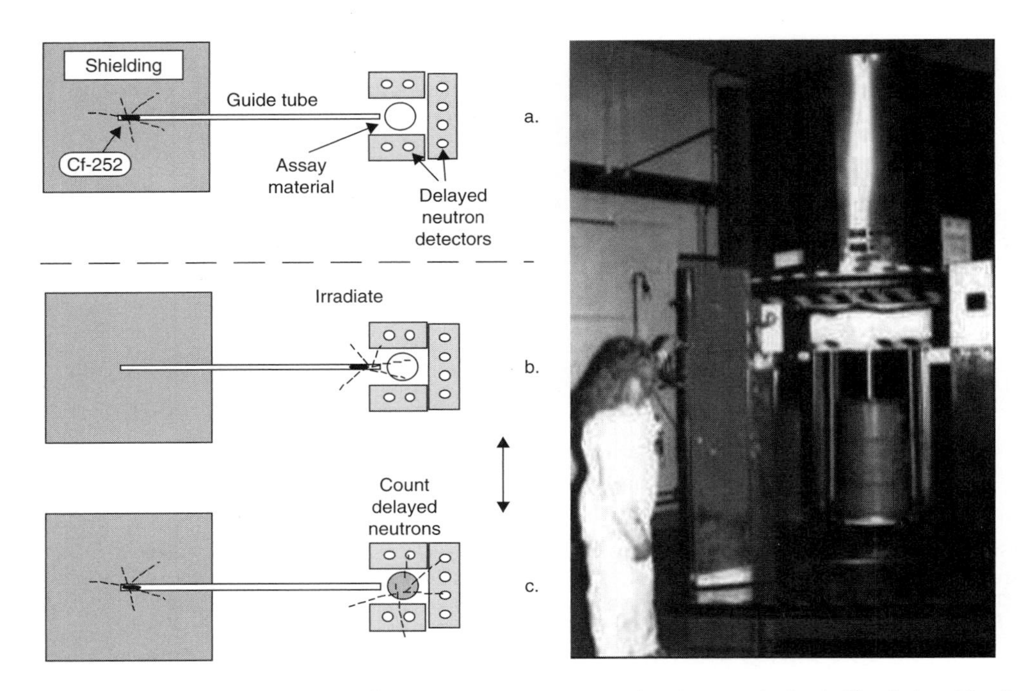

FIGURE 3A.21 Left: schematic of the "shuffler" measurement process. Right: photograph of a shuffler designed for the assay of 208-L drums and cans of U-bearing waste and cans of Pu-bearing pyrochemical salt residues (very high ^{241}Am). The upper, black section shields the source from the neutron detector (lower section) and protects the operator.

counters is the same, as is the control and analysis program, INCC. The collar measurement gives the effective fissile mass per unit length. It can measure fissile material masses (grams) to an accuracy of 2–4% in 1000 s.

Active Delayed-Neutron Assay

When a nucleus fissions, spontaneously or otherwise, there is a prompt emission of neutrons such as are counted in coincidence and multiplicity counters and a delayed emission of neutrons from the decay of certain fission products. These neutrons are delayed by a few seconds to a minute and account for ~1.6% of the total neutrons from ^{235}U fission and ~0.6% from ^{239}Pu fission. They are critically important for the safety and control of thermal reactors. They can also be useful for measuring nuclear materials, especially uranium. A neutron source, generator or isotopic, can induce fissions in uranium; then if the source is shut off or moved away from the uranium, a sensitive detector can count the delayed neutrons from fission. This technique, which is called *delayed-neutron activation analysis* (DNAA), was first studied using Cockcroft-Walton and Van de Graaff accelerators. These accelerators suffered from reliability problems that prompted the development of assay systems that used large ^{252}Cf spontaneous fission neutron sources and rapid source transfer systems to move the neutron source quickly between a large shield and the measurement cavity. A schematic of the "shuffler" process appears on the left side of Figure 3A.21. The ^{252}Cf source (up to 10^{10} n/s) is moved from its shield into the measurement cavity to irradiate the sample. The transfer time is typically ~0.1 s and the irradiation time 1–10 s. The source is then returned to the shield and the delayed neutrons are counted in the neutron detectors (^{3}He and polyethylene) that surround the sample. This process is repeated 30 or more times until the desired counting statistics have been achieved. The rapid motion of the source back and forth and the accompanying sound is the origin of the name "shuffler."[20]

[20] P. M. Rinard, "Application Guide to Shufflers," Los Alamos National Laboratory report LA-13819-MS (Sept. 2001).

Shufflers are usually large and, like other active interrogation systems for 200-L drums, relatively expensive because of the ^{252}Cf source, its drive mechanism, and the required shielding. About 15 are in use throughout the world because they provide the best potential accuracy for cans or drums of uranium-bearing product, scrap, or waste when the uranium particle masses are greater than 1 mg. They have good sensitivity because the delayed neutron signal is directly proportional to the ^{252}Cf strength and the background is typically very low. The calibration of shufflers was originally based on measurements of standards. Today Monte Carlo calculations can create an excellent calibration with only a few standard measurements to benchmark the calculations. For materials with good calibration standards or modeling calculations, shufflers typically provide 1–2% (1σ) accuracy on product or scrap cans and 5 to 50% (1σ) accuracy on waste drums.

The differential die-away technique (DDT) is used by U.S. and European waste generator sites for assay and characterization of transuranic radioactive waste drums or crates prior to disposal.[21] The method uses a pulsed 14-MeV neutron generator to actively interrogate the entire container at a rate of about 100 pulses per second. After each pulse the 14-MeV neutrons scatter, thermalize, and induce fissions in the matrix material. The prompt fission neutrons provide a direct measure of the fissile content of the container and are detected using arrays of bare and cadmium-covered ^{3}He detectors that surround the assay chamber. The term *differential die-away time* was coined because of the large differences between the characteristic lifetimes of the interrogating thermal neutrons and the detected fast neutrons.

DDT systems count neutrons in list mode for roughly 1 to 4 ms after the neutron generator pulse using Pulse Arrival Time recording modules.[21] These modules can also collect delayed neutrons from the induced fissions, starting at roughly 5 ms after the generator pulse. This is part of the same delayed neutron signal used by ^{252}Cf shufflers. In addition to this active interrogation mode, most DDT and shuffler systems also perform passive neutron coincidence measurements with their neutron sources turned off or retracted. Thus both instruments are often called Passive Active Neutron (PAN) systems. Some commercial DDT systems use additional detector packages and collimating materials to provide rough images of the spatial distribution of the fissile material; they are called *imaging PAN systems*, or IPAN.

The high cross-sections for thermal-neutron-induced fissions make DDT systems very sensitive. Detection limits range from a few mg to a few 10s of mg of ^{239}Pu or ^{235}U in a 208-L drum, depending on the matrix, and easily meet the WIPP waste disposal criteria of 100 nCi/g. The strong 14-MeV neutron generators (averaging 10^{8} n/s) also make it possible to perform active assays of remote-handled waste containers despite their high passive neutron background. However, neutron moderation and absorption in matrix materials or self-shielding in lumps of SNM requires the use of matrix correction factors to obtain good measurement accuracy.

The combined thermal-epithermal neutron (CTEN) interrogation system was developed to provide better matrix penetrability and more accurate matrix corrections for waste assays.[22] The CTEN system adds ^{4}He detectors and uses graphite rather than polyethylene chamber walls to detect both thermal and epithermal neutrons. The epithermal neutrons can penetrate further into lumps of fissile material, mitigating the effects of self-shielding, and in some cases can be used to detect self-shielding and provide a correction. Figure 3A.22 shows the crated waste assay monitor (CWAM), which uses CTEN technology to assay large waste boxes.

DDT and CTEN technology is also directly applicable to the detection of potential smuggled SNM hidden in waste containers. A new, highly sensitive solution to this problem is the

[21] N. J. Nicholas, K. L. Coop, and R. J. Estep, "Capability and Limitation Study of the DDT PAN Waste Assay Instrument," Los Alamos National Laboratory report LA-12237-MS (May 1992).

[22] S. G. Melton, R. J. Estep, C. A. Hollas, G. Arnone, G. S. Brunson, and K. Coop, "Development of Advanced Matrix Correction Techniques for Active Interrogation of Waste Drums Using the CTEN Instrument," *Proc. Inst. of Nuclear Materials Management*, Phoenix, AZ, July 20–24, 1997, Los Alamos National Laboratory report LA-UR-97-399.

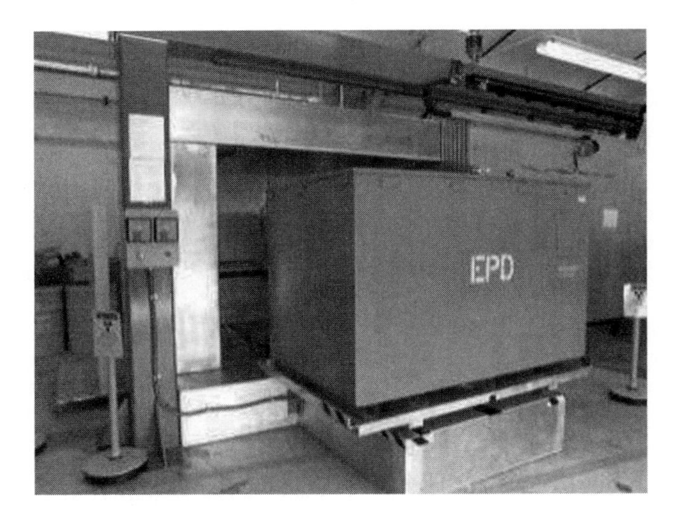

FIGURE 3A.22 The crated waste assay monitor (CWAM) installed at the Y-12 Plant, Oak Ridge, TN.

Table 3A.4 Specific powers of plutonium nuclides and ^{241}Am.

Isotope	mW/g
^{238}Pu	567.0
^{239}Pu	1.9
^{240}Pu	7.1
^{241}Pu	3.4
^{242}Pu	0.1
^{241}Am	114.0
^{237}Np	0.022

Active Interrogation Package Monitor (AIPM).[23] The system is capable of interrogating small packages (~1 m^3) for shielded SNM using both epithermal or delayed neutrons. Based on laboratory tests, the AIPM has a low detection threshold for shielded SNM or, if there is a significant amount of shielding, for detecting the presence of the shielding. The AIPM is being installed at the Y-12 Plant in Oak Ridge, Tennessee, to monitor equipment and supplies leaving a secure nuclear material balance area.

Calorimetry

Plutonium and, to a lesser degree, uranium emit heat from α-particle absorption in the sample. The range of a 5-MeV α particle in condensed matter is less than 10 μm, so the energy from α decay is dissipated in the sample and degraded into heat. Plutonium produces 2–12 W/kg of heat, depending on the isotopic composition. The specific power of the plutonium nuclides is shown in Table 3A.4.

For low burnup plutonium, the principal heat source is ^{239}Pu, but for high burnup the major contributions come from ^{238}Pu and ^{241}Am. Radiometric calorimeters measure the heat

[23]B. D. Rooney, et al., "Active Interrogation Package Monitor," *IEEE Nuclear Science Symposium*, 2, 1027, (1998).

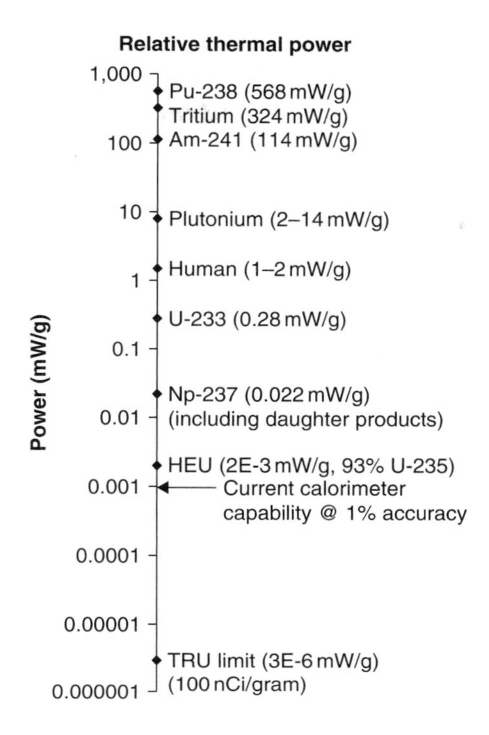

FIGURE 3A.23 The range of thermal power for various nuclear materials and its relation to that of the human body at rest. TRU refers to the burial limit for transuranic waste.

output of nuclear materials, usually plutonium. The new generation of calorimeters is also able to measure kilogram quantities of uranium or neptunium.

Calorimetry was developed many years before atomic and nuclear radiations were known. The calorimetry of nuclear materials began to develop in the middle of the 20th century. Figure 3A.23 shows the range of specific power from nuclear material and its relation to the power of a human body at rest.

Calorimetry is the most accurate and precise nondestructive plutonium measurement technique because the heat measurement is not subject to the matrix problems that affect γ-ray and neutron measurements. Within the U.S. DOE complex, calorimeters are the basis for accountability measurements of most pure and impure Pu metal, oxide, scrap, and residues because of their high accuracy. They are also used to measure ^{238}Pu (e.g., thermo-electric generators for satellites), ^{210}Po, and tritium. The calorimeter's measurement of the item's heat output must be combined with an isotopic analysis, either γ ray or mass spectrometric, to obtain the plutonium mass. Calorimetry is a time-consuming measurement, typically four to eight hours or longer, because of the time required for the measured item and the calorimeter to reach thermal equilibrium. Calibration is usually based on traceable nuclear material standards, although some facilities use electrical current and resistance standards. A calorimetry laboratory may also have secondary heat standards made from ^{238}Pu. Calorimeters are rarely portable, so are not often used by international inspection agencies.[24]

All calorimeters have four common elements: 1) a sample chamber, 2) a well-defined thermal resistance, 3) a temperature sensor, and 4) a constant temperature environment. The

---

[24]D. S. Bracken, R. S. Biddle, L. A. Carrillo, P. A. Hypes, C. R. Rudy, C. M. Schneider, and M. K. Smith, "Application Guide to Safeguards Calorimetry," Los Alamos National Laboratory manual LA-13867-M (January 2002).

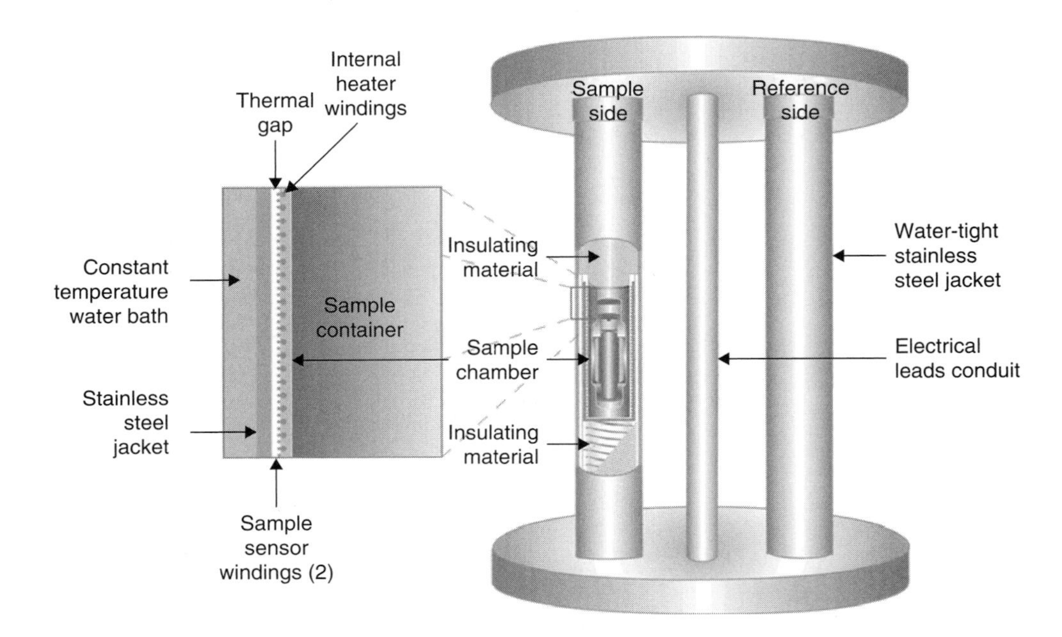

FIGURE 3A.24 Schematic of a twin-bridge, heat-flow calorimeter. The entire calorimeter is immersed in a constant-temperature water bath.

interrelationship of these four components and their hardware determines the type of calorimeter. The most appropriate design for radioactive material is an isothermal, jacketed, heat-flow calorimeter. A heat-flow calorimeter has a sample chamber insulated from a constant temperature environment by a thermal resistance and a means to measure the temperature difference across the thermal gradient produced by the thermal resistance, and thus the heat generated in the sample chamber. When an item is placed in the calorimeter, the temperature gradient changes with time until equilibrium is achieved. The magnitude of the temperature shift determines the thermal power of the item. The curve describing the approach of the temperature difference to equilibrium is a function of several exponentials with different time constants. The time constants are related to the specific heats and thermal conductivities of the matrix material, packaging, and, in some instances, the calorimeter. The type and placement of the temperature sensors, the heat-flow path, and the type of heat sink define the various kinds of calorimeters used for measuring radionuclides. The simplicity of a calorimeter measurement would allow a user to manually collect and analyze data from a digital multimeter. In practice, it is better to have a data acquisition system display results and measurement diagnostics. Figure 3A.24 shows a twin-bridge, heat-flow calorimeter with its basic parts labeled. This entire instrument is immersed in a precisely controlled, constant-temperature water bath.

About 50 heat-flow calorimeters, currently in use in the DOE complex, are based on nickel-wire temperature sensors connected to a Wheatstone bridge. A precision water bath is used to provide a constant temperature heatsink. The measurement chamber has a can that holds the measured item and provides good thermal contact with the chamber wall while preventing any contamination of the inside of the calorimeter. Both the measurement and reference chambers are wound with Manganin wire to provide the internal heater. Two lengths of nickel wire are wound concentrically about the heater windings and serve as two arms of the Wheatstone bridge; the same is done to the reference chamber. The thermal resistance between the sample sensor windings and the water bath is identical to that of the reference sensor and the water bath. The sensitivity of the calorimeter is directly proportional to the thermal resistance of the thermal gap. The insulating material at the top and bottom of the measurement cell is used to force all of the heat radially through the sensing element. The reference bath is controlled to better than $\pm 0.001°C$, which is critical to high-precision, low-power measurements.

FIGURE 3A.25 A twin-bridge, heat-flow calorimeter. The large block is the water bath. The calorimeter can and the insulating plug rest on top of the water bath.

FIGURE 3A.26 Picture of a high-precision, solid-state calorimeter in a water bath. The laptop is used for data acquisition.

Twin-bridge calorimeters usually use large (550–1000 L) water baths. Figure 3A.25 shows a typical twin-bridge, heat-flow calorimeter. In recent years some U.S. and international facilities have also installed air-bath calorimeters, which provide a faster approach to equilibrium but at somewhat reduced measurement precision.

Solid-state calorimeters, developed relatively recently, use thermopiles as heat-flow sensors. A thermopile has numerous thermocouple pairs connected in series. Thermocouples are formed by joining the ends of two dissimilar conductors. A temperature difference between two thermocouple junctions produces a voltage that is proportional to the temperature difference. Figure 3A.26 shows a small-sample, solid-state calorimeter that uses thermopile heat-flow sensors. It was fabricated from commercially available components and makes high-precision measurements of small samples that are comparable to those of much larger calorimeters. With a source power of ~10 mW, equivalent to ~4 g of low-burnup plutonium, the relative standard deviation of six measurements using the solid-state calorimeter system

FIGURE 3A.27 Photograph of the LVC with the calorimeter in the up position to enable loading the 208-L drum.

is 0.11%. The extremely low noise of the heat-flow sensor has a standard deviation of 0.1 to 0.2 μV, allowing for high-precision measurements of items with powers in the sub-milliwatt range. The sensor response to heat is linear. The advantages of thermopile heat-flow sensors compared to Wheatstone bridge sensors include lower cost, wide commercial availability, scalability to any size or shape, insensitivity to mechanical strains, intrinsically low noise, stable baseline (zero power output), increased portability, increased robustness, and no sensor self-heating.

Figure 3A.27 shows another solid-state calorimeter (large-volume calorimeter, or LVC) that is designed to measure 208-L drums. Measurement times are in the range of 12–24 hours because of the time required for the drum to come to equilibrium. The LVC can fill an important gap by providing a measurement capability for drums that cannot be assayed correctly by γ-ray or neutron techniques. The LVC can also be used to provide secondary working standards for those techniques. The drums are placed on a circular insulating plug of extruded polystyrene to prevent heat leakage out the bottom of the calorimeter. The LVC uses two conductive temperature zones heated by silicone rubber-encapsulated wire surface heaters to provide a constant reference temperature to the cold side of the thermopile heat-flow sensors. Temperature control is achieved via servo-controlled feedback loops for each heater. The LVC does not use any water or other neutron moderating or reflecting materials for temperature control. This provides a smaller footprint for facility installation, at the expense of somewhat lower measurement precision.

Calorimetry Precision and Bias

The DOE Calorimetry Exchange (CALEX) Program distributed identical PuO_2 items containing 400 g of plutonium with 5.86% ^{240}Pu by weight to all DOE plutonium facilities. The program tabulates the results from several of these facilities yearly. Calorimeter biases for 23 calorimeters at five DOE facilities are presented in Figure 3A.28. The dashed vertical lines separate the data submitted by each laboratory. These data were collected over a 15-month period starting in October 1993. All measurements have a bias of less than ±0.8%. The average bias is 1.0004 with a standard deviation of the average of ±0.0002. The error expected on a single measurement would be 0.3% one relative standard deviation (RSD) for

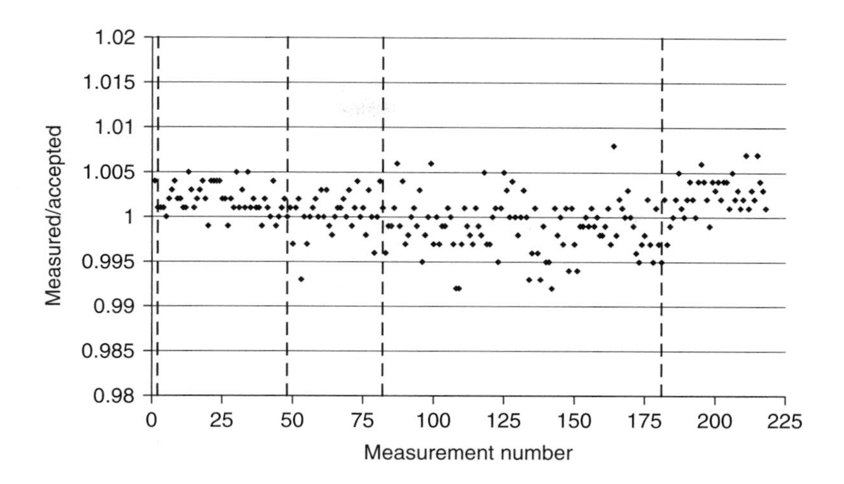

FIGURE 3A.28 Calorimeter measurements of CALEX standards taken over a 15-month period by five DOE laboratories using 23 different calorimeters.

power measurements, and 0.5% to 1% (RSD) for combined power/γ ray isotopics measurements of total Pu mass.

Handheld Gamma-Ray Instruments

With the recent, increased concern over illegal trafficking of radioactive and nuclear materials, radiation dispersal devices, and nuclear terrorism, there has come an increased demand for portable, handheld γ-ray spectrometers for use at border crossings, airports, and seaports to identify and interdict such dangerous materials. Gamma-ray spectrometers are preferred because they are able to identify radionuclides of concern. Neutron counters may also be very useful because of the low neutron rate in the natural background and because they are harder to fool. The very presence of a strong neutron signal may indicate the presence of plutonium. Six years ago there were few portable γ-ray spectrometers, other than health physics dosimeters. Now that a demand exists, there are at least 10 different commercial instruments that use NaI, CdZnTe, and even Ge detectors. Many contain software that can analyze the measured spectra and identify the radioactive nuclides present; this is necessary because the users are largely unfamiliar with γ-ray spectroscopy and spectra. These identification programs are not foolproof, but some are surprisingly effective.

FieldSPEC (ICx Radiation, Formerly Manufactured by Target Systemelectronic gmbh)

The FieldSPEC was developed under the German support program to the IAEA for nuclear safeguards and security purposes. It uses a NaI detector and a small Geiger-Mueller counter that provides the dosimeter response at high dose rates. The NaI detector has a very small ^{137}Cs source that provides a γ-ray peak at 661.6 keV for gain stabilization; this means that the instrument has one fixed energy calibration. The FieldSPEC, shown in Figure 3A.29, is relatively simple to operate, having only three menu-driven push buttons. Its many operating functions include Dosimeter, Source Search, Nuclide Identification, U/Pu Attribute Test, Fuel Assembly/Rod Length Measurement, ^{235}U Enrichment Assay, and full Multichannel Analyzer (1024 channels) capability. The FieldSPEC is a very rugged instrument and weighs only 900g, including four AA rechargeable batteries. The nuclide library contains spectra from almost 80 different nuclides. The FieldSPEC is used extensively by the IAEA (HM-5 is the IAEA name) and various national border control and customs personnel. The FieldSPEC is also available with a CdZnTe detector that provides better energy resolution, albeit with much lower efficiency and sensitivity. The FieldSPEC is also sold under the name identiFINDER.

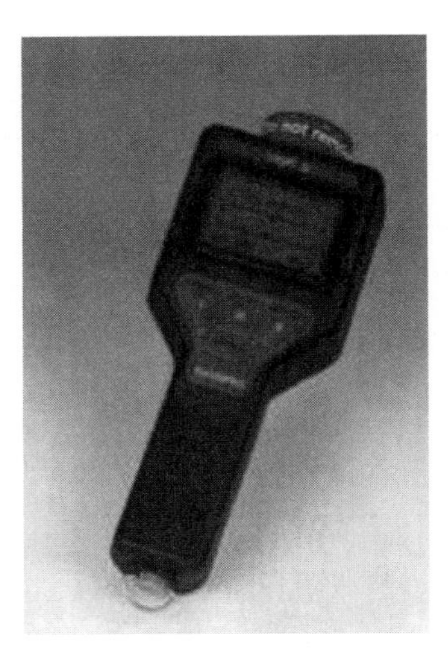

FIGURE 3A.29 The FieldSPEC is a handheld NaI γ-ray spectrometer with digital signal processing, gain stabilization, 8+ hour battery life, and full multichannel analyzer (MCA) capability.

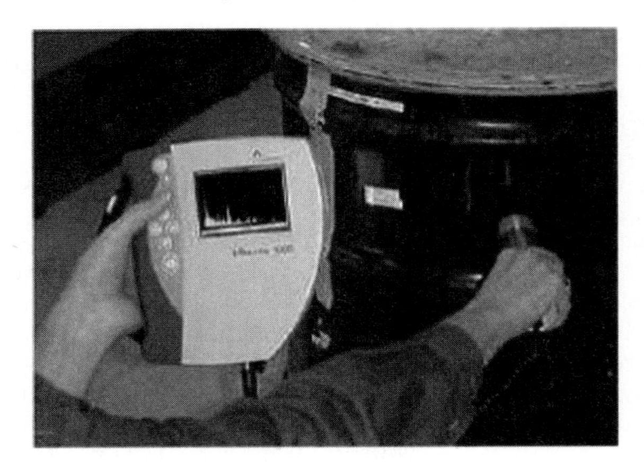

FIGURE 3A.30 The Canberra InSpector 1000 measuring a 208-L drum. The NaI detector fits in the right side of the electronics package.

Canberra InSpector 1000

The InSpector 1000 is similar to the FieldSPEC; it uses a NaI detector with a Geiger-Mueller tube, uses an external ^{137}Cs source for energy calibration, is a full multichannel analyzer (4096 channels), and has some of the same operating functions: Source Finder, Nuclide Identifier, and Spectrometer. Figure 3A.30 shows the InSpector 1000 being used to measure a 208-L drum. The instrument with detector and batteries weighs 1800g. When fully charged, the batteries will run the InSpector 1000 for about 12 hours. The instrument can be operated by a relatively inexperienced person, yet it provides complete spectroscopy functions for a more experienced operator. The instrument has applications in homeland security, customs and border control, health physics, treaty and nonproliferation compliance, monitoring nuclear transportation, and environmental screening.

ORTEC Detective

The Detective has a Ge detector with a Stirling-cycle cooler and takes full advantage of the vastly better resolution of germanium. Unfortunately, the Detective weighs almost 12kg. However, the improved resolution makes the Detective much better able to identify the

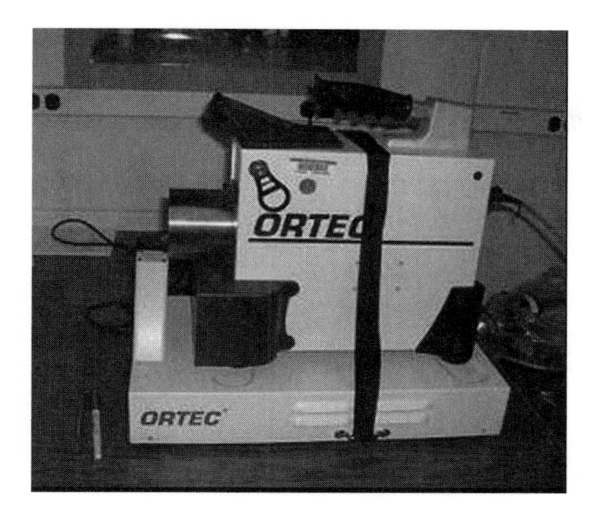

FIGURE 3A.31 The ORTEC Detective is shown mounted on a docking station used for charging, calibration, and cool-down. A small ^{137}Cs source is mounted inside the docking station.

nuclides present. It also provides an enhanced ability to identify mixtures of nuclides, identify nuclides in the presence of high background radiations, and identify nuclides through thick absorbers. Figure 3A.31 shows the Detective in its docking station where it is placed when not in use to charge its batteries.

Another advantage of the Ge detector is its inherent stability. A small ^{137}Cs source is mounted in the docking station to enable the user to occasionally verify the energy calibration. The Detective also has a dosimeter and source search routine like the previous two instruments. Less than 12 hours are required to cool the detector, and the Detective can operate for about three hours on a single charge. The Detective can distinguish between natural uranium, LEU, and HEU. It can also distinguish between reactor-grade and weapons-grade plutonium.

LANL GN5 (Prototype, Portable Ge-Based Spectroscopy System)

Figure 3A.32 shows a prototype Ge-based spectrometer that uses a cylindrical Ge crystal surrounded by an active annular shield of bismuth germanate (BGO) scintillator. The GN5 also includes a small ^{3}He neutron detector. The packaging for this electrically cooled detector is compact, and its battery life exceeds 10 hours. Pulses from the Ge detector are processed in anticoincidence with pulses from the BGO to produce spectra with a suppressed Compton continuum, increasing the sensitivity at low energies. The weight of the new prototype is about 8.6 kg. Target applications include highly portable, low-background, wide-energy-range, γ-ray isotopics for low- to high-burnup plutonium and low- to high-enriched uranium.

General Comments on Nuclide Identifiers

Recent experience shows that ~90% of the false alarms at airports are caused by medical isotopes, e.g., ^{99m}Tc, ^{67}Ga, ^{131}I, etc., in the bodies of travelers. For cargo containers and trucks, the major problem is natural isotopes, e.g., ^{40}K, ^{232}Th, potassium nitrate fertilizers, granite or marble, lantern mantles, and camera lenses. A study of seven nuclide identifiers that used NaI, CsI, and CdZnTe showed that, lumped together, the number of correct identifications, the number of false positives, and the number of false negatives were approximately equal.[25]

---

[25] J. M. Blackadar, S. E. Garner, J. A. Bounds, W. H. Casson, and D. J. Mercer, "Evaluation of Commercial Detectors," *Proc. Inst. of Nuclear Materials Management*, Phoenix, AZ, July 13–17, 2003, Los Alamos National Laboratory report LA-UR-03-4020.

FIGURE 3A.32 The LANL GN-5 prototype instrument is a self-contained compact high-resolution γ-ray spectroscopy system. It incorporates an electrically cooled Ge detector and a BGO anti-Compton annulus for high sensitivity in portable applications.

The expectation is that most users of nuclide identifiers will be personnel such as customs agents, border patrol, and airport police who are relatively untrained in γ-ray spectroscopy. Therefore, the instruments must be easy to operate and must have robust nuclide identification software that can reliably distinguish between, for example, medical isotopes and nuclear materials. The Ge-based instruments have a significant advantage over NaI and CdZnTe because of their far superior resolution. Present Ge instruments are heavy and on the borderline of truly portable instruments. Considerable development is underway now to improve the algorithms used to analyze the low-resolution spectra. An alternative approach is to use the low-resolution identifiers for screening and have a Ge spectrometer available to confirm or verify the results of the first measurement.

Summary

Measurements provide the input data to the nuclear material control and accounting system. Destructive assay usually provides the most accurate measurement, but there are many materials and situations where DA is not possible or reliable. Nondestructive assay techniques and instruments have been developed over the past 40 years to deal with these materials. Analytical chemistry techniques can determine elemental concentration in very pure samples to an accuracy of 0.05 to 0.20%. Mass spectrometry can determine isotopic composition to an accuracy of 0.02 to 0.05%. NDA determines nuclide mass or concentration; x-ray measurements can determine elemental concentration. The NDA techniques covered in this section include γ-ray spectrometry, neutron counting, and calorimetry.

Most of the NDA instruments covered herein are used regularly by the International Atomic Energy Agency. The IAEA designates each instrument with a four-letter acronym and has almost 100 designations in its list of equipment. The Agency also designates software and containment and surveillance equipment with four-letter acronyms. Reference 2 provides a good, elementary discussion of the role of the IAEA in nuclear nonproliferation and the techniques and instruments, DA and NDA, used by Agency safeguards inspectors.[2] Table 3A.5 provides a list of the principal NDA instruments used by the IAEA.

Heavy instruments, such as the HLNC, AWCC, and PSMC, are usually located at important facilities where they are used and rarely shipped to other sites. They are stored under seal or in a sealed room dedicated to IAEA use. Light instruments, such as the HM-5

Table 3A.5 NDA instruments used by IAEA inspectors.

γ	HM-5	Handheld Monitor System Version 5	Active fuel length, complementary access inspections
γ	MMCA	MiniMultichannel Analyzer	General γ-ray spectroscopy, NaI, CZT, Ge
γ	MMCN	MMCA + NaI	Uranium enrichment assay
γ	MMCC	MMCA + CZT	Uranium enrichment assay
γ	MMCG	MMCA + Ge	Uranium and plutonium isotopic composition
γ	IMCA	InSpector Multichannel Analyzer	General γ-ray spectroscopy, NaI, CZT, Ge
γ	IMCN	IMCA + NaI	Uranium enrichment assay
γ	IMCC	IMCA + CZT	Uranium enrichment assay
γ	IMCG	IMCA + Ge	Uranium and plutonium isotopic composition
γ	SFAT	Spent Fuel Attribute Tester	CZT, detect ^{137}Cs from spent fuel underwater
γ	KEDG	K-Edge Densitometer	Plutonium elemental concentration in solutions
γ	ICVD	Improved Cerenkov Viewing Device	Image intensifier, measure Cerenkov light from spent fuel in cooling pond
γ	CBVB	CANDU Bundle Verifier Baskets	Collimated CdTe measures ^{137}Cs or ^{95}Nb for spent CANDU fuel bundles in baskets
γ	CBVS	CANDU Bundle Verifier Stacks	Collimated CdTe measures ^{137}Cs or ^{95}Nb for spent CANDU fuel bundles in stacks
n	HLNC	High-level neutron coincidence counter	NNC for Pu <5 kg in produce cans
n	INVS	Inventory sample coincidence counter	NNC for Pu in <50 dram vials
n	AWCC	Active well coincidence counter	Active NNC for HEU <10 kg in produce cans and research reactor (MTR) fuel assemblies
n	UNCL	Uranium neutron collar	Active NNC for LWR fuel assemblies
n	BCNC	Birdcage neutron counter	NNC for Pu in fast critical assembly plates FCA
n	GBAS	Glovebox assay system	NNC for Pu & MOX in gloveboxes in TRP & RRP
n	PCAS	Pu canister assay system	NNC for Pu <18 kg as input to PFPF & JMOX
n	PSMC	Pu scrap multiplicity counter	NMC for impure Pu in various facilities
n	UFBC	Universal fast breeder counter	NNC for MOX fuel assemblies in U.S. and Japan
n	UWCC	Underwater coincidence counter	NNC to verify Pu in fresh MOX fuel assemblies underwater
n	WCAS	Waste crate assay system	

and the MMCA and IMCA series, are often carried by inspectors from Vienna to the facilities under inspection. With the exception of the most sensitive facilities, nuclear material measurements are only performed during the annual Physical Inventory Verification (PIV).

Most of the measurement techniques and instruments described in this section are designed with the premise of an "honest" operator in mind. An operator who is trying to divert nuclear material and who is intelligent and knowledgeable about the assay techniques may be able to fool the measurement equipment of the IAEA. For example, consider a produce can (diameter ~15 cm and height ~20 cm) declared to have 6.15 kg of UO_2 enriched to 2.3%. Now let the diversion-minded operator fabricate a can with two regions, an inner region that is 10-cm diameter and an outer annulus of 2.5-cm thickness. The operator fills the outer region with 2.3% UO_2 and the inner region with 93% UO_2. Any of the IAEA's γ-ray determinations of uranium enrichment will verify the declared enrichment of 2.3%. ^{235}U γ rays from the center region are shielded by the 2.3% material and are not visible to the γ-ray detector. Assuming a density of 2.0 g/cm,^{3} the inner region would contain ~3.14 kg of weapons-grade uranium that was invisible to the IAEA inspector's instrument. However, if the inspector had access to an AWCC and measured the can in it, he would easily detect the subterfuge. Gamma-ray and neutron assay techniques are complementary, and though one can be fooled in some circumstances, it is difficult to be fooled in combination. This practice is often called *spoofing* and will be covered in more detail elsewhere in this book.

3B

Irradiated Fuel Measurements

Mark E. Abhold

Introduction

As of October 2005, over 440 nuclear power reactors operate worldwide with an installed electrical power generation capacity of about 370 gigawatts of electrical generation power (GW(e)).[1] Roughly 300 research reactors and critical assemblies add to the total of operating reactors.[2] Plutonium contained in irradiated spent nuclear fuel (SNF) from these reactors is discharged at the rate of about 70 to 75 metric tons per year, but only 15 to 20 metric tons of the yearly total is separated in reprocessing plants, leaving the majority of the plutonium in the form of spent fuel assemblies. As of the end of 2003, over 1,300 of the 1,855 metric ton worldwide plutonium inventory was contained in civilian spent fuel, along with about 50 of the world's 1,900 metric tons of highly enriched uranium (HEU, >20% ^{235}U)[3]. In addition, other nuclear materials such as americium and neptunium exist in SNF, and some specialty fuels may also contain thorium and ^{233}U.

Diversion of SNF for chemical reprocessing and extraction of plutonium or other nuclear materials is of particular concern at facilities where spent fuel has accumulated. Diversion could take place by removal of spent fuel from storage with or without substitution of inert or radioactive dummy elements, or by the unrecorded removal of spent fuel from the core with or without replacement by unrecorded fresh fuel assemblies.

Safeguards measures such as radiation monitoring, tags and seals, and video surveillance are often employed to detect possible spent fuel diversion from nuclear facilities. In addition, safeguards measures often require quantitative assay measurements to account for the special nuclear material (SNM) and other nuclear material content. Before irradiation, fresh reactor fuels can easily be characterized using the standard gamma-ray and neutron nondestructive assay (NDA) techniques described earlier. However, directly measurable SNM signatures in irradiated fuels can be completely masked by the intense radiation emitted from fission products that build up during irradiation, thus rendering standard NDA techniques unusable on spent fuel. This section describes nuclear reactor fuels, their radiological characteristics, and the techniques used to safeguard and characterize spent nuclear fuel (SNF).

---

[1]International Atomic Energy Agency Power Reactor Information System, www.iaea.org/programmes/a2/index.html, Oct. 2005.

[2]*Nuclear Research Reactors in the World: Reference Data Series #3*, IAEA Publication IAEA-RDS-3/10, Jan. 1997.

[3]D. Albright and K. Kramer, "Tracking Plutonium Inventories," www.isisonline.org/global_stocks/plutonium_watch2004.html, June 2004.

Nuclear Reactor Designs

This section introduces the most common reactor designs and describes the physical characteristics of their fuels.

Light Water Reactors

Pressurized water reactors (PWRs) and boiling water reactors (BWRs) are collectively known as *light water reactors* (LWRs). These reactors use ordinary water as the moderator and coolant, and their fuel consists of ceramic uranium oxide pellets, enriched in ^{235}U, that are about 1 cm in diameter and 1 to 2 cm long. Hundreds of these pellets are loaded into ~4 m-long fuel pins made of zirconium alloy tubes; these fuel pins are assembled into large square arrays with grid assemblies, spacers, and structural end pieces. PWR assemblies are typically quite large, ranging from arrays with 15 rows and 15 columns of pins (15×15) to 17×17 arrays, and the typical BWR is a 7×7 or an 8×8 array. The core of a large LWR contains hundreds of these arrays, called *fuel assemblies*, placed vertically, typically with a total of 40,000 to 50,000 fuel pins and between 100 and 200 metric tons of uranium oxide.

The initial ^{235}U enrichment typically varies between 3% and 5%. Each assembly spends three to four years in the reactor core burning the ^{235}U, converting some ^{238}U to ^{239}Pu, and subsequently burning both ^{235}U and the converted ^{239}Pu until the remaining fissile material is insufficient to maintain reactor criticality. During operation, the core fuel is not accessible; LWR cores cannot be opened, because the reactor is extremely radioactive and under pressure, so during operation the core fuel load is completely secure. One quarter to one third of the fuel is replaced with fresh fuel every 12 to 18 months. Refueling operations are conducted during reactor shutdown periods, with the core open and the fuel accessible. Loading and unloading are monitored by surveillance cameras to ensure that no diversion takes place during fuel movements to and from fuel storage.

Heavy Water Reactors

Heavy water reactors (HWRs) are designed to use natural uranium (0.7% ^{235}U) because they take advantage of cooling and moderating water made from deuterium, which absorbs fewer neutrons than the hydrogen in ordinary water. The *Canadian Deuterium-Uranium (CANDU) reactor* is a common HWR type that uses fuel similar to LWR fuel except the fuel pins are much shorter at about 50 cm length and are arranged in circular bundles of 28 rods and thus are much smaller and lighter than LWR fuel assemblies. A typical CANDU core has about 4,500 bundles placed horizontally end to end in hundreds of fuel channels.

CANDU reactors are designed for refueling on-line (also called on-load) without shutting the reactor down. About 15 bundles are replaced per day with the spent fuel discharged from one end of the fuel channel and the fresh fuel inserted into the opposite side. On-line refueling has significant operational advantages in that no reactor shutdown is necessary, but, unlike LWRs, the continuous discharge of fuel requires continuous surveillance and monitoring of the core. Like LWRs, the main safeguards concern is the plutonium bred from ^{238}U during reactor operations. The burnup of CANDU reactor fuel is usually 6,500 to 7,500 Megawatt-days per metric ton of heavy metal (MWD/MT), compared to 30,000 to 50,000 MWD/MT and above for LWR reactors. Low-burnup plutonium is more attractive for nuclear weapons use because it contains relatively less ^{240}Pu, an isotope that spontaneously emits neutrons. About 30 CANDU reactors currently operate worldwide, with 10 or more additional CANDU derivatives in India.

Gas-Cooled Reactors

Gas-cooled reactors typically use carbon in the form of graphite as the moderating material instead of hydrogen (in water) because carbon absorbs fewer neutrons than hydrogen.

Some Russian-built carbon-moderated reactors (for example, the RBMK reactors of the type used at Chernobyl) use water confined to pressure tubes as the coolant, but most carbon-moderated reactors use gas as the coolant because gas coolants with very low neutron absorption properties are available. Several gas-cooled, carbon-moderated reactors have been designed that use helium or carbon-dioxide coolant.

An early type of carbon-dioxide, gas-cooled power reactor of interest is the *Magnox reactor*, originally designed and operated in the United Kingdom but exported to and copied by other countries, including North Korea. Magnox fuel is natural uranium metal clad in magnox, a special alloy of magnesium and aluminum; hence the name. Since the fuel is natural uranium, on-load refueling is essential to eliminate frequent refueling outages. Like the CANDU reactor, Magnox fuel burnup is quite low, making the plutonium produced attractive for nuclear weapons use. Magnox reactors have been used for both power and plutonium production.

Another gas-cooled reactor type is the *high-temperature gas-cooled reactor* (HTGR), with fuel in the form of small, highly enriched uranium carbide or uranium oxide particles coated with layers of carbon and silicon dioxide pressed into pellets and assembled into fuel elements. These elements are placed into the core moderator, a massive pile of hexagonal graphite blocks containing passages for helium coolant. This reactor type is of importance because it is capable of breeding ^{233}U by introducing thorium-bearing fuel pellets.

Fast Breeder Reactors

Conventional LWR thermal reactors convert some fertile material (^{238}U) to a fissile material (^{239}Pu) by neutron capture. As the amount of converted Pu builds up in the core, some Pu will fission, releasing additional energy, thus improving utilization of the original fissile uranium. It is possible to increase conversion to the break-even point such that more fissile material is produced than is consumed by keeping the neutron spectrum energetic enough to take advantage of the fact that ^{239}Pu yields many more neutrons from energetic (fast) neutron-induced fission than from fissions induced by thermal (slow) neutrons. When the conversion ratio reaches break-even, the reactor is known as a *breeder reactor*.

The most advanced breeder reactors now in use keep a fast neutron energy spectrum by replacing water as the coolant with a liquid metal such as sodium, because sodium does not moderate neutron energy as efficiently as water and also does not absorb neutrons.

The reactor core of a fast breeder is divided into an active core region and a blanket of fertile material such as ^{238}U. Neutrons leaking from the core are absorbed in the fertile blanket, converting some fraction to plutonium, which is later reprocessed to manufacture fresh plutonium-bearing fuel. The reprocessed fuel is typically a mixture of uranium oxide and recycled plutonium oxide pellets clad in stainless steel.

These reactors are of significance because of their high plutonium production rate and because of the potential to produce weapons-grade Pu with low ^{240}Pu content from the fertile blanket. Fast breeder reactors require continuous radiation surveillance to monitor the production and flow of plutonium-bearing fuels.

Research Reactors

Research reactors have a variety of purposes, including studying interactions between neutrons and matter (where experiments are typically placed outside the reactor, along beam tubes), producing radioactive isotopes for medical or research applications by irradiating materials placed in or near the core, training, and engineering development. A wide variety of research reactors and fuels have been designed with these specialized purposes in mind. Only two examples will be briefly introduced here.

One common research reactor type is the *materials testing reactor* (MTR). MTR fuel is made from 20–93% enriched uranium/aluminum oxide powder or a metal alloy rolled into thin layers and covered with aluminum cladding to make flat or curved plates. The original fissile content of an MTR fuel assembly ranges from about 100 g to 600 g of ^{235}U. When the uranium is in the form of HEU, the fresh fuel is of safeguards concern. MTR fuel is small and light enough to easily move by hand, but spent MTR fuel is extremely hazardous due to its high radiation levels. As part of the Atoms for Peace Program, the United States built MTR reactors in 29 foreign countries.

Another common research reactor type is the *Training, Research and Isotope Production reactor* (TRIGA) produced by General Atomics. TRIGA fuel elements are composed of a ceramic zirconium hydride with 8 to 8.5 weight percent uranium that is nominally 20% enriched. TRIGA fuel elements are either aluminum or stainless steel clad.

Research reactor structure is highly relevant to the application of safeguards. Many research reactors are of the pool type in which the core is visible and accessible for measurements. For these reactors it is possible to perform verification measurements directly on core fuel. However, many high-powered reactors are usually of the tank type with forced cooling, where the core fuel is not visible and not accessible; therefore, in-core verification must rely on reactor instrumentation and operating records.

The U.S. Department of Energy (DOE) Reduced Enrichment for Research and Test Reactors (RERTR) Program is working to convert research reactors from the use of fuels and targets containing high-enriched uranium to the use of fuels and targets containing low-enriched uranium (LEU, $<20\%$ ^{235}U) consistent with the United States' nonproliferation policy goal of minimizing and eventually eliminating the use of HEU in civil programs worldwide.

Naval Reactors

Most nuclear submarines and surface vessels are powered by one or two water-cooled, pressurized water reactors. The fuel is uranium and zirconium or aluminum alloy enriched to between 20% and 93%, with most modern reactors between 20% and 45%. Fully utilized discharged fuel has a high burnup, a very high radiation level, and significant amounts of Pu. Lesser irradiated fuel can retain a large quantity of highly enriched uranium.

Naval reactor fuel can be especially difficult to assay, because it may contain both HEU and Pu as well as large quantities of fission products and neutron-producing actinides.

Irradiated Fuel Characteristics

Reprocessing SNF is difficult and hazardous because special measures are needed to safely handle these highly radioactive materials, including the use of shielded hot-cells or canyons with remote manipulators. The extent of special handling measures depends on the size and weight of the SNF assembly as well as the magnitude of the radiological hazard; therefore, the attractiveness of SNF for use in manufacturing nuclear weapons is a function of not only the amount and type of contained nuclear materials but also its radiological characteristics.

Radiological Characteristics

Fission of a ^{235}U or ^{239}Pu atom creates two medium-mass fission fragment nuclides that are initially rich in neutrons. These fragment nuclides undergo beta decay to approach stability and in the process emit gamma rays. Over 200 isotopes are created by fission, but only about 10 isotopes can be directly measured from their characteristic gamma rays. In addition to the fission product gamma rays, activation of the structural materials in the assembly also produces gamma rays. Uranium present in nuclear fuel captures neutrons creating transuranic nuclides. Many of these nuclides produce neutrons through spontaneous fission or by alpha decay when the emitted alpha particle interacts with oxygen or other low atomic number elements in the fuel in an (α,n) reaction.

High burnup fuel is highly radioactive, making handling outside the spent fuel pool difficult and very hazardous. Burnup (in atom percent) is defined as the number of fissions per 100 heavy nuclei (uranium or plutonium) initially present in the fuel. *Burnup* is often used interchangeably with the term *exposure*, which is defined as the integrated energy released by

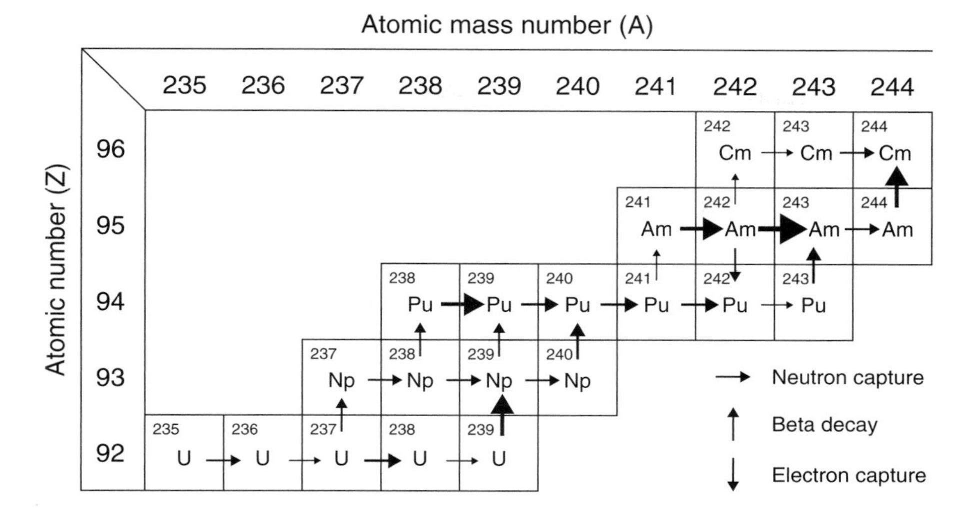

FIGURE 3B.1 The mechanism for the production of plutonium and other actinides in a nuclear reactor. The arrows are sized to give a rough indication of the relative magnitude of each reaction.

fission of the heavy nuclides initially present in the fuel. Exposure has dimensions of megawatt (or gigawatt) days (thermal output of the reactor) per metric ton of initial heavy metal (MWd/MT). One atom percent burnup is roughly equivalent to an exposure of 9.6 GWd/MT; however, the relationship between burnup and exposure changes with the changing ratio of uranium to plutonium fissions as plutonium builds up in the fuel.

Early LWR fuels typically would achieve a burnup of 30,000 MWd/MT, but with improvements to materials and fabrication techniques, modern fuels can achieve 40,000 to 50,000 MWd/MT or more. An LWR fuel assembly irradiated to 40,000 MWd/MT has a radioactivity level exceeding 100,000 Ci, even after cooling for 20 years, and can produce dose rates of many thousands of rem/hour in air—a dose rate that would be lethal in an exposure lasting only a few minutes. Even research reactor fuel can be extremely hazardous to handle. MTR type fuel that is initially irradiated to 60% burnup would have a one-meter dose rate in air that exceeds 100 rem/hr for cooling times up to 20 years.[4] Dose rates that exceed 100 rem/hr are considered to be "self-protecting" by the IAEA because the radiological hazard provides a barrier to theft and misuse and very large and cumbersome shielding casks are needed for safe handling.

Plutonium Production

Figure 3B.1 illustrates the mechanism for the production of plutonium and other actinides in nuclear reactors. The process for breeding ^{233}U is not shown on this figure, but it can be produced by neutron capture in ^{232}Th. The majority of plutonium is produced (bred) by the capture of a neutron in ^{238}U followed by two successive beta decays. One significant variable for Pu production is the initial ^{235}U enrichment.[5] For the same total mass of uranium, higher enrichment produces less Pu as higher enrichments start with relatively less ^{238}U.

[4]R. B. Pond and J. E. Matos, "Photon Dose Rates from Spent Fuel Assemblies with Relation to Self-Protection," Presented at the 1995 International Meeting on Reduced Enrichment for Research and Test Reactors, Paris, Sept. 1994.

[5]M. Swinhoe and M. E. Abhold, "Measurement Techniques for Reactors and Critical Assemblies," in *International Training Course on Implementation of State Systems of Accounting for and Control of Nuclear Materials,* Los Alamos Report, LA-UR-03-2223, 2003.

For plutonium production to take place at an appreciable rate, a high rate of neutron capture is needed, and thus the neutron flux in the reactor must be very high, typically well above 10^{13} n/cm^2/sec. Studies have shown that research reactors operating at more than 25 MW (thermal) are at least theoretically capable of producing a significant quantity of plutonium or ^{233}U (defined by the IAEA as 8 kg) in one year.[6] The actual production rate would depend on the individual reactor design and operating history.

A typical LWR produces about 0.2 gm of ^{239}Pu per MW(thermal)-day of operation, so a large LWR (~3000 MW(thermal)) can produce more than 0.5 kg of Pu per day. A heavy water, graphite moderated, or fast breeder reactor with optimal reflection can produce substantially larger quantities; a rough rule of thumb is up to about 1 gm of ^{239}Pu per MW(thermal)-day.[6]

Spent Fuel Storage

The United States employs a once-through fuel cycle whereby spent fuel from nuclear reactors is stored until it can be disposed of in a permanent high-level waste repository. Reprocessing to recover plutonium from SNF is currently not done in the United States; therefore, significant inventories of SNF are stored at reactors, at dedicated storage facilities, and at various DOE sites. Similar storage facilities are used in countries where reprocessing is employed while the spent fuel is awaiting transfer to the reprocessing facility. A brief description of typical storage facilities follows.

Spent Fuel Pools

A discharged LWR spent fuel assembly generates intense radiation and heat from the decay of fission and activation products. Immediately following discharge, spent fuel is placed into storage racks in a pool of water typically 40 or more feet deep. The pool water is circulated by pumps through heat exchangers, providing needed cooling to remove the decay heat, and in addition to cooling, the water also provides necessary shielding for personnel radiation safety. The storage racks keep the fuel in a safe position, sufficiently separated from adjacent assemblies, to avoid a criticality accident.

Spent fuel assemblies are typically stored for 10 or more years until their heat production rate drops enough to send them to dry-cask storage or to permanent disposal. While in the spent fuel pool, the fuel is accessible for maintenance or measurements. For example, some spent fuel storage facilities have equipment for underwater replacement of failed fuel pins, and measurements can be made by partially lifting assemblies out of the storage rack because the pool is deep enough to allow the assemblies to be raised but still maintain a safe depth for shielding. At the reactor, spent fuel characterization measurements typically take place in the spent fuel storage pool.[7]

Dry-Cask Storage

Most nuclear power plant spent fuel pools were not designed with sufficient capacity to store all the spent fuel generated over the lifetime of the reactor. Therefore, secondary storage capacity is often needed. When a fuel assembly has been sufficiently cooled, it is often moved from the spent fuel pool to a dry storage site, which can be either at the reactor or away

---

[6]W. Theis, "IAEA Safeguards Experience at Research Reactors," in Los Alamos Report LA-10672-C, June 1985.

[7]The security of spent fuel pool storage was recently reviewed by the U.S. National Academies and a finding made that dry cask storage for older, cooler spent fuel has inherent safety and security advantages over pool storage. See *Safety and Security of Commercial Spent Fuel Storage: Public Report*, National Academies Press, ISBN 0-309-09647-2, 2006.

from the reactor. The dry storage site typically consists of a large reinforced concrete pad in a fenced and secured area containing a number of dry-cask storage systems.

Dry-cask storage systems consisting of metal casks, concrete casks, metal canisters housed in concrete modules, and concrete storage vaults made by a variety of vendors have been licensed for use.[8] All these storage systems house multiple spent fuel assemblies, and all are designed to provide containment, radiological shielding, physical protection, and inherently passive cooling of the SNF during normal, off-normal, and accident conditions. Physical access to individual spent fuel assemblies for the purpose of measurements or verification is typically not possible.

Spent Fuel Safeguards

Safeguards measures such as radiation monitoring, tags and seals, and video surveillance are often employed to detect possible spent fuel diversion from reactors, at-reactor and away-from-reactor spent fuel storage facilities, during transportation, in conditioning facilities where fuel assemblies may be disassembled and repackaged, and at the head-end of reprocessing facilities where spent fuel assemblies are chopped and dissolved. Safeguards measures often require quantitative assay to account for the special nuclear material (SNM) and other nuclear material content. Before irradiation, fresh reactor fuels can easily be characterized using the standard gamma-ray and neutron nondestructive assay (NDA) techniques described earlier in this chapter.

U.S. Domestic Safeguards

Recommendations for safeguarding irradiated nuclear fuels have been developed through the Fissile Material Assurance Working Group[9] to ensure that the fissile material content in SNF is validated by direct measurements or other methods. The recommendations include the following:

- Accountability values for nonself-protecting SNF should be based on measured values, substantiated estimates determined from reactor burnup data, age, initial enrichment, or other valid means.
- When technology exists, it should be used to assay SNF and to validate reactor burnup/enrichment calculated estimates.
- Once validated, SNF should be inspected and subject to item accountability checks on a frequency corresponding to routine physical inventories as required by DOE orders.

IAEA Safeguards

Application of IAEA safeguards takes place on the basis of an agreement between the IAEA and the state in which the facility resides. The wide variety of reactor and fuel types implies that the details of the safeguards approach vary from facility to facility, with larger reactors and installations with large amounts of attractive materials tending to have more intensive safeguards. The safeguards approach for spent fuel consists of three major components used in combination:[10]

- *Accountancy.* Reporting by states on the whereabouts of the fissionable material under their control, on the stocks of spent fuel, on the characteristics of the spent fuel, and on the processing and reprocessing of spent fuel.

---

[8]M. G. Raddatz and M. D. Walters, "Information Handbook on Independent Spent Fuel Storage Installations," U.S. Nuclear Regulatory Commission, NUREG-1571, Dec. 1996.

[9]D. W. Crawford, "Safeguards on Spent Fuel and Other Irradiated Nuclear Fuels Destined for Waste Management," *Proc. 40th Annual Meeting of the Institute of Nuclear Materials Management*, Phoenix, Arizona, July 25–29, 1999.

[10]G. Zuccaro-Labellarte and R. Fagerholm, "Safeguards at Research Reactors, Current Practices, Future Directions," *IAEA Bulletin*, vol. 38, no. 4, 1996.

- *Containment and surveillance (C/S).* Techniques such as seals that allow conclusions that no material has been tampered with or film and TV cameras that record any action occurring in a particular area of the reactor or spent fuel storage facility. These systems maintain continuity of knowledge (COK) on the spent fuel from reactor core to the storage pond, and to dry storage if it exists.
- *Inspections.* Site visits by Agency inspectors, checking unattended monitoring system data, seals, and C/S, verifying declarations, and confirming physical inventories of fuel or spent fuel with nondestructive assay measurement techniques.

Nondestructive Assay Measurement Techniques

Characterizing, or assaying, the fissile content of spent nuclear materials through measurements is not only important to safeguards but is needed in support of safety, material management, and disposition activities. For example, characterizing SNF is required for waste acceptance into the DOE Office of Civilian Radioactive Waste Management facilities and for transportation.

The simplest verification measure is visual inspection. Provided with an unobstructed view, typically through shielding water, the safeguards inspector can determine the physical integrity of the assembly or irradiated object but cannot verify whether the object has been irradiated or whether it is a dummy substituted to mask a diversion.

Since DA of irradiated fuels is rarely possible, spent fuel is typically verified using NDA. The type of spent fuel and its characteristics determines the NDA techniques that can be used. After a significant exposure of the fuel in a reactor, the uranium and plutonium signatures are completely masked by radiation from fission products, activated structural components, and transuranic isotopes. Therefore, passive measurements that directly yield the ^{235}U or ^{239}Pu content are not possible. Instead, either indirect signatures or active interrogation must be used to estimate these quantities. For example, indirect signatures based on the emitted radiation can be used to determine the burnup level, and from the burnup, along with other known quantities, the remaining SNM content in the fuel can be estimated through calculational methods.

Safeguards measurement techniques used to verify the presence of irradiated materials and/or directly measure SNM content are described in the following sections.

Cerenkov Radiation

Cerenkov radiation is emitted whenever a charged particle passes through a medium at a velocity exceeding the phase velocity of light in that medium. The bluish glow that can be seen surrounding the core of an operating pool type reactor and can also be seen surrounding spent fuel stored under water are examples of Cerenkov radiation. High-energy electrons, gamma rays, and neutrons from spent fuel are capable of producing Cerenkov light either directly or indirectly as they interact in the water surrounding the fuel, where the phase velocity of light is about 75% of the value in a vacuum. Cerenkov viewing devices (CVD) that amplify the light are used to identify physical characteristics of spent fuel assemblies in storage pools and can also provide an indication that the fuel has been irradiated. Although the absolute Cerenkov light level and its decay in time is related to spent fuel burnup and may provide a possibility of quantitative measurements of spent fuel, in practice CVD is only used to qualitatively verify consistency of the declared burnup and cooling time. This technique is only applicable for wet storage of individual rods or assemblies and is not capable of identifying missing fuel rods or the substitution of dummy fuel rods in an assembly.[11] This is the most common type of verification measurement made on spent fuel.

---

[11] J. R. Phillips, Chapter 18, "Irradiated Fuel Measurements," in D. Reilly, N. Ensslin, H. Smith Jr., S. Kreiner, *Passive Nondestructive Assay of Nuclear Materials*, U.S. Nuclear Regulatory Commission, NUREG/CR-5550, 1991.

FIGURE 3B.2 The FORK detector in use under water in a spent fuel storage pool. The spent fuel assembly has been lifted up from its storage location to allow the FORK detector to be placed in contact with the fuel-bearing middle part of the assembly.

Passive Gamma-Ray and Neutron Total Counting

For cooling times greater than one year, the total gamma-ray and neutron activity is proportional to burnup. The activity can be estimated by measuring the total number of gammas and neutrons emitted from the spent fuel, independent of their energy, at a location near the middle of the assembly. This measurement is usually performed under water in the spent fuel storage pool and requires the assembly to be partially lifted from its storage location. This can be accomplished with the *FORK detector* (FDET),[12,13] a detector commonly used by the IAEA and shown in Figure 3B.2. The FDET measures both gamma rays and neutrons simultaneously using ionization and fission chambers. *Ionization chambers* are sealed detectors containing a gas and two electrodes between which a voltage is maintained. Gamma rays entering the chamber interact with the gas, causing gas molecules to become ionized. The liberated electrons and ions are attracted to the oppositely charged electrodes, causing a signal to be recorded by an external circuit. Fission chambers add the capability to detect neutrons by coating the inside surface with an isotope (typically ^{235}U) that undergoes neutron-induced fission with the release of charged particles that cause ionization in the chamber's gas.

Total neutron measurements have some advantages over total gamma-ray measurements for estimating burnup because neutrons emitted from the spent fuel are not attenuated by the fuel as much as gamma rays and neutron counting can be performed soon after the fuel is removed from the core, whereas gamma-ray measurements cannot because short-lived fission products dominate the gamma-ray emissions. The majority of the neutron emission

---

[12]P. M. Rinard and G. E. Bosler, "Safeguarding LWR Spent Fuel with the FORK Detector," Los Alamos National Laboratory, LA-11096-MS, March 1988.

[13]G. E. Bosler, J. K. Halbig, S. F. Klosterbuer, H. O. Menlove, and J. R. Phillips, "Passive Neutron Measurement Applications for Irradiated Fuel Assemblies," *Trans. Am. Nucl. Soc.* (39) 198:348.

in short-cooled LEU fuel at high burnup comes from ^{242}Cm and ^{244}Cm built in through the mechanism illustrated in Figure 3B.1. These isotopes have half-lives of 163 days and 18.1 years, so the cooling time must be precisely known and corrected for. The initial ^{235}U enrichment and the irradiation history are two other factors that can affect the interpretation of the total neutron counting results. Total neutron counting is not precise enough to identify a single missing pin in a LWR assembly but can easily identify whether half of the assembly is missing and can also identify misdeclared assemblies that were diverted during a refueling outage and replaced with fresh assemblies. The FORK detector is also used occasionally to verify the operating history of LWRs.

Passive Neutron Coincidence Counting

A device for measuring the SNM content in irradiated breeder reactor fuel, the underwater breeder counter (UWBC), has been installed in the BN-350 breeder reactor in Kazakhstan.[14] Unlike LWR spent fuel, this breeder reactor spent fuel has negligible spontaneous fission neutrons from Cm; therefore, the passive neutron signal is primarily from ^{240}Pu, ^{238}Pu, and neutron multiplication within the assembly. Since these fission neutrons are typically emitted in bursts of two or more neutrons at a time, the time distribution of neutron detection events can be used to separate fission reactions from other sources of neutrons such as those produced by (α,n) reactions. The observables of most interest are the passive neutron singles and coincidence counting rates and the gross gamma-ray dose rate in the ionization chamber. The neutron rates are related to the plutonium content, and the ratio of the observed neutron to gamma-ray rates can be used to distinguish between blanket and core assemblies. This device has been used to measure the Pu content of breeder driver assemblies with accuracy better than 10%.

Gamma-Ray Spectroscopy

In spent fuel applications, high-resolution *gamma-ray spectroscopy* is used to identify characteristic spectral lines from fission and activation products. Absolute measurements of the ^{137}Cs content can be performed to get a quantitative measure of the fuel burnup; however, that requires knowledge of the assembly attenuation and the measurement geometry because gamma rays are easily attenuated by the assembly and surrounding water. Spent fuel burnup can also be obtained from the ratio of ^{134}Cs to ^{137}Cs activity with appropriate decay corrections for long irradiation and cooling times. ^{134}Cs is produced by neutron capture on the fission product ^{133}Cs; therefore its production requires two neutron interactions, the first interaction being the neutron that caused the fission. The production of ^{134}Cs is therefore proportional to the neutron flux squared. By dividing the ^{134}Cs concentration by the ^{137}Cs concentration, which is directly proportional to the flux, the ratio is then proportional to the integrated flux and therefore proportional to the burnup. Activity ratios are easier to determine than absolute activity because only the ratio of detector efficiency at the two gamma energies needs to be known. The measurements are usually conducted at poolside with a long sealed tube or air collimator placed into the pool to provide a path for gamma rays from the fuel to stream relatively unscattered to the detector. These measurements require that the measured assembly be lifted partially out of the storage rack and sometimes moved to an empty region of the pool.

The IAEA uses two more gamma-ray systems for attribute verification of spent fuel. These use NaI(Tl) or CdZnTe detectors that have substantially poorer resolution than Ge. The Irradiated Fuel Attribute Tester (IRAT) uses a small CdZnTe encased in a stainless steel cylinder that is approximately 30 cm long and 3 cm in diameter. The cylinder contains the

---

[14] J. P. Lestone, M. E. Abhold, J. Halbig, H. O. Menlove, P. Polk, P. M. Rinard, J. Sprinkle, P. Staples, and R. Holbrooks, "An Underwater Instrument for Breeder Reactor Spent Fuel Assemblies," LANL report LA-UR-98-1588, 1998.

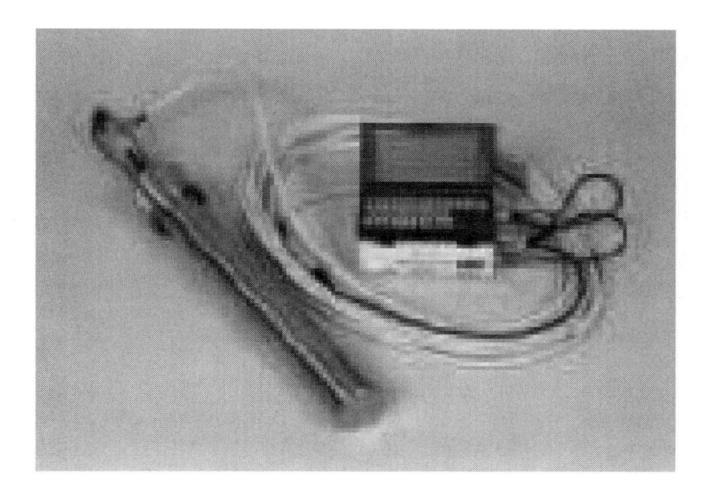

FIGURE 3B.3 The IRAT system used by the IAEA for attribute verification of spent fuel assemblies raised partially out of their storage racks.

detector, preamplifier, shielding, and a collimator. The IRAT is suspended from the spent fuel pond bridge crane and measures the fission product spectrum from a fuel assembly that is partially raised from the storage rack. A multichannel analyzer collects the spectrum, which is checked for gamma rays from fission products such as ^{137}Cs, ^{134}Cs, ^{144}Pr, ^{154}Eu, and others, depending on the cooling time of the assembly. Figure 3B.3 shows the IRAT with a mini-multichannel analyzer (MMCA) and a palmtop computer.

The spent fuel attribute tester (SFAT) comes with either a NaI(Tl) or a CdZnTe detector; CdZnTe is now more common. The principal advantage of SFAT is that it measures fuel assemblies without raising them from the storage racks. Reactor operators are very reluctant to move spent fuel assemblies because of the possibility of damaging them and releasing fuel and fission products into the cooling pond. The detector, preamplifier, and shielding are contained in a watertight stainless steel cylinder. A watertight collimator pipe is attached below the detector housing to restrict the detector from seeing many gamma rays from neighboring fuel assemblies. For fuel cooled less than 4 y, SFAT measures the 2182-keV gamma ray from ^{144}Pr. It measures the 661.6-keV gamma ray from ^{137}Cs for longer cooled assemblies. Because SFAT views the assembly from the top, it also sees a high flux of gamma rays from the activation product ^{60}Co from the steel in the fuel assembly hardware. SFAT is often used if an ICVD is unable to measure Cerenkov radiation, i.e. from low burnup or long cooling time fuel.

High-Energy Gamma Tomography

A tomographic measurement system has been developed and built for cross-sectional viewing of spent fuel assemblies.[15] The tomographic hardware images gamma rays from the fission products ^{154}Eu and ^{144}Pr and can reveal the rod-to-rod distribution of these gamma emitters, thus providing the means to detect single rod diversion scenarios. Gamma-ray tomography can be performed on an irradiated fuel assembly that is partially raised from the storage rack or moved to a measurement position.

[15] F. Lévai, S. Dési, M. Tarvainen, and R. Arlt, "Use of an Underwater Multidetector System for Gamma Emission Tomography of Spent Fuel Assemblies," Proc. 15th ESARDA Symposium, Rome, 1993:387–392.

Active Neutron Interrogation

Direct measurement of the fissile content of irradiated fuel is possible using a large neutron source to induce fissions. The source can be an accelerator, a 14-MeV neutron generator, or an isotopic source such as ^{252}Cf or Am-Be. The source strength must be quite large, on the order of 10^8 to 10^9 n/s, to induce a neutron signal that is comparable in size to the passive neutron yield from the transuranic isotopes in the spent fuel. This technique has found only limited use for specialized measurements, such as the measurement of leached hulls in reprocessing facilities, because large, complex, expensive, and extremely well-shielded instruments are needed to be able to safely handle the neutron sources. In practice, active neutron interrogation systems cannot distinguish between uranium and plutonium.

Active Neutron Coincidence Counting

HEU spent fuels, especially those fuels with enrichments greater than 90%, have very little curium content and therefore produce few neutrons, making active measurements of the fissile content possible with small neutron sources. The active neutron coincidence technique is used by the Research Reactor Fuel Counter (RRFC-II), shown in Figure 3B.4, installed at the U.S. Savannah River Site. The RRFC-II measures the ^{235}U content in MTR-type fuel being returned to the United States.[16] Fuel for MTR reactors was supplied by the United States to foreign nations under the Atoms for Peace Program proposed by President Dwight D. Eisenhower in 1953. In 1996, the United States started a program to take back U.S.-origin foreign research reactor spent fuel to prevent the possibility of the fuel being diverted to

FIGURE 3B.4 The Research Reactor Fuel Counter (RRFC-II) shown at poolside.

[16] M. E. Abhold, M. C. Baker, S. Bourret, W. Harker, D. Pelowitz, and P. Polk, "Second-Generation Research Reactor Fuel Counter," *Proc. of the 42nd Institute of Nuclear Materials Management Meeting*, July 2001.

produce nuclear weapons. Note that this program does not take back LWR fuel. Approximately 13,000 assemblies of material test reactor (MTR) spent nuclear fuel from 29 countries will be returned to the U.S. during the 13 years of the program ending in 2009. The RRFC-II assays the ^{235}U content by first measuring neutrons emitted spontaneously by the fuel, then interrogating the fuel with a small neutron source ($\sim 10^5$ n/s) and analyzing the induced fission reactions in the fuel using coincidence counting techniques. The measurement can be done entirely under water in the spent fuel pool, eliminating the need for costly and hazardous handling operations of spent fuel out of water. The RRFC-II has demonstrated the ability to assay remaining ^{235}U content to within 10%.

Shufflers

Californium shufflers using the delayed neutron technique described in "Nuclear Material Measurement Technologies" have been used in reprocessing plants to assay baskets of leached hulls from chopped-up irradiated LEU fuel rods, HEU spent fuel assemblies, and highly radioactive solutions using neutron sources on the order of 10^{10} n/s. Conceptually, the source strength of a shuffler can be increased to whatever level is needed to produce the desired delayed neutron signal. However, in practice, the intensity of the ^{252}Cf source needed for irradiated fuel can be exceedingly difficult to handle and shield. For example, the shuffler built to assay spent fuel at the Fluorinel and Fuel Storage (FAST) facility weighs over 15 tons to accommodate the necessary shielding. Since high-burnup fuel that is initially low enrichment will have a high curium content, the shuffler's accuracy is limited on fuels of this type.

Summary

The possible types of irradiated fuel assay measurements depend on the physical and radiological characteristics of the fuel, the facilities available for allowing safe access to the fuel, and the cost and complexity of measurement equipment that can be afforded. Table 3B.1 gives a brief summary of the level of verification and the types of nondestructive measurements that are possible.

Table 3B.1 Nondestructive assay techniques and the level of verification possible.

Measurement Technique	Level of Verification	Limitations
Visual inspection	Physical integrity	Must have visual access Cannot differentiate dummy from irradiated assemblies
Cerenkov	Presence of radiation	Must be done under water
Passive neutron	Indication of irradiation exposure level SNM content can be calculated given operator-declared information Can identify diversion of a partial assembly	Access to isolated assembly needed Must have trustworthy operator declarations
Passive gamma	Presence of fission products and actinides SNM content can be calculated given operator-declared information	Access to isolated assembly needed Precise control of measurement geometry needed to correct for shielding effects Must have trustworthy operator declarations
Active neutron	Quantitative measurement of total fissile content	Access to assembly needed Hazardous neutron sources are needed for low initial U enrichments

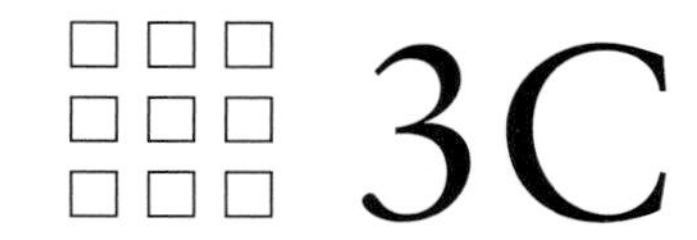

3C Measurement of Nuclear Material Process Holdup

Douglas Reilly

Introduction

The term *holdup* refers to the nuclear material deposited in the equipment, transfer lines, and ventilation systems of processing facilities. Reprocessing, fuel fabrication, conversion, and enrichment require very large facilities that can contain hundreds of kilometers of pipes and ducts, pumps, ovens, centrifuges, filters, diffusers, and so on. During years of operation, significant quantities of uranium and/or plutonium can build up in this equipment. Operators need to know the location and amount of holdup for reasons of accountability, criticality safety, radiation safety, waste management, and efficient plant operation. Sometimes the term *holdup* is also applied to in-process inventory, if this must be known for verification or accountability purposes. Holdup is difficult to measure, and though it is usually a small fraction of plant throughput, it can often amount to many kilograms of nuclear material, and this limits the accuracy of the nuclear material balance within the facility. A diverter could, in principle, remove one or more significant quantities (SQ) of HEU or plutonium and hide the loss in the uncertain material balance caused by holdup deposits within the plant. IAEA safeguards inspectors rarely attempt to measure holdup; although they have participated with Los Alamos and the operator in a holdup measurement campaign at the Ulba Fuel Fabrication Plant in Ust-Kamenogorsk, Kazakhstan. The 1991 *Passive Nondestructive Assay of Nuclear Materials* presents a good summary of holdup measurements.[1]

Holdup measurements must cover a range of material types. Process history determines which materials may be deposited. The range of deposit thickness, presence of different material types (isotopic mixtures), and chemistry influence and complicate holdup measurements. The range of ^{235}U enrichment in some facilities includes depleted (0.3%) up to 97%, and that of ^{240}Pu at other facilities ranges from 2% to 45%. Because the equipment in large facilities is extensive, the total holdup may be large, even if deposit thicknesses are small.

Holdup measurements are usually made using gamma-ray techniques, although neutron measurements are also used. There is some experience with using thermoluminescent dosimeters (TLD) to measure holdup deposits in glove boxes or heavy equipment where it is difficult to insert gamma-ray detectors. Such dosimeters usually receive most of their dose from x-rays or low-energy gamma rays, so the results are more susceptible to attenuation or geometry effects than those obtained with gamma-ray detectors. However, measurement

[1]D. Reilly, N. Ensslin, H. Smith, and S. Kreiner, Chapter 20, "Attribute and Semiquantitative Measurements," in *Passive Nondestructive Assay of Nuclear Materials*, NUREG/CR-5550, March 1991.

Table 3C.1 Common gamma rays for holdup analysis.

Isotope	$E\gamma$ (keV)	Intensity (γ/g-sec)
^{238}Pu	153	5.9×10^6
^{235}U	186	4.32×10^4
$^{241}Pu - ^{237}U$	208	2.04×10^7
^{239}Pu	414	3.42×10^4
^{241}Am	662	4.61×10^5
^{238}U	1001	73

performance can be comparable if the TLDs are carefully calibrated using mockups of the equipment to be measured.[2]

Gamma rays have several advantages over neutrons in measuring holdup, because they are easily collimated, allowing the locations and distributions of deposits to be defined. The gamma-ray peaks confirm the identities of the isotopes present. Multiple isotopes and elements can be measured independently and simultaneously by choosing the detector and peaks appropriately. Shielded gamma-ray detectors and the required electronics can be small and lightweight so that measurements can be performed in locations that are difficult to access.

Gamma-Ray Signatures and Equipment

Faced with a mix of material types for plutonium or uranium, the resolution provided by germanium or Peltier-cooled CdTe should be considered if there are possible biases from spectral interferences. When process knowledge is unable to specify isotopics, these high-resolution detectors may be required for preliminary surveys. When isotopic composition is sufficiently well known and interferences unlikely, even low-resolution scintillators (sodium iodide-NaI, bismuth germanate-BGO) can make useful holdup measurements. Table 3C.1 lists the gamma-ray peaks commonly chosen to measure the nuclides of interest.

If scintillators like NaI or BGO are used, it should be noted that they exhibit a strong gain dependence on temperature. The effective gain of NaI may drop by 1–3% per 10-degree increase in centigrade temperature. A simple and practical stabilization technique is to regularly measure a gamma-ray source to compensate for drift. The 60-keV gamma ray from ^{241}Am ($t_{1/2} = 460$ y) is commonly used as a reference peak.

Figure 3A.4 of the "Nuclear Material Measurement Technologies" section shows the gamma-ray spectrum from low-burnup (93% ^{239}Pu) plutonium measured with four different detectors (NaI, coplanargrid cadmium-zinc-telluride [CPG CZT], Ge, and Peltier-cooled CdTe). The detector most commonly used for holdup measurements is NaI. A NaI thickness of 1.25 cm absorbs 80% of ^{235}U gamma rays at 186 keV. A thickness of 5 cm absorbs 85% of ^{239}Pu gamma rays at 414 keV. The intermediate-resolution CZT is equal in sensitivity to the 2.5-cm-diameter NaI in spite of its limited size. (Cubic crystals as large as 1.5 cm on a side absorb up to 95% and 40% of gamma rays at 186 and 414 keV.)

Interferences can add unwanted counts to the assay peak. Detectors with improved resolution and peak shape reduce bias from interference. The use of Ge detectors is generally not possible because they are too heavy. Recent progress with CPG CdZnTe detectors is favorable for portable gamma-ray measurements.[3] A large CPG CZT detector can

[2]H. Preston and W. Symons, "The Determination of Residual Plutonium Masses in Gloveboxes by Remote Measurements Using Solid Thermoluminescent Dosimeters," AEA report AEEW-R1359, Winfrith, UK (1980).

[3]P. N. Luke and E. E. Eissler, "Performance of the CdZnTe Coplanar-Grid Detectors," *IEEE Trans. Nucl. Sci.*, 43, No. 4, pp. 207–213 (Aug. 1995).

FIGURE 3C.1 The Peltier-cooled CdTe detector is shown measuring Pu isotopics in a glove box.

resolve interfering gamma rays from the ^{232}Th decay chain that appear in recycled uranium. A gamma ray at 238 keV is produced at the end of this decay chain. It is not resolved from the 186-keV gamma in NaI, but it does not interfere in Ge or CZT. Gamma ray peaks from ^{241}Pu-^{237}U (332 keV), ^{241}Am (323–335 keV, 662 keV), and ^{237}Np-^{233}Pa contribute to bias in the NaI assay of ^{239}Pu at 414 keV. Many of these effects are readily addressed with CZT.

The recent availability of Peltier-cooled CdTe detectors with crystals larger than 1 cm^3 has made gamma-ray isotopics for uranium and plutonium truly portable. The earlier figure (Figure 3A.4 in the Measurement section) illustrates the good energy resolution of CdTe. Figure 3C.1 illustrates the compact dimensions of the CdTe detector, shown in use for portable Pu isotopics measurements in a glove box. The capability range of CdTe for isotopic analysis covers 3%- to 30%-^{240}Pu; it also covers ^{235}U from 0.1 to ~80%, and MOX. A 15-min. count with a CdTe detector measures the ^{240}Pu fraction to 2% and the ^{235}U fraction to 3%.

Generalized Geometry Holdup (GGH) Assay Method

Assumptions and Constraints

The *generalized geometry holdup* (GGH) method categorizes each geometry, no matter how complex, as a series of simple point, line, or area deposits.[4] This idea is illustrated in Figure 3C.2. The GGH assay method was developed to simplify the analysis of holdup measurements performed using NaI detectors. It can, however, be applied to any detector. The analysis of holdup data using GGH requires the following constraints:

1. Radiation shielding is used on the back and sides of the crystal.
2. A cylindrical collimator is installed on the front of the crystal.

[4]P. A. Russo, "Gamma-Ray Measurements of Holdup Plant-Wide: Application Guide for Portable, Generalized Approach," Los Alamos National Laboratory report, LA-UR-04-8365, Nov. 2004.

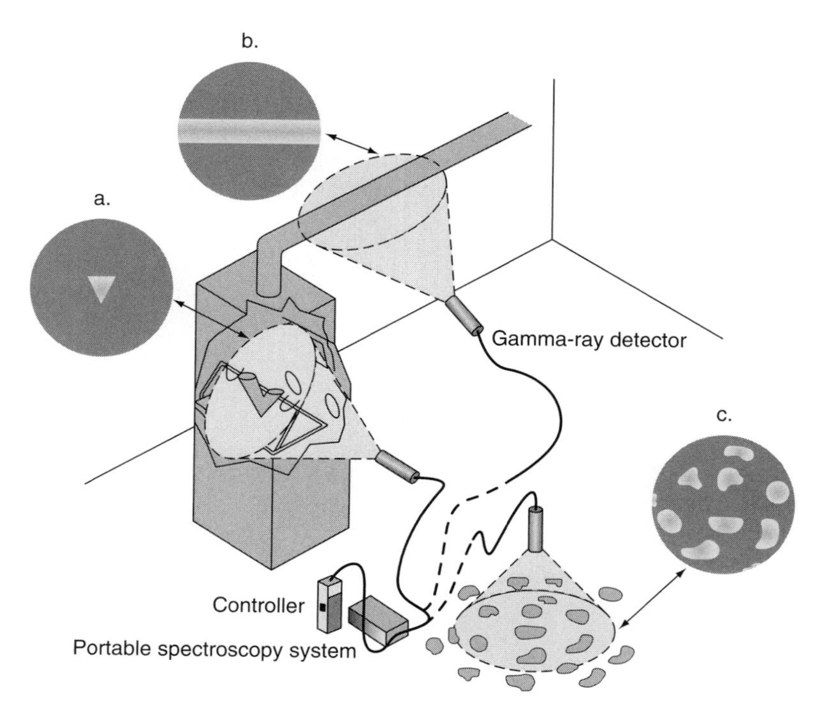

FIGURE 3C.2 Illustration of point (a), line (b), and area (c) holdup deposits.

3. The detector is positioned so that the deposit can be approximated as:
   - A small point at its center, or
   - A narrow, uniform line through its center whose length exceeds its width, or
   - A uniform distribution that fills it (area deposit)
4. Measurements are performed at a known distance r between the detector and the deposit.

Calibration

The calibration of the GGH method determines the relationship between the count rate of the measured gamma ray and the mass of the isotope of interest. Calibration of the assay of a point, line, or area deposit is accomplished with a point source. The response for each gamma-ray peak is measured with this source positioned on the detector axis at a known distance from the crystal. Measurements are also performed with the source displaced at fixed intervals from the crystal axis to obtain the two-dimensional radial response of the detector. These data are used to obtain the calibration for the assay of the specific isotope mass in a point, line, or area deposit.

Performing the GGH Measurement and Assay

Using the GGH method to determine uranium or plutonium holdup requires a portable spectroscopy system and a calibrated detector. Because count times are often very short (5–15 s), the random uncertainty can be large for individual measurements. Propagating the uncertainties of the many measurements to get the total holdup in a piece of equipment greatly reduces the random error.

The initial assay result is the specific isotope mass for a point, line, or area deposit. Three additional corrections are required for equipment attenuation, finite-source dimensions, and the self-attenuation of the deposit. In recent measurements of ^{239}Pu holdup in

FIGURE 3C.3 GGH applied in a uranium facility to measure an overhead duct.

bulk-processing equipment using the 414-keV gamma ray, the equipment attenuation correction factor varied from a low of 1.1 (lead-lined gloves) to a high of 6.2 (steel plates on a glove-box floor).

Holdup Measurement System Examples

The Integrated Holdup Measurement System at the Y-12 HEU plant in Oak Ridge, Tennessee, is a good example of a comprehensive holdup measurement system.[5] Y-12 was constructed during WWII as part of the Manhattan Project. It was originally built to house the Electromagnetic Isotope Separation (EMIS) Program, which produced the HEU for "Little Boy," the bomb dropped on Hiroshima, Japan, on August 6, 1945.

After the war, it was decided that the gaseous diffusion process at K-25 was more efficient, and the EMIS machines, called Calutrons, were removed and Y-12 was converted to process HEU for nuclear weapons. A few Calutrons were saved and used as isotope separators to produce small amounts of pure isotopes of uranium and plutonium.

The HEU facility continues to operate today. To measure HEU holdup within the plant, Y-12 has identified many thousands of measurement points, each indicated by a barcode. Operators carry a small multichannel analyzer (MCA), a collimated NaI detector, and a handheld barcode reader with a data logger/controller. Thousands of locations are measured each month. Data from the data logger is downloaded into a computer running a program called HMS4. This has been used successfully for more than seven years. An extensive study was made of system performance using simulated holdup situations such as pipes, ducts, and V-blenders with known U or Pu sources. Figure 3C.3 shows a technician at Y-12 measuring an overhead duct.

Figures 3C.4 and 3C.5 show other ^{235}U holdup measurements in a uranium processing facility using Ge, CZT, and NaI detectors. Figure 3C.5 shows a very large overhead duct being measured with a portable Ge detector weighing ~10 kg with collimator. Figure 3C.6

---

[5] P. A. Russo, H. A. Smith, J. K. Sprinkle, Jr., C. W. Bjork, G. A. Sheppard, and S. E. Smith, "Evaluation of an Integrated Holdup Measurement System Using the GGH Formalism with the M^{3}CA," published in the *Proceedings of the Fifth International Conference on Facility Operation – Safeguards Interface*, LaGrange Park, IL: American Nuclear Society (1996), pp. 239–248.

FIGURE 3C.4 A large overhead duct is measured from below with a collimated Ge detector.

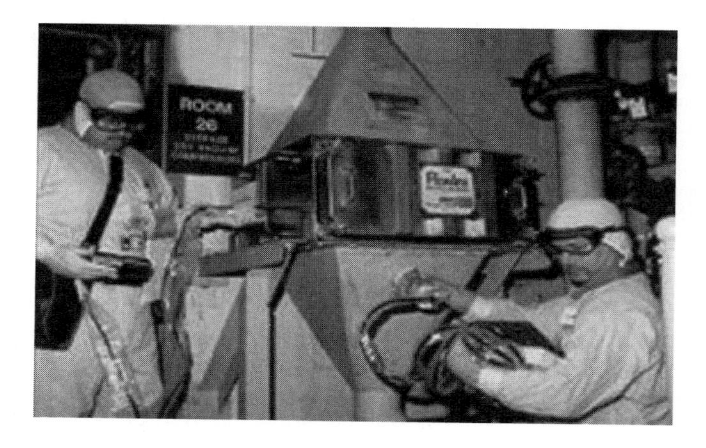

FIGURE 3C.5 Measurements of ^{235}U deposits in a filter system performed with CZT and NaI.

shows CZT and NaI detectors weighing ~1 kg each with collimators. The greater portability of the room-temperature detectors is essential for most holdup measurements.

The Rocky Flats Plant near Denver, Colorado, was built to process plutonium and produce "pits," which are the fission core of thermonuclear weapons. The plant ceased operations in 1993 and has now been dismantled, cleaned, and converted into an environment park. During its operating lifetime (~50 y), Rocky Flats accumulated large quantities of plutonium in the glove boxes, filters, calciners, pipes, and air duct systems of several major processing buildings. This holdup was a significant health and criticality safety concern and at times was a major contributor to the material unaccounted for (MUF) for the facility.

During the decommissioning of the processing buildings, the holdup measurement campaigns were among the largest and most extensive ever reported. The holdup measurement teams pioneered the use of medium-resolution BGO detectors and the use of measurements made with the detectors in contact with pipes or ducts. Although this approach is more susceptible to uncertainties in material distribution than the GGH methodology, it allows routine measurements to be made more quickly. As buildings were decommissioned and the process lines were removed and cleaned out, it was often possible to obtain comparisons

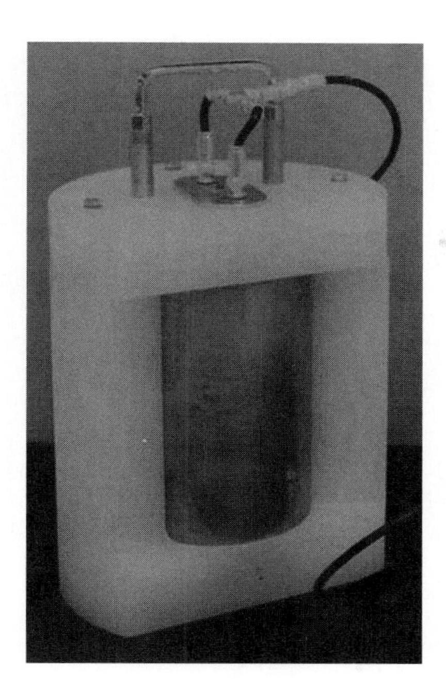

FIGURE 3C.6 Shielded Neutron Assay Probe (SNAP-II) used for U and Pu holdup measurements in heavily shielded situations.

between the measured holdup and cleanout values. The overall results of numerous measurements of extended equipment lines tended to be unbiased.[6]

Neutron Holdup Measurements

Nuclear material processing equipment can be massive and extensive. High-throughput facilities may contain multikilogram deposits. The high attenuation of such equipment and deposits may challenge the capability of gamma-ray holdup measurements. In such cases, the high penetrability of neutrons offers a more reliable option. Neutrons can be detected from pumps, furnaces, and other heavy equipment that is too dense to permit gamma rays to escape. Large polyethylene-moderated ^{3}He slab detectors used to quantify in-process plutonium in glove boxes are a successful tool. Although the spontaneous-fission neutron yield from uranium is low for coincidence counting, the high α,n yield from fluorine enables measurements of uranium deposits using total neutrons from UF_6 and UO_2F_2 in enrichment plants.

Lightweight, directional, portable neutron counters are difficult to design because of the need for a polyethylene moderator surrounding the ^{3}He tubes. However, several reasonably portable detectors have been designed and used for holdup measurements. The original portable counter was the Shielded Neutron Assay Probe (SNAP-II) fabricated in 1975 (see Figure 3C.6). The SNAP-II had two 20-cm-long ^{3}He tubes in a 12.7-cm-diam polyethylene cylinder, wrapped in cadmium and surrounded for 240° by a 5.7-cm-thick directional shield. The intrinsic efficiency of the SNAP-II for fission neutrons was ~17% and it weighed 10 kg. The SNAP-II was used to measure uranium holdup in operating and shutdown gaseous diffusion enrichment plants and plutonium holdup in several scrap recovery facilities.

---

[6]F. W. Lamb, "A Frank Look at Lessons Learned During Holdup Measurements at RFETS: Part 2—Measurements," Rocky Flats Environmental Technology Site report RFP#5560 (April 2005), published in *Nucl. Mater. Manage.* XXXVI, No. 2, pp. 31–34, Winter 2008.

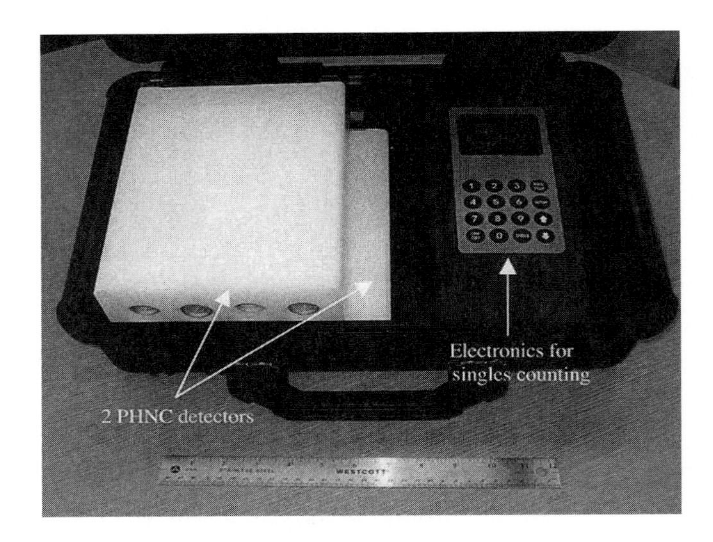

FIGURE 3C.7 Two portable handheld neutron counters (PHNC) and a related electronics package.

FIGURE 3C.8 Wide-area neutron detectors used to measure holdup at gaseous diffusion enrichment plants.

The portable handheld neutron counter (PHNC) shown in Figure 3C.7 is a newer instrument with four ^{3}He tubes. The PHNC has no directionality, but it is more efficient and lighter weight than the SNAP-II. The PHNC is designed for either singles counting of wide-area holdup sources or coincidence counting, with two PHNC slabs, of small containers of Pu materials.[7]

Large slab detectors can provide higher efficiency and better directionality if heavy shielding and collimators are added. Such detectors are too heavy for handheld operation, but they can be moved with carts or can be permanently installed to measure in-process inventory or holdup after cleanup. The slab detectors shown in Figure 3C.8 were used to measure holdup and in-process inventory at uranium enrichment plants.

A new neutron holdup assay method for enrichment facilities, or in any facility with a large distributed volume of material, was recently described.[8] The distributed source-term analysis (DSTA) technique uses Monte Carlo modeling of a centrifuge enrichment cascade

---

[7]H. O. Menlove, "Manual for the Portable Handheld Neutron Counter (PHNC) for Neutron Survey and the Measurement of Plutonium Samples," Los Alamos National Laboratory manual LA-14257-M (Nov. 2005).

[8]D. H. Beddingfield and H. O. Menlove, "A New Approach to Hold-Up Measurement in Uranium Enrichment Facilities," Los Alamos National Laboratory report LA-UR-00-2534, published in *Nucl. Mater. Manage.* XXIX (Proc. Issue) CD-ROM (2000).

hall to derive a calibration curve relating the average neutron count rate to the mass of uranium holdup. Then a portable counter, similar to the PHNC, is used to survey the average neutron count rate in the hall. This approach avoids the high attenuation problems of gamma-ray measurement, the difficulties in measuring individual pieces of equipment to obtain the total holdup, and the long measurements required to assay the entire process line.

Large slab detectors have also been used to measure plutonium holdup in rotary calciners, hydrofluorinators, and other large, highly attenuating items. Two slab detectors, each 50 cm tall by 100 cm wide by 8 cm thick polyethylene with 10 90 cm long ^{3}He tubes, were placed in a rack and moved to various locations around a rotary calciner used to dry plutonium peroxide. The calciner was inside a glove box. The quantity of holdup was sufficiently high to use neutron coincidence counting, so the measurements were less sensitive to background neutrons. The detectors were calibrated with a 2-kg PuO_2 standard. After one of the measurement campaigns, the calciner was cleaned out and the recovered plutonium measured. The holdup assay agreed very well with the recovery value.[9]

Large neutron slab detectors are used in the Plutonium Fuel Production Facility (PFPF) in Tokai-Mura, Japan, to measure holdup in glove boxes inside this automated MOX fuel fabrication facility.[10] These glove-box assay systems (GBAS) are 160 cm high, 100 cm long, and 7.6 cm wide. Each slab contains 20 152-cm ^{3}He tubes. Monte Carlo calculations were used to design the detector and study its response before installation. Six slabs were originally installed in pairs on either side of a glove box. The slabs could be moved remotely to measure different locations on a glove box. A standard matrix of measurement positions was assigned for each glove box and software written to collect, analyze, and combine all the measurements. Measurement data from this system are shared by the IAEA and the facility operator. Experience at PFPF has shown a measurement uncertainty of $\sim$5% for neutron assay and 25–30% for gamma-ray assay.

Accuracy of Gamma-Ray and Neutron Holdup Measurements

The precision or random error can be readily determined for holdup measurements. Because of the many measurements performed, the overall precision is usually of the order of a few percent or less. However, the accuracy or systematic error is very difficult to determine because it is difficult to know the true mass of nuclear material held up in the equipment of a complex facility. Often, the accuracy estimate for a holdup campaign is simply the "best guess" of the measurement team based on judgment and experience. Such estimates are typically in the range of 25–50% or more because of the many unknown factors and assumptions required to calculate the nuclear material mass. In some cases, such as glove boxes, known standards can be introduced and measured in addition to the holdup. In a few cases, an effort was made to clean out and recover the measured material, which was then analyzed destructively and compared with the measured holdup. A complete cleanout is usually difficult and costly, but this is the best way to determine holdup assay accuracy.

In the early 1980s, a holdup measurement campaign was conducted at a shutdown part of the Portsmouth Gaseous Diffusion Plant (PGDP) in Ohio. Gamma-ray measurements were made with a collimated NaI detector and neutron measurements were made using the slab detectors shown in Figure 3C.8. A total of approximately 250 stages (converter, cooler, compressor, and piping) were measured during the campaign. Afterward, three cells (12 stages each) were cleaned out and the uranium recovered. The U was also measured and

---

[9] D. B. Smith, "Safeguards and Security Progress Report, January–December 1985," Los Alamos National Laboratory report LA-10787-PR (March 1987), pp. 9–14.

[10] M. C. Miller, H. O. Menlove, M. Seya, S. Takahashi, and R. Abedin-Zadeh, "Holdup Counter for the Plutonium Fuel Production Facility—PFPF," Los Alamos National Laboratory report LA-UR-90-2312, published in *Nucl. Mater. Manage.*, X (Proc. Issue, July 1990).

Table 3C.2 Evaluation of PGDP holdup assay.

Cell	n kg U[a]	γ kg U[b]	Recovery kg U
A	177	45	120
B	32	3	28
C	29	12	25
Isolated converter	9	10	7

[a]The neutron counters were not well collimated and measured an entire stage and double-counted the cooler.
[b]Gamma-ray measurements covered only the converters.

Table 3C.3 GGH holdup assay evaluation.

	^{235}U[a]	^{239}Pu[a]
Pipe array	0.90	0.72
V-blender	1.22	.02
Al pipe	1.03	.97
Steel pipe	0.97	1.47
Floor spot	0.96	n/a
Duct	1.07	0.96

[a]Number listed is the average ratio of measured U or Pu to the reference value.

recovered from an isolated converter. The results from this activity are summarized in Table 3C.2. Because the gamma measurements only covered the converters, they should only be compared with the neutron assay of the isolated converter. These results are typical of what one finds in such holdup studies.[11]

A six-year study was conducted on the accuracy and precision of holdup measurements using the GGH (gamma-ray assay) approach to measuring simulated holdup situations with well-known nuclear material standards. A series of simulations were fabricated for this study and a training course; they included a pipe array, a steel pipe, an aluminum pipe, a rectangular ventilation duct, a V-Blender, and a contaminated spot on a floor. These were "salted" with U or Pu fuel rods, U metal foils, and small cans of UO_2 or PuO_2. Table 3C.3 summarizes the results of this study, which included measurements made by many people, ranging from students to holdup experts. The results shown here are "best case" vis-à-vis assay accuracy.[12]

[11]D. B. Smith, "Safeguards and Security Progress Report, August 1982–January 1983," Los Alamos National Laboratory report LA-9821-PR (Nov. 1983), pp. 11–17.

[12]P. A. Russo, et al., "Evaluation of the Integrated Holdup Measurement System with the M3CA for Assay of Uranium and Plutonium Holdup," Los Alamos National Laboratory report LA-13387-MS (Aug. 1999).

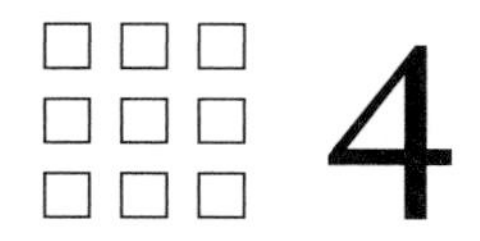

4 Physical Protection

Mary Lynn Garcia

Introduction

A *physical protection system* (PPS) integrates people, procedures, and equipment for the protection of assets or facilities against theft, sabotage, or other malevolent human attacks. The design of an effective PPS requires a methodical approach in which the designer weighs the objectives of the PPS against available resources and then evaluates the proposed design to determine how well it meets the objectives. Without this kind of careful assessment, the PPS might waste valuable resources on unnecessary protection or, worse yet, fail to provide adequate protection at critical points of the facility.

For example, it would probably be unwise to protect a facility's employee cafeteria with the same level of protection as the central computing area. Similarly, maximum security at a facility's main entrance would be wasted if entry were also possible through an unprotected cafeteria loading dock. Each facility is unique, even if performing generally the same activities, so this systematic approach allows flexibility in the application of security tools to address local conditions.

The process of designing and analyzing a PPS is described in the remainder of this chapter. The methodology presented here is the same one used by Sandia National Laboratories in designing a PPS for critical nuclear assets (Williams, 1978). This approach and supporting tools were developed and validated over the past 25 years through research funded by the U.S. Department of Energy (DOE) and development totaling over $200 million.

Although other industrial and governmental assets may not require the highest levels of security used at nuclear weapons sites, the approach is the same whether protecting a manufacturing facility, an oil refinery, or a retail store. The foundation of this approach is the design of an integrated performance-based system. Performance measures (i.e., validated numeric characteristics) for various system components, such as sensors, video, or response time, allow the use of models to predict system performance against the identified threat. This effectiveness measure can then be used to provide the business rationale for investing in the system or upgrade, based on a measurable increase in system performance and an associated decrease in risk to the facility. Looking at system improvement compared to costs can then support a cost/benefit analysis. By following this process, the system designer will include elements of business, technology, and the criminal justice system into the most effective design within the facility's constraints and budget. Before describing this process in more detail, however, it is first necessary to differentiate between safety and security.

Safety versus Security

For the purposes of this chapter, the term *safety* is meant to represent the operation of systems in abnormal environments, such as flood, fire, earthquake, electrical faults, or accidents. *Security*, on the other hand, refers to systems used to prevent or detect an attack by a malevolent human adversary. There are some overlaps between the two: for example, the response to a fire may be the same whether the fire is the result of an electrical short or a terrorist bomb. It is useful, however, to recognize that a fire has no powers of reasoning, whereas adversaries do. A fire burns as long as there is fuel and oxygen; if these elements are removed, the fire goes out. An attack by a malevolent human adversary, on the other hand, requires that we recognize that adversary's capability to adapt and thus eventually defeat the security system.

In the event of a safety critical event such as a fire, security personnel should have a defined role in assisting, without compromising the security readiness of a facility. In this regard, security personnel should not be overloaded with safety-related tasks, because this may increase exposure of the facility to a security event during an emergency condition, particularly if the adversary creates this event as a diversion or takes advantage of the opportunity an event presents. In addition, security personnel may not possess the specific knowledge or training to respond to safety events. For example, in case of a fire, security personnel should not be expected to shut down power or equipment. This task is better left to those familiar with the operation and shutdown of equipment, power, or production lines. Procedures describing the role of security personnel in these events should be developed, understood, and practiced in advance to assure adequate levels of protection and safety.

Deterrence

Theft, sabotage, and other malevolent acts at a facility may be prevented in two ways: by deterring the adversary or by defeating the adversary. *Deterrence* occurs by implementing measures that are perceived by potential adversaries as too difficult to defeat; it makes the facility an unattractive target, so the adversary abandons or never attempts an attack. Examples of deterrents are the presence of security guards in parking lots, adequate lighting at night, posting of signs, and the use of barriers, such as bars on windows. These are features that are often implemented with no additional layers of protection in the event of an attack. Deterrence can be very helpful in discouraging attacks by adversaries; however, it is less useful against an adversary who chooses to attack anyway.

It would be a mistake to assume that because an adversary has not challenged a system, the effectiveness of the system has been proven. The deterrence function of a PPS is difficult to measure, and reliance on successful deterrence can be risky; thus it is considered a secondary function. The deterrent value of a true PPS, on the other hand, can be very high, while at the same time providing protection of assets in the event of an attack. The purpose of this chapter is to describe a process that produces an effective PPS design, validates its performance, and relates the improvement in system effectiveness to the cost. Application of this process allows the design of a PPS that will protect assets during an actual attack as well as provide additional benefits through deterrence.

As more research is done on the measurable and long-term value of deterrents, these may be incorporated into protection system design. To date, however, there is no statistically valid information to support the effectiveness of deterrents. There are, however, studies that indicate that deterrence is not as effective after implementation as is hoped (Sivarajasingam and Shepherd, 1999).

Process Overview

The design of an effective PPS includes the determination of the PPS objectives, the initial design or characterization of a PPS, the evaluation of the design, and, in many cases, a redesign or refinement of the system. To develop the objectives, the designer must begin by gathering information about facility operations and conditions, such as a comprehensive description of the facility, operating states, and the physical protection requirements.

The designer then needs to define the threat. This task involves considering factors about potential adversaries, such as class, capabilities, and range of tactics. Next, the designer should identify targets. Targets may be physical assets, electronic data, people, or anything that could impact business operations. The designer now knows the objectives of the PPS, that is, what to protect against whom.

The next step is to design the new system or characterize the existing system. If designing a new system, people, procedures, and equipment must be integrated to meet the objectives of the system. If the system already exists, it must be characterized to establish a baseline of performance.

After the PPS is designed or characterized, it must be analyzed and evaluated to ensure that it meets the physical protection objectives. Evaluation must allow for features working together to assure protection rather than regarding each feature separately. Due to the complexity of protection systems, an evaluation usually requires modeling techniques. If any vulnerabilities are found, the initial system must be redesigned to correct the vulnerabilities and a reevaluation conducted.

PPS Design and Evaluation Process: Objectives

A graphical representation of the PPS methodology is shown in Figure 4.1. As stated, the first step in the process is to determine the objectives of the protection system. To formulate these objectives, the designer must (1) characterize (understand) the facility operations and conditions, (2) define the threat, and (3) identify the targets.

Characterization of facility operations and conditions requires developing a thorough description of the facility itself (the location of the site boundary, building location, building interior floor plans, access points). A description of the processes within the facility is also required, as is identification of any existing physical protection features. This information can be obtained from several sources, including facility design blueprints, process descriptions, safety analysis reports, and environmental impact statements.

In addition to acquisition and review of such documentation, a tour of the site under consideration and interviews with facility personnel are necessary. These steps provide an understanding of the physical protection requirements for the facility as well as an appreciation

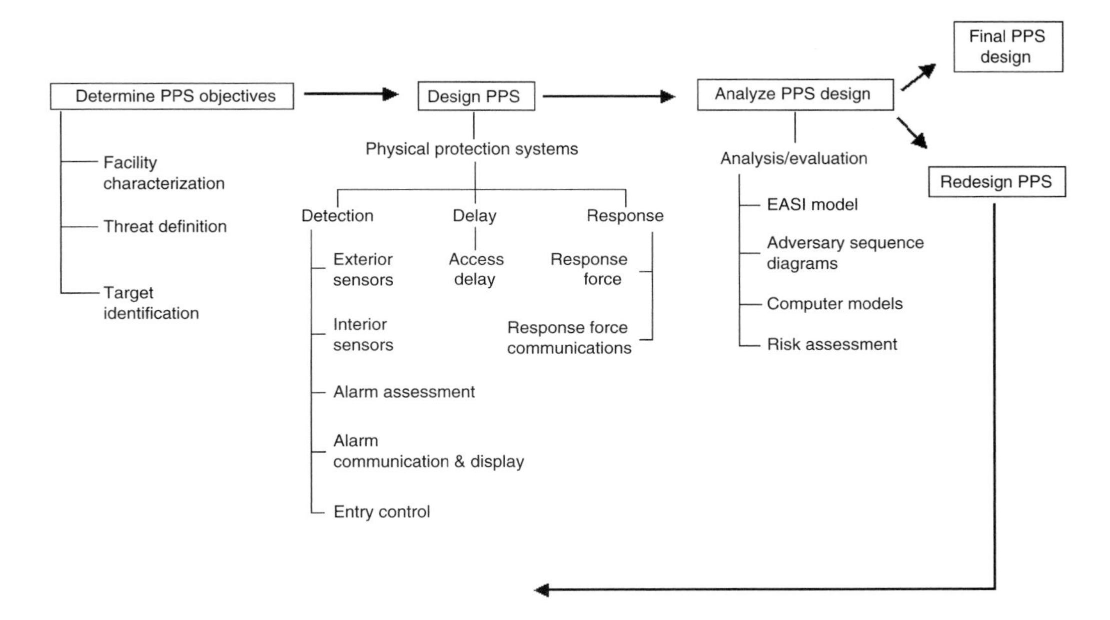

FIGURE 4.1 Design and evaluation process for physical protection systems. The process starts with determining objectives, then designing a system to meet the objectives. It ends with an evaluation of how well the system performs compared to the objectives.

for the operational and safety constraints, which must be considered. Each facility is unique, so the process should be followed each time a need is identified. Compromises must usually be made on all sides so that operation can continue in a safe and efficient environment while physical protection is maintained. Additional considerations also include an understanding of liability and any legal or regulatory requirements that must be followed.

Next, a threat definition for the facility must be made. Information must be collected to answer three questions about the adversary:

1. What class of adversary is to be considered?
2. What is the range of the adversary's tactics?
3. What are the adversary's capabilities?

Adversaries can be separated into three classes: outsiders, insiders, and outsiders working in collusion with insiders. For each class of adversary, the full range of tactics (deceit, force, stealth, or any combination of these) should be considered. Deceit is the attempted defeat of a security system by using false authorization and identification; force is the overt, forcible attempt to overcome a security system; and stealth is any attempt to defeat the detection system and enter the facility covertly.

For any given facility there may be several threats, such as a criminal outsider, a disgruntled employee, competitors, or some combination of these, so the PPS must be designed to protect against all of these threats. Choosing the most likely threat, designing the system to meet this threat, and then testing to verify the system performance against the other threats will facilitate this process.

Finally, target identification should be performed for the facility. Targets may include critical assets or information, people, or critical areas and processes. A thorough review of the facility and its assets should be conducted. Answering such questions as, "What losses will be incurred in the event of sabotage of this equipment?" will help identify the assets or equipment that are most vulnerable or that create an unacceptable consequence.

Given the information obtained through facility characterization, threat definition, and target identification, the designer can determine the PPS's protection objectives. An example of a protection objective might be to interrupt a criminal adversary equipped with hand tools and a vehicle before finished central processing units (CPUs) or microprocessors can be removed from the shipping dock. The process of determining objectives will be somewhat recursive. That is, definition of the threat will depend on target identification, and vice versa. This recursion should be expected and is indicative of the complex relationships among protection system objectives.

PPS Design and Evaluation Process: Design PPS

The next step in the process, if designing a new PPS, is to determine how best to combine such elements as fences, barriers, sensors, procedures, communication devices, and security personnel into a PPS that can achieve the protection objectives. The resulting PPS design should meet these objectives within the operational, safety, legal, and economic constraints of the facility. The primary functions of a PPS are detection of an adversary, delay of that adversary, and response by security personnel (guard force).

Certain guidelines should be observed during the PPS design. A PPS system performs better if detection is as far from the target as possible and delays are near the target. In addition, there is close association between detection (exterior or interior) and assessment. The designer should be aware that detection without assessment is not detection.

Another close association is the relationship between response and response force communications. A response force cannot respond unless it receives a communication call for a response. These and many other particular features of PPS components help ensure that the designer takes advantage of the strengths of each piece of equipment and uses equipment in combinations that complement each other and protect any weaknesses.

PPS Design and Evaluation Process: Evaluate PPS

Analysis and evaluation of the PPS design begin with a review and thorough understanding of the protection objectives the designed system must meet. This can be done simply by checking for required features of a PPS, such as intrusion detection, entry control, access delay, response communications, and a response force. However, a PPS design based on required features cannot be expected to lead to a high-performance system unless those features, when used together, are sufficient to assure adequate levels of protection. More sophisticated analysis and evaluation techniques can be used to estimate the minimum performance levels achieved by a PPS. These techniques include qualitative and quantitative analysis. Systems that are designed to protect high-value critical assets generally require a quantitative analysis. Systems protecting lower-value assets may be analyzed using less rigorous qualitative techniques. To complete a quantitative analysis, performance data must be available for the system components.

An existing PPS at an operational facility cannot normally be fully tested as a system. This sort of test would be highly disruptive to the operation of the facility and could impact production schedules as well as security effectiveness (i.e., create a vulnerability). Because direct system tests are not practical, evaluation techniques are based on performance tests of component subsystems. Component performance estimates are combined into system performance estimates by the application of system modeling techniques.

The end result of this phase of the design and analysis process is a system vulnerability assessment. Analysis of the PPS design will either find that the design effectively achieved the protection objectives or it will identify weaknesses. If the protection objectives are achieved, then the design and analysis process is completed. However, the PPS should be analyzed periodically to ensure that the original protection objectives remain valid and that the protection system continues to meet them.

If the PPS is found to be ineffective, vulnerabilities in the system can be identified. The next step in the design and analysis cycle is to redesign or upgrade the initial protection system design to correct the noted vulnerabilities. It is possible that the PPS objectives also need to be reevaluated. An analysis of the redesigned system is performed. This cycle continues until the results indicate that the PPS meets the protection objectives.

Physical Protection System Design

A system may be defined as an integrated collection of components or elements designed to achieve an objective according to a plan. The designer of any system must have the system's ultimate objective in mind. The ultimate objective of a PPS is to prevent the accomplishment of malevolent overt or covert actions. Typical objectives are to prevent sabotage of critical equipment, prevent theft of assets or information from within the facility, and protect people (executive protection or workplace violence). A PPS must accomplish its objectives by either deterrence or a combination of detection, delay, and response.

The PPS functions of detection and delay can be accomplished by the use of equipment and guards. Facility guards usually handle response. There is always a balance between the use of equipment and the use of guards. In different conditions and applications, one is often the preferable choice. As technology improves, the mix of equipment and guards will change and increase system effectiveness. The key to a successful protection system is the integration of people, procedures, and equipment into a system that protects assets from malevolent adversaries.

Detection, delay, and response are all required functions of an effective PPS. These functions must be performed in this order and within a length of time that is less than the time required for the adversary to complete their task. A well-designed system provides protection in depth, minimizes the consequence of component failures, and exhibits balanced protection. In addition, a design process based on performance criteria rather than feature criteria will select elements and procedures according to the contribution they make to overall system performance. Performance criteria are also measurable, so they aid in the analysis of the designed system.

PPS Functions

The purpose of a PPS is to prevent an adversary from successful completion of a malevolent action against a facility. There are several functions that the PPS must perform. The primary PPS functions are detection, delay, and response. It is essential to consider the system functions in detail, since a thorough understanding of the definitions of these functions and the measure of effectiveness of each is required to evaluate the system. It is important to note that detection must be accomplished for delay to be effective. Remember that the system goal is to protect assets from a malevolent adversary. For a system to be effective at this objective there must be awareness that there is an attack (detection) and slowing of adversary progress to the targets (delay), thus allowing the response force enough time to interrupt or stop the adversary (response).

Detection

Detection is the discovery of an adversary action. It includes sensing of covert or overt actions. The measures of effectiveness for the detection function are the probability of sensing adversary action and the time required for reporting and assessing the alarm. The probability of assessed detection for a particular sensor captures both of these measures. Included in the detection function of physical protection is *entry control*, which refers to allowing entry to authorized personnel and detecting the attempted entry of unauthorized personnel and material. The measures of effectiveness of entry control are throughput, false acceptance rate, and false rejection rate. *Throughput* is defined as the number of authorized personnel allowed access per unit of time, assuming that all personnel who attempt entry are authorized for entrance. *False acceptance* is the rate at which false identities or credentials are allowed entry, and the *false rejection rate* is the frequency of denying access to authorized personnel.

The response force can also accomplish detection. Guards at fixed posts or on patrol may serve a vital role in sensing an intrusion. However, this decision must be carefully considered. Once an alarm is initiated and reported, assessment begins. An effective assessment system provides two types of information associated with detection. This information includes whether the alarm is a valid alarm or a nuisance alarm and details about the cause of the alarm—what, who, where, and how many.

Delay

Delay is the second function of a PPS. It is the slowing down of adversary progress. Delay can be accomplished by personnel, barriers, locks, and activated delays. Response force personnel can be considered elements of delay if they are in fixed and well-protected positions. The measure of delay effectiveness is the time required by the adversary (after detection) to bypass each delay element. Although the adversary may be delayed prior to detection, this delay is of no value to the effectiveness of the PPS because it does not provide additional time to respond to the adversary. Delay before detection is primarily a deterrent.

Response

The *response* function consists of the actions taken by the response force to prevent adversary success. Response can include both interruption and neutralization. *Interruption* is defined as a sufficient number of response force personnel arriving at the appropriate location to stop the adversary's progress. It includes communication to the response force of accurate information about adversary actions and the deployment of the response force. *Neutralization* describes the actions and effectiveness of the responders after interruption. The primary measure of response effectiveness is the time between receipt of a communication of adversary action and the interruption of the adversary action. At sites where there is no immediate response, it is assumed that the asset can be lost and this is an acceptable risk. In these cases, the primary response may be after-loss-event investigation, recovery of the asset, and criminal prosecution.

Deployment describes the actions of the response force from the time communication is received until the force is in position to interrupt the adversary. The effectiveness measure of this function is the probability of deployment to the adversary location and the time required to deploy the response force.

Design Goals

The effectiveness of the PPS functions of detection, delay, and response and their relationship have already been discussed. In addition, all the hardware elements of the system must be properly installed, maintained, and operated. The procedures of the PPS must be compatible with facility operations and procedures. Security, safety, and operational objectives must be accomplished at all times. A PPS that has been well engineered will be based on sound principles, including protection-in-depth, minimum consequence of component failure, and balanced protection.[1]

Design Criteria

Any design must include criteria (requirements and specifications) against which elements of the design will be evaluated. A design process using performance criteria will select elements and procedures according to the contribution they make to overall system performance. The effectiveness measure will be overall system performance.

A feature criteria (also called compliance-based) approach selects elements or procedures that satisfy requirements for the presence of certain items. The effectiveness measure is the presence of those features. The use of a feature criteria approach in regulations or requirements that apply to PPSs should generally be avoided or handled with extreme care. Unless such care is exercised, a feature criteria approach can lead to the use of a checklist method to determine system adequacy based on the presence or absence of required features. This is clearly not desirable, since overall system performance is of interest, rather than the mere presence or absence of system features or components. For example, a performance criterion for a perimeter detection system would be that the system is able to detect a running intruder using any attack method. A feature criterion for the same detection system might be that the system includes two different sensor types.

Performance Measures

The design and evaluation techniques presented in this text support a performance-based approach to meeting the PPS objectives. Much of the component technology material will, however, be applicable for either performance criteria or feature criteria design methods. The performance measures for a PPS function include probability of detection; probability of and time for alarm communication and assessment; frequency of nuisance alarms; time to defeat obstacles; probability of and time for accurate communication to the response force; probability of response force deployment to adversary location; time to deploy to a location, and response force effectiveness after deployment.

Analysis

A PPS is a complex configuration of detection, delay, and response elements. Computerized techniques are available to analyze a PPS and evaluate its effectiveness (Bennett, 1977;

---

[1]Each of these principles is discussed in more detail in Chapter 5 of *Design and Evaluation of Physical Protection Systems*, second edition (Butterworth-Heinemann, 2007), entitled "Physical Protection System Design."

Chapman and Harlan, 1985). Such techniques identify system deficiencies, evaluate improvements, and perform cost-versus-effectiveness comparisons. These techniques are appropriate for analyzing PPSs at individual sites. Also, the techniques can be used for evaluating either an existing protection system or a proposed system design.

The goal of an adversary is to complete a path to a target with the least likelihood of being stopped by the PPS. To achieve this goal, the adversary may attempt to minimize the time required to complete the path. This strategy involves penetrating barriers with little regard to the probability of being detected. The adversary is successful if the path is completed before guards can respond. Alternatively, the adversary may attempt to minimize detection with little regard to the time required. In this case, the adversary is successful if the path is completed without being detected.

The measure of effectiveness used for interrupting an adversary is timely detection. *Timely detection* refers to the cumulative probability of detecting the adversary at a point where there is enough time remaining on the adversary path for the response force to interrupt the adversary. The delay elements along the path determine the point by which the adversary must be detected. That point is where the minimum delay along the remaining portion of the path just exceeds the guard response time. The probability of interruption (PI) is the cumulative probability of detection from the start of the path up to the point determined by the time remaining for the guards to respond. This value of PI serves as one measure of the PPS effectiveness. At high security facilities with an immediate on-site response (often armed), another measure of response is the probability of neutralization (PN), which is defeat of the adversary after interruption.

Physical Protection System Design and the Relationship to Risk

The design and analysis of a PPS include the determination of the PPS objectives, characterizing the design of the PPS, the evaluation of the design, and, possibly, a redesign or refinement of the system. The process must begin by gathering information about the facility, defining the threat, and then identifying targets. Determining whether or not assets are attractive targets is based mainly on the ease or difficulty of acquisition and the value of the asset. The next step is to characterize the PPS design by defining the detection, delay, and response elements. The PPS is then analyzed and evaluated to ensure that it meets the physical protection objectives. Evaluation must allow for features working together to assure protection rather than regarding each feature separately.

The basic premise of the methodology described in this text is that the design and analysis of physical protection must be accomplished as an integrated system. In this way, all components of detection, delay, and response can be properly weighted according to their contribution to the PPS as a whole. At a higher level, the facility owner must balance the effectiveness of the PPS against available resources and then evaluate the proposed design. Without a methodical, defined, analytical assessment, the PPS might waste valuable resources on unnecessary protection or, worse yet, fail to provide adequate protection at critical points of the facility.

Due to the complexity of protection systems, an evaluation usually requires computer modeling techniques. If any vulnerabilities are found, the initial system must be redesigned to correct the vulnerabilities and a reevaluation conducted. Then the system's overall risk should be calculated. This risk is normalized to the consequence severity if the adversary could attain the target. This means that the consequence of the loss of an asset is represented numerically by a value between zero and one, where the highest consequence of loss is represented by one and other lower consequence losses are assigned correspondingly lower values. This method ranks the consequence of loss of assets from unacceptably high down to very low or no consequence. The facility manager is then able to make a judgment as to the amount of risk that remains and whether this is acceptable.

The risk equation used is:

$$R = P_A{}^*[1 - (P_E)]^*C$$

Each term in the equation will be elaborated more fully throughout the text. At this time, it is sufficient to note that the measure of PPS effectiveness, PE, can be related to the probability of attack (P_A) and the consequence associated with the loss (C) to determine risk. In addition, P_E is the product of the probability if interruption (P_I) and the probability of neutralization (P_N), assuming both interruption and neutralization are part of the response.

Once the risk value is determined, the security manager can justify the expenditure of funds based on a scientific, measurable, and prioritized analysis. This information can be presented to executive management of the corporation or facility to demonstrate how the security risk is being mitigated and how much risk exposure remains. The analysis can then form the basis for a discussion on how much security risk can be tolerated or how much to increase or decrease the budget based on risk. This analysis can also serve to demonstrate to any regulatory agencies that a careful review of the security of the facility has been performed and that reasonable measures are in place to protect people and assets. The analysis will allow the facility to state the assumptions that were made (threat, targets, risk level), show the system design, and provide detailed information to support system effectiveness measures.

This process only describes the evaluated risk of the security system and its effectiveness. It should be noted that there are multiple risk areas for a facility or corporation, of which security is only one part. Other areas of risk that need to be considered within the business enterprise include financial risk management, liability risk financing, property/net income financing, employee benefits, environmental health and safety, and property engineering (Zuckerman, 1998). It should be clear that the security program is one that contributes to the bottom line of the corporation, by protecting assets from malevolent human threats. The security manager should be capable of allocating available resources to best protect corporate assets and adjusting resources as required in the face of changing threats. This is the role of the security manager or director in the corporate structure.

Summary

This chapter introduces the use of a systematic and measurable approach to the implementation of a PPS. It emphasizes the function of detection, followed by delay and response, and presents a brief description of the relationship of these functions. Deterrence of an adversary is compared to defeat of an adversary, along with the caution not to rely on deterrence to protect assets. Specific performance measures of various components of a PPS are described, along with how these measures are combined to support a cost/benefit analysis. The process stresses the use of integrated systems combining people, procedures, and equipment to meet the protection objectives. In support of this concept, the difference between safety and security is described to emphasize the difference between accidents or natural disasters and malevolent human attack.

The concepts presented here are somewhat unique in the security industry as a whole but have been demonstrated to be effective in protecting critical nuclear assets for the past 25 years. Although a particular facility may not require the same level of protection or have the same unacceptably high consequence of loss—the loss of a nuclear weapon or material could result in the death of thousands of people, whereas the loss of a piece of jewelry from a retail store is obviously much less—the process described here can still be applied to protect targets against the appropriate threats. Ultimately, this leads to an effective system design that can be used to explain why certain security components were used, how they contribute to the system effectiveness, and how this system mitigates total risk to the facility or corporation.

Further Readings

H. A. Bennett, "The EASI approach to physical security evaluation," SAND Report 760500 1977; 1–35.

L. D. Chapman and C. P. Harlan, "EASI estimate of adversary sequence interruption on an IBM PC," SAND Report 851105 1985;1–63.

J. D. Williams, "DOE/SS Handbooks: A means of disseminating physical security equipment information," *Journal of the Institute of Nuclear Materials Management*, 1978;7(1):65–75.

V. Sivarajasingam and J. P. Shepherd, "Effect of closed circuit television on urban violence," *Journal of Acc Emer Medicine*, 1999;16(4):255–257.

M. M. Zuckerman, "Moving towards a holistic approach to risk management education: Teaching business security management," Presented at 2nd Annual American Society of Industrial Security Education Symposium, Aug. 13–15, 1998, New York, NY.

SAND Reports are available from the National Technical Information Service, U.S. Department of Commerce, 5285 Port Royal Road, Springfield, VA 22161. Phone: 1-800-553-NTIS (6847) or (703) 605-6000; Fax: (703) 605-6900; TDD: (703) 487-4639; Web: www.ntis.gov/help/ordermethods.asp?loc=7-4.0. Or contact the U.S. Government Printing Office, 732 North Capitol St. NW, Washington, D.C. 20401. Phone: (202) 512-1800 or 1-866-512-1800; Fax: (202) 512-2104; email: ContactCenter@gpo.gov; Web: www.access.gpo.gov.

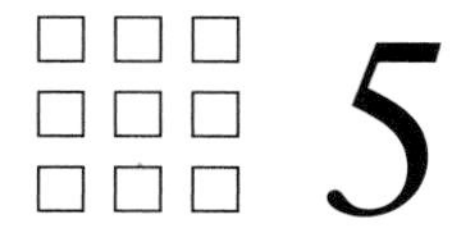

5 International Safeguards Inspection: An Inside Look at the Process

Brian Boyer and Mark Schanfein

Introduction

The International Atomic Energy Agency (IAEA) defines in the model comprehensive safeguards agreement (CSA), INFCIRC/153,[1] the technical aim of safeguards as "the timely detection of diversion of significant quantities of nuclear material from peaceful nuclear activities to the manufacture of nuclear weapons or of other nuclear explosive devices or for purposes unknown, and deterrence of such diversion by the risk of early detection." The whole philosophy underpinning the IAEA safeguards inspection system stems from that statement.

The reader will see in this chapter that the IAEA has built its safeguards system to fulfill this technical aim of safeguards by defining the concepts of significant quantities of nuclear material and timeliness of detection. Hence, the principles behind the IAEA Safeguards Criteria[2] are based on defining a set of guidelines using the concepts of significant quantities and timeliness to allow inspectors to fulfill the technical aims of safeguards for the suite of facilities that the IAEA must inspect. Safeguards inspectors and their management use these criteria and established Agency practices for a facility to lay out an inspection schedule for a material balance period (MBP), which is approximately one calendar year and no more than 14 months long and is the period between two IAEA Physical Inventory Verification (PIV) inspections.[3] The types of and amounts of nuclear materials and their physical forms determine the quantity goals and timeliness goals for a facility, as shown in Table 5.1.

Table 5.1 Definition of significant quantities for IAEA nuclear material types.

Nuclear Material Type	SQ Amount (kg)
Pu (<80% ^{238}Pu)	8 kg Pu
U-233	8 kg ^{233}U
HEU (=>20% ^{235}U)	25 kg ^{235}U
LEU (<20% ^{235}U including natural U and depleted U)	75 kg ^{235}U (or 10 t nat. U or 20 t depleted U)
Thorium	20 t thorium

[1]IAEA, INFCIRC/153 (Corrected), "The Structure and Content of Agreements Between the Agency and States Required in Connection with the Treaty on the Non-Proliferation of Nuclear Weapons," June 1972 (IAEA 1972), p. 1.

[2]IAEA, Safeguards Criteria, 2004 (IAEA, 2004).

[3]IAEA, Safeguards Criteria.

This chapter presents a basic example of an IAEA inspection regime under the traditional CSA at a facility, the activities planned, and the driving philosophies behind it. The reader will be exposed to the concepts of IAEA safeguards in much the same fashion as a neophyte IAEA inspector would be in the three-month Introductory Course on Agency Safeguards (ICAS), albeit in a much condensed and simplified version. The reader should realize that verification of material accountancy in the traditional safeguards is a keystone of IAEA safeguards, and even with the implementation of INFCIRC/540 Additional Protocols[4] and the move to more investigative safeguards as part of the Strengthened Safeguards System,[5] material accountancy will remain the keystone of the safeguards system. Hence, the reader needs to be able to understand traditional comprehensive safeguards as a first step in comprehending the challenges the IAEA safeguards system faces.

The Concept of Significant Quantities of Nuclear Material and Timeliness of Detection

Over the years the IAEA has developed a means of defining the proliferation risk involved in various types, amounts, and forms of nuclear material. The term *significant quantity* (SQ) denotes an amount of a type of nuclear material that can create one nuclear weapon. There is some controversy regarding the definition of the SQ. Part of the assumption in determining the SQ includes process losses associated with the fabrication of a weapon and that the state does not possess a sophisticated nuclear weapons expertise. These amounts stem from estimates of bomb material derived from open sources. In the eyes of the IAEA the nuclear materials and their attendant significant quantities are as shown in Table 5.1.[6]

It should be noted that the isotopic purity of plutonium is not taken into account in defining a significant quantity of plutonium. Hence, although weapon designers would desire plutonium that is purely ^{239}Pu, the IAEA has taken the conservative approach and considered all plutonium capable of being formed into a weapon with the exception of plutonium that is 80% or more ^{238}Pu, which is not considered to be nuclear material in the IAEA definition due to the large heat generation by ^{238}Pu alpha decay. Hence, the IAEA does not distinguish between "reactor-grade" or "weapons-grade" plutonium in its safeguards efforts. However, the Agency does care about uranium isotopic composition focusing on the fissile isotopes ^{233}U and ^{235}U.

Thorium and natural uranium are considered source materials for ^{233}U, ^{235}U, and plutonium. Since it takes time to convert thorium and uranium into weapons-usable materials, the concept of timeliness of detection evolved in the safeguards approaches for nuclear material as defined by the IAEA in Table 5.1. More time and effort should be spent safeguarding material that can be quickly converted into a weapon.

The IAEA established "conversion times" from estimates of the time to convert the different types and forms of nuclear materials. Table 5.2 gives the background on estimates of the conversion times for converting nuclear materials into a weapon. From these conversion times, the IAEA established the timeliness goals stated in Table 5.2.

The IAEA developed this graded safeguards approach to applying manpower and equipment to safeguarding the various forms of nuclear material.[7] Hence, we can see that

---

[4]IAEA, INFCIRC/540 (Corrected), "Model Protocol Additional to the Agreement(S) Between State(S) and the International Atomic Energy Agency For the Application of Safeguards," Sept. 1997 (IAEA, 1997).

[5]Richard Hooper, "The System of Strengthened Safeguards,"*IAEA Bulletin*, vol. 39, no. 4, Dec. 1997.

[6]IAEA, *IAEA Safeguards Glossary, 2001 Edition*, International Nuclear Verification Series No. 3, June 2002 (IAEA, 2002), p. 23.

[7]Don L. Jewell, H. Rod Martin, David D. Wilkey, and Kenneth E. Thomas, "Safeguards Material Attractiveness Level Criteria—History and Prognosis," *Proceedings INMM 41st Annual Meeting*, New Orleans, July 2000.

Table 5.2 Definition of timeliness goals for IAEA nuclear material types.

Nuclear Material	Material Form	Conversion Time	IAEA Timeliness Goals
Pu, HEU or U-233	Metal	Few days (7–10)	1 month
Pure Pu components	Oxide (PuO_2)	Few weeks (1–3)	
Pure HEU or U-233 compounds	Oxide (UO_2)	Few weeks (1–3)	
MOX	Nonirradiated fresh fuel	Few weeks (1–3)	
Pu, HEU or U-233	In scrap	Few weeks (1–3)	
Pu, HEU or U-233	In irradiated fuel	Few months (1–3)	3 months
LEU, Nat U, Dep U, and Th	Unirradiated fresh fuel	Order of 1 year	1 year

the Agency will have much more concern for unirradiated direct-use materials such as highly enriched uranium (HEU) and plutonium, with a goal of verifying the material every month. The IAEA inspects irradiated direct-use material, traditionally power reactor and research reactor spent fuel and irradiated fertile targets, every three months and inspects source material, such as depleted, natural, and low-enriched uranium (LEU) and thorium, once a year.

Timeliness also is contingent on the material amount. In cases where the material quantities at an installation are small (under 1 SQ), the timeliness goals can be relaxed. The IAEA Safeguards Criteria evolved from this emphasis on meeting verification material quantity goals and timeliness goals. The reader should grasp that the Safeguards Criteria drive the traditional comprehensive safeguards down to the aspects of inspection planning and implementation.

Facilities in the IAEA Context

The traditional INFCIRC/153 CSA safeguards focuses on verification of the nuclear materials described previously, specifically in facilities. INFCIRC/153 states that a facility or a location outside facilities (LOF) (facility with 1 effective kilogram of nuclear material or less) where the content of throughput does not exceed 5 effective kilograms of nuclear material shall not have more than one inspection per year. Table 5.3 contains the definitions for an effective kilogram of nuclear material for all the forms of nuclear material.[8] For other facilities with more nuclear material, the number, intensity, duration, timing, and mode of inspections shall be determined by the nominal declared amount of nuclear material in the facility and shall be no more intensive than is necessary and sufficient to maintain continuity of knowledge of the flow and inventory of nuclear material. Hence, the CSA states that the inspection intervals and activities will be of a nature that will not be intrusive and excessive and are based on a graded safeguards concept, as touched upon earlier in this chapter. The Safeguards Criteria[9] give a more detailed description of the allowable activities and categorize facilities as shown in Table 5.4.

A current estimate of the number of each type of facility that is under IAEA safeguards is given in Table 5.4.[10] The Agency classifies the facilities as item or bulk handling facilities. The item facilities have nuclear material that is contained in an "item" form, such as fuel rods and fuel pins; bulk handling facilities have nuclear material that is contained in a "bulk" form, such as UF_6 in cylinders, UO_2 powder, and reprocessed plutonium stored in containers. Item facilities have the advantage to the inspector of having the nuclear material in an integral physical form that will not change. Bulk handling facilities have the disadvantage of having nuclear material in gas, liquid, or powder forms that will be manipulated

[8] IAEA, *IAEA Safeguards Glossary, 2001 Edition*, p. 78.

[9] IAEA, Safeguards Criteria.

[10] IAEA, *IAEA Annual Report*, 2006 (IAEA, 2007).

Table 5.3 IAEA effective kilogram (ekg) definition.

Material Type	Definition of Effective Kilogram
Plutonium	Weight in kilograms
Uranium with an enrichment of 0.01 (1%) and above	Weight in kilograms multiplied by the square of the enrichment of the material
Uranium with an enrichment below 0.01 (1%) and above 0.005 (0.5%)	Weight in kilograms multiplied by 0.0001
Depleted uranium with an enrichment of 0.005 (0.5%) or below	Weight in kilograms multiplied by 0.00005
Thorium	Weight in kilograms multiplied by 0.00005

Table 5.4 Nuclear facilities under IAEA safeguards.

Facility Type (Defined by IAEA Safeguards Criteria)	Approximate Number of Facilities Under IAEA Safeguards Worldwide
1. Light water reactors (LWRs)	180
2. On-load reactors (OLRs)	20
3. Other types of reactors	10
4. Research reactors and critical assemblies (RRCAs)	170
5. Natural and low enriched uranium conversion and fabrication plants	50
6. Fabrication plants handling direct-use material (MOX or HEU)	5
7. Reprocessing plants	10
8. Enrichment plants	20
9. Storage facilities	80
10. Other facilities (~60 other facilities under SGs)	60
11. Locations outside facilities (LOFs)	60–70

chemically or isotopically altered. The inspector will encounter material stored in containers where both the operator and the inspector will always be uncertain of just how much material is present. The operator, whether he is an honest conscientious individual or a devious proliferator, will always have over the course of a material balance period, as stated above to be approximately one year, some material unaccounted for (MUF). MUF is calculated for a material balance area (MBA) over a material balance period (MBP) using the material balance equation, commonly written as:

$$\text{MUF} = (PB + X - Y) - PE$$

where:

PB = Beginning physical inventory
X = Sum of increases to inventory
Y = Sum of decreases from inventory
PE = Ending physical inventory

Because book inventory is the algebraic sum of *PB, PE, X,* and *Y*, MUF can be described as the difference between the book inventory and the physical inventory. (The equivalent term in U.S. domestic safeguards, for both the Nuclear Regulatory Commission and the Department of Energy, is *inventory difference*, or ID, as described in 10CFR 74.4 and DOE

Manual 470.4-6.) For item MBAs, MUF should be zero, and a nonzero MUF is an indication of a problem (for example, accounting mistakes) which should be investigated. For bulk handling MBAs, a nonzero MUF is expected because of measurement uncertainty and the nature of processing bulk materials. The operator's measurement uncertainties associated with each of the four material balance components are combined with the material quantities to determine the uncertainty of the material balance.[11]

Hence, the large enrichment plants and large reprocessing plants with their large throughputs of material in a year will find it statistically difficult using the best measurement techniques to obtain a MUF that is smaller than a significant quantity of nuclear material. This is a major challenge for the IAEA and its inspectors to verify nuclear material in such facilities. Hence, because of the need for experience and more in-depth training for bulk handling facilities, training new inspectors touches generally on bulk handling facilities but focuses on item facilities. An inspector without experience in bulk handling will usually find that it takes almost two to three years to grasp the rudiments of planning and executing a large PIV at an enrichment plant or reprocessing plant. This chapter focuses on the ubiquitous LWR, which is the focus of ICAS. At the end of the ICAS, the class will perform a mock inspection at an LWR.

Basic Goals of IAEA Safeguards at LWRs

An LWR contains two types of nuclear material: LEU and plutonium. As noted previously, it is an item facility so that all nuclear material is an item form. The fresh fuel rods contain the unirradiated LEU, and the core fuel and spent fuel rods contain the irradiated LEU burned by and Pu produced by the fission process. Although a fuel rod will change material composition during the fission process, the uranium and Pu stay contained in the fuel rod. The LWR in this example is a pressurized water reactor (PWR) with LEU fuel and no MOX with a yearly refueling cycle.

The inspector must understand the type of nuclear material and the operations at the facility. The state is obligated to provide to the Agency a Design Information Questionnaire (DIQ) prior to the operation of the plant.[12] The Agency then will do Design Information Verification (DIV).[13] The state has a continuing obligation to provide the Agency with updates to the DIQ. For example, an operator may upgrade the thermal power of an LWR during its operational life by replacing steam generators. The Agency, by doing a DIV, will verify that the operator has done this activity.

An inspector should understand the types of and uses for the nuclear material located in the facility, the operations associated with the nuclear material, and the material quantity and timeliness goals. Starting with a new LWR, the first material to be introduced will be the fresh LEU fuel. This material will have a goal quantity of a significant quantity that is 75 kg of ^{235}U in the LEU fuel and timeliness of one year (Table 5.1). Hence, until the reactor begins to operate, the LWR will need to be inspected yearly.

When the reactor begins to operate with the uranium fissioning, the fresh fuel now becomes core fuel. This material is now seen as irradiated direct-use material since the fission process will convert ^{238}U into Pu. A significant quantity of Pu equals 8 kg elemental Pu. Irradiated Pu has a timeliness of three months. Hence, the inspector needs to inspect the LWR on a quarterly basis, with a yearly PIV at the refueling. When the core is refueled for the first time, there will then be irradiated direct-use material in the form of spent fuel removed to cool in the spent fuel pond. The operator inserts new fresh fuel into approximately one third of the core. The inspector will be responsible for verifying the new fresh fuel, the remaining core fuel, and the spent fuel. The next section describes how the state and

---

[11]IAEA, *IAEA Safeguards Glossary, 2001 Edition*, p. 55.

[12]IAEA, *IAEA Safeguards Glossary, 2001 Edition*, p. 26.

[13]IAEA, *IAEA Safeguards Glossary, 2001 Edition*, p. 27.

the operator designate the material balance areas (MBAs)[14] and the key measurement points (KMPs)[15] of the reactor for the Agency and the Agency's safeguards approach to LWRs.

The MBA and KMP Safeguards Concepts Applied to LWRs

INFCIRC/153 states that the aforementioned design information shall not only identify the features and nuclear material relevant to safeguards, as discussed previously, but shall "determine *material balance areas* to be used for Agency accounting purposes and to select those strategic points which are *key measurement points* and which will be used to determine the nuclear material flows and inventories." INFCIRC/153 states that the MBA size should be "related to the accuracy with which the material balance can be established." The use of "containment and surveillance to help ensure the completeness of flow measurements and thereby simplify the application of safeguards and concentrate measurement efforts at key measurement points" should be pursued. Furthermore, to respect the sensitivity of a proprietary process, "a special material balance area around a process step involving commercially sensitive information may be established." INFCIRC/153, para. 46, shows how the Agency intends to implement safeguards using the design information from an operator to negotiate specific MBAs and KMPs in a facility to enable the IAEA to get the information needed to verify the facility's declarations and to protect the operator's sensitive information.

This balancing of access and protecting sensitive technologies and industrial processes is one of the challenges of safeguards, especially where an operator may have a technical process that he does not want to reveal to a rival and lose his crucial competitive edge. This is a serious consideration with respect to enrichment, reprocessing, and fuel fabrication processes and industrial operations where market shares could be won or lost if the IAEA inspectors performed industrial espionage. Hence, the IAEA has the concept of "Safeguards Confidential" information. The IAEA will honor the confidentiality of the information the state and the operator provide to the Agency. Inspectors are made aware of this obligation not to divulge this information during their Agency careers and beyond.

The IAEA and the state must negotiate subsidiary arrangements that include a general part and a facility attachment for each facility under safeguards in the state.[16] Paras. 39 and 40 of INFCIRC/153 describe the subsidiary arrangements. The Agency and the state shall make subsidiary arrangements that specify the necessary details to permit the Agency to fulfill its procedural responsibilities under the CSA both effectively and efficiently. The general part applies to all common nuclear activities of the state concerned. A facility attachment contains specific provisions necessary for safeguards implementation at a facility.

The subsidiary arrangements include the results of the examination of the design information. From the subsidiary arrangement negotiations and the design information, the IAEA and the state create the facility attachment, which includes facility specific provisions necessary for safeguards implementation. Figure 5.1 shows the legal structure of comprehensive safeguards, including the relationship of the subsidiary arrangements and facility attachments.

The state utilizes a system called the States' Systems of Accounting and Control (SSAC) to organize and transmit reports, to handle the negotiations with the IAEA, and to interface with the Agency regarding international safeguards implementation in the state.[17] Although the quality of the SSAC does not determine the how the IAEA draws its safeguards findings and conclusions, an effective SSAC smoothes the implementation of IAEA safeguards.

The IAEA considers LWRs to be Type I or Type II for safeguards purposes, shown in Figures 5.2 and 5.3, respectively. PWRs, such as this chapter's example generic PWR, are usually

[14]IAEA, *IAEA Safeguards Glossary, 2001 Edition*, p. 46.

[15]IAEA, *IAEA Safeguards Glossary, 2001 Edition*, p. 47.

[16]IAEA, *IAEA Safeguards Glossary, 2001 Edition*, p. 10.

[17]IAEA, *IAEA Safeguards Glossary, 2001 Edition*, p. 28.

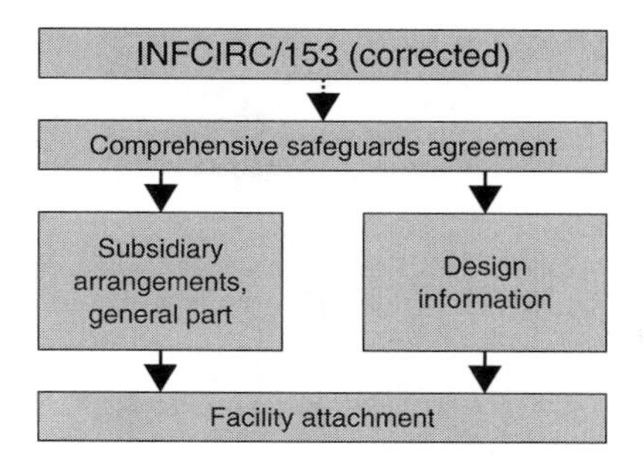

FIGURE 5.1 Legal structure of safeguards.

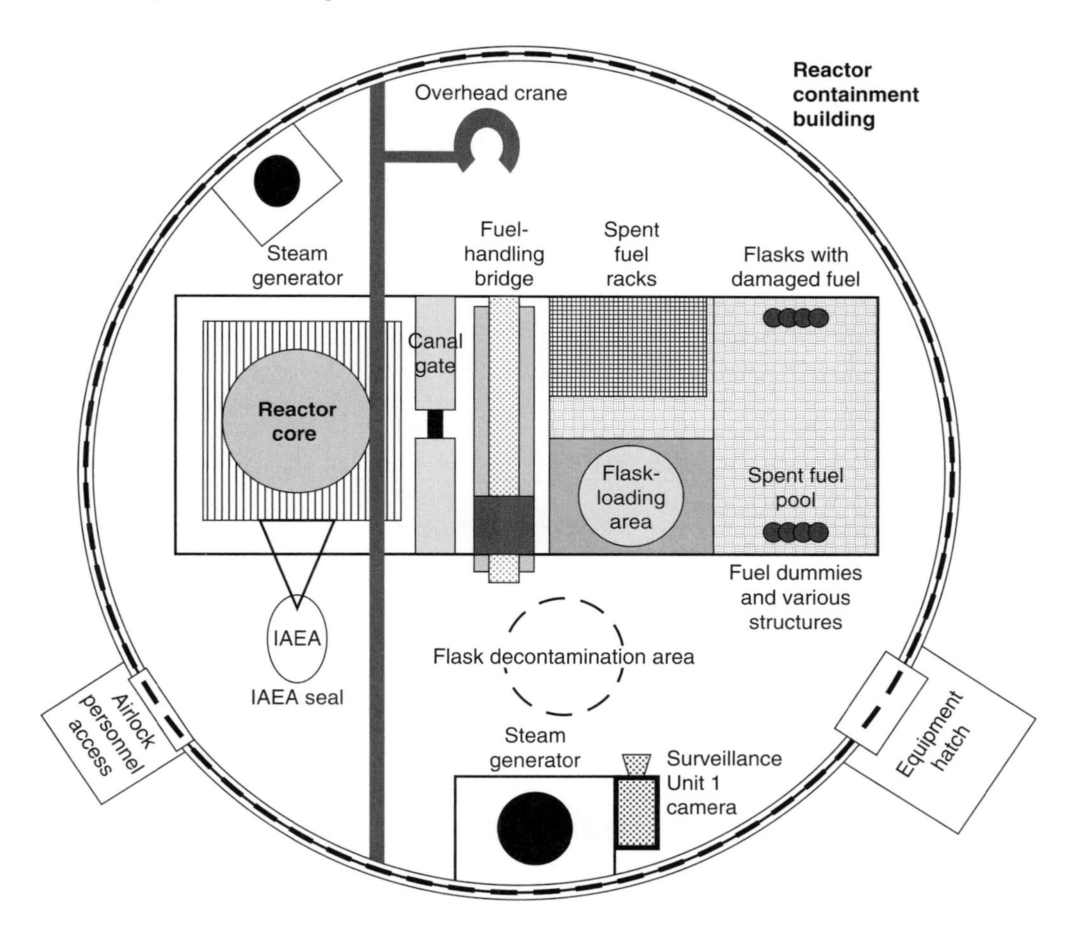

FIGURE 5.2 LWR of the Type I Configuration: Spent fuel pond inside of containment.

Type II with the spent fuel pond outside the containment. These features, which are a basic part of the design information, will determine how the IAEA will create a safeguards approach.

The IAEA, in setting up the MBA and KMP structures in a facility, attempts to take into account the safeguards concerns and possible diversion scenarios in the facility. In an LWR without MOX (Mixed Oxide) fuel, nuclear fuel containing uranium and plutonium, such as this chapter's example PWR, the diversion scenarios exist as listed in Table 5.5. It should be noted that the use of MOX fuel complicates safeguards at a reactor since the MOX will have a timeliness of one month (Table 5.1), forcing the IAEA to inspect the reactor on a

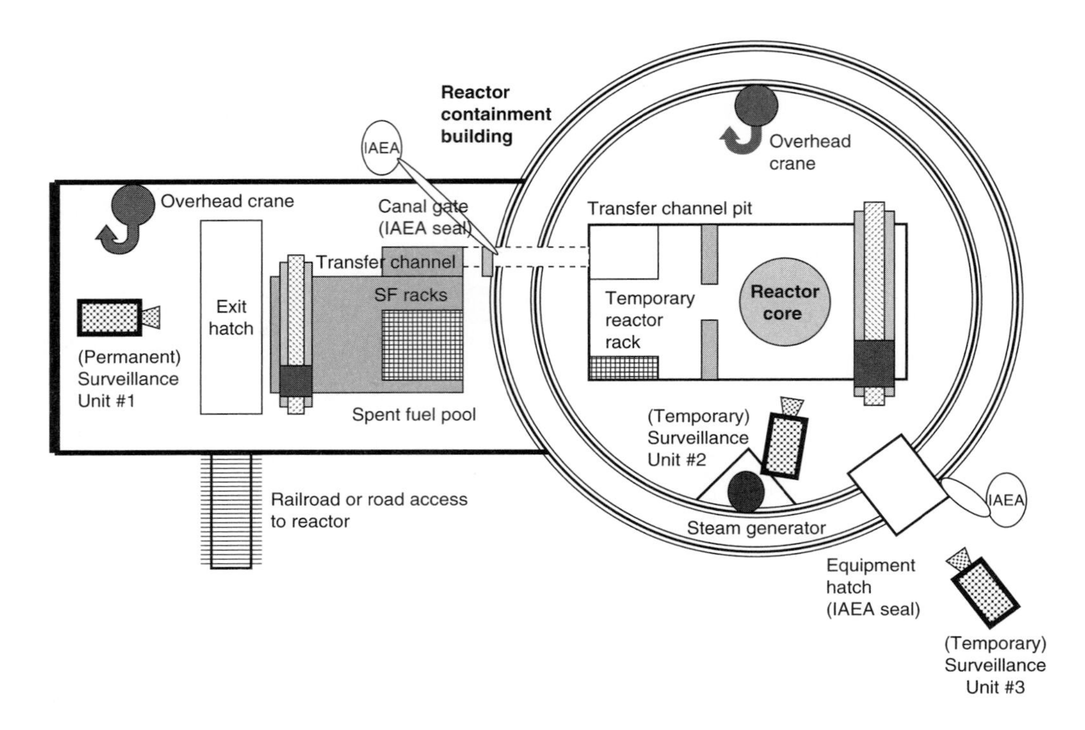

FIGURE 5.3 LWR of the Type II configuration: Spent fuel pond outside of containment.

Table 5.5 LWR diversion scenarios: PWR without MOX.

Diversion	Method	Timing/Location
LEU fresh fuel diversion	Substitution of dummy element for actual element	After fresh fuel verification, prior to core loading
Spent fuel assembly diversion	Substitution of dummy element for actual element	From reactor pool, SF pool, or SF transfer cask
Spent fuel pin diversion	Substitution of dummy element for actual element	From SF pool or SF transfer cask
Unreported Pu production	Insertion of fertile targets for irradiation in core fuel — PWR guide tubes or burnable poison rod	From reactor pool, SF pool, or SF transfer cask

monthly basis and to worry about nuclear material with a high strategic value to a potential proliferator.

A PWR as an item facility, unlike more complicated reprocessing and enrichment plants, has only one MBA, which simplifies accountancy. The PWR's KMPs aim at giving the IAEA the appropriate access to make the measurements needed to verify that the diversions in Table 5.5 can be detected. The example PWR will have the following KMP structure. It should be noted that a flow KMP is a KMP in which material passes into and out of the facility, and an inventory KMP is a KMP in which material is stored or used in the facility:

Flow KMPs

KMP 1. Receipts of nuclear material (nominally fresh LEU fuel)
KMP 2. Nuclear loss and nuclear production for core fuel discharged

where:

Nuclear loss = The reduction in uranium occurring from burnup of fuel
Nuclear production = The production of plutonium from neutron capture in ^{238}U
KMP 3. Shipments of nuclear material (nominally spent LEU fuel to dry storage or reprocessing)
[KMP 4. Exemption/deexemption of nuclear material, accidental gain/loss, etc.]
Inventory KMPs
KMP A. Fresh fuel storage (LEU fuel)
KMP B. Reactor core (LEU fuel and plutonium)
KMP C. Spent fuel pond (spent LEU fuel containing uranium and plutonium)
KMP D. Any other locations of nuclear material

The concept of flow KMPs focuses on shipments (spent fuel) and receipts (fresh fuel) of nuclear material at the plant and the nuclear loss of uranium created by fission and the nuclear gain of plutonium formed by neutron capture in ^{238}U and decay to ^{239}Pu. The inventory KMPs keep stock of the location of the nuclear material in the facility so as to delineate fresh fuel, core fuel being burned, and spent fuel from each other. The inspector's main job in CSA safeguards is to be the material accountant who must verify that the material is where the operator declares it exists and that all items maintain their integrity.

Inspection Frequency at LWRs

For the IAEA to be able to verify the operator's declaration of the nuclear material locations and shipments from and receipts to the facility, the state, the operator and the Agency must agree to facility access. As described previously, the negotiations that created the facility attachment provide the legal framework for Agency access to the facility.

In an LWR under traditional INFCIRC/153 safeguards, the Agency completes a yearly PIV and three interim inspections for timeliness and inspections of shipments of spent fuel in casks to dry storage. The number of inspections is determined by the type of material and movements of material at the facility. As described earlier, this chapter's example PWR has LEU fuel, which has a timeliness goal of one year. Hence, the yearly PIV suffices to verify the LEU fresh fuel on a timely basis. Once the reactor begins operation, the LEU fuel begins to fission and creates plutonium. The reactor now has irradiated direct-use material, plutonium, which has a timeliness goal of three months. Hence, the IAEA must inspect the reactor every three months to verify the correctness of the declaration with respect to the plutonium at the reactor. When the reactor is refueled, the old core fuel now migrates to the spent fuel pool. The spent fuel must also be verified on a quarterly basis exactly as the core fuel.

The Agency performs three basic activities to verify the completeness and correctness of the operator's declaration. The inspector checks the reactor's nuclear material accountancy (accounting and operating records), verifies the material itself and the items' integrity by visual and nondestructive assay (NDA) techniques, and uses containment and surveillance to check that the nuclear material and the reactor are not being diverted and misused, respectively. Material accountancy is basically examining the books that contain the ledger describing the nuclear material at the facility. In an LWR, the operator also provides the Agency with maps of the core and the spent fuel pool. These maps delineate the location of every fuel assembly in these areas. The inspector will then attempt to verify the accountancy by visually counting and identifying the fuel elements and applying various NDA techniques using a random sampling plan designed to give the appropriate confidence level for the desired probability of detection. The confidence level is a limit set around a measured value or estimate that expresses a degree of confidence with regard to the "true" value of the measured or the estimated amount.[18] NDA techniques take advantage of the radiation emitted by nuclear

[18] J. L. Jaech, *Statistical Methods in Nuclear Material Control*, TID-26298, Technical Information Center, Oak Ridge, Tennessee (1973).

material and can, under the correct conditions, provide qualitative (what nuclear material is present) and quantitative (how much nuclear material is present) information.

In the majority of facilities, it might not be practical or possible to verify all nuclear material.[19] In addition, based on INFCIRC/153, the IAEA must try to minimize its interference in the facility's operations. The approach taken by the IAEA to this challenge is to statistically analyze the timeliness and quantity goal requirements for a facility and create a sampling plan that randomly verifies a statistically relevant amount of material.[20] The Agency creates a stratum of nuclear material, a grouping of items and/or batches with similar physical characteristics (e.g. isotopic composition), to facilitate this statistical sampling.[21] The Agency then uses the following formula[22] to calculate the total number of samples, n, in each stratum of nuclear material to be verified:

$$n = N\ (1 - \beta^{1/d})$$

where:

N = The number of items in the stratum
β = The nondetection probability
d = [M/x], the number of defects in the stratum rounded up to the next integer
M = The goal amount
x = The average nuclear material weight of an item in the stratum

Since the IAEA must consider all states potential adversaries, it is critically important that the facility operator have some level of uncertainty as to which items the Agency will select for verification. By using this random approach, even the IAEA inspector does not know for certain which items will be selected. There will be no human-based pattern that an adversary could use to their advantage. Therefore, this places the operator or adversary at great risk for detection should they choose to tamper with or substitute an item. At the same time, by using a scientific-based approach to sampling, which includes random selection, the number of SQs present, and an acceptable risk factor, the IAEA can draw defensible safeguards conclusions.

LWR PIV Inspection

The facility's physical inventory is determined by the operator as a result of a physical inventory taking (PIT) and is reported to the IAEA in the physical inventory listing (PIL). The physical inventory is verified by the IAEA during the physical inventory verification (PIV) inspection.[23] The Agency and the operator center the yearly PIV around the refueling, which in most plants has been a yearly occurrence. The IAEA will be present during the core opening to verify the new core and the new spent fuel pool configuration. In facilities on a stretched-out refueling schedule, a closed core PIV is done as best as possible for timeliness sake, and the core is verified during the refueling period. A PIV at a PWR, used as an example in this chapter, has three distinct phases, denoted as the pre-PIV, the PIV activities, and the post-PIV.

During the pre-PIV the inspector must go to the reactor and prepare the facility for the PIV following the Safeguards Criteria for LWR inspections as the ruling guide. The inspector must verify the fresh fuel that the plant received since the last PIV. The inspector will visually inspect, count, and check the serial numbers stamped on the fuel and perform NDA.

---

[19]IAEA, *IAEA Safeguards Glossary, 2001 Edition*, p. 81.
[20]Jaech, *Statistical Methods in Nuclear Material Control.*
[21]IAEA, *IAEA Safeguards Glossary, 2001 Edition*, p. 53.
[22]IAEA, *IAEA Safeguards Glossary, 2001 Edition*, p. 78.
[23]IAEA, *IAEA Safeguards Glossary, 2001 Edition*, p. 54.

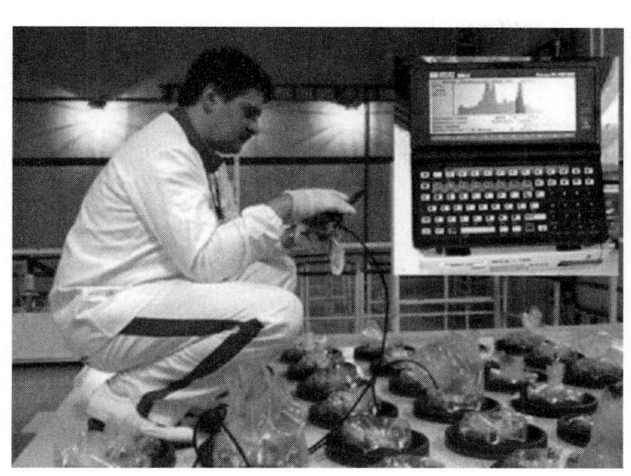

a.

b.

FIGURE 5.4 Using CdZnTe detector to measure uranium spectrum in fresh LEU fuel.

For fresh fuel, the Agency does a "gross defects" test, which entails testing to see if the fuel assembly does contain uranium.

The concept behind these Agency tests is based on the approaches an adversary might take to divert nuclear materials. Such approaches include abrupt diversion of an SQ within the timeliness goal period versus protracted diversion of a SQ within a MBP and complete versus partial removal of nuclear material from an item under any of these scenarios. (Note: A *partial defects* test would test to see whether ~50% of the assembly's nuclear material as declared is present, and a *bias defect* tests to see whether, within a small range of uncertainty, all the assembly's nuclear material as declared is present.) For fresh fuel, the Agency can use a CdZnTe detector to search for the characteristic 185 KeV ^{235}U gamma spectrum peak.

As shown in Figure 5.4, the inspector in the left-hand photograph, squatting among fresh LWR fuel (VVER 440 fuel in this case), is holding the mini multichannel analyzer (MMCA) connected to the CdZnTe detector, which is inserted carefully in a space between the fuel pins in a manner shown in the right-hand photograph of Figure 5.4. The inspectors observe the 185 KeV ^{235}U gamma peak and can attest that the declared LEU fresh fuel really contains uranium.

A reactor, such as the example PWR in this chapter, has containment and surveillance measures to maintain what the IAEA calls *continuity of knowledge* (CofK). Once the Agency has verified nuclear material, it must either maintain a constant vigil over that material to assure that it can detect any tampering with the material or reverify the nuclear material on a required frequency. Reverification can be both costly and time consuming for both the IAEA and the operator.

Since human surveillance would be rather intrusive to the operator and expensive for the Agency to maintain a constant vigil over all the material in the world, safeguards have developed a portfolio of containment and surveillance measures. For the example PWR, both containment and surveillance measures would be in place.

Referring to Figure 5.3 for locations, IAEA tamper-indicating devices (TID), metal "E-Cup" seals (shown in the left-hand photograph in Figure 5.5), seal the huge equipment access hatch on the containment dome where, during refueling, equipment is moved in and out of the reactor hall and the gate that separates the spent fuel pool from the reactor core's pool. These measures provide tamper indication if the operator has opened the reactor hatch or the canal gate to access the core fuel to divert the nuclear material. The design information

a.

b.

FIGURE 5.5 Attached IAEA "E Cup" seal and current IAEA DCM-14 camera surveillance system.

provided by the operator enables the IAEA to create this safeguards approach by allowing the Agency to understand where the diversion pathways exist in the plant.

Of course, if the operator removes core or spent fuel from the spent fuel pond out the containment hatch, he has a clear path to misuse the material as he pleases. If he removes the core fuel into the spent fuel pool, the Agency has surveillance (Figure 5.3) of the spent fuel pool and any diversion access path large enough to remove the core and spent fuel from the spent fuel pool hall. Agency surveillance (shown in the right-hand photograph in Figure 5.5), at present, consists of digital cameras able to capture images in small enough time intervals to detect a diversion and to be able to store this information on a media that the inspector can access during the PIV and quarterly inspections. Hence, the Agency maintains CofK of the core fuel and spent fuel by C/S measures over the course of the year. If the C/S measures fail, causing what the IAEA calls an "anomaly," the inspector may be able to recover the CofK in a timely manner if he detects the failure in an interval shorter than the timeliness goal.

It should be noted that the mere detection of an anomaly is not an indication of a diversion. Power failures at the plant causing a loss of lighting for the cameras or an error by an inspector causing a camera or the control unit to fail are not seen as diversion attempts. However, if a pattern of power outages causing loss of power to the cameras, an operator bumping the camera out of alignment, or an operator breaking Agency seals occurs over a number of years, the IAEA has cause for further investigation.

During the pre-PIV, the inspector must alter the C/S environment to allow the operator to perform the refueling and to still keep the CofK. In a PWR, the inspector will remove the seals on the containment hatch and the canal gate to allow the operator to move old core fuel from the reactor to the spent fuel pond and bring new fresh fuel into the reactor. It should be noted that no C/S measures are practiced on the fresh fuel. Once verified by the Agency, the inspectors will see the fresh fuel again in the reactor core during the PIV, where other measures will be used to reverify the fresh fuel's presence in the core. The inspector will install a temporary surveillance camera in the reactor hall with a backup unit capable of working under low- or no-light conditions outside to cover activity through the hatch. The inspector will then depart to return for the PIV proper.

The operator will want to plan the actual IAEA PIV and operator PIT as efficiently as possible. In today's nuclear power industry the power plants meticulously plan and implement the entire refueling operation. A utility loses money every minute the plant is shut down. Hence, the inspector will arrive at a facility and be on call to perform the verification of the core fuel and spent fuel at the precise time the operator wishes. One must not forget that safeguards should not be a major hindrance in the facility operation.

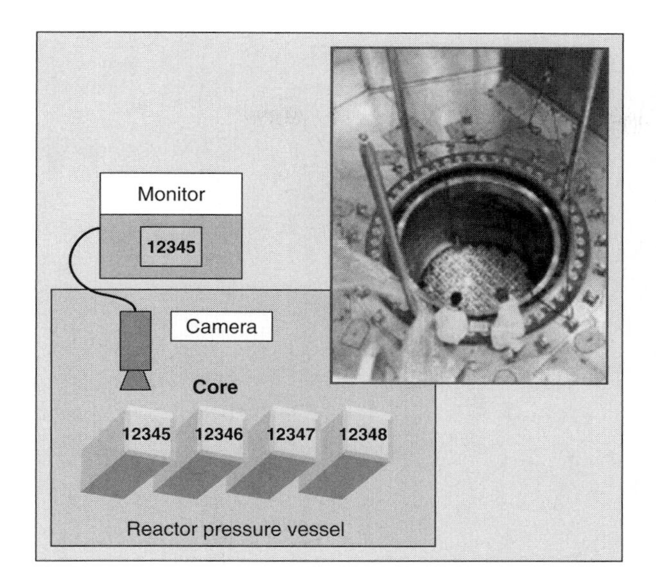

FIGURE 5.6 Use of UWTV camera to verify LWR core fuel.

The inspection team will arrive at the facility and receive the accountancy documents from the operator. The PWR accounting records include the General Ledger (with material accountancy summaries), fuel history cards and fuel assembly certificates for being able to track and identify fuel items, and an itemized list of the fuel assemblies located at the reactor. The operating records can include the power histogram and estimates of burnup, the all-important core and spent fuel pond maps with assembly locations, and cask shipment and crane movement information (important in drawing safeguards observations from images from the surveillance cameras).

The inspector will then proceed to verify the core fuel, as shown in Figure 5.6, by item counting of the core fuel from the core barrel edge and using the operator's underwater TV (UWTV) camera system to check off that the serial numbers of the fuel assemblies match the declared locations on the core map. Since the canal gate seal is not in place, the operator can shuffle items between the core and spent fuel ponds without the Agency's knowledge. Hence, the Agency must verify the spent fuel pool or pools immediately or they cannot assure that the absence of substitution of items in the spent fuel pool for items in the core has occurred between the time of the core verification and the spent fuel verification. For example, if the inspector finishes the core verification late in the day and the inspector and operator agree to continue with the spent fuel pool verification in the morning, the operator has time to reshuffle the core and possibly bring in uranium targets for irradiation and for unreported plutonium production or substituting spent fuel items in the spent fuel pond with spent fuel from the core and removing spent fuel from the facility without the inspector being able to see the state of the spent fuel pool and whether there are suspicious items in the pool.

The inspectors now reconcile their observations of the spent fuel pond with the SF core map and the assembly burnup data. The standard technique is to use the Improved Cerenkov Viewing Device (ICVD) to observe the blue Cerenkov glow emitting from the spent fuel assemblies (Figure 5.7, left-hand graphic). The Cerenkov glow appears in water originating from fission products in the spent LWR fuel. Cerenkov light is seen mainly in the ultraviolet region with peak intensity in the 300 nm region. The light can be seen as a faint blue glow passing through the water, with hotter assemblies being brighter than longer-cooled assemblies.

The ICVD is simply a more sensitive or improved device based on the concept of a night-vision device focused on detecting the visible band of Cerenkov radiation by amplifying the Cerenkov light intensity and discriminating the Cerenkov light from other light sources in

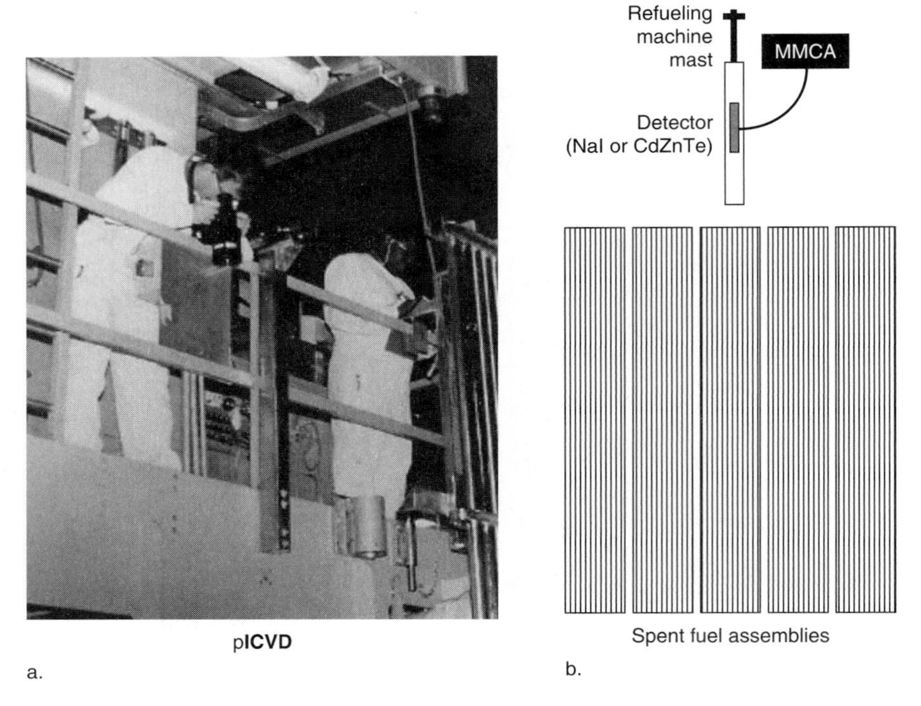

FIGURE 5.7 ICVD and SFAT used to verify spent fuel ponds.

the spent fuel pond area. Improvements to the design extend the device's ability to detect spent fuel with longer cooling times that consequently have reduced Cerenkov light emission.[24]

The inspectors can verify the SF pond using a random sampling plan and target in certain assemblies in certain rows to verify and gain the desired probability of detection. However, inspectors find it easier and faster to verify the entire pond by having the operator move the SF pond bridge row by row, with one inspector performing a sweep of each row moving along the railing, observing the Cerenkov glow with the ICVD. Moving the bridge and locating the exact position of a desired fuel assembly would take longer time. Meanwhile, another inspector will have the core map, which a well-prepared team can have color-coded by burnup so that assemblies with faint or no glow can be reconciled with a recorded low burnup and/or long-term cooling time and will check off each assembly as verified or not. A well-trained inspector can determine whether the Cerenkov glow is from a real spent fuel assembly being observed or from a near-neighbor spent fuel item illuminating a dummy element.

If the team finds assemblies that cannot be verified by ICVD or are questionable, they can resort to the use of a SF attribute tester (SFAT) (Figure 5.7, right-hand graphic).[25] The SFAT can be carefully swung into the water from the bridge and the inspectors measure the gamma spectrum from the fuel assembly. The MMCA connected to the CdZnTe detector, as indicated in Figure 5.7, will provide to the inspector a ^{137}Cs peak indicative of fission products if the fuel assembly is a genuine spent fuel assembly and not a dummy element. Generally, the combination of both ICVD and SFAT should be sufficient to verify all but the most long-termed cooled SF that has a low burnup.

---

[24]Canadian, Finnish and Swedish Support Programmes, *Cerenkov Viewing Devices for Spent Fuel Verification at Light Water Reactors*, 1993 (AECL Research, 1993).

[25]Antero Tiitta, Matti Tarvainen, Serhii Iievlev, Alexander Dvoeglazov, Valerij Bytchkov, Dorel Popescu, Young-Gil Lee, Michio Hosoya, and Valery Goulo, "VVER-1000 SFAT: Final Report on Task FIN A 1073 of the Finnish Support Programme to the IAEA Safeguards," Finnish Centre for Radiation and Nuclear Safety (STUK), Helsinki, 2002.

Once the inspectors have verified the core fuel and spent fuel ponds, they can attend to servicing the surveillance systems in the facility by checking that they have not been tampered with (seal replacement) and accessing and retrieving image data from the system. At this time the inspectors may be able to service and remove the temporary cameras in the reactor hall (and outside the containment) and replace the seals on the containment hatch and the canal gate. Usually the operator will want to have the canal gate open for some time after the PIV. Hence, the inspectors may have to schedule a separate post-PIV inspection to replace the canal gate seal. However, if the next interim inspection is within the time frame for completion of the post-PIV activities, a separate inspection may be avoided.

Generally, the facility at the PIV, especially a PWR, will not have any nuclear material outside of the core fuel and spent fuel to verify. However, there may be extra fresh fuel stored at the facility. With a multiunit power plant, especially Russian VVERs (VVER is the Soviet [and now Russian Federation] designation for light water pressurized reactor; in Western countries, PWR is used as the acronym), there may be fuel for another unit with a different MBA in the MBA for the plant being inspected. The inspector must be judicious in making sure he inspects all declared material and performs all designated activities. However, in the generic PWR of this chapter's example, all material is assumed to be in the core or in the spent fuel pond and has been inspected.

The inspectors must now make sure that they have checked the accountancy and power records and reconciled all errors encountered before leaving the plant. It is far better to reconcile any problems before they become anomalies to be resolved by the Agency and the state.

When the inspectors return to Vienna, they must perform some very important tasks to complete the work of the inspection. They review the surveillance data and reconcile it with the operator's fuel cask and crane movement records. This chapter assumed that for the example PWR the SF is not shipped off-site. In many facilities the operator ships SF to Away From Reactor Storage (AFRS) or puts the SF in dry storage casks and sends these casks to a dry storage site on or off the site. The inspector must verify that these movements occurred and in many cases must verify the casks and seal them with IAEA seals prior to shipment.

The inspectors must also turn into the seal verification group in Vienna any IAEA seals that they removed from the facility for verification. The IAEA seals verification group checks that the seals match the Agency records and the seals working papers from the inspectors and that the seals were not tampered with or inadvertently damaged by the inspector or operator, and that the knots the inspectors tied to secure each seal's wire inside the seal prior to capping the "cup" on the seal are intact. (It should be noted that the ability to tie a perfect square knot is an essential inspector skill!) The inspectors gather up all this input from the surveillance review, the seals verification team, and their working papers from the inspection and complete the inspection report. It should be noted that there are timelines and deadlines for the inspectors to complete surveillance reviews and get the report done so as to obtain the timeliness goals for the facility. When the inspectors complete all the required activities for the PIV, including post-PIV activities, the IAEA can close the material balance for the year and draw safeguards conclusions.

Interim Inspections for Timely Detection in LWRs

As described earlier, the IAEA does timeliness inspections at facilities due to the fact that some materials have shorter timeliness periods than the material balance period or that shipments and receipts occur at the facility and need a timely verification. For the example PWR in this chapter, three interim inspections are required. These are to verify the core fuel and spent fuel (irradiated direct use material: Pu in fuel assemblies). The inspectors will travel to the reactor and perform some of the tasks described in the PIV, but not all of them, due to the fact that the operator should not transfer nuclear materials between the SF pond and the core between core refuelings.

Therefore, the inspectors must verify that the core fuel has not been tampered with. The inspectors replace IAEA seals on the containment hatch door and the canal gate. The removed seals are sent to the IAEA Vienna office for verification. If the seals have not been tampered with, the inspectors have successfully verified the core fuel.

The inspectors must also verify the spent fuel pond. Knowing that the canal gate seal was intact allows the inspectors to verify that no items passed from the SF pond to and from the core. If the surveillance also shows no items removed from the spent fuel pond, the inspectors can verify that the spent fuel pond is as declared. However, if the surveillance images cannot conclusively show that there could not have been a SF assembly removed from the SF pond, whether by camera failure, lighting failure, image storage corruption, or other problems, the inspectors must reverify the SF pond. If they find this out in Vienna after ending the inspection, they must schedule an inspection to "recover" the SF inventory by reverifying the pond as in the PIV before the timeliness period expires.

The inspectors will once again have to look at the accountancy records, but the nuclear material status should not change until the next pre-PIV inspection, when new fresh fuel arrives at the plant. However, the movements of casks and cranes and power history will be needed to verify operational parameters of the plant with respect to surveillance images and calculating the "unreported plutonium production," respectively. Once again, the inspectors will return to Vienna and have tasks to complete prior to finishing the report. They must get the seals verified and review the surveillance images. Again, they must get these activities and the report done in a timely fashion to obtain the safeguards timeliness goal.

Review of Safeguards at an LWR: Generic PWR

This chapter has given the reader a glimpse into the process of the IAEA inspection regime for a sample generic PWR without MOX, which falls in the agency LWR facility category. As one can see, lots of details entwine the IAEA safeguards inspection regime. In moving from the INFCIRC/153 CSA proscriptions for creating the facility attachment, all of which set the legal and operational bounds for the state and the IAEA, one sees that many decisions and negotiations must occur among the state, the operator, and the IAEA.

At the level of performing the inspection, the inspector must manage a myriad number of small details in every step of the verification process to complete an inspection successfully. Timely planning by the inspector with his operations division management and coordination with IAEA support staff that provide NDA instruments, install and service surveillance equipment, handle IAEA seals, monitor radiation exposure and health, and handle travel and visa arrangements makes for success as an inspector. When an inspector looks at this job, he sees a lot of small details that may seem trivial. However, the best inspectors realize that for the IAEA to complete its verification mission of fulfilling the technical aims of safeguards, the attention to details is significant.

SUMMARY

The reader should also realize that at other facilities, especially bulk handling facilities such as enrichment and reprocessing plants, the complexity of the verification work increases dramatically. Furthermore, with states implementing the Strengthened Safeguards System with the Additional Protocol (INFCIRC/540), the inspector is in the process of transitioning from the traditional "accountant" as described here to an "investigator" who has the duty of verifying not only the "correctness" of the state's declaration but the "completeness."[26] Hence, the inspector not only verifies that all declared nuclear activities are in order with the state's treaty obligations, but the inspector must prove the negative: that there are no undeclared nuclear activities occurring in the state.[27] Most inspectors, while finding that this new tasking adds more burdens to their job, enjoy the challenge of having to investigate a state's nuclear program in detail as part of their role of providing the world confidence in a state's intention to operate a peaceful nuclear program.

---

[26] J. Vidaurre-Henry, A. Osipov, and P. Rodriguez, "The Enhanced Curriculum for Strengthened Safeguards System Training," *Proceedings INMM 46th Annual Meeting*, Phoenix, Arizona, July 2005.

[27] J. Carlson, "Safeguards in a Broader Policy Perspective: Verifying Treaty Compliance," INMM/ESARDA, Santa Fe, New Mexico, Oct. 30–Nov. 2, 2005.

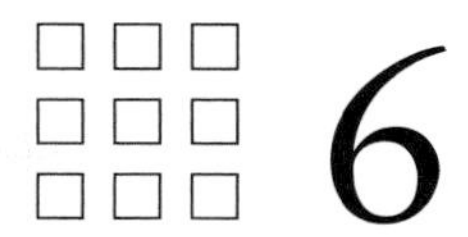

6 International Atomic Energy Agency Unattended Monitoring Systems

Mark Schanfein

Introduction

The International Atomic Energy Agency (IAEA) relies heavily on the use of unattended monitoring systems (UMS) to provide continuous monitoring of nuclear facilities around the world. The states possessing these nuclear facilities permit such monitoring to allow the IAEA to confirm that nuclear material in these facilities is not being diverted from peaceful to military uses. This monitoring is an important tool of international safeguards and helps states meet their obligations under the Nuclear Nonproliferation Treaty (NPT). Several states also use UMS to meet domestic legal and regulatory requirements for accounting for nuclear materials and operating nuclear facilities safely and securely.

There are currently over 100 UMS worldwide, with an average of 10 new systems installed per year. The primary overall goal for these systems is to never lose safeguards significant data under even the most challenging infrastructure and operational environments. The growing reliance on UMS and the stringent data-gathering goals demand that these systems have high reliability. Other issues for monitoring systems include the integrity of hardware and software, the interplay among worldwide vendors, the flexibility for systems upgrades, the ease of implementation and configuration, and operator training.

This chapter introduces current UMS as deployed by the IAEA as well as the goals, benefits, challenges, and financial issues for such systems. The UMS technologies the IAEA uses for international safeguards are always in flux as the next generation is designed, tested, and implemented. Nevertheless, the basic principles do not change.

Background

The concept of unattended monitoring systems is not new, neither in concept nor in implementation. The first example is the use of film cameras to monitor spent fuel ponds in reactors. In the 1970s, the IAEA relied on twin Minolta film cameras for this monitoring effort. The cameras had a fixed interval of 20 minutes based on the operational time for moving a spent fuel cask, and they required regular film changes. The effort to review these images was quite problematic. Black-and-white images of a spent fuel pond with very little activity and no way to rapidly advance to images of interest challenged the most astute viewer to maintain attention through the entire review process. Usually inspectors would look at thousands of images of the same scene.

With the advent of integrated circuitry and computers, this field has seen a revolution in capability. One of the first implementations of a modern distributed UMS took place at the Darlington CANDU reactor in Canada in the late 1980s. A Los Alamos National Laboratory

(LANL)-developed system was installed to monitor the discharge of spent fuel from this on-line reactor. Even today this system continues to operate with high reliability. The only recent upgrade was to the computer data collection system because the old PCs were obsolete, making them difficult to repair or support new devices such as data storage.

What Is an Unattended Monitoring System?

An unattended monitoring system, or UMS, is any automated monitoring system comprising a single set or multiple sets of sensors designed so that it can maintain continuity of knowledge about the content and location of all nuclear material of interest in a facility 24 hours a day and 365 days a year. The concept of continuity of knowledge (CofK) can take many forms, from simply tracking spent fuel bundles or assemblies to performing a quantitative analysis of accountable nuclear material in cans of mixed oxide fuel. The intent is that the system can provide the necessary assurance for the IAEA to draw rapid, comprehensive, and definitive conclusions that nuclear material is not being diverted from peaceful use. This directly relates to two specific IAEA criteria and two scenarios: goal quantity and conversion times, and abrupt and protracted diversion.

As an example, the IAEA considers 8 kgs of plutonium and 25 kgs of highly enriched uranium as quantities of interest to a diverter and therefore these have been selected as detection goal quantities for the IAEA safeguards system. Conversion times are the estimated times for a diverter to convert these quantities into a nuclear weapon. These conversion times are based on the form of the material; conversion times are shorter for pure metals and oxides (weeks) compared to spent fuel and less pure forms (months, up to a year). This is because highly enriched uranium and plutonium in metal form are much more readily usable as the core of a nuclear weapon. Extracting this metal from spent nuclear fuel requires a difficult and time-consuming industrial process.

The two high-level diversion scenarios consider the complete diversion of a goal quantity in a short time (abrupt) versus a series of small diversions that lead to a goal quantity over a long time (protracted). All these factors taken in combination lead to the basis for the detection sensitivity of the IAEA safeguards system and the time periods during which definitive safeguards conclusions must be drawn.

A UMS has the following basic characteristics:

1. It is a system that automatically monitors the flow of nuclear materials 24 hours a day, 365 days a year, without the need for human interaction.
2. It may use a variety of sensors such as radiation, pressure, temperature, flow, vibration, optical, and electromagnetic fields to collect qualitative or quantitative data.
3. It is permanently installed in a nuclear facility.
4. It is computer-based for data retrieval, either on-site or remotely.
5. All components are in tamper-indicating enclosures to ensure data authenticity.

The IAEA defines the word *remotely* to indicate when the data goes from a computer server in a monitored facility to some remote location. This could be somewhere in the facility in an inspector data review room, to an IAEA field office such as Toronto or Tokyo, or to the headquarters in Vienna.

Regarding item 5 above, the IAEA must be able to independently verify its conclusions regarding the nuclear material in a facility. At the same time, the IAEA must consider that every state is a potential adversary with the intention to divert nuclear material to a weapons program. As such, there is always the potential threat that, to conceal such a diversion, a state might try to alter the data the IAEA is collecting. There are several steps that the IAEA takes to prevent an adversary from succeeding in this effort and to assure that the data is authentic. One is to design all data-recording enclosures so that the IAEA can detect any unauthorized tampering. The IAEA has no ability to prevent tampering, because their inspectors are not constantly in the facility nor do they control access to the data-recording equipment. Therefore, it is critical that they have the ability to detect it. Specific examples of tamper-indicating enclosures are presented in Figure 6.1. Detection of data tampering would be grounds for further action by the IAEA with respect to the state operating the facility.

Why Does the International Atomic Energy Agency Use UMS?

Prior to the use of UMS, the routine inspection of safeguarded nuclear facilities was periodic, relying on inspectors to visit at a specified frequency, perform specific activities, and draw timely conclusions based on the data collected.

This approach had a number of problems. Because it was periodic, the information an inspector collected at a facility was no longer current once the inspector finished. This also meant that significant effort could be required to reestablish this information across the inventory during the next inspection. Requiring physical presence to carry out inspections also meant that the operator had to support every inspector's visit with personnel resources and that facility operations were interrupted to allow an inspector access. This placed a heavy burden on both the IAEA and the facility operator. Some facilities, such as reprocessing, on-load reactors, and hot cells, which present radiation hazards to personnel, precluded inspector presence due to safety concerns. Even in the case where inspectors could have complete access, it was not economically feasible for a facility that operated continuously, or for the IAEA, to provide around-the-clock safeguards using inspectors and the facilities' support staff. In fact, it was the design of fully automated facilities that really drove the advancement into modern unattended safeguards systems as the only rational solution to the challenge of effective IAEA monitoring.

The benefits of UMS are as follows:

1. Provides the highest level of safeguards assurance currently available to the IAEA through continuous routine monitoring of activities in nuclear facilities.
2. Minimizes impact on the facility by allowing uninterrupted facility operation.
3. Minimizes the impact on the IAEA by reducing inspector visits and inspection resources (including the high cost of worldwide travel).
4. Reduces radiation exposure to personnel and can operate in radiation areas too dangerous for humans.

A Balanced Approach to IAEA Safeguards Compliance

To fulfill its mission as a neutral arbiter of compliance with the Nonproliferation Treaty (NPT), the IAEA needs the ability to draw independent, verifiable conclusions on a state's nuclear facilities. However, the process of inspections and monitoring also has to meet other conditions of the safeguards agreements between the IAEA and the state. These conditions are found in "IAEA Information Circular (INFCIRC)/153: The Structure & Content of Agreements Between the Agency & States in Connection with the NPT." Part I, Paragraph 4, of INFCIRC/153 states that safeguards agreements shall be implemented in a manner designed:

a. To avoid hampering the economic and technological development of the State [...] in the field of peaceful nuclear activities, including international exchange of nuclear materials;
b. To avoid undue interference in the State's peaceful nuclear activities, and in particular in the operation of facilities; and
c. To be consistent with prudent management practices required for the economic and safe conduct of nuclear activities.

This instruction to the IAEA stressed the need to strike a balance between the ability to reach verifiable conclusions and the need to take the operational realities of nuclear facilities into account. It also gave the state and the facility operator, who has the deepest knowledge of the facility, the right to negotiate efficient safeguards while the IAEA maintained its right to independent verification.

Cost Advantages of UMS

One of the primary advantages of a UMS is the cost savings it allows for safeguards inspections. This includes not only cost-effective safeguards at facilities but also a reduced burden on the IAEA safeguards budget. The use of UMS technology enables the IAEA to require fewer on-site safeguards inspections without sacrificing the capability for independent verification. It therefore saves the IAEA resources that otherwise would be spent on inspector travel and labor to visit various facilities around the world.

Worldwide Deployment of UMS

The push to deploy UMS started in the late 1980s and has rapidly advanced along with technological leaps in the capability and reliability of hardware, firmware, and software. From 2000 to 2004, the IAEA installed an average of 10 new systems per year. By August 2004, 90 UMS systems were installed at 44 facilities in 22 different countries. Of the 90 systems installed, 79 are radiation based, five are thermo-hydraulic based, and six are process-monitoring based. By the end of 2006 it was expected that about 25–30 additional UMS would be installed around the world. Most of the UMS installations in the past two years have been in the Japanese Rokkasho Reprocessing Facility.[1]

Key Characteristics of UMS

The IAEA's commitment to implement unattended monitoring systems for nuclear safeguards around the world is a daunting task. The consequences of failure are technical, economic, and political. On the technical front, loss of confidence in an UMS would create the need to reestablish the knowledge of the inventory and location of the nuclear materials at a facility by means which have traditionally been both more costly and operationally disruptive. The negative political consequences could vary widely. They would include questions about the reasons for failure and whether or not this created an opportunity for undetected diversion of nuclear material. In cases where the UMS fails but no safeguards violation occurs, the impacted state would likely lodge complaints to the IAEA about the additional costs it will face due to a failure of IAEA equipment. This, in turn, can raise questions about the IAEA's competence in accomplishing its mission. Should the case arise where a diversion is discovered by other means despite the apparent proper functioning of an UMS, this would severely damage the technical reputation of the IAEA and question the wisdom of IAEA reliance on UMS. Therefore, the IAEA's UMS Unit established two primary goals in order of priority: no loss of safeguards significant data and assurance that the data is authentic.

The key in the first goal is the phrase *safeguards significant*. It is not possible to assure that all UMS components will work all the time. Instead, the emphasis is on providing fault-tolerant systems that can continue to meet IAEA needs in maintaining knowledge of the nuclear material in a facility in spite of some equipment failures. Therefore, when the IAEA uses systems that are designed to be robust and reliable while eliminating or minimizing single points of failure through redundancy, components can fail without causing a loss in confidence in the data and without jeopardizing the IAEA's ability to draw verifiable safeguards conclusions.

Authentication of data is a tremendous challenge. Since the IAEA must assume that every state could be a potential adversary, it goes to great lengths to protect its data from the point of origin in the sensor through data analysis. Nevertheless, data authentication is of secondary importance to data loss. Authentication must not jeopardize the reliability of the data collection scheme.

UMS Design Considerations

Considering the primary goal, "No loss of safeguards significant data," some key UMS design approaches include the use of high reliability and redundant critical components and reduced use of low-reliability components. Securing reliable components can be a great challenge in

---

[1]See Chapter 9, "Case Study: Safeguards Implementation at the Rokkasho Reprocessing Plant," by Susan Pickett for a detailed analysis of the UMS at the Rokkasho Reprocessing Facility.

the UMS area, primarily due to the low production unit volume for some key components. Specialty components such as data generators will only be made by small vendors, so reliability will always be tested to assure performance. The current design philosophy at the IAEA strives to use as many high-volume, commercial off-the-shelf (COTS) products as possible. This helps not only with reliability but also with costs. Fortunately, there are many high-reliability components available commercially, including sensors, batteries, cables, software, air conditioners, uninterruptible power supplies, industrial PCs and servers, operating software, connectors, encryption and wireless hardware and software, and cabinets.

Using independent redundant components or "backup" systems to monitor the same event is another primary design approach to prevent the loss of data. Defense in depth by layering both the sensor and data collection systems is another fault-tolerant approach. Some approaches include the use of signal splitters. The sensor could be a gas-filled tube for neutron detection, which has demonstrated reliability over many decades, unlike the data generator, which may not be so reliable and will have its signal split and sent to two different data generators. Therefore, a single sensor can be used in conjunction with two data generators (radiation data generators can support multiple sensors) to assure robust data collection if one generator fails.

The use of uninterruptible power supply (UPS) to assure uninterrupted data collection even after a loss of mains power is an obvious application to UMS. It is even more crucial considering that the infrastructure in many countries around the world is such that loss of power or out-of-tolerance power levels of varying duration are routine events. It is the practice of the IAEA to negotiate with the facility to obtain Class 2 power from the facility. (Class 1 supports the facility's safety system. Because of its importance, nothing else is added that might risk its integrity.) This is their backup power system; obtaining power for IAEA systems from it will help assure reliability for the IAEA system. Besides a main UPS that is deployed in an electronics rack, the same defense in depth is applied where other critical components, such as data generators, will have battery backup internally.

Employing multilayer security is the second goal, and it reflects all the activities that the IAEA performs to secure its data. These include mechanical approaches such as secured housings that hold sensors and all other associated equipment used to collect data, such as electronics cabinets, cabling, and junction boxes. It also includes electronic approaches such as data authentication and encryption.

The primary objective for any UMS is to reliably collect safeguards information without an inspector's presence, on a continuous basis, and to accomplish the following:

- Verify flow and inventory of nuclear materials
- Minimize intrusiveness on operator
- Reduce IAEA and operator manpower requirements
- Decrease radiation exposure to IAEA inspectors and facility operators
- Standardize hardware and software for the IAEA, to minimize maintenance and training

In beginning the design of a UMS, the following design considerations are investigated for each application:

- *Cost/benefit.* A cost/benefit analysis is made, comparing inspector days in the field without a UMS and with a UMS.
- *Reliability and stability.* Have the UMS components demonstrated a reliability of at least 150 months mean time between failures (MBTF) for the requested application? Can it perform reliably within the available infrastructure?
- *Meet operation's user requirements.* Can the UMS meet the performance and functionality required by the IAEA Operation Division, whose inspectors will use the data from the UMS?
- *Operator-provided equipment.* Is there operator equipment that could be used jointly by the IAEA and the operator while maintaining the independent verification capability the IAEA requires?
- *Authentication requirements.* Can the UMS be secured to assure the IAEA that the data is authentic?

- *Early involvement of the IAEA in planning stages.* Allows integration of facility-specific safeguards features into final plant designs, attaining the most cost-effective design prior to facility construction, avoiding the high cost and restrictions associated with facility retrofits.
- *Longevity.* Is the system designed to last at least 10 years?

Authentication

Authentication is defined as all measures taken to ensure that the safeguards measurement systems collect and provide authentic data. This is the broader category that also includes encryption. For the classical application of electronic authentication and encryption on cryptographic modules, the IAEA follows U.S. Federal Information Processing Standard-140 (FIPS-140).[2] This standard specifies the security requirements for a cryptographic module used within a security system protecting sensitive information in computer and telecommunications systems. Due to the nature of its work, the IAEA must monitor activities by leaving its equipment in facilities in states that are potential adversaries; this represents a great challenge for UMS authentication. It should be noted that as a cost savings measure, some installed systems are "joint use" or jointly used by the State Inspectorate and the IAEA and therefore jointly specified—the State often has equal responsibility and also often funds the development and installation. However, independence and strong authentication of the IAEA portion of instrument hardware, data collection, and data extraction remain the same.

Some examples of the steps the IAEA has taken to ensure that its data is authentic include the following security methods:

- *Software controlled.* IAEA software is not accessible by states, nor are systems shipped with hard drives in them. This eliminates the opportunities a state could have to examine the software during the process of shipment, customs clearance, and receipt at a facility.
- *Tamper-indicating enclosures.* This includes all external housings and shipping containers.
- *Containment/surveillance (C/S) on detector head and electronics.* Containment devices such as seals and surveillance as with optical sensors can be used to protect any enclosure (seals at any time and surveillance during operation).
- *Visual inspection of components and cables.* Since the IAEA cannot prevent tampering, it is critical that inspections identify potential tampering. Currently, this capability is limited to visual techniques; however, many other techniques are applicable.
- *Efficiency check with normalization source.* It is common practice to use an IAEA source that is stored at a facility under IAEA seal to calibrate equipment that the IAEA will use for any measurements.
- *Supervision of maintenance.* In cases where the IAEA must use local companies to perform maintenance or repair on IAEA equipment, all such work is done under the supervision of the IAEA.
- *Cross-correlation with other safeguards measures.* As difficult as authentication can be for the IAEA, the use of multiple sensors on time-correlated activities increases the difficulty for an adversary to compromise such a system.
- *Use of unique data signature on all digital data.* The IAEA maintains a cryptographic standard for all digital data and communications. Currently, this standard requires 128-bit encryption algorithms.

---

[2] http://csrc.nist.gov/cryptval/140-2.htm (June 2007).

- *Encrypted data transmission between cabinets and for remote monitoring.* Data encryption is used both on- and off-site for all data. (Special consideration for each state is also taken into account for off-site data, as appropriate.)
- *Uninterruptible power supplies (UPS).* Maintaining power is key for not only uninterrupted data collection but for data protection as well. A layered approach is used.

Tamper-Indicating Features

Tamper-indicating features are integrated into all detection mechanisms, both electrical and mechanical. Although not separately categorized, various design features are used to minimize or eliminate access points by an adversary.

At the most basic hardware level, the IAEA uses simple mechanical approaches to eliminate easy tamper opportunities. Consider the standard 19-inch Rittal industrial rack and enclosure that are used to house IAEA instrumentation. Figure 6.1 shows details of this enclosure. Figure 6.1a shows the application of an IAEA tamper-indicating metal seal that is applied to the locking mechanism for the cabinet. Figure 6.1b shows the locking mechanism in the open configuration; this mechanism moves full-length door pins so that edges of the door cannot be pried open. Figure 6.1c shows the use of internal hinges as external hinge pins are readily removed. The standard U.N. blue paint coating used on all IAEA equipment is applied using a powder process to make touching up evidence of tampering more challenging.

The concept of "protected enclosure" reaches out to sensors as well. Figure 6.2 shows a picture of the entrance gate monitor used to measure fresh MOX fuel prior to entry into the reactor core. The application of the standard IAEA metal seal to detect attempts to access the upper portion of the neutron tube electronics can be seen. This neutron collar uses coincidence counting to assay the plutonium content of each assembly.

Figure 6.3 shows the IAEA tamper-indicating conduit used to protect unauthenticated cabling that connects sensors to the electronic equipment in the IAEA instrument enclosures.

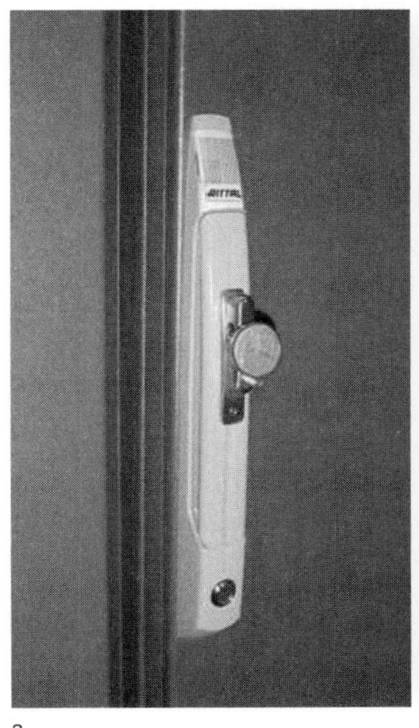
a.

b.

c.

FIGURE 6.1 Cabinet close-ups.

FIGURE 6.2 Entrance gate monitor (ENGM) detector.

FIGURE 6.3 Tamper-indicating conduit.

This attempt to protect the cabling is problematic. Some facilities can have kilometers of cabling, including many portions that must go through wall penetrations that cannot be accessed. The addition of this stainless steel bellows-type conduit is expensive and must be visually inspected periodically by an inspector to detect tampering. This is a very impractical requirement, since some areas cannot be accessed due to operational and radiation constraints, let alone the task of examining the conduit in a thorough enough manner to actually detect tampering. This is clearly an area in need of applying new technical means to either authenticate all cabling or detect electronic tampering.

Security Solution: Virtual Private Network

A *virtual private network* (VPN) is a private data network that uses the public telecommunications infrastructure, maintaining privacy through the use of a tunneling protocol and

FIGURE 6.4 AP-10.

security procedures. The idea of the VPN is to give the IAEA the same capabilities at much lower cost by using the shared public infrastructure rather than a private one. The IAEA uses a VPN device inside the IAEA tamper-indicating cabinet to transmit data both between cabinets and remotely from a facility through various available mediums such as the Internet, satellite, the public switched telephone network (PSTN), and digital subscriber line (DSL).

Wireless Solution

A particularly costly activity in any nuclear facility is the installation of cabling. Installing tamper-indicating conduit further increases this cost. In addition, often there are situations where the IAEA must install monitoring equipment on movable platforms. Taking a wireless data transmission approach is quite attractive for these reasons. The IAEA has installed wireless systems using the Alvarion AP-10 indoor wireless hub (~US$1,000) shown in Figure 6.4 and the SA-10 station adaptor (~US$500) for the end-user computer. This system operates on 10Base-T Ethernet using RJ-45 connectors with data rate up to 3 Mbps and a range up to 150 m (500 feet).

It should also be pointed out that there will always be one-of-a-kind systems that will not warrant the time and effort to migrate such unique limited-use systems to the standard software platform.

Collect Software is the automated software application the IAEA uses in a local cabinet's computer system to collect data from sensors at the facility being monitored. This could be a large distributed sensor system or one that is very small (on the order of just a couple of sensors). The primary function is polling of data from data generators. A data generator is the first electronic device that receives the sensors' input and, if required, digitizes the signal. The IAEA's approved software is called Multi-Instrument Collect (MIC). MIC was designed by Los Alamos National Laboratory under the auspices of the United States Support Program. This application has the following functions:

- Data collection (can support ~40 data generators)
- Startup service (automatically starts up during the PC's startup routine)
- File transfer service (can provide automatic file transfer to an archive)
- Delete files (can automatically clear data storage space as desired)
- Binary files to text (can convert error code to readable text)
- Debug tool (a self-diagnostic tool to discover/resolve problems)
- Tracker (transmission of state of health)

- File copy routine (can provide automatic copies of data as desired)
- Display instrument messages (allows user to view messages from polled instruments)

IAEA inspectors use review software to analyze and draw conclusions about the data collected at a facility. In general, this software is used in an attended mode. That is, it takes an inspector to use this software for the analysis effort. It is also important to note that, in general, the IAEA does no real-time data analysis. That is because the IAEA's inspection methodology is based on drawing timely, but periodic, conclusions based on the type of nuclear material, its form, its quantity, and the estimated time for conversion into a weapon.

The IAEA's Current Data Collection

The IAEA's current data collection standard is called Integrated Review Software (IRS). IRS was designed by Los Alamos National Laboratory for the Rokkasho Spent Fuel Storage Facility and funded by the Japanese Nuclear Material Control Center (NMCC).[3] It was further developed to integrate INCC for the Japanese Plutonium Fuel Production Facility (PFPF) with funding provided by Japan Nuclear Fuel Cycle (JNC) Development Institute. This application has the following functions:

- *Allows for review of data outside active area.* Inspectors are encouraged to minimize their time in a radiation area. Therefore, the IRS software is usually on a PC in an office area controlled by the IAEA.
- *Event screening by threshold settings.* A sort on events detected by a sensor can be screened first using threshold settings to eliminate events that are not of safeguards significance.
- *Graphical display.* Plots the sensors' events against time for a visual display.
- *Data analysis.* Utilizes algorithms to assess events.
- *Campaign management.* Allows for data partition in line with the operator's declaration of activities.
- *Mark of assays by time correlation.* Allows for specific selection of an event based on time.
- *Transfer of marked data to software for analysis.* Allows the use of more sophisticated analysis algorithms in other modules to assess the events.

This modular piece of software allows any vendor to add capability to its suite of analysis tools. The current suite of review tools includes:

- *RAD.* Radiation Review (graphical plot of radiation signals against time).
- *DVR.* Digital Video Review (display of video images against time).
- *PR.* Position Review (display of GPS data on a two-dimensional map).
- *INCC.* IAEA Neutron Coincidence Counting (INCC-quantitative analysis of neutron data).
- *ISO.* Plutonium Isotopic Review (quantitative analysis of isotopic composition).
- *OP.* Operator Review (operator's declaration of events/activities).
- *IR.* Integrated Review (comparison engine that uses defined limits to match the operator's declaration against the IAEA data).

---

[3] M. E. Abhold, S. E. Buck, Y. Yokota, et al., "Integrated Monitoring and Reviewing Systems for the Rokkasho Spent Fuel Receipt and Storage Facility," in *Proceedings of the 39th Annual Meeting of the Institute of Nuclear Materials Management* (INMM), July 1998.

Position Review Module

The Position Review (PR) module displays GPS latitude and longitude position data of a vehicle carrying nuclear material on an area map. The vehicle movement can be played like video. This use of a global-positioning satellite sensor for the purposes of tracking the movement of nuclear material is one of the more modern wireless approaches, the use of which will continue to spread as an effective and important capability to support the IAEA mission.

Integrated Review Module

The Integrated Review (IR) module works in association with the RAD, INCC, and OP modules to compare the operator's declaration of activities with those independently collected by the IAEA. Data disagreements between these comparisons are displayed by the IR module in contrasting colors in the data line on the computer screen to show the IAEA inspector whether there is agreement or disagreement. This top-level comparison is based on input from the two IAEA data collection modules: the Radiation Review module (RAD) and the IAEA Neutron Coincidence Counting (INCC) module against the operator's declaration (OP).

Hardware Standards

A similar learning process has taken place with hardware. Originally, entire UMS cabinets were designed by vendors. However, it quickly became clear that these specialty vendors needed to focus only on safeguards-unique hardware, whereas the remaining components would be commercial off-the-shelf (COTS) units. This would give the IAEA the best combination of dedicated devices and cost-effective commercial components, providing maximum flexibility by using a building-block approach to UMS.

System Components

Computers

One of the least reliable components used in a UMS is the computer. The idea of a standard model is impossible to maintain since the component technology used in PC manufacturing changes so rapidly that having two PCs with the same name and model number might not mean that the internal components remain the same; therefore, reliability always remains an issue. The IAEA has used a variety of hardware and software configurations to assure higher reliability. This method is based on the following basic approaches: minimize or eliminate moving parts (by using passive cooling and solid state drives); use redundant components such as a Redundant Array of Independent Disks (RAID), multiple independent cooling fans, or a failover box that starts up a second PC; and use an independent watchdog that can restart a PC.

A fault-tolerant system should be able to run without a collect PC. One of the approaches the IAEA uses is the Intelligent Local Operating Node (ILON). This device can maintain certain critical functions that allow a distributed sensor network to continue to collect data without the collect PC. In addition, these instruments will collect data even given total failure of both the collect computer and the ILON. However, certain critical functions, such as triggering, will be disabled without the ILON. This particular device is undergoing an upgrade to meet the IAEA's Ethernet standards and to employ strong authentication. It has the following characteristics and functions:

- *Open topology and cabling.* This allows maximum versatility for facility upgrades that require additional monitoring sensors that were not planned at the time of installation due to new facility capability or new IAEA requirements. The new upgraded ILON will be Ethernet compatible.
- *Instrument or collect function.*
- *Time synchronization.* This can keep all the data generators on the same clock setting.

- *Triggering (direct and indirect).* This is a critical function that allows sensors to be combined to obtain the highest level of safeguards assurance. One example is a radiation sensor triggering a camera.
- *Authentication.* The current ILON has weak 32-bit wrapper authentication; the upgrade will bring this to 128-bit.
- *Watchdog.* If a data generator fails to check into the ILON, it can send a hard reset command.
- *Functions independent of the collect computer that is used to collect sensor data from the data generators.* This is the key. It operates at the same independent level as a data generator, using batteries as necessary.
- *Narrow bandwidth.* The current ILON has a limited bandwidth of 78 kb/s since it was not intended for digital images; the upgrade will bring this up to 100 mb/s.

Surveillance Data Generator

The Digital Camera Module 14 (DCM-14) was designed under the auspices of the German Support Program. The module is commonly combined with a CCD camera, as shown in Figure 6.5.

The DCM-14 also has a rotating buffer memory for up to eight images. This plays a key role when triggering is used for safeguards applications. The camera is set at a regular "heartbeat" where an image is taken at a designated frequency. As an untriggered image, this new image is placed in the rotating buffer as the new #1 image and the previous images rotate through the buffer with the last image, old #8 being deleted. When a trigger takes places, such as from a radiation sensor, the DCM-14 will dump the rotating buffer to the permanent data storage as pre-event images. It will then go into a triggered mode with the subsequent post-trigger images at some set interval that will also be dumped to permanent data storage. In this way, a complete cycle of images is available to fully define the triggering event.

The DCM-14 has the following capabilities:

- Digital image
- Scene change detection
- Image compression

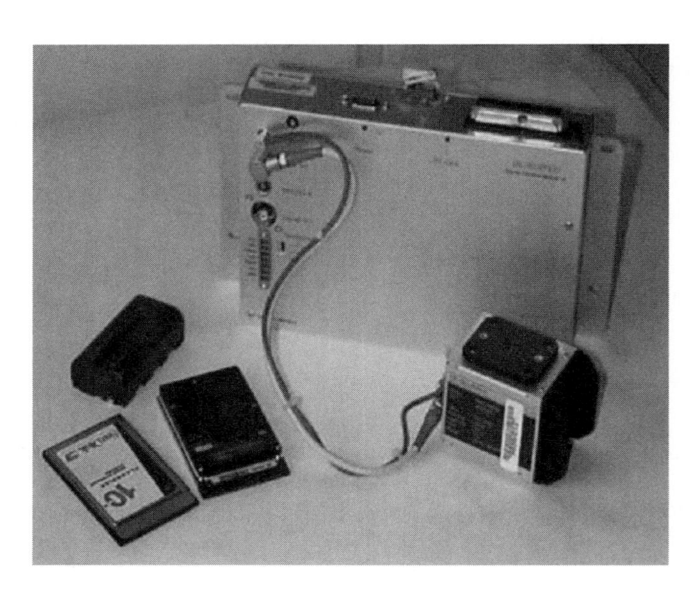

FIGURE 6.5 DCM-14 with CCD.

- Image/data authentication
- Image/data encryption (triple DES algorithm)
- Power management (minimization mode when on battery power)
- Battery backup (about two or three days, depending on image frequency)
- External triggers
- On-board 100 days' data storage (in removable Flash cards)
- State of health

Since the module can store up to 100 days of images, this allows a safeguards inspector on a 90-day inspection cycle to recover all images by removing the Flash cards if the collect computer has failed. The internal battery is another fault-tolerant approach, since the main cabinet will have a "smart" UPS that will extend maximum life to the data generator by shutting down the collect computer; the UPS batteries are dedicated to the data generators. If the UPS life is exceeded, the data generators' internal batteries further extend the life of the module. All these layers of power are intended to carry the IAEA systems through the majority of power outages without loss of data.

Radiation Data Generators

These data generators follow the same functionality as the surveillance generator, the difference being the capability to support radiation sensors and transmit triggers. The two primary radiation data generators are the LANL-designed miniature gamma-ray and neutron detector (MiniGRAND) developed under the auspices of the United States Support Program and the BOT Engineering-designed Standalone Autonomous Data Acquisition Module (ADAM) developed under the auspices of the Canadian Support Program. Both radiation data generators are undergoing upgrades.

The MiniGRAND supports three pulse channels and two current channels, allowing a broad assortment of radiation sensors to be attached. It is shown in Figure 6.6. Besides the MiniGRAND board stack, the lower board is an ILON. The Standalone ADAM is shown in Figure 6.7. This data generator supports eight pulse channels, but an adaptor is under development to allow current-based sensors as well. The twin removable memory cards and its Ethernet capability can be seen in the photo.

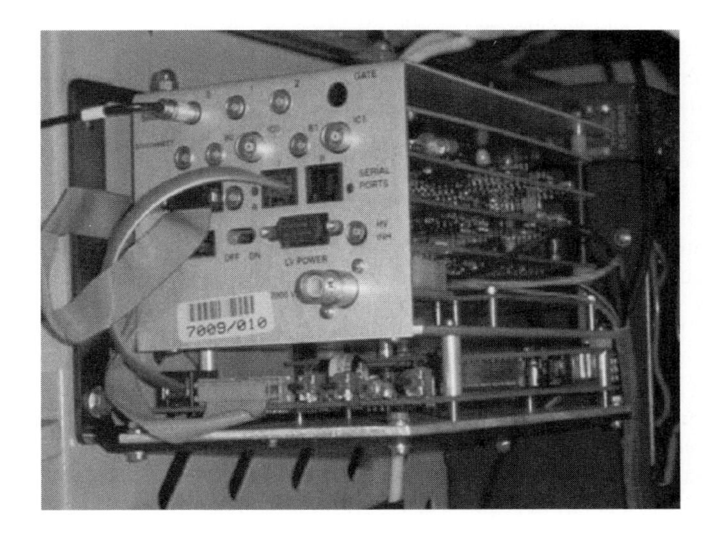

FIGURE 6.6 MiniGRAND and ILON.

FIGURE 6.7 ADAM.

Dedicated Simple Systems

VXI-Based Flow Monitor

In contrast to the modular units mentioned, the IAEA has had complete UMS developed by a single vendor. One example is the VXI-Based Flow Monitor (VIFM), which was designed in the last decade by BOT Engineering, under the auspices of the Canadian Support Program, specifically to monitor spent fuel bundles from CANDU type reactors. The primary specification the IAEA established required the use of the industrial VXI architecture while the manufacturer was free to design the rest of the system. This rack-based unit with expandable slots and backup battery can be seen in Figure 6.8.

This unit is used to monitor core discharges from the reactor face (Core Discharge Monitor, or CDM), and bundles (Bundle Counter, or BC) as they are moved into the spent fuel pond. It has proven to be very reliable. It uses paired ADAM data generators providing full data generator backup and up to eight SOLGEL batteries providing up to 90 days of operation without mains power. The main problem with the unit has to do with the custom design of nearly all components, including the collect computer. Upgrades have been costly and they take a long time. This experience helped the IAEA focus on using COTS where it makes sense.

The VIFM is a qualitative system that can count items but not assay nuclear material quantities. Figure 6.9 shows a typical spectrum from the CDM sensors for a CANDU 600. In this display, we have counts along the ordinate axis (traditional y-axis) and time along the abscissa axis (traditional x-axis). The gamma signal is represented in white and the neutron in black. An algorithm is applied that can count bundle movements using this peak structure. The algorithm is designed to detect off-normal responses. To understand the radiation profile shown in Figure 6.9, a short review of CANDU reactor operations is in order.

FIGURE 6.8 VXI-Based Flow Monitor.

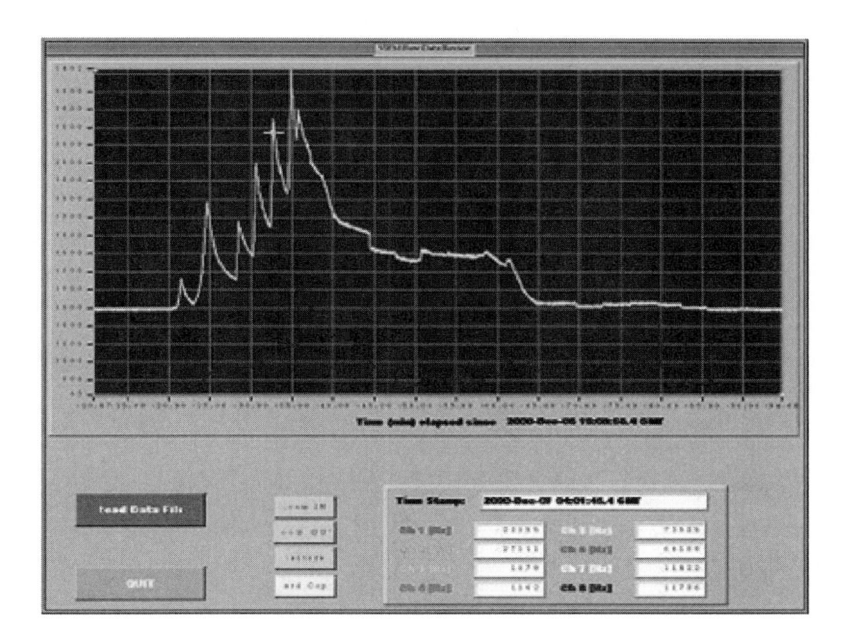

FIGURE 6.9 Core Discharge Monitor.

This reactor uses natural uranium for fuel and uses heavy water to provide the necessary moderation to sustain a nuclear reaction. The fuel channels are aligned horizontally in the reactor. In this on-load power reactor (the category of power reactors that can be refueled during operation), approximately 15 to 23 bundles per full power day are replaced on a daily basis to refuel the reactor. Each spent fuel bundle is tracked because it will contain plutonium once the fission process has taken place. Therefore, the CDM detectors are only on the reactor face from which spent fuel will be removed. This means that the IAEA must be able to conclusively count and maintain surveillance on the nearly 3,000 spent bundles removed from the reactor each year.

The process of removing and replacing fuel bundles follows a specific sequence. Two fuel-handling machines are aligned on the same fuel channel from opposite faces of the reactor. One fuel-handling machine has four pairs of fresh fuel bundles (eight total) loaded in its rotating cylinders (much like the barrel on a revolver-type handgun) for insertion into the reactor (the unmonitored reactor face). The other fuel-handling machine is empty, ready to receive the spent fuel bundles that will be removed. Focusing on the gamma (white) spectrum of Figure 6.9, the first peak represents the removal of the channel plug from the reactor face where the spent fuel will be removed.

The second peak represents the removal of the radiation shield plug, now allowing direct access to the fuel. The next four peaks represent the removal of four pairs of spent fuel bundles, representing eight bundles from the reactor.

The monitoring system also measures radiation using the bundle counter. This counter uses three solid-state gamma-ray detectors located above the tray mechanism that transfers the spent fuel bundles from the reactor hall to the spent fuel storage pond. The response from these three detectors is represented by color-coded sequential response graphs. A fourth sensor is located at a point where it can detect the transfer of the two bundles into the spent fuel pond. The first three sensors are used to verify the transfer of the two bundles from the reactor core face onto the transfer tray. They are located longitudinally along the axis of the tray transfer mechanism, with the first two sensors over the final resting position of the second bundle and the third over the final resting position of the first bundle. Because of the forward location of the first two sensors, they view the spacing between the two bundles, which is evident by dips in response on the graphs.

The response from the third sensor that is located toward the rear of the transfer tray only shows the first bundle; therefore, this sensor never sees this gap, as evidenced by the lack of a response dip. The specific locations of these three sensors were chosen to detect all possible scenarios of diverting bundles from the tray.

The VIFM system, as is true for most UMS, uses counting algorithms to aid the inspector in drawing conclusions from the data collected by these automated systems. In the normal first screen display, the total counts of bundles from the CDM and BC systems are displayed. Under conditions of normal operation, the two bundle counts must match. In this case, the inspector can consider his inspection of the data as complete. There would be no need to look at the detailed response peaks, as shown in Figure 6.9. A mismatch in count would indicate an anomalous condition. It would then be up to the inspector to determine the cause of this discrepancy. Under a condition that requires detailed investigation, the inspector can then "drill down" to a lower-level screen in the software to see the response peaks for each event.

A mismatch does not immediately cause the inspector to suspect potential diversion. In fact, discrepancies are not that uncommon because problems may occur during operations at a facility. As part of the operator's obligation to the IAEA, an operator declaration of activities for the period covered by the inspection is made. It is in this declaration that the IAEA inspector will most likely find mention of some difficulties during operations that may be detected by the UMS.

Advanced Thermohydraulic Power Monitor

Although the majority of UMS use radiation sensors for safeguards on nuclear facilities, the IAEA deploys a wide range of sensors. One nonradiation sensor system is the Advanced Thermohydraulic Power Monitor (ATPM). This system was designed specifically to meet the safeguards challenges associated with plutonium production in research reactor fuel. Unlike commercial power reactors that operate at full power in a fixed-core configuration for extended periods of time as part of a state's base power grid, research reactors typically operate for short two- to three-week periods at varying power levels and can have the flexibility to change the core configuration. In addition, spent fuel is routinely replaced in commercial power reactors on a routine basis (typically every 12 to 14 months in LWRs and daily in on-load reactors such as the CANDU), allowing for direct monitoring and measurement. Due to the periodic nature of operation in research reactors, a single core loading of fuel will last many years.

The challenge to the IAEA is finding a way to accurately determine plutonium production in the fuel in the core throughout the lifetime of a research reactor. Since the IAEA has access to the detailed design of each facility (through the design information verification process), knowledge of power can be used to calculate plutonium production in the core of the reactor. This is the basis for the ATPM approach.

Figure 6.10 shows a schematic representation of the ATPM system design. The data collection side of the system is located in the radioactive cold portion of the facility. On the "Hot Area" portion of the schematic, two redundant sets of sensors are mounted on the hot and cold sides of the primary core cooling loop. Each set contains a resistive type temperature sensor (T) and an ultrasonic flow monitor (F). The output from these sensors provides a velocity measurement on the cooling loop's water and the temperature drop as heat is removed from the loop. The data from these sensors, in combination with the required reactor design information, are used to calculate reactor power output. This calculation is plotted against the unit's internal clock. In addition to being used as a direct comparison with the operator's declaration on reactor operations, this information can be used with predictive reactor operation codes, such as ORIGEN, to calculate plutonium production.

Figure 6.11 shows an IAEA engineer installing an ultrasonic flow sensor on the primary core cooling pipe. The long frame that is strapped onto this cooling pipe is used to secure the ultrasonic sensors at the required angle for this specific pipe diameter. The engineer has one sensor in his hand and is securing it on the frame. A matching sensor will be installed at the far end of the frame (not visible in this image). Note the two threaded rods that are mounted perpendicular to the frame and extend away from the cooling pipe. The first is located immediately adjacent to the engineer's left hand. The second is in the center of the frame. A third rod is on the other end, not visible in this picture. This photo shows the difficult work environment often faced by the UMS staff. Once the sensors are secured, a tamper-indicating cover is applied to

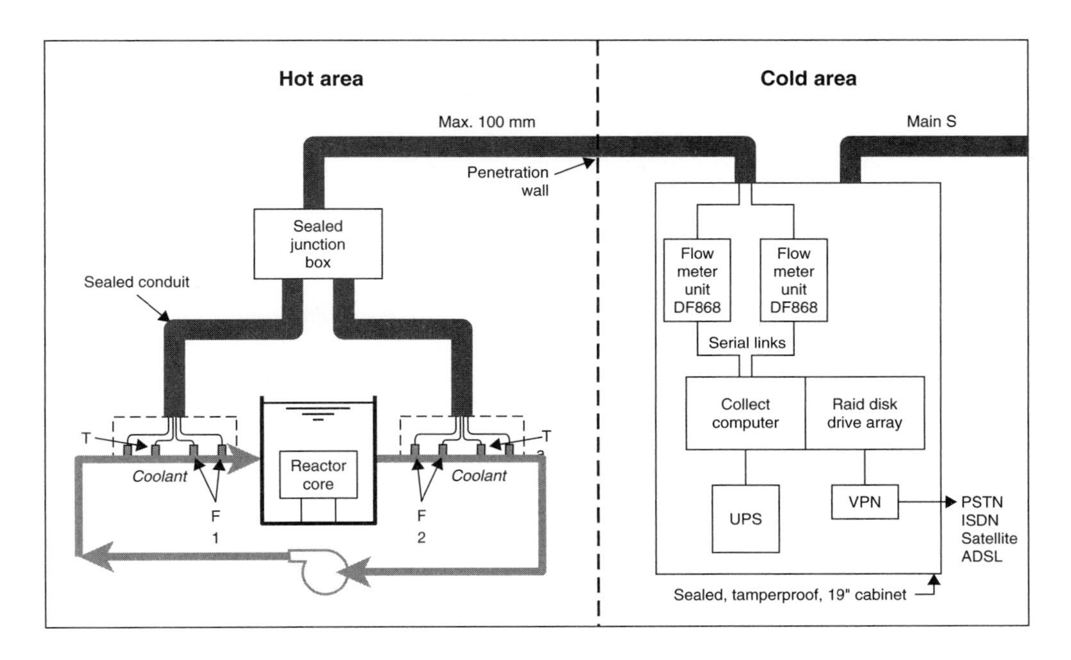

FIGURE 6.10 ATPM system diagram.

FIGURE 6.11 Flow sensor installation.

the sensor frame that slips over these rods. Nuts are tightened down on the threaded rods securing the cover. Tamper-indicating wire is then pulled through the holes in each threaded rod; the wire is tied off at the center rod, awaiting the application of a metal seal by the IAEA inspector.

Dedicated Complex Systems

The previous systems were defined as "simple" systems solely from the perspective of the nuclear material they were designed to safeguard. Both the VIFM and the ATPM were designed to provide safeguards on an item facility. In an item facility, all nuclear material is permanently sealed within a container that has no access point and structurally never changes during the entire time at the facility. Nuclear fuel contained in fuel rods that are mounted in assemblies fits this definition. In a bulk facility, nuclear material can be accessed directly and it might change form. Examples include gases, powders, solutions, and metals. Many of these forms are found in enrichment, reprocessing, and fuel fabrication facilities. Access to bulk materials presents a much greater safeguards challenge for the IAEA since small portions of material can be diverted.

One of the greatest facility challenges is safeguarding a reprocessing facility. The Plutonium and Uranium Extraction (PUREX) process is one example. At a reprocessing facility that uses the PUREX process, nuclear material progresses through the following different forms: feed material (spent fuel, which contains irradiated nuclear material with actinides and fission products), processed material (input liquids, separated liquids, undissolved solids, low-level and high-level liquid and solid wastes), and product material (separated oxides and mixed oxides). High radiation levels dictate that most of these activities be controlled remotely. Good safety and business practices demand that such facilities be fully automated. This is the case at the Japanese Rokkasho Reprocessing Facility (RRP).[4]

Figure 6.12 shows a picture of the Improved Plutonium Canister Assay System (iPCAS). This is a dedicated quantitative system that is fully integrated into the operational flow of the facility. It is owned by the Japanese safeguards authority and used by the IAEA. It was designed by Los Alamos National Laboratory (LANL) and represents one of 10 systems from

[4]See Chapter 9, "Case Study: Safeguards Implementation at the Rokkasho Reprocessing Plant," by Susan Pickett in this volume.

FIGURE 6.12 iPCAS side view.

FIGURE 6.13 iPCAS top view.

LANL that are deployed throughout the flow stream at RRP. iPCAS was designed to precisely measure three cans totaling 36 kg (18 kg of Pu and 18 kg of U) of mixed oxide product ($(Pu/U)O_2$) from the reprocessing activities that will provide the feed material for a fuel fabrication facility currently under construction.[5]

One unique aspect of this neutron coincidence and gamma isotopic counting system is the use of unmoderated He3 tubes to correct for moisture content. The inner unmoderated and outer moderated (high-density polypropylene polyethylene) neutron tubes are shown in Figure 6.13, which shows a top view into the iPCAS counter. Humidity is important because

[5]See M. E. Abhold, and M. C. Baker, "Design of the Improved Plutonium Canister Assay System," Los Alamos National Laboratory, LA-UR-00-5383, Oct. 2000.

the addition of oxygen molecules results in an increase to the noncoincidence neutron background. This in turn would result in an increase in the uncertainty in measuring the coincidence neutron. Using humidity correction eliminates this problem. The spacing of the intrinsic germanium detectors along the vertical axis of the iPCAS allows each gamma detector to view one can and determine the isotopics. Combining the isotopic measurement with the neutron measurement allows for an accurate quantitative determination of plutonium in each can.

Pre-Field Installation Testing

With the primary focus on reliability, the IAEA's UMS Unit spends considerable time assembling and testing complete field-configured systems in the Safeguards Equipment Support Facility (SESF) at IAEA headquarters in Vienna. Although not always achievable, the goal is to test all systems for 90 days prior to installation in the field to ensure that all early failure modes for hardware components and configuration issues have been addressed. The testing protocols include the full range of expected field conditions with one exception, which is testing of the sensors to matching radiation fields. It is not practical to create duplicate radiation environments. Radiation sensors are very robust and reliable. Testing with smaller sources either at the SESF or at the IAEA's Seibersdorf Analytical Laboratory, where stronger sources are available, is sufficient to guarantee performance in the field. Our experience has been that these sensors, if defective, are defective on delivery. Therefore, these simple tests are sufficient to assure reliable performance.

To fully test these units without the actual radiation field present, the IAEA UMS Unit has developed a dedicated signal generator that can duplicate the exact response peak, allowing full testing of the counting algorithms in these instruments. One of these lab-based signal generators is shown in Figure 6.14. Other smaller, portable signal generators are used in the field as part of the setup protocol.

FIGURE 6.14 UMS signal generator for testing radiation-based systems.

The importance of the focus on laboratory testing prior to field installation becomes clear when one considers that the cost of travel alone averages approximately US$5,000 per trip. Whether simple or complex problems surface in the field, trips to repair IAEA safeguards systems are expensive. Due to the distances traveled and the unique nature of the equipment, even simple trips require that a technician or engineer carry sufficient replacement parts and tools to deal with all contingencies; this excess baggage adds considerably to the expense of traveling.

Summary

Over the last two decades, the IAEA has successfully developed and deployed unattended monitoring systems (UMS) that provide continuous safeguards data, automating what once required periodic visits by inspectors. The success of these systems is based on effective design methodologies that emphasize reliability through fault-tolerant designs and taking a component building-block approach to ensure maximum flexibility in meeting the needs at the diverse and complex facilities that comprise the nuclear fuel cycle. This task has been accomplished in a cost-effective manner benefiting both the IAEA and the facility operator. As the use of nuclear power continues to expand across the world, UMS are leading a revolution in safeguards capability so that the IAEA can continue to verify compliance of its member states with their treaty-based obligations in assuring the peaceful use of nuclear technologies.

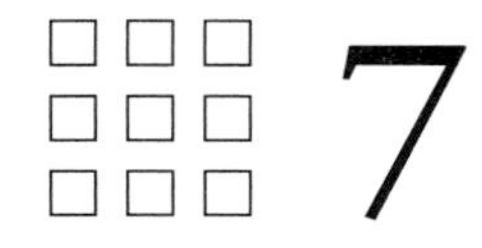

Evaluating International Safeguards Systems

Kory Budlong Sylvester, Joseph Pilat, and Tom Burr

Introduction

The international safeguards system has evolved considerably since the early 1990s. Traditional safeguards measures that focus on monitoring declared materials and facilities have been augmented with new measures aimed at detecting undeclared facilities and activities. How does one integrate such disparate measures into a coherent "state-level" safeguards system and demonstrate that it is both effective and efficient? How can state-specific factors be taken into account? A comprehensive means for assessing safeguards system performance is needed. In this chapter, the system of evaluation developed for traditional International Atomic Energy Agency (IAEA) safeguards based on diversion pathway analysis is described and extended to include detection activities beyond declared sites.

The Need for Evaluation

The evaluation of IAEA safeguards systems serves both an internal and an external function. Evaluation is necessary to provide confidence in the nuclear nonproliferation regime. Inspections produce evidence that states are abiding by their relevant nonproliferation commitments, or alternatively, evidence of noncompliance. For inspections to be relied upon, states must be convinced that the system of safeguards is effective. Without such confidence, the regime is ultimately undermined and its relevance reduced.

This desire for confidence necessitates a degree of transparency in IAEA safeguards design and implementation. Rigorous and transparent testing and demonstration of system performance significantly strengthen confidence in the Agency's safeguards system and the conclusions drawn from the Agency's activities.[1] Detailed reporting of inspection results is also valuable in this context.

Evaluation also assists in the design and implementation of safeguards approaches themselves. This is particularly important as the Agency moves to more flexible implementation of

[1]A comparable analytical process is well established for domestic safeguards. Such analyses include the evaluation of state systems of accounting and control as well as the development of a design basis threat and characterization of physical protection systems in terms of their ability to detect, delay, interdict, and defeat that threat. Although important, this analysis is not within the scope of this chapter.

safeguards under the Additional Protocol.[2] Under so-called "integrated safeguards," measures authorized under traditional safeguards and the Additional Protocol are combined with the objective of producing a more effective and efficient safeguards system.

Evolving Safeguards Objectives

For states with comprehensive safeguards agreements, the IAEA's objectives have been clearly defined in the Agency's Information Circular 153 (INFCIRC/153). The "basic undertaking" of safeguards is to verify that "all source or special fissionable material [...] is not diverted to nuclear weapons or other nuclear explosive devices."[3] Paragraph 28 of INFCIRC/153 further notes that safeguards are to provide "timely detection of diversion of significant quantities of nuclear material from peaceful nuclear activities ..." and "deterrence of such diversion by the risk of early detection."

With the adoption of the Additional Protocol (described in INFCIRC/540), the role of the Agency in providing credible assurance of the absence of undeclared nuclear material and activities was made clear. Conclusions pertaining to the *completeness* of a state's declarations are now included in annual IAEA safeguards implementation reports for states that have concluded an Additional Protocol agreement with the Agency.

So how should the Agency meet its safeguards objectives? What must inspectors do to provide verification, deterrence, and assurance in the completeness of state declarations? These broad safeguards objectives must be cast into tangible guidance for inspectors.

Historically this has been done through the establishment of inspection goals, technical criteria for safeguards implementation, and ultimately, facility-specific safeguards approaches. Inspection goals represent performance targets for IAEA verification activities at a given facility, reflecting the nature of the facility and the nuclear materials present. The quantity component determines the extent of inspection activities at a facility necessary for the IAEA to be able to draw a conclusion that diversion of one significant quantity (SQ) or more of nuclear material over a material balance period has not occurred and that there has been no undeclared production or separation of direct-use material at the facility over that period. The significant quantity depends on material type, e.g., the SQ for Pu is 8 kg. It is loosely taken to mean the amount of material necessary to produce a nuclear explosive device.[4]

The timeliness component of the goal relates to "target detection times" and is dependent on specific nuclear material categories. The frequency of IAEA inspections at declared facilities is typically driven by timeliness goals. The timeliness goal for irradiated direct-use material, such as spent fuel, is three months. It is obviously desirable for detection times to be shorter than the time required by a proliferator to convert nuclear material it obtains into a nuclear weapon. Therefore, there should be a relationship between the timeliness goal and material conversion times.

Historically, safeguards criteria have been established for each facility type—for example, nuclear reactors, which "specify the scope, the normal frequency and the extent of the verification activities required to meet the quantity and the timeliness components of the

---

[2]Model Protocol Additional to the Agreement(s) Between State(s) and the International Atomic Energy Agency for the Application of Safeguards," INFCIRC/540, printed by the IAEA in Austria, Sept. 1997.

[3]"The Structure and Content of Agreements between the Agency and States required in Connection with the Treaty on the Non-Proliferation of Nuclear Weapons," INFCIRC/153 (Corr.), 1972, Paragraph 1.

[4]For the table of significant quantity values used by the IAEA, see IAEA Safeguards Glossary: 2001 Edition, p. 23.

inspection goal."[5] The criteria list a prescriptive set of activities inspectors must complete.[6] If these activities are successfully executed, it can be said that the underlying safeguards objectives have been met.

The inspection goals and criteria-based approach has served traditional, INFCIRC/153-based safeguards reasonably well. However, it has certain limitations and has proven less applicable to modern, INFCIRC/540-type safeguards, which are inherently less prescriptive.

A rigid criteria-driven approach can lead to misplaced priorities. The criteria define the activities for goal attainment, and goal attainment is achieved by meeting the criteria. Goal attainment runs the risk of becoming the end in and of itself rather than fulfilling the functional objectives of safeguards as a whole. Ensuring that planned activities are actually executed is important but leaves open the question of the adequacy of the inspection plan itself.[7]

In examining the criteria, the relationship and relative importance of any given inspection activity to overall system performance can be difficult to discern. What level of confidence should be associated with safeguards conclusions if the inspection goals are attained? How much material has been directly verified—100%? Fifty percent? How strong a deterrent has been provided? Is the probability of timely detection 90% or is it 20%? What is acceptable? These questions are of great interest but are difficult to answer by examining the criteria alone.

Perhaps the biggest criticism of a criteria-driven approach is its limited applicability at the state level. With the adoption of the Additional Protocol, the scope of safeguards has been unambiguously extended to include assurances of the completeness of declarations and the detection of undeclared facilities. Therefore, there is a need to extend international safeguards implementation criteria to meet this requirement.

Rather than a prescriptive, criteria-based approach, development of a clearly defined set of performance objectives would be useful. Instead of specifying a required set of activities, performance goals are identified and multiple means for meeting the goals are allowed. Prospective safeguards approaches are defined and then evaluated to determine whether they collectively satisfy safeguards objectives. The emphasis is placed on meeting functional requirements rather than the means by which they are attained—on results rather than process.

Performance Objectives: Verification, Deterrence, and Assurance

As stated, IAEA safeguards under comprehensive safeguards agreements and the Additional Protocol fulfill three roles for member states: (1) to verify the peaceful use of declared nuclear activities in a state, (2) to deter states from proliferating via a risk of early detection, and (3) to provide assurances of the completeness of a state's declarations.[8] Table 7.1 summarizes the relationships between these objectives and the role of traditional safeguards quantity and timeliness goals in satisfying each objective.

---

[5]IAEA Safeguards Glossary: 2001 Edition, International Nuclear Verification Series, No. 3, International Atomic Energy Agency, Vienna, 2002. p. 25.

[6]These criteria have been revised over time; the latest version was published in 1995.

[7]For example, from a deterrence perspective, whether planned activities/inspections were fully implemented or not in a given year may not be critical. If the Agency could not perform a particular interim inspection for some logistical reason, deterrence objectives may still have been met, even though the timeliness goal was not attained. Focusing exclusively on safeguards implementation misses this point.

[8]The first objective plays an important role in support of international nuclear trade as a state's desire to have confidence that material and/or technology sold abroad is not used to support a nuclear weapons program.

Table 7.1 Comparing safeguards objectives.

	Safeguards Objective	Scope	Role of "Quantity" Goal	Role of "Timeliness" Goal	Primary Safeguards Measures
Verification	Verify that all declared material remains in civilian use	Declared materials	Determines measurement level; establishes target values	Not particularly important; verify inventory with some frequency	Materials accounting augmented by containment/ surveillance (for efficiency)
Deterrence (through risk of timely detection)	Establish the ability to detect all credible proliferation "pathways"	Undeclared actions at both declared and undeclared sites	Not a focus per se, only part of throughput considerations in pathway definitions	A major focus; objective is to detect pathway use prior to path completion	Surveillance/ unannounced inspections/new measures to detect undeclared activities
Assurance of completeness of a state's declarations	Find evidence of use of any plausible proliferation pathways	Undeclared actions at both declared and undeclared sites	Not particularly important	Not particularly important	Information analysis; state-specific review of the IAEA's physical model

Given their differences, it has been proposed that separate performance measures should be established for each safeguards objective.[9] This separation would enable more clear linkage between individual inspection activities and specific safeguards objectives. This context clarifies the application of safeguards and enables more effective and flexible implementation.

In terms of verification, the safeguards system should periodically demonstrate that the declared nuclear material in a state is accounted for in the peaceful nuclear program. Materials accounting would, of course, be the primary tool utilized for this function. Verification objectives would be expressed in terms of fraction of material inventoried and the frequency, with more sensitive materials verified to a greater extent and more often.

In terms of deterrence objectives, the safeguards system should provide confidence that credible proliferation actions, should they occur, would risk detection in a timely manner. For deterrence, possible proliferation actions are the focus of safeguards efforts rather than nuclear material. The scope of actions to be detected (including concealment strategies) should be broad. Certainly it would include material diversion and facility misuse at declared facilities as well as undeclared activities at other locations. Deterrence objectives would be expressed in terms of the detection probability for each action, with more strict detection requirements for more attractive (i.e., quicker, easier, less costly, etc.) proliferation options.

Providing assurance in the completeness of a state declaration shares certain commonalities with deterrence objectives. To provide a level of assurance that all relevant material and activities have indeed been declared, a thorough consideration of possible proliferation activities is needed. When this is combined with an evaluation of all available information and effective investigation activities, greater assurances of the completeness of a state's declaration can be attained. Actively seeking to detect proliferation actions, or a prioritized subset of actions, can enable both deterrence and assurance objectives to be fulfilled.

[9]Kory W. Budlong Sylvester and Joseph F. Pilat, "Performance-Based International Safeguards System," paper presented at the 27th ESARDA Annual Meeting, Symposium on Safeguards and Nuclear Material Management, London, May 10–12, 2005; Los Alamos National Laboratory document LA-UR-04-4259.

Providing assurance of the absence of undeclared activity presents obvious difficulties—one cannot prove a negative. Achieving this performance objective is inherently different. The strength of assurances depends on a variety of factors but will vary with the degree to which proscribed activities have been considered and the degree to which indicators of their presence are sought.

Although deterrence, verification, and assurance are separate goals, they are mutually reinforcing. For some proscribed activities, it may be the case that verifying material is the most effective way of detecting the use of a proliferation "pathway." Conversely, an accurate inventory may rely heavily on assurances that certain concealment activities have not occurred.

Proliferation pathways provide a useful analytical framework for expressing both deterrence and assurance goals. Such pathways represent the minimum set of activities a state must undertake to produce weapons-usable material. The use of pathway analysis in this context would extend the concept of "diversion path analysis," which has been used by the IAEA in the past to define and evaluate safeguards at the facility level, to address state-level considerations.

Proliferation Pathways

The nuclear fuel cycle in all its permutations represents possible technical pathways to the production of plutonium and highly enriched uranium (HEU). If these pathways can be properly characterized, a safeguards system can be measured in terms of its ability to detect their use. At a high level of generality, these pathways are represented in Figure 7.1.

The IAEA's "physical model" describes the entirety of the nuclear fuel cycle. Starting from raw material mining and milling and continuing through conversion to metal suitable for weaponization, the physical model describes the known technical processes for achieving each processing objective. This is the type of information that can be found in many nuclear engineering textbooks on the fuel cycle.

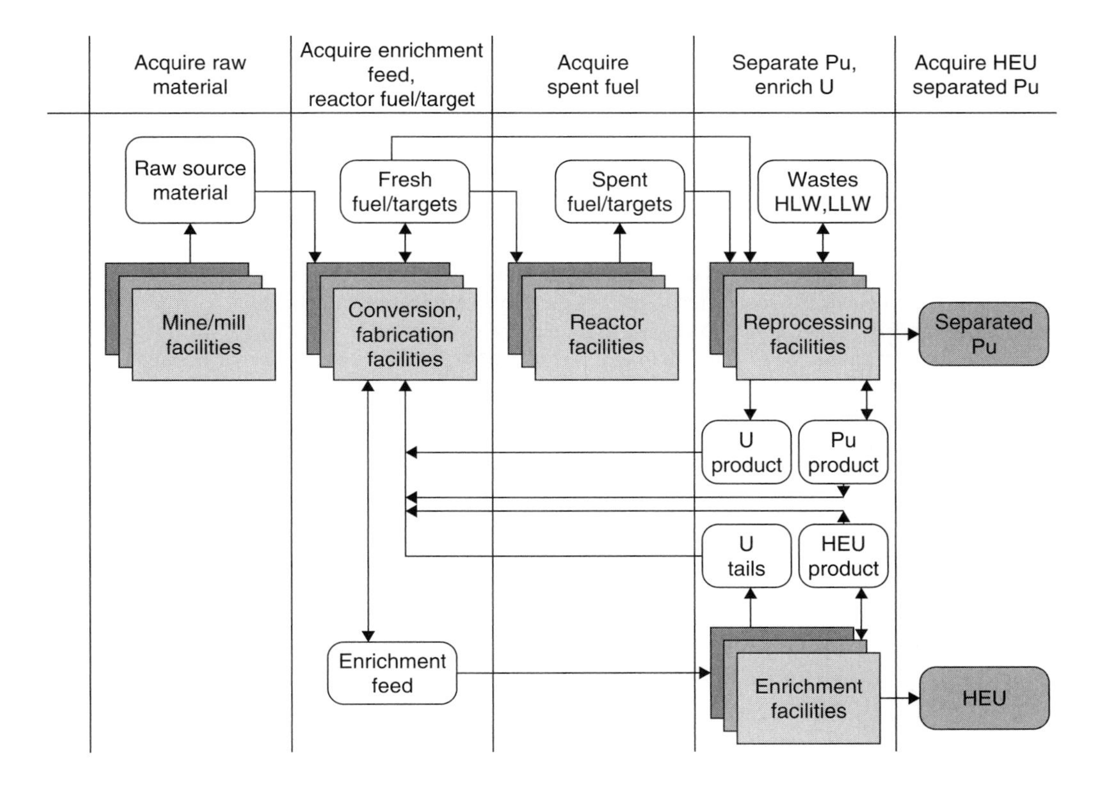

FIGURE 7.1 Proliferation pathways.

Knowledge of the required technical steps for proliferation serves multiple purposes. When inspectors are trying to uncover clandestine nuclear programs, it can tell them what evidence to look for. Linked into ordered steps or pathways, this knowledge can also provide a structured means for evaluating the collection of safeguards measures applied in a given state.

The pathways are actions necessary for a successful proliferation program. As such, safeguards systems can be assessed in terms of their ability to detect and thereby deter the use of any or all of these paths. As we will see, such an approach can provide a context for understanding inspection results as well.

Defining a State-Level Safeguards Approach

Having considered safeguards performance objectives, a set of IAEA activities must be selected that will meet them. The agency has adopted a "state-level" approach to safeguards. In this approach a comprehensive "state evaluation" process is used to plan safeguards activities, assess their results, and determine future activities in an iterative fashion. This has been referred to as an *information-driven approach* to safeguards implementation. For states under so-called "integrated safeguards," the prescriptive safeguards criteria are set aside in favor of a state-specific inspection plan.

Safeguards priorities can be reflected in requirements set for verification, deterrence, and assurance. A greater emphasis on deterrence would mean that requirements for timely detection are set at higher levels. Setting requirements for the assurance of the absence of undeclared activities is difficult. How hard should one look for something that might not be there? Nonetheless, taskings must be developed, and proliferation pathway analysis and the physical model can help structure the process.

Defining safeguards approaches begins with a review of safeguards authorities as derived from various documents. The set of possible inspection "tools" depends on the safeguards agreements into which a state has entered. INFCIRC/153 safeguards agreements, for example, provide for the application of materials accounting as well as containment and surveillance measures.[10] INFCRIC/540 agreements provide new authorities aimed at the detection of undeclared facilities. Of perhaps greatest importance in this context are so-called "complementary access" rights. Together these agreements represent a menu of safeguards authorities and possible inspection techniques.

Designing a safeguards system for the verification of nuclear material is conceptually easy to do. How the material is to be measured, to what accuracy, and how often are the relevant parameters. Designing to meet deterrence objectives is a more difficult task. In this case it is better to think in terms of activities or events the system is designed to detect. Once a standardized baseline of proliferation actions to be detected is established, the ability of various safeguards approaches to detect them can be evaluated.

It is important to note that a consideration of proliferation scenarios is an unavoidable and important step in defining credible safeguards approaches. We will seek to use these scenarios as a baseline against which we can judge the performance of alternate safeguards approaches. Though important, it is not suggested that this is an easy task. To the contrary, describing these scenarios at the appropriate level of detail is difficult and requires an iterative approach.

Analytically, it is useful to categorize proliferation actions by where they are assumed to take place. That is, we will treat separately those actions that take place at specific, declared facilities from those that take place at undeclared facilities. The type of safeguards measures applied as well as the methods used in determining their effectiveness are quite different. It is therefore more convenient to consider them separately.

---

[10]Containment and surveillance (C/S) measures include the use of seals and video cameras to maintain continuity of knowledge for safeguards purposes.

Measures Applied at Declared Facilities

The types of activities at declared facilities that we seek to detect are material diversion and facility misuse.[11] Material diversion is obviously best addressed by direct verification of material inventories. This provides positive confirmation of nondiversion. For misuse activities, it is facility "services" that are targeted by the proliferator—for example, undeclared irradiation of uranium targets in a power reactor. Typically, various surveillance measures are used to "see" these prohibited actions.

For the nuclear material in each facility, it is necessary to consider the safeguard measures to be applied (e.g., item count) and the frequency of their application (e.g., during one interim inspection [IIV] randomly selected between scheduled annual inspections [PIVs]).[12] Similarly, for each area in the facility we must consider the safeguard measures to be applied (e.g., visual observation by the inspector) and the frequency of their application (e.g., during one annual PIV). It is also important to know when any data gathered during an inspection are reviewed, because this information will impact later determinations of timeliness.

Summary tables can be used to extract, organize, and condense the information regarding a safeguards approach. Table 7.2 provides an example of how the previously described information can be summarized for a declared facility. The table describes proposed safeguards measures for application during a PIV at a research reactor with specific areas and materials in the facility.

The designation of "material" and "area" safeguards is intended to be a helpful convention and need not be strict.[13] There can be overlap. For example, a reactor core seal serves the function of safeguarding core fuel but also ensures that undeclared targets for irradiation are not inserted into the core. In the later case it serves an area-monitoring function. What is important is to recognize and identify both types of safeguards and ensure their inclusion in the analysis.

Measures to Detect Undeclared Facilities

A number of tools are available to the IAEA for the detection of undeclared facilities. Under INFCIRC/153, the Agency has the authority to utilize "ad hoc" inspections to verify the completeness of a state's initial declarations. It also has the authority to utilize "special inspections" when a specific concern is raised. This right, however, has been circumscribed in practice because it has been taken to be an *a priori* accusation of noncompliance.

With the completion of the "Strengthened Safeguards" program termed Programme 93 + 2 in 1995, a wider range of measures for detecting undeclared facilities was instituted. Strengthening measures that could be implemented under INFCIRC/153 authorities included expanded declarations, environmental sampling at locations where the Agency had existing access rights, and the use of satellite imagery. The additional information gathered was used to support an improved "information analysis" program for safeguards that seeks to examine information available from all sources to identify inconsistencies in a state's nuclear fuel cycle activities.

New measures that were deemed to require the additional authorities provided for in INFCIRC/540 included expanded declaration requirements, "complementary access" rights, and the right to perform wide-area environmental sampling.

Efforts to detect undeclared facilities are fundamentally different from measures applied at declared facilities. There is no material to "verify," and the area for surveillance can be enormous. Nonetheless, precisely how the measures are applied remains relevant. Details

---

[11]The IAEA's definition of the term "diversion" encompasses the processing of undeclared nuclear material at declared facilities. To more clearly separate activities involving declared material, we refer to the latter as "misuse."

[12]The IAEA acronyms IIV and PIV stand for *interim inventory verification* and *physical inventory verification*, respectively.

[13]*Material safeguards* are applied directly on the declared material itself. The term *area safeguards* applies to those activities with the capability of monitoring a given area in a declared facility, typically for detecting relevant undeclared activities.

Table 7.2 Summary PIV activities for a research reactor.

Area	Area-Related Safeguards	Material	Material-Related Safeguards
Core tank	Visual observation	**Core fuel**	Item count Criticality check
Pool	Visual observation	**Spent fuel**	Item count Gross defect (100%)
Fresh fuel store	Visual observation	**Fresh fuel**	Item count Verify seal Gross defect (100%) Partial defect (50%)
Facility	Review operating records	**All materials**	Review accounting records

regarding how the measures are applied in practice, their scope and frequency of application, and the like must be considered in evaluating their effectiveness.

Proliferation Actions to Be Detected

Because safeguards effectiveness will be measured in terms of detection capabilities, identifying what is to be detected is of paramount importance. As noted previously, defining specific proliferation actions of interest is a challenging task. A very large number of conceivable scenarios can be imagined. Even so, there can be no certainty that *all* possible scenarios (including concealments variants) have been considered. In addition, care must be taken to ensure that only technically credible scenarios are included. However, the caliber of the analysis will be measured in terms of the breadth and depth of the scenarios considered. The more extensive the consideration of scenarios, the more broadly applicable the results.

Since these proliferation actions represent concrete elements of proliferation pathways, we will refer to them as pathway *segments*. A segment describes a discrete, physical action necessary to ultimately produce separated weapons-usable material. Individual segments can be combined to form complete proliferation paths.

Pathway segments can be defined with varying degrees of specificity. What is important is that the chosen level of detail meet the analysis requirements. Each segment presents opportunities for detection. An iterative process may be necessary to establish the appropriate level of analysis and to meet other analysis objectives. Care must also be taken to ensure that the calculus of detection is not made duplicative through improper segment definition.

Similarly, the most attractive alternative from the proliferator's perspective should be chosen for use in the analysis. For example, there are many inferior ways to misuse a facility. The analyst must decide on the most intelligent (which may be the least detectable) manner and use it in the analysis. The specification of pathway segments for use in the analysis can always be revisited and others added if necessary. The objective is to provide a standardized baseline for evaluating various safeguards approaches. If desired, multiple misuse approaches (as in the case where two segments perform equivalent actions) could be assumed and carried forward in parallel in the analysis.

Diversion Segments

Diversion segments describe the undeclared removal of declared nuclear material from a safeguarded facility. A diversion segment must define the specific material involved, how much is diverted, over what period of time, from where, and during what mode of operation. With this information we can determine and properly account for the area and material safeguards that may detect the proliferation action.

As an example, let's consider the case of a reprocessing facility. Generally, at such facilities there are three types, or classes, of proliferation scenarios we are concerned with:

- *Direct material diversions.* These diversions involve simply taking the material present in various stages of the process and removing it from the facility, either in a single, abrupt act or on a protracted basis.
- *Producing a relatively "pure" Pu product prior to removal.* Here *pure* is taken to mean sufficiently free of highly radioactive isotopes such that the Pu could be handled in a glove box rather than a hot cell. Such an action would simplify additional off-site processing but may be readily detectable.
- *Producing output streams from a given unit that are perhaps less detectable (via nondestructive assay, optical surveillance, radiation monitoring, etc.) if diverted.* This effort could involve directing Pu where it is not typically found (e.g., into difficult-to-measure streams or streams that might be difficult to track) or removing signatures (e.g., radiation) from Pu containing streams to avoid detection.

The possible combinations of material quantities and timing involved in such a strategy are numerous. Analytically sound scenarios that can be taken to represent many similar variants should be developed. This technique of using a single segment to represent a class of proliferation actions should be used whenever possible to simplify the analysis. Care must be taken, however, to ensure that the segment used is truly representative.

Misuse Segments

In defining misuse segments we are interested in describing material processing steps that advance a proliferator toward its goal. Misuse segments describe how the services of declared facilities are abused or clandestine facilities are utilized; as such a misuse segment may be implemented at either a declared or an undeclared site. In the definition of a misuse segment at a declared facility, the quantity of material introduced is specified along with where and when it is processed and extracted. For example, there are various ways in which undeclared uranium could be introduced into a declared enrichment facility, processed into highly enriched uranium, and removed from the facility. Each alternative would be represented by a series of misuse segments.[14]

For processing that occurs at a clandestine facility, an additional factor to consider is its assumed location. To capture any differences between the detectability of such facilities when located on a declared site versus off-site, both options should be considered as possible and included in the analysis. Safeguards system performance will vary as inspection authorities differ significantly in each case.

Identifying Applicable Safeguards and Concealment Scenarios

Having identified the activities targeted for detection at a declared or undeclared facility, the next step is to determine which of the proposed safeguards measures "cover" each action.

To cover a pathway segment, a measure must be capable of independently detecting an anomaly, thereby indicating pathway use.[15] For each pathway segment, the safeguards

[14]For segments at undeclared facilities, it is not relevant to specify the area and operating mode for material movements. Misuse, or rather "use," of an undeclared facility to perform a certain function can be treated as single operation. Component acquisition, imports, and the like can be treated as separate segments as well as construction activities. The intent is to examine separately the detectability of indicators related to these activities, separate from those related to facility operation.

[15]It is recognized that the proper functioning of one measure may require other supporting measures (e.g., an effective item count relies on accounting records). What is to be avoided is the listing of these supporting measures as additional, independent opportunities for detecting the segment.

measures that provide an independent opportunity to detect the defined action should be captured in the assessment.

In determining the applicable safeguards, it is necessary to consider (1) whether the measure is capable of "seeing" the described activity and thereby raising an alarm and (2) the resolvability of the anomaly in question.

As a convention, and to simplify initial evaluations, followup actions can be assumed to be entirely effective and timely. However, if there is no possibility of effective followup actions—in other words, if the anomaly cannot be resolved—then detection cannot occur. Examples of this could be where a suspicious container that may contain undeclared material is recorded via surveillance leaving a facility. Future investigation of the container might not produce any evidence of its prior contents.

Concealment Scenarios

The actual performance of an individual safeguards measure in terms of detection will depend on the countermeasures employed by a proliferator. Because the safeguarded entity is fully aware of the detection measures being applied, it will undoubtedly attempt to conceal its actions. To determine the sensitivity of the safeguards system in such a case, concealment scenarios must be considered and system performance measured against them.

This activity is as important as segment definition and no less challenging. Although it is important to consider a wide range of concealment possibilities, as with proliferation scenarios the possibilities should be restricted to those that are technically credible.

In developing concealment scenarios, we must consider (1) how a relevant measure might be completely defeated (i.e., made incapable of producing an anomaly); (2) the difficulty associated with executing the concealment; and (3) what, if any, additional pathway segments would be required to implement the concealment (e.g., fabrication of dummy fuel assemblies).[16]

The assignment of difficulty levels to concealment scenarios can be used in an attempt to capture the resource requirements, technical difficulties, and so on associated with each scenario. In a sense each level represents a different type of proliferator or, alternatively, the level of determination of a particular proliferator. As the determination, resources, and capabilities available to a proliferator increase, more concealment options become available.

Estimating Detection Capability

Each step along a proliferation pathway, in principle, represents an opportunity for detection. The ability of individual safeguards measures to detect each pathway segment can be quantified or estimated. At the next level of analysis, the path level, these segment data can be aggregated to produce path level results. At the state level, safeguards performance across all relevant paths can be examined.

Residual anomalies are those that remain despite the adversary's use of a concealment scenario. These anomalies become the basis of detection and evidence of proliferation. For proliferation utilizing undeclared facilities, the "indicators" of nuclear fuel cycle activities referenced in the IAEA's Physical Model serve as possible anomalies, and the capabilities for detecting these anomalies can be determined through formal expert judgment.[17]

For each safeguards measure we must estimate the detection capability over time. The detection capabilities are estimated in the context of credible concealment scenarios and

[16]For initial evaluations, it is convenient to consider only those concealment approaches or combinations of approaches that are completely effective. While more difficult, it is possible to analyze the partial reduction of detection capability achievable by a concealment. This would represent a higher level of fidelity for the analyses and could be implemented in subsequent analyses.

[17]See: Joseph F. Pilat and Kory W. Budlong Sylvester, "Report on the Illustrative Expert Elicitation to Assess Information Analysis as a Means of Detecting Undeclared Activities," Los Alamos National Laboratory document, LA-UR-01-1890 (Rev. 7/01).

represent point estimates of cumulative detection probability as measured from the time the activity described in the segment is initiated.

Detection at Declared Sites

Calculation of detection probability is straightforward for proliferation actions at declared facilities. The timing and nature of the proliferation action are specified in the segment definition. Similarly, the timing and measurement activities for each inspection are specified in the facility inspection plan. The likelihood of successful detection over multiple inspection periods can therefore be calculated.

Conceptually, to detect a diversion of declared items, such as spent fuel assemblies at a reactor storage pool, three things must happen: The inspector must arrive at the site, he must inspect the portion of the inventory that has been tampered with (assuming an attempt at concealment has been made), and he must use the appropriate safeguards tool to correctly identify the anomaly. Therefore, the probability for detection to occur on any given day (after diversion) can be expressed as:

1. $$P = P_{inspect} \cdot P_{select}(n) \cdot P_{anomaly}(instrument \mid concealment)$$

where:

$P_{inspect}$ is the likelihood that inspection will occur on any given day.

$P_{select}(n)$ is the probability that the relevant material inventory (or item) is selected for measurement, which depends on the sample size (n).

$P_{anomaly}(instrument \mid concealment)$ is the probability that the measurement technique used by the inspector correctly identifies an anomaly leading to detection of diversion, which depends on the effectiveness of the instrument, given the concealment method.

The first term in the equation, ($P_{inspect}$), is driven by the expected frequency of inspection. For example, if random unannounced inspections are planned with a frequency of one per year, $P_{inspect}$ would equal 1/365. It is simply chosen as part of the inspection plan.

The second term of the formula is determined by the measurement plan for material verification. Upon arriving on-site, an inspector does not need to examine every item in an inventory to produce sufficient confidence that material diversion has not occurred. For efficiency purposes, the IAEA uses a sampling approach to verify the operator's inventory declaration.

An estimate of the number of samples (n) to be selected for measurement is:[18]

2. $$n = \frac{1}{2}(1 - \beta^{1/d})(2N - d + 1)$$

where:

N is the number of items in the inventory.

β is the nondetection probability.

d is the number of defects (tampered items) in the inventory.

Replacing β with $1-P_{select}(n)$ and rearranging terms, we find:

3. $$P_{select}(n) = 1 - \left[1 - \frac{2n}{(2N - d + 1)}\right]^d$$

As suggested by Equation 2, it is typical to specify the desired nondetection probability and determine the required sample size. Therefore, the second term of Equation 1 is simply

[18]John Jaech, "Statistical Methods in Nuclear Material Control," Prepared for the Division of Nuclear Material Security, U.S. Atomic Energy Commission, Technical Information Center, TID-26298, 1973. p.321. The referenced expression is an approximation of the hypergeometric probability density function where the allowable number of defects (diversions) in the sample is zero.

chosen. However, the choice does have significant implications in terms of the time required to complete an inspection and the level of inspector effort, which may impose practical limitations.

The third term of the Equation 1, $P_{anomaly}$, is not chosen but rather is fixed by the physical limitations of an inspector's measurement device. The ability of a safeguards instrument to correctly produce an anomaly can be estimated by a variety of means. Empirical verification through testing is of greatest value. When this is not feasible, other means are available. For example, if radiation detection is to be used to identify characteristic signatures, a computer model can be constructed of the nuclear material in its normal and tampered states and combined with a physics-based model of the detection instrument, to determine the probability of alarming. Similar simulations can be developed for other safeguards measures. A library of computed values, or estimates generated by expert elicitation, can be used in assessments of detection probability.

It is instructive to examine the interrelationship between the proliferator's concealment strategy and the inspector's sampling plan and measurement capabilities in determining the overall detection probability.

Consider the case where an inspection is held to verify the nuclear material in the spent fuel pool at a reactor. The inspection plan calls for an item count of the spent fuel assemblies with a fraction of the 1,000 assemblies selected for measurement to guard against individual pin diversion. (It is possible that a proliferator could take apart an assembly, remove a fraction of the pins, and replace them with dummy rods.)

Figure 7.2a shows how the chances of an inspector selecting an assembly that has been tampered with increases with the number of affected assemblies and the sample size. The figure shows scenarios where 4, 10, and 20 assemblies have been modified. The optimal strategy for the proliferator (lowest selection probability) is to tamper with as few assemblies as possible.

However, as more assemblies are tampered with, the effectiveness of the verification technique decreases. Figure 7.2b shows that the best proliferation strategy changes if the ability of the instrument to detect the anomaly is sufficiently reduced (in this case, from 90% to 20%). Ultimately, detection probability is limited by the measurement technique.[19]

Given the probability of detection on any given day (Equation 1), we can utilize the following expression for the probability that detection occurs on Day d:

4. $$P_{detection}(d) = P(1 - P)^{d-1}$$

Equation 4 is the geometric probability distribution and represents the probability that detection occurs on Day d (and not before). As such, the cumulative geometric distribution gives the probability for detection prior (or equal) to Day d:

5. $$P_{detection \leq d}(d) = 1 - (1 - P)^{d}$$

An example calculation for $P_{detection \leq d}(d)$ can be useful for illustrative purposes. Again we consider a pathway segment that describes the diversion of a significant quantity of Pu (8 kg) contained in spent fuel. What is the probability of detection for this event? Within one month? Within three months? How does the detection probability vary over time? Of course the answer depends on the nature of the diversion, the frequency of safeguards inspections, and the verification measures utilized.

Let's assume that the inspection regiment for the spent fuel pool includes two random, unannounced inspections per year. During both inspections, 100% verification of the spent

[19]This is an example in which surveillance measures can be used to compensate. The greater the disassembly activities, the greater probability that review of camera surveillance records will raise questions.

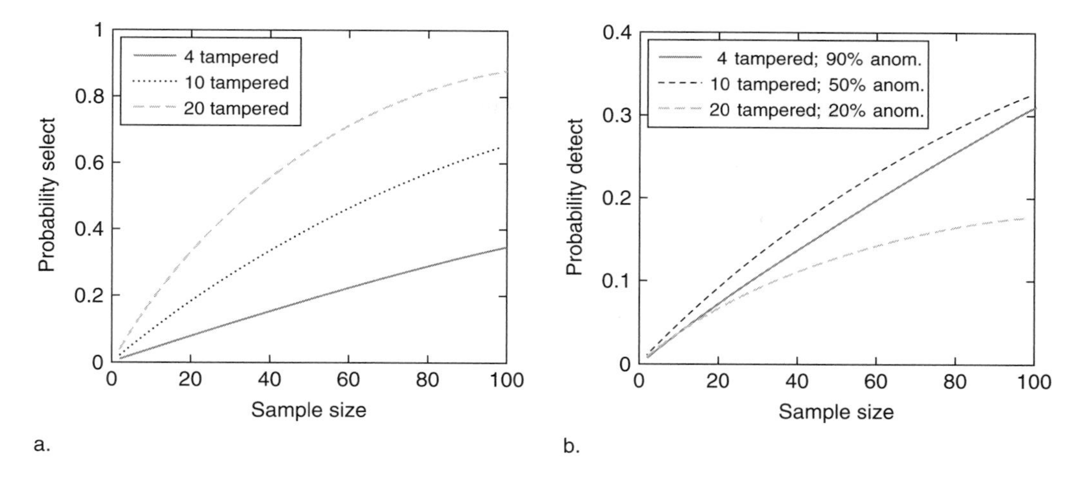

FIGURE 7.2 a. Probability of inspectors selecting a tampered assembly. b. Probability of inspectors detecting a tampered assembly.

fuel inventory for gross defect and 100 assemblies are selected at random for a partial defect measurement.[20] Again we assume there is a total inventory of 1,000 assemblies.

Figure 7.3 shows the increase in detection probability with time for four scenarios. For the scenario where two spent fuel assemblies are diverted without concealment, the probability of detection rises quickly. Timeliness in detection is simply driven by the inspection frequency. If the assemblies are replaced with dummies, performance drops slightly due to reduced effectiveness of the verification technique (assumed to fall to 80% detection probability). As pin diversion strategies are implemented, performance drops off more dramatically as the probability that measurement will detect an anomaly is reduced from the 90% level (when 50% of the pins in an assembly are removed) to 20% (when only 10% of the pins in an assembly are removed).[21]

This reduction can be compensated for with either increased inspection frequency, higher sampling rate, or improved partial defect measurement. Again, this illustrates the point that effectiveness depends on the interplay between choices made by both the inspector and the proliferator.

Off-Site Detection

Estimating detection capabilities for proscribed activities performed at undeclared sites presents a much different situation and a unique challenge. Here the activities to be detected are less specific and the performance of safeguards measures more difficult to estimate.

Expert elicitation is a process used, for example, when physically based data are absent or open to interpretation. The process can be informal or formal. With formal expert elicitations, the quality and accuracy of judgments of knowledgeable people come from the completeness of the expert's understanding of the phenomena and the process used to elicit

---

[20]The specified verification regime introduced two new terms: gross and partial defects. A *gross defect* is one in which the safeguarded item has been "falsified to the maximum extent possible so that all or most of the declared material is missing" [IAEA Safeguards Glossary: 2001 Edition, p. 78]. If no concealment attempt is made, e.g., replacement with a dummy assembly, simple safeguards measures and item count should detect such a defect. *Partial defects*, such as removal of a few pins in the assembly, are more difficult to detect and require more sophisticated inspection equipment, such as a fork detector with ionization chambers to measure the total gamma-ray output and fission chambers to measure the total neutron output.

[21]These values are only notional and do not represent actual measurement capabilities.

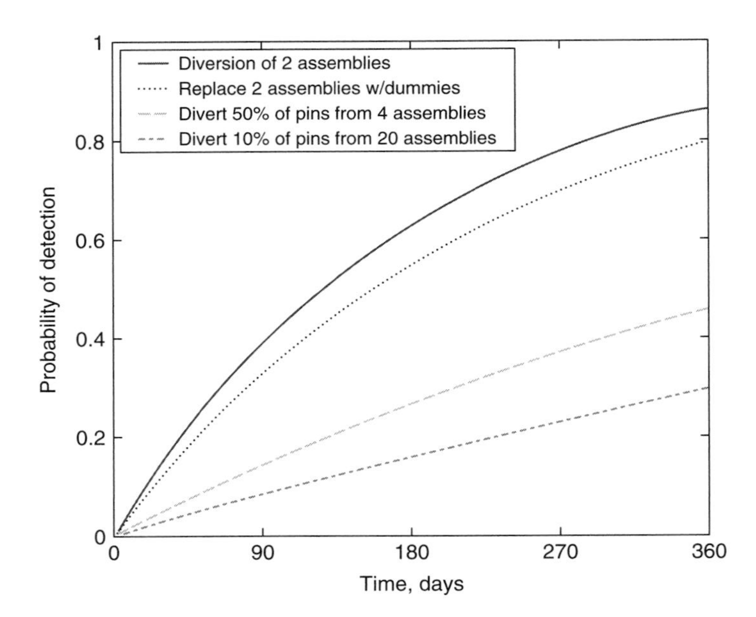

FIGURE 7.3 Cumulative PDF for diversion detection time.

and analyze the data. The use of a more formal process to obtain, understand, and analyze expert judgment provides rigor and transparency in the results. Procedures for formal elicitations can be employed throughout the process to enhance its technical credibility and consistency and its transparency.[22]

For detection at declared sites, virtually any detection capability can be attained with an increase in inspection effort. This is not true for efforts to detect undeclared facilities. Due to the inherent difficulties in producing reliable estimates, bounding assessment can be valuable. Such approaches can be used to assess reliance on assumed capabilities and the impact on overall performance if actual performance does not meet expectations.

Producing Pathway-Level Results

Having completed the segment evaluation, the results must be aggregated to produce pathway-level results. After all, it is the performance of the entire safeguards systems, across all credible proliferation pathways, in which we are most interested.

This aggregation raises a number of methodological questions. The detection capabilities utilized at the segment level may be derived from quite disparate measures and techniques. Some of the detection capabilities will be estimated from sampling plans and others via a formal elicitation process. This fact, along with other considerations such as independence between measures, means that care must be taken during the aggregation process to ensure that system features are not inaccurately characterized. Any interdependencies must be identified and properly addressed.

Assuming independence, the segment-level results can be combined to produce an expression for the detection probability for each path, $P_{path}(t)$, as given by:

6. $$P_{path}(t) = 1 - \prod_{i}(1 - P_{di}(t))$$

---

[22]See: Joseph F. Pilat, Kory W. Budlong Sylvester, and William D. Stanbro, "Expert Elicitation and the Problem of Detecting Undeclared Activities," paper presented at the Institute of Nuclear Materials Management 43rd Annual Meeting, Orlando, Florida, June 23–27, 2002.

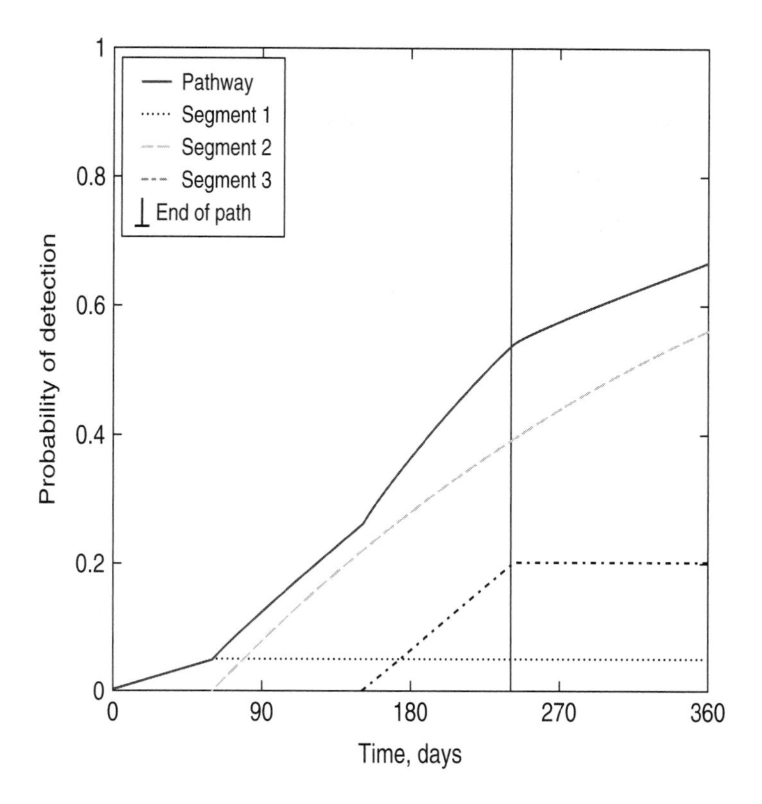

FIGURE 7.4 Cumulative PDF for pathway detection time.

where:

$P_{di}(t)$ is the probability of detecting segment i prior to time t.

(In the evaluation of Equation 6, care should be taken to reconcile segment time [measured in d days] with pathway time t.)

Figure 7.4 shows how the cumulative detection probability varies for a nominal pathway involving three segments: the construction of an undeclared hot cell with reprocessing capabilities, the diversion of spent fuel, and the processing of that fuel to recover 8 kg of Pu. (The detection probability for diversion segment is taken from Figure 7.3 for the case in which 50% of the pins are diverted from four assemblies out of 1,000.)

In addition to the safeguards activities at the reactor pool, IAEA efforts to detect the construction and operation of undeclared facilities also contribute to detection capabilities. Separate estimates of detection probability for these activities were assumed for illustrative purposes. Because the probability of detection prior to pathway completion is of great importance for deterrence, a line has been added to the figure to identify the end of the pathway.

Figure 7.4 illustrates the contributions from efforts to detect each segment to the overall detection probability for the proliferation pathway. Each step along the path creates new opportunities for detection. It is obviously desirable to design safeguards systems that produce high levels of detection probability early in the path. This affords the international community more time to develop alternatives for response.

Evaluating State-Level Safeguards System Performance

Evaluation of safeguards systems must be performed to ensure that inspections and other Agency activities are providing the desired level of verification, deterrence, and assurance of the absence of undeclared activities in a state. Evaluation confirms that expectations from the international community are being met by the safeguards system.

Proliferation pathway analysis is an important means for evaluating safeguards systems. Having a defined, consistent set of paths in which to compare the performance and cost of various safeguards approaches is useful for internal planning purposes alone. Proliferation pathway analysis provides a logical way for assessing together the contributions from traditional and new measures and examining any tradeoffs.

Whether implemented on a subset of proliferation paths or using all credible paths in a state, pathway analyses can be utilized in a variety of ways. Three examples are given here: state-specific safeguards design, generic cost/benefit analyses, and reporting of safeguards implementation.

State-Level, State-Specific Safeguards Design and Optimization

In designing a state-specific safeguards approach, alternate sets of measures can be evaluated for their performance and cost in covering a standard collection of pathways. An iterative process of this nature would facilitate the optimization of a state-specific approach.

For any safeguards approach considered, a number of questions will be important: With regard to pathway coverage, are all credible paths covered? Where are the system's strengths and weaknesses? Is pathway coverage strengthened or reduced compared to alternative safeguards approaches? What is the detection probability prior to the end of a path? Are deterrence objectives being fulfilled? In terms of assurance, is sufficient data being gathered to support a conclusion of the absence of undeclared activities?

Generic Cost/Benefit Analyses

In a state-level evaluation, all routes to weapons-usable material are considered. Nonetheless, when used for comparative purposes, meaningful information can be extracted via the evaluation of a subset of proliferation pathways or even a set of pathway segments.

For example, assume that for a given safeguards approach the continuity of knowledge (via surveillance cameras) on spent fuel is to be replaced with unannounced inspections. Pathway analysis can be used to determine the impact on individual segments (e.g., spent-fuel diversion) and on paths utilizing those segments to determine the net effect. The costs for each approach can also be determined and compared.

Reporting on Safeguards Implementation

For any state-specific safeguards approach, proliferation pathway analysis clearly establishes the set of proliferation paths to be covered and, for each segment in those paths, the measures to be relied on. This represents a set of actions that should be performed and the results reported.

By linking safeguards measures to segments and paths, the purpose of each safeguards measure is clearly specified. Moreover, the importance of each measure in terms of meeting state-level safeguards objectives can also be understood. Any necessary changes in an integrated safeguards approach during the course of a year—for example, for practical purposes—can be tailored to replace the specific purpose of the eliminated measure. If performance standards exist, pathway analysis can be used to verify that the approach was an acceptable substitute and that system performance objectives were attained, both in terms of deterrence and assurance.

The ability to fully specify the required level of verification, deterrence, and assurance the safeguards system must provide, and then demonstrate that this level is being attained, enhances the transparency of Agency activities and increases confidence in the safeguards system. Extending traditional diversion path analysis to proliferation pathways facilitates the implementation of a more dynamic and effective international safeguards system.

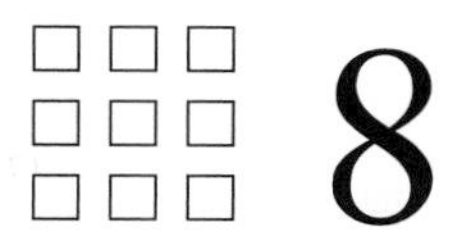

8

Statistical Methods in Nuclear Nonproliferation Activities at Declared Facilities

Tom Burr

Introduction

Nuclear nonproliferation is a multilayered effort that begins with steps to prevent nuclear material that could be used as a weapon from leaving the peaceful energy cycle. In nuclear safeguards at known facilities with declared operations, the main purpose of nuclear materials accounting (NMA) measurements is to confirm the flows and inventory of special nuclear material (SNM) to within relatively small control limits. In addition, containment and surveillance (C/S) is used to try to confirm there has been no diversion of SNM. NMA, C/S, and related nonproliferation topics involve statistical methods that will be described.

In the context of nuclear nonproliferation agreements, facilities that process and/or store SNM are required to perform periodic NMA measurements. In traditional safeguards, a key function of periodic NMA measurements is to confirm the presence of SNM in accountability vessels to within relatively small measurement error. C/S is used as a complementary measure to try to confirm the absence of undeclared flows that could divert SNM for possible illicit use.

International safeguards face several issues that are distinct from domestic safeguards. In the International Atomic Energy Agency (IAEA) context, the entire facility could be involved in a diversion plan and therefore, for example, safeguards components such as gates, vaults, and guards are irrelevant. In domestic safeguards, there is no credible basis for concern that the entire facility might attempt a diversion; therefore, these same components are highly relevant. The commensurate distinctions between IAEA and domestic safeguards will not concern us here, although, for example, the IAEA's need to monitor for possible data falsification and the role of C/S in reducing required verification measurements lead to statistical issues that we will describe.

Statistical hypothesis testing receives considerable attention in NMA, partly because of the appeal of quantified, objective testing.[1] Reference 1 focused on NMA, but as the ensuing discussion indicated, NMA is only one component of safeguards. C/S is another key

[1] T. P. Speed and D. Culpin, "The Role of Statistics in Nuclear Materials Accounting: Issues and Problems," with discussions, *Journal of the Royal Statistical Soc A* 149, Part 4, pp. 281–313, 1986.

component. In addition, in all cases, IAEA must "trust, but verify," by making random confirmatory (qualitative and/or quantitative) measurements of the operator's declared measurements, so the IAEA's measurement error is even larger than the operator's measurement error, making the quantified portions of the safeguards conclusions even less capable.

This chapter focuses on statistical issues in evaluating NMA measurements for safeguards at declared facilities, establishes statistical notation for use throughout this text, and includes related topics involving other lines of defense against illicit nuclear material trafficking, such as portal monitoring at screening locations.

Background

The inventory difference (ID) for SNM at time t is defined as:

$$\mathrm{ID}_t = BI_t + R_t - EI_t - S_t \tag{8.1}$$

where BI is beginning physical inventory, R is receipts, EI is ending physical inventory, and S is shipments (all terms include measurement error).

An example ID sequence is plotted in Figure 8.1. The control limits at $0 \pm 2s_{\mathrm{ID}}$ (s_{ID} is the empirical standard deviation) in Figure 8.1 are static because in this example, we do not have the information needed to adjust them for monthly variations that affect the ID measurement error (such as variation in measurement equipment or throughput). If we know all terms in Equation 8.1 and the measurement uncertainty associated with each measurement system, we apply statistical rules involving the variance of a sum (propagation of variance, or POV) to estimate the measurement error standard deviation of the ID, σ_{ID}. Note that Equation 8.1 is a sum of many terms, some of which have negative signs. We assume that we know the individual terms and have estimates of their associated uncertainties in Equation 8.1.

A good estimate of σ_{ID} is usually all that is required for statistical evaluation of an ID because of the central limit effect, whereby sums of approximately 10 or more random variables (all measurement errors are random at some stage, even the so-called "systematic" errors) will have approximately a Gaussian distribution. Assuming no material loss, the ID has an approximate $\mathrm{N}(0,\sigma_{\mathrm{ID}})$ distribution (denoted $\mathrm{ID} \sim \mathrm{N}(0,\sigma_{\mathrm{ID}})$, where $\mathrm{N}(\mu,\sigma)$ is the normal distribution with mean μ and standard deviation σ. Therefore, to test for SNM loss, the ID can be compared to $k\sigma_{\mathrm{ID}}$ where k is 2 to 3, depending on the desired false alarm probability.

NMA to confirm facility operations involves periodically comparing the latest ID to an estimate of σ_{ID}. To check for abrupt loss, the null hypothesis, H_0: $\mathrm{ID}_{\mathrm{true}} = 0$, is tested versus the alternative H_A: $\mathrm{ID}_{\mathrm{true}} > 1\ SQ$, where SQ is the significant quantity of interest, such as

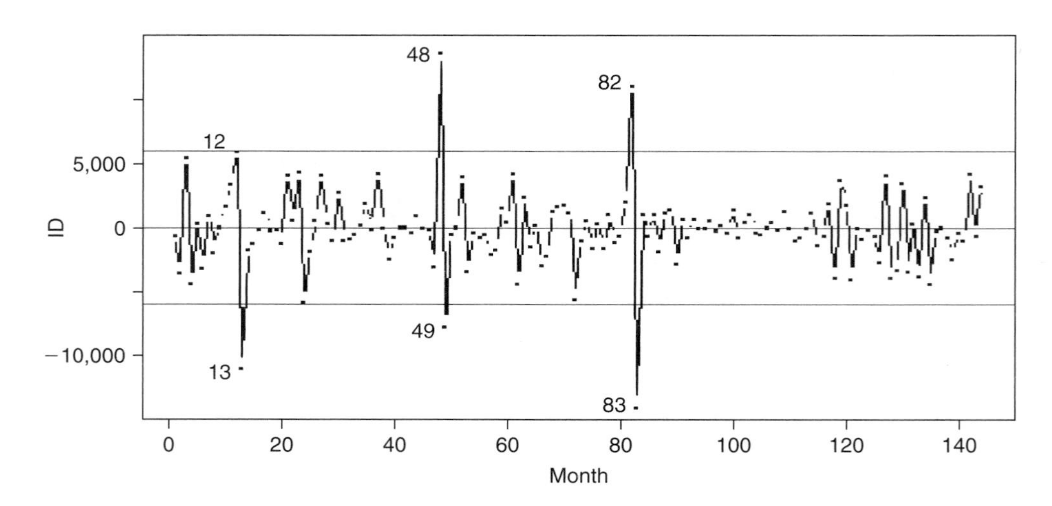

FIGURE 8.1 Monthly IDs over 12 years from a gaseous diffusion facility.

8 kg of Pu. To check for trends in a sequence of IDs, which could indicate small protracted loss, the *n*-by-*n* measurement error covariance matrix Σ_{ID} is used. See the section on sequential testing, and note from Figure 8.1 that successive values can be highly correlated.

In Figure 8.1, apparent losses on periods 12, 48, and 82 are followed by apparent gains on successive periods, and even if these three pairs of large values are removed, a large negative correlation remains between successive IDs. In view of Equation 8.1, where the ending inventory for period t is the beginning inventory for period $t + 1$, this is consistent with the information we have about this gaseous diffusion facility where large SNM (uranium, in this case) inventories are present, so the inventory measurement error will be a large contributor to σ_{ID}, leading to negative serial correlation.

The magnitude of σ_{ID} determines what SNM loss L could would lead to an alarm with high probability. For example, testing only for SNM loss (not gain) with a false alarm probability of $\alpha = 0.05$, the alarm probability $1 - \beta$ is 0.95 for $L = 3.3\ \sigma_{ID}$ (and $1 - \beta > 0.95$ if $L > 3.3\sigma_{ID}$). Usually the safeguards goals include a goal that $1 - \beta$ is at least 0.95 if $L \geq 1$ SQ, which is accomplished if and only if $\sigma_{ID} \leq SQ/3.3$. If $\sigma_{ID} > SQ/3.3$, then either measurement errors should be reduced to achieve $\sigma_{ID} \leq SQ/3.3$ (if feasible) or enhanced C/S is required, but the increased C/S effort level is challenging to negotiate.

Topics covered next include a short review of relevant statistical topics, measurement error modeling, propagation of variance, sequential testing, verification measurements, and difficulties with ID evaluation (such as holdup and poorly measured SNM streams).

Review of Relevant Statistical Topics

The most important statistical concept involved in evaluating ID sequences is variability. Measured values (examples: inventory differences, shipper-receiver differences) are often compared to their nominal values, and decisions must be made whether they are within the estimated variability due to measurement error.

Let X denote a random variable. For example, let $X = 1$ if an item's measured value (M) exceeds its declared value (D), and $X = 0$ otherwise. Assume there are $N = 1000$ measurements and that $\Sigma X_i = 500$, so that the percent with $M > D$ is $p = 0.5$. If we choose an item at random, we will have $X = 1$ with probability (*wp*) 0.5 and $X = 0$ *wp* 0.5. For this review we will consider two questions:

- What is the average value of X?
- What is the variance of the X values?

Solution: We first have to define *average* and *variance*.

Informally, "things vary." More formally, we define one type of variability in the language of random variables as the average squared distance of the random variable from its center (mean or average) value. This is an arbitrary but useful definition (and convenient mathematically).

First, we define average, μ, with respect to (*wrt*) the probability distribution $f(x)$, where $f(x) = 1$ *wp* 0.5 and $f(x) = 0$ *wp* 0.5.

In practice, we must estimate μ, so "it's Greek to us."

$$\mu \equiv \int_{-\infty}^{\infty} x f(x) dx$$

Don't worry if integration isn't your forte, because with our special $f(x)$, in the cases considered here, each integral reduces to a sum (integrals are just generalized sums, after all). Try to work this out (answer follows).

Second, we define variance, σ^2, *wrt* $f(x)$ as follows:

$$\sigma^2 \equiv \int_{-\infty}^{\infty} (x - \mu)^2 f(x)\, dx$$

Note that σ^2 is the *average of* $(x-\mu)^2$ *wrt* $f(x)$.

Again, try to work this out for our $f(x)$ (answer follows)

We will also use the notion of covariance (arises for example when two measurements share a common systematic error) between two random variables. Let f_{XY} denote the joint probability density of random variables X and Y. For example, let X the same as above ($X = 1$ if $M > D$ and 0 otherwise). Let $Y = 1$ if $M >$ True value and $Y = 0$ otherwise. It isn't difficult to see that X and Y are not independent, because when $M >$ True, it is more likely that $M > D$. For this example, we will assume $f(0,0) = P(X = 0$ and $Y = 0) = 0.4$, $f(0,1) = P(X = 0$ and $Y = 1) = 0.1$, $f(1,0) = P(X = 1$ and $Y = 0) = 0.1$, and $f(1,1) = P(X = 1$ and $Y = 1) = 0.4$. *Note:* The mean of Y is $\mu_y = 0.5$, and $\mu_x = 0.5$. We use f_{XY} to define the covariance between X and Y, $\sigma^2{}_{XY}$ as follows (work this out for our f_{XY}).

The rest is easy compared to this:

$$\sigma^2{}_{XY} \equiv \int_{-\infty}^{\infty} (x - \mu_x)(y - \mu_y) f_{xy}\, dx dy$$

Answers for X:

$$\mu = 0.5 \times 1 + 0.5 \times 0 = 0.5$$

Isn't that what you expected if we have half ones and half zeroes?

Okay, this one isn't so intuitive, but try using extreme cases of $p = 0$ or $p = 1$ rather than our middle case of $p = 0.5$. What should σ^2 be if $p = 0$?

$$\sigma^2 = 0.5 \times (1 - 0.5)^2 + 0.5 \times (0 - 0.5)^2 = 0.25$$

The concept: when X is above its mean, is there a tendency for Y to either be above (positive covariance) or below (negative covariance) its mean?

Note: As an alternate solution, consider $0.4 \times 0.4 - 0.1 \times 0.1$. (Why?)

$$\begin{aligned}\sigma^2{}_{XY} &= (0 - 0.5) \times (0 - 0.5) \times 0.4 + (0 - 0.5) \times (1 - 0.5) \times 0.1 \\ &\quad + (1 - 0.5) \times (0 - 0.5) \times 0.1 + (0 - 0.5) \times (0 - 0.5) \times 0.4 \\ &= 0.15\end{aligned}$$

Here's the answer for $\sigma^2{}_{XY}$: Because this text includes statistical concepts for a wide audience, readers are assumed to have a basic familiarity with typical material from an introductory statistics course. We suggest any introductory statistics text that covers variance and covariance, as defined previously, to lay the foundation for statistical issues in NMA.[2,3,4]

Measurement Error Models

All assay measurements involve multiple errors that can have different relative contributions in different contexts. Therefore, measurement error modeling is a large topic that we will briefly discuss here to describe some of the models that are commonly used in NMA.

The terms *random* and *systematic* errors are qualitative terms until we specify a particular measurement system and associated error model. Generally, random errors are unique to each

[2]W. Bowen and C. Bennett, *Statistical Methods for Nuclear Materials Management*, U.S. Nuclear Regulatory Commission, NUREG/CR-4604, 1988.

[3]W. Venables and B. Ripley, *Modern Applied Statistics with Splus*, third edition, Springer: New York, 1999.

[4]M. Neuilly, *Modelling and Estimation of Measurement Errors*, Lavoisier: Paris, 1999.

measurement, whereas a systematic error affects two or more measurements.[5,6,7,8] Sometimes the term *long-term systematic error* is contrasted to *short-term systematic error* (*long-term* implying that all measurements in the campaign share a common systematic error and *short-term* implying that two or more measurements in the time period share that error, but not all do).

An example error model is:

$$M = T(1 + S_{\text{item}} + S_{\text{inst}} + R) \tag{8.2}$$

where M is the measured mass, T is the true mass, S_{item} is the item-specific systematic error (bias), S_{inst} is the measurement instrument specific systematic error (bias), and R is the random error. All errors are random at some stage, which we denote $S_{\text{inst}} \sim N(0,\sigma_{\text{Sinst}})$, for example, and $N(\mu,\sigma)$ is the normal distribution with mean μ and standard deviation σ.

It is often assumed that a new S_{inst} is generated if and only if the instrument is recalibrated, leading to new estimates of calibration parameters. Therefore, $\sigma_{S_{\text{inst}}}$ can be estimated by using calibration data and commensurate results for uncertainties in estimated parameters or by using MC data on standards. Replicate measurements on the same item allow us to estimate the standard deviation σ_R of R. It is always challenging to estimate σ_{Sitem}. In fact, because S_{item} varies from item to item, we would model it as random error at least for the purposes of POV for ID evaluation. However, items having the same characteristics tend to have similar biases, so it is preferred to model S_{item} as being random within a class of items.

To some extent, there has always been an issue regarding how well the relevant properties of the standards match those of the items. This issue is becoming increasingly important in the U.S. DOE complex because of the need to measure scrap, waste, and residues, which can have highly variable material composition. Often S_{item} can be included with R. The implication is that either S_{item} is negligible compared to R or auxiliary methods must be included to measure selected items as part of the MC program and thereby have data to support per-item estimates of systematic error. For example, we might be characterizing a nondestructive (NDA) neutron assay method using Equation 8.2 but with $S_{\text{item}} + R$ redefined to be R.

Provided we have a "gold standard" assay method (with negligible S_{item}) such as calorimetry to occasionally remeasure an item, we can redefine σ_R appropriately and use such item remeasurements to estimate σ_R. The Pu facility at Los Alamos National Laboratory maintains a remeasurement database that provides some data for this purpose. As an important aside, it is necessary to separately estimate σ_S and σ_R in order to estimate σ_B. After estimating σ_B, it is usually acceptable to replace σ_R with $\sigma_{R_{effective}} = \sqrt{\sigma^2{}_R + \sigma^2{}_S}$.

In addition to remeasurement results, calibration data are useful for developing measurement error models by applying results of statistical function fitting (such as regression and least squares fitting of response to predictors). Assay methods are often calibrated on items having accurately known SNM amounts, but the material form of the calibration items might not match the form of test items in all respects, such as density, impurity effects, and source distribution. Therefore, statistical issues are often involved in designing experiments to assess the impact of mismatch between calibration standards and test items and to develop assay methods that measure and correct for mismatches.

---

[5]H. Aigner, R. Binner, and E. Kuhn et al., "International Target Values 2000 for Measurement Uncertainties in Safeguarding Nuclear Materials," *ESARDA Bulletin,* 31, 39–68, 2002.

[6]T. Burr, T. Sampson, and D. Vo, "Statistical Evaluation of FRAM γ-ray Isotopic Analysis Data," *Applied Radiation and Isotopes,* 62, 931–940, 2005.

[7]T. Burr, G. Hemphill, V. Longmire, and M. Smith, "The Impact of Combining Nuclear Material Categories on Uncertainty," *Nucl. Inst and Methods in Physics Research,* A 505, 707–717, 2003.

[8]B. Taylor and C. Kuyatt, "Guidelines for Evaluating and Expressing the Uncertainty of NIST Measurement Results," NIST Technical Note 1297, 1994.

Propagation of Variance

A common way to estimate the variance of a sum of measurements (such as *mass(Uranium)* = Σ(*Volume* $\times$ *Concentration*)) is to apply Equation 8.2 and assume that two measurements have nonzero covariance if and only if they are in the same measurement group and made during the same calibration period. Other models such as additive error models are available.[9] Applying the statistical principle that the variance of a sum of random variable is the sum of all variances plus the sum of all covariances, it is straightforward to derive a useful formula for a given strata with SNM total T:

$$\sigma^2{}_T = T^2(\sigma^2{}_R/n + \sigma^2{}_S) \tag{8.3}$$

where $\sigma^2{}_R$ ($\sigma^2{}_S$) is the sum of all random (systematic) error variances and n is the number of items in the strata. Or, more correctly:

$$\sigma^2{}_T = T^2(\sigma^2{}_R/n + \sigma^2{}_S) + (n - 1)s^2{}_T\sigma^2{}_R \tag{8.4}$$

where $s^2{}_T = \sum_{i=1}^{n}(T_i - \overline{T})^2/(n - 1)$.

Equation 8.4 differs from Equation 8.3 only in the term $\sigma^2{}_R = \sum_{i=1}^{n}(T_i - \overline{T})^2$ which is assumed to be 0 in the Equation 8.3 "stream-average" assumption. Also, if there are recalibrations of all measurements on the same schedule (unlikely, but used for illustration here), with a total of *ncal* recalibrations, then Equation 8.3 is modified to:

$$\sigma^2{}_T = T^2(\sigma^2{}_R/n + \sigma^2{}_S/ncal) \tag{8.5}$$

where again we use the stream average assumption. Generally, instruments will be recalibrated on different schedules, so no simple formulas are available. However, it is often useful to bound the correct solution by assuming two extreme cases: recalibrate each instrument after each measurement (effectively converting all errors to random errors) or never recalibrate. In the simplest case, we apply Equation 8.3 to each strata. Sometimes we must also allow for nonzero covariance among terms in *BI*, *EI*, and *R*, and among terms in *EI* and *S* or *R* and *S*.

Sequential Testing

A typical performance measure for the NMA system of a declared facility is the magnitude of σ_{ID}. Note that because throughput usually increases for longer balance periods, a facility can often use more frequent balance closures (such as weekly rather than monthly) to reduce σ_{ID}. However, this will not improve detection of protracted loss, and because facilities must monitor for abrupt and protracted loss, either sequential testing or at least monitoring IDs for trends is often used.[10]

This section provides a brief review of sequential tests for loss in nuclear materials accounting. Sequential tests include tests for abrupt and protracted loss, and the best sequential test depends on the exact loss scenario. Therefore, in practice it is common to use a

[9] T. Burr, A. Coulter, and J. Prommel, "VPSim: Variance Propagation by Simulation," LA-13382-MS, *Proceedings of the Annual Meeting of the Institute of Nuclear Materials Management*, 1997.

[10] R. Avenhaus and J. Jaech, "On Subdividing Material Balances in Time and/or Space," *Journal of the Institute of Nuclear Materials Management,* 10(3), 24–33, 1981.

sequential test that performs well (although perhaps not as well as the best test for the given scenario) for a wide range of scenarios.

A large abrupt loss leads to a single large ID and large protracted loss leads to multiple large IDs. Assume that the random vector of IDs observed through the nth period, $x = \{x_1, x_2, ..., x_n\}$ is approximately normally distributed with mean μ_n and $n \times n$ covariance matrix Σ_{ID}. If there is no material loss, then $\mu_n = 0$. The covariance matrix, Σ_{ID}, contains the variances of each ID along the diagonal, and the off-diagonal entries are the covariances.

We are concerned with tests for a null hypothesis NH: $\mu_n = 0$ versus an alternative hypothesis: AH: $\mu \neq 0$ with $\sum_{i=1}^{n} \mu_i > 0$.

The best test (the most power to detect loss for a given false alarm rate, in the sense of the Neyman Pearson lemma, depends on the exact form of the alternative hypothesis. We cannot assume we know the exact form of the alternative hypothesis, or there would be no need for a test.[10] Therefore, many tests have been proposed, with each test designed to do well for certain forms of the alternative hypothesis. The reason for assembling many tests is to study their performance over a range of loss scenarios and covariance matrices. In practice, a facility would implement one or a few of these tests, depending on individual circumstances.

Let the false-alarm probability for any one test be denoted α. That is, α is the probability that the test alarms one or more times during the n balance periods when the null hypothesis of zero loss is true.

ID Test

The ID test is the same as a one-at-a-time Shewhart test except there is serial correlation among successive IDs.[11,12,13] This test alarms if $x_i \geq h_i\sqrt{\sigma_i}$ for at least one i in 1, 2, ..., n, where $h = \{h_1, h_2, ..., h_n\}$is selected so that $P(x_i \geq h_i\sqrt{\sigma_i}) \leq \alpha$ for at least one i in 1, 2, ..., n when $\mu_n = 0$.

There are a number of ways to select h. One way is to fix $h_i = h$ for all i and select this threshold via simulation. Analytical methods are also available but are complicated to compute, because Σ_{ID} is not a diagonal matrix in general. Until Σ_{ID} is known, which in practice will not occur until the entire MUF sequence is observed, it is impossible to determine h to achieve the overall false-alarm rate of α.

SITMUF Test

The SITMUF test (MUF is material unaccounted for, which is another name for the ID) is the standardized, independently transformed MUF test.[12,13] This test is based on the unique linear transform of x to y that preserves the time ordering implied in x and with the components of the transformed y being independent, approximately normally distributed random variables with variance 1.

To transform the original MUF sequence x to an independent sequence $y = \{y_1, y_2, ..., y_n\}$, use a well-known linear algebra method (Cholesky decomposition). We assume that the measurement error covariance matrix of x, Σ, is estimated using variance propagation of all key measurements.

---

[11]J. Jaech. *Statistical Methods in Nuclear Materials Control*, U.S. Atomic Energy Commission, 1973.

[12]T. Burr, C. Coulter, E. Hakkila, H. Ai, I. Kadokura, and K. Fujimaki, "Statistical Methods for Detecting Loss of Materials using Near-Real Time Accounting Data," *Proceedings of the Annual Meeting of the Institute of Nuclear Materials Management*, 24, 1995.

[13]R. Picard, "Sequential Analysis of Materials Balances," *Journal of Nuclear Materials Management*, 38–42, 1987.

It is helpful to consider the y sequence as arising from calculating $y_t = x_t - E(x_t|x_{t-1},\ldots, x_1)$, where $E(x_t|x_{t-1},\ldots, x_1)$ is the expected (i.e., average) value of x_t given all previous x values, and then standardizing so that the variance, $\sigma^2_{y_i} = 1$ for all balance periods t. The transformed vector, $\mathbf{y}$, has mean μ_y that depends on the mean μ_n of $\mathbf{x}$. Under the NH, $\mu_y = 0$. Under the AH, $\mu_y \neq 0$. The test alarms if $y_i \geq h_i$ for at least one i in 1, 2, ..., n, where $\mathbf{h} = \{h_1, h_2, \ldots, h_n\}$ is selected so that $P(y_i \geq h_i$ for at least one i in 1, 2, ..., n when $\mu_n = 0$.

Because the y_i are mutually independent, we have $h = z(1 - \alpha)^{1/n}$ where $h = z(1 - \alpha)^{1/n}$ is the $(1-\alpha)$ quantile of the standard normal distribution. To explicitly show the time-order interpretation of the transformed vector, $\mathbf{y}$, write the transformation of the component x_i to y_i as $y_t = (x_t - E(x_t|x_{t-1},\ldots, x_1))/\sigma_i = (x_t - \sigma_{i-1}{}^{T}\, \boldsymbol{\Sigma}_{i-1}{}^{-1}\, \mathbf{x}_{i-1})/\sigma_i$, where we decompose $\boldsymbol{\Sigma}_i$ as $\Sigma_i \begin{pmatrix} \Sigma_{i-1}\sigma_{i-1} \\ \sigma^{T_{i-1}}\sigma_{i,i} \end{pmatrix}$ These two expressions for y_t arise from a standard result for the multivariate normal distribution. The $\mathbf{y} = \{y_1, y_2, \ldots, y_n\}$ sequence is referred to as the standard innovative sequence or the MUF residuals, and testing for loss in this sequence is the same as testing for loss in a sequence of independent normal random variables having mean zero and variance 1. However, because of the transformation from $\mathbf{x}$ to $\mathbf{y}$, any true loss will also be transformed.

Reference 13 has shown that a numerically stable and convenient way to calculate the SITMUF sequence from the MUF sequence is to apply the Cholesky decomposition of $\boldsymbol{\Sigma}$ as follows. For $\Sigma = \mathbf{CC}^{T}$, where $\mathbf{C} = c_{ij}$ is the lower-triangular Cholesky factor, it can be shown that $\mathbf{y} = C^{-1}\mathbf{x}$. The Cholesky decomposition is available in many standard linear algebra libraries such as LinPack.

Other common tests in safeguards that consider more than one period at a time include CUMUF (cumulative MUF) and GEMUF (a likelihood ratio that is equivalent to the Mahalonobis distance from 0 at each time period).

Page's Test Applied to the SITMUF Sequence

An effective sequential test that is often used in nuclear safeguards and elsewhere is Page's test applied to the SITMUF sequence, y.[12,13] Reference 12 examined other sequential tests and found that Page's test applied to the SITMUF sequence is competitive for a wide range of diversion scenarios. Other safeguards studies also evaluated sequential tests for safeguards and for anomaly detection in time series, such as a sequence of MUF values (see Reference 1).

Page's test can be applied to any sequence, but its properties (average run length, for example) are most easily studied if the sequence is independent. It is therefore common to apply Page's test to the SITMUF sequence.

Page's test is like the CUMUF test but restarts the sum at 0 if the sum is negative. The motive for this restart mechanism is to achieve a compromise between the MUF statistic and the CUMUF statistic. Page's statistic applied to the SITMUF sequence, y, is defined as $P_i(y) = \text{maximum}\ (P_{i-1}(y) + y_i - k, 0)$.

The test alarms if $P_i(y) \geq h_i$ for some $i = 1, 2, \ldots, n$, where the h_i are selected to give the desired false-alarm probability, α. The parameter k is a control parameter intended to give the user some control over the size loss that the test is well suited to detect. Generally, smaller values of k are best for detecting small protracted losses, and larger values of k are best for detecting abrupt losses. If we specify that we want good detection probability for a loss of 1σ, then choose $k = \sigma/2$. Because the y sequence is standardized to zero mean, and variance equal to 1, we choose $k = 1/2$.

Verifying Declarations

Verification measurements of items that are randomly selected from various strata (usually available inventory) at the time of physical inventory (*PI*) are compared to their book values. In a processing area that measures each input "batch" and each output "batch," there could be snapshots in time when the in-process batch is difficult to measure. Assume that we have input items that will change material form while in process, possibly distributing some SNM in glove boxes, pipes, furnaces, uncalibrated tanks, and so on. In situations with nonnegligible amounts of SNM in difficult-to-measure for ("holdup"), "data-driven" rules could determine when a process area cleanout is required during PI.

For example, suppose we begin processing in a clean material balance area (MBA).[14] We will consider three cases.

Case 1: After one year of tracking each input and output batch, the cumulative ID (CID) over 200 input and output batches (total of 200 kg input) is 1 kg (0.5%) and the estimated σ_{CID} is 2 kg (ignoring any contribution to σ_{CID} due to in-process material). Engineering estimates of holdup and the maximum capacity for holdup are 0.5 kg and 3 kg, respectively.
Assessment: Because σ_{CID} is 1% of 200 kg (and DOE order M474 suggests that σ_{CID} be as small as reasonably possible or, in any case, 2% or less), it is not yet necessary to perform a cleanout and PI. Also, a holdup measurement (which is time-consuming unless in-line holdup measurements are available) would not yet be required because the 1 kg CID is within expectations for holdup. Reasonable people can negotiate whether 2% is the appropriate target value for σ_{CID} for the particular material types. For example, if 1% is a more appropriate target for this process, then we expect cleanout and PI every one year at this throughput rate.

Case 2: Same as Case 1, but the ID reaches –2 kg within six months.
Assessment: This is an apparent gain of 2/1 = 2 times the estimated σ_{CID} (σ_{CID} accumulates at 2 kg per year), so either the process is producing material (perhaps from recovering in-process material) or σ_{CID} is underestimated. An assessment must be made and could result in measurement studies to better quantify σ_{CID}, or detection of either illicit activity involving replacement of material or faulty accounting.

Case 3: Same as Case 1, but the ID reaches 2 kg within six months.
Assessment: This is an apparent loss of 2/1 = 2 times the estimated σ_{CID}, so either the process is accumulating material (perhaps from buildup of in-process material), σ_{CID} is underestimated, or there is a loss. An assessment must be made and could result in measurement studies to better quantify σ_{CID}, detection of either illicit activity involving loss of material, or faulty accounting. Generally, apparent losses are more closely scrutinized than apparent gains. Should a holdup measurement be required? That would depend on more specific details, but conceivably, if σ_{CID} were allowed to increase to 2 kg to include the uncertainty due to the holdup estimate, then σ_{CID} would be 2% of the six-month throughput of 100 kg, and 2/2 = 1 so that both the CID and CID/σ_{CID} are acceptable. In that case, a holdup measurement need not be required yet.

The "data-driven" rules to determine when a process cleanout and PI are required include (1) σ_{CID} increases to its allowed upper limit (currently 2% of cumulative throughput since the previous cleanout); (2) the ID exceeds 2 or 3 times σ_{CID}; (3) the ID exceeds (by a negotiated amount) what can be expected to be recovered from process cleanout; and (4) it has been more than a negotiated time (five years, for an example) since the previous cleanout.

But when is a PI (without process cleanout) required? Generally, there are two categories of facilities: (1) facilities that rely on activities during PI to get material into measurable form and (2) facilities that evaluate IDs in near real time (NRTA). Those of Type 1 will probably need to continue to do periodic PIs and the associated verification or confirmation measurements of statistical samples of items in various strata. The Los Alamos Plutonium Facility is of Type 2, provided in some cases we assume that material control is adequate to

[14]T. Burr, R. Strittmatter, B. Scott, C. Murdock, and M. Schanfein, "Evaluation of a continuous physical inventory approach for the Los Alamos Plutonium Facility," LA-UR01-3542, *Proceedings of the Annual Meeting of the Institute of Nuclear Materials Management*, 2001.

allow certain shortcuts in assay methods. For example, the input material to one process is Pu metal. This metal is weighed and the weight is multiplied by a historical Pu purity factor to estimate the Pu mass. Strictly speaking, this is not a safeguards measurement because the Pu is not directly measured (but a gross attribute gamma measurement is used to confirm the presence of Pu). If there is a credible substitution scenario that material control procedures failed to detect, the input would be overestimated. Fortunately, the Pu in the final product is assayed, so there is a loss detection capability. We believe it is acceptable to treat the input Pu measurement as though it were a true Pu assay, compare it to the assay of the Pu output, and track the ID of each batch as well as the CID. And generally, there are many facilities that could be granted certain shortcuts in assay methods, provided material control procedures were adequate. When feasible, it is obviously preferred to have complete assay methods for all SNM streams.

In summary, for facilities that do batch tracking or some type of NRTA, it is possible to argue that the PI is required only when a "data-driven" rule as described comes into effect.[14]

Other Purposes of the PI

Because it is desirable to reduce the PI frequency when feasible, it is important to recognize all of the purposes of the PI under current DOE practices. The purposes of the PI are as follows: for some facilities, the PI is the mechanism that allows IDs to be computed; and for all facilities, the PI provides a convenient time for auditors to ensure that the accounting system captures the true picture of SNM in the facility (the audit role to detect data falsification).

Suppose we measure all inputs and outputs for the MBA and do not rely on the PI to compute IDs. How will auditors know that the accounting system captures the true picture of SNM in the facility? In the case of international safeguards as performed by the IAEA, the effort to detect or deter data falsification is substantial. This is because the IAEA must protect against the entire facility or state falsifying data. Generally, the threat of widespread data falsification is thought to be less in the context of domestic safeguards. Nevertheless, the MC&A approaches (such as the "difference statistic" D, which measures the average difference between operator and inspector measurements) that have been in use for many years by the IAEA do provide a logical framework for evaluating verification measurements.

This chapter focuses on domestic safeguards, so verifying facility declarations is not typically as high a priority as it is in international safeguards. Nevertheless, modern domestic safeguards approaches recognize that if IDs are evaluated frequently, the real goal of the PI is not to check for loss but to audit the facility's accounting records for accuracy. Therefore, verification measurements to verify book values of SNM items are required in domestic safeguards, much in the same way that the IAEA's difference statistic is used in international safeguards.

The IAEA's difference statistic D for a given stratum (inputs, for example) is defined as the average difference between operator (o) and inspector (i) measurements for n items, multiplied by the number of items in the strata, N. That is, $D = N\sum_{i=1}^{n}(o_i - i_i)/n$. The IAEA's estimate of the ID is the operator's ID minus D. For example, consider a simple case with one input (x_1) and one output (x_2) measurement by the operator which are both verified by the IAEA (y_1 and y_2, respectively). Then ID – D = $(x_1 - x_2) - \{(x_1 - y_1) + (x_2 - y_2)\} = (y_1 - y_2)$, which is the IAEA's ID. Usually, the IAEA or the domestic safeguards auditor verifies a small random sample from some or all strata (inputs, outputs, inventory) and in the case of the IAEA, the average difference in each strata is multiplied by N. Reference 15 provides variance calculations (and the IAEA maintains several technical manuals regarding D) for several situations to estimate σ_D^2 and $\sigma_{ID-D}^2 = \sigma_D^2 - \sigma_{ID}^2$ (not obvious but true).[15]

---

[15]Picard, R., "Note on the MUF-D Statistic," *ANS Topical Conference, Proceedings Third International Conference on Facility Operations-Safeguards Interface*, American Nuclear Society, Inc., La Grange Park, Illinois: 414–423, 1988.

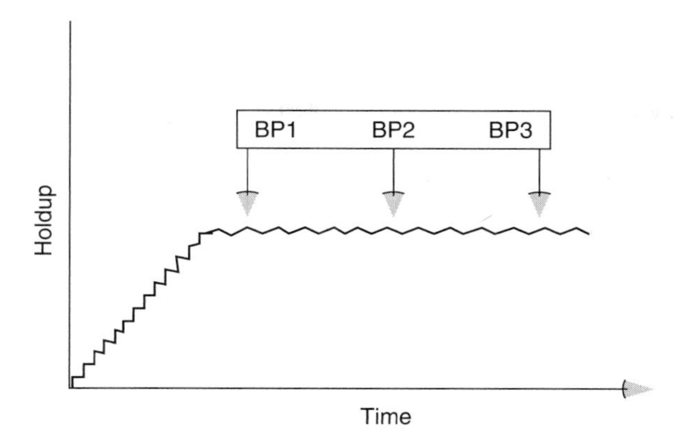

FIGURE 8.2 Simplified model of total holdup versus time.

Difficulties with ID Evaluation

Issues that complicate ID evaluation include (1) the impact of holdup on σ_{ID}; (2) serial correlation in successive IDs; (3) poorly characterized measurement quality leading to poor estimates of σ_{ID}, and (4) time delays in measurement results. Issue 2 was addressed in the section on sequential testing. Issue 3 can be addressed by using a measurement control program that is designed to accurately reflect how measurements are actually performed. An example of Issue 4 is destructive analysis (DA) of grab samples required to assay Pu. Often many days are required for DA lab results, so the ID cannot be computed in a timely manner, perhaps not until many days after a balance period closure. Typically, estimated values based on previous or target concentration values are allowed, but note that this introduces a vulnerability unless Pu presence is at least confirmed using rapid, low-accuracy confirmatory measurements. Generally, real facilities face processing challenges, and the impact of safeguards to operations needs to be as minor as possible. Ideally, safeguards measures are not viewed as pure overhead but can add value by forcing the operator to thoroughly understand and control the process.

We now consider Issue 1 by evaluating the impact of changing material holdup (poorly measured or unmeasured inventory) on ID evaluation. Figure 8.2 illustrates that changing holdup from period to period leads to larger-than-POV-based estimates of σ_{ID}.

Let $\Delta H = \text{Holdup}_i - \text{Holdup}_{i+1}$ be the change in holdup from balance period i to $i + 1$.

The key point in Figure 8.2 is that if holdup is changing (ΔH is nonzero), then instead of estimating the true loss, the ID estimates ΔH. Note that, due to measurement error, the IDs vary randomly around ΔH. This adds uncertainty to the ID equation that must be considered but is often difficult to quantify. Fortunately, NDA measurements are often very effective at estimating changes in holdup and sometimes effective in estimating holdup itself, so NDA measurements play a key role in holdup and ID evaluation.

Related Topics

Probability and statistics are separate but related subjects. Probability involves forward modeling such as specifying an event, associated observables, and probability distributions for those observables that involve both physical and measurement processes. Statistical analysis usually attempts to solve an inverse problem such as: Given the observations, estimate which forward processes are most likely.

We briefly describe three nonproliferation topics associated with declared activities that involve inverse problem solving.

Portal Monitoring

Data from passive radiation portal monitors (RPMs) have been collected at screening locations since 2002.[16] The purpose is to detect potentially harmful radioactive cargo (special nuclear material, SNM) that emits gamma rays (coarsely binned into low- or high-energy counts) and/or neutrons. Each vehicle slowly passes by a set of fixed radiation sensors, resulting in a profile time-series measurement from each sensor. Although this is not a declared facility in the traditional sense, the vehicles declare their cargo, thus making portal monitoring similar to nonproliferation activities at declared facilities.

The basic task with RPMs is to have a high detection probability for threat items and low nuisance and false alarm rates. Because a nonnegligible fraction of nonthreat cargo contains naturally occurring radioactive material (NORM) such as the potassium in cat litter, the majority of alarms are not due to statistical fluctuations but instead are true (nuisance) alarms due to NORM material.[16] Also, because the count criterion leads to many nuisance alarms arising from NORM and because background suppression by the vehicle is smaller for ratios of gamma counts than for counts alone, some systems are including both gamma count and gamma count ratio alarm criteria.[17,18]

Statistical issues include alarm threshold selection and sensor optimization, drifting backgrounds, and pattern recognition methods applied to spectral analysis to distinguish NORM from background and from threats.[19,20]

Solution Monitoring

Solution monitoring (SM) is a form of process monitoring which can be considered to provide a C/S capability. Experience to date has shown that SM is a challenging but useful safeguards measure that contributes to both NMA and C/S. Part of the challenge involves choosing effective evaluations of SM data that avoid data indigestion, enable anomaly detection and/or resolution, and do not burden the operator or the inspector with too many investigations.

Potential benefits of SM include improved abrupt loss detection while controlling for multiple tests, anomaly resolution, measurement error model validation, and data authentication. Data authentication results from the many internal consistency checks that arise when relating level, density, and temperature readings; this makes it very difficult to alter data without being detected.

SM is the nearly continuous monitoring of solutions in all key process tanks. Typically, the level (L) and density (D) of the solution in a tank is obtained by measuring the differences in pressures that are required to bubble air through dip tubes located at various points

---

[16] B. D. Geelhood, J. H. Ely, R. R. Hansen, R. T. Kouzes, J. E. Schweppe, and R. A. Warner, "Overview of Portal Monitoring at Border Crossings," *IEEE Nuclear Science Symposium—Conference Record*, 513–517, 2004.

[17] T. Burr, J. Gattiker, K. Myers, and G. Tompkins, "Alarm Criteria in Radiation Portal Monitoring," *Applied Radiation and Isotopes*, 65, 569–580, 2007.

[18] J. Ely, R. Kouzes, J. Schweppe, E. Siciliano, D. Strachan, and D. Weier, "The Use of Energy Windowing to Discriminate SNM from NORM in Radiation Portal Monitors," *Nuclear Instruments and Methods in Physics Research A*, (2). 373–387, 2005.

[19] L. Pibida, M. Unterweger, and L. Karam, "Evaluation of Handheld Radionuclide Identifiers," *Journal of Research of the National Institute of Standards and Technology*, 109(4): 451–456, 2004.

[20] J. Blackadar, S. Garner, J. Bounds, W. Casson, and D. Mercer, "Evaluation of Commercial Detectors, LAUR03-4020," *Proceedings of the Annual Meeting of Institute of Nuclear Materials Management*, 2003.

in the tank. Temperature (T) is obtained via thermocouples. The (L, D, T) data are collected frequently, perhaps every few seconds or even less. At this stage the data can be analyzed to check their validity, filtered and perhaps compressed, before they are uploaded to some form of real-time database. Various data storage and/or change-detection rules determine how frequently these measurements are archived.[21] The various time histories are then combined with tank calibrations to estimate their associated volume and mass histories. All these histories can then be evaluated by so-called solution monitoring evaluation systems.

Reference 10 considered the effect of analyzing frequent IDs over small material balance areas (individual tanks in this case) and concluded that in the worst-case protracted diversion scenario, less frequent IDs over a single material balance area actually lead to higher detection probability. The proof used the Neyman Pearson lemma from classical statistics and assumed that the diversion was optimally (from the diverter's view) allocated. This worst-case loss vector is proportional to the sum of the rows of the variance-covariance matrix Σ_{ID}.

SM involves tanks and frequent balance closures (each transfer and wait mode in the example that follows). If the worst-case diversion occurred, it is straightforward to prove that the optimal strategy is to compare the total input to tank 1 to the total output from tank 15. This would be classical ID accounting, not SM as we have defined it. However, frequent balance closures around each transfer and wait mode will have very high detection probability against abrupt loss and nearly as high a detection probability against the worst-case loss as an annual ID comparing tank 1 input to tank 15 output.

SM cannot improve protracted loss detection against this worst-case loss vector; however, it dramatically improves loss detection against other protracted loss vectors and against any abrupt loss.[21] In addition, there is the possibility that the bias corrections that become available via SM data can reduce the volume measurement error, thereby leading to improved loss detection against even the worst-case loss vector. There will be many paired comparisons of shipments and receipts between tanks; each of these should agree with the propagated total measurement error, or the error models must be refined.

Any statistically detectable diversion would have to be concealed by replacing the lost mass with proper density solution. The adversary would have to work hard to conceal the diversion (and would probably be discovered at the time that Pu was measured off-line because of out-of-specification chemical species). The same type of calculations would apply if we used an in-tank Pu concentration measurement, and in that case there would be no way for the adversary to conceal a statistically detectable diversion.

Process Monitoring

Process monitoring is a broad term that includes monitoring by radiation detectors, cameras, and monitoring solutions in vessels using pressure-sensing dip tube or other technology. Radiation detectors can monitor either the declared SNM transactions (an item was shipped from Point A to Point B, so the detector should confirm this using detected radiation) or can monitor for undeclared transactions (such as portal monitors do). Smart cameras can save and archive scenes involving declared transactions, watch for undeclared transactions, and alert an inspector to sections in the archive that require human review.

Process monitoring in our context is any type of monitoring such as quality control checks that could provide safeguards assurances. Solution monitoring is an example. It is generally agreed that process monitoring data can and should be a safeguards component. However, facilities do not want to reveal proprietary process information nor to resolve anomalies for safeguards purposes that do not impact process quality. Therefore, facilities must negotiate the type of process monitoring information to be used for safeguards.

---

[21]E.C. Miller and J. Howell, "Tank Measurement Data Compression for Solution Monitoring," *Journal of the Institute of Nuclear Materials Management,* 27(3), pp. 25–32, 1999.

Summary

This overview of statistical methods for nonproliferation at declared facilities included: (1) propagation of variance of algebraic combinations such as products and sums of random variables, such as *mass*(uranium) = Σ (*Volume* $\times$ *Concentration*); (2) measurement error modeling; (3) sequential testing to support near-real-time accounting; (4) specialty topics such as holdup and NDA of heterogeneous material, and (5) other specialty topics such as process and solution monitoring. The chapter briefly mentioned second line of defense screening for illicit nuclear material and associated statistical issues.

At declared facilities, the key statistical concept is variability and the main tool for ID evaluation is σ_{ID}. Statistical methods are used to estimate measurement uncertainties of individual assay methods and to combine these via the ID equation to estimate σ_{ID}. In the broader nonproliferation context, there are many statistical issues in monitoring for undeclared activities.

9

Case Study: Safeguards Implementation at the Rokkasho Reprocessing Plant

S. E. Pickett

Introduction

Rokkasho-mura Reprocessing Plant (RRP) in Rokkasho-mura, Aomori, Japan is the largest commercial nuclear spent-fuel reprocessing plant under International Atomic Energy Agency (IAEA, or also referred to as the Agency) safeguards and the only such facility located in a nonnuclear weapons state. The plant is a key facility in Japan's nuclear fuel cycle policy, which commenced in 1957 with the first Long Term Plan.[1] The Long Term Plan specified a "National Project" to develop a fast breeder reactor, enrichment, and reprocessing. Under this plan, Japan planned to close the nuclear fuel cycle, reprocessing spent nuclear fuel to use the plutonium in fuel for the fast breeder reactor. Although a closed nuclear fuel cycle has not been realized according to the original plan, Japan has completed its reprocessing facility and is moving toward construction of a mixed plutonium-uranium oxide (MOX) fabrication facility (named J-MOX) to fabricate MOX fuel for use in its light water reactors. Commercial operation of this facility is scheduled for 2011.[2]

RRP is designed to reprocess 800 metric tons of spent reactor fuel and to recover approximately 8 metric tons of plutonium annually. The plant will be safeguarded by both the IAEA and by the Japanese government (Ministry of Education, Culture, Sports, Science and Technology, Japan Safeguards Office (JSGO), and the Nuclear Material Control Center (NMCC)), with each drawing independent conclusions.

RRP is illustrative of the technical challenges involved in safeguarding large-scale facilities. With the quantity of throughput of nuclear material at RRP, it is critical that adequate measurement and detection systems are in place to ensure that no special fissionable material is diverted for undeclared purposes.[3] As a non-nuclear weapons state (NNWS) signatory to the Nuclear Nonproliferation Treaty (NPT), Japan has committed to safeguards and verification

---

[1]"Long-Term Program for Research, Development and Utilization of Nuclear Energy," Atomic Energy Commission (AEC) of Japan, Nov. 24, 2000; www.japannuclear.com/files/Japanese%20Government%20Long-Term%20Program%20for%20Nuclear%20Energy.pdf (July 2007).

[2]Akiyoshi Minematsu, "The Current Status of Active Tests at the Rokkasho Reprocessing Plant and the Preparation for MOX Fuel Fabrication Plant," *Book of Abstracts ICAPP*, 2007; www.inspi.ufl.edu/icapp07/program/abstracts/7599.pdf (July 2007).

[3]The term *special fissionable material* means plutonium-239; uranium-233; uranium enriched in the isotopes 235 or 233; any material containing one or more of the foregoing; and such other

at all its facilities to provide confidence that no such diversions occur.[4] In 1999, Japan signed an Additional Protocol (INFCIRC 540), which provides the (IAEA) with a broader set of tools to search for undeclared materials and activities in a state, with an emphasis on improved access to information and physical locations within a state.[5]

The approaches for safeguarding these large-scale facilities bring together advanced technologies and IAEA processes in an integrated system to provide timely assurance that the plant is operated as declared. To provide such assurance, the IAEA draws on technical capabilities and personnel resources including continuous design verification, advanced safeguards technologies, containment and surveillance, inspectors, and data acquisition and analysis systems.

As noted in earlier chapters, the array of technologies must be integrated to support IAEA *safeguards conclusions*, a term that refers to the IAEA's right and obligation to ensure that safeguards are applied, to all source or special fissionable material in all peaceful nuclear activities within the state, under its jurisdiction or carried out under its control anywhere.[6] The IAEA must "be able to draw an overall conclusion that all nuclear material has been placed under safeguards and remains in peaceful nuclear activities or has been otherwise adequately accounted for, the Agency must draw conclusions of both the nondiversion of declared nuclear material and the absence of undeclared nuclear material and activities for the State as a whole."[7] It is these conclusions that provide confidence to the broader international community that the state is not diverting material for nonpeaceful purposes. For states with both comprehensive safeguards agreements (as required by the NPT and defined in IAEA INFCIRC 153) and additional protocols (Additional Protocol, INFCIRC 540) in force, the measures included in the Additional Protocol (AP) strengthen the confidence basis on which the conclusion is drawn.

As noted, to draw these safeguards conclusions, the IAEA relies on verification measures (such as on-site inspections, visits, monitoring, and evaluations). As the facilities under safeguards become more complex and the operating throughputs increase, new technologies can provide assistance to ensure that the facility is being operated as declared and that material is not being diverted. The IAEA continues to strengthen safeguards through development of new guidelines for reviewing states' declarations of nuclear material, new software for processing data more efficiently, new databases for imagery, and development of new safeguards instruments for material measurement.[8] RRP provides an excellent case study of how advanced safeguards technologies are integrated into a large processing facility through a collaborative process among the IAEA, the state (government), and the operator as well as associated technology suppliers. These advances, in containment/surveillance, nuclear material accountancy and data acquisition, can improve the effectiveness and help the IAEA draw safeguards conclusions.

This chapter presents a brief overview of international safeguards and the domestic safeguards system in Japan and then describes some of the safeguards technologies that have

---

fissionable material as the Board of Governors shall from time to time determine; but the term *special fissionable material* does not include source material. Source: Statute of the International Atomic Energy Agency (opened for signature at New York on Oct. 26, 1956; entered into force on July 29, 1957), Article XX: Definitions.

[4]Japan signed the NPT in 1976, thereby accepting IAEA safeguards as stated in Article 3 of the Treaty.

[5]Japan brought the Additional Protocol into force on Dec. 16, 1999, and provided its Protocol Declaration to the Secretariat of the IAEA.

[6]Paragraph 2 of "The Structure and Content of Agreements Between the Agency and States Required in Connection with the Treaty on the Nonproliferation of Nuclear Weapons"; [INFCIRC/153(Corrected)], IAEA, www.iaea.org/Publications/Documents/Infcircs/Others/infcirc153.pdf (July 2007).

[7]Tariq Rauf, "Drawing Safeguards Conclusions," Presentation to the 2004 NPT Preparatory Committee, www.iaea.org/NewsCenter/Focus/Npt/npt2004_ppt_2904.pdf (July 2007).

[8]*Strengthening the Effectiveness and Improving the Efficiency of Safeguards System*, International Atomic Energy Agency, Aug. 17, 2001, GC (45)/23.

been integrated into the RRP to ensure material accountability throughout the plant and assist the IAEA in drawing safeguards conclusions.

International Safeguards

IAEA Goals

The IAEA works to ensure that the state is fulfilling its international obligations to use its nuclear facilities and materials for peaceful uses of nuclear energy (in other words, not use civilian nuclear programs for nuclear weapons purposes) through its safeguards program. Today the IAEA safeguards nuclear material and activities under agreements with more than 140 states.[9]

To fulfill its responsibility and be able to draw safeguards conclusions regarding the activities of a particular state, the IAEA sets goals and subsequent criteria that each state must follow. The IAEA safeguards objectives are for the "timely detection of significant quantities of nuclear material from peaceful nuclear activities to the manufacture of nuclear weapons or other nuclear explosive devises or for purposes unknown, and the deterrence of such diversion by the risk of early detection."[10] Thus the IAEA has set "significant quantity" goals for safeguards, and safeguards criteria are defined such that diversion of a significant quantity of material should be detected within a fixed time and with set detection probabilities.[11] Safeguards technologies have been developed to measure the type and quantity of special fissionable material present in a given process or stage in the facility. In short, the safeguards technologies provide *scientific* measurements to assist the IAEA in verifying a state's declaration, monitoring a state's material and activities, and performing data acquisition and analysis.

In general, IAEA safeguards consist of three main elements:

- Nuclear materials accountancy (NMA)
- Inspection and verification, process monitoring
- Containment and surveillance (C/S) measures

None of these measures alone is capable of meeting all objectives. For example, extensive use of NMA alone is not capable of detecting certain types of facility misuse. Different combinations of NMA, process monitoring, and C/S could potentially satisfy safeguards goals as long as all credible material diversion paths are appropriately monitored.

State System of Accounting and Control

With regard to the state's obligations to the IAEA, the state undertakes steps to implement and carry out its responsibilities under its agreements with the IAEA. The state will thus develop a system to meet its reporting requirements based on the model put forth in the "Structure and Content of Agreements Between the Agency and State Required with the Treaty of Nonproliferation of Nuclear Weapons," INFCIRC 153, concluded in 1972. INFCIRC 153 allows for the implementation of international safeguards, creation of a verification system,

---

[9]*IAEA Safeguards Overview: Comprehensive Safeguards Agreements and Additional Protocols*, www.iaea.org/Publications/Factsheets/English/sg_overview.html (July 2007).

[10]"The Structure and Content of Agreements Between the Agency and States Required in Connection with the Treaty on the Nonproliferation of Nuclear Weapons"; [INFCIRC/153 (Corrected)], p. 9, www.iaea.org/Publications/Documents/Infcircs/Others/inf153.shtml (July 2007).

[11]*Significant quantity*: Warning and verification goals are specified in terms of detection probability within a given amount of time. In both cases, the amount of material to be detected is specified as a *significant quantity* (e.g., 8 kg of Pu, 25 kg HEU, 75 kg LEU). See also, "IAEA Safeguards Glossary, 2001 Edition, International Nuclear Verification Series No. 3," IAEA, www-pub.iaea.org/MTCD/publications/PDF/nvs-3-cd/PDF/NVS3_scr.pdf, 2001.

and development of a system of accounting and control in each state to help provide assurances that all nuclear material is used only for peaceful purposes. This agreement, which is in accordance with the Nonproliferation Treaty, Article III, defines the application of safeguards as well as the rights and responsibilities of the state and the IAEA in the implementation of those safeguards.[12] The state's system is referred to as the *State System of Accounting and Control* (SSAC) and is obligated to account for and control all nuclear material subject to safeguards.

The state will also often undertake separate steps to implement its own safeguards, known as *domestic safeguards*. The overall objective of a national or domestic safeguards system is to prevent the theft or diversion of weapons-usable nuclear material by unauthorized entities, which are assumed to seek the material to inflict serious harm on the interests of that state (vice the international community). In Japan, the domestic safeguards system also ensures that the state submits its required declarations to the IAEA. This section focuses on the Japan SSAC as it pertains to activities and reporting required by the IAEA.

Japan, as a signatory to the NPT, must abide by its obligations under the NPT and Comprehensive Safeguards Agreements. Japan, like other countries, has established an SSAC through which the state carries out its responsibilities to the IAEA. The domestic legal basis for Japan's SSAC is through the "Law Concerning Regulation of Nuclear Raw Materials, Nuclear Fuel Materials and Nuclear Reactors," enacted in 1957. [13]

As illustrated in Figure 9.1, the Domestic Safeguards System in Japan is supervised by the Ministry of Education, Culture, Sports, Science, and Technology (MEXT), which has designated the Nuclear Material Control Center (NMCC) as the Authorized Safeguards Inspection Executing Organization.[14] MEXT is responsible for submitting the required reporting to the IAEA, and the NMCC is the only organization that has been approved by federal law to carry out national safeguards inspections in Japan. NMCC is responsible for inspecting nuclear facilities in Japan, conducting statistical analyses, ensuring measurement accuracy, and adjusting safeguards equipment. It is also responsible for managing the SSAC as well as for compiling the accounting reports and Additional Protocol (AP) declarations, submitted by the facility operator, and submitting them to IAEA (via MEXT). NMCC's activities also include an analysis of samples and research and development in support of safeguards and physical protection technologies.[15]

All Japanese nuclear facilities are required to maintain accounting and operational records, submitting relevant accountancy reports and AP declarations to NMCC. The facilities must also allow national and IAEA inspectors at their facilities in conjunction with the NPT safeguards agreement (INFCIRC/153) and Japan's Additional Protocol (INFCIC/255 add.1).

As with all other nuclear facilities in Japan, Rokkasho Reprocessing Plant adheres to Japan SSAC requirements, providing detailed reports of its material, allowing national and

---

[12]For more information, see INFCIRC/153, www.iaea.org/Publications/Documents/Infcircs/Others/inf153.shtml (July 2007).

[13]U.S. National Academy of Sciences, "Protection, Control, and Accounting of Nuclear Materials: International Challenges and National Programs—Workshop Summary (2005)," Committee on Development, Security, and Cooperation (DSC), p. 41; www.nap.edu/catalog/11343.html (July 2007).

[14]Government of Japan, Ministry for Education, Culture, Sports, Science, and Technology, www.mext.go.jp/english/org/science/37.htm.

[15]U.S. National Academy of Sciences, "Protection, Control, and Accounting of Nuclear Materials: International Challenges and National Programs—Workshop Summary (2005)," Committee on Development, Security, and Cooperation (DSC), p. 42; K.Naito, "Enhanced Cooperation between Agency and Japanese SSAC," presented at ESARDA 5 2006, www.inmm.org/esarda5/data/papers/010_1.6_Naito.pdf.

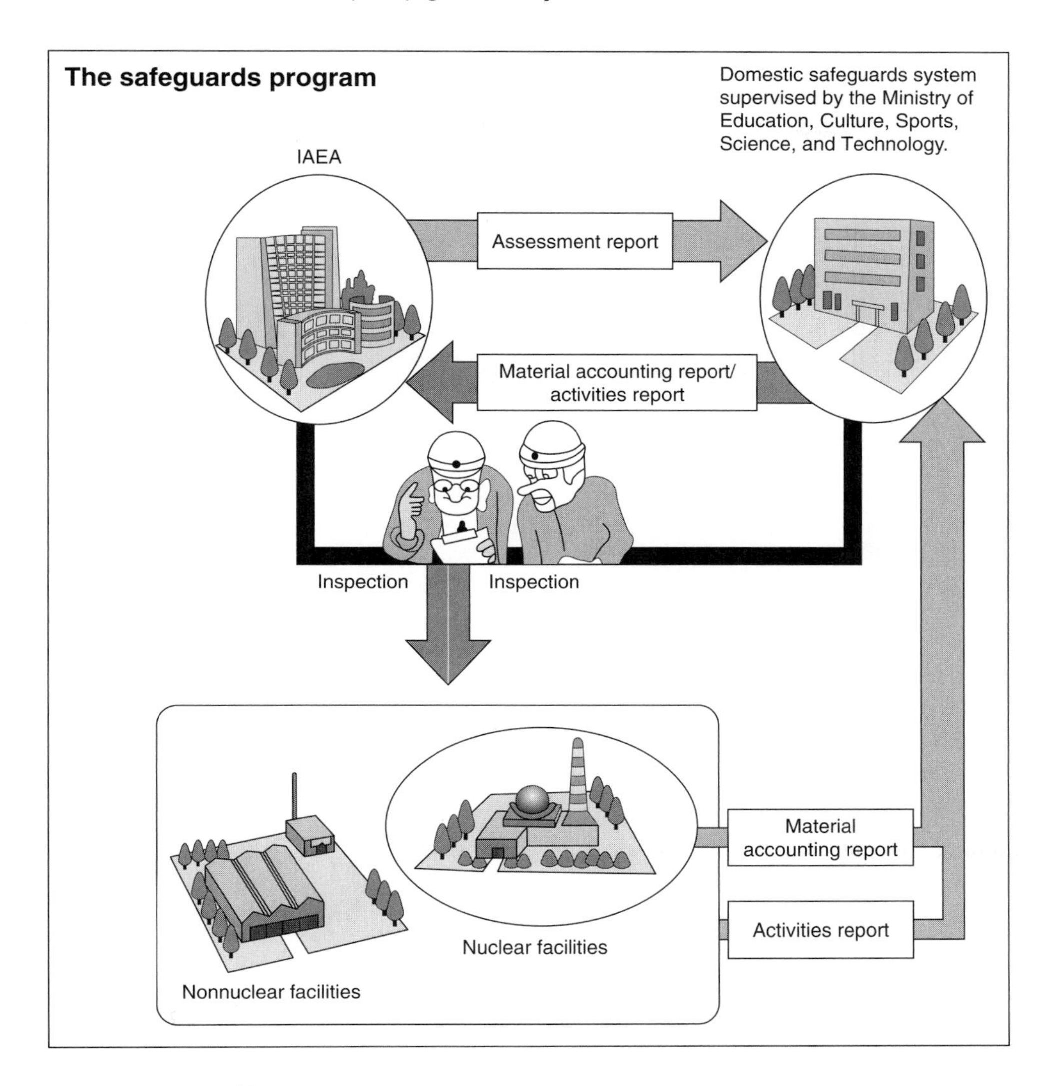

FIGURE 9.1 SSAC process.[16]

IAEA inspectors access, and installing safeguards technologies that help the IAEA monitor and verify activities at the facility.

Reprocessing and Safeguards

The IAEA has successfully implemented safeguards at small and medium-sized reprocessing plants since the 1970s. The new reprocessing plants envisaged in the 1990s were designed to process spent fuel in quantities about four times larger than those of plants built in the 1970s, utilizing more advanced process technology.

In view of the importance of safeguards for such plants, development and testing of new safeguards techniques were pursued intensively by operators and safeguards technology

[16]"Japan's Nuclear Power Program," Federation of Electric Power Companies, Japan, www.japannuclear.com/nuclearpower/nonproliferation (July 2007).

developers. They recognized that a mutual awareness and understanding of these new safeguards techniques would benefit all parties.[17]

Challenges

The RRP, operated by Japan Nuclear Fuels, Ltd. (JNFL), has a design throughput of 800 metric tons of spent fuel per year, with a spent-fuel pool capacity of 3,000 tons. The plant will recover 8 tons of plutonium per year in the form of mixed oxides. This is enough to reprocess the spent fuel produced by 40 reactors at 1,000 MW-class nuclear power stations.[18]

Because the throughputs for reprocessing plants like RRP are large—and much of the operation is automated—material accountancy measures and the uncertainties associated with them are likely to be outside the target values for "conventional" nuclear materials accountancy. Thus the main challenge is to ensure that all the material is accounted for throughout the process, to meet the aforementioned IAEA detection goals.

To satisfy the IAEA goals and maintain effective safeguards assurance that material is not being diverted to illicit uses, this large-scale, high-throughput facility required new measurement capabilities, including higher-precision MOX measurements, measurement on high-dose-rate vitrified canisters, measurements of the head-end fuel assembly feed and leached hulls, and an on-site laboratory for DA measurements.[19]

Facility and Process Description

The process at RRP comprises the cask receipt and storage, the spent fuel storage area, a head-end process, the main process including uranium oxide conversion, uranium/plutonium codenitration conversion process, mixed uranium/plutonium oxide (MOX) storage, uranium oxide storage and waste treatment and storage areas. Figure 9.2 illustrates the operations process. At each process and key measurement area, various safeguards technologies are employed to verify the amount of nuclear material and process.

Fuel assemblies are received from power reactors in shielded transport casks, typically 14 to 20 fuel assemblies per cask. Casks may be stored temporarily in a storage area before transferring into a water-filled pool where the assemblies are removed from the casks under water and are stored, awaiting processing. Fuel assemblies are then transferred individually to the dry head-end area, where they are chopped into short lengths to expose the fuel inside the protective cladding. Chopped pieces fall into nitric acid, where the fuel dissolves, leaving the cladding pieces ("hulls") to be washed and removed from the process as waste. The dissolved fuel solution is clarified in a centrifuge to remove undissolved "fines" before collection in an accountancy tank to be measured for uranium and plutonium content. After this high-accuracy measurement, the solution is fed to the main process. Virtually all the uranium and plutonium from the original fuel assemblies is contained in nitric acid solution transferred to

---

[17]In 1987, the IAEA established extrabudgetary funding to accelerate and broaden these activities, and this initiative led to the formation of the consultative forum for large-scale reprocessing plant safeguards, referred to as LASCAR (short for *large-scale reprocessing*). Between 1988 and 1992 the LASCAR team reviewed safeguards for large-scale reprocessing plants and determined that it is feasible to design effective and efficient safeguards with advanced techniques, including near-real-time accounting of material.

[18]"Reprocessing," Japan Nuclear Fuels Limited (JNFL) Website, www.jnfl.co.jp/english/reprocessing.html (July 2007).

[19]K. Naito, "Safeguards at RRP," presented at the International Nuclear Nonproliferation Science and Technology Forum, Tokyo, May 2006; www.jaea.go.jp/04/np/documents/fr06_naito01_E.pdf.

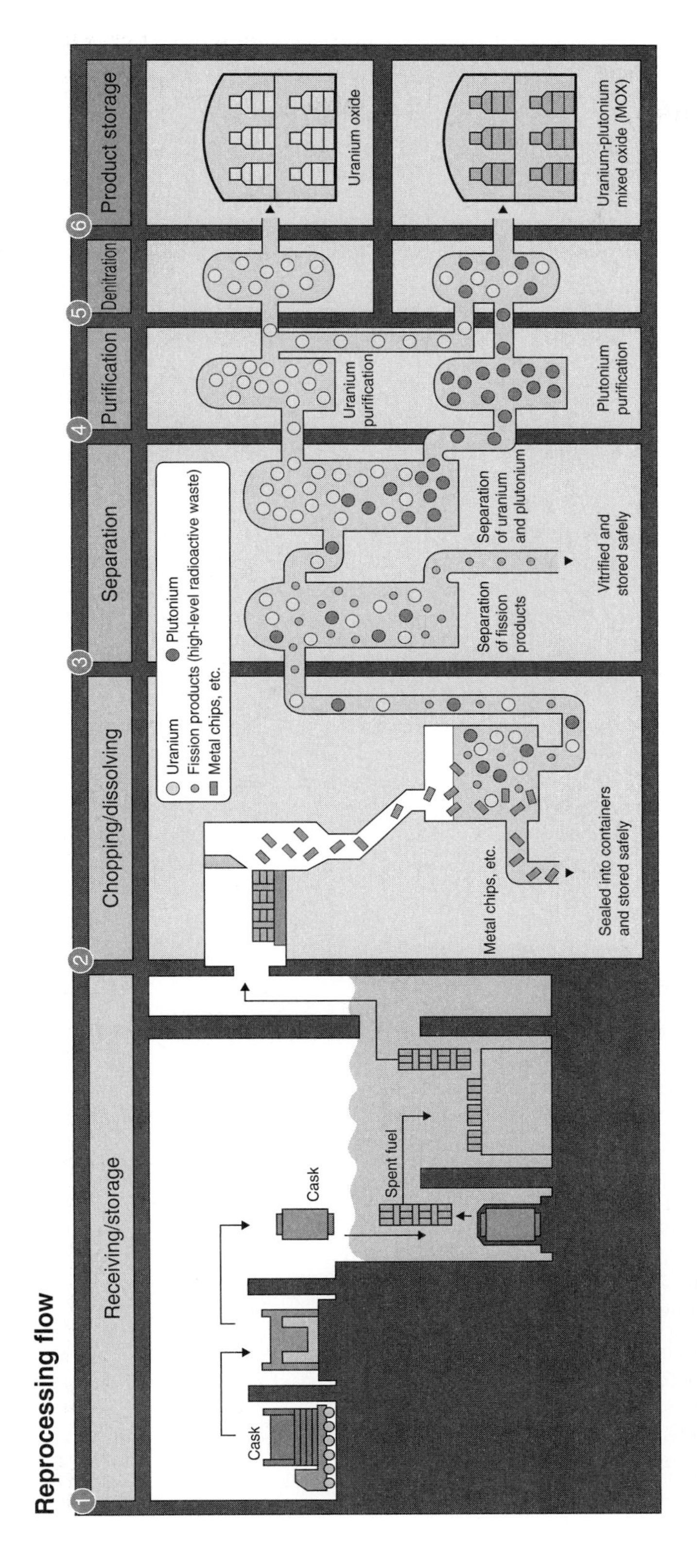

FIGURE 9.2 Reprocessing flow diagram.
Source: Japan Nuclear Fuels Limited.

the main process. Trace amounts are transferred to waste in the washed hulls and, with undisolved "fines," removed in the centrifuge.

The main process employs solvent extraction for the removal of fission products and the partitioning and purification of uranium and plutonium. By adjusting the acidity of the solution, fission products, uranium, and plutonium can be made preferentially to transfer from the aqueous phase to the solvent or vice versa. The main process employs several stages in which aqueous and solvent are mixed and then allowed to separate. In the first stage the fission products are transferred to solvent, leaving the U and Pu in the aqueous phase. In later stages the U and Pu are then purified. To avoid storing separated plutonium, for safeguards purposes, the Pu nitrate is mixed with uranium nitrate in a 50/50 mixture prior to the conversion to oxide. Uranyl nitrate and plutonium nitrate solutions are measured before being transferred to the conversion process. The uranium from the spent fuel is made into UO_3 powder, measured, and transferred to the uranium product storage area. Again, virtually all the U and Pu transferred to the main process are included in the product solutions or powder. The highly radioactive waste from the first separation stage contains about 98% of the fission products. The rest is removed in the U/Pu separation and purification stages. The combined waste streams are concentrated and transferred to a vitrification process.

In the conversion process, mixed U and Pu nitrate solutions undergo a codenitration heat treatment process and are converted into MOX powder for later use in the manufacture of fresh fuel. The MOX powder is collected in sealed cans and weighed on a high-accuracy device. Three cans are placed in a canister and transferred to the product storage area. The canisters are measured with a high-accuracy neutron detectors and have their isotopic composition measured with a high resolution gamma detector. All the Pu, with the exception of very small amounts in wastes, is converted to MOX.

All solid and liquid radioactive wastes are treated and stored in the waste treatment and storage area. This area includes the vitrification process for high active liquid waste. The vitrification process involves mixing concentrated highly active waste with glass in a melter. The waste is incorporated in glass in stainless steel canisters for long-term storage. Lower-activity solid waste is placed in drums or crates for storage.

A characteristic feature of reprocessing plants is that, because of the highly radioactive nature of the material, almost all processes are carried out behind heavy biological shielding, either water or concrete. Operations are carried out remotely, and personnel access to the processing areas is generally not possible.

Safeguards Approach

The approach to safeguarding RRP has involved the state (NMCC and MEXT), the operator (JNFL), and the IAEA. From the initial development of RRP, the state and the operator have worked in consultation with the IAEA to develop an effective safeguards system, employing traditional material accountancy technologies as well as advanced techniques for near-real-time accountancy (NRTA), containment, and surveillance and automated data collection and evaluation.[20]

The IAEA applies a safeguards approach comprising several different techniques in nuclear materials accountancy, process monitoring, inspection and verification, and containment and surveillance (C/S) measures. The combination of information from the various systems taken together allows the IAEA to make technically justifiable safeguards statements for the facility. The safeguards approach comprises:

- Measurement systems that are highly sensitive, reliable, and independent and/or authenticated, namely destructive assay (DA) and nondestructive assay (NDA) technologies.
- Comprehensive design information examination and verification (DIE/DIV) during the design and construction phase as well as verification activities during the

[20]Ibid.

operation of the facility. Such verifications include interim inventory verification (IIV) and physical inventory verification (PIV).
- Measures to provide assurance on facility operations, such as surveillance systems, solution monitoring, and radiation monitors.

The measurement systems incorporate many authentication features such as sealed equipment enclosures, conduits, and transit timing of samples. All material received into the facility as well as all material removed is measured (verification of transfers). Thus a simple account of the quantity of material present is recorded. In a manner similar to financial accountancy, a balance can be quoted for the account at any given time. This is referred to as *book inventory*. The book inventory is determined from the arithmetic accumulation of receipts and issues. Verification of material in the process, either by NDA or DA, is done by measuring the quantity of material physically present at a point in time. The quantity found to be present is referred to as the *physical inventory.*

Additionally, the IAEA will employ short evaluation periods on shipper/receiver differences (SRD) and material unaccounted for (MUF). SRD measures ensure that there is no material unaccounted for between the shipper and receiver of the material. MUF is the term used in nuclear materials accountancy and safeguards as a quality indicator of the control of nuclear materials. It is the physical inventory minus the book inventory. If the value is positive, it will appear that there is a gain of nuclear material; if it is negative, a loss.

To assist in the taking of such inventories, a material balance area (MBA) is identified for a process or part of a process such that all flows in and out are known and well characterized. Additional measurement points, known as Key Measurement Points (KMP), may also be identified based on the type of material, verification approach, and material flows.

The IAEA will also rely on well-trained inspectors, 24-hour inspector presence, full access to facilities and staff, and authenticated unattended measurement systems in continuous operation. There is also integrated data collection and evaluation software as well as the IAEA-JSGO On-Site Laboratory (OSL) that the IAEA can use for destructive analysis sampling.[21]

IAEA Activities

Working with the state and the operator, the IAEA was given access to the facility plans at an early stage to assess the proposed facility design, to ensure that it is compatible with the stated processing techniques and throughputs. During construction, IAEA inspectors verified that the process is built according to the specification.

During the verification process, the IAEA collects the *calibration* data, both independent and operator's data, and subjects them to a series of statistical performance tests. Using these data the IAEA constructs a set of independent calibrations for all relevant vessels (measurement instruments).[22]

The data generated by each safeguards system or technology used by the IAEA need to be *authenticated*. This means that the data themselves and their origin need to be genuine. In the case of IAEA-owned systems, the sensors and the data collection equipment are protected by sealed, tamper-indicating enclosures, including sensor housings, cable conduit, and electronics cabinets.

Key process vessels have measurement systems for level, density, and temperature. This information, combined with the analysis of samples, allows the Agency to follow the

---

[21]S. J. Johnson, et al., "Development of the Safeguards Approach for Rokkasho Reprocessing Plant," IAEA-SM-367/8/01, www-pub.iaea.org/MTCD/publications/PDF/ss-2001/PDF%20files/Session%208/Paper%208-01.pdf.

[22]The IAEA has installed separate, independent measurement systems to allow it to verify the operator's data. Inspectors can take independent samples of materials from the process and analyze them in their separate on-site laboratory. Depending on the stream, the Agency will design a suitable sampling and analysis plan; *the sampling plan is not shared with the operator.*

movement of solutions through the process, to confirm transfers between, and inventories in, MBAs and to confirm the status and presence of material.

Systems of cameras with video recording, radiation detection systems, and seals are implemented to monitor various operations. Particularly in areas where items are handled, such systems are used to detect the potential diversion of material. In key areas, notably product stores, dual systems (dual C/S) will be installed. In dual C/S, two independent systems with no common failure mode (such as a physical seal and a camera system) are installed and analyzed separately.

Rokkasho Reprocessing Plant: Safeguards Technologies

Nuclear Material Accountancy

Nuclear material accountancy technologies have been the fundamental basis of traditional safeguards, providing the necessary information about the material type and quantity at a nuclear facility under IAEA safeguards. Applied to various nuclear facilities and throughout the material processing system, these technologies are used to measure material, whether for SRD measurements, inventory measurements, or others. They remain a critical component in comprehensive safeguards under the NPT as well as for verification activities in states that have signed the Additional Protocol and/or are implementing integrated safeguards approach.

Many of the traditional safeguards technologies, such as neutron measurement instruments (to determine assay of fissionable material) and cameras, have been further developed to address the continuous and often unattended processes inherent to RRP and large-scale processing facilities.

The wide range of advanced nuclear material accountancy technologies, including both DA and NDA, facilitate the inventory and verification procedures. RRP also relies on neutron based (helium-3 tubes and fission chamber) assay systems, some supplemented with high-resolution gamma spectroscopy (HRGS).[23]

Additionally, all RRP NDA systems are equipped with cameras to record and verify the sample ID. This section introduces some of the advanced NDA safeguards technologies installed at RRP, which fall into four general categories:

- Accountability tank samples
- Waste stream assay
- Product assay
- Holdup measurement

Accountability Tank to Denitration

The primary measurement of plutonium concentration in the main process area, from the accountability tank to denitration (see previous diagram), is DA sampling of the liquid stream. The liquid samples from the accountability tank are pneumatically transferred to the OSL for both DA and NDA measurements. Measurements are then taken to provide information on the material from the tanks. The Hybrid K-edge densitometer[24] is used to determine the plutonium and uranium concentrations; curium-244 is measured using neutron counting. The ratio of the curium to the plutonium is then used in the waste assay measurements.

---

[23]The principles of neutron-based methods and HRGS methods are described in detail in Chapter 3 of this volume and in "Passive Nondestructive Assay of Nuclear Materials," by Doug Reilly, Norbert Ensslin, and Hastings Smith; U.S. Nuclear Regulatory Commission, Office of Nuclear Regulatory Research, Los Alamos National Laboratory, NUREG/CR-5550, LA-UR-90-732, Chapters 8 and 13–17.

[24]Ibid, Chapters 9 and 10.

This ratio is an indirect method of determining the mass of plutonium and uranium from an observed curium neutron measurement.[25]

Waste Streams

Numerous systems have been developed to measure the material in the various waste streams at RRP. These systems allow the IAEA and the state to assess and verify the nuclear material, not only at these specific areas but at RRP as a whole. Some of these systems include the Vitrified Canister Assay System (VCAS), Rokkasho Hulls Drum Measurement System (RHMS), the Waste Drum Assay System (WDAS), and the Waste Crate Assay System (WCAS) systems to measure vitrified waste, leached hulls, drummed waste, and crated debris waste, respectively. The RHMS, for example, provides verification of transfers to retained waste of U and Pu in hulls and end-pieces. The RHMS applies the U:Pu:Cm-244 ratios analyzed in the dissolver solution to a passive neutron measurement using Helium-3 detectors located in the operator's active neutron system.

Product Measurements

Measurements are made on the product in two places prior to the blender by the TCVS and on MOX containers before long-term storage by the iPCAS (see Figure 9.3). The Temporary Canister Verification System (TCVS) is an unattended neutron coincidence system designed to measure the plutonium mass in canisters in temporary storage before the blender. iPCAS (the Improved Plutonium Canister Assay System) measures the Pu content and isotopic composition of containers of 36 kg MOX before they are transferred to long-term storage. It is an NMCC-owned system that is used by the IAEA.

FIGURE 9.3 iPCAS with germanium detectors and neutron coincidence counter.

[25]For more information on the curium ratio measurement, see N. Miura and H. Menlove, "The Use of Curium Neutrons to Verify Plutonium in Spent Fuel and Reprocessing Wastes," LA-12774-MS; www.fas.org/sgp/othergov/doe/lanl/lib-www/la-pubs/00285668.pdf, 1994.

Holdup

Another area where NMA technologies are applied is in the glove boxes, where there is a potential for holdup. Measurements on the powder handling glove boxes are made with the Plutonium Inventory Measurement System (PIMS). The PIMS system comprises Helium-3 tubes installed permanently on all glove boxes in the MOX conversion area. PIMS also provides process flow information.[26]

Process Monitoring and Verification

Further assurance that the plant is operating as declared is obtained from process monitoring. Since RRP is primarily an largely automated process with limited inspector access due to the nature of the material being processed, the IAEA has worked with the state and operator to develop process monitoring systems that include advanced monitoring and data acquisition systems. Given the large amount of material processed at RRP, shorter intervals between data acquisition and monitoring are required to ensure no diversion of a significant quantity of material.

The inspectorates (Japanese and IAEA) use a number of monitoring systems on the nuclear material flows, to provide additional assurance that:

- The plant is being operated as declared.
- Possibilities for removal of nuclear material from the process streams are reduced.
- The flow of nuclear material in process can be monitored.
- Verification of the material accountancy system for conventional and near-real-time accountancy (NRTA) in unattended mode is improved or supported.
- Validation of the other safeguards systems, including the automated sampling system, is provided.

At key processes throughout RRP, there are measurement systems for level, density, and temperature. The IAEA collects data from either independent or authenticated operators' systems. NRTA is used to give timely information on the nuclear material balance in the plant and to watch for possible trends. Flow sheet verification is used to give information on the alternative nuclear materials in the plant. This information, combined with the analysis of samples, allows the IAEA to follow the movement of solutions through the process, to confirm transfers between and inventories in MBAs, and to confirm the status and presence of material.

Primary systems for monitoring include:

- *Solution Measurement and Monitoring System (SMMS).* The SMMS/SMS is installed to collect measurements from individual pressure transducers for tank-level and density measurements, temperature sensors, and neutron counters installed on selected vessels in the main process stream through the various facility areas.
- *Integrated Spent Fuel Verification System (ISVS).* ISVS consists of time-synchronized closed-circuit television (CCTV) cameras and radiation detectors. The objective of the ISVS is to maintain continuity of knowledge from the time of discharge from the transport casks into the storage pools, during storage, and eventual movement into the head-end portion of the facility.
- *Integrated Head-End Verification System (IHVS).* IHVS consists of a number of surveillance cameras/radiation detectors (CRDs), including cameras and radiation detectors mounted in the cell walls, ID check cameras, and CCTV units installed in the shearing cell to maintain continuity of knowledge.
- *Plutonium Inventory Measurement System (PIMS).* The objective of the PIMS is to verify the Pu quantity in the glove boxes and monitoring of conversion process operation in unattended mode.

---

[26]T. Iwamoto, "Holdup Measurement in Reprocessing Facility," Institute of Nuclear Materials Management, Holdup Workshop, www.inmm.org/holdup_workshop/2C%20Iwamoto.pdf.

FIGURE 9.4 Camera and radiation detector.

Containment and Surveillance

Containment and surveillance and continuous DIE/DIV inspections help to provide strong assurance that no material is diverted from the plant before the input accountability tank or after the measurement of filled MOX canisters. C/S technologies are implemented to assure the continuing integrity of previously verified materials and confirm the absence of any interference with these materials.[27] Systems of cameras with video recording, radiation detection systems, and seals can be implemented to monitor either process flows or largely static areas. It should be noted that C/S systems are quite labor intensive to review. For large facilities, the number of cameras that require review becomes extensive and at RRP, there are approximately 70 camera systems.[28]

C/S technologies are employed at various points in the facility to ensure that there is no *undeclared* removal of nuclear material. C/S measures are used extensively in the fuel pool, head-end, and product stores. In the spent-fuel storage area, cameras and radiation monitors are used to ensure that there is no undeclared removal of fuel assemblies (or fuel pins) from the storage pool. In the head-end of the plant, cameras, radiation monitors, and solution monitoring are used, together with design information verification, to give high confidence that all the nuclear material that enters the head-end ends up in either the input accountancy tank or in the hulls waste stream.

In other key areas, notably product stores, dual systems (dual C/S) will be installed. In dual C/S, two independent systems with no common failure mode (for example, a physical seal and a camera system) are installed and analyzed separately.

One prime example at RRP is the camera and radiation detector (CRD) for the Integrated Head-End Verification System (IHVS), shown in Figure 9.4. The IHVS is a key component of the safeguards approach, designed to monitor the movement of spent-fuel assemblies and leached hulls through the plant. It is an NMCC-owned system that will be used by the IAEA. The CRD incorporates in one unit a radiation-tolerant video camera and neutron and gamma-ray radiation detectors. Through the use of these sensors and associated

---

[27] M. Zendel, "Experiences and Trends for Safeguarding Plutonium Mixed Oxide (MOX) Fuel Fabrication Plants," *Journal of Nuclear Materials Management*, Feb. 1993.

[28] T. Iwamoto,T. Ebata, K. Fujimaki, and H. Ai, "Establishment of the Safeguards at Rokkasho Reprocessing Plant," Nuclear Material Management Department, Japan Nuclear Fuel Limited (JNFL), Rokkasho-mura, Kamikita-gun, Aomori-ken, Japan; Pacific Basin Nuclear Conference, 2006, www.pacificnuclear.org/pnc/2006-Proceedings/pdf/0610015final00362.pdf.

data analysis software, the direction and speed, as well as the neutron and gamma-ray emission rates, of these highly radioactive materials found in the head end (input) of RRP can be determined.[29]

Summary

The application of international safeguards at a plant such as RRP presents a number of challenges. The measurement uncertainties associated with large throughput can prevent classical nuclear material accountancy reaching conventional target values, and automated operation means that material is not always available for "hands-on" measurements by inspectors. The early and frequent consultation between the operator, the State, and the Agency has helped to identify critical safeguards technology needs and implement safeguards systems in conjunction with the development of the facility. The challenge of automated operation has been met by the installation of high-sensitivity and high-reliability instruments that operate in unattended mode and measure the items in relevant streams. The limitations of conventional nuclear materials accountancy have been supplemented by a number of additional measures. These include the application of containment and surveillance, continuous DIE/DIV during the operation of the plant, continuous inspector presence, and process monitoring. Such systems generate an enormous amount of data, making "intelligent" software for data collection and automated analysis key to safeguarding the facility.

Large-scale facilities require advanced safeguards technologies that will improve the efficiency and effectiveness of safeguards—and continue to provide IAEA a foundation that will allow the IAEA to draw defensible safeguards conclusions. The information from these techniques, taken together with the collaboration of all parties, gives assurance that the Rokkasho Reprocessing Plant is operating as declared and allows the IAEA to make technically justifiable safeguards statements for the facility that extend well beyond the results of traditional nuclear material accountancy.

[29]James Tape, "Personal Views on Integrated Safeguards and the Status of Safeguards R&D in the United States," LA-UR 03-0615, Los Alamos National Laboratory, 2003.

10

Case Study: Nonproliferation Activities at the BN-350 Reactor, Kazakhstan

Mike C. Browne

Introduction

When Kazakhstan declared independence from the crumbling Soviet Union on December 16, 1991, it found itself in possession of a significant fraction of the Soviet nuclear weapons infrastructure. This included 1,040 strategic nuclear warheads mounted on 104 ICBMs, 370 nuclear-tipped air-launched cruise missiles, nuclear material mining and processing facilities, and the largest weapons-testing complex in the world—Semipalatinsk.[1,2] For a brief period, Kazakhstan was the fourth largest nuclear power on Earth.

During the 1990s, Kazakhstan agreed to eliminate its nuclear weapons inheritance. All nuclear warheads and delivery vehicles were transferred to Russia;[3] all ICBM launch silos and a large number of testing facilities at the Semipalatinsk Nuclear Test Site were dismantled.[4] Shortly after declaring independence, Kazakhstan signed the Nuclear Nonproliferation Treaty as a nonnuclear state and completed a safeguards agreement with the International Atomic Energy Agency (IAEA). The country's posture was and continues to be one of nonproliferation and the strictly peaceful use of nuclear technology.

In 1991, the U.S. Congress initiated the Nunn-Lugar Cooperative Threat Reduction (CTR) Program to help former Soviet republics dismantle their nuclear, chemical, and biological weapons stockpiles. As part of the CTR Program, Kazakhstan began working with the United States in 1993 to place its civilian nuclear facilities under IAEA safeguards. In 1994, the U.S. and Kazakhstan governments agreed to secretly airlift approximately 600 kilograms (kg) of highly enriched uranium to the United States, an action referred to as Operation Sapphire.[5]

---

[1]"Politics and Policy," Embassy of Kazakhstan to the U.S. and Canada, Oct. 30, 2006, www.kazakhembus.com/NuclearDisarmament.html.

[2]N.-O. Bergqvist and R. Ferm, "Nuclear Explosions, 1945–1998," FOA-SIPRI User Report, July 2000.

[3]The Nuclear Threat Initiative, "Kazakhstan Nuclear Facilities: Nuclear Weapons," www.nti.org/e_research/profiles/Kazakhstan/Nuclear/4278_4316.html, updated Aug. 2004; Oct. 30, 2006.

[4]Jon Brook Wolfsthal et al., "U.S. Nonproliferation Assistance Program," *Nuclear Status Report: Nuclear Weapons, Fissile Material, and Exports Controls in Former Soviet Union*, vol. 6, 2001.

[5]Nuclear Threat Initiative, "Country Overviews: Kazakhstan: Nuclear Overview," www.nti.org/e_research/profiles/Kazakhstan/Nuclear/index.html, Sept. 2006.

FIGURE 10.1 The BN-350 facility.

Kazakhstan's willingness to disarm and dispense with its high-risk nuclear materials has been attributed to many causes, one of which is the Kazakhs' strong antinuclear sentiment fostered during the Cold War. However, the country has not eliminated its enormous uranium mining, processing, and reactor fuel production capacity. Kazakhstan's uranium-handling infrastructure is too viable an income source to dismantle, so the nation began positioning itself to become one of the world's major suppliers of reactor-grade uranium fuel assemblies.[6] The international safeguards community has applauded Kazakhstan for its positive attitude toward nuclear safeguards.

The BN-350 reactor (see Figure 10.1) was part of a five-reactor series designed, built, and operated as technology demonstration facilities for a Soviet program to develop a multi-use fast breeder reactor. The design of each BN-series reactor includes various experimental technologies and manufacturing techniques.[7] The goal of the program was to develop an economically viable reactor design that would serve a dual purpose—a source of electricity for civilian use and a source of plutonium for the Soviet nuclear weapons program.

The BN-350 reactor cooling system was of the loop type containing liquid sodium coolant, where the primary heat exchangers and pumps are located outside the core containment vessel. The steam generators used to produce electricity from the thermal output of the reactor were required to boil water using the heat contained in the secondary loop's liquid sodium.

The BN-350 was operational from 1972 to 1999. The reactor was designed for a thermal output of 1,000 MW and a resulting electrical output of 350 MW. However, the safety constraints on the plant's operation kept the maximum operational thermal output at 750 MW.[7] The electrical output of the facility was used to power the city of Shevchenko (renamed Aktau in 1992), a city of 143,000 people, according to the 1999 census.[8] Some low-pressure steam was siphoned off at the electrical turbines and used in the adjacent sea-water desalinization facility.

---

[6]Nuclear Threat Initiative, "Kazakhstan: Uranium Mining and Milling," www.nti.org/db/nisprofs/kazakst/fissmat/minemill.htm, Dec. 26, 2000.

[7]Argonne National Laboratory, "Design Description of Soviet Liquid-Metal-Cooled Fast-Breeder Reactors," ANL/SETC-90/1, vol. 1, 1990.

[8]The name of the city was originally Aktau but was changed to Shevchenko by the Soviets in 1964 to honor a Ukrainian poet who had been exiled there. The name was changed back to Aktau in 1992.

The BN-350 required highly enriched uranium in its fresh fuel rods as a result of its fast breeder reactor design. The spent fuel contained significant quantities of plutonium that were intended to be separated and used in the Soviet nuclear weapons industry. The existence of both highly enriched uranium and plutonium in the remaining fresh and spent-fuel assemblies at the BN-350 make it a very important facility for the implementation of nonproliferation safeguards. This is especially true considering the remote nature of the site and its proximity to nations known to have a history of nuclear proliferation activities.

In 1999, the BN-350 reactor was shut down and subsequently decommissioned.[9] At this time there were still 2,900 kg of fresh highly enriched reactor fuel and an estimated 300 metric tons of spent reactor fuel (containing significant quantities of plutonium) in storage at the BN-350 facility.[10] Existing IAEA safeguards were upgraded to monitor the facility during the interim period between decommissioning and the transfer of the fuel away from the BN-350.

In 2001, a joint effort between the Kazakh government-owned National Atomic Company Kazatomprom and the private U.S. Nuclear Threat Initiative was undertaken to remove the fresh fuel from BN-350 and have it down-blended. This project culminated in October 2005, when the last of the fresh fuel was transferred from BN-350 to the Ulba Metallurgic Facility, where it was down-blended to low-enriched uranium reactor fuel.[11]

The remaining proliferation threat at the BN-350 is the ~300 metric tons of spent fuel lying in special storage canisters in the on-site storage pond. The goal is to move the spent fuel to an outdoor intermediate storage site at the BN-350 and eventually to a long-term or permanent storage facility elsewhere.

Unfortunately, the design of the BN-350 facility was distinctly Cold War vintage, with little thought given to nuclear materials control and accounting practices. Implementing robust nuclear safeguards at the facility toward the end of the reactor life cycle was and continues to be a major challenge for the IAEA, considering that the facility was not designed with safeguards in mind.

Overview of Safeguards History and Approach at the BN-350

The initial approach to safeguarding the BN-350 was based on a proven, traditional IAEA approach of inventory verification followed by dual containment and surveillance (C&S). At the BN-350, inventory verification was at first limited to item counting. No comprehensive radiation-based measurements of the quantity of nuclear material occurred initially. The dual C&S was made up of seals and video data. During scheduled inspections, seals were examined, video data were reviewed, and operator declarations were studied.

The first unattended safeguards systems were installed in 1993 under the aegis of the IAEA. This system applied unattended electronics coupled to radiation detectors that utilized ionization chambers for gamma-ray detection and ^{3}He neutron detectors to monitor the loading and discharge of the reactor core.

This was followed by the installation of an unattended domestic safeguards system for the facility operator starting in 1996 and ending in 1998. This system did not involve the IAEA or its inspectors and was intended to allow the Kazakh facility operator to ensure that

---

[9] "Republic of Kazakhstan: Nuclear Power Reactors," Power Reactor Information System, IAEA, April 24, 2005; www.iaea.org/programmes/a2/.

[10] J. P. Lestone, J. M. Pecos, J. A. Rennie, J. K. Sprinkle Jr., P. Staples, K. N. Grim, R. N. Hill, I. Cherradi, N. Islam, J. Koulikov, and Z. Starovich, *Nucl. Inst. and Meth.*, A490, 409 (2002).

[11] "Government of Kazakhstan and NTI Mark Success of HEU Blend-Down Project; Material Could Have Been Used to Make Up to Two Dozen Nuclear Bombs," NTI Press Release, Oct. 8, 2005; www.nti.org/c_press/release_Kaz_100805.pdf.

no material was being diverted. The system gave the operator confidence in the security of the material and provided a development test bed for advanced unattended safeguards equipment that the IAEA would use to perform international monitoring of the facility.

When the decision was made to empty the reactor and package the assemblies into proliferation-resistant canisters, the IAEA required a system to monitor this activity, to have confidence that Kazakhstan was meeting its safeguards obligations. This phase used an unattended system that was similar to the system that was installed for the facility but was completely separate from that system, with enhanced capabilities and a different focus. As the packaging proceeded, two attended systems were used to characterize the assemblies and the canisters. The data obtained by these systems will be used later to ensure that the spent fuel has not been replaced between inspections by other neutron-emitting material, thus allowing a diversion of Pu-bearing spent fuel.

Once the assemblies were packaged, they were placed back in the spent-fuel pond. Monitoring was performed during this time using the same IAEA system that was used for the packaging, but with some additional capabilities.

The initial phase of the long-term storage plan for the fuel involves packaging the canisters into dry storage casks. These casks will be placed temporarily at the site and then eventually moved to a final off-site location. The safeguards system for this process has more stringent requirements and is still in development.

The safeguards approach for the long-term storage process for the BN-350 spent fuel consists of a complementary system of attended and unattended safeguards designed to provide both qualitative and quantitative information. Combined, these systems provide a baseline material characterization that can be used for later recovery of the requisite continuity of knowledge (CofK) as well as providing information that can be combined with the continuous monitoring data to provide both the facility and the IAEA with high confidence that no nuclear material has been removed or diverted.

The following sections describe in detail the application of the attended and unattended safeguards for the BN-350 spent fuel.

Attended Safeguards

The goal of the attended safeguards at the BN-350 is to directly measure and characterize the Pu content of every spent-fuel assembly and enable the subsequent confirmation of this information throughout the safeguards lifetime of this material. The measured Pu content was compared to facility records to establish a total material inventory baseline. Additionally, these measurements are to be used later in the project with supplemental measurements to confirm material presence according to IAEA requirements. To meet these requirements, two detector systems were designed that provided accurate measurement of Pu content in the spent fuel. The detectors were designed to perform measurements on the material at critical stages during the repackaging process so that subsequent measurements could always reconfirm previous measurements as the process was completed.

Accurate nondestructive assay of the plutonium content was possible because, even at the highest fuel burnup levels, the buildup of the curium isotopes remained very low. Thus the dominant source of neutron emissions from the fuel was from ^{238}Pu and ^{240}Pu, with negligible neutron emissions from ^{241,242}Pu as well as the curium isotopes.[10] To determine the plutonium content within each individual spent-fuel assembly, neutron-coincident counting measurements were performed on each fuel assembly while the assemblies were under water in the facility storage pond. These measurements were accomplished by using the Spent Fuel Coincidence Counter (SFCC), which consisted of a single ring of 20 ^{3}He proportional counters, each with a pressure of 4 atm (standard atmospheric pressure at sea level), a diameter of 2.54 cm, and an active length of 30 cm, along with a single ionization chamber embedded in polyethylene, as shown in the schematic drawing presented in Figure 10.2a. To ensure that the SFCC would properly function in the extremely high gamma-ray radiation level that was associated with the spent-fuel assemblies, a 6.8 cm thick lead inner ring was built into the counter, as shown as the black region in Figure 10.2a. The ionization chamber was used to

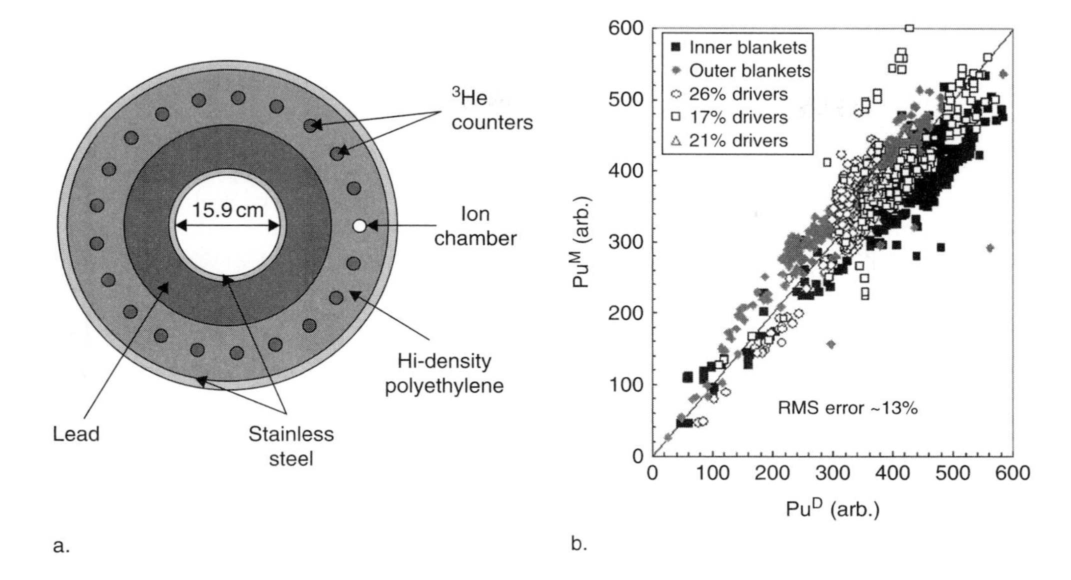

FIGURE 10.2 a. Schematic drawing of the SFCC. b. Measured plutonium masses using a single SFCC measurement at the midplane of an assembly as a function of facility-declared plutonium masses for ~1,600 spent-fuel assemblies. (Taken from the paper by Lestone, et al.)[10]

ensure that the gamma-ray radiation level at the location of the ^{3}He tubes was low enough (<50 Rads/hr) so that the ^{3}He tubes were insensitive to the gamma ray flux. This upper limit of 50 Rads/hr corresponded to a surface contact gamma-ray radiation level from the spent-fuel assemblies as high as10^5 Rads/hr.[10]

To convert the measured singles and doubles (coincidence) neutron rates into the plutonium content of a given assembly, extensive modeling of the spent fuel was performed to determine the relative composition and location of the various neutron-emitting isotopes as a function of the total integrated neutron flux for each of the various BN-350 assembly types. The REBUS-3 fuel-cycle analysis code[12] was used to determine the specifications of the neutron irradiation field to which an assembly was exposed throughout its lifetime, based on an equilibrium cycle that represented typical BN-350 fuel-loading and discharge patterns. In conjunction with the REBUS-3 code, the ORIGEN point depletion code[13] was used to determine the spatial details of the isotopics within each assembly in the reactor.

From this comprehensive modeling of the BN-350 spent fuel, the distribution of plutonium throughout a given assembly as well as the relationships between the plutonium linear density (ρ_{Pu}), the ratio of neutrons from (α,n) reactions to neutrons from spontaneous fission in the assembly (α), and the linear density of the effective ^{240}Pu (ρ_{240}') could be determined for a given assembly, depending on the assembly type and its location within the reactor core. A more detailed description of the extensive modeling of the BN-350 spent fuel can be found in the paper by Lestone, et al.[10]

Based on the results of this modeling effort as well as the extensive testing and modeling of the SFCC detector using the neutron transport code MCNP,[14] an iterative analysis

---

[12]B. J. Toppel, "A user's guide for the REBUS-3 fuel cycle analysis capability," Argonne National Laboratory Report ANL-95/40 (1995).

[13]Oak Ridge National Laboratory, "ORIGEN: isotope generation and depletion code-matrix exponential method," Radiation Shielding Information Center Report CCC-217 (1977).

[14]Judith F. Briesmeister (Ed.), "MCNP™: A General Monte Carlo N-Particle Transport Code, Version 4A," Los Alamos National Laboratory Report LA-12625-M (1993).

procedure was developed based on the technique of neutron coincidence counting corrected for self-multiplication effects.[15] This analysis technique converted the measured singles and doubles counting rates into the linear Pu density within an assembly based on either multiple measurements along the length of a single assembly or a single measurement at the midplane of the assembly as well as a minimal amount of facility declaration information. Figure 10.2b shows the results of the measured plutonium content using a single SFCC measurement at the midplane of the assembly as a function of the facility-declared plutonium content for approximately 1600 spent-fuel assemblies.[10] The root mean squared percentage difference (RMS%) between the measured and the facility-declared plutonium content was 13.4%, with the major contributors to the difference between measured and declared values being the variation of the isotopic relationships on the initial enrichment and assembly position in the BN-350 reactor (which were not considered in the analysis procedure), the variation of the distribution of fissile isotopes within the reactor following different fuel reloadings, and uncertainty in the facility declarations. A smaller RMS% difference between facility-declared and measured plutonium masses of 8.2% was observed for 34 assemblies when multiple SFCC measurements were made along the length of an individual assembly, combined with more detailed spent-fuel assembly knowledge. The sum of the measured plutonium masses for the approximately 1,600 spent-fuel assemblies that were only measured at the midplane of the assembly indicated a −1.9% bias in the measurements relative to the sum of the facility-declared values for these assemblies. For the purposes of the inventory baseline, these measurements were in sufficient agreement with facility declarations. For detailed information on the fuel assemblies and reactor design, see Reference 7.

Upon completion of the initial Pu inventory, the assemblies were packaged into welded steel canisters that held either four or six individual spent-fuel assemblies, depending on the type of assembly. This packaging of the spent-fuel assemblies was performed to improve their stability for long-term storage. In addition, because many of the assemblies had cooled to the point where the level of radioactivity was no longer a theft deterrent, the packaging also decreased the proliferation risk by increasing the radiation dose as well as making movement more physically difficult due to increased bulk and weight.

Although the packaging of the assemblies was essential for their long-term secure storage, a new series of measurements was needed to maintain the capability to verify the presence of the spent-fuel assemblies inside the welded steel canisters. These measurements were performed using the Spent Fuel Attribute Monitor (SPAM), a neutron coincidence counter very similar in design to the SFCC.[16]

The SPAM consisted of 15 ^{3}He proportional counters along with two ionization chambers, as shown in the schematic drawings of SPAM presented in Figure 10.3. As was the case of the SFCC, a 7.62 cm thick inner ring of lead was present in the SPAM to reduce the gamma dose at the location of the ^{3}He counters.

Because each canister contained a mixture of different driver and blanket assemblies from the BN-350 core, no attempt was made to measure the absolute mass of Pu that was present inside the welded steel canister. Instead, the analysis of the neutron coincidence data from the measurements of the welded steel canisters using the SPAM yielded an attribute that was proportional to the total amount of Pu mass that was present inside the canister. This measured canister attribute (CA) was then recorded into a database and stored as a baseline for future verification measurements on that canister that would be performed using the SPAM detector.

---

[15]D. Reilly, N. Ensslin, H. Smith Jr., and S. Kreiner (Eds.), *Passive Nondestructive Assay of Nuclear Materials*, Office of the Nuclear Regulatory Commission, Washington, D.C., NUR-EG/CR-5500 (1991).

[16]Parrish Staples, John Lestone, and David G. Pelowitz, "Spent Fuel Attribute Monitor: SPAM User's Manual," Los Alamos National Laboratory Report LA-UR-01-6610 (2001).

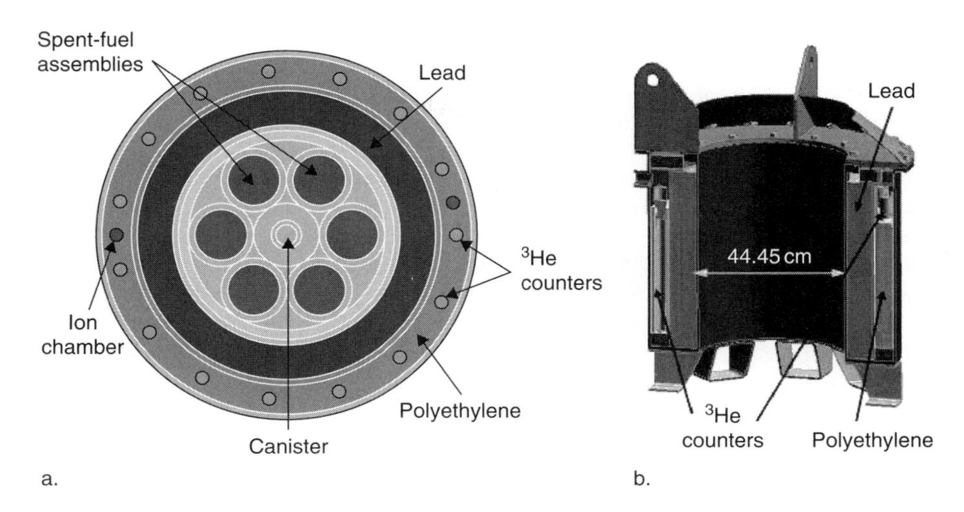

FIGURE 10.3 a. Top view of the SPAM counter with a welded steel canister containing six BN-350 spent-fuel assemblies. b. Side view of SPAM counter.

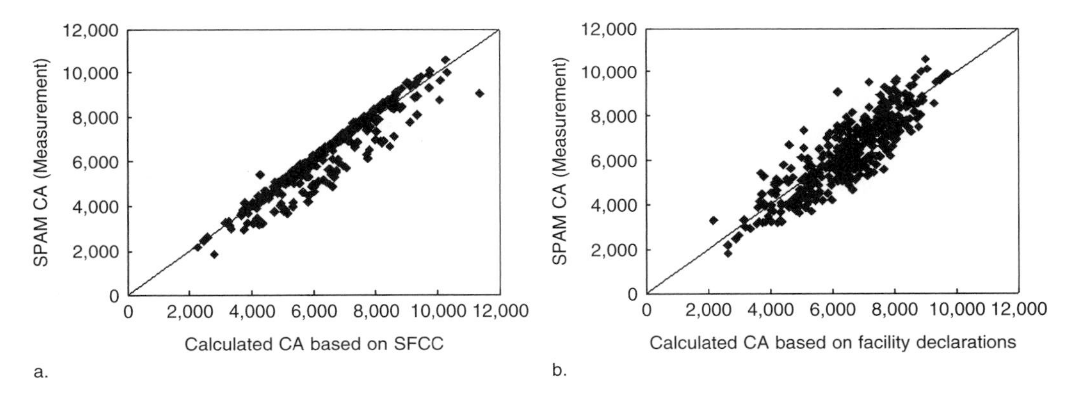

FIGURE 10.4 a. Measured CA value using SPAM for a canister containing BN-350 spent fuel as a function of the calculated CA value for the canister based on the SFCC measurements. b. Measured CA values as a function of the calculated CA value based on the facility declarations.

Figure 10.4a shows the measured CA values using the SPAM as a function of the calculated CA values based on the SFCC measurements of the individual spent-fuel assemblies that comprised a canister for approximately 400 canisters. In general, the data shows an excellent agreement between the measured and the calculated CA values, as evidenced by the fact that nearly all the data points shown in Figure 10.4a lie very close to the straight line corresponding to an exact agreement between the measured and calculated values. An additional comparison between the measured CA values and those calculated from the declared values for the assemblies in a given canister shown in Figure 10.4b demonstrates the traceability of the information from the initial declarations to the SPAM measurements. This comparison between measured CA values and those calculated based on the facility declarations shows a similar correlation between the measured and calculated CA values but with a larger spread in the data than what is observed in Figure 10.4a.

Although ensuring that the CA accurately represents the amount of Pu in a canister is important, the essential component of the SPAM measurements is the precision with which the measurements are made.

The better the precision with which the CA is established for an individual canister, the more sensitive the value of the CA will be to the removal of material from the canister. Because the lower group of data points shown in Figure 10.4a is offset from the straight line due to noise problems with the SFCC measurements rather than any issues with the SPAM measurements, a reasonable method for determining the precision of the CA values is to exclude the approximately 70 data points that are associated with the noisy SFCC measurements. The RMS% difference between the measured and the calculated CA values is then 3.8%.

Unattended Safeguards

Safeguards Instruments and Software

The IAEA visits the BN-350 facility every 90 days to retrieve and review data collected by the Unattended and Remote Monitoring (UNARM) system that is monitoring the spent fuel in temporary storage. Because the monitoring instruments are left unattended for extended periods of time, they must possess capabilities and attributes not normally found in off-the-shelf instruments.

The IAEA has worked extensively with Los Alamos National Laboratory (LANL) and other technical institutions over the past 40 years to develop a suite of instruments that meet their needs. The result has been a family of instruments that are based on the same design principles and share design and operational philosophies and details. The basic radiation signatures utilized by these instruments are described in Chapter 3 on nuclear material measurements, and some of the characteristics of the hardware and software are described in Chapter 6 on unattended monitoring systems.

The instruments and software used at the BN-350 are drawn from this set and are described in the following sections.

BN-350 UNARM System Devices

The Gamma Ray and Neutron Detector (GRAND3) electronics connects to two ionization chambers for measuring large gamma fluxes and to three pulse-counting detectors, such as ^{3}He tubes, for measuring neutrons, or plastic scintillators for measuring gammas. It contains a microcontroller that filters the incoming data, adapts to changing signal levels, detects radiation events, and communicates with a collection computer. It is highly configurable and contains internal storage for recording measurements taken while not connected to the collection computer. The GRAND3 is the size of a briefcase.

The GRAND3 was developed in the late 1980s and still sees considerable use, though it is no longer manufactured. At the BN-350, the GRAND3 was used to monitor core loading and discharge, and it is used to detect illicit movement of materials to and from the hot cell during the temporary storage phase of the disposition program.

The GRAND3 has been superseded by the MiniGRAND, a more modern instrument that is smaller, contains a faster processor and more memory, uses less power, and has enhanced data acquisition capabilities that are a superset of those in the GRAND3. The MiniGRAND is smaller than the GRAND3, at 4 inches × 4 inches × 8 inches, making it easier to integrate into detector systems.

At the BN-350, the MiniGRAND is used in two types of detectors; a fixed-area radiation monitor (FARM) and a characterization radiation monitor (CHARM). A FARM provides a measure of the gross amounts of radiation present in an area of the facility. FARMs typically contain ionization chambers to measure gamma radiation and ^{3}He tubes to measure neutrons. Sometimes other types of detectors are substituted, such as Si for gamma measurements. A FARM cannot supply information about the characteristics of the radiation present, merely the information that certain levels are present.

Characteristics of Instruments Used in UNARM Systems

The following are characteristics of instruments used in UNARM systems:

- *Reliability.* Instruments must not fail during the inspection period, resulting in a loss of data or requiring a visit by an inspector.
- *Low-power operation.* Instruments must be able to operate on their internal batteries for an extended period in the event of a mains power loss. In addition, because the instruments are frequently packaged inside tamper-proof enclosures, low-power operation limits the amount of heat that must be dissipated.
- *Configurability.* Operations, both nominal and of significance to safeguards, vary greatly among different facilities, and instruments must be sufficiently configurable to take such variances into account.
- *Adaptability.* The instrument must be able to recognize events, even in changing radiation conditions.
- *Data reduction.* Over an extended inspection period, instruments may capture large volumes of data, much of which is statistically identical and reflects nominal facility conditions. By "compressing" statistically identical data while keeping all the data during and around a safeguards-significant event, the amount of data that must be saved and reviewed is reduced.
- *Triggers.* The instrument must be able to signal to an external device that an event is occurring.
- *Security.* The integrity and validity of safeguards data must be assured in order for the IAEA to draw their conclusions.

UNARM Instruments

A CHARM is used when the characteristics of radiation, in addition to quantities, must be known. A CHARM contains a MiniGRAND in combination with differentially shielded ionization chambers and ^{3}He tubes. The shielding allows different MiniGRAND channels to detect different amounts of radiation, which allows inspectors to draw conclusions about the type of material present.

The μGRAND (microGRAND) is the next-generation MiniGRAND, at a substantially lower cost. It is still smaller and uses less power than the MiniGRAND and currently contains four pulse-counting channels, though current-sensing capabilities (for use with ionization chambers) can be added in the future. It has the same configurability, autonomous operation, and internal storage required for UNARM system operation. The uGRAND can be used anywhere a MiniGRAND would be used to perform pulse-counting applications.

Though the MiniGRAND-type instrument can make some energy discrimination using shielding, it cannot measure the energy spectrum of the radiation present. In some cases it is necessary to be able to monitor and capture the spectrum of the radiation. This capability is provided by the MiniADC, a small ($4 \times 4 \times 8$ inches) 1,024-channel multichannel analyzer that also contains unattended-monitoring characteristics.

A user can configure a MiniADC to define up to five energy-level regions of interest (ROIs) within which absolute counts can be measured and among which ratios can be calculated; triggers may be configured based on ROI values and ratios. Like the MiniGRAND, the MiniADC incorporates the configurability, autonomous operation, and internal storage required for UNARM system operation. While the MiniADC typically saves only the ROI measurements, the MiniADC can also be configured to save spectra as well when certain events occur.

In addition to radiation data, video surveillance is used as another layer of monitoring. The DCM-14 is a digital camera module used by the IAEA for surveillance. Although it can be configured to operate standalone with configurable image rates and triggers based on motion in a frame, in an UNARM system it is connected to a device that receives event messages from instruments and issues trigger signals to the DCM-14 to make it capture images much more frequently. The DCM-14 timestamps, authenticates, and, optionally, encrypts all images.

An UNARM system frequently consists of instruments, cameras, computers, and other sensors distributed throughout a facility that must communicate with each other. Communications among these devices is effected using an intelligent local node (ILON). The ILON provides the means by which collection computers communicate with instruments, instruments trigger cameras, events are logged, and devices in the system are time-synchronized. The ILON also collects data from electronic seals and provides the failover capability for redundant collection computers.

Each computer, instrument, camera, or other sensor in the system is connected to an ILON, and all ILONs are typically connected via twisted pair, though many other types of physical media (RF, power-line, RS-485, and the like) may also be used. The ILONs communicate with each other using a networking protocol developed by Echelon Corporation that includes authentication on the messages.

The auxiliary communications device (ACD) is an update to the ILON. It contains a more powerful processor, has more internal memory, operates at lower power, and provides stronger authentication for network messages. It is a drop-in replacement for the ILON.

To be able to compare data collected by different instruments, cameras, and other sensors, the devices must be time-synchronized. At the BN-350, this is achieved by connecting an ILON to a Trimble Accutime 2000 GPS antenna. The Accutime 2000 receives time information and passes it to the ILON, which then broadcasts it to the other ILONs in the system, which then synchronize the times in their attached instruments. All devices in the UNARM system, including the collect computer and the video server to which the DCM-14 sends data, may be synchronized in this fashion. The Accutime and ILON may also be used to track the position of a system component that moves from one location to another.

Balanced magnetic switches (BMS) are contactless switches that can be used to detect the opening of a door. The two halves of the BMS are constructed such that the magnetic fields of the two sides are matched, making it exceedingly difficult to defeat the switch. The BMS output signal is fed into the input port of an ILON or ACD that is configured to send event messages when the state of the BMS changes.

The data from the sensors in the UNARM system are ultimately collected and stored on a central collect computer. At the BN-350, the collect computer runs a program called Multi-Instrument Collect (MIC), which polls the instruments for their data and stores the data on a redundant array of independent disks (RAID). The RAID provides data redundancy by distributing the collected safeguards data across a number of disks in such a manner that if one disk fails, no data are lost, and the failed drive data can be rebuilt automatically when a new disk is hot-swapped into the RAID. The RAID also has redundant power supplies and fans and are simultaneously connected to the hot and cold redundant computers in the UNARM system.

At the BN-350, the collect computer is actually a pair of computers, one in operation and the other in cold standby. A failover box is used to control power to the computers. The failover box contains an ILON, which receives heartbeat messages from MIC on the active computer, and power-switching circuitry. If the heartbeat messages cease, the ILON attempts to restart the active computer. If the restart fails, the failover box shuts down the active computer and starts up the cold-standby computer.

In addition to polling the instruments in the UNARM system and retrieving data from them, MIC generates overall state-of-health (SOH) information that provides snapshots in time of the system's operation. A set of MIC-related utilities moves data between computers as needed for archiving and review.

Review is the generic name for a suite of tools used by inspectors to examine the different types of data the UNARM system produces. Data to be reviewed may consist of radiation traces from instruments, images from cameras, position data from a GPS, operator declarations of facility operations, binary trigger data, and other data generated by the UNARM system.

Detailed Descriptions of the Safeguards Systems

Initial Facility Safeguards

In 1996, the U.S. Department of Energy (DOE) funded LANL to design and install the radiation-monitoring component of a safeguards system for the BN-350 facility (see Figure 10.5), called the Rapid Response (RR) System. RR also included physical safeguards that were implemented by Sandia National Laboratory. The intent of the RR system was to allow the operator to be aware of any radioactive material leaving a perimeter defined as the reactor hall, the spent-fuel pond where discharged fuel was stored, and the hot-cell, where materials research was performed. The system was designed to cover entry and exit points as well as out-of-the-way locations where material might be temporarily stored for later diversion.

The radiation component of the system consisted of instruments connected over an ILON network to a collect computer in the Central Alarm Station (CAS). The CAS was manned by security personnel who are trained to take notice of radiation events within the facility. The system also incorporated locking turnstiles at the "clean" entry and exit points of the building as well as an instrument to monitor a "dirty" door.

The instruments installed for the RR system were FARMS and CHARMs, MiniADCs with NaI detectors, DCM-14s, and a sonar device.

FARMs and CHARMs were placed in the hot cell to monitor the movement of assemblies into and out of the room and to characterize the material in the assemblies. Ionization chambers connected to a GRAND3 were installed in tubes that extended beneath the water surface in the spent-fuel pond area, to monitor movement of assemblies through the channel to and from the hot cell; the direction of the assemblies' motion could be determined by examining the radiation traces. Cameras were installed in the spent-fuel pond area to provide video coverage of activities above the pool. In addition, a sonar device was placed into the channel near the ionization chambers to detect any attempts to move assemblies under cover of darkness (blind cameras) or to shield the underwater ionization chambers.

An alarm panel was installed in the spent-fuel control room for a while, but since it alarmed whenever assemblies were moved, which occurred frequently, the operators eventually had it removed. Room 122, the exit from the hot cell to the outside, had two instruments installed: a MiniGRAND and a MiniADC. These instruments triggered colocated cameras. In the hall at the clean entrance to the building, a pair of detectors was installed to provide signals to a MiniADC that was to lock the turnstiles when radiation above a certain threshold was detected.

Two collect computers were placed in the CAS; the computers ran Windows NT and MIC and shared a SCSI connection to a RAID containing 20 GB of storage. The computers were configured identically, even to their Ethernet address, and only one was turned on at a time. In the event of a failure of the operating computer, a failover box turned off the failed computer and turned on the other computer. Because the computers were configured identically and contained identical software, the switchover was transparent to the rest of the system.

The collect computers were connected by Ethernet to two review computers in the engineer's offices. From the review computers, technicians could examine the radiation data as well as images taken by the cameras in the facility.

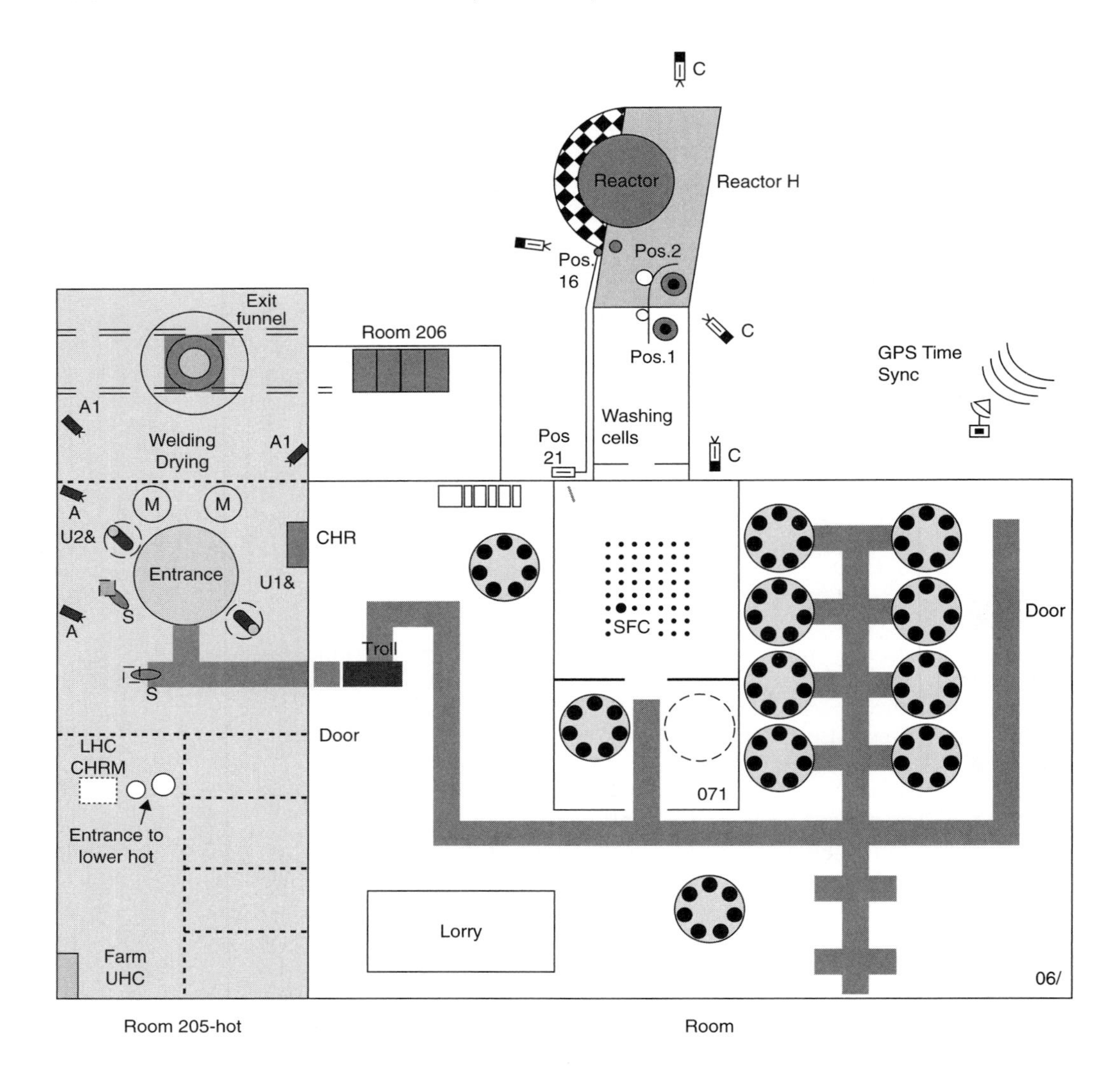

FIGURE 10.5 Schematic representation of the BN-350 facility and the IAEA monitoring equipment (UNARM system).

Packaging

With the advent of packaging the assemblies into canisters, the IAEA contracted with Los Alamos to install a parallel safeguards system that would allow the IAEA to monitor the packaging operation. This system used many of the same components as the RR system but packaged and located them differently.

The biggest difference between the systems was the use of integrated detector packages. In the RR system, the detectors were separated by some distance from the instruments to which they were connected. For example, the detectors in the hot cell were connected to a GRAND3 in the spent-fuel pond area over a distance of about 20 feet. This allowed the possibility of tapping into the detector cables and injecting a false signal to the instrument. To prevent this in the IAEA system, the instruments and detectors were packaged together, with a network node, battery and battery charger, shielding, and moderating material in one enclosure. The enclosure could be sealed by the IAEA to guarantee that no tampering occurred, and the only cables accessible to an adversary were the network cable, which carried authenticated data, and power (see Figure 10.6).

For the unattended components of the system, MiniGRAND instruments in integrated detectors were used, along with DCM-14s, GRAND3s, and ILON network nodes. See Figure 10.5.

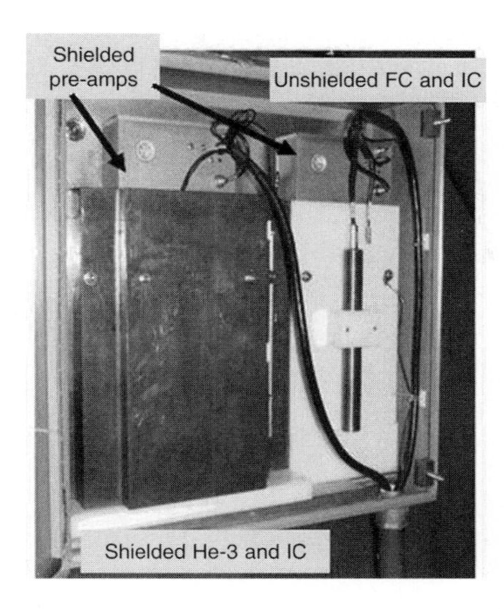

FIGURE 10.6 Example of an integrated detector installed at the BN-350 showing differentially shielded ionization chambers (IC) and fission chambers (FC).

For the attended part of the system, which characterized each assembly before packaging and each canister after packaging, a GRAND3 and intelligent shift register (ISR) were used. The ISR was connected to the spent-fuel coincidence counter for assembly measurement and the SPAM counter for canister measurement.

Integrated detectors were installed on the walls of the hot cell to monitor the assemblies as they were moved to the welding stations and the canisters as they were moved back to the spent-fuel pond. CHARMs were installed near the main material entry and exit points in the hot cell, and a FARM was installed on a wall in the far corner of the hot cell, where it could oversee all activities.

Two more integrated detectors were located beneath the floor of the hot cell, below the water at the material entry point. Each of these contained a MiniGRAND, ionization chamber, and DCM-14 and monitored movement of assemblies and canisters in the area near the entry funnel. See Figure 10.7.

DCM-14s were installed in the spent-fuel pond area, watching the area where the assemblies and canisters were accessed, and in the hot cell, watching the entrance and exit funnels and the welding stations where the assemblies were placed into canisters and then welded shut.

The camera units were configured to take images at regular intervals of 1 to 30 minutes and were triggered by the instruments when the instruments detected radiation events. When triggered, the DCM-14s would capture images at a more rapid rate, 5 to 10 seconds between images, until the event ended. In this way the IAEA could monitor the facility constantly at low time resolution when nothing of interest, from a radiation perspective, was happening and then monitor much more closely when a radiation event was occurring.

Another integrated detector containing a ^{3}He slab detector and an ionization chamber was installed in Room 122 to monitor the movement of material from the exit funnel. This detector is intended to discriminate between legal declared movements and unauthorized movements of neutron-emitting material.

A lower hot cell lies beneath the main hot cell, accessed by a small port in the main hot cell. To prevent material from being surreptitiously dropped into this area and retrieved later, a miniGRAND with an ionization chamber and ^{3}He tube was installed to monitor the lower hot cell.

All radiation data, triggers, and time-synchronization messages traveled over the ILON network that connected all instruments, cameras, and computers. An ILON with a Trimble

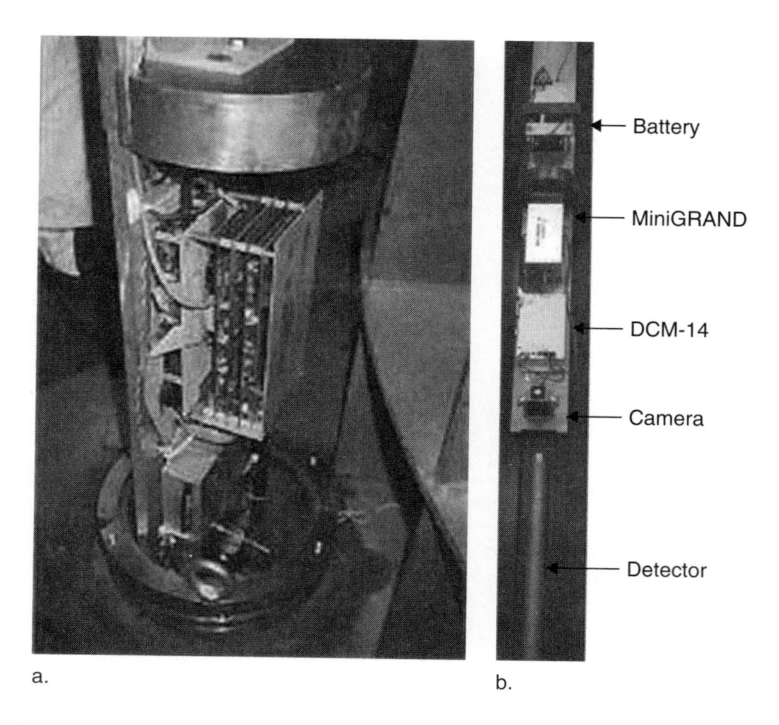

FIGURE 10.7 a, b. Integrated underwater detector installed at the BN-350 facility.

Accutime GPS receiver was installed on the top of the building to receive time information to keep all the nodes, instruments, computers, and cameras synchronized.

As with the RR system, two collect computers were used in a similar failover configuration, sharing a 110 GB RAID and configured identically. These were located in a secure area near the spent-fuel pond, accessible only to IAEA personnel. This location also held the video servers, power conditioning and uninterruptible supplies, a review computer, and backup hardware.

Short-Term Storage

The system the IAEA used to monitor the wet storage phase consisted of the same system as used for packaging, with the addition of an integrated detector containing a Si detector and MiniGRAND to monitor a hatch that provides an exit from the spent-fuel pond.

The IAEA maintains a secure office in the clean part of the facility. A review computer in the office is connected via VPN-protected Ethernet to the system in the secure area near the spent-fuel pond. This connection allows the IAEA to retrieve and review the safeguards data from the office without having to enter the dirty areas of the facility, as well as to dial in from Vienna to collect SOH information about the monitoring system. Only SOH data are transferred remotely to Vienna. This capability allows them to remotely diagnose problems with the system and send a properly equipped technician when necessary.

Long-Term Storage and Permanent Disposition

Long-term storage of the spent fuel will be in dual-use casks constructed of concrete and steel. They are called *dual-use* because the casks are used for both storage and transportation. At the time that this case study was written, the long-term storage phase had not yet begun. The safeguards plan for this phase is currently an extension of the installed system with the capability of providing radiation monitoring of each individual cask. In addition, attended measurements are planned that enable reverification of the material in each cask.

The casks will initially be located near the BN-350 but will ultimately be moved to another location. Secure, safeguarded permanent disposition of the fuel is currently being studied, but no decisions on the exact configuration for safeguards or the location of the long-term storage site have been made.

Summary

Safeguards at the BN-350 facility continue to offer unique nonproliferation challenges and opportunities. Started by the collapse of the Soviet Union and subsequent creation of and support from the Kazakhstan state, safeguards activities by the International Safeguards Community continue to protect the material remaining at the BN-350. This effort represents one of the largest and most enduring programs dedicated to the safe and final disposition of a large quantity of highly attractive nuclear material. The cooperative approach involving multiple countries and agencies has fostered development of new safeguards techniques and equipment that have been utilized in other applications around the world. Though the project is not yet complete, its continued success is paramount to the goals of nuclear nonproliferation and countering nuclear terrorism.

PART II

Detecting Nuclear Proliferation and Verifying the Elimination of Nuclear Weapons Programs

11 Using Open Sources for Proliferation Analysis

Richard Wallace and Arvid Lundy

Introduction

Open-source analysis is an essential component of nonproliferation analysis and a vital tool for detecting undeclared nuclear activities. In some cases, open-source analysis can provide the first clue that a state might be pursuing a nuclear weapons program counter to its treaty obligations and public declarations. Open sources include all information generally available to the public. Basic open-source analysis resembles traditional research as conducted by scholars, economic analysts, or legal investigators. National intelligence agencies, law enforcement organizations, nongovernmental organizations (NGOs), and international treaty-monitoring entities such as the International Atomic Energy Agency (IAEA) also carefully examine a broad range of open-source information to meet their needs.

Open sources are defined to include the following:

- Publicly available information (information such as found on the Internet or provided by NGOs, companies, the news media, and governments)
- "Fee-based" information such as found in published scientific and technical literature or subscription databases
- Information that is normally only made available on request or to specific individuals, including:
  - Company financial reports
  - Conference information (participant lists or paper titles, abstracts, or full text)
  - Internal publications of various organizations
  - Internal travel reports
  - Technical cooperation summaries
  - Unpublished scientific papers
  - Patent applications

Open sources do *not* include information that is legally protected, classified, or restricted in distribution (unless it becomes available to the public by some means, in which case it requires particularly careful validation by independent sources).

Open-Source Analysis

Open-source analysis is an important component of all-source analysis. *All-source analysis* literally means the use of all sources available to the analyst or investigator. This could include classified information generated by a state's intelligence organization or other legally protected sources (such as proprietary trade information or the confidential results of IAEA safeguards inspections that are available only to IAEA analysts). The integration of open-source analysis with intelligence data in conducting all-source analysis yields powerful advantages. For example, combining Website information containing commercial overhead imagery with government statements and scientific literature can lead to relationships and clues that are compelling evidence of proliferation activities.

The five greatest challenges to using open-source information to answer a specific nonproliferation question include the following:

- *Scarcity of information.* Sometimes there is little information available on a particular individual, organization, or activity. Example: Tightly controlled information on possible North Korean uranium enrichment activities.
- *Information overload.* In other cases, the vast amount of information available on a particular topic requires advanced analysis techniques and careful selection to concentrate on the highest priority, most reliable information. Example: The nuclear fuel cycle infrastructure in Japan.
- *Validation.* Open-source researchers must remain aware that inaccurate and deliberately false or misleading information is common. Many sources have an established political agenda and look for facts and conclusions that support their point of view.
- *Language barriers.* The most detailed information is often found in the native language of a country, which might not be widely spoken by analysts. Such situations also complicate forming effective information search strategies. In addition, summaries based on translations are seldom as reliable as original-language articles.
- *Information analysis.* Collecting, organizing, determining associations, tracking, and drawing conclusions from a wide variety of information types can be a daunting task. Example: Given the information on the Iran nuclear program, is it most likely purely civilian, or is it partially a military program? The analysis also includes a determination of the reliability of the available information.

Open sources can be numerous and continually changing (particularly Internet information). Most information is textual, with some graphical content (such as images that can include both satellite and ground photographs, organization charts, process flow sheets, site diagrams, infrastructure schematics, or building blueprints). Sources exist in multiple formats and languages, with varying levels of detail and accuracy. Open sources alone are unlikely, in isolation, to produce definitive proof of undeclared proliferation related activity, although this occasionally happens, as in the case of Iran.[1] Open-source information tends to be indirect and circumstantial, so a wide range of sources must be scanned for multiple independent types of evidence. The key results from careful analysis of open sources are usually identification of interests; names and locations of people, projects and organizations; patterns; and connections.

The following sections discuss various aspects of open-source analysis in more detail. The specific sources and examples mentioned are representative of those that were found to be useful for analyzing proliferation behavior. However, the rapidly evolving nature of the Internet, published scientific literature, and technologies to make information available prevents providing up-to-date lists of references, Websites, and tools. The mention of an information source is not necessarily an endorsement of the quality or accuracy of the information

---

[1]Violations of Iran's safeguards agreement (related to uranium centrifuge enrichment and heavy water production) were alleged in public announcements of an Iranian dissident group, verified by commercially available overhead imagery, proven by subsequent IAEA inspections, and finally admitted by Iran. (See IAEA DG report to BOG, GOV/2004/83, Nov. 15, 2004, and David Albright, "Iran at a Nuclear Crossroads," ISIS report, Feb. 20, 2003.) Also see Chapter 12 in this volume, by Frank Pabian.

found there. Open-source researchers must remain aware that inaccurate and deliberately false or misleading information is common. Diligent cross-checking and vetting of information is required to conduct high-quality analysis.

A scholarly approach to open-source (and all-source) analysis can be categorized into the following phases:

1. *Define the work and research plan.* Identify goals and hypothesis against which to test information and to guide searches for additional information. Identify counter-hypotheses and the evidence that would confirm or refute them.
2. *Collect data.* Gather sufficient information to test hypotheses, and develop methods for cross-checking and validating to obtain the most relevant, usable, and reliable data from a large supply. Finding unexpected data often leads to work plan revisions and additions.
3. *Process data.* Some data may need to be sorted, translated, structured, or formatted to make it available for quantitative assessments or for further analysis using additional software tools.
4. *Analyze data.* Look for relationships between information (people, research and development projects, organizations, places, and events). Sometimes visual or tabular representations of the data can be useful. Identify important observations/findings.
5. *Draft report and obtain thorough peer review whenever possible.*
6. *Report.* Finished reports or findings should be thorough, clear, concise, based on directly available information, and defensible.

The approach described above is somewhat idealistic in that it does not consider the requirement of resource and time limitations (imposed perhaps by political events or a publication deadline). Such realities often require careful consideration regarding which of the preceding steps might be compressed. The most effective nonproliferation analyses are often generated by analysts who are able to work in consultation with others or in teams for a sustained period on a limited subject. In-depth expertise in political matters for the country or geographical region; the technical areas examined (particularly when analyzing science and technical literature); and the information search and analysis tools and techniques are all essential components of strong nuclear nonproliferation analysis rarely found in a single individual.

Proliferation Pathway Analysis

One analytic approach to investigating whether a country is conducting undeclared nuclear weapons development activities is to consider the nuclear fuel cycle and weaponization activities that are required to successfully produce and deliver a nuclear weapon, and then conduct an analysis of the state in question to determine which components are present or missing. These key areas can then be targeted for deeper investigation. In many cases, the critical technologies will be:

- Fissile material production and handling
- Uranium enrichment and facilities with isotope separation capabilities
- Weapons-usable plutonium production reactors
- Plutonium separation and purification (reprocessing), and metallurgy technologies
- Criticality and health physics
- Weaponization
- Electronic fire-sets, fusing/detonation, high explosives testing, modeling, delivery vehicle development, and so on

The type of R&D information that might be most useful would be experimental studies in fields related to fissile material production, weaponization, and relevant facilities and activities that are not publicly declared or acknowledged by the state. The analyst is looking primarily for trends and patterns in R&D, not just topical research. This implies the need to build databases of topics, authors, affiliated individuals, and institutions and to look for relationships and patterns over time.

Integrating Multiple Data Sources

Analysis often proceeds by pulling on a lose string to unravel a problem. As a hypothetical example, a news story might mention a company that was accused of or sanctioned for selling dual-use commodities without a license. Further research into the export control records might identify the selling company and alleged purchasing entity. Investigation of the selling company could identify an individual whose background includes a technical education at a particular institution (and perhaps a patent on relevant technology). Research into papers published in the relevant field by that institution at the appropriate time could lead to discovery of another person from the alleged sanctioned purchasing country who attended the same department at the same time (perhaps as a coauthor) as the individual from the selling company. A search of science and technology (S&T) literature published by the identified individual from the purchasing country could lead to an association with another research institution or facility that has previously been suspected of involvement in a nuclear program. Satellite imagery of this facility might reveal evidence of proliferation activities. This chain illustrates how analyzing a broad range of research topics, S&T literature, individuals, companies, or institutions using a variety of sources can lead to a reasonable picture of a potential proliferation activity.

Search Strategies

Searching the Web

An incredible amount of information is now available on the Internet. Some contemporary (2007) Web search engines[2] include Google,[3] Yahoo!, Ask,[4] and Live[5] (formerly MSN Search), but many people use aggregate search tools (such as Dogpile and Metafind) to simultaneously search a variety of search engines. Vivisim/Clusty[6] will automatically determine themes within the search results and cluster results in self-defined categories.

Some search engines like Kartoo[7] will provide novel graphic displays showing relationships between search results (see Figure 11.1.). Most analysts use a variety of search tools, since the proprietary index and Web crawler engines for each search tool work in different ways and often ferret out different information. Many of these search engines also provide access to many "images" if searched under that heading or one for "news photo."

The Internet search field is evolving rapidly, with many mergers and popular search sites switching their underlying engines. Sites such as Search Engine Watch[8] and Search Engine Showdown[9] try to report on new developments, offer search strategies and suggestions, provide tests and reviews of the various engines, and provide charts showing their features and index sizes. For example, some search engines do not provide full Boolean search capabilities (in particular, no OR function), only a few offer proximity searching (e.g., *nuclear* within five words of *weapon*), and some offer archival searching (searching on stored images of Web pages that no longer exist). In addition, the engines vary in terms of how

---

[2]A *search engine* is an application that allows users to submit a query (such as "country *x* uranium enrichment") that is then matched against a database of information and references and the resulting relevant references presented to the user for further examination.

[3]www.google.com, accessed May 5, 2007, for example, claimed to index over 8 billion Internet pages in 2005.

[4]www.ask.com, accessed May 5, 2007.

[5]www.live.com, accesses May 5, 2007.

[6]www.vivisimo.com or www.clusty.com, accessed May 5, 2007.

[7]www.kartoo.com, accessed May 5, 2007.

[8]http://searchenginewatch.com, accessed May 5, 2007.

[9]www.searchengineshowdown.com, accessed May 5, 2007.

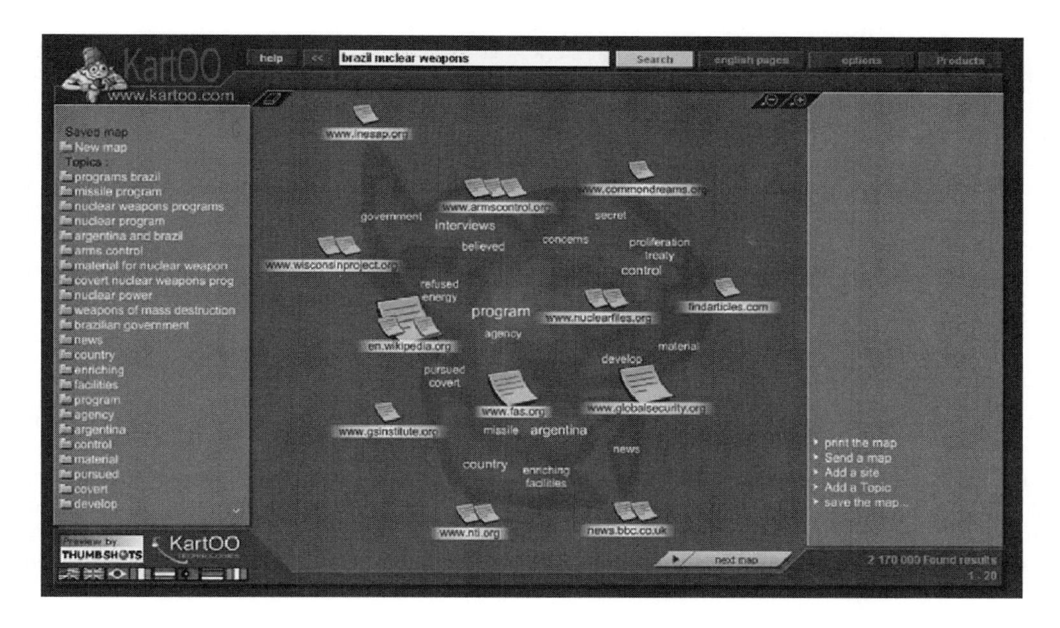

FIGURE 11.1 Example of simple Kartoo search for "Brazil nuclear weapons."

much of a Web page or file is indexed; for example, some only index the first 100–120 kb of each Web page or file.

Some analysts use software such as Copernic Agent Pro[10] to conduct a metasearch of predefined sites and search engines. The sources and keywords can be stored for future use, and useful sites can be collected, indexed, and shared among a group of analysts in the same organization.

Special mention should be made regarding the use of an Internet-based, community-created consensus encyclopedia such as Wikipedia. In early 2007, there were over 75,000 active contributors working on more than 5,300,000 articles in more than 100 languages.[11] As of April 2007, there were 1,752,545 articles in English; every day hundreds of thousands of visitors from around the world make tens of thousands of edits and create thousands of new articles to enhance the amount of knowledge held by the Wikipedia encyclopedia. The Wikipedia editorial page provides the following thoughts on the benefits and limitations of such a resource:

> *Wikipedia is written by consensus—an approach that has its pros and cons. Censorship or imposing "official" points of view is difficult to achieve and almost always fails after a time. Eventually for most articles, all notable views become fairly described and a neutral point of view reached. In reality, the process of reaching consensus may be long and drawn-out, with articles more fluid or changeable for a long time while they find their "neutral approach" that all sides can agree on. Reaching neutrality is occasionally made harder by extreme-viewpoint contributors. Wikipedia operates a full editorial dispute resolution process that allows time for discussion and resolution in depth, but also permits months-long disagreements before poor quality or biased edits will be removed forcibly.*

Studies suggest that Wikipedia is broadly as reliable as *Encyclopedia Britannica*, with similar error rates on established articles for both major and minor omissions and errors.

---

[10]www.copernic.com, accessed May 5, 2007.

[11]www.wikipedia.com, accessed May 5, 2007.

There is a tentative consensus, backed by a gradual increase in academic citation as a source, that it provides a good starting point for research, and that articles in general have proven to be reasonably sound. That said, articles and subject areas sometimes suffer from significant omissions, and although misinformation and vandalism are usually corrected quickly, this does not always happen. (See, for example, the incident in which a person inserted a fake biography linking a prominent journalist to the Kennedy assassinations and Soviet Russia as a joke on a coworker; the joke went undetected for four months. Afterward the prankster said he "didn't know Wikipedia was used as a serious reference tool.") Therefore, a common conclusion is that it is a valuable resource and provides a good reference point on its subjects, but like any online source, *unfamiliar information should be checked before relying upon it.*[12]

Developing and Using Local Databases

A key open-source analysis technique is the creation and maintenance of customized, subject-relevant local databases. This is the digital equivalent of building files on a certain subject for later use. Mining local data sets can be an effective first step in researching a new question or topic. Many organizations and individuals create local databases of their search results. Some routinely download articles that meet specified criteria from news feeds and store these in a local database to facilitate future searches or analysis of these selected data with advanced software tools. Individual analysts will eventually build up substantial amounts of information, including finished analysis reports as well as processed versions of the data that were used to generate them. Frequently, a future question will involve the same subject that was researched earlier. Local databases can be maintained and are not composed of ephemeral or perishable data such as data from the Web.

Some large organizations use a document storage-index-search-retrieval tool to build local databases. These tools place data from a wide variety of external sources (S&T literature, news feeds, Web searches, summary articles, printed reference summaries) into a consistent data format that can be searched simultaneously with a single advanced query. Verity's Topic or Knowledge Navigator K2[13] system is an example of a system that can accept news feeds, manual data loads, and other information; add metadata (for example, evaluations of the validity of the source); then organize it into an shared data warehouse so that the data can be searched by anyone having the proper network access. This is a complex system that may require a full-time administrative staff. Similar systems with simple features are available from other vendors.

A variety of organizations are now producing desktop search software that indexes and rapidly searches a particular desktop computer or network file system for relevant information. One of the earliest effective systems of this type was dtSearch,[14] which currently claims to be able to index over a terabyte of information and then search that information in typically less than a second. Although somewhat expensive, dtSearch can create indexes that can be used by different individuals, which allows data and indexes to be stored on a central network file system and then searched independently by several analysts.

Some analytical tools, such as Visual Analytics' Visual Links,[15] i2's Analyst Notebook,[16] and other analytic applications that perform link analysis, require the underlying data to be in relational database tables. This is natural for law enforcement agencies that deal with names, organizations, telephone numbers, bank accounts, and bank transactions but is more challenging if your data are in unstructured text files or S&T citations (with text abstracts). Various software solutions are attempting to deal with unstructured text, including the AskSAM[17]

[12]http://en.wikipedia.org/wiki/Wikipedia:About, accessed May 5, 2007.
[13]www.autonomy.com, accessed May 5, 2007.
[14]www.dtsearch.com, accessed May 5, 2007.
[15]www.visualanalytics.com, accessed May 5, 2007.
[16]www.i2inc.com, accessed May 5, 2007.
[17]www.asksam.com, accessed May 5, 2007.

unstructured database system, several systems for entity extraction, the Attensity[18] solutions that extract full relational data from unstructured text files, and Inxight SmartDiscovery,[19] which not only does entity extraction but also handles categorization and relationships. As the volume of data increases, a data processing step (such as automatic translation, entity extraction, or unstructured to relational data transformations) is becoming an important step between data collection and analysis.

When searching for individual authors, the most obvious caveat is to check for alternate spellings. In addition to trying full names or initials, variant spellings can be a serious problem. Some languages with non-Latin characters can be transliterated in a wide variety of ways. If one is careful in conducting name searches, sometimes a careful S&T literature search for an author can track someone from his thesis institution (frequently not in his home country) through several intermediate positions to his final position, including a year or two sabbatical outside his home country. Often the affiliated institution or department will have a bio or list of papers for its staff members. It is also becoming more common for scientists, even in Third World countries, to maintain individual Websites listing their CV and major accomplishments. These can often be found through a general Internet search once you have a few key pieces of information about the person or his research interests.

Avoiding Misinformation from the Internet

The primary issue with Internet information is credibility. Numerous examples of high-profile Internet inaccuracies and deliberate spoofs exist. Numerous sites provide incorrect or incomplete information on a range of topics.

Some deceptions are deliberate. For example, in February 2003 the Website of *Computerworld* magazine published a story claiming a radical Islamic group was behind the "Slammer worm" attack that clogged the Internet. The next day the story was retracted after it was learned that one journalist had deceived another. Dan Verton had based his article on an email interview with a person he identified as "Abu Mujahid," a member of Pakistan-based Harkat-ul-Mujahadeen. But Mujahid was really Brian McWilliams, 43, a freelance journalist in Durham, New Hampshire, who has written for Salon.com and Wired News. McWilliams said he had duped Verton because he wanted to teach reporters "to be more skeptical of people who claim they're involved in cyberterrorism." The fiasco is a good reminder of the risks involved in relying on email interviews and the importance of verifying sources.[20]

In another alleged hoax case, fake photographs claiming to show the tsunami that struck Indian Ocean shorelines in December 2004 fooled many media outlets, including the *Times of India*, the *Calgary Herald*, and Channel Nine and Sky News in Australia. The photos, allegedly of the tsunami, were actually from a 2002 tidal bore (a predictable periodic rise in a river flow) in China.[21]

Other sources of error may include omission, translation, or mutation. Important caveats or qualifications may be omitted (for example, in the 2002 U.S. Intelligence reports on Iraq weapons of mass destruction) and guesses may become "common knowledge" facts. A remarkable study in translation was once provided when an Iranian news agency reported on an IAEA inspection in Farsi, which was translated into Japanese, then picked up by a Russian news agency, and finally translated back into English by FBIS. The inspector's name (which was Scandinavian) was almost unrecognizable. This is a particular danger when non-Western characters are used by the original or intermediate sources. In addition, a story tends to be distorted more each time it is repeated.

---

[18]www.attensity.com/www/, accessed May 5, 2007.

[19]www.inxight.com, accessed May 5, 2007.

[20]Dan Verton, "Journalist perpetrates online terror hoax," *Computerworld*, Feb. 6, 2003, accessed April 2007 at www.computerworld.com/securitytopics/security/cybercrime/story/0,10801,78238,00.html.

[21]Snopes.com article, accessed April 2007 at www.snopes.com/photos/tsunami/tsunami1.asp.

Any information obtained through the Internet should be carefully evaluated to determine its origin and the motivations that the author may have been under when posting that information. It is best to try to identify the original (primary) source of information rather than use secondhand reports and to examine the details to look for pieces of information (names, dates, places, activities) that can be verified through further searches or by contacting primary sources. (For example, prior to running the photos, nobody called the media outlets in the purported Thailand location of the tidal wave described previously.) The reputation and previous track record of the publishing site or organization can help with the evaluation, but these are not foolproof. One can often look for common threads in *multiple* sources, especially if they are truly *independent*, to find reliable facts.

Generally, primary sources will tend to be more accurate than secondary ones. It is usually worthwhile to expend the effort to identify the primary original source or statement. Minimize the amount of translation done on the primary source. An author who has direct knowledge of an event tends to be more accurate, as does one who understands the subject. For example, nuclear trade publications tend to be more accurate in reporting nuclear materials trafficking incidents or nuclear proliferation issues than do the general media, owing to the greater specific technical knowledge and topic experience of trade writers. Another common example is the press conference. If possible, it is better to obtain a transcript of the relevant press conference rather than rely on a media summary of it. Often the context of a statement, the bias of a reporter, or the phrasing of the question that led to a particular answer influences how one interprets the statement.

Balancing Effort with Deadlines

Knowing when to quit can also be problematic. When searching for a specific type of information, you can seldom be sure that you have located every possible reference. At some point, you may continue to obtain more general information on the topic, but less of it is relevant to your search. (A search on a person may begin to exclusively provide more detailed information on an unrelated pop singer with the same name, for example.) And, of course, if there is no information on your given topic, you could search forever for it. Given the specific topic at hand, its importance, and other aspects of the situation, you might want to define a search strategy that, when exhausted, means you will stop looking and go on to another project.

Deep Web Sources of Additional Information

The information available on the Web that is indexed by general search engines such as Google is information that resides in permanent pages that are directly accessible by typing in the server and address of these files (the *http://www...* address). Such Internet files are sometimes referred to as the *surface Web*. Other information available through the Internet is isolated behind network firewalls or contained in searchable database systems (such as library holdings) that require authorized users to establish a user account and log in to conduct specific searches using database-specific search protocols and terms. This additional data is called the *deep Web*.

Deep Query Manager (DQM) is a software system for searching the deep Web.[22] Search tools such as DQM tend to be expensive, but they provide the ability to simultaneously search thousands of databases that are unreachable through conventional Web searches. These products use the unique format required by each database, collate the results, apply user-defined filters to eliminate unwanted information, and present the results in an understandable format. Both the queries and the results can be stored for future use or further processing and culling. On the other hand, you could find that your particular interests can be served by accessing a small set of specific databases without specialized software.

[22]Michael K. Bergman, "Comprehensive Overview of BrightPlanet's Technology," white paper published by BrightPlanet, Dec. 2004; accessed in April 2007 from www.brightplanet.com.

In either case, you should be aware of the sites that contain information of interest to you but that might not be indexed by surface Web search engines.

Sources for Data Collection

Country Profiles

A *country profile* is a brief overview of a state, including its history, culture, current issues and near-term outlook, geography, people, government, economy, communications infrastructure, transportation systems, military capabilities, transnational issues and agreements (including perceived national threats), and nuclear infrastructure. Country profiles are produced by many governments, businesses, and NGOs. Such profiles often contain information relevant to an analysis of nuclear activities. Specific sources for this information include the *CIA World Factbook*,[23] which provides a basic free overview of most states, and the Economist Intelligence Unit,[24] which provides a free "country profile" but charges a fee for in-depth analysis for nearly 200 countries.

Typical information for nuclear nonproliferation might include whether the state has specific security concerns and the nature of those concerns and whether its neighbors have nuclear weapons. The nature of a state's political system and the role of military force in its strategy and history are also important. Its diplomatic behavior, particularly how committed it is to the nonproliferation regime (treaties, agreements, participation in regional or international nonproliferation activities, NPT standing, and so forth), is another key issue. Often the information in country profiles can be combined with recent information from news media and other open sources to determine what nuclear technologies a state is importing or developing and how these might correspond to the country's nuclear power plans or undeclared pursuit of nuclear weapons.

Such analysis of a state's nuclear-relevant industrial infrastructure might help provide evidence that it is developing sensitive nuclear program capabilities, such as uranium conversion, uranium enrichment, heavy water production, or plutonium extraction from irradiated fuel. Similarly, the presence of advanced high explosives research, development and testing, detonation physics, or hydrodynamic shock wave modeling program, combined with the acquisition of dual-use advanced machine tools or metal forming capabilities, could be an indicator of efforts to build nuclear explosive devices.

Some countries have abandoned nuclear weapons programs before such weapons were produced or deployed; however, it is important to understand what capabilities might have been preserved from the historical programs and to closely examine the current state of those relevant capabilities as well as the organizations, entities, and technically competent personnel who developed them. Information could also be analyzed to assess the probability of any undeclared fissile material remaining from a previous program.

More detailed information specifically concerning a state's (or even a terrorist group's) nuclear programs and activities is often available from official government publications or the Internet. Some of the more reliable Internet information is available from academic institutions or internationally respected NGOs, such as the Nonproliferation Program at the Carnegie Endowment for International Peace, the Center for Nonproliferation Studies (CNS) at the Monterey Institute of International Studies (MIIS), the Nuclear Threat Initiative (NTI), the Institute for Science and International Security (ISIS), and Global Security (particularly noted for its analysis of satellite imagery). However, the information from even these sources must be validated wherever possible. One source that provides thorough information on countries' nuclear capabilities and missile systems is *Deadly Arsenals: Tracking Weapons*

[23]See https://www.cia.gov/cia/publications/factbook/.

[24]See www.eiu.com.

of Mass Destruction, from the Carnegie Nonproliferation Project.[25] Annual reports from a facility, company, or regulatory agency might provide useful information regarding current activities and processes as well as future plans possibly related to undeclared activities.

Information produced by NGOs with specific political agendas, such as the Federation of American Scientists (FAS)[26] and the Wisconsin Project on Nuclear Arms Control,[27] or by political organizations such as the National Council of Resistance of Iran (NCRI) can be extremely valuable but should be carefully validated. The analyst must keep the goals, motivations, and potential biases of organizations providing information in mind. For example, the NCRI is a revolutionary group that has been listed by the U.S. State Department as a terrorist organization; it advocates the overthrow of the ruling regime in Iran. The NCRI revealed at a press conference in Washington, D.C., on August 14, 2002, the locations of two top-secret nuclear sites in Iran. The NCRI subsequently revealed the uranium enrichment assembly and test facility at Kalaye Electric Co. in Tehran. The two sites, Natanz and Arak, were being constructed without the knowledge of the IAEA, as was the prototype centrifuge assembly and test facility at Kalaye Electric. Failure to notify the IAEA of these projects violated Iran's NPT obligations.[28]

Natanz and Arak were designed to produce enriched uranium and heavy water (the latter for use in an adjacent possible plutonium production reactor now under construction), respectively. Enriched uranium and plutonium are the two key fissile materials necessary for the production of nuclear weapons. Iran did not acknowledge nuclear activities at these sites until satellite imagery acquisition and further investigation by the IAEA made the nature of these facilities unmistakable. The specific geospatial information on the location of the sites in Iran was only available in the original NCRI transcript. In a textbook example of open-source analysis, it was the careful reading and interpretation of all information presented by NCRI to precisely locate the facilities, combined with the use of commercial satellite imagery to help characterize their function, which eventually led the IAEA to formally investigate Iran's undeclared activities.[29]

News Media

News media reports (newspapers, newswires, radio, and TV) are often valuable because reporters sometimes have sources that differ from others available to a scholar or researcher. However, such reports must be validated against other reports of the same event or activity or other corresponding information. It is best to identify the original source for a summary article and carefully examine this original (in the original language, if possible), as well as understand the authors' motivation and historical reliability. One of the most comprehensive news feeds is Factiva[30] (the Reuters/Dow Jones news service), which is fee-based. There are also many free Internet sources, such as sites from the BBC, CNN, *Time*, and other major news agencies.

The amount of information reported daily by global media outlets is enormous. A number of commercially available software packages, often called *news aggregators*, can monitor news feeds, filter them to your specifications (usually using a combination of keywords that you define), and then notify your computer in real time when a new story appears

[25]Joseph Cirincione, Jon Wolfsthal, and Miriam Rajkumar, *Deadly Arsenals: Nuclear, Biological, and Chemical Threats*, second edition (Washington, D.C.: Carnegie Endowment for International Peace, 2005).

[26]www.fas.org, accessed May 5, 2007.

[27]www.wisconsinproject.org, accessed May 5, 2007.

[28]*Implementation of the NPT Safeguards Agreement in the Islamic Republic of Iran*, IAEA GOV/2004/83, 15 Nov. 2004.

[29]Chapter 12 provides a detailed account of this open-source analysis case.

[30]www.factiva.com, accessed May 5, 2007.

that meets your interests. A few examples of the many available news aggregators in 2006 include Botbox,[31] Novobot,[32] and Radio UserLand.[33]

The Foreign Broadcast Information Service (FBIS)[34] is a U.S. government operation that translates the text of daily news media and journalism broadcasts as well as government statements from non-English sources around the world. This is an invaluable resource for non-English news. However, users must belong to an organization that is authorized to use the FBIS service.[35] For those who do not have access to FBIS, the World News Connection[36] is a more generally available U.S. government resource that contains many of the FBIS-translated articles.

Industry-focused technical journals and newsletters are another category of open-source publications that often cover nuclear items in more detail than do general news media. Such publications include the *World Nuclear Association*, *Nuclear Fuel*, and *Nucleonics Week*. Some European import/export trade information is available through Comext (the European Commission Database on External Trade).

For weapons and delivery systems, there are several well-known publications. These include the *Aviation Week* family of publications[37] and *Jane*'s.[38] Jane's Information Group is a world-leading provider of intelligence and analysis on national and international defense, security, and risk developments. Some of the information and consulting subjects covered by *Jane's* family of publications include:

- Country-by-country internal and external security and threat assessments
- Defense news and analysis
- Orders and formations of worldwide armies, navies, and air forces
- Military systems and equipment (including specifications)
- Worldwide geopolitical intelligence and news analysis
- Terrorism intelligence, news and assessment services

Jane's online service is subscription based and provides access to over 200 information sources. However, the Website also provides some information for free.

Image Sources

There are an increasing number of sources for images (photographs, drawings, maps, and video) that can be useful in assessing a state's nuclear activities. Relevant images may include photographs (ground or aerial) of facilities, equipment, and even notable individuals within the government or research establishments of a given state. One source for images is Google. Another is Getty Images.[39] Some NGO sites such as ISIS and Global Security.org specialize in open-source satellite imagery that can be extremely valuable in assessing nuclear infrastructure. For example, Global Security has a wonderful page comparing before, during, and after construction images of several Iranian nuclear facilities.[40] The Satellite Imagery chapter of

---

[31]www.botbox.com, accessed May 5, 2007.

[32]www.proggle.com/novobot/index.shtml, accessed May 5, 2007.

[33]http://radio.userland.com, accessed May 5, 2007.

[34]www.opensource.gov, accessed May 5, 2007.

[35]Susan B. Glasser, "Probing Galaxies of Data for Nuggets: FBIS Is Overhauled and Rolled Out to Mine the Web's Open-Source Information Lode," *The Washington Post*, Nov. 25, 2005, p. A35.

[36]http://wnc.fedworld.gov.

[37]See www.aviationweek.com/aw/, accessed May 5, 2007, for a description.

[38]http://www.janes.com/, accessed May 16, 2007.

[39]http://editorial.gettyimages.com/, accessed May 5, 2007.

[40]www.isis-online.org/images/iran/lavizanshian_joint.html, accessed April 2007.

this volume provides more information on the use of open-source overhead imagery for nonproliferation analysis and reporting.

Google Earth[41] has received wide acclaim for revolutionizing the use of satellite imagery. Many sophisticated nuclear facility analyses have been done using the free version alone, and the fee-based version offers added capabilities and higher-resolution images. One novel concept is that the vast numbers of Internet users suddenly have the ability to search for obscure details in areas that were formerly reserved for a small cadre of professional satellite imagery analysts. Text-searchable Weblogs have emerged that capture many observations of interest to nuclear nonproliferation analysts.

Most NGOs provide full country locator maps on their Websites. However, more precise maps are available from *Microsoft Encarta World Atlas*, which can locate sites by geocoordinates. Web searches of a specific location with the word *map* will usually provide additional map links. The Perry-Castañeda Library Map Collection at the University of Texas at Austin[42] is a great source of detailed foreign city street maps. Other sources of hardcopy maps for purchase that are not digitally available online is Omni Resources[43] or Maps2Anywhere.[44] These vendors can provide access to some rather esoteric maps that include foreign geologic and geophysical maps.

Business or Facility Information

Company and facility profiles are similar to country profiles and provide some history, current management team, contact information, location(s), products, capabilities, activities, number of employees, main customers, and so on. Some examples of the type of sources available for such information are discussed here.

One good source for basic commercial company information is the fee-based KOMPASS database.[45] In 2007, KOMPASS claimed to index 2.2 million companies in 70 countries, referenced by 54,000 product and service keywords, 822,000 trade names, and 4.2 million executive names. Business searches can be focused on specific countries or regions (although coverage is not uniform around the globe). One could search, for example, for all companies in a specific country that produce permanent magnets or maraging steel (components of interest for uranium enrichment centrifuge construction) or nuclear reactor containment vessels.

Another specialty fee-based service is LexisNexis, which provides legal, news, public records, and business information, including tax and regulatory publications in online, print, or CD-ROM formats. LexisNexis Company Dossier provides basic and in-depth business information on more than 35 million global companies.[46] Clients of this service tend to be law firms or large corporations that investigate companies or individuals.

Host governments often provide official information on industrial and scientific institutions. A state's atomic energy organization is particularly useful for information on "declared" nuclear facilities and its publications often include helpful images. Sometimes such publications are available from government Websites, but (especially for less developed countries) often you must find the information in a technical library (the British Library has a particularly good collection and has interlibrary loan agreements with many organizations) or by contacting the organization directly to request printed information.

Be aware, however, that these are not necessarily "independent sources of information" compared to a state's public declarations. Other government-related sources include records of

---

[41]http://earth.google.com/, accessed May 5, 2007.

[42]www.lib.utexas.edu/maps/, accessed May 5, 2007.

[43]www.omnimap.com, accessed May 5, 2007.

[44]www.maps2anywhere.com, accessed May 5, 2007.

[45]www.kompass.com, accessed May 5, 2007.

[46]www.lexisnexis.com, accessed May 5, 2007.

government hearings or actions. (For example, the U.S. Department of Commerce publishes information on violations of export laws and lists of entities that are blacklisted for such previous violations.) In addition, some countries occasionally publish findings by their intelligence organizations in the open literature—for example, the CIA's periodic nonproliferation report to Congress or the French government report on Iran to the nuclear supplier's group.

Other sources of information on specific facilities may include the following:

- *Operator information.* The facility or organization itself may have literature describing its activities (for example, the Bhabba Atomic Research Centre in India Website[47] contains information, photos, and publications on many of its facilities, capabilities, and activities). These might be available on the Internet, as print publicity or recruiting brochures, or as annual reports to funding agencies or stockholders. If the facility provides information in both the local language and English, the local language pages are generally far more detailed and informative than the English pages. For more obscure facilities for which only a name may be available, searching for the telephone number, address, or the email address of a facility manager could be more successful than just searching by name. This is because facility names are sometimes concealed by governments or exist in many variations that can be confusing.
- *Environmental/community groups.* A number of local and national environment or special interest groups keep close watch on facilities in their area. For example, the Websites of the Alliance for Nuclear Accountability[48] and Citizen Alert focus on U.S. nuclear facilities. Similar groups exist in other countries and often have some information on nuclear activities.
- *Companies that have been investigated for violations of export control laws.* The customs services, judicial authorities, or export control administrations of several nations make their list of such companies public. News stories on these companies and their violations can sometimes be found in the general media via Web searches or FBIS. Other important export control information is compiled by the Monterey Institute for International Studies' Nonproliferation Program in a monthly publication called the *International Export Control Observer.*[49]

Information on Individuals

Talented individuals are a critical element of any successful nuclear program. The establishment of the nuclear fuel cycle and the construction of nuclear weapons require highly skilled scientists, engineers, and administrators. Therefore, information on the skills and activities of individuals can often provide insights on national efforts in the nuclear sphere and the degree to which progress has been made. Finding the names of relevant individuals and information regarding their activities is a challenging task.

As mentioned, official scientific organizations, government agencies, and universities sometimes provide the names, titles, and professional contact information for leading individuals. Universities often encourage their faculty and staff to create personal Websites with CVs that list education, employment history, research interests, affiliations, and publications.

Unfortunately, general Internet searches by individual name are seldom successful unless the individual is a notable public figure, has a fairly unique name, or has published extensively in a narrow topical area. Government officials or leading business executives can sometimes be found in KOMPASS or another geographically specific business directory. Professional or scientific personnel are frequently members of local, national, or international

[47]www.barc.ernet.in, accessed May 5, 2007.

[48]www.ananuclear.org, accessed April 15, 2007.

[49]http://cns.miis.edu/pubs/observer/index.htm, accessed May 5, 2007.

professional societies or organizations. Occasionally, country- or university-specific databases of thesis topics will lead to information about specific authors.

Analyzing S&T Capabilities

Usefulness of S&T Capability Analysis

Analysis of the technological capabilities of a country suspected of having or known to have a nuclear weapons program should focus on the capabilities required to conduct a nuclear weapons program and its associated production of weapons-grade nuclear materials. Such analysis helps determine:

- The credibility of the development effort
- How long it might take the effort to succeed
- What expertise and commodities are apt to be most sought by the program
- What technological routes are most apt to be followed
- What organizations are likely to be involved

S&T analysis can provide a window into the general relevant technical capabilities and strengths of a country, specific areas of concentrated research, activities at specific organizations or R&D facilities, individuals who are prominent in various areas and who might warrant further individual study, and relationships between people, institutions, and research groups (both domestic and international relationships).

S&T analysis is particularly helpful in looking for patterns in research that are inconsistent with the public declarations of a state. For example, a search of mechanical engineering conference papers in a given country may indicate research on maraging steel or 7075 aluminum. These materials are used in centrifuge uranium enrichment systems. Questions would be raised if the authors of such papers were from a state that has claimed it has no capability or interest in the field of uranium enrichment. On the other hand, technically advanced countries will often have legitimate reasons for pursuing R&D on sensitive technologies (for example, similar aluminum tubes are used for both centrifuge enrichment equipment and conventional rocket motor bodies). Additional research is needed to determine whether publications themselves are in fact evidence of undeclared nuclear activities.

Conducting this type of S&T capabilities analysis can have other side benefits for nonproliferation and counterproliferation efforts. For example, a technological capabilities analysis can provide information useful to crafting more effective economic sanctions and export controls designed to hinder a nuclear weapons program. In addition, if other means fail and a decision is made to use military force to counter the nuclear weapons development program, detailed knowledge of a state's nuclear and industrial infrastructure can aid military planning.

Factors Affecting S&T Capability Analysis

All countries that have developed nuclear weapons have involved many of their top scientists and engineers as well as key industries in the effort. Chemical engineering, mechanical engineering, numerical modeling, and project management capabilities may be as important to a proliferant as nuclear engineering or nuclear physics knowledge. Materials science, metallurgy, electrical engineering, high explosive materials, and precision machining all play an important part as well. As a result, a survey of technological capability must analyze technology in various separate sectors and the links between them:

- Academic and civil or governmental research laboratory capabilities
- Industrial capabilities including R&D, manufacturing, and import capabilities
- Military R&D, manufacturing, and testing capabilities
- Delivery vehicle R&D, manufacturing, testing, and deployment

Characteristics of the subject country will affect both the ease and the value of technological capability analysis. In very underdeveloped countries, technology capability analysis may be done quite quickly because there are so little data to analyze, especially if the subject state is a closed society with tight information control. An underdeveloped country with relatively low technological competence can easily encounter roadblocks that can slow or stop a nuclear program. However, accurate analysis of technological capability in these countries can also be very difficult. If the country possesses a small cadre of highly competent individuals, the program can advance more rapidly than expected. North Korea is perhaps an example of this.

Highly developed countries have often mastered all technology necessary for a nuclear weapons program or could very quickly do so. They might even have stocks of nuclear materials. In such cases analysis of proliferation risk is most dependent on assessment of motivation and internal decision-making processes. However, even in these cases certain technological capabilities and activities may indicate early efforts on a nuclear weapons program. Some examples of capabilities that would be useful to the development of nuclear weapons include:

- The uranium (U) spectroscopy studies needed to obtain data for uranium Atomic Vapor Laser Isotope Separation (AVLIS)
- Separation of actinides from irradiated fuel or targets in a research reactor
- Research on beryllium-neutron reaction properties
- Advanced manufacturing processes for heavy metal spherical parts
- Experimental or computer modeling studies of shock-wave compression of spherical metal shapes
- Extremely high-density, high pressure equation of state studies
- Hydrodynamic test sites for military shaped charge munitions development
- Flash radiography
- High-speed photography (mechanical or electronic)
- Pin diagnostics
- Precise timing signal generation and recording, etc.
- Hot-cell facilities for handling and processing radioactive materials
- Magnetic materials technology (especially for high-strength permanent magnets)

Expertise Required

Ideally a technology analysis team would include people with hands-on backgrounds in relevant nuclear fuel-cycle technology (the complete process for both HEU and plutonium production), and nuclear weapons design and production, including preparation of fissile metals and shaped explosives. For example, it may be important to know that research on the use of TBP (tri-n-butyl phosphate) and kerosene (or dodecane) as a solvent in a mixer-settler to extract metals from industrial waste streams is directly relevant to the PUREX process for extracting plutonium from irradiated reactor targets. Similarly, research on certain atomic spectra from uranium is relevant to the Atomic AVLIS uranium enrichment system. Another good indicator of potential AVLIS work is the specific use of copper-vapor lasers in atomic physics and chemistry research. The forgoing are a few examples of hundreds of relevant technical relationships. Finally, the team members should be familiar with open-source data collection on the countries of concern and have relevant language capabilities.

Sources for S&T Literature

One of the best resources for nuclear-related S&T publications is the IAEA's International Nuclear Information System (INIS), which is available for free to member states and for a small fee to others.[50] This system is unique in that it utilizes offices in many different nations around the world to review and collect nuclear-related S&T information from their

[50]www.iaea.org/inisnkm, Aug. 2007.

geographical area. The abstracts are translated into English, and pointers are provided to the location of full-text articles, which may be in English or in a local language.

Because of this indexing method, many local publications (local research reports, small-scale publications, conferences, university theses, and the like) are included that would not be indexed by large commercial publications databases. On the other hand, the subject matter is limited to directly nuclear-related items (the main purpose of the database is to share nuclear technology with developing countries), and many of the local articles are only available in the local language and by special request to the coordinating organization for INIS input in that country.

Commercial S&T database systems have a broader topical coverage than INIS, although not the depth in local publications. Some of the most useful commercial databases include Thomson DIALOG,[51] which is expensive but comprehensive; ISI's Web of Science (WOS)[52]; and Elsevier's Science Direct,[53] although there are many others that might be more suitable for particular applications, such as Google Scholar. Google Scholar can search by keyword or author but does not yet have an option to explicitly search by affiliation.

Each service has its own set of advantages and disadvantages. For example, DIALOG is extremely comprehensive (including the ISI databases and patent information) but can be expensive unless the user is meticulous in defining the search string. In addition, DIALOG only indexes the affiliation of the first author on a paper, so searches by affiliation or country will miss papers where the first author is not from the organization or country of interest. WOS is limited to abstract searches unless the user invests in an expensive full-text option; however, it does index the affiliation of every author. Science Direct also mostly indexes abstracts but its scope is limited to Elsevier publications.

A federated search tool can greatly expedite S&T search and analysis. Such tools allow a single complex query to be directed simultaneously to several different databases at different locations (such as, perhaps, INIS, WOS, and Science Direct), gather the results, and present them in an organized fashion. An ideal tool would then extract the major components (subject, authors, affiliations, date, and so on) and allow the analyst to sort or display various combinations of parameters to look for trends or relationships.

Catalogues of Proliferation Relevant Technologies

One useful approach is to use a recognized catalog of nuclear weapon and nuclear fuel cycle technologies and open-source research methods to search for those technologies that the subject country possesses. Looking specifically at nuclear-related technology, there are a number of technology "catalogs" that can be useful:

- The Nuclear Section of the U.S. Department of Defense's Militarily Critical Technologies List (MCTL), especially earlier editions, which are considerably more detailed than later editions[54]
- The Nuclear Suppliers Group (NSG) export control guidelines supplemented by special attention to three areas:
  - Fluorine chemistry expertise, especially on an industrial scale (critical for all UF_4 and UF_6 based uranium enrichment processes)
  - Computerized numerical modeling capabilities, especially those involving shock wave physics, hydrodynamics, and radiation transport (normally Monte Carlo based)

[51]www.dialog.com, accessed May 5, 2007.

[52]http://scientific.thomson.com/products/wos/, accessed May 5, 2007.

[53]www.sciencedirect.com, accessed May 5, 2007.

[54]Currently the old 1998 MCTL edition is posted on the Federation of American Scientists Website; see www.fas.org/irp/threat/mctl98-2/p2sec05.pdf (July 2007).

  - Capability in precision high explosive work such as the development of shaped charge munitions and precision timed initiators
- The NSG *Guidelines for Nuclear Transfers* (INFCIRC/254, Part 1), which deals with especially designed or prepared equipment for use within the fuel cycle[55]
- *Guidelines for Transfers of Nuclear-Related Dual-Use Equipment, Materials, Software, and Related Technology* (INFCIRC/254, Part 2), which deals with dual-use items for both fuel cycle and weaponization activities[56]

Academic and Research Output

A rough measure of a country's scientific productivity can be made by looking at the total scientific paper production within the country. Scientific paper output can be assessed through major scientific databases (Engineering Index, Inspec, or ISI Proceedings and ISI SciSearch, etc.). This is done by entering only the country name in the institution field, entering a date range, and noting the number of papers found. This process can be repeated for individual years to find whether production is increasing or decreasing. In countries with small scientific establishments it can be useful to read the title of every paper published over a given time period. Often relevant papers appear that would not have been located when keywords were used in narrowing searches.

By carefully selecting subject headings for searches, an estimate of competence in particular relevant areas can be made. The number of papers found is not apt to be very meaningful unless compared with numbers from other countries. For example, comparing numbers from Israel, Turkey, Pakistan, Iran, and Libya would be more meaningful than the number from any single country. Looking year by year and comparing results by country may also be meaningful. For example, the total production of scientific papers by Iran has increased every year for 10 years, and the rate of increase has been larger than any other country in the world. China has the second highest rate of increase (Lundy, 2006, unpublished). Publication policies or restrictions must also be considered in making comparisons.

Looking at the publication record for relevant technologies will usually provide the name of the institutions that produced the papers. These institutions frequently have Websites that can be researched for more specific information. Research organizations often give a mission statement and list all their publications. Professors in research universities frequently have Websites with listings of their research interests, names of the graduate students they supervise, and complete listings of their papers. Academic departments may give histories of their development and list the areas they excel in. Sometimes the department's laboratory equipment is listed. Studying coauthors' credits will indicate which organizations work together in particular areas. Graduate students also often have Websites that can be very informative about ongoing research and connections with other organizations. Sometimes it's important to look beyond paper titles and abstracts and find a real expert in the particular subject matter to analyze full text papers.

Absence of any papers in a particular proliferation relevant area can also be significant. If other papers in related areas and other evidence of scientific or industrial activities indicate there is likely expertise in the missing area, one might suspect that publication in that particular subject area is being censored or controlled, possibly to hide a clandestine research program. Other tips for effective researching of scientific literature and documentation include:

- The vocabulary used with scientific and engineering subjects varies by country or region and the analysts must be able to recognize various descriptions for the same area of technology.

[55]The guidelines are available on the IAEA Website but are most easily found at the NSG's own site, www.nsg-online.org/guide.htm (July 2007).

[56]Ibid.

- Publication analysis may reveal links to military R&D organizations and foreign researchers, possibly indicating where foreign acquisition of technology could be occurring.
- Look at Websites of professional science and engineering organizations within the country. For example, the presence of a fluorine chemistry society or a computer society with a section devoted to applied numerical modeling would indicate some capabilities in these areas.
- Look at a country's entry into international science, math, or engineering contests. Most of these contests are for undergraduate-level students. The Association for Computing Machinery (an international organization) runs an annual programming contest, for example. How has the country performed in such competitions relative to other countries?
- Study the attendance of scientists from particular nations at international meetings, to gain insights on their research activities. Knowledge of the papers presented and meetings attended by such individuals could be useful. Examples of the types of meetings that are of interest include:
  - The Information Exchange Meeting on Actinide and Fission Product Partitioning & Transmutation
  - The International Detonation Symposium
  - The International Autumn Seminar on Propellants, Explosives, and Pyrotechnics (always in China)
  - The International Workshop on Separation Phenomena in Liquids and Gases

Industrial Capabilities

The NSG Guidelines (INFCIRC/254, Parts I and II) mentioned earlier are excellent starting points to evaluate a country's industrial and technological capabilities. How many of the commodities listed in INFCIRC/254 can the country produce or otherwise acquire in the required quantities? Is it skilled in utilizing the commodities for a range of technical and industrial purposes? The NSG listings were developed by very knowledgeable experts from NSG member countries. The parameters given represent realistic values for needed commodities that might be controlled. Certainly in many cases commodities with lesser specifications might be successfully used in proliferation activities, but these commodities are judged to be more widely available in the international marketplace and thus relatively uncontrollable. Also, in some cases technology advances may offer other routes to implementing required technology, but the NSG lists have not been updated to reflect this. Obviously many other commodities are also needed for a nuclear weapons program, but again, due to wide availability and wide usage, these other commodities do not appear on the NSG control lists.

One can search for specific categories of products listed by the NSG both via the Internet and in industrial directories. For example, Kompass (www.kompass.com) is one directory that allows searching simultaneously by commodity and by country. Another directory focused especially on mechanical and electrical products, Global Spec (www.globalspec.com), calls itself an engineering search engine. It is very good on application information and identifying leading companies for various commodities but weak on searching out small companies in lesser-developed countries.

You can also find a number of specialized regional or country directories on the Internet. For example, the Iranian Isfahan University of Technology Chemistry Department maintains a listing of chemicals manufactured in Iran along with links to the manufacturer's Websites.[57] Examples of more general national listings are the Federation of Malaysian Manufacturers (www.fmm.org.my/index.asp) or Iran Access (www.iranaccess.com). Usually it is necessary to look at individual companies' sites (if they exist) to determine whether the commodity listed meets the specific requirements as stated in the NSG guidelines.

---

[57]See www.iut.ac.ir/department.php?deptname=Chemistry%20Dept. and http://chem.iut.ac.ir/alprolist.htm (June 2007).

Considerable technical expertise, especially in applied engineering, might be needed to make accurate assessments of the appropriateness of specific commodities for use in nuclear development programs. Significant help may be available on Websites of leading manufacturers of the commodities or their competitors. These manufacturers often have extensive tutorial information relative to specifying and applying their products. In addition, the sales engineering or application engineering staffs of leading companies are often willing to discuss their assessments of smaller producers in countries of concern; most companies monitor competitors closely. Searching an online global business-to-business marketplace like Alibaba.com (www.alibaba.com) or Business OnLine (www.b2b-bestof.com) may also be useful to see both what commodities the country is trying to buy and what commodities it is offering for sale.

In countries where there are stock markets, detailed information on capabilities of particular companies can sometimes be found in their filings with the stock exchange. For example, when it was revealed that Scomi Precision Engineering Sdn. Bhd. (SCOPE) in Malaysia had built centrifuge parts that ultimately ended up in Libya, looking at a recent initial public offering prospectus of its parent company on the Kuala Lumpur Stock Exchange (available free on the Internet) gave many details. All the key manufacturing equipment at SCOPE was listed, its manufacturing rates, its principal raw material needs and where the material was obtained, its principal customer in Dubai (Gulf Technical Industries LLC), its number of employees, its key managers and their backgrounds, and some of its history. (This was all in image format on the stock exchange server, so it was not found by search engines other than by the name of the parent company, Scomi Group Berhad.)

Other potential resources include:

- *Regional or international trade fairs.* These can provide useful information. An example is Metal Asia (www.bmp.mta-asia.com), a regional trade show for the precision engineering industries in Asia that provides Web listings on all the companies exhibiting. BizTradeshows (www.biztradeshows.com) provides a directory of most but not all trade shows by region, city, and focus.
- *Government Websites.* Look for a country's Department or Ministry of Trade and Industry or equivalent (Department of Commerce in the United States) Website. These government departments often give good overviews of their country's industrial capabilities and goals.
- *Patent databases.* These can sometimes give useful information on areas of industrial strengths in more developed countries. In general, though, it is difficult to draw many conclusions from patent information. Some countries emphasize patents, others don't, and using technology in a weapons program certainly doesn't depend on holding patents on the technology.
- *Marketing reports by commercial organizations.* These are another source of industrial capability information, particularly in developed countries that often produce excellent surveys of particular industries on a global scale with country-by-country comparisons. The high cost of these reports (often over US$1,000) make their use difficult for some proliferation researchers.

Large industrial projects in a country can sometimes aid the acquisition of other technological skills. A large commercial nuclear power plant like Bushehr in Iran doesn't itself add many technological capabilities to a nuclear weapons program, but the frequent technical interactions between Russian and Iranian participants in this project create opportunities for the acquisition of additional expertise more relevant to a weapons program. A different example would be the U.S. allowing and facilitating the production of M1 tanks in Egypt for the Egyptian military. This surely has resulted in the transfer of significant production knowledge that can be used more generally in Egyptian industry, in both civilian and weapons applications. Moreover, nearly all countries must import certain commodities for high-technology efforts, so easy access to major world trading centers like Dubai, Singapore, or Hong Kong can effectively improve a country's industrial capabilities.

Technological capability is also highly dependent on project management skills. These are very difficult to assess except by looking at the success or failure of other technological programs within the country. Sometimes discussions with engineers who have worked in the country can give helpful insights.

Finally, don't overlook more general country assessments and the specialized work of NGOs that has already been done. The *CIA World Factbook* (https://www.cia.gov/library/publications/the-world-factbook/index.html) is always a good starting place. Talk to people from the country or travelers to it, or people who have worked in the country. Get a geographical sense of the country by continually locating key facilities on maps and Google Earth or other overhead coverage. Take advantage of the assessments done by other organizations.

Assessing Military R&D Capabilities

In some countries, the military operates large research establishments. These are often hard to get much information about. Often the production of high explosives needed in implosion system development will reside in military facilities as well as any hydrodynamic test facilities. Jane's (www.janes.com/index.shtml) sometimes can provide useful information on a country's military R&D programs.

Usually it is useful to determine, as much as possible, the military involvement in a nuclear weapons program as well as the organizational structure of the program and its fit within governmental structures. This is outside S&T capability analysis per se, but clues might come up based on associations of authors from different institutions on individual papers. In addition, in some developing countries, defense plants also produce civil goods and there is considerable information available on their capabilities. (Egypt has this situation for example.)

Nuclear Delivery Vehicle Capabilities

Assessing efforts and capabilities for the manufacture of nuclear weapons delivery systems can sometimes provide vital clues on the nuclear weapons development program itself. The principal nuclear delivery vehicles are usually considered to be aircraft, ballistic missiles, and cruise missiles. It is possible that other low-technology methods of delivery (truck, boat, or secret assembly from smuggled parts in the target city) might also be used effectively for a single strike or a few strikes.

The Missile Technology Control Regime (MTCR) defines both ballistic and cruise missiles to be of concern for nuclear delivery if they can carry a 500 kg or greater load a distance of 300 km or greater.[58] Because of the limited accuracy of long-range ballistic missiles (especially early development models), they can be a strong indicator of interest in nuclear weapons because arming them with a conventional warhead has limited military utility.

Weight distribution and shape of the payload in missiles is usually critical to their flight performance, so there would necessarily be some coordination between nuclear weapons designers and missile designers. Missile designers would need to know size, shape, and weight distributions of the warhead, and weapons designers would need to know the environmental envelope of the missile—accelerations, vibration spectrum, temperature extremes, and pressures.

In addition, if solid propellant missiles are being pursued, there may be an overlap in technologies to develop reliable solid propellant and those needed for reliable high explosives for nuclear weapons implosion systems. Solid propellants are often based on the same high explosive compounds used in nuclear weapons, and in the case of solid propellants, there is great concern knowing what conditions might result in their high-order detonation.

---

[58]Text of the Missile Technology Control Regime guidelines can be found at www.mtcr.info/english/guidelines.html (Aug. 2007).

Conversely, for nuclear weapons high explosives, there is great interest in how to obtain reliable, rapid initiation of detonation. Thus some testing facilities could easily be shared between a solid propellant missile program and a nuclear weapons implosion program.

In the case of aircraft delivery, if available aircraft have adequate payload capability, the fitting of a nuclear weapon to the aircraft is apt to be less technically demanding than missile delivery. However, the weapon designers would still need the environmental envelope and other characteristics (speed, range, altitude capability, navigation systems, penetration aids, and the like) of the aircraft that influence its feasibility as a nuclear delivery system. In the cases of all delivery vehicles, the safing, arming, and firing systems will require coordination between vehicle experts and weapons experts.

Nuclear-capable delivery vehicles might be acquired by foreign purchase, purchase of foreign design and in-country production, or by in-country development and production. If the vehicles are acquired by foreign purchase, obviously the weapon designers will have additional constraints to design the weapon to fit the vehicle. In the case of in-country development, much might be learned about the general technological capabilities and project management skills through information about the vehicle tests, if available. Missile test firings (but not necessarily their results) nearly always become known. In some cases, like that of North Korea, the country is an exporter of missiles and missile technology but apparently is not nearly as advanced in its nuclear weapons technology.

Additional Open-Source Analysis Software Tools

Tools to Identify and Visualize Relationships

Government agencies, businesses, and industry sometimes collect and store enormous amounts of information; however, only a small amount of this data is actually used due to its complexity and volume. Examples of such data include descriptive data such as name, phone number, address, affiliation, Social Security number (SSN), bank account numbers, and transaction data such as phone calls, bank transfers, and credit card purchases. Some software tools will allow such data to be automatically grouped into relational categories that the software discovers by analyzing the data. A graphical representation of relationships discovered in the data can be made available to the human analyst. Analyst's Notebook and Visual Links are two common software packages that are used to discover, analyze, and display such links.[59]

The example shown in Figure 11.2 displays information on two individuals using the same SSN. They are unlikely to be related because of the wide distance between addresses, so this is an example of a situation that probably warrants closer scrutiny. Similar situations might arise in nonproliferation studies by linking the same individual to different companies or different companies to the same address or telephone number. Such analytical techniques would be helpful in uncovering the procurement activities of the A. Q. Kahn procurement network.

Another example might involve starting with an open-source reference to an individual and organization and finding links to other banks or organizations. Such link analysis could show that one person is a key individual for the organization. See Figures 11.2 and 11.3.[60]

The visual diagrams showing links between identified entities provided by tools such as Analyst Notebook, IBM's Entity Analytic Solution (EAS),[61] and Visual Analytics' Visual Links can be effective; however, they are complex systems to learn, require structured

[59]See references 60 and 61.

[60]www.visualanalytics.com/products/visuaLinks/images, accessed May 5, 2007, used by permission of Visual Analytics.

[61]www-306.ibm.com/software/data/db2/eas/relationship, accessed May 5, 2007.

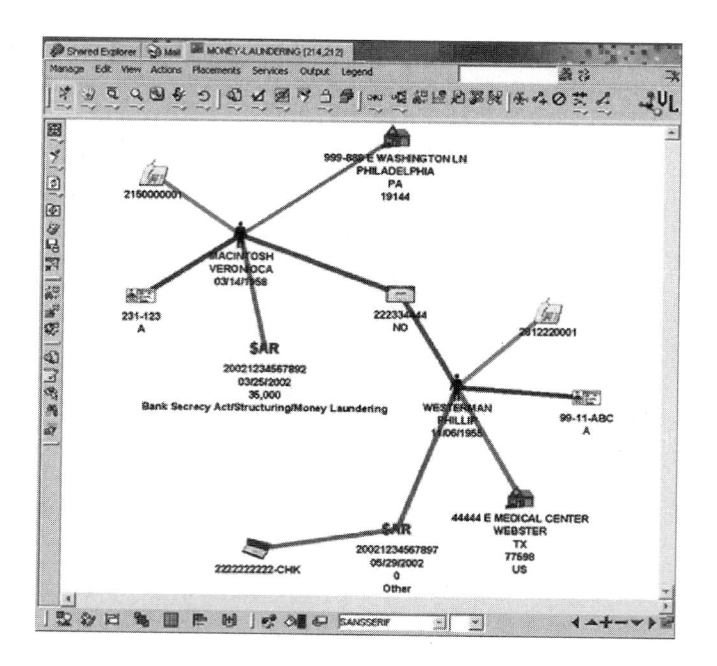

FIGURE 11.2 View of link analysis software tool.

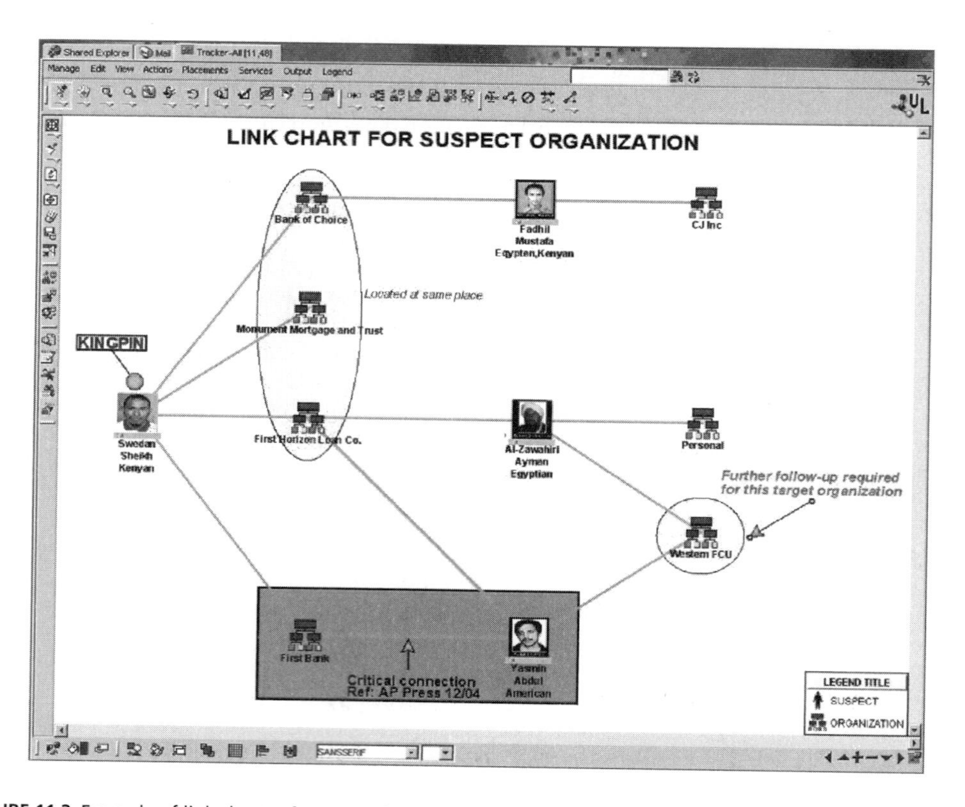

FIGURE 11.3 Example of link chart software tool.

(and sometimes proprietary) database formats, and require extensively trained and experienced analysts for effective use.

Desired Software Tools

Many software packages have been mentioned here as examples of currently available tools and capabilities. Now we describe some ideal capabilities that would be helpful for the nonproliferation analyst.[62] In many cases, some versions of such tools are beginning to become available, but often their accuracy, comprehensiveness, maintainability, cost, speed, and/or ease of use has not yet reached a point where they are routinely used by many analysts.

An important issue for many analysts is the ease of both installation and configuration of a software tool and minimal training time for effective use of the tool. Some analysts do not have the full-time support of an information processing department or programmers to convert large data sets into formats convenient for further analysis. In addition, many excellent software tools have been introduced to analysts but not used often or effectively because the analysts simply do not have the weeks of time needed for training and gaining experience with these newer tools.

Computerized Language Translation and Search

Automatic language translation has been an elusive goal for many years. However, the capabilities of such systems are gradually improving as new algorithms are developed and as computers increase in capability. What is really needed for nonproliferation analysis is a tool that can accept an English-language query, translate the query into effective target keywords in a foreign language, conduct a search for relevant sources in that language, and then translate the results back into English for presentation to the analyst.

Some such tools are being developed, but the automatic translations are still limited in accuracy and comprehensiveness. Unfortunately, such tools are most needed for languages that are most difficult to translate into English. Many Western publications, especially S&T literature, are already available in English; however, many publications in Arabic, Farsi, Korean, Chinese, and Japanese are not currently translated. Better tools for the English-speaking analysts to access these data are needed. It should be kept in mind, however, that the optimum means for assessing foreign-language S&T literature for proliferation analysis is to have analysts that are both language trained and knowledgeable in the technical aspects of nuclear proliferation.

Entity Extraction

Much nonproliferation open-source data is in unstructured text or graphical formats. Extracting specific entities from these data and transforming the data into a universal storage format (perhaps XML?) or a relational data format that can be used for further processing is a rapidly evolving field. Again, some of the issues involve recognition of a complete set of entities, language differences, and the complexity of software installation and use. In addition, some of these systems are currently prohibitively expensive for many small analytical groups.

Better tools are needed to automate the extraction of "relational events" from free-form text—not only who or what but why, when, where, and how. A solution would enable analytical processing by automating the transformation of written language into structured, relational data. The result would allow dramatically faster and more comprehensive detection of trends, anomalies, patterns, and linkages. Once extracted in this structured form, the information can be pushed downstream to feed virtually any system that processes relational tables. Such tabular data can be stored in a shared data warehouse repository for data mining

---

[62] R. K. Wallace, "Improving Detection through Improved Collection/Evaluation of Open Source Information," INMM/ESARDA Workshop, Santa Fe, New Mexico, Oct. 30–Nov. 2, 2005.

or link analysis purposes. For example, these data can be input directly to a link analysis tool such as the previously mentioned EAS, Visual Links, Analyst Notebook, or SmartDiscovery for further analysis, or they could be sent via an automated email alert to an analyst. Currently, however, these tools are rather complex and expensive enterprise-based solutions. A wider variety of more user-friendly entries in the field of entity extraction and categorization of unstructured text data is needed.

Eliminating Duplicates

Many different news media might pick up the same story from multiple sources. Sometimes these accounts contain useful complementary information; however, they often just repeat the same information. An S&T search using several different bibliographic databases may contain a large number of duplicate references, with each source providing only a small (but perhaps important) number of unique entries. The data collector or analyst needs an automated method to eliminate such duplicates. This capability is needed for both structured (S&T references, entities extracted, tabular data) and unstructured (text reports) data.

Expanded Data Sources

Although not technically a software tool, new databases of information related to international nuclear activities available through open sources could greatly assist the job of tracking proliferation. Databases that are needed include detailed information on global imports and exports (particularly export applications that are denied by virtue of nuclear or duel-use regulations), data on criminal records, and expanded data on businesses in non-Western countries.

Summary

Open-source analysis techniques are a powerful tool for detecting, characterizing, and assessing undeclared nuclear activities. A wide variety of sources are available, including global news media, scientific literature, government reports, trade publications, geographic maps, technical drawings, photographs, and satellite imagery. Each type of information, when assessed in relation to the others, can lead to a more integrated understanding of a nation's nuclear activities. Open-source analysis using automated tools on large sets of data sources requires the careful formulation of queries and search strategies as well as understanding the origin and quality of sources.

Moreover, a broad technical and general expertise is needed by analysts and analyst teams to perform effective assessments of a nation's technological capability for the purpose of detecting and monitoring nuclear weapons development efforts. Technology assessments must be conducted using integrated information from all the other open-source techniques to develop a comprehensive understanding of a state's proliferation activities and intentions. This type of analysis requires sustained resources, excellent research skills, and multidisciplinary teams. It also requires the proper recording and citation of sources and methods to allow others to follow your analysis and support or challenge its conclusions. Finally, analyses should include peer reviews and caveats reflecting the levels of confidence in the reliability of sources and conclusions.

Acknowledgments

The authors would like to acknowledge helpful discussions with John Lepingwell and Victor Braguine of the IAEA. Particularly useful reviews of this text were contributed by Professor Alexander H. Montgomery of Reed University, Dr. Lynn Eden, and Mr. Lewis Franklin of the Center for International Security and Cooperation (CISAC) at Stanford University.

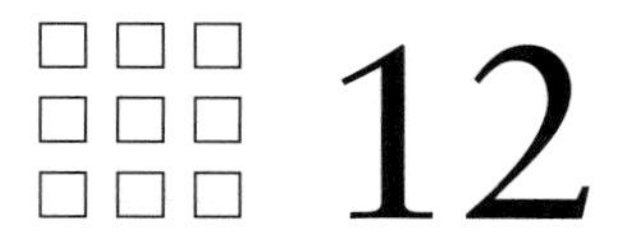

12 Commercial Satellite Imagery: Another Tool in the Nonproliferation Verification and Monitoring Toolkit

Frank Pabian

Introduction

This chapter highlights the evolutionary and revolutionary role that commercial satellite imagery (as another independent "open source") is playing on the international stage in providing a heretofore unimaginable basis for greater global transparency and the way it has helped, and will continue to help, detect and monitor undeclared nuclear-related facilities and activities. In that role, such imagery serves as an enabler of transparency for nation states previously lacking their own "national technical means" (NTM, a euphemism for "spy satellites"), sometimes even being referred to as a "Poor Man's NTM,"[1] but also, perhaps more significantly, for multinational organizations, nongovernmental organizations, the news media, academics, or even just plain individual interested parties or hobbyists. The opportunity now exists for anyone with relatively modest computer access (and a credit card for supplemental imagery purchases when necessary) to view, identify, and monitor nuclear facilities and associated activities that have the potential to help threaten international peace and security and that might otherwise have remained clandestine.

Although commercial satellite imagery can have myriad applications covering a whole spectrum of subjects of interest to humanity (and is yet another example of the *democratization* of information in the Internet age[2]), this chapter focuses on the imagery's capabilities to serve the interests of nuclear nonproliferation. In this chapter, this author also hopes to provide the reader with a greater appreciation and understanding of some of the methodologies

---

[1]Lewis Dunn and Marjorie Robertson, *Satellite Imagery Proliferation and the Arms Control Intelligence Process*, Science Applications International Corp., April 22, 1997, as quoted by Lt. Col. Larry K. Grundhauser, USAF, "Sentinels Rising: Commercial High-Resolution Satellite Imagery and Its Implications for US National Security," *Airpower Journal*, Winter 1998, p. 70; www.airpower.au.af.mil/airchronicles/apj/apj98/win98/grund.pdf.

[2]Chris Dibona, "Widely Available, Constantly Renewing, High-Resolution Images of the Earth Will End Conflict and Ecological Devastation As We Know It," http://edge.org/q2007/q07_7.html#dibona.

associated with the interpretation of commercial satellite imagery as well as the utility and limitations of such imagery for detecting undeclared nuclear activities.

Because it is impossible within this short chapter to school the reader in the varied and complex aspects of imagery analysis and associated image processing and remote sensing[3] technologies, it is this author's intention to provide the reader with at least the means to pursue the subject in greater depth by way of selected references along with an original exemplar illustration (along with some "how to" and "what is possible" information in the Appendixes on this book's companion Website). In that exemplar, a veritable "textbook case," we will see how freely available commercial satellite imagery can yield new, previously unavailable information on a clandestine, allegedly nuclear-related, facility, as a followup to open-source "tip-off" information.

Commercial Satellite Imagery: A New Basis for Modern Geospatial Awareness and Global Transparency

Only quite recently have the hardbound printed-paper atlases and topographic-sheet maps that for centuries had been the mainstay of geographical awareness and navigation become (for the most part) only collectible relics of a bygone era. At their best, such maps (no matter how accurate) were effectively only artistic symbolic renderings that could never match the detail, completeness, and timeliness currently available with modern *geographical information systems* (GIS)[4] combined with the publicly accessible digital overhead imagery archive. Past limitations to map completeness were largely due to physical space limitations, hence an emphasis was placed on urban developed areas (often ethnocentrically biased), with rural or undeveloped areas generally given less attention. Other omissions, however, were more the result of state censorship, particularly in less open societies, to limit enemy navigation and to prevent the unauthorized detection and identification of sensitive facilities or activities (and even in open societies such activities persist; see Figure 12.1[5]).

Until the advent of commercial satellite imagery, the status quo of remoteness providing "security through obscurity" had proved to be quite effective for concealing undeclared facilities and activities (except against the few nation states having the requisite enormous intelligence collection and analysis capabilities to overcome such obstacles). This is largely because, as recently as 1999, acquisition and use of high-resolution satellite imagery—and the searching for, identifying, and characterizing clandestine facilities and activities—was the

---

[3]For the purposes of this study, *imagery analysis* is defined as "the process by which humans and/or machines examine photographic images and/or digital data for identifying objects and judging their significance," and *remote sensing* is "the measurement or acquisition of information of some property or object or phenomenon by a device that is not in contact with the object or phenomenon." (Source: the *Manual of Photographic Interpretation*, American Society of Photogrammetry and Remote Sensing.)

[4]"A GIS is a computer system capable of capturing, storing, analyzing, and displaying geographically referenced information; that is, data identified according to location. Practitioners also define a GIS as including the procedures, operating personnel, and spatial data that go into the system. The power of a GIS comes from the ability to relate different information in a spatial context and to reach a conclusion about this relationship." (Source: http://erg.usgs.gov/isb/pubs/gis_poster/.)

[5]An apparent Swedish government attempt to censor its national overhead imagery archive was recently exposed using Google Earth through its employment of commercial satellite imagery from DigitalGlobe.

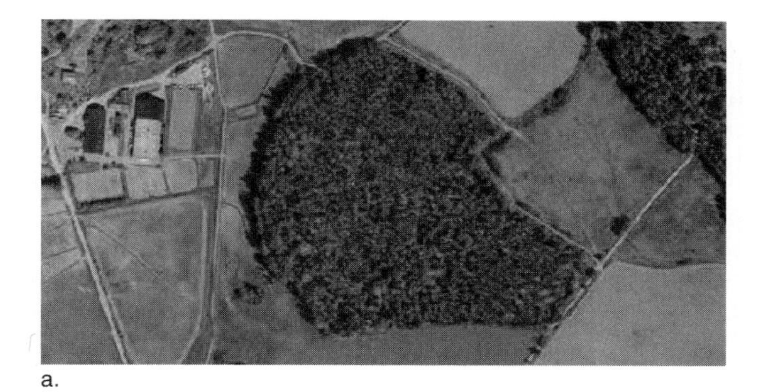
a.

b.

FIGURE 12.1 On April 7, 2006, bloggers revealed that Sweden's GIS agency had censored aerial images to conceal its spy headquarters. Figure 12.1a shows the Swedish government mapping service image of the area; Figure 12.2b is a Google Earth image. Even open societies have facilities that they would prefer to remain clandestine, and even on Google Earth (and Virtual Earth). The U.S. Naval Observatory (which includes the U.S. Vice President's residence) in Washington is currently obscured through pixilation. Most recently, Google Earth reportedly agreed to pixilate sensitive sites in India and France. In such cases, therefore, rather than "data without borders," we may increasingly find "pixels with prior-restraint."

sole purview of superpower national governments that kept their capabilities and knowledge classified and under extremely tight security controls.

Now, however, there has been a true paradigm shift in the quality and quantity of geospatial data readily available to the public. Not only are some of those former NTM historical archives being declassified and made accessible for public review, but more current high-resolution satellite imagery is now either freely available over the Internet or through direct purchase through commercial vendors.[6] It is available to anyone who has access to a typical home computer linked to the Internet. Foremost among the growing family of freely available GIS tools (also known as *applications*) and commercial satellite imagery data sets that now provide the basis for our new "Digital Earth" are these: Google Earth

[6]"Private eyes in the sky," *The Economist*, May 4, 2000, www.economist.com/displaystory.cfm?story_id=333111; and Patrick Clarke, "Commercial Satellite Imagery Matures as an Asset," April 23, 2004, www.military-geospatial-technology.com/article.cfm?DocID=461.

(http://earth.google.com), Virtual Earth[7] (http://maps.live.com), NASA's World Wind[8] (http://worldwind.arc.nasa.gov), USGS Global Visualization Viewer (http://glovis.usgs.gov), TerraServer (www.terraserver.com), SkylineGlobe (www.skylineglobe.com), Yahoo Maps Beta (http://maps.yahoo.com/beta/#env=F), GeoFusion (www.geofusion.com), Flash Earth (www.flashearth.com), or France's developing Geoportail (www.geoportail.fr).

These can all be augmented by additional separate imagery data purchases, if so desired. Some of these GIS tools and satellite image data sets can also be cross-linked (or augmented) with global positioning systems (GPS) for improved geospatial accuracy related to enhanced personal navigation. (As of this writing, it is already an additional option for use with enabled cell phones and for an in-vehicle navigation system from at least one automobile manufacturer.) This new capability has created a new era of empowerment whereby anyone so inclined can conduct highly sophisticated investigative surveillance, assessments, and reporting of clandestine facilities and associated activities anywhere of their choosing.[9]

In this chapter we review some specific examples of how this process can play out, using nothing more than a Wi-Fi (wireless) or broadband access-enabled notebook computer while sipping your favorite drink at your local coffee shop, or perhaps while under the shade of a palm tree at some tropical island resort. (And for stay-at-homes, in the near future it will very likely be possible to "fly and zoom" around a newly discovered clandestine facility in

---

[7]Google Earth (basic) was offered as a free application in June 2005; by mid-2007 it had recorded over 250 million activations. It is PC based and was the first such data set to provide seamless infinite zoom (from global "synoptic" views down to individual houses) and infinite pan, along with high-resolution terrain drape capability for three-dimensional (3-D) perspective visualization, free measurement tools, and a free 3-D building and modeling tool (Sketch-up). Three-D "fly-through" videos of your favorite locales are also possible with the for-fee version, Google Earth (Pro). Interestingly enough, Google Earth was based on a product called Keyhole, which was previously owned by a company by that same name that was founded by In-Q-Tel, a Central Intelligence Agency-created Silicon Valley "seed" company. Virtual Earth, a free Microsoft product (derived from GlobeView 3-D Viewer by GeoTango, now Microsoft owned), which debuted on November 7, 2006, is Web browser based (Explorer only at present) and has the advantage of already built and "textured" (photorealistic, not plain gray as had earlier been provided by Google Earth) for selected large population cities on a list to be continually expanded to around 3,000 over the next five years. SkylineGlobe seems to have the best "Gazetteer" (place name list) and offers the advantage of overlaying live video feeds directly onto the static base imagery. This author is a proponent of all such systems because they offer unlimited new and exciting ways to view Earth. It should also be noted that there now appear to be two globally significant competitive consortia forming as a result of ongoing negotiations of long-term contracts. Those consortia could consist of GeoEye, Microsoft, and Yahoo versus DigitalGlobe, Google, and the National Aeronautics and Space Administration (NASA).

[8]On Dec. 18, 2006, NASA announced a joint partnership with Google Earth. See www.nasa.gov/centers/ames/news/releases/2005/05_50AR.html and http://earthissquare.com/2006/12/18/google-and-nasa-joining-up-to-release-well-world-wind/.

[9]One very clear example just came to this author's attention regarding the discovery by a Google Earth Community member, Ken Grok, from Germany, who located a previously unheralded site in China that takes the form of a huge outdoor terrain model that completely replicates a huge area of disputed border between China and India. The outdoor terrain model was clearly constructed for some type of military training purposes and was not necessarily something that China would want to have publicized. See: Lester Haines, "Chinese black helicopters circle Google Earth Mystery military project wows the crowd," *The Register*, July 19, 2006, www.theregister.co.uk/2006/07/19/huangyangtan_mystery/, accessed 19 July 2006.

3-D space on a large-screen high-definition television [HDTV] with the aid of the latest video game controller.)

The utility of commercial satellite imagery, and its impact on global transparency,[10] has grown in direct proportion with the quality (or resolution[11]) of that imagery—in other words, the better the quality, the more useful it is. Publicly available satellite imagery has rapidly improved with time after it first became available in 1972 (see box). As of this writing, the best-resolution (best possible from nadir) commercial satellite imagery is now 41 centimeters (~16 inches) in panchromatic (*pan*, or black and white) and 2.44 meters (~8 feet) multispectral (MS).[12] By way of illustration, it should now be possible to "identify a spare tire on a medium-sized truck"[13] or even identify a basketball hoop within the context of an outdoor basketball court. At this writing, the commercial satellite constellation dedicated to the high resolution or "military" market (sub-one-meter resolution) consists of five systems (three from two separate U.S. firms, one from Israel, and one from Russia).[14] Table 12.1 shows what is currently available (in standard black type) with what is also planned (boldface type). (*Note:* Imagery sold to commercial customers will be resampled to 0.5-meter resolution. Both DigitalGlobe's and GeoEye's current operating licenses with the National Oceanographic and Atmospheric Administration (NOAA) do not permit the commercial sale of imagery below 0.5-meter resolution. It is not clear if or when such a limitation might be rescinded.)

The Table 12.1 list represents only satellites that acquire imagery from multiple bands, usually four, with three (red, green, and blue) taken in the *visible* and one from the near-infrared (*nonvisible*) part of the electromagnetic spectrum. These MS bands generally offer lower resolution as a result of engineering concerns with respect to signal to noise than is true for panchromatic. However, one advantage that they do offer is that if the three visible bands are combined such that they are reproduced as red, green, and blue, respectively, they can be combined such that they artificially recreate a composite "natural-appearing" color image. That resulting relatively low-resolution color image can then be enhanced with a higher-resolution panchromatic overlay. The resulting "fused" product (four bands total) is referred to as *pan-sharpened MS* and is generally the best for infrastructure analysis and visualization purposes.

Other imaging satellites, which are not the central focus of this review, can acquire images from the nonvisible portions of the electromagnetic spectrum, including the microwave

---

[10]Some have argued that global transparency has its downside. For more on that subject, see Kristin M. Lord, *The Perils and Promise of Global Transparency: Why the Information Revolution May Not Lead to Security, Democracy, or Peace*, 2006, ISBN10: 0-7914-6885-2; and Alasdair Roberts, *Blacked Out: Government Secrecy in the Information Age,* 2006, ISBN-10: 0521858704.

[11]*Resolution* "generally refers to the size of the smallest object that can be distinguished in an image from its surroundings. Higher-resolution images enable imagery analysts to detect and identify smaller objects." (Source: Baker et al, *Commercial Observation Satellites*. See also Table A in the Appendix, available on this book's companion Website.)

[12]W. E. Stoney, *ASPRS Guide to Land Imaging Satellites,* Updated for the NOAA Commercial Remote Sensing Symposium Key Trends and Challenges in the Global Marketplace, Sept. 12–14, 2006, www.asprs.org/news/satellites/satellites.html; see also www.licensing.noaa.gov/Optical_Remote_Sensing_Satellites_8-9-06%20without%20sensors.xls.

[13]According to the *National Imagery Interpretability Rating Scale* (NIIRS, high 6), a methodology established by the U.S. National Geospatial Agency (NGA) to define and quantify the relative quality of overhead imagery for interpretation purposes.

[14]W. E. Stoney.

Table 12.1 Current and planned high-resolution commercial imaging satellites.

Optical Land Imaging Satellites by Best Resolution

Satellite	Country	Launch	Pan Res.m	Ms Res.m	Swath Km
Very High Resolution (.41 To 1 Meters)					
GeoEye-1	***US***	**08/22/08**	**0.41**	**1.64**	**15**
WorldView-1	*US*	09/18/07	0.5		16
WorldView-2	***US***	**mid-2009**	**0.5**	**1.8**	**16**
QuickBird-2	US	10/18/01	0.6	2.5	16
EROS B1	*Israel*	04/25/06	0.7		7
EROS C	***Israel***	**mid-2009**	**0.7**	**2.5**	**16**
Pleiades-1	***France***	**07/01/08**	**0.7**	**2.8**	**20**
Pleiades-2	***France***	**07/01/09**	**0.7**	**2.8**	**20**
IKONOS-2	*US*	09/24/99	1.0	4	11
OrbView 3	*US*	06/26/03	1.0	4	8
Resurs DK-1 (01-N5)	Russia	06/15/06	1.0	3	28
KOMPSAT-2	Korea	07/28/06	1.0	4	15
IRS Cartosat 2	India	01/10/07	1.0		10

Note: OrbView 3 has been inoperable since April 23, 2007.

Source: W. E. Stoney, *ASPRS Guide to Land Imaging Satellites,* Updated for the NOAA Commercial Remote Sensing Symposium Key Trends and Challenges in the Global Marketplace, Sept. 12–14, 2006; www.asprs.org/news/satellites/satellites.html.

region (radar imagery) and thermal infrared bands. Each of those remote imaging sensor systems has its own unique value that can certainly add to the overall body of knowledge about clandestine sites and their associated activities, particularly when used to augment visual spectrum imagery. However, to date, they have simply not been as readily accessible by the public, and such imagery generally requires a far greater degree of expertise and training to exploit adequately.[15] The advantages and disadvantages of each sensor type, beginning with the visual bands, are provided in Table B in the Appendix (see preface for companion site information).

Along with the improved *quality* of the data now available there has been a concomitant increase in the *quantity* of data available. Currently, there are 35 separate commercial imaging satellite systems in Earth orbit from 17 different nations having resolutions of 39 meters or better. The number of satellites is expected to roughly double by 2010. For a complete list of all satellites, their capabilities, and their relative footprints ("swath" or spatial/areal coverage), see www.asprs.org/news/satellites/satellites.html.

[15]Italy's upcoming radar constellation, called COSMO-SkyMed, is a dual-use system, so data distribution will have some limits. Four satellites will be launched between 2007 and 2009 that will provide day/night all-weather X-band SAR imagery at resolution down to 1 meter over a 10 km by 10 km area, depending on mode (1, 3, 15, 30, 100 m GSD options available, with swaths from 10 to 100 km). Products will include polarized SAR and interferometric images or digital topography. The system can acquire 1,800 images per day. Revisit time can be as little as 12 hours. These specs come from a brochure, so some caution is needed in taking them literally; some combinations or locations may not be possible. Also, not all plans are realized, but it looks powerful on paper and, if true, could prove quite useful in conjunction with electro-optical imagery.

Commercial Satellite Imagery: A Chronology

Commercial satellite imagery had its roots in the U.S. government-funded Earth Resources Survey Program that was established in 1965 under the auspices of NASA. That program was an effort to apply space technology to help monitor, evaluate, and solve such growing global problems as environmental degradation due to urban growth and pollution, deforestation, desertification, and so on. It also was intended to provide a more accurate basis for mapping to aid in exploring for increasingly scarce mineral and fossil resources and for inventorying agricultural development. The program led to the launch in July 1972 of Landsat (originally dubbed the Earth Resource Technology Satellite, or ERTS) which, via its four-band multispectral scanner (MSS), had an effective resolution of 80 meters, which was sufficient to barely detect only the largest known cultural features such as cities and connecting freeways.

Resolution improvements were implemented in the early 1980s in which the effective resolution was improved via a seven-spectral band scanner, called the Thematic Mapper, which was the prime instrument on Landsats 4 (1982), 5 (1984), 6 (this satellite failed to attain orbit during launch and thus has never returned data), and 7 (1999). That newer system was capable of resolving objects as small 30 meters. Landsat 7 carries a single sensor, the ETM+, which has a panchromatic ("black and white") band that achieved 15-meter resolution (from Nicolas M. Short, Sr.'s, *The Remote Sensing Tutorial*, http://rst.gsfc.nasa.gov/Intro/Part2_20.html). However, the Landsat system's best resolution capabilities had already been eclipsed in 1986 with the successful launch by France of the SPOT-1 satellite that had a 20-meter resolution in multi-spectral (MS) bands and 10-meter in panchromatic (SPOT-5 now both PAN and MS at 2.5 meters resolution).

In 1995, India took the lead with its IRS-1C satellite that provided a 5-meter resolution panchromatic capability. In 1999, Space Imaging (now owned by Geo-Eye; see www.geoeye.com) launched its Ikonos-2 satellite, which provided a 1-meter resolution and 4-meter MS, followed in 2001 by DigitalGlobe (formerly Earthwatch) launching Quick-Bird-2, which had similar capabilities but which was flown at a lower orbit, enabling a resolution of 0.61-meter panchromatic and 2.44 meter MS (see www.digitalglobe.com). In 2003, OrbImage (now Geo-Eye) launched OrbView-3, which has capabilities similar to Ikonos-2. Most recently, Digital Globe successfully launched and is now operating its WorldView-1 satellite, providing startling clear 0.5 meter resolution panchromatic imagery (see www.digitalglobe.com/index.php/86/WorldView-1). For graphical resolution comparisons, see www.fas.org/irp/imint/imint_101.htm.

An informative update on the status of emerging European capabilities in this field and their impact on a future world in which "there is no place to hide" and including plans for a 0.5-meter radar imaging satellite can be found in Theresa Hitchens, "European Eyes in the Sky," *Imaging Notes*, Fall 2006, pp. 20–24.

Source: www.imagingnotes.com/go/article_free.php?mp_id=70&cat_id=14&PHPSESSID=6ae5ea4d7ea3fd46498f9f9ca332be13

Imagery Analysis: The Process

Commercial satellite imagery, by itself, is nothing more than compendium of raw data in the form of picture elements (pixels) that is of little use without the addition of interpretation and analysis. The two keys to remember are (1) commercial satellite imagery is most often acquired from a perspective that is most generally quite unfamiliar to most people, being from an overhead, nearly vertical, view, and (2) commercial satellite imagery never comes with "labels." Imagery analysis is the process of deriving those labels and the process of determining their significance, both of which "add value" to that raw data. The process can involve at least two strategies, either alone or in conjunction with each other. The first is *direct recognition*, which is identifying a nuclear reactor by the presence of a signature containment dome adjacent to a cooling system. The second is by *inference;* for example, the presence of a security fence around a facility might be inferred even when not directly observed due to differential vegetation growth across a fenced boundary that might arise from animals grazing outside such a perimeter but not inside.[16]

Imagery analysis of clandestine nuclear sites is not something that can automatically be done simply because the necessary tools are now readily available. It takes years of experience and skill, diligently working with those tools with creativity and imagination, to use them effectively and to maximum utility. Furthermore, technical training is necessary for an in-depth understanding of the nuclear fuel cycle and weaponization infrastructures to correlate what is observed on imagery with what is known about a nation's overall program (including the political context).

Moreover, a working knowledge of common industrial processes and infrastructure is necessary to provide a sound basis for quickly sifting "the wheat from the chaff." Some industrial facilities with no connection to nuclear or military production can appear to have some of the same features as facilities that are of proliferation concern (i.e., aluminum plants could be easily be mistaken for uranium enrichment plants). Most modern industrial and chemical plants require multiple utilities services, heat transfer in the form of cooling towers or ponds, ventilation stacks, waste ponds, and the like that can be identical to those found in use by the nuclear industry. On a more subtle scale, a working knowledge of the culture, regional economy, and general level of industrial development in the country that is under observation can be very helpful in quickly identifying an unknown facility or object. Here's a simple example: Oil-fired or coal-fired brick-making kilns (factories) are common in many less developed parts of the world and less common in more developed areas, such that an uninitiated imagery analyst might easily mistake one as something of significance (see Figure 12.2). It is therefore necessary that the analyst, while examining the fine details of an image, always keep the big picture in mind regarding the geographic location of the facility of interest and its cultural context.

Volumes have already been written on generic imagery interpretation and analysis, which address the various techniques and methodologies to identify objects from imagery (interpretation) and how to derive new information (analysis) in the process. The Bibliography on this book's companion website provides further in-depth reading on these two subjects. Nonetheless, for this chapter, we will briefly cover several of the main points.

One knowledgeable practitioner succinctly describes this process of imagery interpretation as enlisting the basic "Five Ss": size, shape, shadow, shade, and surrounding objects.[17] An analyst's ability to correctly identify objects on imagery and determine their significance requires, at minimum:

- *Size.* An awareness of the scale of the imagery being studied and hence the true (and relative) sizes of the objects that are imaged
- *Shape.* The physical characteristics of the objects (i.e., cultural or "manmade" objects are commonly distinguished by angularity and are often comprised of geometric shapes rather than the random natural features)

[16]Nicolas M. Short, Paul D. Lowman, Jr., Stanley C Freden, and William A. Finch, *Mission to Earth: Landsat Views of the World,* National Aeronautics and Space Administration, 1976, p. 8.

[17]Dr. William C. Green, California State University San Bernardino, Department of Political Science, *Strategic Intelligence Syllabus*, PSCI 621, Topic 7, Imagery Intelligence IMINT; see http://dcr.csusb.edu/psci/green/621SYW04.pdf.

FIGURE 12.2 Numerous, innocuous (and air-polluting) oil-fired brickmaking kilns are present on this image that are a common feature in the developing world, but not necessarily a familiar one to most nonproliferation imagery analysts. Also be aware that, although nuclear facilities frequently contain ventilation stacks that are never intended to exhaust "smoke," a smokestack may be present at, or near, a nuclear facility in association with a supporting thermal (conventional) power plant or steam plant.

- *Shadows.* The sun angle and its orientation relative to the viewer providing information on both true and relative heights and silhouette shapes of the imaged objects
- *Shade.* The tonal brightness and contrast of the objects, both individually and in relation to their surroundings (assuming panchromatic, but the color capability of most commercial satellite imagery is an additional bonus)[18]
- *Surroundings.* The context and setting of the objects in the scene being studied, including both topographic and geographic associations

However, further review of the process by others reminds us that there are even more factors that come into play in image interpretation, such as determining an identification from an image that forms the necessary ground work for subsequent analysis as follows:

- *Signatures.* In general, cultural or manmade features have certain consistent common functional characteristics that help separate them from their surroundings

---

[18]With the advent of computer-based digital image processing such as is now possible with software tools like ENVI and Adobe Photoshop combined with fast processors and the requisite Random Access Memory (RAM), contrast stretching and other enhancements are readily possible to highlight features for better visualization that might otherwise be lost, such as objects hidden within shadows. However, the analyst is cautioned always to ensure the integrity of the raw data. Any manipulation (for example, to add or remove data such that fake objects are created or real objects disappear) is anathema and will in any case ultimately be found out thanks to the frequency of coverage from multiple independent satellites. The topic of image processing is an entire discipline in itself and is beyond the scope of this chapter. For further study, see, for example, Chris McGlone, with Edward Mikhail and James Bethel, *Manual of Photogrammetry*, fifth edition, 2004, ISBN 1-57083-071-1.

and uniquely identify them. Here, the spatial arrangements of many key elemental objects form a whole that is unique, or nearly so. Pattern and texture both come into play. For example, all nuclear reactors require some type of cooling system, therefore cooling towers are an example of one key element that (though not a unique identifier by themselves) when taken together with other key signature elements, generally form a consistent pattern that can be used to uniquely identify such a facility. Spray ponds, another type of cooling methodology, are readily distinguished from their surroundings by textural differences. The "association" of multiple key signature elements leads ultimately to identification.[19] This is the basis for *image interpretation keys*, or guidebooks, that provide details on the signatures helpful in making such identifications.[20] But be forewarned, "signature suppression," such as the concealment effort employed by Iran to inhibit functional identifications at the Natanz uranium enrichment facility (see Figure 12.3), is an ever-present possibility and should always be taken into consideration when conducting any imagery-based site evaluation. (See the Bibliography on this book's companion Website under "Camouflage, Concealment, and Deception.")

- *Time.* Temporal changes determined from monitoring a site with multiple images acquired at different times (regardless of the interval that can vary from minutes to years) provide insights on functionality, operational history, expansions to the facility, construction pace, and the like.

Some could argue that *imagery interpretation* and *imagery analysis* are two different forms of *imagery exploitation* that are mutually exclusive in their requirements with respect to skills, knowledge, and abilities. This case can be made by saying that interpretation merely answers the *objective* questions of what, when, or where because it only provides "identifications, basic descriptions, and limited information." The argument would go that imagery analysis involves a higher level of deduction and relies heavily on what is generally termed *convergence of evidence*, which incorporates multiple pieces of information deriving from multiple identifiable imagery signatures, as well as *collateral information* derived from other information sources beyond the imagery alone (and which can also include the analyst's personal subject area knowledge and experience). Although it might not be explicitly stated, if we accept that definition, imagery analysis provides the answers to the more *subjective* questions of why, how, and what is the significance.

Regardless of how one wants to define imagery analysis, with respect to issues of nuclear nonproliferation verification and monitoring, every imagery analyst should aspire to have the ability to provide "value-added" answers to all the above questions. The key to success is the ability to discern, to the greatest degree humanly possible, what is significant on the imagery and why. Moreover, imagery analysis, because it is always somewhat subjective (an art as much as a science[21]), should almost always include caveats such as "possible" or "probable," depending on the level of confidence the analyst assesses to be appropriate.[22]

---

[19] *Manual of Photographic Interpretation*, second edition, American Society of Photogrammetry and Remote Sensing, 1998, ISBN 1-57083-039-8, pp. 56–62.

[20] Defense Mapping Agency, *Photo Interpretation Student Handbook: Module 2: Cultural Features* (vols. 1 and 2), April 1996, available from the U.S. Government Printing Office (in particular, vol. 2, pp. 435–451 pertaining to the nuclear fuel cycle).

[21] Dino A. Brugioni, "The Art and Science of Photoreconnaissance," *Scientific American*, March 1996, pp. 78–85.

[22] There is an old adage in this business: "If it looks like a duck, walks like a duck, and quacks like a duck, it is probably a duck." When there is similar reinforcing "convergence of evidence," the analyst is generally safe is assigning a high probability to his or her conclusions.

The Utility of Commercial Satellite Imagery for the Detection of Clandestine Nuclear Facilities

Clandestine facilities, by their very definition, are not intended to be readily detectable or identifiable. History has shown, however, that many such facilities have nonetheless come to light as a result of detection and correct identification by outside observers through a variety of means. Those means often relied heavily on the use of satellite imagery. Among the examples from the past, the detection of the Kalahari nuclear test preparations in South Africa by the former Soviet Union in the late 1970s and the detection of the clandestine reprocessing plant in North Korea by the United States in the early 1990s are notable. Those cases involved the use of classified imagery from entirely state-controlled systems either alone or possibly in some combination with information derived from some other sources.

Some of those formerly highly classified (covert or "spy") satellite systems have now been declassified along with sample images and associated derived analytical reporting and documentation that reveal amazing insights into this previously tightly controlled technology and tradecraft. Specific examples pertinent to detecting undeclared nuclear activities can be found in *Corona: America's First Satellite Program*, published by the Central Intelligence Agency (CIA) History Staff.[23] That volume describes how, in August 1964, imagery analysts working for the CIA, using classified satellite imagery, were able to correctly locate China's then clandestine nuclear weapons test site and correctly identify and describe, *in advance*, preparations for an atmospheric nuclear test in October 1964 (China's first ever). That report was particularly salient at this time, since very similar discoveries of similar sites are now being made with commercial satellite imagery.

In late 2006, independent imagery analysts working for the *New York Times* (and at least one nongovernmental organization or NGO) were able to follow up on open-source press leads to correctly locate North Korea's clandestine underground nuclear test site near Kilju *in advance* of the first test that occurred there on October 9, 2006.[24] That find is particularly significant when one considers that North Korea is one of the most secretive and least known countries on Earth. Even the South Korean newspaper *Hankook Ilbo*

---

[23]Kevin C. Ruffner, editor, CIA Cold War Records, *Corona: America's First Satellite Program*, Center for the Study of Intelligence, 1995. Available through Military Reference Branch (NNRM), Textual Reference Division, National Archives and Records Administration, Washington, D.C. 20408. See in particular CIA/NPIC, Photographic Intelligence Report, "Uranium Ore Concentration Plant, Steiu, Romania," Dec. 1961, pp. 157–168; Regional Nuclear Weapons Storage Site near Berdichev, USSR," May 1963, pp. 169–174; and Special National Intelligence Estimate 13-4-64, "The Chances of an Imminent Communist Chinese nuclear Explosion," Aug. 26, 1964, pp. 237–246.

[24]Douglas Jehl and David E. Sanger, "North Korea Nuclear Goals: A Case of Mixed Signals," *New York Times,* July 24, 2005, www.nytimes.com/2005/07/25/politics/25korea.html?ex=1279944000&en=2c0ad62d6b979e6b&ei=5088&partner=rssnyt&emc=rss; John Pike, www.globalsecurity.org/wmd/world/dprk/nuke-test.htm; and William B. Scott, David A. Fulghum, and Michael Bruno, "Nuclear Poker," *Aviation Week and Space Technology*, Oct. 9, 2006. Also noteworthy is that after the test, the *New York Times* developed a 3-D visualization graphic derived from Google Earth and made with DigitalGlobe satellite imagery that verified that the identified test site is in close proximity to the USGS determined seismic epicenter; see http://graphics8.nytimes.com/images/2006/10/10/world/1010-web-KOREA.jpg and Megan Kuhn, "GeoEye Spies Off-Limit Sites," *Leesburg Today*, Nov. 2006; www.leesburg2day.com/articles/2006/11/02/loudoun_business/biz68geoeye110106.txt.

claimed that, "Finding the test site beforehand would be akin to finding a needle in the Han River."[25]

Other revelations of clandestine nuclear sites and/or related clandestine nuclear activities have involved the use classified satellite imagery that was declassified on a case-by-case basis when the United States government felt it necessary to further the cause of nuclear nonproliferation (or other national security objectives[26]). One prominent case was in 1992, when the United States revealed the existence of two clandestine and undeclared North Korean radioactive waste sites by providing:

> the IAEA with satellite images (not made public, hence classified) showing two structures that had not been listed in the DPRK's Initial Report. Both were the type of facility in which nuclear waste is customarily stored. It was clear that the DPRK authorities had attempted to disguise the function of the two facilities by planting trees and using other camouflage.[27]

However, now that unclassified sub-one-meter resolution imagery is commercially available, such releases are largely unnecessary. According to several reports, the U.S. government has used commercial satellite imagery to press the case against Iran's clandestine nuclear weapons program using commercial satellite imagery to show a consistent pattern of concealment and deception associated with Iran's nuclear facilities that are inconsistent with a purely peaceful nuclear power program.[28]

The *Modus Operandi*

The detection of clandestine nuclear facilities through the use of commercial satellite imagery is most heavily dependent on the more readily observable signatures associated with large fissile material production facilities. As was described by Demetrius Perricos from the International Atomic Energy Agency (IAEA):[29]

> for large scale uranium enrichment plants (gaseous diffusion, EMIS, aerodynamic, gas centrifuge, etc.) these include, inter alia, large production halls, large electrical switchyards, heat transfer/cooling systems (cooling towers, ponds, outfalls) large process ventilation systems including vent stacks.

---

[25]Jeffrey Lewis (who also created a Google Earth placemark for the site), www.armscontrolwonk.com/1225/more-north-korea-nuclear-testing-rumors.

[26]Jeffrey T. Richelson, editor, "Eyes on Saddam: U.S. Overhead Imagery of Iraq," *National Security Archive Electronic Briefing Book No. 88*, April 30, 2003; www.gwu.edu/~nsarchiv/NSAEBB/NSAEBB88/w.

[27]"The DPRK's Violation of Its NPT Safeguards Agreement with the IAEA," excerpt from *History of the International Atomic Energy Agency,* by David Fischer (published by the IAEA, 1997); www.iaea.org/NewsCenter/Focus/IaeaDprk/dprk.pdf.

[28]Dafna Linzer, "U.S. Deploys Slide Show to Press Case Against Iran," *Washington Post,* Sept. 14, 2005, p. A07, www.washingtonpost.com/wp-dyn/content/article/2005/09/13/AR2005091301837.html; Jacqueline W. Shire, "U.S. Briefing on Iran Alleges Pattern of Concealment, Deception," ABC News, Sept. 14, 2005, http://abcnews.go.com/WNT/International/story?id = 1127021. (The slides are downloadable. Similar but more current slides are also available on a State Department Website: http://vienna.usmission.gov/media/speeches/files/iranpdf.pdf and http://vienna.usmission.gov/media/speeches/files/doeiranprogram.pdf.)

[29]Demetrius Perricos, "Production of HEU: Some Verification Aspects," *Proceedings from the Fissile Material Cut-Off Seminar in Stockholm*, June 1998, pp. 130–139.

Other authors (working for the IAEA) have put together a comprehensive but nonetheless succinct primer on the use of satellite imagery in assessing various portions of the nuclear fuel cycle.[30] There they state:

> The nuclear fuel cycle and related activities, such as heavy water production and nuclear research centres, present a diverse set of physical features to an imagery analyst. Not only are there major differences in the characteristics of each type of activity with respect to others (e.g., enrichment versus reactors versus reprocessing), but the results are highly dependent on whether the activity is being assessed at the laboratory, pilot, or commercial scale. Further, for some fuel cycle activities there are different technologies that could be associated with a single type of activity. Each of the technologies may present entirely different opportunities for the use of satellite imagery.

Some of these activities, such as involving gas centrifuges (together with laser and chemical enrichment processes), can be conducted successfully at smaller scales in such a way as to be less obtrusive and therefore more difficult to detect from overhead imagery alone.

For plutonium production, a reactor is the most likely means of generation, and the reprocessing of the spent fuel from such a reactor requires a significant chemical processing line located within a hot cell facility that is also served by radioactive waste treatment and storage facilities. Even more than with uranium enrichment processes, there is also a need for significant process filtering and ventilation. Large vehicles for fuel transfer may also be detectable but are not necessary if the reactor core is directly connected to a hot cell facility.

For both fissile material production routes that can lead to nuclear weapons proliferation, we might also expect that there will be some type of substantial physical security associated with such facilities. These most often can take the form of multiple security perimeters, guard towers, security checkpoints, external or remote personnel vehicle parking areas, passive air defenses (camouflage, concealment, deception, and anti-aircraft barriers, static wires, barrage balloons, etc.) and active air defenses (AAA guns and missile batteries). The sites for such clandestine facilities are often chosen in relatively remote terrain that also provides some natural passive air defense and ground-level visual concealment, and, increasingly, such facilities are being built underground to inhibit overhead detection and evaluation and to provide additional physical defense.

It should be remembered that there can always be an exception to the rule, but these signatures are generally representative of what one can expect to observe in some form in association with clandestine fissile material production facilities. Most if not all of these security signatures are readily detectable at current and projected commercial satellite imagery resolving capabilities. Those signatures provided the basis for some interesting discoveries very early on, even with only relatively low-resolution commercial satellite imagery.[31] It is instructive to review two of the very earliest works in this regard.

Perhaps the first time commercial satellite imagery was used to verify the location and layout of a clandestine facility capable of producing fissile material was by Vipin Gupta in 1992.[32] In that study, Dr. Gupta located and described a Chinese-constructed nuclear

---

[30]K. Chitumbo, S. Robb, and J. Hilliard, "Use of commercial satellite imagery in strengthening IAEA safeguards," Section 3, *Commercial Satellite Imagery: A tactic in nuclear weapon deterrence,* Bhupendra Jasani and Gotthard Stein (editors), Springer-Praxis Publishers, ISBN 3-54042-643-4, 2002.

[31]Tomas Ries and Johnny Skorve, *Investigating Kola: A Study of Military Bases Using Satellite Photography*, Norwegian Institute of International Affairs, Brassey's Defense Publishers, 1987, ISBN 0-08-034755-X; William A. Kennedy and Mark G. Marshall, "A Peek at a French Missile Complex," *Bulletin of Atomic Scientists*, Sept. 1989, pp. 20–23; and Joshua Handler, "Lifting the Lid on Russia's Nuclear Weapon Storage," *Jane's Intelligence Review*, Aug. 1999, pp. 19–23.

[32]Vipin Gupta, "Algeria's Nuclear Ambitions," *Jane's International Defense Review*, vol. 25, 4/1992, pp. 329–331.

research reactor facility at Ain Oussera in Algeria that had remained hidden from public scrutiny for at least five years, until early 1991, when it was first publicly revealed that U.S. NTM satellites had imaged the then nearly complete facility. Armed only with the name Ain-Oussera mentioned in those reports, Dr. Gupta ordered some commercial satellite imagery (including 80-meter Landsat and 10-meter SPOT) from which he was able to not only detect the presence of the site but determine the precise geographic coordinates. Dr. Gupta also found that the facility included 12 major structures located within an octagonally shaped area of 4.5 square kilometers surrounded by multiple security perimeters. He also described how a single road and multiple powerlines served the facility and that the facility was sufficiently close to a military airfield to have air defense cover.

Then, in 1998, David Albright published a SPOT image of the site of a reported clandestine plutonium production center located near Khushab, Pakistan.[33] That site, too, had remained out of public consciousness through much of its early construction phase until Mark Hibbs first brought it to light in an article in October 1994.[34] Because of the descriptive information provided in that article, Mr. Albright, using that SPOT image, was able to detect the center and identify two major facilities (one of which he identified as the "probable plutonium production reactor" and the other a "possible nuclear materials processing center") as well as two housing/support areas. The image also provided locational references to the towns of Khushab and Sargodha as well as to the Jhelum River and an intersecting canal. Higher-resolution imagery, once available, provided more detailed information about this clandestine site. The function and purpose of the possible nuclear materials processing facility was then made possible. That facility was subsequently identified as a heavy water plant in March 2000.[35] Most recently, in July 2006, Mr. Albright, using commercial satellite imagery once again, reported having identified a second plutonium production reactor under construction at Khushab. That discovery and subsequent reporting has led to several reactions by concerned governments (according to media reports) and is a clear case in which such finds by NGOs or others using commercial satellite imagery can have profound international impact at the highest levels.[36]

Case Study: Looking at Iran—The Visual Evidence of a Pattern of Deception and Denial Under the Nonproliferation Treaty

Since August 2002, numerous public allegations of covert nuclear facilities and associated clandestine activities in Iran were subsequently verified by onsite United Nations inspections. Iran was forced to admit that it had engaged in deliberate denial and deception with respect to its nuclear energy program over a period dating back to 1987 and conducted undeclared experiments with fissile materials in violation of its obligations under the Treaty on the Nonproliferation of Nuclear Weapons (NPT). Nonetheless, despite those admissions, Iran remains steadfast in its claim that it has had no interest in acquiring nuclear weapons and that its nuclear intentions are purely peaceful. When viewed in totality, the accumulated body of evidence raises serious questions about the validity of Iran's claim of having only peaceful intentions. That body of evidence includes:

- The extensive clandestine laboratory and construction work that Iran illegally conducted for many years relevant to fissile material production that was

[33]David Albright, Frans Berkhout, and William Walker, Stockholm International Peace Research Institute (SIPRI), *Plutonium and Highly Enriched Uranium 1996 World Inventories, Capabilities and Policies* (Oxford University Press, 1997), see Figure 9.1, p. 280.

[34]Mark Hibbs, "Bhutto may finish plutonium reactor without agreement on fissile stocks," *Nucleonics Week*, Oct. 1994.

[35]www.isis-online.org/publications/southasia/ikonoskhushabheavyh2o.html.

[36]Brahma Chellaney, "Engaging India to contain it," *Japan Times*, Nov. 9, 2006; http://search.japantimes.co.jp/cgi-bin/eo20061109bc.html.

acknowledged only after a government opposition group, the National Council for the Resistance of Iran (NCRI), publicly revealed the activity.

- The inconsistent and inaccurate statements explaining its nuclear activities, even after being challenged with contrary evidence.
- The omissions and delays in responding to requests for information and inspections by the IAEA.
- The continuing *reactive transparency* (as opposed to *proactive* transparency) by Iran. It was the dissident group, the NCRI, *not* Iran, that first revealed the existence of such nuclear production-related sites as Natanz, Arak, Kalaye Electric, Lashkar-Abad, and Ardekan.[37]
- The extremely suspicious nuclear facility construction, alterations, and razing activities (such as at the Physics Research Center [PHRC] at Lavizan) that have been detected on commercial satellite imagery and verified by IAEA inspections. In the Lavizan case, prior to its razing, the NCRI had alleged various WMD development activities were being conducted there. A few months later, before the IAEA could inspect it, the facility disappeared. The only explanation given by the Iranians was that the land was needed for a park. (See more on this site in Appendix B on this book's companion Website.)
- The continued obstructionism with ongoing IAEA investigations regarding alleged clandestine activities that can only be explained as having nuclear weapons applications that include, for example, the machining of enriched uranium into hemispheres.[38]
- The continuing prevention of IAEA interviews of key named, Iranian personnel whom the NCRI alleges are involved with the military nuclear program.

Some of the strongest evidence supporting the view that Iran is engaged in strategic deception has been gathered through analysis of publicly available commercial satellite imagery of nuclear facilities throughout Iran. This compelling visual evidence, combined with the results of environmental sampling and reports derived from on-site inspections conducted by the IAEA, have raised serious doubts regarding Iranian declarations of peaceful nuclear intent. Furthermore, multiple inconsistencies exist in Iranian statements and behavior with respect to recent information on its nuclear activities and its obligations under the NPT. Together, these sources of data and analytical approaches offer a powerful body of forensic science for conducting inquiries into problems of treaty verification and compliance. This multidisciplinary approach to assessing state actions and compliance with treaty obligations is readily available to states, international organizations, and even individuals.

As a nonnuclear weapons state party to the NPT, Iran has pledged under Article II of the treaty not to manufacture or otherwise acquire nuclear weapons or other nuclear explosive devices and not to seek or receive any assistance in the manufacture of nuclear weapons or other nuclear explosive devices.[39] It also is obligated to declare its nuclear activities and materials, place its nuclear facilities under safeguards, and permit IAEA inspections to verify these commitments. For more than a decade, the United States and several other nations have contended that Iran is acting in violation of these obligations and is making efforts to develop nuclear weapons. For example, in February 1996, Director of Central Intelligence

[37]"Iran opposition's report on past allegations," updated Nov. 21, 2005; ww.nci.org/05nci/11/Iranian-oppositions-status-report.htm.

[38]*Implementation of the NPT Safeguards Agreement in the Islamic Republic of Iran*, report by the Director General, GOV/2006/53 (paragraph 14), Aug. 31, 2006; www.iaea.org/Publications/Documents/Board/2006/gov2006-53.pdf.

[39]The Treaty on the Non-Proliferation of Nuclear Weapons (NPT), www.state.gov/t/np/trty/16281.htm#treaty.

John M. Deutch stated, "Iran is now developing its nuclear infrastructure and the means to hide nuclear weapons development."[40]

A major breakthrough supporting suspicions regarding Iran's nuclear intentions came on August 14, 2002, when the NCRI opposition group held a press conference in Washington, D.C., and revealed the existence of two "top secret" nuclear facilities at Natanz and Arak.[41] The first facility, near Natanz, located about 100 miles north of the central Iranian city of Esfahan, has now been confirmed to be Iran's large-scale uranium enrichment plant that employs gas centrifuge technology illicitly obtained from Pakistan. Iran never declared the construction of this facility, begun in 2000, to the IAEA or to the international community until well after it was first identified by the NCRI. In fact, prior to December 2002, the official explanation for this facility was that it was to be an innocuous agricultural research center for desert eradication. According to the NCRI, a front company named Kala-Electric, with headquarters located in Tehran, had specifically been created for the project. Interestingly, at the time the Iranians claimed that company was to have been only engaged in "watch-making" (subsequently proven to be a false cover story).[42]

Several notable characteristics of the Natanz plant are clear in satellite pictures taken at various times in the construction process.[43] First is the sheer size of the facility. According to the findings of David Albright and Corey Hinderstein:

> Analysis of the images reveals that the two underground centrifuge cascade halls have a combined area of over 60,000 square meters, estimated to be able to hold about 50,000 centrifuges total (Note: recent reporting says 100,000[44]), where each centrifuge requires on average roughly one square meter of floor space. Substantially more centrifuges could be located in the halls, particularly if the centrifuges are packed more tightly or stacked vertically. If each centrifuge has an enrichment capacity of up to 5 separative work units (SWU[45]) per year, the total capacity of this facility when finished is estimated to be up to 250,000 SWU per year.[46]

---

[40]John M. Deutch, Worldwide Threat Assessment Brief to the Senate Select Committee on Intelligence by the Director of Central Intelligence, Feb. 22, 1996.

[41]"Information on Two Top-Secret Nuclear Sites of the Iranian Regime's Nuclear Program (Natanz and Arak)," National Council of Resistance of Iran, U.S. Representative Office, originally revealed Aug. 2002, updated Dec. 2002; www.iranncrfac.org/Pages/Dossiers/Weapons%20of%20Mass%20Destruction/nuclear/WMD_new_informaion_nuclear.htm.

[42]See *List of Revelations on Iran's Nuclear and WMD Activities by the Iranian Opposition since 2002*, item #3, www.nci.org/06nci/01-31/Revelations.htm.

[43]About one month prior to the NCRI revelations about Natanz and Arak in which the NCRI speaker responded to a request from the audience (asking for proof using satellite imagery) with "anyone can get satellite photos," the following article was published: Bryan Bender, "Commercial Satellites to Enhance WMD Detection," *Global Security Newswire,* July 3, 2002; www.globalsecurity.org/org/news/2002/020703-eye1.htm.

[44]Ali Akbar Dareini, Associated Press, "Iran Set on Expanding Nuclear Program," *Washington Post*, Oct. 1, 2006; www.washingtonpost.com/wp-dyn/content/article/2006/10/01/AR2006100100488.html.

[45]*Separative work unit (SWU)* is the standard measure of enrichment services. The effort expended in separating a mass F of feed of assay x_f into a mass P of product assay x_p and waste of mass W and assay x_w is expressed in terms of the number of separative work units needed, given by the expression $SWU = WV(x_w) + PV(x_p) - FV(x_f)$, where $V(x)$ is the "value function," defined as $V(x) = (1–2x)\ 1n((1–x)/x)$. (Source: The Energy Information Administration; www.eia.doe.gov/glossary/glossary_s.htm.

[46]David Albright and Corey Hinderstein, "The Iranian Gas Centrifuge Uranium Enrichment Plant at Natanz: Drawing from Commercial Satellite Images," Institute for Science and International Security (ISIS), March 14, 2003; www.isis-online.org/publications/iran/natanz03_02.html.

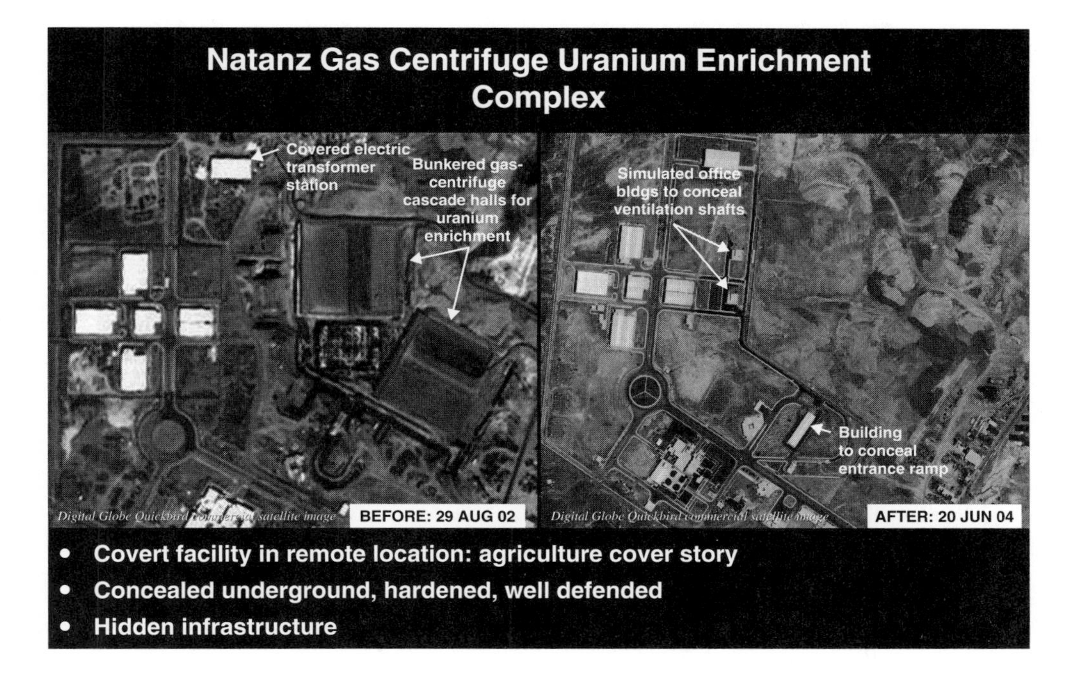

FIGURE 12.3 An illustration of the utility of commercial satellite imagery to reveal a clandestine facility that was never intended to be known by the outside world: the Natanz gas centrifuge uranium enrichment complex. A covert facility in a remote location, this complex had an agricultural cover story. The actual facilities were concealed underground, hardened, and well defended, with infrastructure hidden. Figure 12.3a, taken August 29, 2002, shows cascade halls at top right, which, as Figure 12.3b shows, were no longer visible on June 20, 2004. The Natanz complex offers several examples of "signature suppression."

Albright and Hinderstein go on to conclude:

> This capacity is far larger than required for a nascent nuclear weapon program alone, supporting Iran's statement that the facility is aimed at producing low enriched uranium for nuclear power reactors. Nonetheless, such a facility could use a relatively small fraction of its capacity, say 10,000 SWU per year, to make enough highly enriched uranium for three nuclear weapons a year, while using the remaining capacity to produce low enriched uranium.

The second notable characteristic is the rapidity with which the site was constructed. Figure 12.3 shows the dramatic changes over the two years following its initial discovery. Had it not been captured on overhead imagery during construction, the Iranians would very likely have been successful in hiding the existence of the enormous underground facility destined to become the primary gas centrifuge uranium enrichment plant in Iran (as well as one of the largest known cut-and-cover, hardened, underground complexes worldwide). The third notable feature is that it is now clear that the original plans were devised such that no one would have suspected that the facility would be for any other purpose than for agricultural research on antidesertification. Nonetheless, in the event that those concealment and deception efforts were to fail such that the true nature of the site was discovered (as is now the case), the fallback plan was to have the site hardened against any potential attack (a form of passive defense, which after exposure has been supplemented by active defenses that include antiaircraft artillery).

Among the concealment and deception techniques employed at Natanz are the roofing and disguising of a large high-voltage electric substation (built to appear as an innocuous workshop building), with all subsequent electrical connections placed underground.

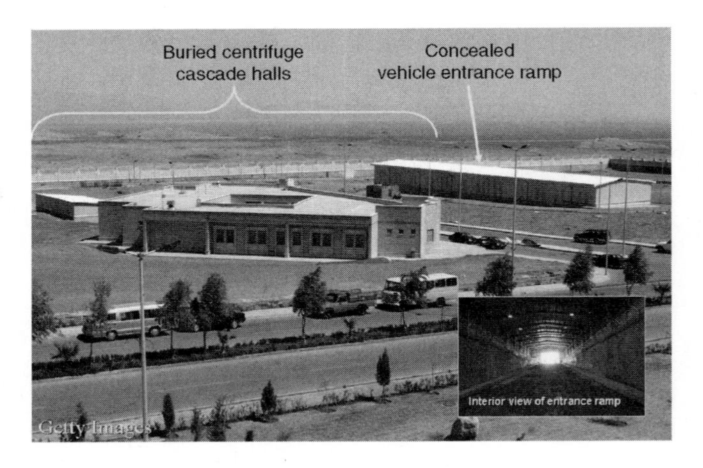

FIGURE 12.4 This is a ground view of the buildings at the Natanz uranium enrichment facility in Iran, showing, in the absence of prior knowledge of the site layout and function, how difficult it would have been to assess the true nature of the facility after its completion. (Copyright Getty Images.)

Two other similarly disguised buildings conceal the presence of personnel accesses and likely ventilation and cooling systems. The sole vehicle access to the underground cascade halls (via a looping ramp that provides additional passive defense) is concealed by an innocuous warehouse building "shell" (see Figure 12.4). Note how the cascade halls were cleverly placed outside the innermost security wall to help further disguise their existence. However, once the location and true function of the site was made public by the NCRI and confirmed by IAEA through onsite inspection, these passive defense measures were quickly supplemented with active defense measures that include multiple rings of antiaircraft artillery (and probably missile) batteries.

In March 2005, Iranian President Mohammad Khatami visited the underground uranium enrichment complex near Natanz and brought media personnel with him in an effort to lend credence to the claim that Iranian intentions are purely peaceful. One likely unintended consequence of that visit was that the ground photos of the visit only served to reinforce how well the original cover story was crafted—such that even the most astute observer visiting the site would be oblivious to the true nature and scope of the facility had it not been revealed earlier.

Following Up on Open-Source Leads Using Google Earth and Commercial Satellite Imagery: A Data Fusion Exemplar

The following study illustrates the utility of commercial satellite imagery in combination with Google Earth for verification of open-source provided information on an alleged clandestine nuclear related facility in Iran. Not only is the exercise successful in confirming the allegations of a clandestine underground missile complex (alleged to have nuclear weapons-associated capabilities),[47] but it exemplifies the resultant synergy that can occur when

[47]"Iran Building Nuclear-Capable Missiles in Underground Secret Tunnels," Statement by Alireza Jafarzadeh, President, Strategic Policy Consulting Inc., at the National Press Club, Washington, D.C., Nov. 21, 2005, www.nci.org/05nci/11/PC-Transcript.htm; and Statement by Alireza Jafarzadeh, President, Strategic Policy Consulting, National Press Club, presented at the Joint Press Conference of Nuclear Control Institute and Iran Policy Committee, Sept. 16, 2005, www.nci.org/05nci/09/PressConfText-16Sept05.htm.

multiple open sources are used in combination, sometimes referred to as *data fusion*. That synergy provides new, unique, value-added information (never before published correlations and derivations pinpointing the location, orientation, and geospatial layout of a clandestine underground production facility) that to date has remained otherwise publicly unknown and still unreported anywhere in the media. The exercise goes beyond that discussed in earlier work[48] using commercial satellite imagery alone by unequivocally confirming the *bona fides* of the source material and therefore providing a sound basis for determining the validity of the accompanying allegations.

The study also illustrates the utility of Google Earth as an unsurpassed visualization and pre-inspection planning tool by allowing for the creation of compelling perspective views (and fly-around videos with a purchasable upgrade) when used either alone or in conjunction with separately purchased or downloaded imagery. Finally, Google Earth is an extremely cost-effective means, heretofore unavailable to parastatal entities such as the IAEA, for independent global broad area search applications by its Satellite Imagery Analysis Unit, created in 2000. The reader is cautioned, however, to the fact that inherent limitations remain for these remote sensing tools when they are used in the absence of such detailed collateral information.

Figure 12.5 is a composite of images from two separate press briefings presented at the National Press Club in 2005 to provide never before revealed details about an underground facility alleged to be associated with Iran's development program for nuclear-capable ballistic missiles. Four posters were shown (each individually downloadable[49]), including a ground photo, two engineering drawings, and a map locating the site. The transcript described the facility as being built with North Korean assistance and located at the base of Kuh-e Barjamali peak east of Tehran under the auspices of Hemat Industries. The first step that had to be undertaken in any followup was determining the precise location of Kuh-e Barjamali peak. Figure 12.6 shows how that peak was quickly located using Google Earth (after clicking on the geographic features layer that shows topographical features in green). By searching another useful Internet database, GeoNames.org, confirmation of the location was obtained. See Figure 12.7.

The next step was to search all available Web information about Hemat Industries, ballistic missiles in Iran, and so forth. Figures 12.8 and 12.9 show that there is indeed information already available that links this site to ballistic missiles and, perhaps most important, to ballistic missiles of North Korean design.

Figures 12.10 and 12.11, illustrate the utility of Google Earth for visualization purposes. They show how, despite the initial lack of a high-resolution imagery for the area under study, it is still possible to place separately purchased imagery over that same terrain base using the Google Earth tool: ADD: IMAGE OVERLAY. For more on what is possible with Google Earth in this regard, the tool offers its own online tutorials. Many useful blogs offer up-to-date information. They also offer user insights that are unavailable anywhere else.[50]

Figures 12.12 and 12.13 illustrate how, using another Google Earth tool called RULER (see the Tools tab), one can measure the distance between tunnel entrances 1 and 3. It also

---

[48]Frank Pabian, "The Utility of Commercial Satellite Imagery for the Detection of Clandestine Activities for FMCT Verification and Monitoring," *Proceedings: FMCT verification- Detection of clandestine activities*, Swedish Defense Research Establishment (FOA), June 20–22, 1999.

[49]www.nci.org/05nci/11/Map-underground-warhead-factories.htm, www.nci.org/05nci/09/nuclear%20tunnel.htm, www.nci.org/05nci/09/tunnel%20engineering%20plans.htm, and www.nci.org/05nci/09/boiler%20and%20ventilation%20system%20plans.htm.

[50]See, for example, www.gearthblog.com, www.ogleearth.com, www.viavirtualearth.com, and www.armscontrolwonk.com. (With respect to the last blog, whose focus is particularly on nonproliferation, see Jacqueline Shire, "Blogging for Arms Control," ABC News, Oct. 11, 2005, http://abcnews.go.com/US/LooseNukes/story?id=1200881&page=1).

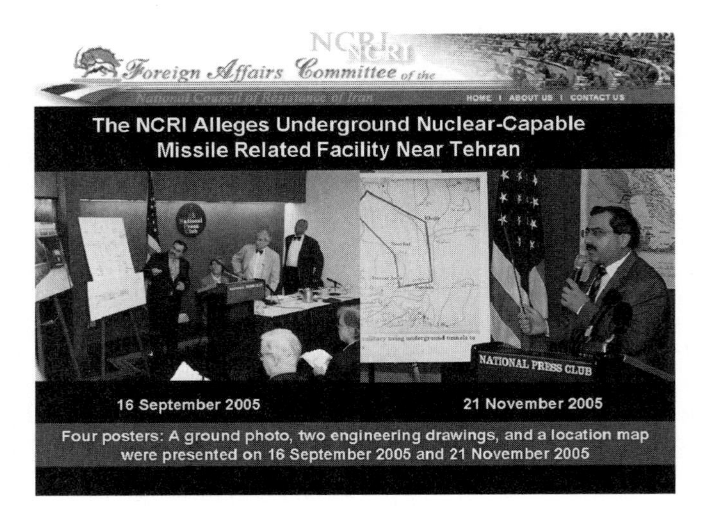

FIGURE 12.5 An affiliate of the National Council for the Resistance of Iran (Iran Policy Committee), together with the Nuclear Control Institute, held two separate press conferences in Washington D.C. in 2005 to reveal details about an underground facility in Iran. The facility was alleged to have been built with North Korean assistance for the purposes of supporting the development of nuclear-capable ballistic missiles. As of this writing, no news media outlet, academic researcher, or NGO has yet published a followup on this information to check on its validity or verify its source's bona fides. (*Sources*: www.iranpolicy.org/images/stories/91605Nuclear/nuclear9.jpg and www.iranpolicy.org/images/stories/Nov21event/img_9526.jpg.)

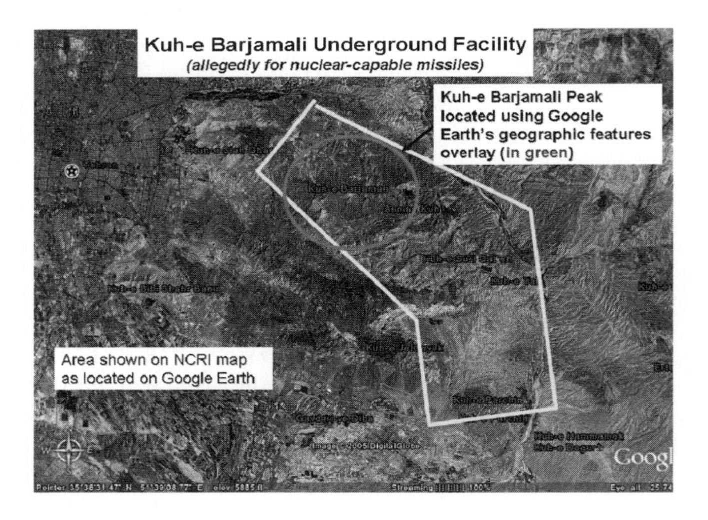

FIGURE 12.6 Area shown on map on poster at November 21, 2005, press conference, overlain on Google Earth. (*Source*: www.nci.org/05nci/11/Map-underground-warhead-factories.htm.)

shows how it is possible to monitor subtle changes over time, in this case over three years. Note the growth of the landscaping vegetation.

Figure 12.14 is a closeup of the main engineering drawing presented at the September 16, 2005, press conference. Cross-section views of the tunnels are shown as well as some detailed dimensions in the plan view. Orientation with the overall site plan is shown as well in the lower-left corner (repeated in larger scale in Figure 12.15). From that overview, combined

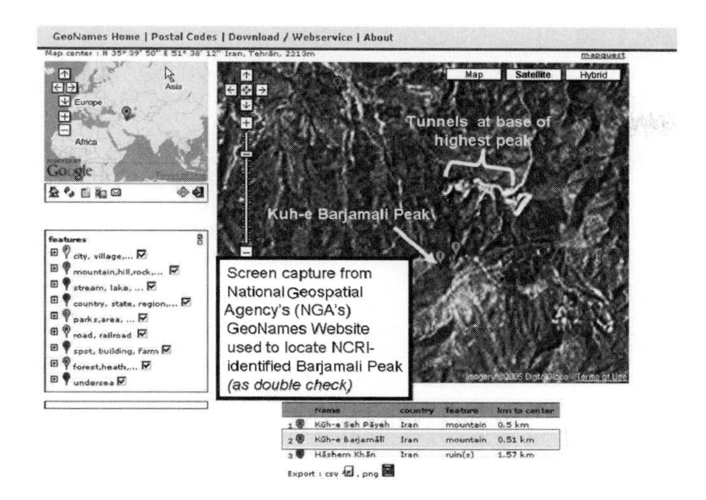

FIGURE 12.7 Confirmation of the peak being correctly located. Interestingly, the tunnels were already visible at the base of the mountain. (*Note:* This image has since been replaced with a higher-resolution one courtesy of Google Earth.)

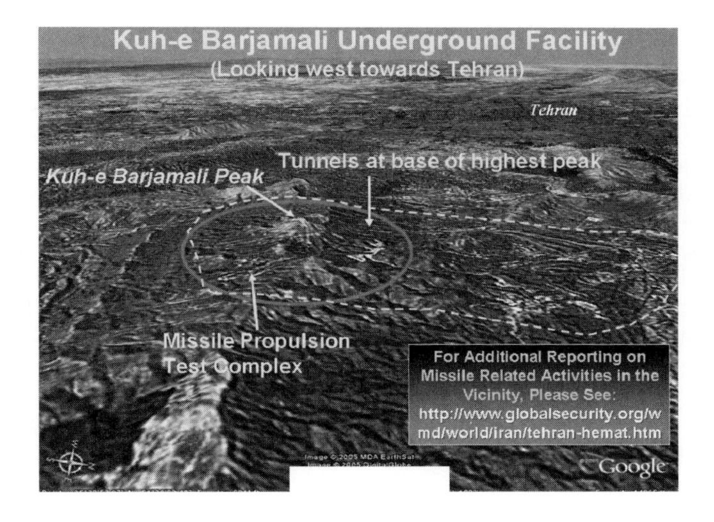

FIGURE 12.8 Perspective visualization view of Kuh-Barjamali site using Google Earth, looking west toward Tehran. Again, interestingly, the site does correlate with a previously reported ballistic missile complex described (in reporting in referenced Globalsecurity.org web link) as being subordinate to Hemat Industries group. (*Source*: www.geonames.org/maps/showOnMap?q=kuh-e%20Barjamali+country:IR.)

with the dimensions shown in the detail section, it is possible to scale off the distance between tunnel entrances 1 and 3. The scaled dimensions match perfectly with those derived by measurement using Google Earth's RULER tool, thereby lending credence to the veracity of the reporting.

Nonetheless, though the design and location of the tunnel entrances do match perfectly in orientation and dimensions, the buildings (rectangles) do not make a good match with the imagery, and neither does the supporting road network outside the entrances. It appears that the overview drawing is somewhat stylized and does not take into account the topography

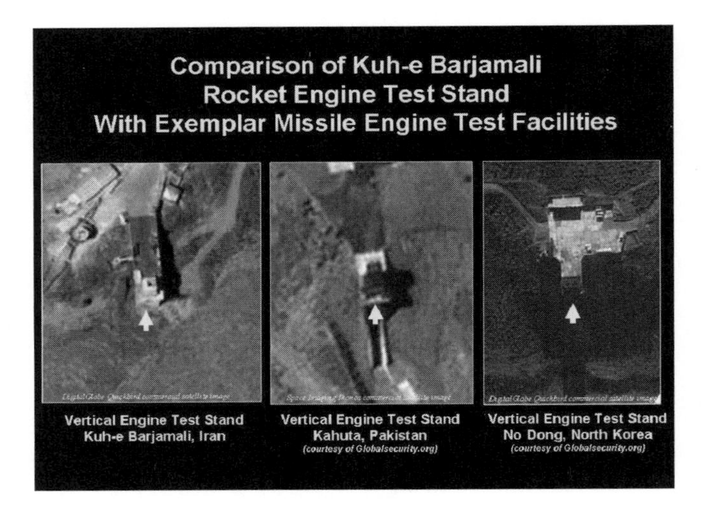

FIGURE 12.9 This missile propulsion test complex (of apparent North Korean design) supports NCRI allegations of both North Korean involvement and ballistic missiles in association with the Kuh-e Barjamali underground facility. (*Source*: www.globalsecurity.org/wmd/world/iran/tehran-hemat.htm.)

of the site as opposed to the reality that we see on the imagery, where the road network and building layouts conform more closely to the hillside topography and its contours.

Figure 12.16 shows how the engineering drawing plan views match up with the imagery when scaled and oriented properly. Figure 12.17 is extremely important because not only does it provide the basis for establishing the bona fides of the sourcing, it shows that the source was aware of a fourth tunnel entrance (identified by caption only as "Nuclear Tunnel"), even though that tunnel entrance was not included in the engineering drawings. We can therefore be quite confident that the engineering drawings are authentic and not created by someone just looking at the imagery.

Figure 12.18 reiterates that correlation between the ground photo and what is observable on commercial satellite imagery, which, together with the help of the enlargement and further study, yields the identification of "probable cooling units," and that they are evidently necessary for heat transfer from the underground facility. The reader should also be aware that every aspect of this figure was generated entirely from what is currently available and downloadable from the Internet. All that was done to the originals was a little enlargement, rotation, cropping, and brightening of the image. You could do this at home!

Analytical Findings from This "Textbook Case"

Integrating, or *fusing*, commercial satellite imagery data together with other information sources can yield new insights. As this study has shown, commercial satellite imagery is a unique and invaluable tool for following up on open-source leads regarding undeclared and clandestine nuclear-related facilities and their associated activities, particularly when there is such an abundance of locational information (maps, drawings, photos). Together with newly available commercial satellite imagery data sets and visualization tools such as Google Earth, it is now much easier to assess the credibility and value of those open sources and their allegations. Although, in this case, others had previously described the NCRI claims of tunnels being associated with North Korean design ballistic missile facilities, the new details of the interior of the underground production facility were entirely new and determined to be credible.

Because of this entirely independent and original study, we were able to glean new information not previously possible about one of Iran's most highly secure (multiple perimeter

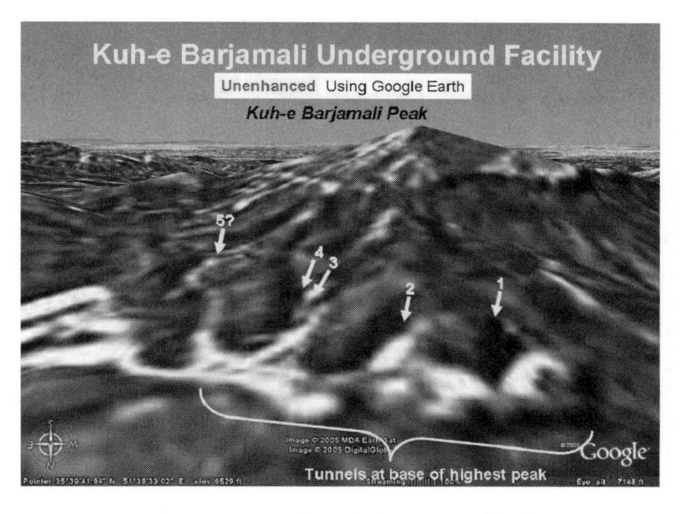

FIGURES 12.10 (top) and **12.11** (bottom) A comparison of a typical low-resolution commercial satellite image provided by Google Earth, as viewed with the Terrain layer clicked "on" and rotated for visualization looking south (Figure 12.10), with an enhanced version (Figure 12.11) that overlays a separately purchased commercial satellite imagery using Google Earth's "add an image" feature. (*Note:* After these slides were created, Google Earth updated its base imagery of this area to include free high-resolution imagery so that a separate imagery purchase is no longer necessary.)

fences and checkpoints are clearly visible) clandestine weapons of mass destruction (WMD) facilities, based on the association with ballistic missile development and testing. We learned that the engineering drawings provided by the NCRI, while very accurate in dimension and plan view with regard to three of the tunnel entrances, did not reflect the presence of a fourth tunnel (along with a possible fifth tunnel), even though the ground photo presented in association with the drawings was subsequently discovered to be none other than that fourth tunnel, the "Nuclear Tunnel." Because we were able to make that correlation, it also became possible for the first time to identify the object outside of the fourth tunnel as probably being a cooling unit, and we could confirm that vehicle access to the tunnel is possible underneath that structure. With Google Earth we gained a much better understanding of the physical setting of this facility (it is set at the base of a large mountain) than had been heretofore possible without that capability to visualize the site in perspective view. Although it is still not possible to verify whether these tunnels have a "nuclear" association, their

FIGURES 12.12 (top) and **12.13** (bottom) An illustration of what is possible with Google Earth tools for measuring to verify dimensions provided on the engineering drawings from the NCRI (they perfectly match) and to show how commercial satellite imagery can be used to monitor changes over time at such a site.

connection with a nuclear-capable ballistic missile delivery system (Shahab-3/No-Dong or North Korean design variant of the Soviet SCUD) does justify additional investigation by those who are tasked with such responsibilities.

Errors in Interpretation Can and Will Happen

Even the best imagery analysts can make mistakes. Among the keys for best avoiding such mistakes is to be as knowledgeable as possible of the subject area; to be aware of the cultural context, the chronological setting, and history of developments associated with the facilities and activities under scrutiny; and always to solicit peer review whenever possible. Earlier in this section, examples were given on how clandestine nuclear test sites were correctly located and identified with satellite imagery in advance of nuclear testing in China and North Korea. Although those are each clear examples of success, other examples would have to be described as less so.

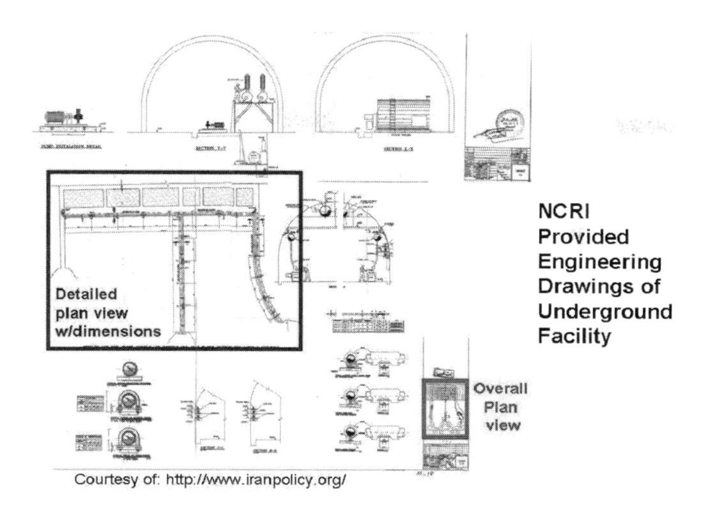

FIGURE 12.14 Engineering drawing provided at September 16, 2005, briefing, downloadable in high resolution from www.nci.org/05nci/09/tunnel-engineering-plans-hr.gif. The larger highlighted area is a dimensioned drawing of the three tunnels while the smaller highlighted area is the overall site plan of the production area (without dimensions).

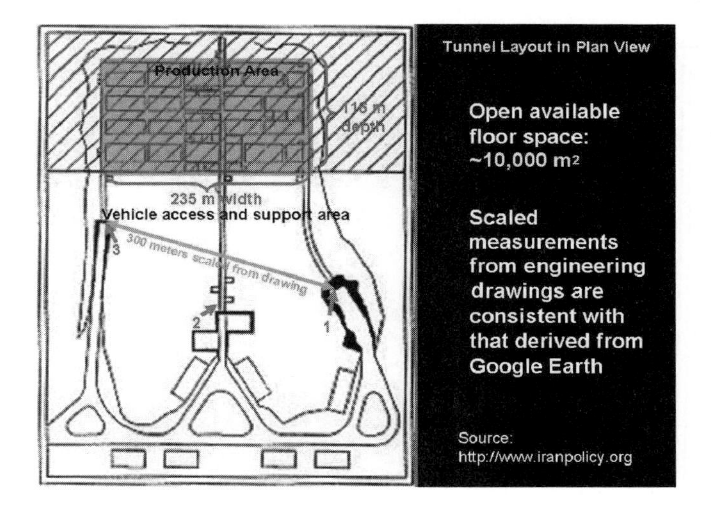

FIGURE 12.15 The overall site plan with dimensions scaled from the smaller area highlighted in Figure 12.14. The dimensions scaled from that drawing between the two tunnel entrances match that measured for the same two tunnels on imagery using the Google Earth measurement tool, showing that the drawings are consistent in dimensions and layout with that provided by the Iran Policy Committee.

The following are two brief examples in which imagery was misinterpreted in attempts to locate and describe clandestine nuclear facilities. The first example was published in the May 25, 1998, issue of *Newsweek* magazine in which a Russian commercial satellite image was shown of an area near (but outside and not including) the location of the 1998 Indian nuclear tests, and, despite being many kilometers away from that test site, was nonetheless labeled as "Ground Zero." The image that was shown, however, was actually a closeup of what appeared

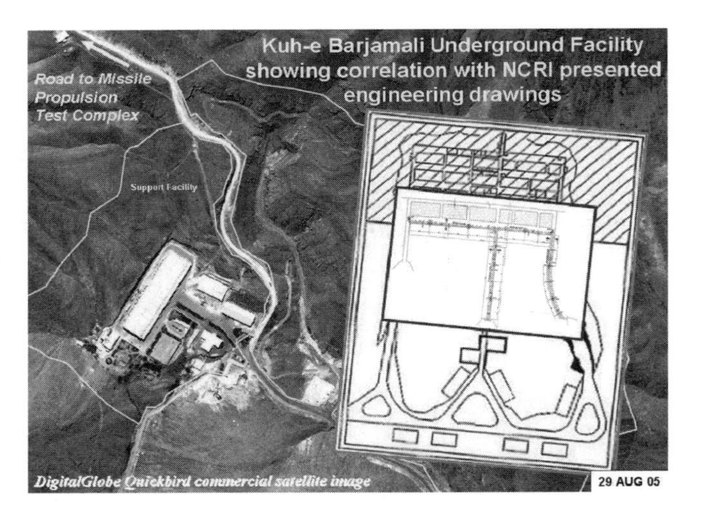

FIGURE 12.16 This figure shows the engineering drawings as correctly scaled, oriented, and overlain upon the commercial satellite image. (Engineering drawings from www.nci.org/05nci/09/tunnel-engineering-plans-hr.gif.)

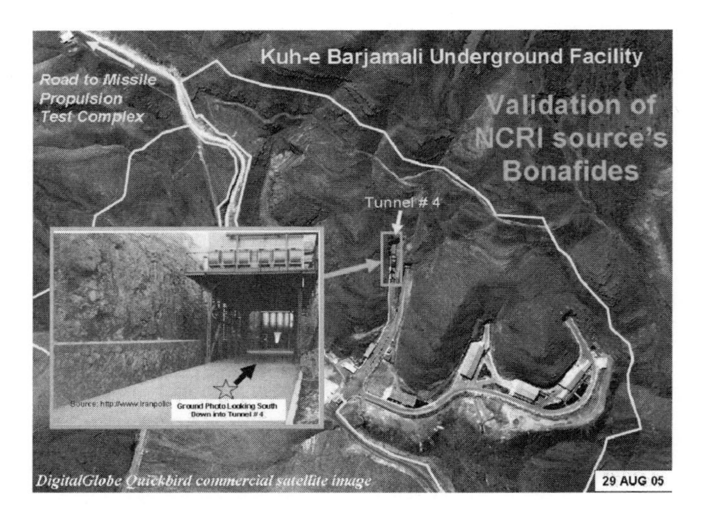

FIGURE 12.17 This figure shows the entrance to the fourth tunnel (which, despite being labeled "Nuclear Tunnel," was not included in the engineering drawings, so there is more to the underground area than those drawings show). The ground photo (inset) provided at the joint briefing held by the Nuclear Control Institute and the Iran Policy Committee provides the necessary proof of the source's bona fides, because it could have been obtained only by someone with access to the site; only someone allowed inside the innermost security perimeter of this clandestine underground ballistic missile related complex could have taken the photo. (*Source*: www.nci.org/05nci/09/nuclear%20tunnel.htm.)

to be an ornate agricultural study area, and the alleged "Ground Zero" was nothing more than a livestock pen. *Newsweek* printed a retraction several weeks later.[51]

[51]See Yahya A. Dehqanzada and Ann Florini, *Secrets for Sale: How Commercial Satellite Imagery Will Change the World,* Carnegie Endowment for International Peace, 2000, p. 24; also see www.ceip.org/files/projects/tcs/remotesensingconf/powerpoint/Livingston/sld001.htm. (*Note:* Interestingly, if you go to that site now using Google Earth, it has been labeled by

FIGURE 12.18 This figure is a closeup of the feature (believed to be cooling units on a framework) that is located outside the fourth tunnel ("Nuclear Tunnel"), compared with the ground photo provided by the Iran Policy Committee. (*Sources*: www.nci.org/05nci/09/nuclear%20tunnel.htm and www.globalsecurity.org/wmd/world/iran/images/dg_23may02_tehran-hemat_28.jpg.)

The second example had to do with claims in the British press of finding a clandestine underground nuclear test site in Iraq.[52] The alleged 4-kilometer-long tunnel described in the reporting as leading to the site under Lake Rezazza was subsequently claimed to have been identified on commercial satellite imagery. However, the feature identified as a tunnel was nothing more than a dry wadi (river bed) that had been cross-diked to retain water for subsistence farming/grazing near freshwater springs (the dikes were misinterpreted as security gates for the "tunnel"). Moreover, an alleged railway line, said to support the site, was nonexistent (it was only a large radius paved road).[53]

Summary

In the field of nonproliferation, commercial satellite imagery has been shown in a number of cases to be very effective in detecting, evaluating, and monitoring clandestine nuclear facilities and activities. Now that imagery from commercial imaging satellites is becoming more ubiquitous (and in many cases, available cost-free through publicly available data sets and visualization platforms such as Google Earth), it can be expected to increasingly provide a timely, accurate reference to support, supplement, and/or enhance nuclear-related ongoing treaty monitoring and verification activities. It should increasingly be viewed as another important tool in the larger information "toolkit" used to support the nuclear nonproliferation

---

someone, going by the name Yaar, as "Bhaadariya ji Maharaj & Shakti Mandir" and is accompanied by this text: "Mata ji's Temple rising in middle of the desert is an amazing place. Bhaadariya ji Maharaj lives here." See http://bbs.keyhole.com/ubb/showthreaded.php/Number/478888.)

[52]See: www.globalsecurity.org/wmd/library/news/iraq/2001/stirevnws01015.htm.

[53]Larry O'Hanlon, "Seismic Sleuths," *Nature*, 411, pp. 734–736 (June 14, 2001); www.nature.com/nature/journal/v411/n6839/full/411734a0.html.

regime through the detection and assessment of clandestine facilities. It should not be viewed as a panacea, or as a standalone substitute for other information, but rather as complementary to other collateral information in the overall verification and monitoring process.

Given the inherent limitations of analytical resources applied to such an effort, detection of clandestine facilities historically has been largely dependent on prior reporting that substantially narrowed the initial focus of the search. In most instances, therefore, commercial satellite imagery will likely not be the "sole" means of detection of clandestine fissile materials production facilities and activities. However, when it is used to corroborate other geographically specific information made available from other sources, there is now no question as to its value in providing the means of their detection.

Regarding the new visualization tools such as Google Earth, they should be relied on as a starting point for further investigations—in other words, when cued to a specific area, see what imagery is freely available via those tools, then if necessary, follow up with commercial imagery archive search and acquisitions, then analyze the imagery to verify details (locational, spatial, temporal, etc.). These tools are the best available means for global visualization because they incorporate detailed 3-D terrain layers allowing for compelling perspective views (and video "fly-arounds" for slight additional cost). They can also serve as everyman's "broad area search tool" in that they allow for independent manual detection of new construction, utility and transportation networks, and high-security exclusion areas that could potentially lead to the detection and subsequent identification of a clandestine nuclear facility (with more current, higher-resolution imagery coverage being added all the time).

Because of their widespread availability and use, these new visualization tools, which incorporate commercial satellite imagery, provide a never before possible basis for virtual global transparency. That global transparency is now being made manifest by a growing virtual (and cost-free) cadre of imagery analysts who are increasingly capable of finding clandestine facilities anywhere on Earth.[54] There is also an ever-growing global community of users eager to post and share their "discoveries" online in blogs,[55] user communities,[56] and Wikis[57] (but because these users are unskilled and much of their work is not well vetted, always be cognizant that the expression "all that glitters is not gold" is just as true for imagery analyses). As a new member of that growing community, you, too, will have the power to make such discoveries.

Epilogue

A final note: While becoming both optimistic and enthusiastic about commercial satellite imagery as a new means for global transparency, we must never lose sight of the fact that

[54]Michael F. Goodchild, "Citizens as Sensors: The World of Volunteered Geography," National Center for Geographic Information and Analysis and Department of Geography, University of California Santa Barbara, CA, USA (2007; www.ncgia.ucsb.edu/projects/vgi/docs/position/Goodchild_VGI2007.pdf).

[55]With respect to blogs, the best that I've found include: "IMINT and Analysis," http://geimint.blogspot.com/; and http://ddanchev.blogspot.com/2006/07/open-source-north-korean-imint.html with its included links.

[56]The community of users can be found at various sites on the Internet, including: http://bbs.keyhole.com/, www.gearthblog.com/, www.ogleearth.com/, http://googleearthuser.blogspot.com/, http://google-latlong.blogspot.com/, http://viavirtualearth.com/, http://virtualearth.spaces.live.com/, http://earthissquare.com/, http://wikimapiablog.blogspot.com/, and www.virtualglobes.org/blog/ (particularly useful for its cross-link directory to many other blogs).

[57]An excellent example is Wikimapia, http://wikimapia.org/, which can be downloaded as a Google Earth layer.

such imagery is really no more than just another information source. Although satellite imagery can certainly provide a unique and much sounder basis for public dialogue, debate, and, hopefully, negotiation with respect to nuclear nonproliferation between the peoples of this world, it is not without its limitations. Commercial satellite imagery can only bring to light, and help us to better comprehend the symptoms of nuclear proliferation problems (in other words, illicit nuclear activities). It is only a messenger and can in no way address the *raison d'être* or root of those problems. Nonetheless, it is this author's sincere hope that commercial satellite imagery will increasingly raise the salience and public consciousness of the existence of these problems such that it will both stimulate and better equip public leaders to effectively deal with those problems, sooner rather than later, by peaceful means.

13 Nuclear Test Monitoring

Loren Byers

Introduction

Though its capabilities have been called into question, the verification system embodied in the Comprehensive Nuclear Test-Ban Treaty (CTBT) currently meets or exceeds its design requirements. From a technical perspective, therefore, verification does not represent an insurmountable impediment to treaty implementation. As a monitoring system with global coverage, the system is also capable of detecting nuclear tests conducted by states or other actors that are not parties to the treaty. This chapter describes some of the history of nuclear test-detection efforts and the technologies used to verify the CTBT.

History

Indian Prime Minister Jawaharlal Nehru is credited with first proposing, in 1954, a worldwide ban on all nuclear test explosions. It took nearly a decade before limited steps were taken toward this goal: In 1963, the Partial Test Ban Treaty (PTBT), which banned nuclear tests in the atmosphere, underwater, and in space, was signed, though neither France nor China, both nuclear weapon states, became signatories.

The next major nuclear arms-control achievement was the conclusion and signing of the Nuclear Nonproliferation Treaty (NPT) in 1968, which, in its preamble, recalls language from the PTBT, calling for negotiations of a treaty to discontinue *all* nuclear weapons test explosions.

Though the issue continued to be raised, it was not until 1991 that the state parties to the PTBT held a conference to discuss a proposal to expand the treaty to ban all nuclear weapons tests. With strong support from the U.N. General Assembly, negotiations for a comprehensive test-ban treaty began in 1993, and over the next three years, the treaty text and its two annexes were drafted.

Three issues in particular were sources of controversy:

- A U.S. proposal to include a provision enabling a state to leave the treaty after 10 years without any need to provide justification
- Whether the treaty should allow low-threshold nuclear tests or ban all nuclear tests
- Whether the treaty should become legally binding after a certain number of states had ratified it or whether ratification by specific states would be required

Eventually, the first issue was withdrawn, a complete ban was agreed to, and entry into force was made conditional to the ratification by the 44 states operating nuclear power or research reactors at the time.

Increased Pressure to Conclude a CTBT

The decision taken by the 1995 NPT Review and Extension Conference to extend the treaty indefinitely included, as a key requirement, the negotiation of a comprehensive test-ban treaty no later than 1996. Negotiation of the treaty—at times contentious—took place in 1995 at the United Nations Conference on Disarmament (CD), and this deadline was met. Australia presented the CTBT to the U.N. General Assembly, which adopted it by an overwhelming majority (158 to 3, with 5 abstentions), on September 10, 1996. The treaty was opened for signature on September 24, 1996, and was signed by 71 states, including the five nuclear weapons states. As of May 2008, 178 states have signed the Treaty and 144 have ratified it. Of these, 41 of the 44 Annex 2 states—those that possessed nuclear research or power reactors in 1996—have signed the Treaty, but only 35 have ratified it. The nine remaining states are China, Egypt, India, Indonesia, Iran, Israel, North Korea, Pakistan, and the United States.

During treaty negotiations, two issues proved difficult to resolve. The first was the means of verification, with some states favoring national technical means (NTM) as the primary verification mechanism and others favoring the creation of a treaty-based verification organization. The second issue was the requirement that the 44 Annex 2 states ratify the treaty before its entry into force. In the end, compromises were reached on both points.

The positions of the Annex 2 states that have not yet ratified the treaty are summarized here:

- *China.* Although a vocal supporter of the CTBT, China has been slow to pass the treaty through its bureaucracy for ratification. China's hesitance may be connected to calculus that includes the possible future need to test new weapons for its arsenal and the fact that the United States has not yet ratified.
- *Egypt.* Egypt is a vocal advocate for nonproliferation, especially with regard to a Middle East Weapons of Mass Destruction Free Zone (MEWMDFZ), and frequently reiterates the call for Israel to join the nuclear nonproliferation regime by renouncing nuclear weapons and signing the NPT. Egypt's delay in ratifying the CTBT is tied to this issue. It is unclear whether Egypt would ratify the CTBT after Israel did or if it would continue to call for Israel to join the NPT as well.
- *India.* India is unlikely to ratify the CTBT before Pakistan and China do. Although there is no overt indication, it may also be holding back because of a possible need for another series of nuclear tests.
- *Iran.* Iran is unlikely to embrace ratification of the CTBT under its current leadership, which is unwilling to resolve suspicions of a clandestine nuclear weapons program. Iran would also likely insist on Israel's ratification as a precondition.
- *Israel.* Although a signatory, Israel is not likely to ratify the CTBT while it believes the region is unstable, while there are others in the region rattling nuclear sabers, and without a seachange in its policy of nuclear "ambiguity."
- *Pakistan.* Pakistan's position is primarily related to India's nuclear posture; the country will likely delay ratification out of a need to maintain parity with India in the quality and quantity of its nuclear arsenal.
- *North Korea.* The likelihood that the Democratic People's Republic of Korea (DPRK) will ratify the treaty in the near future is uncertain. Despite its bellicose underground test of a nuclear weapon in 2006, the DPRK has agreed to the shutdown of its plutonium production capabilities and is actively negotiating within the six-party framework for the complete elimination of its nuclear weapons program. This process envisions the DPRK ratifying the CTBT at a future date.
- *United States.* Opponents of CTBT ratification in the U.S. Congress, to support their position, questioned the effectiveness of the CTBT verification regime. However, this appears to be a hollow argument, given the volume of credible analysis to the contrary and particularly in light of the solid performance of the CTBT verification system following the 2006 DPRK test. More likely, those who oppose ratifying the CTBT support the current administration's position on modernizing the U.S. nuclear arsenal and anticipate a possible need for further nuclear tests to certify new warhead designs before they are accepted into the stockpile. It is unclear how a new administration will proceed.

Verification

Under Article I of the CTBT, member states agree not to carry out or participate or assist in carrying out any nuclear explosion and to prohibit and prevent any nuclear explosion anywhere under their jurisdiction or control. To assure that treaty obligations are being met, the CTBT outlines an international global verification system—the first such treaty to do so.

The International Monitoring System (IMS) will be the responsibility of the Comprehensive Test Ban Treaty Organization (CTBTO) once the treaty enters into force. The CTBTO will be headquartered in Vienna, near the International Atomic Energy Agency (IAEA).

Article IV of the Treaty and the Protocol establish a global verification regime with the capability to monitor for nuclear tests conducted in any terrestrial environment. The Treaty stipulates that the monitoring system will be capable of detecting, locating, and identifying explosions down to less than 1 kiloton TNT equivalent, detonated anywhere within the atmosphere, under water, or under ground. The monitoring system is to comprise four separate but interdependent elements:

- The International Monitoring System (IMS)
- Consultation and clarification
- Onsite inspections (OSIs)
- Confidence-building measures (CBMs)

The IMS will comprise a network of 321 monitoring stations and 16 radionuclide laboratories, located in 89 countries.[1] The system will include four types of monitoring technology: seismic, hydroacoustic, infrasound, and radionuclide. When it is complete, there will be 50 primary and 120 auxiliary seismic monitoring stations, which will be able to detect nuclear explosions and distinguish them from other manmade or natural seismic events. The IMS will include 60 infrasound and 11 hydroacoustic stations designed to detect and identify the acoustic signals of nuclear explosions in the atmosphere and in the oceans. In addition to these three waveform-monitoring systems, the IMS will include 80 atmospheric radionuclide-sampling stations and 16 associated analytical laboratories.

The IMS is designed to provide uniform global coverage. Due to global geography, most of the hydroacoustic stations are located in the oceans of the Southern Hemisphere, whereas more seismic stations are located in the Northern Hemisphere, owing to its larger landmasses.

The establishment of the IMS poses engineering challenges unprecedented in the history of arms control, with many stations located at remote sites and in harsh environments. Though some of the stations already existed when the IMS was planned, most have had to be newly constructed or have had to undergo substantial upgrades. In many cases, the political challenges associated with the cooperation and communication necessary to establish and maintain the IMS have proved no less demanding. Furthermore, as time passes, older stations that were incorporated into the IMS, and the earliest stations to be installed, are increasingly in need of upgrades and maintenance. Once established and certified that they meet all technical requirements, the monitoring stations will be operated by local institutions under contracts with the Technical Secretariat.

As of the end of 2006, 193 facilities, including nine radionuclide laboratories, had been certified, 244 stations had been installed and substantially met specifications, 40 stations were under construction or under contract negotiation, and 95 stations and four radionuclide laboratories had contracts for operation and maintenance. In addition, approximately 190 stations were configured in the International Data Center (IDC) operating system. The Provisional Technical Secretariat (PTS) for the CTBT Organization (CTBTO) anticipates that approximately 90% of the IMS network will be installed by mid-2008.[2]

Data collected by the IMS will be transmitted to the IDC, based at the headquarters of the PTS in Vienna, for analysis, distribution, and archiving. The data will be made readily available to all member states, which will allow those with few or limited technical means to

---

[1]The locations of the proposed stations are listed in Annex 1 of the Treaty's Protocol.

[2]"IMS Network Status," *CTBTO Spectrum*, Issue 9, Jan. 2007, p. 12.

participate more fully in the verification and enforcement of the treaty. The IDC will process the data and produce bulletins of detected events.

The test explosion conducted in the DPRK on October 9, 2006, demonstrated that the verification system will work as intended. Seismic signals were detected at more than 10 IMS primary seismic stations, from which the PTS was able to confirm the event and determine an approximate location within two hours. On October 11, the PTS released a bulletin confirming the location and time of the event, and—after experts' analysis of the original data, plus the data from one additional primary and many auxiliary seismic stations—reducing the uncertainty in the location to less than 1,000 square kilometers, the maximum allowed for an OSI under the Treaty. This was achieved with less than 60% of IMS stations operating and with the IDC's data processing systems and procedures still under development.

The consultation and clarification elements of the Treaty outline procedures for state parties to resolve differing interpretations of the data and determine possible instances of noncompliance. In addition, individual states' national technical means (NTM) are expected to complement the verification regime, as is explicitly provided for in the Treaty. Finally, CBMs will contribute to effectiveness of the verification monitoring system. Some examples of CBMs mentioned in the Treaty include notification regarding planned large chemical explosions, inviting observers to the sites of such explosions, and assisting in the calibration[3] of the IMS network.

Based on the system's current performance, the 1-kiloton minimum detection level specified in the Treaty will be exceeded in most if not all regions. Even partially completed, the IMS has already detected explosions well below 1 kiloton. For example, seismologists in the United States analyzed 76 seismic events that occurred between January and May 2004 from an area of Wyoming. The authors compared their experimental seismographic classifications with actual blast records from one of the largest mines in the area and found that they had successfully identified 74 out of 76 seismic events, or 97%.[4] Seismic signals from large-scale, delay-fired mine explosions—sequentially detonated explosives in multiple boreholes—can be successfully distinguished from nuclear explosions as well.

Seismic Monitoring

Due to the likelihood that a nuclear explosive test would be conducted underground, seismic monitoring plays a leading role in the IMS. Seismic monitoring includes the steps of detecting the seismic signals, separating the seismic signals that are associated with a particular source from the background, estimating the location and size of the event, and identifying the seismic source (earthquake, explosion, or the like).

The challenge for the IMS seismic network is to distinguish nuclear explosions from other frequent events, such as earthquakes and large chemical explosions (associated with mining operations, for example). Worldwide there are more than 200,000 earthquakes and mining-related explosions per year of similar magnitude to a small nuclear explosion. This task is made possible because the seismic signals from earthquakes and explosions differ. The energy released by sudden movements in the Earth's crust produce two types of elastic body-waves. Because they have the highest velocity, the first to arrive at a seismic station are P-waves (*primary* or *pressure* waves). P-waves cause particles to move, or compress, parallel to the direction of travel of the wave energy. S-waves (*secondary* or *shear* waves), which travel more slowly, move as transverse waves, with motion perpendicular to the direction of wave propagation. An explosion produces relatively little S-wave energy compared to that produced by an earthquake, so the waveforms recorded by seismometers are different. Figure 13.1 shows the seismograms from a small earthquake and a chemical explosion recorded at the same seismic station. As this example shows, given good data, distinguishing between an explosion and an earthquake is relatively straightforward.

---

[3]Incorporating seismic travel-time corrections for specific locations based on the seismic data from events of known location, magnitude, and time.

[4]Ari Hartmann, "Seismic Signature Helps Reveal Nuclear Tests," *Geotimes*, June 5, 2007.

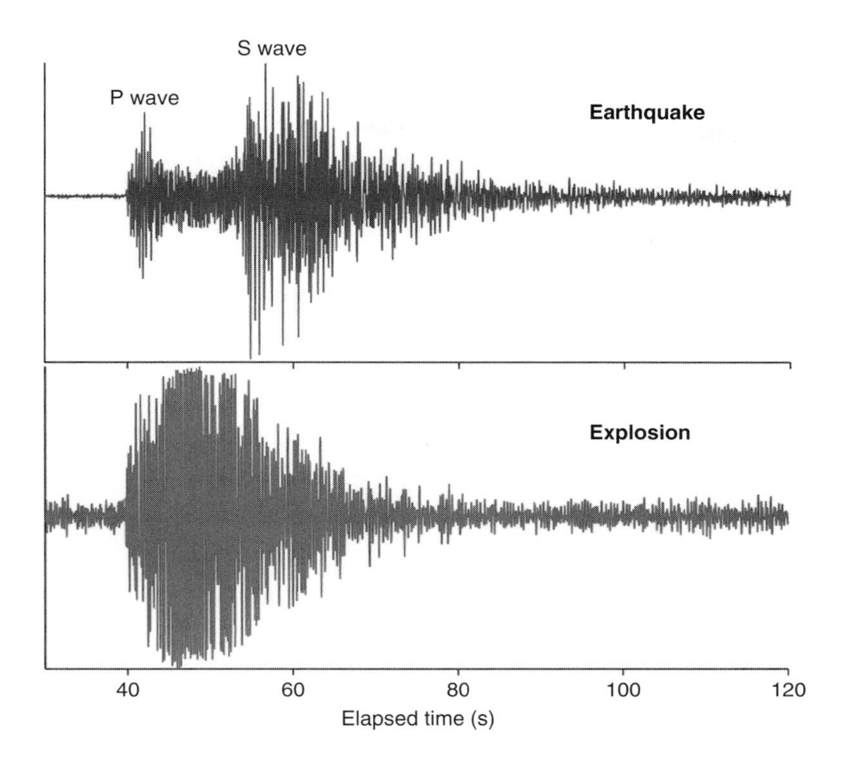

FIGURE 13.1 Seismograms of an earthquake and a chemical explosion, recorded at the same station.[5]

The global network of 50 *primary* stations in 34 countries will use two basic kinds of equipment:

- Three-component (3-C) broadband seismometers, which measure ground motion in three directions (one vertical and two horizontal, east-west and north-south).
- Arrays, which combine 3-C systems with a cluster of between nine and 25 narrowly spaced, vertical-component, short-period seismometers, arranged over an area of up to 500 km^2. (Because of its geometry, an array can optimize the signal-to-noise ratio, and the direction and distance can be determined more accurately using data from an array. Approximately half the proposed stations will be arrays.)

In addition to the primary seismic network, there will be an *auxiliary network* comprising 120 stations in 61 countries. This network will mainly augment the IMS stations that are part of countries' existing seismic monitoring networks. The main purpose of the auxiliary network is to provide additional data when needed, to improve event detection, localization, and characterization. The primary network stations will send data continuously to the IDC; the auxiliary network will operate continuously but will transmit data only for requested time periods.

The IMS currently has a detection threshold that exceeds the 1-kiloton capability, which corresponds to around 4.0 on the Richter scale, for virtually all of Eurasia and North America. In certain areas of particular interest, such as the Nevada and Novaya Zemlya test sites, the IMS seismic network has a detection capability typically below 2.5 on the Richter scale.

In addition to the IMS seismic network, hundreds of institutions and organizations worldwide operate seismic-monitoring networks of various types. Much of North America, Europe, the western Pacific, and parts of Central Asia, northern and southern Africa, and the

[5]Recorded by the Los Alamos Seismic Network, Los Alamos National Laboratory.

Middle East are now being monitored closely for earthquake activity down to magnitude 3.0 on the Richter scale or lower by organizations whose data and analysis are openly available.

The Treaty specifies that the role of the IMS and the IDC with regard to seismic-source characterization is only to provide "expert technical advice [...] to help [...] identify the source of specific events." The ultimate responsibility for identification belongs to the member states.

Hydroacoustic Monitoring

By monitoring the world's oceans, the IMS hydroacoustic network will take into account over two-thirds of the earth's surface. Underwater explosions produce such powerful broadband acoustic signals that the network, when complete, will be able to detect any manmade explosion in any ocean. The network will include 11 stations: six hydrophone stations in the Indian, Atlantic, and Pacific Oceans and five T-phase seismic stations (see following discussion) located on steep-sloped islands in the Caribbean, northern Pacific, and Atlantic. With the certification, on June 8, 2007, of the hydroacoustic station on Wake Island, located in the western Pacific, 10 of the 11 stations comprising the worldwide hydroacoustic network are in operation.[6] The locations selected for the hydroacoustic network were chosen to achieve maximum coverage of ocean areas as well as for considerations related to ease of construction and operation.[7]

A hydroacoustic station consists of a sensitive hydrophone suspended from a moored buoy or anchored to the sea floor. The signal from the hydrophone is relayed by cable to a shore station, from which the data are transmitted to the IDC. When the shore stations are located on small islands (which most will be), hydrophones will be deployed on opposite sides of each island, to receive sound from all directions. The hydrophones will monitor low frequencies (1 to 100 Hz), which is optimal for detecting explosions at large distances.

Only a small number of stations are required because acoustic energy propagates so efficiently in water. If there is no substantial land mass or extensive area of shallow water between the explosion and the detector, hydrophones can detect acoustic signals in the oceans at distances of more than 15,000 km from their source. Even nuclear explosions on small islands have been detected by hydrophones thousands of kilometers away. As an example, a 200-ton underwater explosion is so large that, provided there is no bathymetric barrier (a large area of shallow water or an island, for example), it will be detectable by at least one IMS hydroacoustic station. This sensitivity also means that the hydroacoustic monitoring system will detect an explosion in shallow water or even a near-surface atmospheric test over water.

The hydroacoustic network is particularly effective at discriminating between explosions and other phenomena, such as undersea volcanoes and earthquakes, because the signals from underwater explosions have a characteristic waveform and have much more energy at higher frequencies. Also, hydroacoustic signals from seismic events have a more gradual onset and decay over a longer period of time.

One of the reasons that even small explosions—only a few kilograms in yield—can be detected from great distances in the oceans is that there is little attenuation as sound waves travel through the oceans' *sound fixing and ranging channel* (SOFAR). The SOFAR acts as an acoustic waveguide, restricting sound to depths at which energy loss due to scattering or

---

[6]"Braving the Storms: Constructing Hydroacoustic Station HA11 at Wake Island," www.ctbto.org, July 6, 2007.

[7]Three of the hydroacoustic stations that existed prior to the Treaty will be upgraded: Wake Island (northern Pacific Ocean), Ascension Island (mid-Atlantic Ocean), and Queen Charlotte Island (west coast of Canada). The other stations will be new: Cape Leeuwin (Australia), Crozet Island (southwest Indian Ocean), Chagos Archipelago (northern Indian Ocean), Guadeloupe (Caribbean), Clarion Island (west coast of Mexico), Juan Fernandez Island (Chile), Flores Island (northern Atlantic Ocean), and Tristan da Cunha (southern Atlantic Ocean).

refraction from the sea surface or the sea floor is minimized.[8] The region of peak SOFAR transmission is around 1,000 m in tropical waters. Hydrophones will be placed at a depth near the region of peak SOFAR transmission.

Hydrophones are sensitive, but they are expensive to install and maintain. An alternative to a hydrophone is a high-frequency (0.5 to 20 Hz) seismometer, located near a steep shoreline. Such a seismometer responds to T-phase seismic waves—compression waves generated by conversion of the incident sound at the land boundary. T-phase seismometers are not as sensitive as hydrophones, but they are much cheaper to install and maintain.

Infrasound Monitoring

Any explosion in the atmosphere generates a pressure wave that radiates in all directions and attenuates as it travels away from the source. The IMS infrasound network will have 60 stations in 35 different countries that will use sensitive microphones to detect very low-frequency sound waves, which can travel long distances in the atmosphere, by measuring small changes in pressure. The data from these stations can be used to locate atmospheric explosions and distinguish them from natural phenomena such as volcanoes, storms, or wind and from other man-made phenomena such as rocket launches or supersonic aircraft. The infrasonic signal from an above-ground 1-kiloton explosion can be detected at a distance of a few thousand kilometers. Atmospheric explosions of 0.5 kilotons can be detected for most of the planet, but the detection limit is as low as 0.3 kilotons for large continental areas and 0.1 kilotons for a few locations.

Some infrasound stations consist of seven-microphone arrays, but most have an array of four microphones placed at the center and apices of an equilateral triangle, with sides whose length is 1–3 km. The array configuration provides information about the direction of the signal source.

Atmospheric Radionuclide Monitoring

The IMS radionuclide network will comprise 80 radionuclide-sampling stations, located in 39 countries. All the other IMS components provide real-time monitoring; radionuclide monitoring requires a delay for sampling and analysis. In its favor, the results are definitive: the fission products produced by a nuclear explosion, and their isotopic ratios, are unambiguous and well characterized. Atmospheric and underwater explosions release radionuclides as particulates and noble gases (xenon and krypton); underground explosions vent a smaller but still significant fraction of the noble-gas radionuclides into the atmosphere. Without extraordinary mitigation efforts, the fission-product radionuclides from a nuclear explosion will be released into the atmosphere.

Radionuclide sampling can confirm a nuclear event but cannot pinpoint the location with exactness. By contrast, the waveform networks described previously can detect and locate suspicious events but they cannot provide definitive evidence of a nuclear test. Combining the radionuclide and meteorological analysis with the waveform data from the same area and time, however, the results would be definitive. Furthermore, in principle, even with a carefully devised evasive scenario, a nuclear event would likely be detected by the radionuclide monitoring system alone.

---

[8]Acoustic waves bend, or refract, toward the area of minimum sound speed, which is directly related to the temperature and density of seawater. Sound waves become "channeled" as they are refracted downward as their speed decreases due to lower water temperature near the surface, then upward as the density increases with depth (and pressure). This phenomenon allows low-frequency sound waves to travel thousands of meters without significant energy loss. The depth of the SOFAR channel varies with the temperature, density, and, to some extent, salinity of the water. It is deepest in the subtropics and comes to the surface in high latitudes, where the sound propagates in the surface layer.

In addition to the radionuclide sampling stations, there will be 16 associated Secretariat-certified laboratories that will analyze filters from the stations (and, should an OSI take place, samples taken by inspectors). All the radionuclide network stations will include high-volume air samplers, which draw air through high-efficiency filters to collect particulates. These filters are analyzed at one of the certified laboratories, using high-resolution gamma-spectroscopy. Half the stations will also have lower-volume cryogenic samplers to collect noble-gas samples for beta- and gamma-analysis of xenon isotopes.

Because this monitoring method is passive—the air must move to the stationary detector—a relatively dense network of detectors is necessary for it to be effective. Meteorological monitoring must also be integrated into the process to provide information about wind patterns from the days and weeks preceding a detected event.

For radionuclide monitoring there is a 50% probability of detecting a 1-kiloton explosion within five days; the probability increases to 90% within 10 days.

International Data Center

As part of normal operations, the IDC will provide system monitoring (of the IMS stations, communications, and IDC processing systems); data handling (acquisition, authentication, quality control, and storage); special data management (radionuclide laboratory reports, data from CBMs, consultations and clarifications data, OSIs, and NDCs); and expert analysis. Member states may also request technical assistance, special IDC software, and other IDC support. The IDC has been providing IMS data and IDC products to member states through secure signature accounts since early 2000.

The IDC receives several gigabytes of data from the IMS daily. The IDC processes and analyzes these data and makes the first automated products—referred to as IDC Standard Products—available to member states within two hours.

One of the IDC Standard Products is the Standard Event List 1 (SEL1), which includes the preliminary locations of events from which signals have been detected by at least two IMS primary seismic stations—mostly earthquakes or mining-related chemical explosions. SEL1 is prepared automatically, around the clock, and is issued for every 20-minute interval of time. It typically includes over 100 events every day.

The Standard Products are reviewed by IDC analysts, who then prepare and release the Reviewed Event Bulletin (REB). The REB contains all those events which have been detected at IMS seismic, hydroacoustic, and infrasound stations and which meet specific quality criteria (referred to as the *event definition criteria*) for a given day. The REB may include data from both primary and auxiliary seismic stations.

Communications

The IDC will be linked to every IMS station via the Global Communications Infrastructure (GCI), which transmits data using a network of three geosynchronous satellites. The IDC will also be linked to the national data centers of signatory countries. Five GCI hubs, which are connected via terrestrial links to the IDC, have been installed, and GCI terminals have been set up at 46 IMS stations, NDCs, and development sites. In September 1998, the CTBTO and the international partnership of Hughes Olivetti Telecom Ltd. signed a $70 million contract to design, install, manage, operate, and maintain the GCI through 2008. The network's very small aperture terminals[9] (VSATs) will have the capacity to transfer up to 11.4 gigabytes of data per

[9] A very small aperture terminal (VSAT) is a two-way satellite-communications ground station with a dish antenna that is smaller than 3 meters. VSAT is commonly used in communications by businesses (for credit card transactions or corporate communications, for example) and in broadband Internet and other communications in remote or rural locations. Nearly all VSAT systems are IP-based.

day among the monitoring stations, the IDC, and the signatory states. As of September 1999, seven VSATs were operating at seven IMS stations and five national data centers.

Confidence-Building Measures

Member states consent to pursue CBMs as one of their CTBT obligations. Specifically, each state agrees to voluntarily notify the PTS when any single chemical explosion of 300 metric tons, or .3 kilotons, or more of TNT-equivalent explosives is going to take place. They may also invite observers to the sites of such explosions. Analysis of the data from these known large explosions assists in the calibration of IMS stations. In the context of CBM activities, Kazakhstan, Russia, and the United States have carried out a number of information exchanges and calibration explosions.

Consultation and Clarification

The consultation and clarification component of the verification regime encourages state parties to attempt to resolve, either among themselves or through the CTBTO, ambiguous events *before* requesting an OSI. A state party must provide clarification of an ambiguous event within 48 hours of receiving such a request from another state party or the Executive Council.

Onsite Inspections

Each state party has the right to request an OSI based on IMS data or NTM. Requests for OSIs must be approved by at least 30 members of the Treaty's 51-member Executive Council. The Executive Council must act within 96 hours of receiving a request for an inspection, during which time the Technical Secretariat can make preparations to deploy an inspection team reasonably quickly, if an inspection is approved. The Technical Secretariat will not maintain a standing group of inspectors; instead, an OSI team would be selected from a group of designated and trained experts who are on standby. In the event that the Executive Council finds an OSI request to be frivolous or abusive, it may impose punitive measures on the requesting state that could include financial compensation for Technical Secretariat expenses and suspension of the state's right to request future inspections.

An OSI request must include the time of the ambiguous event, the approximate geographical coordinates, the estimated depth, the proposed boundaries of the area to be inspected (less than 1,000 square kilometers), the state party or state parties to be inspected, and all evidence on which the request is based.

Activities conducted as part of an OSI may include:

- Visual and photographic inspection
- Measurement of aftershocks (to more definitively distinguish between earthquake and explosion and to help determine more precisely the event's location)
- Radiation measurements
- Environmental sampling
- Measurement of radioactive noble gases (especially Ar-37 and Kr-85) or debris
- Observation of surface changes due to spallation[10]
- Locating human artifacts characteristic of test activity
- Active seismology
- Imaging with ground-penetrating radar
- Drilling

---

[10]Spallation of the ground surface occurs above an underground explosion when, as a result of the shock wave generated by the explosion, surface materials are thrown upward and then fall back to the surface.

Considering the options available to inspectors, it would be difficult for a violator to anticipate ways to conceal *all* potential evidence. The first phase of an inspection, lasting 25 days, is to include less intrusive techniques. A second phase, which would immediately follow the first phase unless terminated by the Executive Council, would last 35 days and apply more intrusive measures. Refusal to allow inspection on the part of a country that had tested clandestinely, in the face of strong evidence of a test, would likely be seen as a tacit admission of a violation. Many arms-control experts believe that the right, under the treaty, to conduct OSIs may be a deterrent, regardless of whether the inspection actually takes place.

Summary

If the CTBT enters into force, it will have one of the most sophisticated verification regimes in the history of arms control. After the 2006 nuclear test explosion in the DPRK, the location, yield, and type of fissile material used in the explosion were identified from seismic and radionuclide data of the kind that will be collected by the CTBT's verification system. This fact indicates that when completed, the CTBT IMS will be capable of meeting or exceeding the performance parameters specified for it by the Treaty. Correspondingly, the likelihood that a future nuclear detonation could be concealed from a completed IMS, coupled with NTM, and, if necessary, an OSI, is very, very small.

To maintain and improve the CTBT verification system's current capabilities, work needs to be done on calibration of the seismic network, and aging equipment and infrastructure need to be upgraded as necessary. However, these are technical problems with known solutions. The greatest remaining challenge to enforcing a global nuclear test ban is finding the political will and diplomatic compromises to achieve entry into force of the CTBT.

Further Readings

"Benefits to the U.S. Derived From the CTBT International Monitoring System," Fact Sheet released by the Bureau of Arms Control, U.S. Department of State, Washington, D.C., Oct. 8, 1999.

"Braving the Storms: Constructing Hydroacoustic Station HA11 at Wake Island," www.ctbto.org, July 6, 2007.

"Capability to Monitor the Comprehensive Test Ban Treaty," Joint Statement by the American Geophysical Union (AGU) and the Seismological Society of America (SSA) of Sept. 1999, reaffirmed Dec. 2003.

"CTBT Facts and Fiction," Fact Sheet released by the Bureau of Arms Control, U.S. Department of State, Washington, D.C., Oct. 8, 1999.

"Comprehensive Test Ban Treaty: Now or Never," *The Acronym Institute for Disarmament Diplomacy*, Acronym Report No. 8, www.acronym.org.uk/acrorep/a08ass.htm, Oct. 1995.

"Final Report of the Independent Commission on the Verifiability of the CTBT," www.ctbtcommission.org.

Richard L. Garwin and Frank N. von Hippel, "A Technical Analysis of North Korea's Oct. 9 Nuclear Test," *Arms Control Today*, www.armscontrol.org/act/2006_11/NKTestAnalysis.asp, Nov. 2006.

David. Hafemeister, "Effective CTBT Verification: The Evidence Accumulates," *Verification Yearbook 2004*:29–44.

Ari Hartmann. "Seismic Signature Helps Reveal Nuclear Tests," *Geotimes*, June 5, 2007.

Wolfgang Jans, "Hydroacoustic CTBT Monitoring of the World Oceans: Strengths and Weaknesses," Talk given at the Vertic Seminar during the Article XIV Conference, 21–232, Sept. 2005.

Wang Jun, "CTBT Verification Regime: Preparations and Requirements," Disarmament Forum, Onsite Inspections: Common Problems, Different Solutions, 1999, no. 3: 39–44.

Jungmin Kang and Peter Hayes, "Technical Analysis of the DPRK Nuclear Test," *CISAC*, Policy Forum Online 06-89A, www.nautilus.org/fora/security/0689HayesKang.html, Oct. 20, 2006.

Daryl Kimball, "Keeping Test Ban Hopes Alive: The 2005 CTBT Entry-into-Force Conference," *Disarmament Diplomacy*, Issue No. 81, Winter 2005.

Ben Mines, "The Comprehensive Nuclear Test Ban Treaty: virtually verifiable now," Verification Research, Training and Information Centre (Vertic), Vertic Brief 3, April 2004.

Miles A. Pomper, "Test Ban Infrastructure: A Concrete Reality," *Arms Control Today*, Oct. 2004.

"Potential Civil and Scientific Applications of the CTBT Verification Technologies," CTBTO Preparatory Commission, Oct. 2002.

David Ruppe, "U.S., India Differ on Nuke Test Moratorium Language," Global Security Newswire, April 25, 2006.

"Seismic Verification of Nuclear Testing Treaties," U.S. Congress, Office of Technology Assessment, OTA-ISC-361 (Washington, D.C.: U.S. Government Printing Office, May 1988).

Statement by the Chinese Delegation on the Nuclear Disarmament and Reduction of the Danger of Nuclear War at the First Session of the Preparatory Committee for the 2010 NPT Review Conference, Ministry of Foreign Affairs of the People's Republic of China, www.chineseembassy.org/eng/wjb/zzjg/jks/kjfywj/t317959.htm, May 11, 2007.

"Technical Issues Related to the Comprehensive Nuclear Test Ban Treaty," Committee on Technical Issues Related to Ratification of the Comprehensive Nuclear Test Ban Treaty, Committee on International Security and Arms Control, National Academy of Sciences, National Academy Press, Washington, D.C., 2002.

"The Global Communications Infrastructure and the International Data Centre," CTBTO Preparatory Commission, *Basic Facts: Booklet 4*, 2001.

"The Global Verification Regime and the International Monitoring System," CTBTO Preparatory Commission, *Basic Facts: Booklet 3*, 2001.

Rodney W. Whitaker, "Infrasonic Monitoring." Los Alamos National Laboratory LA-UR-95-2275, 1995.

Ray Willemann, "The Role of Non-IMS Stations in Explosion Monitoring," Talk at the Vertic seminar during the CTBT Article XIV Conference, 2003.

Jennifer Yauck, "Put to the Nuclear Test: Seismology and the International Monitoring System," *Geotimes*, March 2007.

John J. Zucca, "Forensic Seismology Supports the Comprehensive Test Ban Treaty," www.llnl.gov/str/Zucca.html.

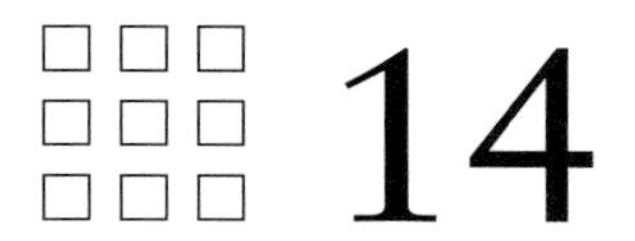

Evaluating Nonproliferation Bona Fides[1]

Amy Seward, Carrie Mathews, and Carol Kessler

Introduction

Anticipated growth of global nuclear energy in a difficult international security environment heightens concerns that states could decide to exploit their civilian nuclear fuel cycles as a means of acquiring nuclear weapons. Such concerns partly reflect a fundamental tension in the Treaty on the Nonproliferation of Nuclear Weapons (NPT). On one hand, Articles II of the NPT clearly prohibits nonnuclear-weapon states party from acquiring nuclear weapons. On the other hand, Article IV of the NPT confers the "*inalienable right*" of parties to the treaty to "develop research, production and *use* of nuclear energy *for peaceful purposes*," and indicates all parties shall cooperate to "facilitate [...] the fullest possible exchange of equipment, materials and scientific and technological information for the *peaceful uses* of nuclear energy" and "cooperate in contributing [...] to the further development of the applications of nuclear energy *for peaceful purposes*." The tension arises because some nuclear technologies can be used for either peaceful or nonpeaceful purposes and the NPT does not provide a distinction in technology transfers or development of these more sensitive dual-use technologies such as enrichment and reprocessing. But once these technologies are in hand, a country is just steps—significant ones, but only a few—from being able to produce nuclear material suitable for weapons.

Such a path to the acquisition of nuclear weapons is more than a theoretical possibility. As NPT parties, Iraq, North Korea, and Libya all used ostensibly peaceful nuclear activities as a basis for establishing clandestine nuclear weapons programs. Thus the international community may legitimately seek ways to determine whether a state has good faith commitments ("*bona fides*") to develop only peaceful applications of nuclear energy; in other words, whether the state is exercising its Article IV rights in a manner that is consistent with its full NPT obligations. Such a determination of a country's bona fides may be relevant, for example, before exporting nuclear equipment, material, or technology to that country.

Such a determination will be more persuasive and defensible if it is not made ad hoc but rather on the basis of criteria that are as objective as possible and consistently applied. The purpose of this chapter is to propose such a set of criteria for gauging a country's peaceful uses bona fides. These criteria or indicators are then applied to the cases of Iran and Brazil.

---

[1]This chapter summarizes the PNNL publication, Morris et al., "Peaceful Uses Bona Fides: Criteria for Evaluation and Case Studies," PNNL SA-16641. The authors wish to acknowledge the contributions of Danielle Peterson, Chris Ajemian, and Fred Morris to this research and chapter.

Peaceful Use vs. Peaceful Purpose

Article IV of the NPT refers to both peaceful *purposes* and peaceful *use*. Peaceful *use* is straightforward: Is the state using nuclear equipment, material, and technology solely for civilian applications, or is the state instead misusing its nuclear program to build a weapons capability? Peaceful *purpose*, on the other hand, suggests that the state's *intent* be considered: What is the state's purpose in acquiring or indigenously developing nuclear equipment, material, or technology and will it remain peaceful over the lifetime of their nuclear program, or does the state intend to use such material, equipment, or technology for weapons purposes, perhaps at some time in the future?

For example, a state could construct an enrichment facility as part of a civilian nuclear power program but with the intent of later using it as a hedge against a rival's threats or as a critical component of a nuclear weapons program. If a state has reactors for power generation and uranium is being enriched for fuel for use in those reactors, it can be argued that the enrichment plant meets the *peaceful-use* criterion. However, if a country with one or two reactors and negligible domestic uranium deposits seeks to invest in a uranium enrichment facility, this may lead to reasonable questions about the country's intent in constructing the facility. That is, if the economic justification for the enrichment facility appears weak and there is no other compelling rationale for it, such as extensive plans for future nuclear power generation, it could be concluded that the intended use may not be peaceful. Under the proposed methodology, reaching this judgment would require careful evaluation of other peaceful-use indicators to see if they add clarity as to whether to judge such activities as peaceful or not peaceful.

Pursuant to NPT Article IV, nonnuclear weapons states (NNWS) are obligated to limit their pursuit of nuclear energy exclusively to peaceful endeavors, leaving no room for a state to start pursuing nuclear energy for peaceful use with the purpose of subsequently using the material, equipment, or technology thus acquired as a means of developing nuclear weapons. Through the safeguards system established to implement Article III, the NPT empowers the International Atomic Energy Agency (IAEA) on behalf of the international community to examine the purpose and operation of a NNWS nuclear program. If and when doubt arises based on IAEA safeguards conclusions, the NPT further provides justification for states and the IAEA to suspend support or assistance to the state in question until the IAEA can verify the state's peaceful intent.

Despite Iraq being a signatory to the NPT as a NNWS, it became apparent in 1991 that the country had violated its Article II and III obligations. As a result, the IAEA member states agreed to the establishment of the Additional Protocol (AP) to IAEA safeguards agreements. The AP established stiffer inspection and reporting requirements, providing the IAEA with the ability to evaluate the correctness *and* completeness of a state's declaration of nuclear material and activities. This new capability gave the IAEA the tools and authority to check for undeclared nuclear material or facilities, which had not been true under previous safeguards arrangements. To the extent that it is adopted by member states, the AP enhances safeguards coverage significantly and increases other member states' confidence in conclusions drawn by the IAEA regarding the absence of undeclared nuclear material and activities. Implementation of an AP in a state reduces the probability that its nuclear programs may be redirected to weapons purposes without detection by the IAEA.

In a positive sign of the evolution in thinking that is reflected by the AP, Malaysia submitted a working paper at the 2005 NPT Review Conference on behalf of the Non-Aligned Movement (NAM) that, while focusing on the right to peaceful use, recognized that the parties' "inalienable right to develop research, production and use of nuclear energy for peaceful purposes" should be "*in conformity with articles I, II, and III of the Treaty ....*"[2] The statement was significant because the NAM had supported the position that Article IV's inalienable right to nuclear technology meant any nation had the right to any peaceful nuclear technology, equipment, or materials. In this statement, the NAM acknowledged the conditional link

---

[2]2005 Review Conference Working Paper by the Group of Non-Aligned States Parties to the NPT, NPT/CONF.2005/WP.20, para. 1.

between NPT Articles II and III and Article IV, thus accepting limits on the exercise of the right to peaceful nuclear technology.

Indicators of Peaceful Uses

Indicators of both peaceful and nonpeaceful use were identified through this methodology to provide a systematic means of evaluating a state's longer-term peaceful-use intent or bona fides. The first four tables present four categories of indicators:

- A state's nonproliferation credentials
- Its fulfillment of Article III obligations and its commitment to transparency
- The coherence of its nuclear program
- Its degree of geopolitical integration and overall record of international cooperation

Although no single indicator (with the exception of conducting weaponization activities) is decisive, applying the complete set of indicators to a state's actions can provide an assessment of the probability that a state's nuclear program will remain peaceful or not. Tables 14.1–14.4 list the indicators and the metrics selected for the bona fides evaluation and

Table 14.1 Peaceful-use indicators (nonproliferation credentials).

		Nonpeaceful-Use Indicator	Peaceful-Use Minimum Indicator	Elevated Peaceful-Use Indicator
Nonproliferation Credentials	Metric	Threatened or actual withdrawal from NPT	Ratification and full adherence to NPT	Declared policies supporting NPT regime; enables NPT compliance by other countries
	Metric	No signature to any nonproliferation treaty	Participation in regional safeguards regimes if applicable	Member NWFZ; ratified other nonproliferation treaties such as CTBT

Table 14.2 Peaceful-use indicators (fulfillment of Article III obligations).

		Nonpeaceful-Use Indicator	Peaceful-Use Minimum Indicator	Elevated Peaceful-Use Indicator
Fulfillment of Article III Obligations	Metric	Safeguards agreement not in force; subsidiary arrangements not in place	Safeguards agreement and subsidiary arrangements in force	AP and subsidiary arrangements to AP (as applicable)
	Metric	Noncompliance with safeguards agreements; confirmed undeclared nuclear facilities or materials	Full compliance with safeguards agreements; no undeclared nuclear facilities or materials	No undeclared facilities or materials; confirmed correct and complete expanded declaration per AP
	Metric	Interference with, suspension of, or prevention of safeguards inspections	Full cooperation in conduct of safeguards inspections	Complementary access per AP
	Metric	Inadequate or poorly functioning SSAC[1]	SSAC satisfies IAEA guidelines	Exemplary SSAC
	Metric	Exports of EDP[2] equipment or materials not under safeguards (noncompliance with NPT III.2)	Exports of EDP equipment or materials under safeguards (compliance with NPT III.2)	Export information per AP provided to IAEA

[1]State system of Accounting for and Control of Nuclear material.

[2]Especially designed or prepared (EDP).

Table 14.3 Peaceful-use indicators (coherence of nuclear energy program).

		Nonpeaceful-Use Indicator	Peaceful-Use Minimum Indicator	Elevated Peaceful-Use Indicator
Coherence of Nuclear Energy Program	Metric	Development, acquisition, or plans for sensitive nuclear facilities lacking reasonable economic or energy security justification	Current and planned nuclear fuel cycle facilities have reasonable economic or energy security justification	Agreement to forego enrichment and reprocessing and other sensitive facilities or implement multilateral ownership/control of facilities
	Metric	Operating research reactor(s) fueled with HEU and no willingness to decommission or convert to LEU; retains sensitive nuclear facilities	Commitment to decommission or convert and safeguard HEU research reactors to LEU; agreement on multilateral ownership/control of sensitive facilities	No research reactors operating with HEU; agreement to close sensitive nuclear facilities
	Metric	Fresh or spent HEU fuel in storage with no willingness to plan or arrange for return to supplier	Commitment or arrangement for return of fresh or spent HEU to supplier	No fresh or spent HEU in storage; active arrangements for HEU return to supplier
	Metric	Some fuel cycle activities lay solely on weapons development pathway	No fuel cycle activities lay solely on weapons development pathway	

Table 14.4 Peaceful-use indicators (geopolitical cooperation).

		Nonpeaceful-Use Indicator	Peaceful-Use Minimum Indicator	Elevated Peaceful-Use Indicator
Geopolitical Cooperation	Metric	Country located in area of regional or interstate instability or perceives its position in regional hierarchy as untenable	Country located in area of moderate to high regional or interstate stability or views its position as acceptable in regional hierarchy	Country located in area of high overall stability or perceives its position as satisfactory in the region hierarchy
	Metric	Few to no international treaty obligations or commitments	Member of major international nuclear treaties (Nuclear Safety Convention; CPNMM; Spent Fuel and Nuclear Waste Convention; Nuclear Terrorism, Assistance and Notification Conventions)	Member of other multilateral nonproliferation mechanisms (Proliferation Security Initiative; Nuclear Suppliers Group)

these are applied to case studies for Brazil and Iran in Tables 14.5 and 14.6. These are followed by a comparative analysis.

Nonproliferation Credentials

This category focuses on a state's adherence to, participation in, and leadership of nonproliferation regimes. The minimum standard for a state to demonstrate the peaceful purposes of

Table 14.5 Peaceful-use indicators, Brazil.

		Nonpeaceful-Use Indicator	Peaceful-Use Minimum Indicator	Elevated Peaceful-Use Indicator
Nonproliferation Credentials	Metric	Threatened or actual withdrawal from NPT	Ratification and full adherence to NPT	Declared policies supporting NPT regime; enables NPT compliance by other countries
	State action/ activity		**Brazil ratified the NPT in 1997; Brazil's nuclear materials were determined in 2005 to have remained in peaceful activities**[1]	**Quadripartite/ABACC agreements require close Brazilian-Argentine and IAEA cooperation for full-scope safeguards compliance**
	Metric	No signature to any nonproliferation treaty		Member in regional nuclear weapons-free zone if applicable; ratified other nonproliferation treaties such as Comprehensive Test-Ban Treaty
	State action/ activity			**Party to regional Treaty of Tlatelolco (and amendment); party to CTBT**
Fulfillment of Article III Obligations	Metric	Safeguards agreement not in force; subsidiary arrangements not in place	Safeguards agreement and subsidiary arrangements in force	AP and subsidiary arrangements to AP (as applicable)
	State action/ activity		**Quadripartite/ABACC provides for full-scope IAEA safeguards; subsidiary arrangements in place**	**AP is under consideration by Brazil**
	Metric	Noncompliance with safeguards agreements; confirmed undeclared nuclear facilities or materials	Full compliance with safeguards agreements; no undeclared nuclear facilities or materials	No undeclared facilities or materials; confirmed correct and complete expanded declaration per AP
	State action/ activity		**Full compliance with Quadripartite/ ABACC agreements; extensive recent negotiations regarding Resende enrichment facility successfully completed**	
	Metric	Interference with, suspension, or prevention of safeguards inspections	Full cooperation in conduct of safeguards inspections	Complementary access per AP
	State action/ activity		**Past disagreement with IAEA over protection of centrifuge enrichment technology at Resende is now resolved; IAEA is now confident that it has necessary information to make safeguards conclusions about he facility**	

(Continued)

Table 14.5 (*Continued*)

		Nonpeaceful-Use Indicator	Peaceful-Use Minimum Indicator	Elevated Peaceful-Use Indicator
	Metric	Inadequate or poorly functioning SSAC	SSAC satisfies IAEA guidelines	Exemplary SSAC
Fulfillment of Article III Obligations	State action/ activity			**Highly effective SSAC per Quadripartite/ABACC agreements**
	Metric	Exports of equipment or materials not under safeguards (noncompliance with NPT III.2)	Exports of equipment or materials under safeguards (complies with NPT III.2)	Export information per AP provided to IAEA
	State action/ activity		**Brazil is a Nuclear Supplier Group member; Brazil chaired NSG plenary in 2006**	
	Metric		Research base supporting technical studies ensuring safety of nuclear program; nuclear operator training; nuclear regulatory body	Sponsorship of international forum, meetings or workshops promoting nuclear safety and/or nonproliferation training and cooperation
	State action/ activity		**Brazil has five nuclear research centers; the Angra power plant simulator has provided operator training for countries such as Spain, Switzerland, Germany, and Argentina**[2]	**Brazil hosted the IAEA's 2002 Conference on Safety Culture in Nuclear Installations**[3]
Coherence of Nuclear Energy Program	Metric	Development, acquisition, or plans for sensitive nuclear facilities lacking reasonable economic or energy security justification	Current and planned nuclear fuel cycle facilities have reasonable economic or energy security justification	Agreement to forego enrichment and reprocessing and other sensitive facilities or implement multilateral ownership/control of facilities
	State action/ activity		**Brazil's stated plans for power generation and sale of enriched fuel meet output estimates for its Resende enrichment facility**	**Brazil's spent fuel reprocessing capability has been shut down; no active commercial reprocessing of spent reactor fuel**
	Metric	Operating research reactor(s) fueled with HEU and no commitment to decommission or convert to LEU; retains sensitive nuclear facilities	Commitment to decommission or convert and HEU research reactors to LEU; agreement on multilateral ownership/ control of sensitive facilities	No research reactors operating with HEU; agreement to close sensitive nuclear facilities
	State action/ activity			**No research reactors operating with HEU**

	Metric	Fresh or spent HEU fuel in storage with no plans or arrangement for return to supplier	Commitment or arrangement for return of fresh or spent HEU to supplier	No fresh or spent HEU in storage; active arrangements for HEU return to supplier
	State action/ activity		**HEU fuel assemblies returned to U.S. under the Foreign Research Reactor Spent Fuel Acceptance Program**[4]	**All spent HEU fuel has been returned to the U.S.; only other HEU was received from China and has been down-blended to LEU just under 20%**[5]
	Metric	Fuel cycle activities lay solely on weapons development pathway	Fuel cycle activities do not lay solely on weapons development pathway	See minimum peaceful-use indicator
	State action/ activity		**Once secret weapons program discontinued; light water reactors; enrichment facility under full-scope safeguards**	**Former prototype naval propulsion program redirected to possible peaceful applications for small power plants**
Geopolitical Cooperation	Metric	Country located in area of regional or interstate instability	Country located in area of moderate to high regional or interstate stability	Country located in area of high overall stability
	State action/ activity		**Former nuclear competition with Argentina has subsided**	**Some degree of low-level regional conflict and some internal but strong economic and cultural ties across region**
	Metric	Few to no international treaty obligations or commitments	Member to major international nuclear and nonproliferation treaties	Member to other multilateral nonproliferation mechanisms (Proliferation Security Initiative; Nuclear Suppliers Group)
	State action/ activity		**Party to Nuclear Safety Convention; CPNMM; Spent Fuel and Nuclear Waste Convention; Nuclear Terrorism, Assistance and Notification Conventions**	**Nuclear Suppliers Group member**

[1]IAEA Safeguards Statement for 2005, p. 1, para. 2; p. 11, "List of States: The 77 states listed in paragraph 2 are ... Brazil ...."

[2]IAEA Brazil Country Profile 2003; www-pub.iaea.org/MTCD/publications/PDF/cnpp2003/CNPP_Webpage/countryprofiles/Brazil/Brazil2003.htm.

[3]IAEA International Conference on Safety Culture in Nuclear Installations 2002, Rio de Janeiro, Brazil: Announcement and Call for Papers; www.iaea.org/worldatom/Meetings/2002/infcn97.shtml.

[4]www-pub.iaea.org/MTCD/publications/PDF/te_1508_web.pdf.

[5]www.isis-online.org/global_stocks/end2003/civil_heu_watch2005.pdf.

Table 14.6 Peaceful-use indicator, Iran.

		Nonpeaceful-Use Indicator	Peaceful-Use Minimum Indicator	Elevated Peaceful-Use Indicator
Nonproliferation Credentials	Metric	Threatened or actual withdrawal from NPT	Ratification and full adherence to NPT	Declared policies supporting NPT regime; enables NPT compliance by other countries
	State action/ activity	**NPT withdrawal threatened since 2005**		
	Metric	No signature to any nonproliferation treaty; or statements denying NPT validity	Effective participation in NPT nonproliferation regime	Member regional nuclear weapons free zone; ratified Comprehensive Test-Ban Treaty or other nonproliferation treaties
	State action/ activity			
Fulfillment of Article III Obligations	Metric	Safeguards agreement not in force; subsidiary arrangements not in place	Safeguards agreement and subsidiary arrangements in force	AP and subsidiary arrangements to AP (as applicable)
	State action/ activity		**Safeguards agreement and subsidiary arrangements in force; AP signed, not ratified; until February 2006 AP in effect as though in force**	
	Metric	Noncompliance with safeguards agreements; confirmed undeclared nuclear facilities or materials	Full compliance with safeguards agreements; no undeclared nuclear facilities or materials	No undeclared facilities or materials; confirmed correct and complete expanded declaration per AP
	State action/ activity	**Failure to report new facility construction of uranium conversion facility at Esfahan; undeclared uranium enrichment and plutonium separation efforts revealed by IAEA since 2003; undeclared import and use of natural uranium in 1991 and 1993; undeclared import of UF6**[1]		
	Metric	Interference with, suspension of, or prevention of safeguards inspections	Full cooperation in conduct of safeguards inspections	Complementary access per AP

	State action/ activity	**Failure to provide one-year multiple reentry visas to IAEA inspectors; inspector access prevented, postponed or restricted; AP suspension has limited inspector access to Iran's enrichment facilities**		
	Metric	Inadequate or poorly functioning SSAC	SSAC satisfies IAEA guidelines	Exemplary SSAC
Fulfillment of Article III Obligations	State action/ activity	**Failure to report nuclear material, its processing and use, and to declare the existence of fuel cycle facilities; Iranian Nuclear Regulatory Authority (INRA) maintains SSAC**		
	Metric	Exports of EDP[2] equipment or materials not under safeguards (noncompliance with NPT III.2)	Exports of EDP equipment or materials under safeguards (complies with NPT III.2)	Export information per AP provided to IAEA
	State action/ activity			
	Metric		Research base supporting technical studies ensuring safety of nuclear program; nuclear operator training; nuclear regulatory body	Sponsorship of international forum, meetings or workshops promoting nuclear safety, training and cooperation
	State action/ activity		**Atomic Energy Organization of Iran (AEOI) National Nuclear Safety Department (NNSD); Iran Nuclear Regulatory Agency (INRA)**	
Coherence of Nuclear Energy Program	Metric	Development, acquisition, or plans for sensitive nuclear facilities lacking reasonable economic or energy security justification	Current and planned nuclear fuel cycle facilities have reasonable economic or energy security justification	Agreement to forego enrichment and reprocessing and other sensitive facilities or implement multilateral ownership/control of facilities
	State action/ activity	**Fuel cycle infrastructure development largely not justified by existing or planned nuclear power generation; ongoing suspension of international nuclear cooperation (Russia; IAEA) reinforces lacking justification**		
	Metric	Operating research reactor(s) fueled with HEU and no commitment to decommission or convert to LEU; retains sensitive nuclear facilities	Commitment to decommission or convert and HEU research reactors to LEU; agreement on multilateral ownership/control of sensitive facilities	No research reactors operating with HEU; agreement to close sensitive nuclear facilities

(Continued)

Table 14.6 *(Continued)*

		Nonpeaceful-Use Indicator	Peaceful-Use Minimum Indicator	Elevated Peaceful-Use Indicator
	State action/ activity	**New planned facilities include heavy water research reactor that is proliferation prone; enrichment plant construction continues despite international concern**		
	Metric	Fresh or spent HEU fuel in storage with no plans or arrangement for return to supplier	Commitment or arrangement for return of fresh or spent HEU to supplier	No fresh or spent HEU in storage; active arrangements for HEU return to supplier
	State action/ activity			
	Metric	Fuel cycle activities lay solely on weapons development pathway	Fuel cycle activities do not lay solely on weapons development pathway	
	State action/ activity	**Production of uranium targets; unresolved questions over two military sites that may house undeclared nuclear activities: Parchin and Lavizan; *Green Salt Project* high explosive testing and missile reentry vehicle design may have military nuclear dimension**		
Geopolitical Cooperation	Metric	Country located in area of regional or interstate instability	Country located in area of moderate to high regional or interstate stability	Country located in area of high overall stability
	State action/ activity	**High regional instability; Geopolitical isolation**		
	Metric	Few to no international nuclear treaty obligations or other international commitments	Member to major international nuclear nonproliferation treaties (Nuclear Safety Convention; CPNMM; Spent Fuel and Nuclear Waste Convention; Nuclear Terrorism, Assistance and Notification Conventions)	Member to other multilateral nonproliferation mechanisms (Proliferation Security Initiative; Nuclear Suppliers Group)
	State action/ activity	**Not member of most of IAEA-related international conventions**		

[1]Representative examples; not an inclusive list.
[2]Especially designed or prepared (EDP).

its nuclear program is ratification and full adherence to the NPT, and participation in a bilateral or regional safeguards regime if applicable. States may demonstrate a greater level of nonproliferation commitment by working with other countries to facilitate their compliance with the NPT, through membership in regional nuclear weapons-free zones (NWFZ), and through assuming leadership roles in nonproliferation. Conversely, states that are not parties to the NPT or that threaten withdrawal from the NPT undermine their stated commitments to peaceful uses.

Brazil

Brazil joined the NPT in 1997. The IAEA Safeguards Statement for 2005 concluded that Brazil as a state party, though without an Additional Protocol (AP), is a state whose "declared nuclear material remained in peaceful activities."[3] Its joint system of nuclear materials accounting and control with Argentina through the Brazilian-Argentine Agency for Accounting and Control of Nuclear Materials (ABACC) is a unique, two-state regional safeguards/nonproliferation regime employing a rigorous state system of accounting for and control of nuclear materials (SSAC) in both countries.[4] The subsequent quadripartite agreement (signed by Brazil, Argentina, the IAEA, and the ABACC) provides for full-scope IAEA safeguards on Argentine and Brazilian nuclear materials, full rights over any proprietary technology developed by both countries, and nuclear energy for the propulsion of submarines.

Iran

Unlike Brazil, Iran signed the NPT on the day it opened for signature, July 1, 1968, and ratified the Treaty on February 2, 1970. However, Iran has threatened withdrawal from the NPT on numerous recent occasions since its nuclear program came under renewed international scrutiny in 2002. Iran is a signatory to the Comprehensive Test Ban Treaty (CTBT) and a state party to the Partial Test Ban Treaty (PTBT). Iran does not participate in any regional nonproliferation regimes, since the Middle East has not negotiated a NWFZ. Iran has on several occasions expressed willingness to participate in a regional weapons-free zone.

Fulfillment of NPT Article III (Safeguards) Obligations

States' commitments to peaceful intent are indicated by concluding safeguards agreements and their subsidiary arrangements with the IAEA, complying fully with those safeguards agreements by declaring all nuclear material and activities, developing an adequate State System of Accounting for and Control of Nuclear Material (SSAC), fully cooperating with IAEA inspections, and establishing nuclear export and import policies that are in compliance with Article III.2 of the NPT.

A more advanced standard of demonstrating peaceful purposes involves a country's adoption and implementation of an AP, including provision for a complete and timely declaration of all nuclear sites, materials, and equipment to the IAEA, as well as transfer of information to the IAEA relating to exports of certain nuclear materials and equipment. Maintaining a highly effective SSAC and working to establish superior transparency in a nuclear program by volunteering for visits by an IAEA SSAC Advisory Service (ISSAS) mission further reinforces confidence in peaceful intent. Additionally, transparency can be demonstrated by accepting visits of the IAEA such as an IAEA Operational Safety Review Team

---

[3]IAEA Safeguards Statement for 2005, p. 1, para. 2; www.iaea.org/OurWork/SV/Safeguards/es2005.pdf.

[4]See INFCIRC/435, "Agreement of 13 December 1991 between the Republic of Argentina, the Federative Republic of Brazil, the Brazilian-Argentine Agency for Accounting and Control of Nuclear Materials and the International Atomic Energy Agency for the Application of Safeguards," March 1994, www.iaea.org/Publications/Documents/Infcircs/Others/inf435.shtml.

or International Physical Protection Advisory Service or visits of international organizations such as the World Association of Nuclear Operators (WANO).

Transparency of a state's activities provides significant confidence in its peaceful intent. When South Africa decided in 1994 to dissolve its nuclear weapons program and joined the NPT as an NNWS, its high level of cooperation with the IAEA and its subsequent ratification of an AP were examples of the kind of transparency that create international confidence. Today South Africa is considered to have a fully peaceful nuclear program.

Brazil

Brazil has not concluded an Additional Protocol with the IAEA but is considering doing so, and its nuclear nonproliferation policy has come a very long way in the last 15 years. Its secret nuclear weapons program that it publicly revealed of its own accord in 1990 is now ended and it has publicly committed to peaceful use. Brazil's Quadripartite Agreement with Argentina provides for full-scope IAEA safeguards on Argentine and Brazilian nuclear installations. ABACC provides Brazil with a complete and effective SSAC. Brazil's two-year-long negotiations with the IAEA over the safeguards approach at its Resende enrichment facility did raise concern as to its overprotectiveness or uncooperativeness with the IAEA. However, the safeguards approach was agreed with the IAEA, and Brazil has committed to adhere to the safeguards requirements despite its concerns over the high industrial cost of protecting proprietary technology. As for trigger list export and import control, Brazil is a member of the Nuclear Suppliers Group (NSG) and assumed its chairmanship in January 2006.[5] It is a member of the Missile Technology Control Regime and has submitted timely UNSC resolution 1540 reports.

Iran

Iran has been slow and in many cases unwilling to provide the IAEA with information concerning its nuclear activities that would validate its commitments to NPT Article II obligations. Moreover, Iran has assumed an uncooperative and often confrontational posture in its conduct vis-à-vis the IAEA.

In its safeguards conclusions, the IAEA has not determined that Iran is pursuing a weapons program, yet it cannot verify that Iran's nuclear program is of an entirely peaceful nature. Iran has a comprehensive safeguards agreement in place with the IAEA,[6] yet there have been many instances of noncompliance with this agreement. Iran "failed in a number of instances over an extended period of time to meet its safeguards agreements with respect to the reporting of nuclear material, its processing and its use, as well as the declaration of facilities where such material has been processed and stored."[7] Furthermore:

> Many aspects of Iran's fuel cycle activities and experiments, particularly in the areas of uranium enrichment, uranium conversion and plutonium research, had not been declared to the Agency in accordance with Iran's obligations under its Safeguards Agreement.[8]

Iran's "many failures and breaches"[9] of its obligations to comply with its safeguards agreements have prompted its case to be referred to the Security Council, which imposed sanctions in December 2006.

---

[5]Steve Kidd, "A Latin Nuclear Revival?" *Nuclear Engineering International*, Jan. 26, 2006; www.neimagazine.com/story.asp?sectionCode=147&storyCode=2034782.

[6]View full text at www.iaea.org/Publications/Documents/Infcircs/Others/infcirc214.pdf.

[7]www.iaea.org/Publications/Documents/Board/2005/gov2005-67.pdf.

[8]Ibid.

[9]Ibid.

Iran has subsidiary arrangements to its safeguards agreement and accepted modifications to these arrangements proposed by the Agency requiring Iran to inform the Agency of new nuclear facilities and modifications to existing facilities through the provision of preliminary design information as soon as the decision to construct, to authorize construction, or to modify has been taken, and to provide the Agency with further design information as it is developed.[10] However, there are instances of violations in this area, such as in March 2005, when Iran failed to report to the Agency in a timely manner certain underground excavation activities that were already under way in December 2004 at the uranium conversion facility at Esfahan. Although Iran submitted the necessary design information in December 2004, Iran should have provided such information to the Agency at the time of the decision to build, not after groundbreaking.[11]

Moreover, in 2006 Iran stated in a letter to the Agency that "according to comprehensive Safeguards Agreements, Iran was not required to report to the Agency information on P-2 centrifuge drawings and the handful of rotor tubes (domestically made), since neither construction of a nuclear facility nor nuclear material was involved." The IAEA responded that "if, as Iran states," no construction decision was taken nor nuclear material introduced, neither would have to be reported. The Agency's response points to what has become a central issue for the international community and IAEA: the uncertainty regarding Iran's intent in developing its nuclear activities.

When Iran agreed to sign the Additional Protocol in December 2003 and allow its provisional entry into force, it allowed inspections to begin that provided greater access to its facilities. Subsequent inspections revealed a great deal of new information—"almost two decades' worth of undeclared nuclear activities"—including uranium enrichment and plutonium separation efforts. Iran's continued cooperation with the IAEA under an Additional Protocol would have been essential to clearing up questions regarding Iran's nuclear program and building confidence in its nuclear intentions. However, on February 6, 2006, Iran informed the Agency that "[o]ur commitment to implementing safeguards will only be based on the NPT safeguards agreement" and that "all voluntarily nonlegally binding measures including the provisions of the Additional Protocol and even beyond that will be suspended."[12] Iran's suspension of the Additional Protocol is hindering the Agency's ability to "assess fully Iran's enrichment related research and development activities, including the possible production of centrifuges and related equipment."[13] As of January 2008, Iran has not indicated it is willing to resume application of an Additional Protocol with the IAEA.

Coherence of a State's Nuclear Energy Program

Indicators in this category are intended to answer the questions: Are elements of the current and planned nuclear program logical, are they economically and technically consistent, and are they aligned with the stated purpose? The fundamental indicator in this category deals with the presence or absence of sensitive nuclear materials or facilities which could be used for weapons-related activities.

States pursuing activities that can directly support weapons development, as indicated in Figure 14.1, raise an immediate flag about the state's peaceful purposes commitment and should involve enhanced transparency to enable verification of any claim of peaceful use.

Other indicators in this category on nuclear program coherence are reflected by choices made in designing a nuclear program. Positive indicators include procuring reactors powered

[10]www.iaea.org/Publications/Documents/Board/2003/gov2003-40.pdf.

[11]www.iaea.org/Publications/Documents/Board/2005/gov2005-67.pdf.

[12]"Implementation of the NPT Safeguards Agreement in the Republic of Iran: IAEA Board of Governors Report by the Director General," GOV/2006/15, Feb. 27, 2006.

[13]www.iaea.org/NewsCenter/News/2006/dgbriefsboard.html.

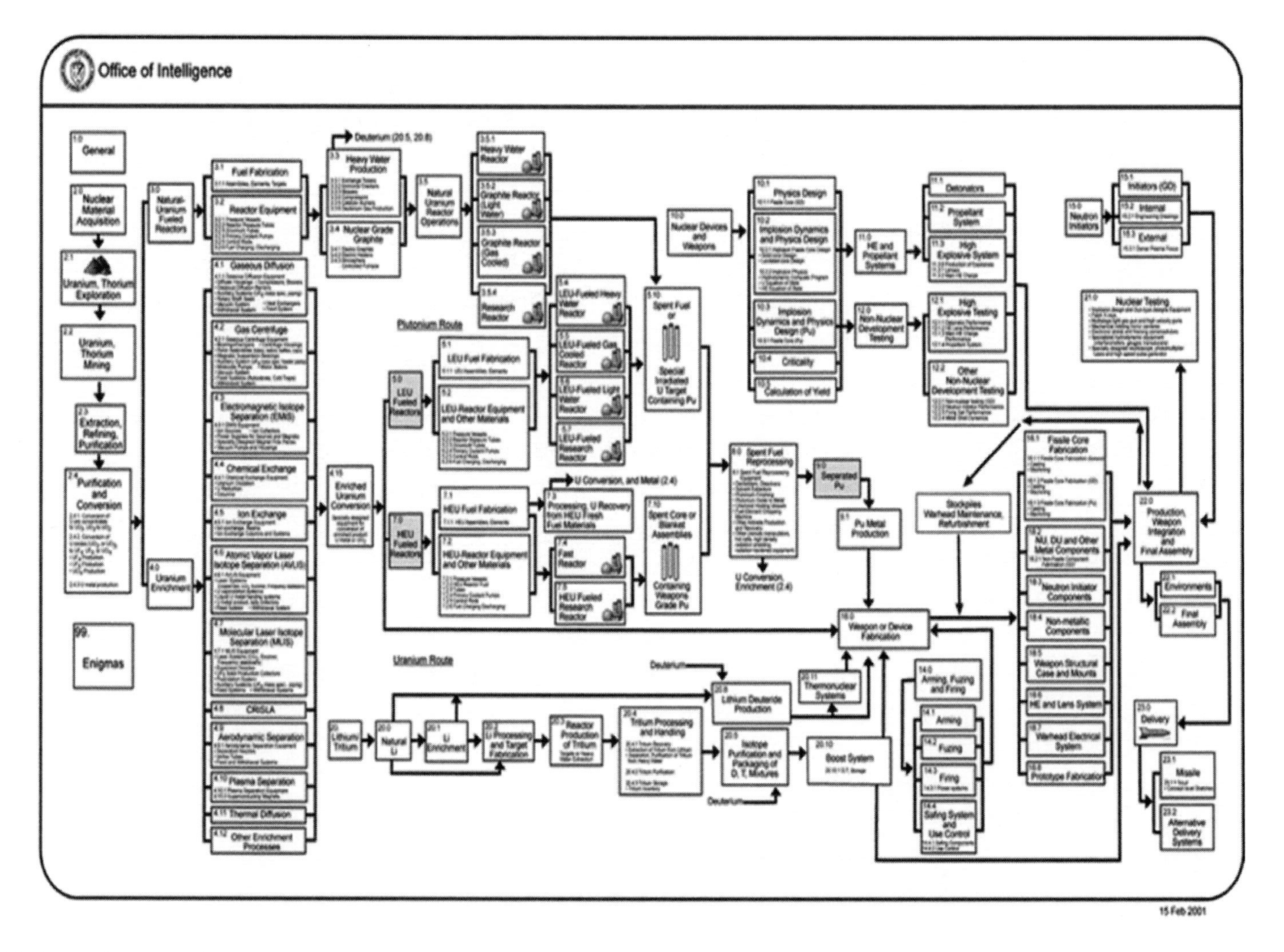

FIGURE 14.1 Nuclear fuel cycle and weapons development process.

by low-enriched uranium, seeking enrichment and reprocessing services from others rather than investing in national capabilities, leasing nuclear fuel, abstaining from the use of highly enriched uranium (HEU) fuel, and repatriating any existing HEU (either spent or fresh) to the country of origin. On the other hand, a country with one to a few nuclear reactors, embarking on development of enrichment technology raises concerns. Such an investment may not be economically justified since enriched fuel could be more cheaply procured from international market sources. Investment in such a facility does not appear to be part of a coherent nuclear program. A state's choice to develop sensitive enrichment and/or reprocessing facilities will nearly always raise serious concern if the state is unable to articulate a reasonable economic, technical, or energy security justification for such investment.

A state may choose to share its facility ownership with another state or accept multilateral investment in or oversight of its nuclear facilities to help increase confidence that a facility will remain exclusively for peaceful use. In 2005, an expert group convened by the IAEA positively evaluated the concept of multilateral ownership of nuclear fuel cycle facilities for such purpose.[14] Multilateral ownership, such as exists with EURODIF and URENCO, increases the transparency of the facility operations which increases the difficulty for any member state to turn the facility to nonpeaceful use. A state's acceptance of nuclear fuel cycle services such as fuel leasing and/or return of spent fuel to the supplier and/or its decision to forego enrichment and reprocessing plants of its own volition are significant indicators of peaceful use. However, further evaluation is needed to determine the practicality of fuel leasing, particularly how long there will be sufficient capacity in states currently possessing enrichment and fuel fabrication capabilities to meet demand from nuclear power generation. It is believed that current enrichment and fuel suppliers are spread sufficiently across the political spectrum that they can credibly assure customers against fuel supply disruption for political reasons (rather than for legitimate proliferation concerns).

Brazil

Brazil's rationale for developing domestic enrichment capability is to add value to its large uranium resources for export, eliminate the cost of overseas enrichment services for domestic reactor fuel supplies and establish the capability to support possible future nuclear-powered naval vessels. Brazil is the world's fifth-largest country and has extensive developed and undeveloped hydroelectric power and natural uranium reserves. Hydroelectric power plays the paramount role in the Brazilian electricity system, whereas thermal power plants (both conventional and nuclear) are minor contributors.[15] The country's economy and demand for electricity are growing and its two operating reactors provide only 4% of its electricity supply.[16] By enriching domestically, Brazil expects to save at least $12 million per year. This is not much compared with the approximately $180 million investment Brazil Nuclear Industries (INB) is making at its Resende enrichment plant, not including operational costs.[17] However, Brazil is considering a program to finish the 1,300 MW Angra 3 reactor and build another 1,300 MW plant and two more Brazilian-designed 300 MW plants. Ownership of a larger fleet of reactors makes Brazil's enrichment investment more logical and economic.

In addition, Brazil hopes to participate in the $5-billion-a-year global nuclear fuel market in the future.[18] About 90% of the world's nuclear power plants depend on foreign enrichment

---

[14]*Multilateral Approaches to the Nuclear Fuel Cycle: Expert Group Report to the Director General of The International Atomic Energy Agency,* 2005, p. 6.

[15]IAEA Brazil Country Profile 2003.

[16]Steve Kidd.

[17]Erico Guizzo, "How Brazil Spun the Atom," Spectrum Online (IEEE), March 2006, www.spectrum.ieee.org/mar06/3070.

[18]Ibid. See quotes by Samuel Fayad Filho, director of nuclear fuel production at Nuclear Industries of Brazil; www.spectrum.ieee.org/mar06/3070.

services to get their fuel. Demand for enriched uranium over the next two decades and rising uranium prices could also add to Brazil's justification for investment in Resende.[19] More recent estimates of output at the Resende facility vary between 200,000[20] and 300,000 SWU annually.[21] Current goals are to produce 20 to 30 metric tons of enriched uranium per year, or about 60% of domestic fuel requirements, by 2008 or 2009, possibly reaching 100% by 2010.[22] In all, rising electricity demand, adequate natural uranium reserves, the possibility of constructing several more power plants, and a stated desire to sell nuclear enrichment internationally provide for the coherency sought by this methodology in Brazil's nuclear program.

Iran

Iran denies it is creating a nuclear weapons program and maintains that as a party to the NPT, it has the right to pursue and develop any nuclear energy technology for peaceful purposes. Given the pace of Iran's economic development and population growth and its increasing energy consumption and dwindling fossils resources, the Iranian government maintains that "[d]iversification—including the development of nuclear energy—is the only sound and responsible energy strategy for Iran."[23] Iran's economy is still dependent on oil revenue, and the government asserts that it "can't allow the ever increasing domestic demand [to] affect the oil revenues from the oil export."[24] Tehran reasons that if it can use nuclear power to fulfill its domestic energy needs, it can export more oil and generate more foreign currency revenue.[25] To this end, Iran is vigorously pursuing its nuclear energy program.

Iran's argument for energy independence is weakened by the fact that its energy investments are skewed to a nuclear program, which does not accord with its domestic resource endowments or its near-term energy needs. Iran's own uranium resources are not enough to supply a self-sufficient fuel cycle, and the cost of indigenous fuel manufacture (including uranium enrichment) far exceeds the price at which fuel could be purchased on the open market. Iran's stated goal of energy independence "also is not due to shortages in the electricity sector as much as in the transportation sector. Its goals could be pursued much more effectively through any number of projects formulated to efficiently use natural gas and/or increase its refinery production."[26] Currently, the natural gas sector is so poorly managed that a significant portion of its annual gross production is flared at the wellhead, and due to limited refinery capacity and low production at existing refineries, Iran imports some 40% of its gasoline needs.[27]

Despite international concerns about the Iran's nuclear program, the country continues to pursue sensitive nuclear technologies, including uranium conversion; enrichment, including

---

[19]IAEA, INFCIRC/640, "Multilateral Approaches to the Nuclear Fuel Cycle: Expert Group Report submitted to the Director General of the International Atomic Energy Agency," p. 49, para. 129; www.iaea.org/Publications/Documents/Infcircs/2005/infcirc640.pdf#search=%22IAEA%2C%20INFCIRC%2F640%22.

[20]Steve Kidd, "A Latin Nuclear Revival?" *Nuclear Engineering International*, Jan. 26, 2006; www.neimagazine.com/story.asp?sectionCode=147&storyCode=2034782.

[21]IAEA Brazil Country Profile 2003.

[22]Erico Guizzo, "How Brazil Spun the Atom," Spectrum Online (IEEE), March 2006, www.spectrum.ieee.org/mar06/3070.

[23]The government is planning for the depletion of fossil resources "within two to five decades."

[24]www.iran-embassy.org.in/More_events.asp?ID=759.

[25]http://newsvote.bbc.co.uk/mpapps/pagetools/print/news.bbc.co.uk/2/hi/middle_east/4688984.stm.

[26]Thomas W. Wood, Matthew D. Milazzo, Barbara Reichmuth, and Jeffery Bedell, "The Economics of Energy Independence for Iran," *Nonproliferation Review*, vol. 14, no. 1, March 2007.

[27]Iran possesses some 90 years' worth of oil at current domestic production rates and some 220 for natural gas reserves. Ibid.

centrifuge research and development; metal purification and casting; and work on a heavy water reactor. The proliferation sensitivity of uranium enrichment and a heavy water reactor in particular and Iran's weak track record with regard to safeguards and transparency make these technological pursuits cause for more concern to the international community.

Geopolitical Cooperation

Regional and international political integration and cooperation are also important indicators of a state's willingness to maintain its commitment to international norms such as nonproliferation. If governments have good track records with respect to implementation of international treaties, most importantly of the NPT, this can provide reassurance to others of their peaceful intent. For example, it is hard to imagine proliferation concerns with respect to Belgium or Canada, which have long and strong histories of abiding by international nonproliferation commitments. This is not to say that a state's adherence to treaty commitments or cooperation through other political mechanisms can be conclusive indicators. These are factors to be considered together with other indicators. Yet, the degree of international political cooperation by a state can help refine an analysis developed from the preceding criteria. For example, the international community has found Brazil's plan to develop commercial-scale enrichment facilities much less objectionable than similar plans announced by Iran. One notable difference between the two countries is that Iran has chosen by its policies and demeanor, relative political isolation. Brazil, however, is an active player in many international venues and appears more concerned with behaving in a manner that is aligned with global norms, such as complying with its NPT obligations; indeed, Brazil held the chair of the Nuclear Suppliers Group (NSG).

Brazil

Brazil is an active participant in many international efforts such as the Generation IV International Forum and International Project on Innovative Reactors (INPRO) programs,[28] multiple treaties, and organizations. Moreover, Brazil gave up its nuclear weapons program voluntarily and formed a partnership with Argentina, its former enemy, and the IAEA.

Iran

Iran is a member and cofounder of the United Nations, the Nonaligned Movement (NAM), the Organization of the Islamic Conference (OIC), and the Organization of the Petroleum Exporting Countries (OPEC). Iran has not signed nor is it party to several major international treaties in the nuclear field, including the Nuclear Safety Convention, the Convention on the Physical Protection of Nuclear Material, the Joint Convention on the Safety of Spent Fuel Management and on the Safety of Radioactive Waste Management, the International Convention for the Suppression of Acts of Nuclear Terrorism, the Convention on Assistance in Case of a Nuclear Accident or Radiological Emergency, and the Convention on Early Notification of a Nuclear Accident. This track record does not support Iran's claims of a peaceful nuclear program.

Iran was recently elected vice chair of the United Nations Disarmament Commission and has solid support from much of the developing world in its standoff with the United States over its uranium enrichment program. During its last summit in Havana, Cuba, all 118 NAM member countries declared support for Iran's nuclear program for civilian purposes in their final written statement.[29]

---

[28]IAEA Brazil Country Profile 2003.

[29]www.globalsecurity.org/wmd/library/news/iran/2006/iran-060917-irna01.htm.

Iran's location in Eurasia and its vast energy reserves—in particular, its large petroleum supply, its large population and economic output, its growing military power, and its regional influence—provide it significant geostrategic influence. These factors make it a key regional and international player. However, Iran's actions internationally have resulted in its being criticized for its role in destabilizing Iraq, and in its being declared a state sponsor of terrorism, particularly in its support of Lebanon-based Hezbollah, both of which contribute heavily to the lack of international support for its nuclear program by many in the international community.

Although many Iranians would like to improve relations with the international community, this does not appear to include the leadership which currently control Iran's foreign policy. The installation of a government under President Mahmoud Ahmadinejad in 2005 has led Iran to a confrontational strategy over its nuclear program and in its dealings with the IAEA. In sharp contrast to former President Mohammad Khatami, Ahmadinejad has returned to the fiery rhetoric of the Khomeini era, advocating a clash of civilizations between the Islamic world and the West. In a September 2006 speech at the United Nations, Ahmadinejad warned "foreign governments against meddling in Iranian affairs."[30] A month later, he verbally attacked Israel by quoting Khomeini: "Israel must be wiped off the map."[31]

Analysis of Case Studies: Brazil and Iran

In the preceding tables, Brazil and Iran's peaceful-use *bona fides* were evaluated according to our methodology. These two countries were selected for case studies because each chose to pursue development of a uranium enrichment capability. The analysis of their peaceful-use *bona fides* sheds light on why in Brazil's case, its actions have eventually been accepted by the international community, whereas in Iran's they have not.

As shown in the tables, Brazil and Iran maintain very different relationships with the international community in terms of the level of confidence in the peaceful nature of the respective nuclear programs. Such confidence reflects keenly on perceptions about each country's future nuclear intentions.

Brazil's peaceful-use bona fides appear strong. In all four criteria, Brazil meets either the minimum or higher standard for peaceful use. After revealing and disavowing its clandestine nuclear weapons program, Brazil joined and has adhered to the NPT. It has not concluded an Additional Protocol but is considering doing so. It has engaged in recent, extensive safeguards negotiations with the IAEA over its enrichment facility at Resende, which were resolved to mutual satisfaction. Brazil appears to have a coherent, transparent nuclear power program and is well integrated into important international nonproliferation treaties and organizations. Brazil held the Chair of the Nuclear Suppliers Group and has aspirations for exporting nuclear services in the future.

Conversely, although Iran is a party to the NPT, it does not meet the minimum peaceful-use standards across the four criteria. Although the IAEA has not conclusively determined that Iran is pursuing a weapons program, it cannot verify that Iran's nuclear program is of an entirely peaceful nature. Iran has been cited for numerous violations of its safeguards agreement, has consistently been less than forthright about its nuclear activities, has made contradictory and untrue statements regarding its nuclear activities, has blocked IAEA inspections, has made inflammatory statements regarding the nonproliferation regime, and has on several occasions threatened to withdraw from the NPT, all of which indicate Iran's peaceful intent is suspect.

---

[30]www.un.org/webcast/ga/60/statements/iran050917eng.pdf.

[31]Ibid.

Summary

The methodology for assessing a state's "peaceful-use" bona fides presented in this study provides a mechanism for identifying states that may be of potential proliferation concern. As an increasing number of states turn to nuclear power to meet energy diversity or security interests and/or to offset global climate change, the importance of identifying states that may be using—or intending to use—peaceful nuclear technology as a cover for nuclear weapons development is magnified.

As stated in Article IV of the NPT, the NNWS possess an "inalienable right to develop research, production and use of nuclear energy for peaceful purposes." The exercise of this right to peaceful nuclear technology is contingent, however, upon meeting all relevant NPT commitments, especially, the Article II commitment to not manufacture or otherwise acquire nuclear weapons and the Article III obligation to accept IAEA safeguards. That all states recognize and accept these obligations is crucial to the effectiveness of the nonproliferation regime. The safeguards mechanism of the NPT empowers the international community, through the IAEA, to verify the peaceful intent of a NNWS nuclear program. The Treaty provides justification for states to suspend support or assistance to the state in question when doubt arises as to its peaceful intent.

Iran consistently claims that its nuclear program is of entirely peaceful intent and alleges discrimination by those that seek to limit its nuclear fuel cycle. This argument strikes a chord of sympathy among NNWS frustrated over a perceived lack of access to nuclear technology. Tension between NWS and NNWS can be overcome by development of more proliferation-resistant nuclear technologies that enable broader dissemination of nuclear technology with lesser proliferation concerns. Methodologies such as this one that apply a set of straightforward measures to assess and understand a state's nuclear program, may help distinguish countries of potential concern. Taken separately, most of the indicators presented in this chapter provide singular glimpses of certain aspects of a state's nuclear program. Taken together, however, these indicators present a comprehensive picture of a state's credentials relevant to assessing its peaceful intent.

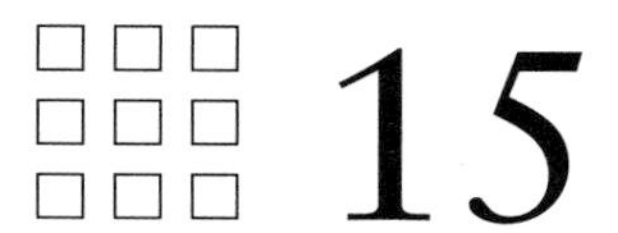

15

Dismantling Nuclear Weapons Activities: Politics and Technology

James E. Doyle

Introduction

It is well known that many more nations have the financial and technical capabilities to produce nuclear weapons than have done so. Only the United States, Russia, China, Great Britain, France, India, Pakistan, Israel, and North Korea are believed to possess nuclear arms. The vast majority of states, including dozens that could build nuclear arms, have chosen not to do so.

Less well known is the fact that of those nations exercising nuclear restraint, many have made prior efforts toward acquiring nuclear weapons or taken concrete steps to support a nuclear weapons option. Moreover, South Africa, Ukraine, Kazakhstan, and Belarus temporarily possessed nuclear weapons and then gave them up. The historical record thus makes clear that states can move in either direction along a continuum bound at one end by a decision not to acquire nuclear weapons and at the other by the possession of a fully functional nuclear arsenal. In considering whether or not to seek nuclear weapons, states take other factors into account, such as economic and environmental concerns, nonproliferation norms, and relations with allies, before deciding to create their own nuclear arsenals. When the strategic and political situations facing states change, their decisions regarding weapons programs can be delayed, cut back, or canceled. As discussed in detail in Chapter 18, Libya provides the latest example of a nation reversing its efforts toward acquiring nuclear arms.

The renunciation of nuclear weapons by such states and their corresponding actions to confirm their nonnuclear weapons status provide examples of *nuclear rollback*. Nuclear rollback occurs when a nation with a nuclear weapons program cancels that program and gives up some of the tools needed to acquire a nuclear weapons capability and/or accepts the emplacement of additional barriers to going nuclear. Rollback thus involves a voluntary decision by either a potential proliferator or a state with nuclear weapons to give them up.[1]

[1]However, a strategy of nuclear rollback could include the objective of inducing states to consent to maintaining a nonnuclear status after their nuclear weapons capability was eliminated by force, as in the case of Iraq. For reasons discussed in detail in this chapter, nuclear rollback could also be termed *denuclearization*. Both terms refer to nuclear weapons programs, however, not civilian nuclear power.

There are many possible reasons for nations to change their policies with respect to nuclear weapons. Changes in the security environment, the political strategy of the national leadership, or economic concerns can all influence decisions on nuclear arms development. The technical resources, expense, and time required to develop a functional nuclear arsenal also present challenges to a nation in its pursuit of nuclear weapons. To overcome these challenges, states must create and maintain political coalitions that marshal human, material, and financial resources toward the objective of building nuclear weapons. Such coalitions encounter constraints such as competition with other national objectives for scarce human and financial resources, international regimes that deny access to key technologies and materials, and domestic and international political opposition. To be successful in securing the resources they require, nuclear weapons development programs must be either effectively insulated from these challenges or politically defended and legitimated in spite of them.[2] This insulation and legitimization must be sustained for a period of many years, along with material support, in order for states to succeed in building nuclear weapons.

Over time these common difficulties faced by all states attempting to acquire nuclear arms create opportunities for U.S. and international nonproliferation policy. Policies and diplomatic actions can be created that influence the way a proliferating state evaluates the costs and benefits of acquiring nuclear arms. The international community can offer a state positive inducement for abandoning its pursuit of nuclear arms. In some cases the long-term implementation of nonproliferation diplomacy, using a diverse set of policy tools, can help persuade a nation to abandon or verifiably suspend a nuclear weapons program.

It must also be recognized that some states will be extremely resistant to any penalties or positive inducements that seek to hinder their nuclear weapons development efforts. Clear examples of this dynamic are provided by India, Pakistan, and North Korea. In these states the incentives to acquire nuclear weapons were very strong and stemmed from deeply held security concerns and issues of nationalism. These states were willing to take many years, endure international sanctions, and keep key elements of their efforts secret to successfully develop nuclear weapons. It will likely require a fundamental improvement in the respective security situations of these states for them to even consider abandoning nuclear arms. It is too soon to tell if the Six-Party diplomatic process with North Korea can achieve this outcome. In the case of India, even this improvement probably would not suffice, because nuclear weapons are seen as a symbol of India's emerging great-power status.

Another important point is that there is no standard package of incentives and disincentives that can influence the nuclear weapons aspirations of all states. Each nation makes a calculation of its interests based a broad range of domestic and external considerations that change over time. Therefore, the same policy tools and incentives that might induce one state to refrain from developing nuclear arms might not be effective elsewhere. Nuclear rollback is likely only when domestic and external conditions converge on the wisdom of nuclear restraint.[3]

In recent years a greater recognition has developed among scholars and policymakers of the need to address a complex range of incentives and disincentives for nuclear proliferation within a given state.[4] An attempt to influence the balance of incentives and disincentives for proliferation within a state can be made by employing different categories or types of diplomatic and policy action. One level of nonproliferation policy is strong and continuing support for international nonproliferation norms as embodied by the NPT, efforts at nuclear

---

[2]Etel Solingen, "The Political Economy of Nuclear Restraint," *International Security*, vol. 19, no. 2 (Fall 1994), pp. 136–139, and Steven Flank, "Exploding the Black Box: The Historical Sociology of Nuclear Proliferation," *Security Studies*, vol. 3, no. 2, pp. 259–294.

[3]Ariel Levite, "Never Say Never Again: Nuclear Reversal Revisited," *International Security*, vol. 27, no. 3, Winter 2002–03, pp. 59–88.

[4]The most comprehensive analysis is Mitchell Reiss, *Bridled Ambition: Why Countries Constrain Their Nuclear Capabilities*, Washington, D.C.: Woodrow Wilson Center Press, 1995.

Table 15.1 Types of nonproliferation policy.

Type of Policy	Nonproliferation Policy Tool
Policies that promote nonproliferation globally	Supporting nuclear, chemical, biological, and missile arms reductions and deemphasizing nuclear weapons Supporting international nonproliferation norms, comprehensive nuclear test ban and fissile material cutoff Support for international nonproliferation export controls
Policies promoting nuclear nonproliferation that need to be tailored for specific countries	Addressing security concerns Creating economic incentives Influencing fuel cycle economics Technical and financial assistance Diplomatic assistance or intervention
Policies not directly related to nonproliferation that have positive proliferation consequences	Support for economic liberalization Support for democracy

arms reduction, comprehensive nuclear test ban, fissile material production ban, and nuclear export controls. There is general consensus that this is one type of policy action that helps discourage nuclear proliferation in all states.[5]

A second category of nonproliferation tools consists of actions and initiatives designed to reduce incentives and increase disincentives for proliferation in a specific state or region. These policies address a range of political, economic, and military factors specific to the target states. Finally, there is a broad set policies designed to achieve central objectives of U.S. foreign policy, such as the spread of democracy and liberal economic systems, the success of which can help discourage states from acquiring nuclear arms or increase the likelihood that they will give them up.[6] The three types of nonproliferation policy action and some examples of policy tools that correspond with each are summarized in Table 15.1.

Policies aimed at inducing a state to give up its nuclear weapons or efforts to acquire them may take many years to achieve results, if they are successful at all. However, the typically long period it takes to build nuclear weapons creates the opportunity to influence domestic conditions (leadership, political orientation, security situation) as well as external ones that might change a state's nuclear course. If a state decides to eliminate its nuclear weapons development efforts or accept more rigorous international safeguards, careful thought must go into selecting effective technical and administrative measures that provide confidence that the decision has been implemented.

Obtaining such confidence may take many months or years and require extensive inspections and/or the implementation of technical safeguards measures on both retired and enduring nuclear assets within the state. One major factor that will determine the scope of effort required to confirm the rollback of a nuclear weapons program will be the size and level of advancement of the program. For example, if North Korea should in the future

[5]Thayer, "The Causes of Nuclear Proliferation ... ," pp. 506–508.

[6]For an argument that the spread of democratic governance deters nuclear proliferation, see Glen Chafetz, "The End of the Cold War and the Future of Nuclear Proliferation: An Alternative to the Neorealist Perspective," *Security Studies*, vol. 3/4 (Spring/Summer 1993), pp. 127–158. For the view that economic liberalization can decrease proliferation incentives, see Etel Solingen, "The Political Economy of Nuclear Restraint," *International Security*, vol. 19, no. 2 (Fall 1994), pp. 162–169.

declare a willingness to abandon its nuclear weapons infrastructure and rejoin the NPT, the IAEA will have to verify the elimination of North Korea's stocks of military plutonium, its plutonium production capability, its nuclear weapons assembly and testing facilities, and any uranium enrichment facilities that have a weapons-related function. This would likely be a very complex and time-consuming process requiring intrusive on-site inspection and significant decommissioning of several large contaminated nuclear facilities.

The other major factor determining the difficulty of verifying the elimination of a nuclear weapons program is the degree of cooperation demonstrated by the government that has agreed to the elimination process. In the case of South Africa (described in Chapter 16), the IAEA was able to verify in a very brief time the elimination of a program that had produced six workable nuclear weapons and spanned more than 15 years. The South African government was very forthcoming and transparent regarding all information on its past program, including technical aspects of the facilities that produced the weapons and detailed accounting of the quantities of nuclear materials that were produced and then eliminated. It allowed IAEA inspectors sufficient access to facilities, documentation, and personnel. These actions, in addition to facilitating the dismantlement of capabilities for the production of nuclear weapons, were also evidence of South Africa's political commitment to become a nonnuclear weapons state.

By contrast, Iraq's behavior following its defeat in the 1991 Gulf War and passage of several United Nations Security Council Resolutions mandating the elimination of its nuclear weapons programs was uncooperative. Information provided by the Iraqis was often tardy, incomplete, confusing, and suspect. IAEA inspectors were taken to facilities and then denied entry or told to wait for hours. Iraqi cooperation with IAEA and United Nations inspection efforts was so inconsistent that these organizations could not solidly dispel suspicions (later disproved) that Iraq was reconstituting its nuclear weapons program. This suspicion was one of the pretexts for the U.S. invasion of Iraq in March 2003.[7]

The examples of South Africa and Iraq should be kept in mind as national security strategists and the IAEA consider their approach to possible future operations to verify the elimination of a nuclear weapons program. Clearly, a key explanation of the contrasting experiences was the motivation or willingness on the part of the South African and Iraqi governments to eliminate their programs. In the case of South Africa, the national leadership had made the decision that foreswearing nuclear weapons and joining the NPT was in the national interest and then forged a policy to achieve that objective. The elimination of Iraq's nuclear program was imposed after its defeat in war. The political context therefore has great influence over the course of future elimination operations.

In cases where motivations for elimination and the degree of cooperation are weak, it is likely that more comprehensive and intrusive technical measures will be required to develop confidence that nuclear weapons capabilities no longer exist. This could mean that some attempts at nuclear rollback will entail long diplomatic and technical processes involving phased reciprocal actions. The agreed framework for the denuclearization of the Korean peninsula had these characteristics and ultimately collapsed due to dissatisfaction by both sides that commitments were being met in a timely and transparent manner.[8] The Joint Statement of the Fourth Round of the Six-Party Talks, signed in Beijing on September 19, 2005, by North Korea, South Korea, the United States, China, Russia, and Japan, outlines a similar process for denuclearization. Despite the fact that North Korea again permitted the monitored disablement of its plutonium production reactor and reprocessing plant by early 2008, the degree of success for the Six-Party process has yet to be determined.

---

[7]Malfrid Braut-Hegghammer, "Rebel Without a Cause: Explaining Iraq's Response to Resolution 1441," *The Nonproliferation Review*, vol. 13, no. 1, March 2006, pp. 17–34.

[8]Jonathan D. Pollack, "The United States, North Korea, and the End of the Agreed Framework," *Naval War College Review*, vol. LVI, no. 3, Summer 2003; www.nwc.navy.mil/PRESS/Review/2003/Summer/art1-su3.htm (Jan. 2007).

Chapters 16, 17, and 18 describe in detail some examples of nuclear rollback. They provide background on how states changed their policies regarding nuclear weapons and describe the measures they adopted to give the international community confidence in their decisions. The chapters also provide, whenever possible, a description of the inspections, measurements, and safeguards activities that were utilized to confirm the elimination of a nuclear weapons program. This background should be useful in crafting future nonproliferation policies and selecting the appropriate technical approaches for verification and safeguarding in states that abandon efforts to acquire nuclear arms.

16
South Africa

Sara Kutchesfahani and Marcie Lombardi

Introduction

South Africa is the only country known to have manufactured nuclear weapons and then dismantled them. This reversal became known as *nuclear rollback*. Nuclear rollback occurs when a nation unilaterally and voluntarily relinquishes its nuclear weapons program.

South Africa's nuclear weapons program was not actually publicized until 1993, two years after South Africa signed the NPT as a state without nuclear weapons capabilities. The construction and dismantlement of South Africa's nuclear weapons program was classified as top secret, and many relevant documents pertaining to its weapons program are still classified to this day. South Africa can already serve as a role model for other nations that are considering abandoning their nuclear weapons efforts, and if South Africa chooses to declassify more information about the dismantlement, it would be easier for countries to learn about the process and follow the South African example.

The South African model has already been followed in Sweden, Argentina, and Brazil, and it is hoped it can be repeated again with today's nuclear weapons challengers: North Korea and Iran. This chapter first analyzes why South Africa pursued the bomb; second, it provides details of South Africa's nuclear weapons program; third, it discusses how South Africa eliminated its weapons program; fourth, it analyzes why South Africa gave up the bomb; and finally, it discusses how South Africa can serve as a nuclear rollback model to North Korea and Iran.

Why Did South Africa Pursue the Bomb?

As noted in the chapter introduction, many other states have initiated efforts to acquire nuclear weapons and thus abandoned them. The history of rollback in each of these cases is unique. In the case of South Africa, the country's decision to pursue a nuclear weapons program cannot be tied to one single reason alone but to a number of possible factors. Political motivations, security issues, technological momentum, and regional dominance all played a role in the country's decision. Furthermore, the availability of natural uranium resources, the U.S.-lead Atoms for Peace Program of 1953, prosperous uranium business, and the U.S. Plowshare Peaceful Nuclear Explosive (PNE) Program also contributed to South African motivations. Finally, South Africa's strong relationship with nuclear weapons states like the United States and the United Kingdom and its nuclear collaboration work with Israel were all factors in the South African proliferation puzzle.

Since the 1950s, South Africa's policy of racial separateness, or *apartheid*, aroused international criticism and led to the nation's political isolation. For example, after the Sharpeville shootings in March 1960, when South African police fired on a crowd of unarmed blacks protesting against apartheid, the United States supported a U.N. Security Council Resolution deploring the policies and actions of the South African government. In April 1961 South Africa was forced to withdraw from the British Commonwealth, and 24 African states introduced a U.N. General Assembly Resolution urging the severance of diplomatic ties to the new Republic of South Africa.

In 1964 the United States, the United Kingdom, and other Western governments joined a voluntary embargo on arms sales to South Africa and later supported a mandatory U.N. arms embargo in 1977. Over time, South Africa was also excluded from a number of important world bodies, including the Food and Agriculture Association, the International Labor Organization, the Economic Commission for Africa, and the World Health Organization. It was also temporarily denied participation in the International Atomic Energy Agency (IAEA). This political isolation influenced South Africa's national security situation. By the mid-1970s it was clear that South Africa had no real political or military allies in the international community. Throughout the 1970s and 1980s international pressure against the apartheid system continued to mount and included arms embargoes, trade bans, and moves to withdraw investment from the South African economy. All these activities added to the white regime's sense of isolation and insecurity.[1]

In addition to this political ostracism, South Africa faced regional instabilities and conflicts in the 1970s that created the perception of direct threats to South African security. In 1974, the Portuguese regime fell, triggering a withdrawal from its African colonies, Mozambique and Angola. In January 1975 talks aimed at achieving a peaceful transition to a new political order in Angola broke down and three rival liberation movements began a civil war. South Africa, with U.S. covert support, intervened militarily against the Soviet-backed *Movimento Popular de Libertacao de Angola* (MPLA), which had taken control in Luanda by providing assistance to the rival liberation groups *Frente Nacional de Libertacao de Angola* (FNLA) and *Uniaco Nacional para a Independencia Total de Angola* (UNITA). Due to public condemnation and Congressional action, however, the United States ended its covert assistance activities in Angola in December 1975. This intervention caused hostility between Pretoria and the MPLA and led to the eventual arrival of 50,000 Cuban troops, which were airlifted to Angola by the Soviet Union. By the spring of 1976, the South African Defense Force (SADF) began a phased withdrawal from Angola.

Other events brought forces hostile to the apartheid regime closer to its borders in the 1970s. On South Africa's northeastern border, a pro-Soviet, Marxist-Leninist regime assumed power in Mozambique, and white rule ended in Zimbabwe (formerly Rhodesia). In 1975 the United Kingdom terminated the 1955 Simon's Town Agreement for bilateral South Atlantic naval defense. To the northwest, the movement for the independence of Namibia (formerly South African-controlled South West Africa) caused further regional instability, with black nationalist forces becoming increasingly organized militarily and equipped with weapons from outside groups. In 1977, the United States, Britain, France, Canada, and West Germany formed a "contact group" to negotiate Pretoria's withdrawal from Namibia, as required by a U.N. resolution, and to manage that country's transition to independence.

In addition to political isolation and regional instability, South Africa became subject to sanctions on its nuclear dealings with other states. These sanctions were primarily motivated by the domestic policies of the South African government. For example, the United Kingdom allowed an agreement for nuclear cooperation between South Africa and its Atomic Energy Authority to expire in the early 1970s. In 1975 the United States refused to export enriched uranium for the Safari-1 research reactor and by the late 1970s was refusing to honor an agreement to provide enriched uranium fuel for South Africa's Koeberg power reactors.[2] Also in the mid-1970s, German firms curtailed cooperation with South Africa in the development of uranium enrichment technology.[3] In June 1977 South Africa was expelled from the IAEA's Board of Governors and in 1979 was denied participation in annual IAEA General Conferences.

---

[1]For a summary of South Africa's political isolation, see Robert Scott Jaster, *The Defense of White Power* (New York: St. Martin's Press, 1989), pp. 42–43.

[2]J. D. L. Moore, *South Africa and Nuclear Proliferation* (New York: St. Martin's Press, 1987), pp. 97–102.

[3]David Fischer, *Stopping the Spread of Nuclear Weapons: The Past and the Prospects* (London: Routledge, 1992), p. 52. See also Zdenek Cervenka and Barbara Rogers, *The Nuclear Axis: Secret Collaboration between West Germany and South Africa* (London: Julian Friedman, 1978).

The combination of a deteriorating security situation, political isolation, and external constraints on its nuclear development provided the motivations for a South African nuclear weapons development effort. Security concerns seem to have been the strongest of these three motivating factors. South Africa's National Party, which assumed control of the government in 1948, has a long tradition of xenophobia. During the Cold War the National Party touted global communism as the principal external threat, claiming that this threat would inevitably materialize in the form of a Soviet-led conventional military attack on South Africa.[4] In 1972 the Commandant-General of the SADF, Admiral Bierman, asserted: "In the final analysis it is a prerequisite for the successful defense of the Southern hemisphere that the deterrent strategy based on nuclear terror and the fear of escalation should also be applicable in the region."[5]

The events of the early 1970s galvanized South Africa's sense of vulnerability and most likely triggered the decision to seek a nuclear weapon capability. The National Party described the constellation of economic, political, and military pressures facing the country as a "total onslaught" against South Africa by the forces of international communism. A "Total National Strategy" was needed to mobilize and coordinate South Africa's resources to meet this onslaught.[6] This strategy included decisions taken in the mid-1970s to double the size of the SADF and triple the defense budget. The total national strategy also included a decision to develop nuclear weapons. At the time the South African government may have viewed nuclear weapons as improving the country's security in several ways: by making the country seem impregnable because it held a devastating weapon of last resort, by bolstering the morale of the Afrikaner community, and by serving as an equalizer in the case of a large-scale conventional military attack.[7]

Disincentives against the overt acquisition of nuclear weapons also played a role in determining South African behavior. Overt acquisition of nuclear arms could erode Western tolerance for the South African regime, leading to increased external political pressure and possibly to direct diplomatic opposition or support for South Africa's regional adversaries. In response to Pretoria's acquisition of the bomb, Western nations might impose strict trade embargoes and pass legislation requiring the withdrawal of investments from the South African economy. Other economic sanctions such as the freezing of assets were also possible. The foreign trade that South Africa was able to sustain provided capital and materiel important to the maintenance of its armed forces. Given the fact that South Africa was involved in several conflicts on its borders, a complete cutoff of trade could eventually weaken the SADF and undermine South African security.

If South Africa had revealed its plans to develop nuclear arms, one of the first reactions of the international community would likely have been increased efforts to constrain South Africa's nuclear development. The reaction would have jeopardized plans for the Koeberg nuclear power plants and the commercial enrichment plant. Also, several states that purchased South African natural uranium, even after the approval of U.N. resolutions condemning apartheid, would probably have canceled these transactions.

Because South Africa's domestic policies already faced strong international criticism, diplomatic interactions with other states would become even more difficult if its nuclear ambitions became public. African states in particular might demand that the Western powers impose strengthened sanctions against South Africa or ask for direct assistance against a racist government that possessed weapons of mass destruction. It is likely that

---

[4]Moore, p. 52, and Kenneth L. Adelman, "The Strategy of Defiance: South Africa," *Comparative Strategy*, vol. 1, 1978, p. 35.

[5]Daryl Howlett and John Simpson. "Nuclearisation and Denuclearisation in South Africa," *Survival*, vol. 35, no. 3 (Autumn 1993): pp. 154–155.

[6]Mitchell Reiss, *Bridled Ambition: Why Countries Constrain Their Nuclear Capabilities*, Washington, D.C., Woodrow Wilson Center Press, 1995, p. 9.

[7]Michele A. Flournoy and Kurt M. Campbell, "South Africa's Bomb: A Military Option?" *Orbis*, vol. 31 (Summer 1988): pp. 398–399.

Table 16.1 Details of South Africa's nuclear weapons program.

Key South African Sites	Facilities
Pelindaba	AEC site housing Safari-1 research reactor, a hot cell complex, a waste disposal site, and conversion and fuel fabrication facilities. It was an assembly facility, and it was here where the nuclear weapons were made. At Building 5000 complex (isolated buildings at Pelindaba), the development and assembly of nuclear explosives were carried out.
Pelindaba East (Valindaba)	Site of AEC pilot-scale uranium enrichment facility (Y-Plant), closed in 1990.
Vastrap	Site of two nuclear explosive test shafts in the Kalahari Desert, built in the 1970s; filled with concrete in 1993.
Circle Facilities (Advena)	ARMSCOR facility used in the 1980s/early 1990s for design, manufacture, and storage of nuclear weapons.
Somchem	A military facility involved in the development and manufacture of explosives and propellants.

an overt nuclear weapons program in South Africa would have incurred opposition from the Soviet Union and China as well.[8] Finally, South Africa was aware that a declaration that it possessed nuclear weapons, or intended to do so in the near future, would have raised the possibility that a rival African state such as Nigeria might seek its own nuclear deterrent.

Nuclear Weapons

South Africa's first nuclear weapon was completed in November 1979 (see Table 16.1 for full details of the country's weapons program). For the next 10 years weapons were built at an average rate of about one every 18 months. All weapons were similar in design to the "Little Boy" atomic bomb that the United States dropped on Hiroshima and contained highly enriched uranium (HEU) as the fissile material. The two subcritical halves of each gun-type weapon were stored in separate vaults, and elaborate security precautions were devised to prevent unauthorized use.[9]

As South Africa's bomb-building expertise improved, the design of each weapon was modified slightly to enhance safety and reliability.[10] As Lt. Colonel Roy E Horton notes, "ARMSCOR invested heavily in refining and qualifying the various parts of the weapon with an emphasis on safing and arming features. While the gun-type design had the advantage of not using explosives, there were still considerable challenges to prevent accidental detonation if the weapon was dropped. ARMSCOR engineers developed a unique means of physically preventing an accidental detonation prior to final arming, but the mechanical devices involved took several years to qualify and eventually proved extremely difficult to

[8]George Barrie, "South Africa," in *Nonproliferation: The Why and the Wherefore*, Jozef Goldblat, ed. (London and Philadelphia: Taylor and Francis, 1985), pp. 154–155.

[9]"In the gun-assembly technique, a propellant charge propels two or more subcritical masses into a single supercritical mass inside a high-strength gun-barrel-like container"; U.S. Congress, Office of Technology Assessment, *Technologies Underlying Weapons of Mass Destruction*, OTA-BP-ISC-115 (Washington, D.C.: USGPO, 1993), p. 174.

[10]Mark Hibbs, "South Africa's Secret Nuclear Program: From a PNE to a Deterrent," *Nuclear Fuel*, May 10, 1993, p. 5.

maintain.[11] A total of six weapons were completed, and a seventh was under construction at the time the program was canceled.

Uranium Enrichment and Plutonium

By the time South Africa decided to end its nuclear weapons program, it had gained experience with much of the nuclear fuel cycle. The pilot enrichment plant at Valindaba, named the Y-Plant, first produced a small quantity of HEU in January 1978. By the time the Y-Plant stopped enriching uranium in 1989, South Africa had increased its annual output to roughly 100 kilograms of HEU, enough material for two crude nuclear devices per year.

South Africa also constructed a larger-scale semicommercial uranium enrichment plant at Valindaba, which began operation in 1987. This plant had the theoretical capacity to produce 1,500 kilograms of weapon-usable HEU per year, enough for 30 weapons annually. South Africa has claimed and the IAEA has confirmed, however, that the semicommercial enrichment plant was not used to produce any nuclear weapons material. The plant was closed in early 1995. (*Note*: The Y-Plant did produce the enriched uranium for the weapons; the Z-Plant did not.)

An early interest in producing plutonium was canceled after South Africa realized that natural uranium-fueled heavy water reactors were not competitive with light water reactors and that development of a plutonium production capability would reduce the resources available for uranium enrichment.[12] South Africa has no large-scale plutonium reprocessing facilities. It does, however, have a hot-cell complex that is capable of reprocessing small amounts of plutonium from spent fuel stored near its research reactor and power reactors. South Africa continued to evaluate methods to produce and recover plutonium and to produce tritium, a radioactive gas used to boost the explosive power of nuclear weapons. These efforts focused on the design of a 150-megawatt pressurized water research and development reactor to be built at Gouriqua, near Mosselbay in Cape Province. This program ended in 1989 or 1990 and construction of the proposed reactor never went beyond the site preparation phase.[13]

Peaceful Nuclear Explosives

During the late 1960s and early 1970s, several countries were pursuing the possibility of using nuclear explosive devises for civil applications. In the United States this was called the *plowshare project*. These applications included dams, canals, and mining, aided by nuclear explosions.[14] In 1969 the AEB established an internal committee to investigate the economic and technical aspects of using PNEs in mining.[15] In early 1974, a report was prepared that concluded that the development of nuclear explosive devices for peaceful uses was feasible. The then South African Prime Minister John Vorster approved the program, and funds for the development of the Kalahari nuclear test site were allocated. However, due to the world's reaction to India's 1974 "peaceful" test, the program was kept top secret.[16]

---

[11]Lt. Colonel Roy E. Horton, "Out of (South) Africa: Pretoria's Nuclear Weapons Experience," *USAF Institute for National Security Studies*, Aug. 1999, p. 10.

[12]David Albright, "South Africa's Secret Nuclear Weapons," ISIS Report 1, no. 4 (May 1994), p. 4.

[13]*Ibid*, p. 12.

[14]Richardt Van Der Walt, Hannes Steyn, and Jan Van Loggerenberg, *Armament and Disarmament* (New York: iUniverse, Inc., 2005).

[15]David Albright, "South Africa and the Affordable Bomb," *Bulletin of the Atomic Scientist*, July/Aug. 1994, vol. 50, no. 04, www.thebulletin.org/article.php?art_ofn=ja94albright.

[16]Waldo Stumpf, "Birth and Death of the South African Nuclear Weapons Programme," Presentation given at the conference "50 Years After Hiroshima," organized by USPID (Union Scienziati per il disarmo) and held in Castiglioncello, Italy, Sept. 28–Oct. 2, 1995, www.fas.org/nuke/guide/rsa/nuke/stumpf.htm#02, p. 3.

Testing Infrastructure

South Africa built and maintained the infrastructure needed to test the reliability of its nuclear weapons design, even though it is believed to have never conducted a nuclear test. Between 1975 and 1978 a nuclear weapons test site was constructed at the Vastrap military base in the Kalahari Desert in northwestern South Africa. Three test shafts were drilled. One hit unfavorable geological conditions and flooded, but the other two were completed.[17]

In 1977, South Africa planned a fully instrumented "cold test" of its nuclear device using a depleted uranium core to show the behavior of uranium metal under the conditions expected when exploding a nuclear weapon. Some observers claim that a "true" nuclear test using an HEU pit was planned for the following year.[18] Preparations for the test were discovered by the Soviet Union and the United States in August 1977. After receiving demarches from Moscow, Washington, and Paris, South Africa sealed the test shafts with concrete and abandoned the site. Pretoria also gave assurances that no nuclear explosive tests would take place in South Africa.[19]

In 1988, South African technicians reexamined the Kalahari test site to determine how long it would take to prepare for a nuclear test. They checked the condition of one of the test shafts by pumping out the water and lowering a specially designed probe. The shaft was still intact and other preparations for a nuclear test would take a week or two.[20] In late 1989, the testing site was completely abandoned. In June 1993, under IAEA supervision, South Africa filled in the test shafts.

Because it did not conduct any nuclear test explosions, South Africa had to verify the effectiveness of its nuclear weapon design through a series of tests on its components and subsystems. In 1979 the job of building additional nuclear devices was given to ARMSCOR (state-owned Armaments Corporation), the South African arms manufacturing agency. ARMSCOR took a much more systematic approach to testing the reliability and safety of the nuclear weapons design. Using methodologies required to test and develop conventional high explosives and battlefield weapons for SADF, ARMSCOR eventually qualified the nuclear weapons design in terms of its internal ballistics and mechanical arming and safing operations. ARMSCOR engineers also studied failure modes and effects and conducted criticality analysis under a range of postulated storage, delivery, and accident scenarios to ensure the safety of the design.[21]

Research, Development, and Weaponization

Beginning in the late 1970s, South Africa conducted research and development studies on implosion devices, thermonuclear explosive technology, and ballistic missile delivery systems. Advena Central Laboratories (also known as the "Circle" facility), commissioned in 1981, was to be South Africa's sole facility for manufacturing implosion-type nuclear weapons and integrating them with ballistic missiles. Advena had an array of manufacturing capabilities for advanced nuclear weapons designs. For example, at Advena, ARMSCOR technicians built an advanced laboratory featuring flash X-ray analysis and ultra-high-speed photography (up to 20 million frames/second) for recording detonation phenomena. This equipment was for implosion-type nuclear weapons research and development program. Other projects related to implosion devices included the development of insensitive high explosives, which are less prone to accidental ignition than conventional high explosives, and efforts to build miniaturized neutron generators.[22]

---

[17]Reiss, *Bridled Ambition*, p. 10.

[18]*Ibid.*

[19]Moore, *South Africa and Nuclear Proliferation*, p. 112.

[20]Frank V. Pabian, "South Africa's Nuclear Weapon Program: Lessons for U.S. Nonproliferation Policy," *The Nonproliferation Review*, vol. 3, no. 1 (Fall 1995), p. 9.

[21]Albright, "South Africa's Secret Nuclear Weapons," pp. 7–10.

[22]*Ibid.* Also see "Evidence Builds of Advanced Weapons Work by South Africa," *Nucleonics Week*, Jan. 20, 1994, pp. 5–7.

Research on the use of tritium to boost the explosive yield of South Africa's gun-type nuclear device was conducted by the AEB. Despite this research and some investigations into the future production of tritium, no decision to build a boosted device was ever taken. One reason may have been the lack of facilities to handle tritium at Advena.

Many circumstances led to South Africa achieving its status without general knowledge, including secret help from other countries. An example is seen in a 1973 letter from a German ambassador to Pretoria, suggesting that since "both telex and telephone communication to South Africa is periodically monitored [...] confidential communications [could be sent through the] Embassy utilizing ... cipher facilities."[23] Usually, countries that discreetly cooperated in the transfer of technology, know-how, or materials to South Africa had an interest in the country's uranium ore reserves or some other benefit that South Africa would provide in return.

How Did South Africa Eliminate Its Weapons Program?

In September 1989, South African State President F. W. de Klerk called for an investigation to completely dismantle South Africa's nuclear weapons program; his aim was for South Africa to join the NPT as a state without a nuclear weapons capability. Due to internal and external political factors, it was decided that South Africa's nuclear weapons program would not be publicized until after the country signed the NPT. This meant that the dismantlement program started off being classified as top secret, with many relevant documents still classified to this day. South Africa signed the NPT in 1991, and two years later, in 1993, President de Klerk declared in a speech to the South African Parliament that "at one stage, South Africa did, indeed, develop a limited nuclear deterrent capability."[24]

De Klerk appointed a steering committee of senior officials of the AEC, ARMSCOR, and the South African Defense Force to work on the dismantlement program. Although de Klerk called for the investigation in September 1989, actual written confirmation was not received until February 26, 1990—the official date of implementation of the termination of South Africa's nuclear weapons program. The committee was provided with the following brief:

- To dismantle the six completed gun-type devices at ARMSCOR under controlled and safe conditions
- To melt and recast the HEU from these six devices as well as the partially completed seventh device and return it to the AEC for safekeeping
- To decontaminate the ARMSCOR facilities fully and to return severely contaminated equipment (such as a melting furnace) to the AEC
- To convert the ARMSCOR facilities to conventional weapons and nonweapons commercial activities
- To destroy all hardware components of the devices as well as technical design and manufacturing information
- To advise the government of a suitable timetable of accession to the NPT, signature of a Comprehensive Safeguards Agreement with the IAEA, and submission of a full and complete national initial inventory of nuclear material and facilities, as required by the Safeguards Agreement
- To terminate the operation of the Y-Plant at the earliest moment[25]

---

[23]Z. Cervenka and B. Rogers, *The Nuclear Axis, Secret Collaboration Between West Germany and South Africa*, NY Times Book Co., 1978, p. 373.

[24]Adolf von Baeckmann, Gary Dillon, and Demetrius Perricos, "Nuclear Verification in South Africa," *IAEA Bulletin*, vol. 37, no. 1, 1995, www.fas.org/news/safrica/baeckmann.html, p. 4.

[25]Waldo Stumpf, "Birth and Death of the South African Nuclear Weapons Programme," Presentation given at the conference "50 Years After Hiroshima," organized by USPID (Union Scienziati per il disarmo) and held in Castiglioncello, Italy, Sept. 28–Oct. 2, 1995, www.fas.org/nuke/guide/rsa/nuke/stumpf.htm#02, p. 7.

IAEA Involvement

Soon after its accession to the NPT in July 1991, South Africa signed a comprehensive safeguards agreement with the IAEA in September 1991 (INFCIRC/394). The first IAEA inspections in South Africa took place in November 1991, two years before de Klerk's admission that South Africa had developed and dismantled a "limited nuclear deterrent capability." In the two years prior to the admission, IAEA activities included the examination of contemporary operating and accounting records and analysis of the nature and quantity of nuclear material.[26]

De Klerk's disclosure prompted the IAEA to increase its safeguards team in South Africa by including nuclear weapons experts. Over the next five-month period, the bolstered IAEA team carried out inspections at a number of facilities and locations that had been declared to have been involved in the former nuclear weapons program. The objectives of these inspections were to:

- Gain assurance that all nuclear material used in the nuclear weapons program had been returned to peaceful usage and had been placed under IAEA safeguards
- Assess that all nonnuclear weapons-specific components of the devices had been destroyed; that all laboratory and engineering facilities involved in the program had been fully decommissioned and abandoned or converted to commercial nonnuclear usage or peaceful nuclear usage; that all weapons-specific equipment had been destroyed and that all other equipment had been converted to commercial nonnuclear usage or peaceful nuclear usage
- Obtain information regarding the dismantling program, the destruction of design and manufacturing information, including drawings, and the philosophy followed in the destruction of the nuclear weapons
- Assess the completeness and correctness of the information provided by South Africa with respect to the timing and scope of the nuclear weapons program and the development, manufacture, and subsequent dismantling of the nuclear weapons
- Consult on the arrangements for, and ultimately to witness, actions at the Kalahari test shafts to render them useless
- Visit facilities previously involved in or associated with the nuclear weapons program and to confirm that they are no longer being used for such purposes
- Consult on future strategies for maintaining assurance that the nuclear weapons capability would not be regenerated[27]

The IAEA Unravels the Scope of South Africa's Nuclear Weapons Program

The IAEA team was able to document the timing and the scope of South Africa's nuclear weapons program as a result of direct South African cooperation. Cooperation included official documents, program records, and information obtained through interviews with principal personnel at the various facilities and locations involved in the nuclear weapons program. Based on information collected by the IAEA and presented in the *IAEA Bulletin*, volume 37, number 1, of 1995, Table 16.2 represents a summary of the timing and scope of the South African nuclear weapons program as uncovered by the IAEA.[28]

---

[26] Adolf von Baeckmann, Gary Dillon, and Demetrius Perricos, "Nuclear Verification in South Africa," *IAEA Bulletin*, vol. 37, no. 1, 1995, www.fas.org/news/safrica/baeckmann.html, p. 1.

[27] *Ibid*, p. 4.

[28] *Ibid*, pp. 5–6.

Table 16.2 South Africa's nuclear weapons activity uncovered by the IAEA.

Date	Activity Uncovered by IAEA
1969–1979	All R&D work on nuclear explosive devices was done by the South African Atomic Energy Board, the forerunner of the AEC. This work resulted in the production of a "nondeliverable demonstration device." Throughout the program, it was never converted to a deliverable device, even though it had the capability.
1979	The responsibility for the nuclear weapons program was transferred to ARMSCOR, while AEC was made responsible for the production and supply of HEU and for theoretical studies and some development work in nuclear weapons technology. ARMSCOR's principle nuclear weapons activities were carried out in the Circle facilities, located about 15 km away from the AEC's establishment at Pelindaba.
1981	By this time, the South African nuclear weapons program involved the development and production of a number of deliverable gun-assembled devices; lithium-6 separation for the production of tritium for possible future use in boosted devices; studies of implosion and thermonuclear technology; and research and development for the production and recovery of plutonium and tritium.
1985	The South African government decided to limit its nuclear weapons program to the production of seven gun-assembled devices; to stop all work related to possible plutonium devices; and to limit the production of lithium-6. However, the government allowed further development work on implosion technology and theoretical work on more advanced devices.
1987	The first qualified production model of a deliverable device was completed.
1989	The government decided to stop the production of nuclear weapons. However, by this time, four further qualified deliverable gun-assembled devices had been completed and the HEU core and some nonnuclear components for a seventh device had been fabricated.
1990	Preparation for South Africa's accession to the NPT: de Klerk's steering committee to dismantle the country's nuclear weapons was convened: *inter alia*, all existing nuclear devices were to be dismantled and the nuclear materials were to be melted down and returned to the AEC.

The IAEA Verifies South Africa's Dismantlement

Once the nuclear weapons program was dismantled, South Africa had the IAEA verify that everything was destroyed. This was done so that South Africa would have the IAEA's official endorsement and could sign the NPT as a state without any nuclear weapons. To help verify the process, South Africa gave the IAEA unprecedented access to the facilities and records involved with the program. In addition, the country gave the IAEA extensive production data from the Y-Plant so that it could reproduce the day-to-day operations and form accurate estimates of the materials produced.

Then, after two years of investigation, the IAEA concluded that the quantities of material declared by South Africa were reasonable and could be independently confirmed to within one significant unit. The IAEA considers one significant unit as 25 kilograms of HEU. The IAEA believed that they would have detected that much material if South Africa had tried to hide it or export it illegally.[29] South Africa acceded to the NPT in 1991, and the finalizations of the dismantlement process were completed by 1992. All the declared material was then put under IAEA safeguards. Table 16.3 compares the declarations made by de Klerk to the activities verified by the IAEA, as documented in *IAEA Bulletin*, volume 37, number 1, of 1995.[30]

[29]Helen Purkitt and Stephen Burgess, *South Africa's Weapons of Mass Destruction* (Bloomington & Indianapolis: Indiana University Press, 2005), p. 127.

[30]Adolf von Baeckmann, Gary Dillon, and Demetrius Perricos, "Nuclear Verification in South Africa," *IAEA Bulletin*, vol. 37, no. 1, 1995, www.fas.org/news/safrica/baeckmann.html, pp. 6–7.

Table 16.3 Tracking IAEA verification of South African nuclear disarmament.

De Klerk's Brief (1989–1990)	Verified by the IAEA (1991–1995)
Dismantle the six completed gun-type devices at ARMSCOR under controlled and safe conditions.	By the time of the IAEA team's visit in April 1993, the dismantling and destruction of weapons components and the destruction of the technical documentation had been nearly completed. Dismantling records concerning the HEU components of the weapons were available. They provided sufficient detail to enable the ARMSCOR data to be correlated with the corresponding data in the nuclear material accountancy records maintained by the AEC.
Melt and recast the HEU from these six devices as well as the partially completed seventh device and return it to the AEC for safekeeping.	The IAEA team audit of the associated records indicated that all of the HEU the AEC provided to the nuclear weapons program had been returned to the AEC and was subject to IAEA safeguards at the time the safeguards agreement entered into force.
Decontaminate the ARMSCOR facilities fully and return severely contaminated equipment (such as a melting furnace) to the AEC.	The IAEA team carried out an audit of the records of the transfer of enriched uranium between the AEC and ARMSCOR/Circle. As a result of this audit, the team concluded that the enriched uranium originally supplied to ARMSCOR/Circle had been returned to the AEC and was subject to IAEA safeguards at the time the safeguards agreement entered into force.
Convert the ARMSCOR facilities to conventional weapon and nonweapons commercial activities.	The IAEA team found no indication to suggest that there remained any sensitive components of the nuclear weapons program that had not been either rendered useless or converted to commercial nonnuclear applications or peaceful nuclear usage.
Destroy all hardware components of the devices as well as technical design and manufacturing information.	The equipment used for uranium metallurgy at ARMSCOR/Circle had been returned to the AEC at the end of the program. The whole uranium metallurgy process area at ARMSCOR/Circle had been dismantled and decontaminated. The machine tools used for manufacturing the HEU and high explosives components had been decontaminated and are now available for commercial nonnuclear applications.
Advise the government of a suitable timetable of accession to the NPT, signature of a Comprehensive Safeguards Agreement with the IAEA, and submission of a full and complete national initial inventory of nuclear material and facilities, as required by the Safeguards Agreement.	South Africa acceded to the NPT on July 10, 1991. INFCIRC/394: Agreement between South Africa and the IAEA for the application of safeguards in connection with the NPT signed on September 16, 1991.
Terminate the operation of the Y-Plant at the earliest moment.	The Y-Plant was closed down on February 1, 1990.

Implementation of the Dismantlement Plans

After the 1992 presidential election, some representatives from the de Klerk government tried to sell some of the HEU stockpile to Clinton political appointees. Initially the political appointees did not immediately act on the offer from the de Klerk government. Eventually the U.S. tried to reopen negotiations about the sale of the HEU, but the South African government was no longer interested in making a deal. Political issues and rumors surrounding the HEU and its possible sale to other countries made South Africa feel apprehensive about continuing any more "behind the scenes" negotiations. For the time being, South Africa claimed to have just kept its entire HEU stockpile. Then, following the 1994 South African elections, most African National Congress (ANC) members of the South African government decided that it would be best to keep all the HEU for use in the SAFARI-1 reactor. Their reasoning was that if South Africa kept the HEU, they could increase its value by using it to create other isotopes or use it for other commercial irradiation services. The leftover HEU is subject to IAEA inspections.

The IAEA only looks beyond the date of South Africa's accession to the NPT (after 1991), so the HEU that receives inspection is only that which was claimed by South Africa after the dismantlement occurred. It is unknown how much HEU was created by South Africa. The South Africans could have claimed all that they produced, or they could have hidden some of it or sold it to another country. In addition to only being able to inspect the HEU that is claimed, the IAEA can only inspect weapons facilities that are claimed. Since the IAEA was not involved in the dismantlement of South Africa's nuclear weapons program, there will probably always be questions about whether all the HEU was taken into account.[31]

South Africa also had to destroy all the technology involved in making the nuclear weapons during the dismantlement period. First, the Y-Plant at Valindaba was dismantled and decontaminated. This was then scheduled to be turned into a civilian production plant. Another uranium enrichment plant at Pelindaba East as well as the Advena complex, where the completed nuclear warheads were kept for storage, were decommissioned and decontaminated. South Africa then claims that all the weapons designs and all documentation pertaining to the weapons were destroyed. This is another difficult part of the dismantlement process to investigate and verify. South Africa could very well have destroyed all the documentation pertaining to its nuclear weapons, but it is difficult to believe that something did not slip by during the process. The test shafts at Vastrap were filled with concrete as well so that no underground testing could continue to take place. ARMSCOR filled the test shafts under IAEA supervision in July 1993.[32]

In addition, the "know-how" of the nuclear weapons program had to be terminated. This is inherently a very difficult task because one cannot take away prior learning. South Africa tried several different methods to try to limit the spread of "know-how" from the nuclear weapons program. First, South Africa attempted to offer people in the program alternative employment so that they would not be unemployed and tempted to sell information on how to build a nuclear weapon. They also offered early retirement for people involved in the programs, and they attempted to reskill the workers so that they could find work in another industry. Before people were terminated from their jobs, they were also debriefed and remotivated. In special cases where the termination of jobs caused financial hardship, the security followups were said to take place.[33] This logic was followed by some U.S. assistance programs to Russia and former Soviet states.

---

[31]Helen Purkitt and Stephen Burgess, *South Africa's Weapons of Mass Destruction* (Bloomington & Indianapolis: Indiana University Press, 2005), pp. 127–129.

[32]"South Africa's Nuclear Weapons Program: Building Bombs," *Nuclear Weapons Archive Website*, Sept. 2001: http://nuclearweaponarchive.org/Safrica/SABuildingBombs.html.

[33]Richardt Van Der Walt, Hannes Steyn, and Jan Van Loggerenberg, *Armament and Disarmament* (New York: iUniverse, Inc., 2005), p. 100.

Why Did South Africa Give Up the Bomb?

In contrast to the cases of rollback in Sweden, Argentina, and Brazil, there was no public debate on nuclear weapons in South Africa. The government's nuclear weapons development program was carried out in nearly total secrecy. In fact, there seems to have been little public awareness anywhere of the nation's nuclear option until 1977, when the international press reported the discovery of South African preparations for a nuclear test. Even the South African Foreign Ministry was not fully informed of the nation's nuclear status.[34] Another unique feature of this case is the speed and decisiveness with which the South African government made its decisions—first to build nuclear arms and then to dismantle them and join the NPT. It is notable that once the decision to develop nuclear weapons was taken, it took only five years to acquire them.

South Africa's decisions to acquire and then abandon nuclear arms were based on the worldview of a very small group of actors and were executed without wider consultation. The primary motivation for building nuclear arms was to improve the security of the state against the perceived threat of a Soviet-backed attack by neighboring Marxist regimes. Because of its racial policies, South Africa had no regional or international allies. Its leaders believed that a South African capability to use nuclear weapons would increase the chance that the United States would intervene on their nation's behalf during a military crisis.

The decisions to build and then renounce nuclear weapons were made by a small number of South African leaders. The way in which these individuals rationalized their decisions is therefore key to explaining South African behavior. Unfortunately, many of these individuals are still in the South African government and consider their deliberations regarding nuclear arms to be secret. Moreover, the government has declared that all documents relating to the nuclear decisions were destroyed. Although this is unlikely, it places constraints on decision makers that might be forthcoming with any remaining documents, or even with their own recollections of the events. As the key events recede into the past, it becomes less likely that a thorough historical analysis of South Africa's nuclear decisions will ever be written.[35] Until such a history is available, one can only speculate as to why South African leaders came to believe it was logical and reasonable to forswear nuclear arms.

Security Concerns

The South African government's decision to abandon its nuclear weapons program was probably due in large part to the lessening of external security threats. Two events that caused sharp reductions in South Africa's perception of the threat were the Angolan/Namibian peace settlement and the collapse of the Soviet Union. On August 5, 1988, a negotiated ceasefire took effect between the parties fighting over Namibian independence. U.S.-mediated negotiations among Angolan, Cuban, and South African representatives led to the signing on December 22, 1988, of an accord that linked South African acceptance of U.N.-supervised elections in Namibia to the phased withdrawal of Cuban troops from Angola. By late 1989 the last Cuban troops had left Angola. Elections were held in Namibia in November 1989, and Namibia became independent on February 16, 1990.[36]

The collapse of the Soviet Union's conventional power projection capabilities in 1989 and 1990 essentially eliminated the possibility that that nation would intervene directly in a Southern African conflict.[37] It also eliminated the chance of any large-scale Soviet military

[34]Renfrew Christie, "South Africa's Nuclear History," paper presented at the Nuclear History Program Fourth International Conference, Sofia-Antipolis, Nice, France, June 23–27, 1993.

[35]Author's (Jim Doyle's) conversation with David Albright, Jan. 1996.

[36]Howlett and Simpson, "Nuclearisation and Denuclearisation in South Africa," p. 161.

[37]Melvin Goodman, ed., *The End of Superpower Conflict in the Third World* (Boulder, CO: Westview Press, 1992), pp. 1–16.

assistance to pro-Communist states in the region. Without such assistance, even the combined forces of South Africa's regional adversaries could not pose a serious military threat to Pretoria.

In the context of this changed security environment some South African officials began to doubt the logic of the nuclear strategy. Jeremy Shearer, a Foreign Ministry official with knowledge of the nuclear program, admitted in September 1993 that he and others in government worried that if South Africa exercised its nuclear option in a military crisis it might invite combined opposition from Washington and Moscow. Instead of prompting the United States to intervene on Pretoria's behalf, this result would increase the chances of a strategic defeat of South Africa, an outcome the nuclear strategy was intended to avoid.[38]

Domestic Politics

From the standpoint of domestic politics, the primary reason that South Africa gave up nuclear weapons was that once their security justification was no longer compelling, their continued possession was incompatible with the government's vision of political reform and its strategy for ending South Africa's pariah status.[39] To rejoin the community of nations in good standing and end economic sanctions against it, South Africa had to remove suspicions surrounding nuclear weapons status. In short, the government decided that its nuclear status created obstacles to achieving its main domestic political objectives, including an end to civil unrest, transition to majority rule, and revitalization of the economy.[40]

A related domestic political motivation for terminating the nuclear weapons program may have been to allay the suspicions of the ANC that the national party leadership was prepared to use nuclear weapons in whatever fashion they could to retain their grip on the government of South Africa. In essence, the decision to abandon nuclear arms could have been seen as one way to improve the image and political position of the National Party, however marginally, among a broader cross-section of the South African population prior to a transition to majority rule.[41] Another possibility is that once domestic political reform was recognized as inevitable, the South African government may have been concerned that nuclear weapons or nuclear weapons material might fall into the hands of a new ruling political faction that could use them to advance extremist objectives. For example, the decision to dismantle the nuclear weapons program may have been motivated by the possibility that an ANC-led government might transfer any remaining weapons-grade uranium to Libya, Cuba, Iran, or the Palestine Liberation Organization (PLO) to pay off old political debts. It has also been suggested that the white regime was motivated simply by the prospect that if they did not act, nuclear weapons would be inherited by a black majority government.[42]

South Africa's nuclear rollback was also due in part to changing personnel in key bureaucratic positions. Before the decision to eliminate the nuclear program was made in November 1989, two of its strongest proponents, Wally Grant and A. J. A. "Ampie" Roux, had retired from the AEB.[43] By September 1989, others closely associated with the nuclear

---

[38]Albright, David. "How South Africa Abandoned Nuclear Weapons." Washington, D.C.: The Henry L. Stimson Center, Occasional Paper 25, 1995.

[39]Waldo Stumpf, "South Africa's Nuclear Weapons Program: From Deterrence to Dismantlement," *Arms Control Today* (Dec. 1995/Jan. 1996), p. 6; Reiss, p. 20.

[40]"De Klerk Discloses Nuclear Capability to Parliament," FBIS-AFR-93-056, March 25, 1993, pp. 5–9.

[41]David Albright and Mark Hibbs, "South Africa: The ANC and the Atom Bomb," *Bulletin of the Atomic Scientists*, 49 (April 1993): p. 33.

[42]Reiss, *Bridled Ambition*, p. 20.

[43]Wally Grant and A. J. A. "Ampie" Roux were strong supporters of South Africa's uranium enrichment program, and they were instrumental in persuading the government to develop nuclear weapons. See Reiss, *Bridled Ambition*, p. 20.

program, such as Prime Minister P. W. Botha, who in July 1979 endorsed a decision to direct ARMSCOR to build seven nuclear weapons, and Magnus Malan, chief of the defense force, had either left government or lost influence. Some observers also claim that by 1989 other senior officials at the AEB seriously questioned the value of the nuclear stockpile and supported the dismantlement decision.[44] The most significant personnel change was of course the election of F. W. de Klerk as President in September 1989. Immediately upon taking office, de Klerk let it be known to his cabinet that he wanted to make South Africa a respected member of the international community and that he believed political reform and accession to the NPT as a nonnuclear weapons state were prerequisites for this outcome.[45]

An economic reason for nuclear rollback was simply to transfer the industrial, scientific, and financial resources that had been devoted to the nuclear program to other uses.[46] For example, South African officials had decided that some of the HEU from the weapons program will be used in the Safari-I research reactor for the commercial production of medical isotopes such as molybdenum-99.[47] Such explanations are consistent with South Africa's broader efforts to privatize its government-controlled industries and convert defense production to commercial uses.[48]

The real payoff economically, however, was expected to be realized with the lifting of international economic sanctions in response to South Africa's nuclear rollback. A lifting of sanctions and access to capital markets was needed to finance the large public works projects that were envisioned as a key element in an economic reform program that would eventually provide improved jobs, services, and a higher standard of living to the black majority. This is discussed in greater detail in the following section.

International Politics and Economics

In the late 1980s, South Africa's regional security situation was improving, but it still faced domestic political turmoil, international ostracism, and economic recession. The de Klerk government recognized certain linkages between these problems and devised a strategy to address them. For example, to end civil unrest, South Africa had to accept majority rule and improve the economic conditions of the black population. Correspondingly, to finance democratic reforms and economic restructuring, South Africa needed to regain its access to international lending institutions and attract foreign investment.

South African access to Western capital markets had largely been closed since the mid-1980s. With the passage of the Comprehensive Anti-Apartheid Act in October 1986, the U.S. Congress imposed broad economic sanctions on South Africa. Also in the wake of South Africa's declaration of a state of emergency in 1985, which allowed greater repression of civil disturbances, anti-apartheid economic sanctions were imposed by the European Community, the British Commonwealth nations (with Great Britain abstaining), and France. The French

[44] *Ibid*, p. 20. Reiss cites J. W. De Villiers, a weapons designer who later became president and then chairman of the board of South Africa's AEC, and Waldo Stumpf, who is the current CEO of the AEC.

[45] Albright, "How South Africa Abandoned Nuclear Weapons."

[46] Estimates of the financial cost of the nuclear weapons program range from $300 million to $2.5 billion.

[47] J. W. de Villiers, Roger Jardine, and Mitchell Reiss, "Why South Africa Gave Up the Bomb," *Foreign Affairs*, vol. 72, no. 5 (Nov./Dec. 1993): p. 107. See also "South Africa AEC Head Says Stockpile of HEU Will Be Maintained for Safari," *Nuclear Fuel*, Aug. 15, 1993, p. 6.

[48] See, for example, Andre Buys, "The Conversion of South Africa's Nuclear Weapons Facilities," *Bulletin of Arms Control*, Council for Arms Control and Center of Defense Studies at London University, no. 12, Nov. 1993.

sanctions included a partial nuclear trade embargo.[49] These actions and the large-scale withdrawal of investments from the South African economy by European companies after 1985 had a significant negative impact on the South African economy. For example, at the beginning of 1986 there were 250 American companies with investments in South Africa worth approximately $2.5 billion. By the middle of 1991 there were only 125 companies, with investments worth about $1 billion.[50] Also, beginning in the mid-1980s, these sanctions imposed a ban on borrowing from the IMF and Export-Import Bank.

Without access to international credit South African businesses could not finance planned expansion. For example, the Electrical Supply Commission of South Africa (Eskom) had plans to bring electric power to 20 million South African blacks who lacked it and to expand electrical supply to neighboring nations. Because of sanctions, however, Eskom could not obtain loans for this project. In 1990 Eskom officials advised South Africa to join the NPT while its managers toured Europe and the United States to make their case for easing trade sanctions and to get financing for their expansion plans.

A removal of economic sanctions was also necessary to allow normalization of nuclear trade relations with the West. By the late 1980s sanctions had taken a toll on South Africa's nuclear industry. The longstanding U.S. refusal to supply fuel for the Koeberg reactors, triggered by Congressional opposition to apartheid and South Africa's refusal to allow safeguards on the Y-Plant, was very expensive for South Africa. As a result of this embargo South Africa was compelled to construct its own semicommercial enrichment plant to meet the needs of its reactors. In addition, the Comprehensive Anti-Apartheid Act contained new restrictions on U.S. imports of uranium from South Africa. Most states had been unwilling to curtail these imports because of the reliance of their own commercial nuclear fuel industries on South African uranium. The Act also prevented the United States from supplying South Africa with any nuclear technology or services until it became a party to the NPT. This resulted in the eventual cancellation of a technical services contract between South Africa and Westinghouse for the Koeberg reactors.

As mentioned, in 1977 South Africa was removed from its seat on the IAEA Board of Governors and replaced by Egypt. Beginning in 1979 it was prevented from participating in annual IAEA general conferences, and in 1982 it was barred from receiving technical assistance from the Agency. Efforts by factions of the IAEA membership continued throughout the 1980s to expel South Africa from the agency. These actions prevented South Africa from influencing the proceedings of this important institution and lowered its prestige in international nuclear industry circles. South Africa's active participation in the IAEA had been valuable as well for the access this provided to the latest information on nuclear technology and to agency technical assistance programs.[51]

When South Africa acceded to the NPT in July 1991, President de Klerk asserted that joining the treaty "will facilitate the international exchange of nuclear technology, which is not only important for the maintenance and further development of South Africa's own nuclear program, but will also be to the benefit of its neighboring states and the international nuclear community."[52] The prospect of renewed nuclear cooperation with the West was therefore an element of South Africa's overall strategy of reform and a key factor in its nuclear rollback decision. On July 11, 1991, the day after South Africa signed the NPT, President Bush lifted most economic sanctions imposed by the 1986 Comprehensive Anti-Apartheid Act. Sanctions imposed by other Western states have eased as well. South Africa

---

[49]Leonard S. Spector, *The Undeclared Bomb*, (Cambridge, MA: Ballinger, 1988), p. 292.

[50]Thomas L. Friedman, "Bush Lifts a Ban on Economic Ties to South Africa," *New York Times*, July 11, 1991, p. A1.

[51]Moore, *South Africa and Nuclear Proliferation*, pp. 106–107.

[52]David Ottaway, "South Africa Agrees to Treaty Curbing Nuclear Weapons, *Washington Post*, June 28, 1991, p. A25.

can now import fuel and services for the Koeberg reactors, it can export its uranium without restriction, and it can market its nuclear services such as the technology used to blend-down HEU for use as fuel in commercial power or research reactors. South Africa's decision to join the NPT as a nonnuclear weapons state has also made possible its renewed participation in the IAEA.

Aftermath: Impact of the Dismantlement on South Africa's Politics, Development, and Economy

Since dismantlement of its nuclear weapons program, South Africa has been focusing more on its nuclear power program, which generates more than 1,800 MWe, and covers 6% of the nation's total electric power consumption. Considering the fact that the existing nuclear plants at Koeberg-1 and Koeberg-2 will be inadequately serving the growing power demand, South Africa has been seeking an international partner to receive a more economical, more efficient, and safer reactor (4,000 MWe) by 2010.[53] Occasionally, South Africa receives special privileges because of its unique experience with nuclear dismantlement. For example, in the 1995 NPT Conference, South Africa was chosen as a "chief mediator" between the nonaligned movement and the nuclear weapons states. South Africa was instrumental in the discussions resulting in the adoption of a set of Principles and Objectives for Nuclear Non-Proliferation and Disarmament.[54] Furthermore, in April 1996 the African Nuclear Weapon-Free Zone treaty was named after Pelindaba of South Africa (the Pelindaba Treaty).

The IAEA has already accepted South Africa as a nuclear weapons-free country, and it also believes that the country has faithfully adopted the INFCIRC/153 safeguarding agreement. To reinforce its commitment to the NPT, the South Africa Parliament in return passed the Non-Proliferation of Weapons of Mass Destruction Act, which committed South Africa to abstaining from the development of nuclear weapons. As a consequence, the country lost its space defense program. Financial consequences of the nuclear dismantlement vary from the loss of 15,000 job opportunities in the defense industry to the loss of 7,000 jobs as a result of the scaling-down of the nuclear industry. Over all, ARMSCOR lost its military influence and faced great losses of many of its engineers and entrepreneurs.[55]

"Know-how" of the nuclear program, as well as the knowledge of the buildings where sensitive materials are kept, is another difficult aspect of dismantlement. For example, the lack of unemployment benefits enticed "16 nuclear weapons and ballistic missile employees threatened to sell secrets about the nuclear program" if their former company, ARMSCOR, would not give them each a million dollars.[56] There was also the worry of theft, as in a case in which two workers were fired and had to be monitored closely after it was learned that they were trying to steal nuclear material.

Remaining Proliferation Concerns

When it comes to technology, dual-purpose technologies and devices, though decontaminated and strictly devoted to be used for nonmilitary purposes, still remain one of the major

---

[53]"Nuclear Power in South Africa." Uranium and Nuclear Power Information Centre Briefing Paper #88, Aug. 2006; www.uic.com.au/nip88.htm.

[54]"South Africa Nuclear Overview," NTI, Feb. 2006; www.nti.org/e-research/profiles/SAfrica/Nuclear/index.html.

[55]Richard Van Der Walt, Hannes Steyn, and Jan Van Loggerenberg, *Armament and Disarmament.*

[56]*Ibid.*

sources of nuclear proliferation concerns. For instance: the SAFARI-1 research reactor with an average power emission of 20 MW is one of the largest research reactors in the world. Unlike ordinary research reactors, it could be used to generate nuclear power. The spent fuel, rich in plutonium and other radioactive materials, would only need to go through an additional reprocessing plant for plutonium extraction before becoming a highly concentrated weapons-grade material. According to the ISIS's annual nuclear material inventory stocks, South Africa uses 5.8 tons of plutonium for nuclear power and has between 430 and 580 kgs of HEU in stock, both under strict safeguards of the IAEA.[57]

Can South Africa Serve as a Nuclear Rollback Model to North Korea and Iran?

The South African example set a precedent in the case of nuclear rollback because soon after South Africa dismantled its nuclear weapons program, Argentina, Brazil, and Sweden followed suit. Today's nuclear challenges include North Korea and Iran, but it is doubtful that these two countries will follow the South African model.

When South Africa admitted to its nuclear weapons program, there was a change in the political landscape as the country moved away from apartheid, which influenced the South African decision. The leaderships of Iran and North Korea are stable and are not likely to change anytime soon. It is unlikely that Iran will move away from its Islamic Republic status, and North Korea is set in its Communist ways. To some extent, both Ahmadinejad and Kim Jong-Il, respective leaders of Iran and North Korea, enjoy popular support at home, and both leaders relish the thought of being a thorn in the side of the United States. With North Korea having already conducted a nuclear test[58] and the international community imposing a number of sanctions on Iran over its dubious nuclear power program, it therefore seems very unlikely that either of these two nations will be following the South African example.

However, there is one identifiable trait that paints North Korea and Iran with the same brush as South Africa during its apartheid days in the 1970s and 1980s. Today, North Korea and Iran are both perceived as pariah states and are both internationally isolated (North Korea more so than Iran), just as South Africa was in the 1970s and 1980s. Due to its decision to abolish the practice of apartheid, South Africa was able to move away from its international isolation. Furthermore, the South African government embarked on a series of confidence-building measures, notably its public announcement that it had relinquished its nuclear weapons program, which was soon after followed and verified by IAEA inspections. South Africa should therefore be an example for North Korea and Iran to follow since these two nations are to some extent in the same position that South Africa was 20 to 30 years ago. Soon after South Africa's nuclear weapons program disclosure, South Africa and 42 other African states signed the African Nuclear Weapons-Free Zone Treaty (the Treaty of Pelindaba). Should Iran and North Korea choose to follow the South African model, maybe the world can see two further nuclear weapons-free zones: one on the Korean Peninsula, the other in the Middle East. A way to see this move would be marshalling the international resources to verify a willingness to end a nuclear weapons program, because much could be determined by the public diplomacy of the state and how it handled disarmament.

[57]ISIS Online, "Global Stocks of Nuclear Explosive Materials: Summary Tables and Charts," July 12, 2005, Revised Sept. 7, 2005, www.isis-online.org/global_stocks/end2003/summary_global_stocks.pdf.

[58]Even though the North Korea reactor and reprocessing plant have already been disabled, although not much more permanently than they were under the 1994 framework, it is still too soon to predict what might happen in the immediate future.

Summary

South Africa is the only country to have built up a nuclear weapons program and then completely and voluntarily dismantled it. Information about South Africa's acquisition of weapons, technology, and know-how is very valuable in studying the possibility of a nonnuclear weapons state becoming a proliferator. Although the technology of basic nuclear weapons and uranium enrichment is of a very high level, we learn that it is still within the reach of a reasonably advanced industrialized country and is, therefore, not an unachievable barrier. This is even more realistic when the country is not necessarily seeking the latest technology but simply to acquire nuclear weapons capabilities.

The disarmament of a nation is a huge task to attempt, for all parties involved. South Africa had less than seven completed weapons at most. A nation with hundreds of nuclear weapons would be much more difficult. Also, there is no true way to verify whether or not the details of the dismantlement were carried to completion. This reverts to problems of the NPT and the IAEA; the processes will only work if the participating countries are honest and do not withhold any information. There is a chance that South Africa may still have nuclear capabilities or at least the possibility of proliferating again. There is no concrete proof of the complete destruction of technology; additionally, the shredding of documents cannot be verified. This issue is further exaggerated by the fact that a majority of the process remains classified. Finally, there exists the reality that knowledge cannot be erased from one's mind, independent of delegated new tasks. It is also difficult to accurately monitor workers once they have been debriefed and exit the program. There must be laws that clearly prohibit proliferation activities, and there must be means to enforce these laws. Following the key people involved in the weapons program is an obligation of the state. Nevertheless, the case of South Africa is very helpful in learning about the process of nuclear disarmament, and because it has already served as an example for other countries to follow, it is hoped that this history will continue to serve as a role model for countries such as North Korea and Iran.

17 Argentina and Brazil

James E. Doyle

Introduction

From the late 1960s through the 1980s Argentina and Brazil made efforts to create nuclear power infrastructures that could provide the materials for nuclear weapons. There is little evidence, however, that either of these two countries ever made the decision to go forward with the construction of a nuclear weapon. This is clearly true in the case of Argentina. In Brazil it seems that efforts were made to investigate nuclear weaponization but that these efforts were terminated before a decision to complete a nuclear explosive device was taken.[1] In addition, neither Argentina nor Brazil appears to have assessed seriously the role of nuclear arms in their respective national security strategies.

Argentina and Brazil generated suspicions regarding their nuclear intentions by constructing facilities that could produce bomb-grade nuclear materials and refusing to accept international safeguards on all their nuclear activities. Until the early to mid-1990s, both states also refused to bring fully into force the terms of the Treaty of Tlatelolco, which establishes a nuclear weapons-free zone in Latin America, or to join the Treaty on the Nonproliferation of Nuclear Weapons (NPT). In addition, throughout the 1970s and 1980s there was little public accountability of the nuclear programs in Argentina and Brazil. This is especially true in the case of the secret nuclear activities of the Brazilian military. Finally, political and military leaders in both countries often declared their right to explore a nuclear weapons option and to be free to develop the capability for so-called peaceful nuclear explosions (PNEs).[2]

---

[1]There is no evidence of weaponization in Argentina. In 1967 Brazil's National Council for Nuclear Energy (CNEN) commissioned a study of the feasibility of building an atomic bomb. See H. Jon Rosenbaum, "Brazil's Nuclear Aspirations," in *Nuclear Proliferation and the Near-Nuclear Countries*, Onkar Marwah and Ann Schulz, eds. (Cambridge, MA: Ballinger, 1975). In 1978 Brazil launched a secret nuclear development program that proceeded in parallel to its acknowledged civil nuclear program. For some members of the Brazilian military who participated in this program, the objective was to develop the capability to construct a nuclear explosive device. See Michael Barletta, "The Military Nuclear Program in Brazil," CISAC, Stanford University, 1987; http://iis-db.stanford.edu/pubs/10340/barletta.pdf (Jan. 2007). In addition, former Brazilian Minister of Science and Technology José Goldemberg has stated that he believes a nuclear explosive would have been designed by the Brazilian Air Force at the Aerospace Technology Center near São Paulo. See David Albright, "Brazil Comes in From the Cold," *Arms Control Today*, Dec. 1990, p. 13.

[2]In Brazil such statements started with General Artur da Costa e Silva, head of the military government in 1966. See Michael J. Siler, *Explaining Variation in Nuclear Outcomes Among Southern States: Bargaining Analysis of U.S. Nonproliferation Policies Toward Brazil, Egypt, India, and South Korea* (Ph.D. dissertation, University of Southern California, May 1992), p. 163.

The efforts by Argentina and Brazil to maintain a nuclear option were motivated only in part by security concerns. Motivation stemmed primarily from the perceived domestic political benefits of maintaining independent nuclear postures, a belief in the economic benefits of advanced nuclear technology, and the desire for international prestige. The view that nuclear energy development was an important determinant of overall economic and technological advancement and the corresponding belief that international controls on nuclear activities would constrain such advancement were particularly strong in both states. The political leadership in both nations rejected the NPT as discriminatory and opposed efforts by the advanced nations to impose nuclear supplier guidelines that constrained Argentina and Brazil's ability to acquire modern nuclear technology.

Moreover, Argentina and Brazil derived political benefits from supporting one another's decisions to remain outside the global nonproliferation regime. These benefits included a tacit mutual approval of their self-proclaimed right to develop nuclear explosive technology and the creation of an informal agreement to defy the nonproliferation regime. This agreement to maintain common policies toward the NPT guaranteed that neither state could be singled out by the international community for refusing to join the treaty. Ironically, this strategy for mutual opposition to the NPT evolved into a mechanism for more substantive bilateral nuclear cooperation and eventual integration into the international nonproliferation regime.

Over time Argentina and Brazil came to see the disadvantages of a policy designed to preserve a nuclear weapons option and the benefits of forswearing that option and joining the international nonproliferation regime. Argentina signed the NPT in 1995 and Brazil joined in 1997. Both countries have brought the Treaty of Tlatelolco fully into force on their territories, renounced their right to conduct PNEs, and are strengthening their nuclear export controls. Moreover, they have created a joint system of inspections of all their nuclear facilities that includes accepting full-scope IAEA safeguards. Finally, both states have canceled plans to build reprocessing plants and have scaled back uranium enrichment capabilities. In short, they have accepted political barriers to acquiring nuclear weapons.

The role played by external actors in this case of nuclear rollback was limited. Nonetheless, several important elements of U.S. and multilateral nonproliferation policy helped shift the balance of proliferation incentives and disincentives and thus made a contribution to nuclear rollback in South America. These include providing continuing support and leadership for the international nonproliferation regime, efforts to harmonize the export control policies of nuclear suppliers, promotion of liberal economic reforms, and the maintenance of a nonproliferation dialogue with Argentina and Brazil. In addition, the United States played a more active role after the two countries reached a 1985 agreement on bilateral nuclear cooperation. Specifically, the United States took a leading position in the provision of technical and financial support for nuclear safeguards development. In this respect, the case of Argentina and Brazil contains some lessons for ongoing or future efforts to facilitate the rollback of nuclear proliferation in other countries.

Motivations for Acquiring the Capability to Build Nuclear Weapon

Beginning in the 1950s, Argentina and Brazil, like many other states, viewed nuclear energy as having great potential for economic and scientific development. The development of

---

For additional statements, see "Navy Minister Says Country Could Build a Nuclear Bomb," FBIS/LAT, June 29, 1981, p. D2; and "Army Minister said to Favor Building Atomic Bomb," FBIS/LAT, Sept. 4, 1985, p. D2. For similar statements by Argentine leaders, see Joseph Pilat and Warren Donnelly, *An Analysis of Argentina's Nuclear Power Program and Its Closeness to Nuclear Weapons* (Washington, D.C.: Congressional Research Service [Dec. 2, 1982]), pp. 19–36.

nuclear energy was seen as way to reduce reliance on foreign energy supplies.[3] These two countries also witnessed the power of nuclear weapons in World War II. Both countries began nuclear energy research programs in the 1950s and by the end of the 1960s had planned to acquire nuclear infrastructures that could produce weapons-usable fissile materials.[4] One of the primary motivations in both countries for the development of a nuclear weapons option was simply to hedge against the possibility that many states would acquire them and the consequence that states that did not would be relegated to second-class status in world affairs.[5] Argentina and Brazil had aspirations to join the ranks of the highly industrialized nations, and they believed that an unconstrained nuclear energy sector and nuclear weapons might be needed to achieve this goal.[6]

In addition, the two countries were suspicious of one another. Traditionally, Argentina and Brazil have been regional rivals competing for political and economic leadership in Latin America. Throughout the period during which initial decisions on nuclear matters were made, the possibility of armed conflict between the two states remained the focus of military planning in both. This mutual distrust and competition were present in the early nuclear programs of both states. For example, there was speculation in Brazil and elsewhere in the early 1950s that Argentina was conducting experiments aimed at producing a nuclear explosion.[7] This may have contributed to Brazil's decision in 1953 to send Admiral Alvaro Alberto, president of the country's National Research Council, to West Germany to obtain gas centrifuge technology for uranium enrichment.[8]

Although these early attempts to achieve advances in nuclear technology were unsuccessful, they established a lasting pattern whereby Brazil sought to match or better Argentine nuclear accomplishments.[9] Moreover, the two states have always been attuned to the military potential of their nuclear developments. For example, Argentina's 1968 decision to purchase a West German heavy water reactor using natural uranium fuel for its first nuclear power plant raised Brazilian concerns that this reactor type was chosen for its ability to produce more weapons-usable plutonium outside of international safeguards than could a light water reactor using enriched uranium.[10]

Although the chance of military conflict between Argentina and Brazil has traditionally been low, there have also been territorial disputes and periods of increased regional tension. During the 1970s Brazilian economic and military assistance to Bolivia, Paraguay, and Uruguay resulted in Argentine perceptions of an increased capability by Brasilia to project

---

[3]This was particularly the case for Argentina and Brazil after the 1973 oil embargo. See Wolf Grabendorff, "Brazil," and Antonio Sanchez-Gijon, "Argentina," in *A European Non-Proliferation Policy*, Harald Muller, ed. (Oxford: Clarendon Press, 1987), pp. 323–400.

[4]For Argentina, see Daniel Poneman, "Nuclear Proliferation Prospects for Argentina," *Orbis* 27 (Winter 1984): pp. 853–880. For Brazil, see David J. Myers, "Brazil: Reluctant Pursuit of the Nuclear Option," *Orbis* 27 (Winter 1984): pp. 881–911.

[5]Ibid.

[6]John R. Redick, Julio C. Carasales, and Paulo S. Wrobel, "Nuclear Rapprochement: Argentina, Brazil, and the Nonproliferation Regime," *The Washington Quarterly* 8 (Winter 1995): p. 110.

[7]These fears proved unfounded because Ronald Richter, the scientist leading the experiments for Argentina, was revealed as a fraud. None of Richter's experiments ever resulted in a nuclear explosion. See Daniel Poneman, *Nuclear Power in the Developing World* (London: Allen and Unwin, 1982), pp. 68–70, and John R. Redick, "*Nuclear Illusions: Argentina and Brazil*," Occasional Paper 25, The Henry L. Stimson Center (Dec. 1995): p. 2.

[8]Myers, "Brazil's Reluctant Pursuit," p. 883.

[9]Ibid.

[10]Myers, "Brazil's Reluctant Pursuit," p. 889, and Norman Gall, "Atoms for Brazil, Dangers for All," *Foreign Policy*, no. 23 (Summer 1976): pp. 183–184.

power and influence throughout these so-called buffer states. In turn, military planning in Brasilia has focused at times on thwarting Argentine desires to regain the "lost territories" of the former Spanish colonial viceroyalty of the Rio de la Plata.[11] These territories include the present-day states of Paraguay and Uruguay and include parts of Bolivia and Brazil. These concerns were increased by Argentina's 1982 attempt to reoccupy the Malvinas/Falklands Islands, territory it had not controlled for 150 years. This action created doubts in Brazil regarding Argentina's commitment to regional stability.[12]

A final motivation for both Argentina and Brazil to develop the capability to manufacture fissile material and thus preserve a nuclear weapon option was economic. Nuclear energy development was seen as a way to gain experience with engineering and construction technology that would be useful in other sectors of the economy. Acquisition of the full nuclear fuel cycle was needed to avoid reliance on foreign suppliers, add value to natural and processed uranium, and create an indigenous nuclear industry capable of exports. Like South Africa, Argentina and Brazil argued that some of their uranium enrichment and spent-fuel reprocessing activities would be closed to international inspection to protect industrial secrets.[13]

Reservations and Constraints

Argentina and Brazil also faced reservations, constraints, and disincentives to acquiring nuclear weapons. First, the strategic rationale for either country to develop nuclear weapons was never compelling. Argentina and Brazil were wary of one another, but like most Latin American countries, their security concerns were primarily internal. Throughout the 20th century, war between the two was never likely. In fact, the benefits of good bilateral relations, trade, and technical exchange have often been acknowledged by governments in both capitals. Nor did either country face a serious threat from outside the region.

Second, given their rivalry, it was clear to both that neither country would allow itself to be left behind in a race to develop nuclear weapons if one began in earnest. Brazil's superior wealth meant that, over time, it could commit greater resources than could Argentina to a nuclear weapon program. However, Brazil could foresee the dangers in an arms competition that could result in its weaker rival acquiring a weapon that has been called the "ultimate equalizer."[14]

Even before either country had mastered the technology to produce weapons-usable nuclear material, they had reached an initial agreement for peaceful nuclear cooperation. In May 1980 Brazil's military leader, João Figueiredo, visited Buenos Aires and signed an agreement between the two national nuclear commissions that included joint research and development on nuclear power reactors, exchange of nuclear materials, uranium prospecting, and the manufacture of fuel elements.[15] In November 1985, the two states signed the Argentine-Brazilian Joint Declaration on Nuclear Policy (the Declaration of Iguazu) that reemphasized their mutual commitment to develop nuclear energy for exclusively peaceful purposes,

---

[11]Frank D. McCann, "Brazilian Foreign Relations in the Twentieth Century," in *Brazil in the International System: The Rise of A Middle Power*, Wayne A. Selcher, ed. (Boulder, CO: Westview Press, 1981), p. 6.

[12]Myers, "Brazil's Reluctant Pursuit," p. 881.

[13]Spector, *Nuclear Proliferation Today*, pp. 195–269.

[14]Mitchell Reiss, *Bridled Ambition: Why Countries Constrain Their Nuclear Capabilities*, Washington, D.C.: Woodrow Wilson Center Press, 1995, p. 52.

[15]Foreign Broadcast Information Service (FBIS), *Nuclear Development and Proliferation* (June 25, 1980), pp. 4–16. Also see Leonard S. Spector with Jacqueline R. Smith, *Nuclear Ambitions: The Spread of Nuclear Weapons, 1989–1990* (Boulder, CO: Westview, 1990), pp. 388–389.

to promote close cooperation in the nuclear field, and to coordinate activities to surmount increasing obstacles to obtaining nuclear equipment and materials.[16] The continued application of nuclear energy for military purposes by either state was clearly inconsistent with the intent of these agreements.

In addition, both nations' nuclear programs faced varying degrees of domestic opposition.[17] The cost, impact on the environment, and political implications of the respective nuclear programs were criticized in certain public and official circles. For example, the foreign ministries in both states were sensitive to the diplomatic costs of their nations' independent nuclear policies and were concerned that Argentine-Brazilian nuclear ambitions might inadvertently spark proliferation in other neighboring states. Financial officials and private corporate interests also recognized the independent nuclear policies as an impediment to foreign investment and trade.[18] Members of the scientific elites as well believed that the two countries' rejection of the international nonproliferation regime was foreclosing access to advanced Western technology and undercutting the objectives of economic development and modernization.[19] Finally, environmentalists and local officials were often opposed to nuclear activities, especially those that had military potential. The election of civilian governments in both countries in the 1980s made it easier for these groups to voice their opposition concerning the nuclear programs.

Pursuit of a Nuclear Weapons Option

In 1990 and 1991, when Argentina and Brazil made firm political and legal commitments providing reassurance to the international community that they would not develop nuclear weapons, they both possessed advanced nuclear infrastructures with military potential. This section provides some history on the nuclear programs of both nations.

Argentina

In Argentina, the National Commission for Atomic Energy (CNEA) was organized in 1950 to conduct the nation's nuclear program. The country's first research reactor, RA-I, was built in 1953 under the Atoms for Peace Program and used heavy water imported from the United States. By the 1960s, Argentina could build its own research reactors and had mastered nuclear fuel element processing. In the 1970s Argentina constructed two reactors northwest of Buenos Aires: a West German-built natural-uranium heavy water power reactor at Atucha (Atucha I) and a Canadian heavy water power reactor at Embalse. The existence of plentiful supplies of natural uranium in Argentina made heavy water reactors a logical choice because this design would lessen Argentina's reliance on foreign suppliers for enriched uranium. Plutonium could be also separated from the spent fuel used in these reactors. The Argentine program

---

[16]See Julio Cesar Carasales, "A Unique Component of the New Argentine-Brazilian Relationship: Nuclear Cooperation," in *Averting a Latin American Nuclear Arms Race,* Paul L. Leventhal and Sharon Tanzer, eds. (New York: St. Martin's Press, 1992), and John R. Redick, "Argentina and Brazil: An Evolving Nuclear Relationship," Occasional Paper Seven (Southampton, U.K.: Program for Promoting Nuclear Non-Proliferation, 1990).

[17]For information on domestic opposition in Brazil, see Michael J. Siler, "Explaining Variation in Nuclear Outcomes Among Southern States: Bargaining Analysis of U.S. Nonproliferation Policies Toward Brazil, Egypt, India, and South Korea" (Ph.D. dissertation, University of Southern California, May 1992), p. 140.

[18]Paulo Wrobel, "Brazil-Argentina Nuclear Relations: An Interpretation," unpublished draft prepared for the Rockefeller Foundation, Oct. 1993, pp. 26–27.

[19]See "Physicist Warns Against Objectives of FRG Nuclear Deal," *FBIS Latin America* (May 21, 1979), p. D1, and Redick, "Nuclear Illusions," pp. 42–45.

also included uranium production and nuclear fuel manufacturing facilities. In March 1976, CNEA announced plans for a third nuclear power plant, Atucha II, and a commercial heavy water plant that would eliminate the need for overseas suppliers for the Atucha I and Embalse plants. The Atucha II reactor contract was awarded to West Germany and the heavy water plant contract to the Swiss firm Sulzer Brothers.[20]

By the early 1990s several facilities in Argentina had the potential to be used to produce material for nuclear weapons. Argentina had constructed a gaseous diffusion enrichment facility in the Andean resort village of Bariloche—called the Pilcaniyeu facility—and a demonstration plutonium reprocessing unit at Ezeiza, near Buenos Aires. In addition, Argentina had an experimental pilot-scale heavy water facility situated near the Atucha I and II nuclear power stations in Buenos Aires province. Finally, there were unsubstantiated claims regarding a possible Argentine nuclear test site in Patagonia.[21]

The Ezeiza reprocessing facility, begun in 1978, was planned to be in operation by the early 1980s, providing Argentina direct access to weapons-usable plutonium. But economic and technical problems combined to delay the project and construction has been suspended since 1990. The Ezeiza plant was subject to international safeguards only when it reprocessed safeguarded spent fuel (which was the only type of spent fuel Argentina had at the time because all of its nuclear reactors were safeguarded). If it had been completed as planned, the reprocessing facility was expected to extract enough plutonium for one or two nuclear weapons per year.[22]

Construction of the Pilcaniyeu enrichment plant was also begun in 1978. The project was kept secret for five years by Argentina's military government and was revealed only weeks before civilian president Raul Alfonsin's inauguration in 1983. On November 10, 1983, Castro Madero, head of CNEA, announced; "Argentina has successfully demonstrated the technology for the enrichment of uranium."[23] Theoretically, the plant had the potential to enrich enough weapons-grade uranium for four to six nuclear bombs per year.[24] However, Argentine scientists claimed that Pilcaniyeu was designed to enrich uranium to only 20% U^{235}, which is not considered to be weapons-usable. The plant reportedly began enriching uranium to 20% U^{235} in 1988.[25]

Although Argentina in the early 1990s had come close to completing a nuclear infrastructure that could begin producing weapons-usable fissile material, it never achieved an actual capability to do so. The Pilcaniyeu uranium enrichment ran into technical and financial difficulties and never produced weapons-usable HEU.[26] It now operates under international safeguards.

In August 2006, Buenos Aires announced a major nuclear initiative worth $3.5 billion to finish its third nuclear reactor plant (Atucha II), restart a heavy water production plant in Neuquen Province, and conduct feasibility studies for construction of a fourth reactor at Embalse. It also plans to resume nuclear enrichment activities at the Pilcaniyeu complex using a gaseous diffusion-based enrichment technology.

[20]Leonard Spector, *Nuclear Proliferation Today*, p. 210.

[21]John R. Redick, "Argentina, Brazil, and the Nuclear Non-Proliferation Regime," address to the Council on Foreign Relations, Feb. 14, 1994, p. 3.

[22]Spector, *Nuclear Proliferation Today*, pp. 197, 204–5, 218.

[23]Reiss, *Bridled Ambition*, p. 47.

[24]Spector with Smith, *Nuclear Ambitions*, pp. 228, 388, 391.

[25]David Albright, "Bomb Potential for South America," *Bulletin of the Atomic Scientists* (May 1989): p. 16.

[26]Leonard S. Spector and Mark G. McDonough, *Tracking Nuclear Proliferation: A Guide in Maps and Charts, 1995* (Washington, D.C.: Carnegie Endowment for International Pace, 1995), p. 147.

Brazil

In 1955, after unsuccessful early efforts to obtain uranium enrichment technology from West Germany, Brazil signed an agreement for nuclear cooperation with the United States under the Atoms for Peace Program. In 1956 Brazil's National Atomic Energy Commission (CNEN) was created. Brazil also signed a nuclear cooperation agreement with France in 1967 and established a relationship with the West German Nuclear Research Center in 1969.[27] Unlike Argentina, Brazil maintained two parallel nuclear programs. The first was a publicly acknowledged nuclear energy program managed by the state-owned Brazilian Nuclear Corporation (Nuclebras) and subject to IAEA safeguards. The second was a secret, unsafeguarded program run by the Brazilian military to acquire the means to produce weapons-usable fissile materials and enrich uranium for naval propulsion reactors. It should be noted, however, that although Argentina did not have two separate programs, its nuclear program in the 1960s and 1970s was also run by the military and not subject to international safeguards.

One of the original components of Brazil's civilian nuclear energy program was a turnkey power station (Angra I), purchased from the U.S. firm Westinghouse in 1971. Brazil turned to West Germany, however, for a massive purchase of nuclear technology after the United States and Canada insisted on Brazil's acceptance of full-scope safeguards as a condition for further nuclear supply.[28] Brazil's June 1975 "nuclear deal of the century" with West Germany was a multibillion-dollar agreement for the largest transfer to date of nuclear technology to a developing country. It encompassed all aspects of the nuclear fuel cycle, including uranium exploration, fuel fabrication, two 1,250 megawatt power reactors, an industrial-scale uranium enrichment plant, a pilot reprocessing plant, and nuclear waste storage. The deal, which was to be implemented over a 15-year period, also included an option for six additional power reactors.

The German deal raised suspicions concerning Brazil's nuclear intentions. It appears that West Germany was chosen as a nuclear supplier because it did not require Brazil to accept IAEA safeguards at all its nuclear facilities. In addition, the agreement included the transfer to Brazil of enrichment and reprocessing technology. This technology could be transferred to military-run facilities to produce weapons-usable fissile material.

Brazil's unsafeguarded, military-directed nuclear program began in the mid-1970s and received the support of CNEN. This program relied on indigenously developed technology as well as personnel and technology transferred from the foreign-supported civilian program. By 1988 it included a laboratory-scale gas centrifuge uranium enrichment plant at the navy-run Research Institute on Nuclear Engineering (IPEN) near São Paulo. The first modules of an industrial-scale enrichment plant were also under construction at the Aramar Research Facility in Ipero. A laboratory-scale reprocessing facility had been constructed at IPEN as well and plans were under way for a graphite plutonium-production reactor at an army-run facility near Rio de Janeiro. Furthermore, in 1984–1985 the Brazilian military had prepared deep shafts for a possible nuclear testing program in northern Brazil.

---

[27]Edward Wonder, "Nuclear Commerce and Nuclear Proliferation: Germany and Brazil, 1975," *Orbis* (Summer 1977): p. 287.

[28]It also seems that Brazil may have turned to Germany because of doubts over U.S. ability to continue supplying fuel for Angra I and another planned reactor. In July 1974, the U.S. Atomic Energy Commission (AEC) retroactively classified as "conditional" the enrichment contacts for 45 foreign reactors, including two in Brazil. This was done because the AEC projected at the time that demand was exceeding its enrichment capacity. See U.S. Congress, Senate Committee on Foreign Relations, International Organization and Security Agreements, testimony by Myron B. Kratzer, Acting Assistant Secretary, Bureau of Oceans and International Affairs, State Department, hearing on July 22, 1975; and Gall, "Atoms for Brazil," pp. 163–166.

Like Argentina, Brazil is not believed to have ever produced any significant quantities of weapons-usable nuclear material, despite its extensive efforts to do so.[29] The military's laboratory-scale reprocessing facility at IPEN might never have gone beyond experimental operation, and its enrichment facilities at IPEN and the Aramar Research Facilities never overcame the technical obstacles needed to produce HEU.[30] Both facilities are now under international safeguards. President Jose Sarney's claim in September 1987 that the IPEN facility had conducted the successful laboratory-scale enrichment of uranium appears to have referred only to minute amounts of uranium enriched to 3–4% U-235, material that is considered unsuitable for nuclear weapons.[31]

Currently, Brazil mines uranium, which is shipped to foreign countries for conversion and enrichment, and returned to Brazil, where it is fabricated at the civilian enrichment plant at Resende into fuel for its two nuclear power reactors. When completed, a uranium enrichment plant under construction at Resende will allow the country to make its own low-enriched uranium fuel for its nuclear power industry. The plant initially will produce 60% of the nuclear fuel used by Brazil's two operational nuclear power reactors. Brazil has indicated that eventually it hopes to produce sufficient fuel for its reactors and for export.

Brazil's new Resende centrifuge enrichment facility was formally opened on May 6, 2006, but the plant is not forecast to be fully operational until 2010. Negotiations with the IAEA took over two years to enact mutually satisfactory safeguards for Resende that protect Brazilian proprietary interests in its enrichment technology.[32] Brazil claims it will save money by enriching its uranium domestically rather than sending it overseas to Urenco, the European enrichment consortium. It also claims its enrichment facility is 25% more efficient than those in France or the United States. In addition to domestic energy production, Brazil hopes in the future to participate in the $5-billion-a-year global nuclear fuel market.[33] About 90% of the world's nuclear power plants depend on foreign enrichment services to get their fuel. Demand for enriched uranium over the next two decades could justify Brazil's investment in the capability.[34]

Domestic Determinants of the Rollback Decisions

Many factors, both domestic and international, play a role in explaining the decisions by Argentina and Brazil to renounce nuclear weapons. Most officials from these countries who participated in these decision and scholars who have studied the events agree that domestic factors within the two countries and the evolution of their bilateral relationship provide the most convincing reasons for the outcome. The actions and policies of international community also played a role, but one that became more significant after the two nations were already on a course to become nonnuclear weapons states. Some of the main domestic causes of this outcome are discussed in the following sections.

---

[29]Spector and McDonough, *Tracking Nuclear Proliferation*, pp. 153–157.

[30]Some "hot cells" at IPEN may have operated briefly. Author's conversation with Lewis Dunn, March 1997.

[31]Reiss, *Bridled Ambition*, p. 56.

[32]Steve Kingstone, BBC News, Sept. 6, 2006; http://news.bbc.co.uk/2/hi/americas/4981202.stm.

[33]Erico Guizzo, "How Brazil Spun the Atom," Spectrum Online (IEEE), March 2006; see quotes by Samuel Fayad Filho, director of nuclear fuel production at Nuclear Industries of Brazil; www.spectrum.ieee.org/mar06/3070.

[34]IAEA, INFCIRC/640, p. 49, para. 129; www.iaea.org/Publications/Documents/Infcircs/2005/infcirc640.pdf#search=%22IAEA%2C%20INFCIRC%2F640%22.

Bilateral Nuclear Cooperation

A key feature of the decisions to foreswear nuclear arms in Argentina and Brazil is the high degree of bilateral cooperation between the nuclear policies of the two states. This cooperation was possible because Argentine and Brazilian security concerns about each other were never overriding. This fact weakened national security arguments for the development of nuclear weapons. Instead, both countries saw an opportunity to improve their security and economic prospects through a reduction in the tensions produced by the nuclear competition.[35]

Ironically, bilateral nuclear cooperation also evolved from mutual opposition to the nuclear nonproliferation regime. During the completion of negotiations for a Latin American nuclear free zone (the Treaty of Tlatelolco) in 1966 and 1967, Argentina and Brazil jointly opposed any prohibitions on peaceful nuclear explosions (PNEs) and differed with other regional states on issues such as the transportation of nuclear weapons through the zone, the entry-into-force process, and treaty reservations.[36] Ultimately the two states refused to bring the Treaty of Tlatelolco fully into force on their territories until 1990.

As time progressed and nuclear cooperation deepened, it appears that Argentina and Brazil became more concerned with avoiding the constraints of the international nonproliferation regime than with one another's nuclear energy development programs.[37] This view is supported by Argentina's support in the mid-1970s of Brazilian efforts to import reprocessing and enrichment technology from Germany. This support was reciprocated by Brazil in the late 1970s, when Washington objected to Argentina's efforts to buy a third power reactor and heavy water production facilities without accepting full-scope safeguards.[38] This mutual support in opposing supplier restrictions extended the pattern of political cooperation between Argentina and Brazil for maintaining independent nuclear policies.

In the late 1960s and early 1970s the nuclear policies of Argentina and Brazil converged against what they saw as an unjust nuclear order imposed by the nuclear weapons states.[39] For example, the two states opposed the NPT as unequal and discriminatory, objecting in particular to the prohibition on PNEs and the lack of binding security guarantees to non-nuclear weapons states. Both nations also rejected the efforts of the Nuclear Suppliers Group to restrict nuclear exports, a move they saw as threatening their independence and development objectives. In 1977 U.S.-Brazilian relations deteriorated over U.S. attempts to dissuade West Germany from selling Brazil reprocessing and enrichment technology.

Shared hostility to the international nonproliferation regime provided incentives for increased bilateral cooperation on nuclear policy. In January 1977, the Argentine and Brazilian foreign ministries issued a joint communiqué stressing the importance of nuclear policy cooperation and the initiation of technological exchanges between the two countries' respective nuclear energy commissions. The exchanges between CNEA and CNEN under this agreement provided the original foundation for later development of the joint Argentine-Brazilian Accounting and Control System (SCCC) and its administrative body, the Brazilian-Argentine Agency for Accounting and Control of Nuclear Materials (ABACC). This joint

---

[35]Julio C. Carasales, "The Argentine-Brazilian Nuclear Rapprochement," *The Nonproliferation Review* 2 (Spring/Summer, 1995).

[36]Redick, *Nuclear Illusions,* p. 17.

[37]An alternative view is that neither country had confidence that the nonproliferation regime could halt the nuclear ambitions of its rival. If so, the objective of avoiding restrictions on one's own nuclear program would be more important than having ineffective curbs applied to one's rival. In this case it would be reasonable to oppose all supplier restrictions.

[38]Virginia Gamba-Stonehouse, "Argentina and Brazil," in *Security with Nuclear Weapons? Different Perspectives on National Security*, Regina Cowen-Karp, ed. (Oxford: Oxford University Press for Stockholm International Peace Research Institute, 1991), pp. 229–256.

[39]Redick, *Nuclear Illusions*, p. 19.

communiqué was followed in May 1980 by an agreement between CNEA and CNEN on a wide range of joint projects, which included research and development on experimental and power reactors, exchange of nuclear materials, uranium prospecting, and the manufacture of fuel elements.[40] Under the agreement, Argentina leased uranium concentrate to Brazil and sold zircalloy tubing for nuclear fuel elements. Brazil, in turn, manufactured part of the pressure vessel for Argentina's Atucha II nuclear power reactor.[41]

The ABACC system of nuclear materials accounting and control that ultimately resulted from bilateral nuclear cooperation with Argentina is a unique, two-state regional nonproliferation regime employing a rigorous state system of accounting and control (SSAC) over its nuclear materials. The subsequent quadripartite agreement (signed by Brazil, Argentina, the IAEA, and ABACC) provides for full-scope IAEA safeguards of Argentine and Brazilian nuclear installations, full rights over any proprietary technology developed by both countries, and nuclear energy for the propulsion of submarines.

Political Rapprochement and Confidence Building

This deepening of nuclear cooperation between the two countries was facilitated by a warming political relationship and the resolution of key bilateral disputes. In October 1979 Argentina, Brazil, and Paraguay signed the Rio de la Plata agreement that resolved a dispute over the use of water resources and the construction of a hydroelectric dam on the Parana River that flows from Brazil into Argentina.[42] This agreement eased bilateral tensions that had persisted throughout the 1970s over exploiting the fertile Rio de la Plata basin, which lies astride the two states. The agreement marked the beginning of an improved phase of Argentine-Brazilian relations. In May 1980, Brazil's military leader, João Figueiredo, became the first Brazilian president to visit Buenos Aires in 40 years. During this visit the nuclear agreement between CNEA and CNEN was finalized.

Despite the improved political relationship, the two states encountered problems in implementing the May 1980 nuclear agreement. This was due in part to suspicions that remained between the two regarding their regional intentions and their nuclear programs. Argentina's 1982 occupation of the Malvinas/Falklands Islands led Brazil to proceed cautiously in dealing with Argentina's military regime. Moreover, in December 1983, Argentina informed Brazil that the Pilcaniyeu facility was capable of enriching uranium, a fact that raised concern in Brazil.[43]

Another breakthrough that improved political relations and nuclear cooperation between Argentina and Brazil was the emergence of civilian governments in the mid-1980s. In Argentina, military defeat in the Malvinas/Falkland Islands war with Great Britain led to the October 30, 1983, election of President Raúl Alfonsin. In 1984, an economic crisis in Brazil forced the military government to step down and permit civilian elections. Brazil's new president-elect Tancredo Neves met Alfonsin in February 1985. The two leaders promised to revive nuclear cooperation and to work toward the goal of mutual inspections of each other's nuclear installations.[44]

Unfortunately, Brazilian president-elect Neves died before taking office in March, and his successor, José Sarney, did not support the proposed nuclear inspection arrangement. Sarney nevertheless met with Alfonsin in November 1985 and signed the Declaration

---

[40]The text of the agreement can be found in *FBIS Nuclear Development and Proliferation*, June 25, 1980, pp. 4–16. Also see John R. Redick, "The Tlatelolco Regime and Nonproliferation in Latin America," *International Organization* 35 (Winter 1981): pp. 130–131.

[41]Spector with Smith, *Nuclear Ambitions*, pp. 388–389.

[42]Redick, *Nuclear Illusions*, p. 20.

[43]Reiss, *Bridled Ambition*, p. 54.

[44]Richard Kessler, "Argentina, Brazil Agree to Mutual Inspection of Nuclear Facilities," *Nucleonics Week*, March 14, 1985, p. 14.

of Iguazu. This declaration reemphasized the two nations' mutual commitment to develop nuclear energy for exclusively peaceful purposes. It also established a Joint Committee on Nuclear Policy to continue the bilateral dialogue on nuclear matters.[45] This committee, composed of the foreign ministers, officials from CNEA and CNEN, and industry representatives, became a mechanism for continuous contact on nuclear policy and nonproliferation issues.

In July 1986 Alfonsin and Sarney signed a major trade agreement committing their nations to the phased elimination of trade barriers and the creation of a Southern Cone Common Market (MERCOSUR). This agreement included plans for cooperation on nuclear safety in the event of an accident, and a subsequent presidential meeting in December 1986 resulted in agreements for joint research for breeder reactors and the development of safeguards techniques.[46]

In December 1986, Brazil allowed Argentine nuclear officials to visit the laboratory-scale facility at IPEN, where the navy secretly conducted research on both uranium enrichment and reprocessing outside of international safeguards.[47] In advance of a September 1987 public announcement of the successful operation of the uranium enrichment facility at IPEN, Brazilian President Sarney sent a letter to President Alfonsin of Argentina notifying him of the upcoming public announcement. This action reciprocated the prior notice that Argentina had given Brazil regarding its enrichment plant in 1983. These actions and the Presidents' previous meetings led to an invitation from President Alfonsin to President Sarney to visit Argentina's unsafeguarded Pilcaniyeu enrichment facility. The visit, which took place in July 1987, was an important confidence-building measure and prompted discussions on regularizing the process.

The process of reciprocal inspections and confidence building continued with a second visit by Argentine officials to Brazil's IPEN facility in April 1988 and Brazilian officials' visit to Argentina's Ezeiza pilot reprocessing facility in November 1988. Once again, the process was boosted by domestic political developments. In August 1989, Argentina's new President, Carlos Menem, met with Sarney in Brazil to agree on additional measures of nuclear cooperation and to intensify bilateral political and economic coordination. In response to Congressional pressure in Brazil, President Sarney had combined the official and secret military nuclear programs under a reorganized CNEN that reported directly to the office of the Presidency.[48] In December, 1989 Fernando Collor de Mello was elected president in Brazil and replaced Sarney in March 1990.

There were now popular presidents in Argentina and Brazil who were committed to economic reform, increased trade and foreign investment, reduction of the military's influence, and the exclusively peaceful use of nuclear energy. The two presidents saw nuclear cooperation as a way to accelerate the bilateral political and economic coordination that was already under way. On November 28, 1990, Presidents Collor and Menem signed the landmark Joint Declaration of Common Nuclear Policy at Foz de Iguazu. Both countries pledged to use nuclear energy only for peaceful purposes, create a formal system of bilateral inspections, forsake the right to conduct peaceful nuclear explosions, and adhere jointly to the Treaty of Tlatelolco. They also pledged to develop mechanisms for the acceptance of full-scope IAEA safeguards.[49]

---

[45]This committee evolved into the Permanent Standing Committee and later into the Commission of ABACC. The committee had three subgroups to deal with technical cooperation, foreign policy coordination, and the legal and technical aspects of nuclear cooperation.

[46]Redick, *Nuclear Illusions,* p. 21.

[47]Richard Kessler, "Sarney Visit to Pilqaniyeu Was Key to Reciprocal Inspections," *Nucleonics Week,* July 23, 1987.

[48]Redick, *Nuclear Illusions,* p. 23.

[49]Redick, *Nuclear Illusions,* p. 24.

This declaration was implemented at a July 1991 foreign ministers meeting in Guadalajara, Mexico. The Guadalajara Accord established the Joint System of Accounting and Control of Nuclear Materials (SCCC), the purpose of which was to verify that no nuclear materials were diverted for military purposes. To implement this control system, the accord created the Brazilian-Argentine Agency for Accounting and Control of Nuclear Materials (ABACC), which was modeled on the multipartite inspection system set up by the European Atomic Energy Community (Euratom).[50] ABACC began operations in July 1992 and initially monitored nuclear installations in Argentina and Brazil which were not under IAEA safeguards. Although these bilateral safeguards arrangements went a long way toward demonstrating that Argentina and Brazil were no longer seeking a nuclear weapons option, they were not yet sufficiently integrated with the IAEA safeguards system to convince major nuclear suppliers such as the United States, Canada, and Germany to lift nuclear export controls.

To satisfy these concerns and fulfill the pledge regarding full-scope safeguards made at Foz de Iguazu, Presidents Collor and Menem flew to Vienna in December 1991 to sign a Quadripartite Agreement among Brazil, Argentina, ABACC, and the IAEA. Under this agreement the two countries affirmed that international safeguards would apply "on all nuclear material in all nuclear activities within their territories [...] for the exclusive purpose of verifying that such material is not diverted to nuclear weapons or other nuclear explosive devices." The Quadripartite Agreement was ratified by Argentina and Brazil and entered into force on March 4, 1994. All nuclear activities in Argentina and Brazil would now be under international safeguards. In addition, by the end of May 1994, both countries had ratified a revised Treaty of Tlatelolco in which the IAEA played a larger role.[51]

Domestic Politics

Nuclear rapprochement between Argentina and Brazil was facilitated by domestic political changes within each country. Newly elected civilian governments needed to assert control over their respective military institutions. One method for doing this was to establish greater control over their nuclear activities. If the new governments were to convince their own populations and the world that they indeed had a firm grip on political power, they needed to be seen as having total control over national programs as crucial as nuclear research and development. One way to gain better control over nuclear activities was to publicize them. Once exposed, nuclear programs would have to justify economic and political costs, just like other government activities, and compete with other government priorities for resources.

On taking office in 1983 in Argentina, President Alfonsin sought to distinguish his government from previous military regimes. He opposed the pursuit of nuclear weapons and committed his government to ending the nation's diplomatic and economic isolation. One of the reasons that Alfonsin suggested to Brazilian President Sarney in 1985 that they create a bilateral nuclear inspection system was to symbolize that he was in full control of Argentina's nuclear program and, by extension, of the military and the country.[52]

A similar pattern developed in Brazil, albeit more slowly. Beginning in 1986, after the Brazilian press exposed military preparations of a deep shaft for possible nuclear explosive tests at Cachimbo in western Brazil, steps were taken within the Sarney government to gain greater control of the clandestine nuclear program and guarantee that it was devoted to peaceful purposes. After his December 17, 1989, election, President Fernando Collor moved aggressively to establish civilian control over all nuclear activities. Collor replaced Nazare

[50]Reiss, *Bridled Ambition*, p. 61.

[51]"Brazil, Argentina, and Chile Bring into Force the Treaty for the Prohibition of Nuclear Weapons in Latin America and the Caribbean," *U.S. Arms Control and Disarmament Agency Fact Sheet*, ACDA Office of Public Information, June 3, 1994.

[52]Reiss, *Bridled Ambition*, p. 57.

Alves, who had coordinated the parallel military nuclear program, as head of CNEN, and informed military leaders that he was going to dismantle the military nuclear program. In summer 1990, Collor brought members of the press to the suspected test site at Cachimbo to witness its official closing. Finally, in an address to the U.N. General Assembly in September 1990, Collor renounced Brazil's right to conduct PNEs.[53] This action removed an official rationale for the parallel program and a key issue of disagreement between Brazil and the international nonproliferation regime.[54]

The move toward greater openness and accountability regarding nuclear programs that was promoted by civilian governments in Argentina and Brazil increased the domestic political acceptability of changes in the overall nuclear posture. In Brazil, public awareness of the activities at Cachimbo and related revelations in 1987 and 1989 of secret bank accounts used by previous military governments to import equipment for the secret nuclear program, led to a new constitutional provision that limited all nuclear activities to peaceful purposes and subjected them to approval by the Congress.[55] The cost of the nuclear programs, particularly the naval nuclear propulsion program, also came under attack. In Argentina, the military's influence declined markedly after the Malvinas/Falklands war. President Alfonsin used this opportunity to reduce government spending on the nuclear program, to ease an economic crisis.[56]

Of course, major changes in the nuclear policies of Argentina and Brazil could not have been made without the acquiescence of the military. Fortunately, military figures in both countries began to realize the opportunity costs of maintaining independent nuclear programs. Resources devoted to such programs would not be available for other priorities such as improving conventional military forces. The arrival of civilian governments and economic difficulties in the mid-1980s led to reductions in defense budgets. As a consequence, military officials questioned the logic of investing in a nuclear capability that made little strategic sense in a regional context.[57]

Another key domestic political development that contributed to nuclear rollback in both countries was the rising influence of the foreign ministries in government decisions on nuclear matters.[58] The Declaration at Foz de Iguazu created a bilateral working group for nuclear cooperation. This working group had three subgroups to deal with technical cooperation, foreign policy coordination, and the legal and technical aspects of nuclear cooperation. Staff for these working groups was established in the Argentine and Brazilian foreign ministries. This action accelerated and institutionalized bilateral nuclear cooperation and transferred nuclear expertise to this area of the bureaucracy. It also placed the source of action on nuclear matters in the bureaucratic sector of the government that was most aware of the political and economic costs of remaining outside the international nonproliferation regime.

Economics

Economic considerations played a large, if not central, role in the decisions of Argentina and Brazil to forswear nuclear arms and join the international nonproliferation regime.[59]

---

[53]David Albright, "Brazil Comes in From the Cold," *Arms Control Today*, Dec. 1990, pp. 13–16.

[54]Reiss, *Bridled Ambition*, p. 59.

[55]Redick, Carasales, and Wrobel, "Nuclear Rapprochement," p. 113. Also see Jean Krasno, "Brazil's Secret Nuclear Program," *Orbis* (Summer 1994): p. 430.

[56]Paulo Wrobel, "Brazil-Argentina Nuclear Relations," pp. 12–14.

[57]Conversation with John Redick, University of Virginia, Charlottesville, Virginia, Feb. 1996.

[58]Ibid. Also see Redick, *Nuclear Illusions*, pp. 42–43.

[59]Paulo Wrobel, a Brazilian scholar, believes that the desire to remove barriers to advanced technology to achieve economic development goals was the primary motivating factor in Latin American nuclear rollback. See Wrobel, "Brazil-Argentina Nuclear Relations," p. 4.

Over time support for maintaining a nuclear weapons option was eroded by growing recognition of the high costs of the nuclear programs and the denied access to advanced technology that resulted from refusals to accept safeguards on sensitive nuclear activities. Economic development driven by trade, increased foreign investment, and advanced technology became a national priority in both countries. This priority undercut the view that a nuclear weapons option should be maintained despite the political and economic consequences.[60] National nuclear programs could be successfully used as vehicles for industrial and technological development only if they did not trigger international sanctions that blocked access to technology or foreign investment. It became increasingly difficult for Argentina and Brazil to avoid such penalties without renouncing nuclear arms and accepting full-scope safeguards on their nuclear activities. So economics, a clearly international activity, influenced thinking on the nuclear issue due to its domestic impact in Argentina and Brazil.

A similar logic prevailed in the bilateral context. Argentina and Brazil came to see suspicions regarding one another's nuclear programs as an obstacle to bilateral and regional technical cooperation in nuclear energy. Both countries had high expectations that such cooperation would yield economic benefits, including growth in other industrial sectors and increased trade.[61] The diplomatic process that started with agreements on nuclear cooperation in 1980 led to greater interaction on commercial relations. In late 1985, the two sides announced the Argentine-Brazilian Integration and Cooperation Program (PICAB), which facilitated Argentina's reintegration into the regional and international community after the Falklands/Malvinas war. The 1986 Act for Argentine-Brazilian Integration expanded bilateral trade relationships that were initiated in the 1985 Foz de Iguazu declaration.

In the late 1980s governments in Argentina and Brazil realized that this strategy of using nuclear cooperation to promote economic integration and expansion would be more likely to succeed if the impediments to cooperation that were imposed by secret nuclear activities and mutual suspicions were removed. This view provided motivation for a series of joint declarations on nuclear policy and mutual reciprocal inspections of sensitive nuclear facilities that began in December 1986 and culminated in the 1990 Joint Declaration of Common Nuclear Policy. This policy contained the mutual renunciation of nuclear arms and formalized the inspection process.

As mentioned, however, the common nuclear policy still did not go far enough to satisfy the major nuclear suppliers. Because Argentina and Brazil did not accept full-scope safeguards until the Quadripartite Agreement of 1991, they could not purchase nuclear technology or services from states that required such acceptance as a condition of supply. The United States and several other industrialized nations refused to sell certain specialized products such as computerized precision machine tools, electronic components, and supercomputers.[62] Not only did this practice slow the nuclear program and increase its costs, it led to increasing opposition to the independent nuclear policy from finance ministry officials,

---

[60]Etel Solingen, "The Political Economy of Nuclear Restraint," *International Security* 19 (Fall 1994): pp. 126–169, and Etel Solingen, "Macropolitical Consensus and Lateral Autonomy in Industrial Policy: The Nuclear Sector in Brazil and Argentina," *International Organization* 42 (1993): pp. 263–298. Also see Wrobel, "Brazil-Argentina Nuclear Relations," pp. 6, 11, 27.

[61]Specifically, it was thought that nuclear development would stimulate growth in metallurgy, chemistry, mineralogy, welding, nondestructive testing, quality assurance methods, and the application of improved industrial standards. See the chapter on "Industrial and Economic Benefits of Latin American Nuclear Cooperation" in *Averting a Latin American Nuclear Arms Race*, Leventhal and Tanzer, eds., pp. 76–144.

[62]"Import of Nuclear-Related Material Difficult," in Joint Publications Research Service (JPRS), Sept. 12, 1991, p. 15. Despite the restrictions imposed by most suppliers, Argentina and Brazil both acquired some nuclear-related technology by using clandestine means and offering false assurances on intended end uses.

scientists, and businessmen seeking improved access to foreign high technology for economic development.[63]

Supplier restrictions imposed because of Argentina and Brazil's nuclear stance also hindered their development goals in other industrial sectors. U.S. technology denied for nonproliferation reasons slowed the development of Brazil's Satellite Launching Vehicle and increased its costs. For example, Colonel Antonio Carlos Pedrosa, former director of Brazil's Space Activities Institute, estimated in November 1988 that U.S. restrictions on technology sales increased the costs of the satellite launcher program from $6 million to $14.4 million.[64]

Although some advanced nations were willing to export nuclear-related technology to Argentina and Brazil despite their refusal to renounce a nuclear weapons option, these states became less willing to do so over time. The best example is Germany, which signed the 1975 nuclear deal with Brazil. After years of not requiring its nuclear technology customers to accept full-scope safeguards as a condition of supply, Germany announced a change to this policy at the 1990 NPT review conference. The new German policy was not to initiate any new nuclear cooperation agreements with states that did not accept full-scope IAEA safeguards and to require this acceptance from all states with which it did cooperate by 1995.[65]

International Factors Influencing the Argentine and Brazilian Rollback Decisions

Argentina and Brazil were motivated to renounce their respective quests for a nuclear weapons option by the changing nature of their bilateral relationship and by the mutual realization that continued ambiguity regarding their nuclear ambitions would entail political and economic costs. In short, political elites in both states concluded that the benefits to remaining outside the nuclear nonproliferation regime did not outweigh the costs. This shift in attitude occurred over a period of approximately 15 years and was influenced by many factors, the most salient of which have been discussed here.

International and U.S. nonproliferation policy played a minor role in influencing the shift in Argentina and Brazil from pursuit of a nuclear weapons option to full membership in the nuclear nonproliferation regime. This was done through policies that directly and indirectly buttressed domestic political forces within Argentina and Brazil that favored rejection of nuclear weapons and highlighted the potential benefits of joining the international nonproliferation regime. Some of these external influences on the rollback decisions are discussed in the following sections.

Increasing International Pressure

A general strengthening of the international nonproliferation regime corresponding with the end of the Cold War and the resulting political pressures also could have contributed to nuclear rollback in Argentina and Brazil. Reductions in U.S.-Soviet nuclear arsenals and discussions between Presidents Ronald Reagan and Mikhail Gorbachev embracing the goal of eliminating nuclear arms represented progress by the superpowers at that time toward meeting their obligation under Article VI of the NPT.[66] This weakened traditional anti-NPT arguments, made frequently by Argentina and Brazil, against an imbalance of obligations and behavior between nuclear weapons states and nonnuclear weapons states.

---

[63]Redick, *Nuclear Illusions,* p. 43.

[64]"U.S. Policy Hinders Satellite Program," in *FBIS Latin America*, November 9, 1988, pp. 2–3.

[65]Reiss, *Bridled Ambition*, p. 63.

[66]Article VI of the NPT obligates the nuclear weapons states to seek the elimination of nuclear arms. U.S.-Soviet nuclear arms reductions included those resulting from the Intermediate-Range Nuclear Forces Treaty (INF), START I, and the Presidential Nuclear Initiatives of 1991–1992.

In addition, the French, Chinese, and Ukrainian decisions in the 1980s and early 1990s to adhere to the NPT strengthened the nonproliferation regime and isolated countries like Argentina and Brazil. This increased political isolation came at the same time when both Argentina's and Brazil's profiles on the international scene were in decline. In the 1970s their close ties with the advanced industrial world and their economic growth established them as leaders of the developing world. Their independent stance on nonproliferation issues and the global economic order also made them strong members of the nonaligned movement. These relationships attenuated the negative political consequences of remaining outside the NPT. The 1980s economic decline and, in the case of Argentina, the conflict with the U.K. in the Falklands resulted in a lowered international stature for Buenos Aires and Brasilia. The end of the Cold War and the collapse of the Soviet Union undercut the ideological basis for the nonaligned movement and severely weakened that organization. Under these new circumstances Argentina and Brazil were left without alternative political relationships that limited the effects of international nonproliferation pressures. Therefore, the effect of these pressures intensified in the late 1980s and became a factor that contributed to the process of nuclear rollback.

Export Controls

Other nations first blocked exports in the 1950s to prevent the acquisition of enrichment or reprocessing technology by Argentina and Brazil. For example, U.S. and British officials intervened in 1954 in Antwerp and other European ports to prevent the transfer of ultracentrifuges that had been acquired in West Germany by a Brazilian admiral.[67] After the NPT came into force in the early 1970s, U.S. export controls were designed to encourage Argentina and Brazil to accept safeguards on all their nuclear activities. When they refused, U.S. export controls had the additional objective of denying technology or materials that could be used to expand unsafeguarded nuclear activities in these two states.

The United States refused uranium enrichment technology requested by Brazil in connection with the Angra I nuclear power station purchased from Westinghouse in 1971. The decision not to transfer this technology was made because Brazil had not joined the NPT. Subsequent U.S. efforts to convince Germany and the Netherlands not to transfer enrichment or reprocessing technology to Brazil as part of the 1975 deal were unsuccessful. However, U.S. diplomacy did succeed in persuading the Germans and Dutch to require stronger bilateral and trilateral safeguards (with the IAEA) on declared Brazilian nuclear activities.[68]

With the passage of the U.S. Nuclear Nonproliferation Act of 1978 (NNPA) the United States banned the export of enriched-uranium fuel to countries refusing full-scope safeguards. This meant that after 1980 it would renege on its 1973 contract to supply Brazil's Angra I reactor with LEU fuel. In 1981 Brazil contracted with the West German/British/Dutch enrichment consortium URENCO as an alternative supplier of fuel for Angra I. The NPPA also blocked the previously agreed supply of U.S. LEU for Argentina's research reactor. However, the Carter Administration approved the transfer of this fuel to Argentina in June 1980, in part due to Argentine threats to buy LEU from the Soviet Union.[69]

After the 1975 Brazilian/German deal and implementation of the NNPA, a key dimension of U.S. strategy to control nuclear exports to Argentina and Brazil was its leadership in creation of the Nuclear Suppliers Group and its lobbying within that organization for a harmonization of export control policy among the leading nuclear suppliers. It was within this forum that the United States began to pressure Germany and others to demand full-scope

[67]Redick, *Nuclear Illusions*, p. 6, and Barletta, "Military Nuclear Program in Brazil," p. 5. The centrifuges were ultimately delivered to a Brazilian university research center.

[68]David Binder, "U.S. Wins Safeguards in German Nuclear Deal with Brazil," *The New York Times*, June 4, 1975, p. A16.

[69]Redick, *Nuclear Illusions*, p. 5.

safeguards as a condition of supply.[70] In fact, the United States did convince West Germany in 1979 to require full-scope safeguards for a proposed sale of a power reactor and heavy water plant to Argentina. However, Bonn withdrew its requirement for full-scope safeguards after it decided to supply a reactor only.[71] Despite these exceptions, in the late 1980s it became clear that Germany, France, and the United Kingdom were in the process of joining the United States and Canada in requiring full-scope IAEA safeguards as a condition of supply.[72]

West Germany announced at the NPT Review Conference in August 1990 that it would require full-scope safeguards as a condition for future nuclear exports. Existing nuclear deals would have to be renegotiated by 1995 to conform to this new policy. This shift in German policy had implications for both the Argentine and Brazilian nuclear programs but would have the greatest impact on nuclear activities in Brazil. At the time of the announcement, Brazil's Angra II power station, which was over 80% complete, and Angra III, which had not yet been canceled, required German technology to complete. Moreover, in October 1993, when German Foreign Minister Klaus Kinkel visited Brazil to encourage its ratification of the Quadripartite safeguards agreement, Brazil was negotiating with Germany for an additional $750 million in financial support to complete the Angra II reactor and had plans to discuss future financing for the Angra III plant.[73] As discussed earlier, Germany's decision to condition future exports on the acceptance of full-scope safeguards provided strong incentives for Brazil and Argentina to ratify and implement the Quadripartite Agreement.

In summary, international nuclear and dual-use export controls imposed on Argentina and Brazil ultimately slowed the completion of nuclear projects and raised their costs. This outcome increased the domestic constraints faced by proponents of the unsafeguarded nuclear programs in both countries and made them more difficult to justify in light of other national priorities.

Support for Economic Liberalization

Over time the promotion of economic liberalization policies and the links between investment, technology transfer, and nonproliferation strengthened constituencies within Argentina and Brazil that supported nuclear safeguards and opposed the development of a nuclear weapons option.[74] The members of these constituencies included business interests, officials from the Foreign and Finance Ministries, research scientists, and university officials who had frequent contact with international financial and commercial interests. These groups embraced liberal strategies of economic development through free trade and international investment as a central national objective and the primary means to improve their domestic economies. This objective conflicted with the state-financed maintenance of unsafeguarded nuclear activities and the economic penalties imposed as a consequence by members of the international nonproliferation regime. For example, to relieve its debt crisis in 1982 and 1983, Brazil was persuaded to implement economic liberalization measures advocated by the United States and the International Monetary Fund. The measures included reductions

---

[70]Leslie H. Gelb, "Nuclear Nations to Tighten Export Controls," *The New York Times*, July 6, 1984, p. A1.

[71]Switzerland supplied the heavy water plant, also without requiring full-scope safeguards. See "Canada, FRG Dispute Safety Standard on Reactor sale to Argentina," in *FBIS/NDP*, Dec. 10, 1979, p. 18. Also see John M. Geddes, "Swiss, Germans Ignore U.S. Objections, Sell Nuclear Technology to Argentina," *Wall Street Journal*, June 16, 1980.

[72]Reiss, *Bridled Ambition*, p. 70.

[73]Reiss, *Bridled Ambition*, p. 63.

[74]Policies of economic liberalization generally include a reduction of state control over markets and barriers to trade, an expansion of private economic transactions and foreign investment, and the privatization of public sector enterprises. See Etel Solingen, "The Political Economy of Nuclear Restraint," *International Security* 19 (Fall 1994): p. 137.

in state-financed infrastructure development projects, including a 40% cut in the Nuclebras budget in 1983.[75]

Advocacy of and assistance to Argentina and Brazil for economic liberalization policies had its greatest effect on nuclear decision making in the late 1980s and early 1990s. In 1990, both Presidents Menem of Argentina and Collor of Brazil implemented radical economic liberalization policies designed to reduce inflation, balance state budgets, privatize public services, attract foreign investment, and renegotiate foreign debt payment.[76] In Argentina, President Menem placed the nuclear program under the control of the director of planning. This office, with advice from large Argentine corporations and joint ventures, coordinated the privatization of some nuclear activities and the closing of sensitive nuclear facilities.[77] In Brazil, the Foreign and Economic Ministries successfully lobbied the House of Deputies to approve the Quadripartite safeguards agreement in September 1993.[78] At the end of 1990 and just days before President George H. W. Bush was to visit Argentina and Brazil to promote the idea of a hemispheric free-trade zone, the two states had signed the Joint Declaration of a Common Nuclear Policy at Foz de Iguazu that committed them to accepting international safeguards on all nuclear activities.[79] A strong incentive for acceptance of IAEA safeguards was the expectation that such an agreement would facilitate access to advanced technology and create a favorable climate for foreign investment and economic assistance.[80]

Maintaining a Constructive Nonproliferation Dialogue

Succeeding U.S. administrations have worked to maintain a constructive nonproliferation dialogue with Argentina and Brazil despite political acrimony over the NPT, the Nuclear Suppliers Group, and the constraints of the NNPA. Even at one of the lowest points of U.S.-Brazilian relations following Washington's failed attempt to convince Germany not to transfer reprocessing or enrichment technology as part of its 1975 deal with Brasilia, Vice President Walter Mondale stated publicly that there were no major obstacles to "excellent" relations between the two countries.[81] The Reagan Administration, in a similar attempt to avoid a potential impasse in nuclear relations, permitted the retransfer to Argentina from Germany of 143 tons of U.S.-origin heavy water in 1982 for use in a safeguarded Argentine reactor.[82] In March 1988, Ambassador-at-Large Richard Kennedy visited Argentina to discuss nuclear matters and reach agreements on exchanging information on nuclear safety issues and increasing Argentina's participation in a U.S. program to develop reduced enriched uranium fuels for research and test reactors.[83] Ambassador Kennedy continued the dialogue with

---

[75]David Myers, "Brazil," in *Limiting Nuclear Proliferation*, Jed Snyder and Samuel Wells, Jr., eds. (Washington, D.C.: The Wilson Center, 1985), p. 135; and Scott Tollefson, "Nuclear Restraint: Argentina and Brazil," Paper Presentation, Atlanta, Georgia: International Security Studies Section of the International Studies Association Annual Meeting, Nov. 2, 1996, pp. 35–36.

[76]Etel Solingen, "Macropolitical Consensus and Lateral Autonomy in Industrial Policy: The Nuclear Sector in Brazil and Argentina," *International Organization* 42 (1993): pp. 263–298.

[77]JPRS, Aug. 21, 1991, p. 5.

[78]Solingen, "The Political Economy of Nuclear Restraint," p. 162.

[79]Shirley Christian, "Argentina and Brazil Renounce Atomic Weapons," *The New York Times*, Nov. 29, 1990.

[80]For a statement of this expectation, see U.N. Conference on Disarmament, CD/PV.610, Feb. 6, 1992.

[81]Charles A. Kraus, "Mondale Shuns Atom Dispute in Visit to Brazil," *The Washington Post*, March 23, 1979, cited in William H. Courtney, "Brazil and Argentina: Strategies for American Diplomacy," in *Nonproliferation and U.S. Foreign Policy*, Joseph A Yager, ed. (Washington, D.C.: The Brookings Institution, 1980), p. 384.

[82]Milton Benjamin, "U.S. to Allow Argentine Nuclear Aid," *The Washington Post*, Aug. 18, 1983.

[83]Richard Kessler, "Kennedy Visit Prompts Hopes of U.S.-Argentine Nuclear Thaw," *Nucleonics Week*, March 17, 1988, p. 7.

Argentina and Brazil during the George H. W. Bush Administration, traveling to Buenos Aries in August 1989 to advocate the acceptance of IAEA safeguards and to propose U.S.-Argentine joint development of inherently safe reactor designs.[84]

By providing a constant reminder of the bargain that was available to them, this dialogue may have contributed to Argentina and Brazil's ultimate decisions to roll back the ambiguity surrounding their nuclear programs and accept international inspections. In general, the United States used this dialogue to propose cooperative activities that would benefit Argentina and Brazil and to communicate the specific changes in their ambiguous nuclear postures that were required to initiate them.[85] It also helped encourage the shift toward Argentine-Brazilian nuclear transparency. This was particularly true during the period 1988–1994, when the United States used this dialogue to influence the evolution of the landmark agreements that were signed in 1990 and 1991 and to promote their ratification. Beginning in late 1988 and 1989, the United States began broadening its nonproliferation dialogue with Argentina and Brazil to include potential access to technology such as supercomputers, nuclear safety equipment, environmental monitoring techniques, and satellites in exchange for further nonproliferation commitments.[86] The result was a series of agreements with the United States that provided greater access to technology for Argentina and Brazil, removed export restrictions previously imposed upon them, and allowed them greater participation in international export control regimes. These events proceeded in parallel with Argentine and Brazilian acceptance of international inspections of their nuclear facilities and greater political commitments against nuclear proliferation.[87]

Technical Assistance for Nuclear Safeguards

U.S. and international technical assistance for nuclear safeguards in Argentina and Brazil helped define the technical mechanisms for bilateral nuclear cooperation and thus contributed indirectly to the decision to eventually accept full-scope international safeguards. In addition, the willingness of the United States and others to increase safeguards assistance has been essential to the successful implementation of the Quadripartite Agreement. The prospect of acquiring advanced safeguards techniques such as nuclear materials measurement

---

[84]Richard Kessler, "Kennedy Discourages Argentine Idea for Non-IAEA Safeguards," *Nucleonics Week*, Aug. 31, 1989, pp. 12–13.

[85]This was the basic goal of a nonproliferation strategy of "constructive engagement" as practiced by Ambassador Kennedy and others. See Spector, *Nuclear Proliferation Today*, p. 216.

A key part of this strategy, especially during the Reagan and Bush administrations, was to renew and expand military assistance to Argentina and Brazil. This policy may have been instrumental in easing regional security concerns, thus undercutting the military rationale for nuclear weapons. It also demonstrated U.S. willingness to improve relations with these two states across a broad front. See Peter Clausen, *Nonproliferation and the National Interest: America's Response to the Spread of Nuclear Weapons* (New York: Harper-Collins, 1993), pp. 176–177.

[86]For example, the Bush Administration supported the sale of supercomputers to Brazil in 1990. However, to overcome the opposition of the U.S. Congress to the sale, Brazil sent two government ministers, including José Goldemberg, who directed Brazil's nuclear program as Minister of Science and Technology, to reassure Washington regarding Brazil's nonproliferation commitments. The sale was ultimately approved. See Albright, "Brazil Comes in From the Cold," p. 16.

[87]In August 1991 NASA signed an agreement with the Argentine Space Research Commission for the joint development of Argentina's first satellite. For this and other examples of the give and take that occurred according to the nonproliferation bargain discussed above, see James S. Tomashoff and Lewis A. Dunn, "Latin America Nonproliferation Game Plan," prepared for the Volpe National Transportation Support Center by Science Applications International Corp., Jan. 4, 1995, pp. A1–A9.

equipment, remote monitoring, and environmental monitoring systems that can have other industrial or scientific applications also supplied some motivation for expanding the bilateral safeguards proposal to allow for IAEA inspections as well.[88]

Since 1976 Brazilian officials have participated in IAEA safeguards courses taught in the United States at National Laboratories.[89] The United States has supported the safeguards agreements that Argentina and Brazil have had with the IAEA since the 1960s and 1970s which covered their declared nuclear facilities even when neither state was a party to the NPT. These agreements derived from both states' nuclear cooperation agreements with the United States and other nuclear suppliers such as Germany, Canada, and Switzerland.[90] Even though this participation did not involve the nuclear facilities of most concern to the U.S. at the time, it did introduce Argentine and Brazilian officials to nuclear safeguards techniques that would eventually be adopted in their concept for a comprehensive safeguards regime.

For example, beginning in the late 1980s, U.S. safeguards experts began suggesting that a possible model for the evolution of an Argentine-Brazilian nuclear safeguards system was provided by the European Atomic Energy Community (Euratom).[91] This approach was eventually taken with the formation of ABACC, which allowed the integration of the Argentine-Brazilian bilateral safeguards into the IAEA system.[92] In this way U.S. technical support for safeguards in South America provided practical examples of mechanisms that could be used to provide mutual confidence that nuclear facilities were not being used for the development of nuclear arms. This international technical support also helped demonstrate to Argentina and Brazil that IAEA safeguards could be implemented without being so intrusive or disruptive that they negatively impacted the operation of the nuclear facilities.

After the 1991 formation of ABACC, the U.S. Department of Energy began direct assistance for the design of a safeguards regime for Argentina's gaseous diffusion enrichment plant at Pilcaniyeu.[93] An agreement between the U.S. Department of Energy and ABACC was signed on April 14, 1994. This framework agreement established the legal basis for a wide array of cooperative ventures between DOE (and the U.S. national laboratories) and ABACC to enhance ABACC's competence to administer nuclear safeguards at facilities in Brazil and Argentina. Washington arranged for U.S. experts to lecture at ABACC safeguards courses for regional inspectors, supported training for ABACC at U.S. national laboratories, and has provided ABACC with equipment for nondestructive analysis (NDA) of nuclear materials.[94] In addition, U.S. voluntary funding of the IAEA's Program of Technical Assistance to

---

[88]*ABACC News*, Brazilian-Argentine Agency for Accounting and Control of Nuclear Materials, Jan./April 1996, p. 4.

[89]J. Rundo, *Technical Assistance to Brazil: A Summary Prepared for the International Nuclear Technology Liaison Office*, Division of Educational Programs, Argonne National Laboratory, Feb. 1988, p. 8.

[90]Marco A. Marzo, Alfredo L. Biaggio, and Ana C. Raffo, "Nuclear Co-operation in South America: The Brazilian-Argentine Common System of Safeguards," *IAEA Bulletin*, March 1994, p. 30.

[91]See William A. Higinbotham and Helen M. Hunt, "Some Examples of Multilateral Safeguards Agreements," in *Averting a Latin American Nuclear Arms Race*, Leventhal and Tanzer; also see Redick, ed., *Argentina and Brazil*, p. 13.

[92]Jose Goldemberg and Harold A. Feiveson, "Denuclearization in Argentina and Brazil," *Arms Control Today*, March 1994, pp. 10–14.

[93]Tom Zamora Collina and Fernando de Souza Barros, *Transplanting Brazil and Argentina's Success*, ISIS Report, vol. 2, no. 2 (Washington, D.C.: Institute for Science and International Security, Feb. 1995), p. 6.

[94]For a description of nuclear materials measurement technologies, see Chapters 3A, 3B, and 3C.

Safeguards (POTAS) supports the purchase of advanced safeguards monitoring equipment by ABACC.[95]

The role played by international technical safeguards support has clearly been more significant in the period following the signing of the December 1991 Quadripartite Safeguards agreement. The cooperative nature of this assistance has been important in easing the concerns held by some who initially opposed the acceptance of international safeguards, fearing that they would be too intrusive or expensive.[96] It has also been instrumental in vindicating those who argued that accepting international safeguards would have tangible benefits in terms of increased scientific exchange and access to technology. As discussed previously, the belief that this bargain was in the interests of Argentina and Brazil was a primary motivating force behind the nuclear rapprochement. In this way technical safeguards assistance helped facilitate Argentine-Brazilian integration into the international nuclear safeguards regime during a time when that process was still vulnerable to domestic opposition.

Unofficial Contacts (Track II Diplomacy)

Another element of nonproliferation policy that appears to have had some influence in the case of Argentine-Brazilian nuclear rollback is governmental support for unofficial contacts by nongovernmental organizations (NGOs) that promoted nonproliferation objectives in the region. Within the global community of NGOs and parts of the U.S. government, such contacts are referred to as *Track II diplomacy*.[97] In the case of Argentina and Brazil, U.S. NGOs provided both technical and organizational assistance to groups favoring greater civilian controls over nuclear programs, including the application of technical safeguards at nuclear facilities.

One example of this policy dimension was the collaboration between the Commission for Nuclear Questions of the Brazilian Physical Society (BPS) and the Non-Proliferation Project of the Federation of American Scientists (FAS).[98] Beginning in 1988 the FAS gave technical advice to the BPS commission on nuclear safeguards, noting in particular how uranium enrichment plants designed to produce 20% enriched uranium could be modified

---

[95]In 1993, the President's Section 601 Report to Congress noted that " ... DOE is assisting Argentina, Brazil and the ... (ABACC) in developing nuclear safeguards capabilities in preparation for the application of full-scope safeguards. Specific activities include preparation of inspector and operator training programs, joint development of safeguards systems for uranium enrichment plants and participation in nondestructive assay technology development and evaluation programs and in chemical assay intercomparison experiments."

[96]*Gazeta M ercantil,* in JPRS-TND-91-014 (Sept. 12, 1991), p. 11, and *Gazeta Mercantil,* in JPRS-TND-93-024 (July 27, 1993), p. 14.

[97]In some of these activities there is a high degree of cooperation and coordination between NGOs and government officials. Government officials usually must approve the participation of U.S. National Laboratory personnel in such activities and they are often invited to amend or approve the content of presentations given by such nongovernmental officials. Sometimes government officials from participating states attend Track II activities in an unofficial capacity. This coordination is usually intended to ensure that U.S. government positions are properly represented in these unofficial forums. Sometimes such forums are used intentionally by government officials to raise potential policy initiatives on a trial basis. U.S. government coordination of and influence over Track II activities is strongest in cases where the U.S. government has provided financial support for the activity. There is a growing consensus that such contacts have advantages for advancing a nonproliferation agenda, especially when official government-to-government contacts are deadlocked for political reasons.

[98]David Albright and William Higinbotham, "FAS Nonproliferation Experts Provide Technical Assistance to Brazilian Physical Society," *F.A.S. Public Interest Report*, vol. 42, no. 2 (Feb. 1989), pp. 3–4.

so that they could produce uranium of much higher enrichments. At the time, the BPS was advocating an inspection system controlled by the Brazilian Congress for the unsafeguarded nuclear facilities in Brazil. FAS representatives conducted a series of workshops in December 1988 at the University of Rio de Janeiro on nuclear safeguards and government oversight of nuclear programs. These workshops were attended by a Brazilian Congressman and resulted in an initiative to bring Brazilian officials to the U.S. to learn about U.S. Congressional oversight of civil and military nuclear programs. Some FAS staff then traveled to Argentina to meet with members of the Argentine Physics Association who had also formed a committee to their increase civilian oversight of nuclear activities in their country.

A second important Track II activity occurred in October 1989 in Montevideo, Uruguay. This was a conference organized by the Nuclear Control Institute of the United States and financed by the Ford Foundation. Three former directors of the Argentine CNEA and many important Argentine and Brazilian officials, including José Goldemberg, who would later oversee Brazil's nuclear program as Minister of Science and Technology, attended the conference. The American participants included Nuclear Control Institute staff, former government officials, industry representatives, academics, and U.S. National Laboratory personnel. Topics discussed included bilateral and international safeguards, the economic advantages of nuclear cooperation, and the international nonproliferation regime.[99]

The direct influence of these and similar activities on Argentine and Brazilian decision making was limited and difficult to specify. However, it is likely that these exchanges had a positive influence on the evolution of Argentine and Brazilian thinking on the mechanics of joining the nonproliferation regime. The contacts between FAS and BPS, for example, produced a proposal for an independent Brazilian nuclear inspection organization modeled partially on the U.S. General Accounting Office and the Nuclear Regulatory Commission.[100] Unofficial contacts also suggested to Argentine and Brazilian officials that safeguards procedures already developed through the multilateral Hexapartite Safeguards Project could provide a model for safeguards at Brazil's sensitive Aramar enrichment plant. This collaborative project designed safeguards for gas centrifuge enrichment plants in Europe and Japan that would not reveal commercial secrets to IAEA inspectors.[101] A safeguards system similar to this type was eventually applied to the Aramar facility.

In addition, the Track II process allowed the transfer of information on safeguards and bureaucratic mechanisms for oversight of nuclear activities to Argentine and Brazilian officials in a manner that did not weaken them politically. Thus officials who eventually supported the acceptance of IAEA safeguards safely gained information important to advancing their agenda against domestic opposition. The information gained by these individuals through Track II activities in 1988 and 1989 enabled them to more rapidly implement a safeguards system after the decisions to do so were taken in 1990 and 1991.

Summary

Argentina and Brazil were primarily motivated to accept international safeguards on all their nuclear activities by the mutual realization that continued ambiguity regarding their nuclear intentions would hinder the achievement of national goals such as modernization and technological development.[102] The causes of this change in thinking were primarily but

---

[99]Leventhal and Tanzer, *Averting a Latin American Nuclear Arms Race*, pp. 68–74.

[100]David Albright, "Brazil Comes in From the Cold," p. 15.

[101]Ibid. The United States, West Germany, Britain, the Netherlands, Japan, Australia, the IAEA, and Euratom participated in the Hexapartite Safeguards Project.

[102]For example, in justifying ratification of the Quadripartite Safeguards Agreement, Mario Cesar Flores, former secretary for strategic affairs in Brazil, said, "[E]ither Brazil modernizes and its internationalizes or it loses its place in the world and in history." See José Goldemberg and Harold A. Feiveson, "Denuclearization in Argentina and Brazil," *Arms Control Today*, March 1994, p. 14.

not exclusively domestic. A sharp improvement in bilateral relations, the transition to more transparent, democratic governments in both states, and the adoption of liberalizing economic policies are probably the three factors that had the greatest influence on changes in nuclear policy.

Other, more internationalist factors also came into play, however. Concern for image and the role that both countries portrayed in the region and the world influenced the nuclear policies, as did the desire to avoid export controls that would hinder economic and technological growth. The existence of the Euratom regional model for a nuclear safeguards system and the technical and administrative assistance offered by outside nations facilitated the creation of ABACC and its eventual integration with the IAEA safeguards system. Finally, sustained diplomatic efforts on the part of several states, including the United States, to encourage membership for Argentina and Brazil in the NPT as nonnuclear weapons states, played a minor role in the outcome. In summarizing the influence of international factors on nuclear rollback in Argentina and Brazil, three factors stand out: the maintenance of a nonproliferation dialogue, the use of export controls, and, related to the latter, the use of economic leverage provided by advanced technology and other economic incentives. Support for the nonproliferation regime was a fourth aspect of international policy that was instrumental.

It has been 15 years since Argentina and Brazil joined the international nonproliferation regime, and their international behavior and the operation of their nuclear programs has been consistent with their nonproliferation obligations. Exchange of technical information on nuclear energy and nuclear safeguards has continued and intensified between the two nations and with other advanced nuclear states such as the United States, Japan, and the European Union. Both states are strong advocates of nonproliferation norms and are vocal critics of the acquisition of nuclear weapons by additional states and the failure of the nuclear weapons states to eliminate their arsenals.

The conditions that led to the renunciation of nuclear arms in Argentina and Brazil do not exist in other states such as North Korea, Iran, India, Pakistan, or Israel. In some cases the security concerns in these states are too great to give up nuclear weapons or the option to acquire them. In addition, domestic and regional political conditions make nuclear weapons appear vital to state interests. Nevertheless, these factors change over time and many of the dynamics that influenced nuclear nonproliferation in South America may eventually play a role elsewhere. Certainly the international community is making serious and sustained efforts to change the perceived balance of incentives and disincentives for nuclear arms in North Korea and Iran and is employing many of the political and economic strategies that have influenced nuclear proliferation behavior in other regions.

The cases of Argentina and Brazil do provide insights that can inform nonproliferation efforts directed at other states and regions. This is because all states face some of the same challenges when striving to create and sustain a nuclear arsenal. The particular set of issues and conditions, especially the unique characteristics of the relevant national leaderships, must be taken into account, even though some of the fundamental dynamics are universal.

Nonproliferation and international safeguards efforts in South America must be maintained. As both states acquire more advanced nuclear infrastructures, additional measures will be required to maintain confidence that nuclear technologies are not being used for military purposes. For example, Brazil is constructing industrial-scale uranium enrichment facilities that could be operational by 2010. This plant uses proprietary technology, and Brazil denied full access to IAEA inspectors for several years, claiming that industrial secrets had to be protected until the details of access could be negotiated. As of early 2007 this issue seemed to have been resolved to the satisfaction of the IAEA and regional states. Future challenges of this type can be expected and will require the further evolution of ABACC and its relationship to the IAEA.

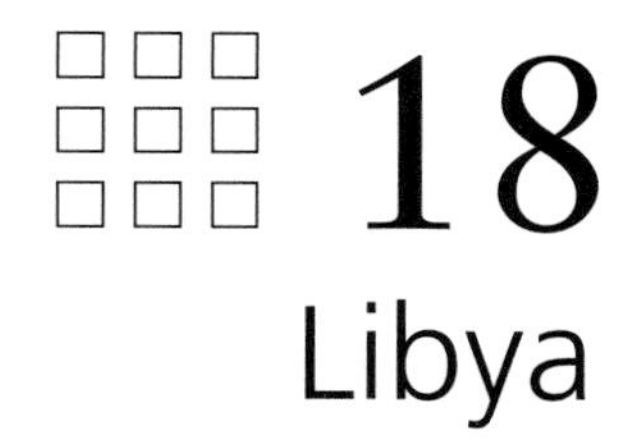

18 Libya

Wyn Bowen

Introduction

In December 2003 the Libyan regime of Col. Muammar Qadhafi publicly announced that it had opted to give up its nuclear weapons program following secret negotiations with the U.K. and U.S. governments. In doing so, Libya committed itself to dismantling the program in a transparent fashion, including multilateral verification by inspectors from the International Atomic Energy Agency (IAEA).[1] The country's pursuit of a nuclear weapons capability had been widely suspected since the 1970s despite the regime's status as a nonnuclear weapons state under the Nuclear Nonproliferation Treaty (NPT), which Libya ratified in 1975.

This chapter examines Libya's pursuit and subsequent abandonment of the quest to develop a nuclear weapons capability. In exploring the origins and evolution of any nuclear weapons program, it is essential to consider both the underlying political intent and the technical efforts implemented to develop the capability itself. Although these two elements should not be seen as mutually exclusive, they are considered here in turn. The first section of the chapter includes an assessment of the likely motives that underlay Libya's nuclear aspirations from 1969, when Qadhafi came to power, through the end of 2003. The second and third sections examine Libya's efforts to acquire the requisite capability for manufacturing nuclear weapons from 1969 through late 2003. The fourth section examines the subsequent cancellation of the program, including a discussion of the possible motivating factors that prompted Libya's nuclear reconsideration. The final section looks at the political and technical actions taken to foreswear this option, to create confidence that the program had indeed been dismantled, including a discussion of the role of the United States, the United Kingdom, and the IAEA.

Nuclear Intentions, 1969–2003

It is important to note from the outset that the Libyan regime adopted a perennially ambiguous position on the nuclear question. On one hand, Qadhafi frequently declared his country's peaceful intentions in the nuclear field. In 1979, for example, he stated that his country had "signed all agreements on the nonproliferation of nuclear weapons" and wanted to reduce its oil dependence by finding "alternative sources of energy including atomic sources," and that Libya was a victim "of the story that we want to build an atom bomb," which he described as "not true."[2] On the other hand, Qadhafi also publicly espoused the pursuit of nuclear

[1]"Libyan WMD: Tripoli's statement in full," BBC News Online, http://news.bbc.co.uk/2/hi/africa/3336139.stm (Dec. 20, 2003).

[2]Youssef M. Ibrahim, "Libya and the World: Interview with Colonel Qaddafi," Dec. 10, 1979, *Survival*, 22/2 (March/April 1980), pp. 80–82.

weapons, primarily in the context of an "Arab bomb." In 1976, for example, he said, "We will have our share of this new weapon."[3] Several years later, in 1987, he said, "The Arabs must possess the atomic bomb to defend themselves until their numbers reach one billion, until they learn to desalinate seawater, and until they liberate Palestine."[4]

Regime Security and Survival

Despite or because of this ambiguity, it was widely suspected that Libya became interested in acquiring a nuclear weapons capability in the early 1970s, and it appears that this desire intensified over the next two decades. The security and survival of the Qadhafi regime itself were evidently of pivotal importance to Libya's nuclear pursuits, and to a lesser extent but still important, it appears that Qadhafi also used the nuclear issue for political propaganda, to further his position and influence within the Arab and Muslim worlds. This was reflected most notably in his frequent rhetoric for the development and acquisition of an Arab bomb to offset the strategic imbalance with Israel.

After taking power, Col. Qadhafi's principal interest became maintaining the security of his regime in the face of internal and external threats and challenges. Externally, the focus was on deterring interference in Libya's affairs by neighbors and states from further afield, such as the United States and Israel. During the 1970s and 1980s the need to deter intervention was particularly important, given the regime's radical approach to external relations, which encompassed attempts to destabilize neighboring states, the sponsorship of international terrorism, anti-Israel rhetoric and activities, and the pursuit of weapons of mass destruction (WMD). It is within this context that Libya's nuclear program must be viewed, because the inherent deterrent value of these weapons meant that their acquisition, or at least the creation of the perception that the regime was seeking to acquire them, appeared to constitute a major part of Qadhafi's approach to deterrence. Despite the acquisition of sophisticated conventional weapons systems from the Soviet Union and France, Libya's shortcomings in this area, based primarily on its limited manpower resources, meant that seeking nuclear and other WMD, notably chemical weapons, was the only serious option if the regime's goal was to deter external interference.

More specifically, Israel and the United States featured prominently in Libya's nuclear calculations. On top of the Qadhafi regime's support for radical Palestinian groups, Israel became concerned during the 1970s about reports that Libya wanted to obtain a nuclear capability.[5] The response was to make it clear to Tripoli that the Israel Defense Force (IDF) could hit targets across the Arab world, including Libya.[6] The Israeli Air Force's destruction of the Osiraq research reactor in Iraq in 1981 and its targeting of the Palestinian Liberation Organization (PLO) headquarters near Tunis in 1985, when its aircraft crossed through Libyan airspace, demonstrated that the Israelis possessed the resolve to undertake offensive military action if a situation was perceived to warrant it.[7]

Obviously, Israel's nuclear weapons and delivery systems were also seen to pose a threat to Libya. For example, the Libyan foreign minister, together with the Syrian and

---

[3]Craig R. Black, *Deterring Libya: The Strategic Culture of Muammar Qadhafi*, Counterproliferation Paper no. 8, USAF Counterproliferation Centre, Air War College, Air University, Maxwell Air Force Base, Alabama, October 2000, p. 6, www.au.af.mil/au/awc/awcgate/awc-cps.htm (April 21, 2005).

[4]Leonard S. Spector and Jacqueline R. Smith, *Nuclear Ambitions: The Spread of Nuclear Weapons, 1989–1990* (Boulder: Westview Press, 1990), pp. 175–185. See also Frank Barnaby, *The Invisible Bomb: The Nuclear Arms Race in the Middle East* (London: I. B. Taurus & Co. Ltd, 1989), p. 150, and Craig R. Black, *Deterring Libya*, p. 6.

[5]Jacob Abadi, "Pragmatism and Rhetoric in Libya's Policy Toward Israel," *The Journal of Conflict Studies*, 20/1 (Fall 2000), p. 93.

[6]*Ibid.* p. 93.

[7]*Ibid.* p. 87.

Iranian foreign ministers, stated in 1985 that their countries would seek nuclear weapons to counter the Israeli nuclear threat.[8] Two years later Qadhafi stated, "The Arabs have the right to manufacture nuclear weapons and to acquire the atomic bomb to defend their existence. After all, their enemy possesses this weapon, and atomic bombs are now found in the Middle East."[9] The Israeli angle persisted into the early years of the 21st century. For example, the Libyan leader said in 2002, "We demanded the dismantling of the weapons of mass destruction that the Israelis have; we must continue to demand that. Otherwise, the Arabs will have the right to possess that weapon."[10]

Growing antagonism between Libya and the United States in the 1980s following the arrival of the Reagan Administration in office, which was based on Tripoli's involvement in international terrorism and its hostile actions against U.S. citizens and interests, also factored Washington into the regime's nuclear calculations. For example, in a 1990 reference to the American attack on Libya in April 1986, a response to Libya's terrorist activities, Qadhafi noted, "If we possessed a deterrent—missiles that could reach New York—we would have hit it at the same moment. Consequently, we should build this force so that they and others will no longer think about an attack. Whether regarding Libya or the Arab homeland, in the coming twenty years this revolution should achieve a unified Arab nation [...] This should be one homeland, the whole if it, possessing missiles and even nuclear bombs. Regarding reciprocal treatment, the world has a nuclear bomb, we should have a nuclear bomb."[11]

Regime security and survival were central, therefore, to understanding the rationale behind Libya's nuclear weapons program. However, the Qadhafi regime's nuclear rhetoric, notably its ambiguity over peaceful versus military intent, contributed significantly to the reluctance of most countries to supply Libya with nuclear technology and assistance. Doubts about Libya's true intentions ultimately put a major brake on what the country was capable of achieving in both the legitimate and clandestine nuclear fields. At one level this involved nuclear-exporting countries exercising increased self-restraint in their dealings with Libya, given the regime's opaque intentions in the nuclear field. At another level it also involved the frequent application of pressure from Washington on other countries, notably in Western Europe, to exercise such restraint. Interestingly, as shall be demonstrated later, regime security was also central to understanding Libya's decision to forego its nuclear program in 2003.

Nuclear Capability: 1970s and 1980s

Although the main focus of Libya's efforts in the nuclear weapons field was on acquiring from abroad the technology and know-how to develop its own program, the Qadhafi regime did try during the 1970s to completely cut the corner by purchasing a full-up weapons capability off the shelf. This effort reportedly involved approaches to several countries, including China,[12]

---

[8]Mari Peled, *Ha'aretz* (Hebrew), Sept. 9, 1985, pp. 61–63 (via the Nuclear Threat Initiative Nuclear Database, www.nti.org).

[9]Leonard S. Spector and Jacqueline R. Smith, *Nuclear Ambitions*, pp. 175–185.

[10]"Libya, Syria, Cuba Need Scrutiny for Weapons Programs, U.S. Says," Address by Under-Secretary of State for Arms Control and International Security John R. Bolton to the Heritage Foundation, Washington, D.C., May 6, 2002, International Information Programs, U.S. Department of State, http://usinfo.state.gov/regional/nea/text/0506.htm (Sept. 3, 2004).

[11]Leonard S. Spector and Jacqueline R. Smith, *Nuclear Ambitions*, pp. 175–185.

[12]John Wright, *Libya: A Modern History* (London and Canberra: Croom Helm, 1983), pp. 201–219; John K. Cooley, *Libya Sandstorm* (London: Sidgwick & Jackson, 1983), pp. 229–239; Lewis A. Dunn, *Controlling the Bomb* (New Haven and London: Yale University Press, 1982), pp. 14–15, 30–31, 50–51; Mohamed Heikal, *The Road to Ramadan* (London: Collins, 1975), pp. 76–77; Anthony Cordesman, *Weapons of Mass Destruction in the Middle East* (London: Brassey's UK, A RUSI Study, 1991), pp. 151–153.

France,[13] India,[14] and the Soviet Union.[15] Not surprisingly, these endeavors failed to produce results. Efforts to develop a domestic program were also unsuccessful in the end despite the investment of not insignificant time and resources and Libya's conduct of activities in contravention of its Safeguards Agreement, which the IAEA failed to detect at the time. In examining how the program evolved, it makes sense to initially consider the period prior to the involvement of the A. Q. Khan proliferation network during the latter half of the 1990s.

In the 1970s and 1980s, Libya sought to acquire from overseas the foundations of a "civil" nuclear program, including uranium exploration, conversion and enrichment capabilities, research and power reactors, and a plutonium reprocessing/separation capability. During this time companies and governments in numerous countries were approached in the quest for relevant technology, infrastructure, and technical expertise. These countries included, but were not confined to, Argentina, Belgium, Brazil, Egypt, France, India, and Pakistan, although most of Libya's approaches did not result in significant material gains. The one real exception, however, was the Soviet Union, which provided Libya with some significant assistance from the late 1970s into the 1980s. The Soviet Union supplied Libya with the 10 MW research reactor (the IRT-1) at the Tajoura Nuclear Research Centre (TNRC), which it also helped to construct. The reactor came on-line in 1981. The TNRC subsequently became the focal point of Libya's covert work on plutonium separation, uranium conversion, and enrichment during the 1980s. The IAEA does not appear to have detected any safeguards violations at this time. Moreover, it is not known whether the Agency had specific suspicions or unanswered questions, given the regime's ambiguity on the nuclear issue.

Uranium

Libya's failure to discover much in the way of domestic and exploitable deposits of uranium resulted in the Qadhafi regime seeking source material from beyond the country's borders. Notably, it involved obtaining 2,263 tonnes of yellowcake from two producers in Niger between 1978 and 1981; the total imported uranium amounted to 1,587 tonnes in 6,367 containers.[16] This quantity could potentially have formed the basis for the production of enough enriched uranium for tens of nuclear weapons, if not more. Libya actually received 587 of these tonnes before its Safeguards Agreement with the IAEA entered force in July 1980,[17] which meant that it did not have to declare this acquisition. There has been speculation in open sources that Libya retransferred some of the uranium to Pakistan, potentially as much as 450 tonnes, prior to 1982.[18] Indeed, it was widely reported in the 1980s that, in exchange for uranium from Niger for Pakistan's clandestine enrichment program and potentially financial assistance for its nuclear weapons program, Libya hoped to be provided with technology and assistance related to weapons development, particularly in the enrichment

[13]Shyam Bhatia, *Nuclear Rivals in the Middle East* (London and New York: Routledge, 1988), pp. 64–71.

[14]Clyde R. Mark, *CRS Issue Brief for Congress: Libya*, Congressional Research Service, The Library of Congress, Washington, D.C., April 23, 2002, p. 4.

[15]Shai Feldman, *Nuclear Weapons and Arms Control in the Middle East* (Cambridge, MA: MIT Press, 1997), pp. 63–65.

[16]Director General, International Atomic Energy Agency, *Implementation of the NPT Safeguards Agreement of the Socialist People's Libyan Arab Jamahiriya*, May 28, 2004, Annex 1, p. 2.

[17]Director General, Implementation of the NPT Safeguards Agreement, May 28, 2004, p. 2.

[18]Anthony Cordesman, *Weapons of Mass Destruction in the Middle East*, pp. 151–153; Frank Barnaby, *The Invisible Bomb*, p. 104; Federation of American Scientists, "WMD Around the World: Libya"; Leonard S. Spector and Jacqueline R. Smith, *Nuclear Ambitions*, pp. 175–185; W. P. S. Sidhu, "Pakistan's Bomb: A Quest for Credibility," *Jane's Intelligence Review*, June 1996, p. 278.

and reprocessing areas.[19] Since Libya gave up the nuclear option in late 2003, no further information has been forthcoming on the Pakistani relationship during this period as a result of IAEA investigations; the only Pakistan link highlighted has been related to the A. Q. Khan network from the mid-1990s onward.

Plutonium

Beyond the acquisition of yellowcake, Libya conducted experiments at the TNRC in the 1980s to separate plutonium from uranium targets irradiated in the IRT-1 reactor. These illicit experiments were hidden from the IAEA and constituted part of the regime's exploration of the plutonium route to nuclear weapons. Between 1984 and 1990, 38 of several dozen small uranium oxide and uranium metal targets (on a gram scale) that Libya had fabricated were irradiated in the IRT-1. Each target contained around 1 g of uranium, and radioisotopes were extracted using ion exchange or solvent extraction methods in hot cells at the radiochemical laboratory next to the reactor. According to IAEA reporting in 2004, very small amounts of plutonium were separated from "at least two of the irradiated targets."[20] This evidently constituted a safeguards violation, although it was not picked up by the IAEA at the time. The regime tried to acquire separation technology from Argentina in the 1980s, and a Libyan request for a small-scale "hot-cell" facility was declined by the Argentineans under U.S. pressure.[21]

It is unclear whether a specific decision was taken by the regime at that time to seek a nuclear weapons capability as opposed to simply developing some of the underlying technical wherewithal to provide options down the line. Indeed, the Libyans made only limited achievements in the field of plutonium separation, and they came nowhere near to producing the quantity that would have been required for a weapon. Moreover, the size of the IRT-1 reactor meant that this would not have been a large enough source for a weapons program. Although the Libyans did try, during the 1980s, to purchase a 440 MWe power reactor from the Soviet Union that would reportedly have been of sufficient scale to produce 70 kg of plutonium a year, negotiations in this area ultimately came to nothing[22] and the project did not develop beyond the feasibility and design development stage.[23] Moscow appears to have backed away from the power reactor negotiations in part because of proliferation concerns, although other factors may have included anxiety about Libya's ability to pay the

---

[19]Shyam Bhatia, *Nuclear Rivals in the Middle East*, pp. 64–71; Leonard S. Spector and Jacqueline R. Smith, *Nuclear Ambitions*, pp. 175–185; "WMD Around the World: Libya," Federation of American Scientists, www.fas.org/nuke/guide/libya/index.htm; Library of Congress, *Libya: A Country Study*; John K. Cooley, *Libya Sandstorm*, pp. 229–239; Anthony Cordesman, *Weapons of Mass Destruction in the Middle East*, pp. 151–153.

[20]Director General, International Atomic Energy Agency, *Implementation of the NPT Safeguards Agreement of the Socialist People's Libyan Arab Jamahiriya*, Feb. 20, 2004, p. 6.

[21]"Annex 8: Nuclear Infrastructures of Argentina and Brazil," *Nuclear Technologies and Non-Proliferation Policies*, Issue 2, 2001, http://npc.sarov.ru/english/digest/digest_2_2001.html.

[22]Henry S. Rowen and Richard Brody, "Nuclear Potential and Possible Contingencies," in Joseph A. Yager, ed., *Nonproliferation and U.S. Foreign Policy* (Washington, D.C.: The Brookings Institution, 1980), p. 208.

[23]Project Number LIB/9/005, Siting of Nuclear Power Plant, first approved 1983, completed Dec. 23, 1985. The project involved Libya's National Scientific Research Council. See Department of Technical Cooperation, International Atomic Energy Agency, www-tc.iaea.org/tcweb/projectinfo/default.asp; Project Number LIB/4/005, Nuclear Power Plant, first approved 1984, cancelled June 30, 1986. The project involved the National Scientific Research Council, Department of Power. See Department of Technical Cooperation, International Atomic Energy Agency, www-tc.iaea.org/tcweb/projectinfo/default.asp.

projected US$4 billion cost[24] and, potentially, Libyan concerns about the safety standard of Soviet technology;[25] the Chernobyl accident had of course occurred in April 1986. Libya also sought technical assistance from Belgium, Bulgaria, and Yugoslavia for its ultimately aborted power reactor project.[26] In the final analysis it was primarily concerns about Libya's motives that halted this particular project.

Conversion and Enrichment

During the 1980s Libya conducted undeclared work related to uranium conversion and enrichment. Between 1982 and 1992 the Libyans worked at the TNRC in conjunction with a German engineer on a gas centrifuge design that he had procured.[27] No uranium was ever enriched, however, and the Libyans did not introduce any UF6 for testing purposes, despite a centrifuge reportedly operating at one point. According to IAEA reports in 2004 that documented Libya's past safeguards violations, centrifuge components from this initial period included a "small number of unfinished, maraging steel cylinders" with diameters the same as "the more advanced L-2 centrifuges" given to Libya in 2000 (discussed later), although their origin remains unknown.[28] Libya also succeeded in procuring centrifuge-related technology during this period, including the reported acquisition of a specialized furnace from Japan in 1985 and vacuum pumps from Europe.[29] A "foreign expert" also supplied two mass spectrometers in the early 1980s.[30] With Argentina Libya may also have discussed, but without any results, the acquisition of enrichment-related technologies.[31]

The Libyans did realize more success in the conversion field, including experiments at the TNRC and the acquisition of a modular uranium conversion facility (UCF). Undeclared but small-scale experiments were performed at the TNRC during the period 1983–1989 using 34–39 kg of feed material taken from the imported yellowcake stored at Sabha. The experiments were designed to develop expertise in the dissolution of yellowcake, the purification of uranium

---

[24]"Libya Abandons Plans for First Unit," *Nuclear Engineering International*, April 1986, p. 6; "Moscow Retreats from Libyan Nuclear Scheme," *MidEast Markets*, Aug. 3, 1987; *Nuclear Engineering International*, Dec. 1987, cited in "Other Proliferation Developments," *PPNN Newsbrief*, no.1, March 1988, p. 2, www.ppnn.soton.ac.uk/nb01.pdf; K. D. Kapur, *Soviet Nuclear Non-Proliferation Diplomacy and the Third World* (New Delhi: Konark Publishers PVT Ltd., 1993), p. 148; "Atomstroyexport: Background Brief," JSC Atomstroyexport, Russia, www.atomstroyexport.ru/eng/history.

[25]Ann MacLachlan and Mike Knapik, "Belgium and Libya Will Sign an Agreement on Nuclear Cooperation," *Nucleonics Week*, vol. 25, no. 21, May 24, 1984, p. 5; Federation of American Scientists, "WMD Around the World: Libya"; Frank Barnaby, *The Invisible Bomb,* pp. 98–99.

[26]Office of Technology Assessment, Technology Transfer to the Middle East, p. 380; Ann MacLachlan and Mike Knapik, *Belgium and Libya,* p. 5; Federation of American Scientists, "WMD Around the World: Libya"; Leonard S. Spector and Jacqueline R. Smith, *Nuclear Ambitions,* pp. 175–185; "Soviets Draw Back From Helping Libyan Program," *Nuclear Engineering International*, Dec. 1987, p. 27; Frank Barnaby, *The Invisible Bomb,* pp. 98–99.

[27]Director General, IAEA, *Implementation of the NPT Safeguards Agreement*, May 28, 2004, Annex 1, p. 5; Douglas Frantz and Josh Meyer, "For Sale: Nuclear Expertise," *Los Angeles Times*, Feb. 22, 2004.

[28]Director General, IAEA, *Implementation of the NPT Safeguards Agreement*, May 28, 2004, Annex 1, pp. 5–6.

[29]Frantz and Meyer, "For Sale."

[30]Director General, IAEA, *Implementation of the NPT Safeguards Agreement*, May 28, 2004, Annex 1, pp. 5–6.

[31]"Annex 8: Nuclear Infrastructures of Argentina and Brazil," *Nuclear Technologies and Non-Proliferation Policies,* Issue 2, 2001.

solutions, and the manufacture of uranium tetrafluoride (UF4) and uranium metal. The Libyans succeeded in manufacturing uranyl nitrate, uranium dioxide (UO2) and uranium trioxide (UO3), UF4, and uranium metal.[32] The modular UCF was procured from "a Far Eastern country" thought to be Japan, although upon receipt in 1986 it subsequently remained in storage and unpacked until 1998,[33] and it was not capable of producing UF6. However, the equipment arrived without assembly, operating instructions,[34] or a fluorination module. The Libyans also benefited from working alongside colleagues from a nuclear weapons state, presumed to be the U.S.S.R., from 1983 to 1986. Moreover, scientists from Libya studied fluorine chemistry during the mid-1980s "in an East European country,"[35] possibly Yugoslavia.

The Libyans were also turned down on several fronts in terms of acquiring conversion-related technology and expertise. The IAEA did not fulfil a request from the Libyans in the early 1980s for assistance in uranium fluoride production.[36] Moreover, in the early to mid-1980s a West European company, Belgonucleaire, opted not to provide Libya with a pilot conversion plant at Sabha and related laboratories at the TNRC after the United States applied pressure on the government in Belgium to stop any negotiations.[37] During the same period, negotiations also failed with a nuclear weapons state, assumed to be the Soviet Union, regarding the acquisition of a UCF capable of producing 120 tonnes of natural UF6 per year. Nevertheless, 100 kg of yellowcake was shipped to this country by Libya in 1985, "in connection with the possible construction in Libya of a uranium conversion facility," and approximately 39 kg of UF6, 6 kg of U3O8 (yellowcake), 6 kg of UO2, and 5 kg of UF4 (all masses refer to the uranium content) were subsequently sent back in February 1985. The IAEA has since verified a 2004 Libyan claim that these compounds were never actually used as sample materials for conversion work.[38]

During the 1980s, then, Libya actively explored the plutonium- and uranium enrichment-based routes to acquiring fissile material. Some limited progress was made, but major achievements were not forthcoming due to the unwillingness of most nuclear suppliers to trade in nuclear technology with Libya, primarily because of proliferation concerns. Indeed, the 1970s and 1980s were typified by the country's reliance on acquiring technology and assistance from abroad, a situation that persisted into the 1990s and beyond.

Nuclear Capability: The 1990s and Beyond

The Libyan nuclear weapons program was reportedly reinvigorated in the mid-1990s and the A. Q. Khan proliferation network played a pivotal role in supplying technology primarily related to centrifuge enrichment as well as weapon designs and manufacturing instructions. As Ambassador Donald Mahley, the senior U.S. WMD representative in Libya in early 2004, has noted, without this network's support the nuclear threat Libya posed would have been significantly constrained, "if not thwarted altogether."[39] Unlike the previous illicit activities,

---

[32]Director General, IAEA, *Implementation of the NPT Safeguards Agreement*, May 28, 2004, pp. 3–4.

[33]*Ibid.* pp. 4–5; "Japanese Parts Used in Libya's Nuke Program," *Herald Asahi*, March 13, 2004.

[34]*Ibid.* pp. 4–5; "Japanese Parts Used in Libya's Nuke Program," *Herald Asahi*, March 13, 2004.

[35]*Ibid.* pp. 3–4.

[36]*Ibid.* pp. 3–4.

[37]*Ibid.* p. 4; Spector, *Going Nuclear*, p. 157; Spector and Smith, *Nuclear Ambitions*, pp. 175–185; Cordesman, *Weapons of Mass Destruction in the Middle East*, pp. 151–153.

[38]Director General, IAEA, *Implementation of the NPT Safeguards Agreement*, May 28, 2004, pp. 3–4.

[39]Ambassador Donald Mahley, "Dismantling Libyan Weapons: Lessons Learned," *The Arena*, no. 10, Nov. 2004, p. 5.

which were conducted primarily at the TNRC, the country's renewed efforts in the enrichment field occurred at locations away from Tajoura for concealment purposes.[40]

Centrifuge Enrichment

Once Libyan intelligence initiated contact with A. Q. Khan in 1997, the Libyans began to receive centrifuges, UF6, weapon designs, training opportunities abroad, specialized equipment such as flow-forming machines, and sensitive materials like maraging steel, all via the network. In the process it appears that Libya could have spent well over US$100 million and potentially up to US$500 million.[41] The network was notable for its complex and transnational nature wherein centrifuges were designed in one country, the components were manufactured in a second country, and shipments occurred via a third country, with delivery to a fourth country on a "turnkey" basis, with no clarity about the end user.[42] (See the chapter on the A. Q. Khan proliferation network for more on this topic.) This meant that once the Libyans had placed orders with Pakistani contacts, these individuals approached middlemen, who then approached suppliers, who did not know the true end user, to manufacture the requisite items before transferring them on to Libya.[43] In the process the network capitalized on weak export control systems in countries such as the United Arab Emirates (UAE) and Malaysia.[44]

The activities of the Malaysian company Scomi Precision Engineering (SCOPE) are perhaps the most notable in relation to the network's dealings with Libya. A. Q. Khan's business associate, B. S. A. Tahir, established contact with Scomi in early 2001 and subsequently negotiated a deal that December for the production of thousands of 14 types of centrifuge components. SCOPE had been established as a subsidiary of the Malaysian company Scomi to work on the contract, although it was later cleared of knowingly participating in proliferation-related work.[45] The SCOPE workshop was equipped with modern milling, turning, and tooling machines procured from France, Japan, Taiwan, and the United Kingdom,[46] and the network reportedly obtained approximately 300 tonnes of aluminium tubes through a Singapore-based subsidiary of a German company; the tubes were machined at SCOPE and sent off to Libya in four batches via a trading company in Dubai, UAE, during the period

---

[40]"Unclassified Report to Congress on the Acquisition of Technology Relating to Weapons of Mass Destruction and Advanced Conventional Munitions: 1 January Through 30 June 2001," www.cia.gov/cia/reports/721_reports/jan_jun2001.htm; statement by Spector, Deputy Director, Center for Nonproliferation Studies, Monterey Institute of International Studies, before the Subcommittee on International Security, Proliferation, and Federal Services of the U.S. Senate Committee on Governmental Affairs, June 6, 2002 www.senate.gov/~govt-aff/060602specter.pdf.

[41]Bill Gertz, "Libyan sincerity on arms in doubt," *Washington Times*, Sept. 9, 2004; Royal Malaysia Police, "Press Release by Inspector General of Police in Relation to Investigation on the Alleged Production of Components for Libya's Uranium Enrichment Program," Feb. 20, 2004, www.rmp.gov.my/rmp03/040220scomi_eng.htm; "Libyan sincerity on arms in doubt," *Washington Times*, Sept. 9, 2004.

[42]Ian Traynor in, "Nuclear chief tells of black market in bomb equipment," *Guardian*, Jan. 26, 2004, p. 14; Anwar Iqbal, "Khan network supplied N-parts made in Europe, Southeast Asia," *Dawn* (online edition), Oct. 14, 2004, www.dawn.com/2004/10/14/top9.htm.

[43]"A. Q. Khan & Libya," GlobalSecurity.Org.

[44]Stephen Fidler and Mark Huband, "Turks and South Africans helped Libya's secret nuclear arms project," *Financial Times*, June 10, 2004, p. 11.

[45]Raymond Bonner, "Did Tenet Exaggerate Malaysian Plant's Demise?" *New York Times*, Feb. 9, 2004, p. 4; *Agence France-Presse*, "Libyan nuclear workers trained in Malaysia: official," May 29, 2004, via Channel News Asia, Singapore.

[46]Bonner, "Did Tenet Exaggerate Malaysian Plant's Demise?" p. 4.

December 2002 through August 2003.[47] The UAE was evidently exploited for transhipment purposes because of its status as a major re-export center.[48] It should be noted that Turkish workshops also manufactured components for Libya's centrifuge program, including motors and the frequency converters. In this instance, subcomponents were acquired from Europe and other places for assembling in Turkey and then shipping to Libya via the UAE using false end-user paperwork.[49] Training opportunities were also arranged in Malaysia on the operation of quality-control machines,[50] in the UAE,[51] in Spain on operating a precision lathe,[52] and reportedly in another African country, possibly South Africa, to examine a complete set of supporting equipment for 10,000 centrifuges ordered from the network.[53]

Despite such acquisitions and assistance, however, Libya's progress was significantly constrained by problems involving planning and personnel, and the network itself was not the most reliable source of high-quality nuclear merchandise. Nevertheless, a quick run through the achievements during this final period demonstrates the depth of the proliferation problem posed by the A. Q. Khan network.

Libya obtained P-1 and the more advanced P-2 types of centrifuge via the network.[54] Work was conducted on P-1 centrifuges, which incorporate aluminium rotors, at a facility in Al Hashan on the edge of Tripoli. The machines, received directly from Pakistan, were used to train Libyan personnel. In all, 20 complete P-1 centrifuges, probably retired from Pakistan's enrichment program, were procured, along with majority of the parts for a further 200 machines, with the exception of magnets and aluminium rotors, which were obtained from another part of the network. Libya also received frequency converters and systems needed for process gas feeding and withdrawal.[55] The initial test of a single preassembled P-1 machine had taken place by October 2000[56] and two successful high-speed tests of P-1 centrifuges were later conducted in the period May–December 2002, although the Libyans have stated that uranium was not introduced.[57]

---

[47]*Ibid.* p. 4; Bill Gertz, "Libyan sincerity on arms in doubt," *Washington Times*, Sept. 9, 2004

[48]"UAE Economic Overview and Guide to Doing Business," U.K. Trade and Investment, https://www.uktradeinvest.gov.uk.

[49]Anwar Iqbal, "Khan network supplied N-parts made in Europe, Southeast Asia"; Fidler and Huband, "Turks and South Africans helped Libya's secret nuclear arms project," *Financial Times*, June 10, 2004, p. 11.

[50]*Agence France-Presse*, "Libyan nuclear workers trained in Malaysia: official," May 29, 2004, via Channel News Asia, Singapore.

[51]Bonner, "The Two Faces of Nuclear Suspect in Malaysia," *International Herald Tribune*, Feb. 20, 2004, p. 1.

[52]Elizabeth Nash, "Spanish Firms in Secret Arms Trade to Libya," *Independent*, Feb. 12, 2004.

[53]Director General, IAEA, *Implementation of the NPT Safeguards Agreement*, May 28, 2004, Annex 1, p. 6; Fidler and Huband, "Turks and South Africans helped Libya's secret nuclear arms project," p. 11.

[54]David Albright and Corey Hinderstein, "Libya's Gas Centrifuge Procurement: Much Remains Undiscovered," Institute for Science and International Security, March 1, 2004, www.isis-online.org/publications/libya/cent_procure.html; Sammy Salama and Lydia Hansell, "Companies Reported to Have Sold or Attempted to Sell Libya Gas Centrifuge Components, Issue Brief," NTI Database, March 2005, www.nti.org/e_research/e3_60a.html.

[55]Director General, IAEA, *Implementation of the NPT Safeguards Agreement*, Feb. 20, 2004, p. 5; Iqbal, "Khan network supplied N-parts made in Europe, Southeast Asia."

[56]Director General, IAEA, *Implementation of the NPT Safeguards Agreement*, Feb. 20, 2004, p. 5; Iqbal, "Khan network supplied N-parts made in Europe, Southeast Asia."

[57]Director General, IAEA, *Implementation of the NPT Safeguards Agreement*, May 28, 2004, Annex 1, p. 5.

By the end of 2000 several centrifuge cascades had started to be installed at Al Hashan. This work progressed to the point at which, by April 2002, a nine-machine cascade had reportedly been set up under vacuum.[58] A vacuum is required to prevent "air friction on the rotor which would cause convection currents" but also vibration, thereby allowing the centrifuge to successfully separate uranium isotopes.[59] Indeed, a great deal can be learned about centrifuges by simply operating them in a vacuum and without introducing UF6. Information acquired in this way can relate to "the life expectancy and durability of key mechanical components, the failure of materials, the effects of vibrations, electric power requirements [...] a detailed understanding of the different ways that centrifuges can fail, and information needed for the development of more advanced centrifuge systems."[60]

By April 2002, a 19-machine cascade had also been set up with 10 rotors installed, although not under vacuum, and a 64-machine cascade with process equipment was ready to be installed.[61] For security purposes, however, all the machines and equipment were removed from Al Hashan in the spring of 2002 and moved to Al Fallah.[62] It is unclear what this security issue involved, but it might have reflected a growing concern about being discovered at a time when the Bush Administration was becoming increasingly vocal and bellicose vis-à-vis the perceived threat posed by Iraq's weapons programs in the run-up to the 2003 war to unseat Saddam Hussein. Subsequent environmental testing in 2004 performed by the IAEA at Al Hashan revealed low-enriched uranium (LEU) and highly enriched uranium (HEU) contamination on the floor of the P-1 test area, on centrifuge and crashed rotor parts, on feed and takeoff systems, and on a mass spectrometer used during the tests. The IAEA subsequently determined that at least one P-1 casing had previously been in Pakistani service until 1987 and that the contamination in Libya was similar to that found at the place of origin.[63]

Libya also received from Pakistan, in September 2000, two complete P-2 centrifuges incorporating rotors made of maraging steel.[64] Early deliveries of components for an additional 10,000 P-2 machines also started to arrive in Libya in late 2002 from other parts of the network.[65] Investigations performed by the IAEA in 2004 subsequently found the two machines and some of the P-2 components to be contaminated with HEU, again probably because of previous applications in the Pakistani enrichment program.[66] Nevertheless, the two centrifuges were reportedly not in a workable state.[67] Moreover, although a large number of P-2 parts had been procured by late 2003, none of the rotating components was

---

[58]Director General, IAEA, *Implementation of the NPT Safeguards Agreement*, Feb. 20, 2004, p. 5.

[59]"Centrifuge process to separate the isotopes of uranium hexafluoride," U.S. Patent issued on May 23, 1995; www.patentstorm.us/patents/5417944-description.html.

[60]This assessment was taken from a report by the IAEA in relation to Iran's centrifuge program and cited in an article by David Albright and Jacqueline Shire, "Better Carrots, Not Centrifuges: Why Iran Must Halt Enrichment and How the U.S. Can Make It Happen," Institute for Science and International Security (ISIS), July 10, 2006, www.isis-online.org/publications/iran/iranissuebrief.pdf.

[61]Director General, IAEA, *Implementation of the NPT Safeguards Agreement*, Feb. 20, 2004, p. 5.

[62]*Ibid.* p. 5.

[63]Director General, IAEA, *Implementation of the NPT Safeguards Agreement*, May 28, 2004, Annex 1, p. 5.

[64]Director General, IAEA, *Implementation of the NPT Safeguards Agreement*, May 28, 2004, Annex 1, pp. 5–6.

[65]Director General, IAEA, *Implementation of the NPT Safeguards Agreement*, May 28, 2004, Annex 1, pp. 5–6.

[66]*Ibid.* Annex 1, pp. 5–6.

[67]Albright, President and Founder, Institute for Science and Security, "International Smuggling Networks: Weapons of Mass Destruction Counterproliferation Initiatives," statement before the U.S. Senate Committee on Governmental Affairs, June 23, 2004; www.senate.gov/~govt-aff/index.cfm?Fuseaction=Hearings.Testimony&HearingID=185&WitnessID=673.

included. The components that had been delivered were later found by IAEA inspectors in January 2004 but in boxes that had been unopened, seemingly validating the Libyan regime's claim that no P-2 centrifuges had been put together or tested.[68] Significantly, many of the P-2 components acquired via the network were scratched and so could not have been used successfully in an operating centrifuge.

Beyond importing centrifuges and associated components, Libya also went about acquiring a domestic manufacturing capability in this area, with the primary focus in a machine shop at Janzour (referred to as Project 1001).[69] The A. Q. Khan network was reportedly approached to set up and obtain equipment for the workshop, including lathes and flow formers.[70] One news report claimed that the workshop was designed for the replacement of broken imported centrifuges and, possibly, to expand the total number of machines available.[71] The IAEA was told in 2004 that a large amount of high-strength aluminium as well as maraging steel and been acquired for the workshop.[72]

Conversion and UF6

Libya did not succeed in producing UF6 during the final stages of its enrichment program, although in 1998 the UCF modules procured in the 1980s were taken to a suburb of Tripoli (Al Khalla), assembled, and subsequently cold tested in early 2002 without uranium feedstock. The modules were moved once again later that year to another suburb of Tripoli (Salah Eddin) for "reasons of security and secrecy."[73] Despite this failure to produce UF6 indigenously, Libya did procure such material via the A. Q. Khan network, asking for some 20 tonnes but receiving only two small cylinders with natural and depleted uranium in September 2000 and a larger one with 1.7 tonnes of LEU (enriched to around 1% U235) in February 2001.[74] There has been speculation about the source, and it has been suggested that North Korea could have been the originating point, with the material transferred by a company in Pakistan via the UAE to Libya, although no evidence exists in open sources that Pyongyang knew about the final delivery point.[75] The North Korean theory is based on tests performed on the material at Oak Ridge National Laboratory, which determined the origin "with near certainty" based on the elimination of other potential sources. If correct, this finding would confirm North Korea's possession of an enrichment program.[76]

---

[68]Director General, IAEA, *Implementation of the NPT Safeguards Agreement*, May 28, 2004, Annex 1, pp. 5–6.

[69]Director General, IAEA, *Implementation of the NPT Safeguards Agreement*, Feb. 20, 2004, pp. 6–7.

[70]Joby Warrick and Peter Slevin, "Probe of Libya Finds Nuclear Black Market," *Washington Post*, Jan. 24, 2004; Broad, Sanger, and Bonner, "A Tale of Nuclear Proliferation: How Pakistani Built His Network," *New York Times*, Feb. 12, 2004, p. A1; Owen Bowcott, John Aglionby, and Traynor, "Atomic Secrets: Businessman Under Scrutiny 25 Years Ago After Ordering Unusual Supplies," *Guardian*, March 5, 2004.

[71]Iqbal, "Khan network supplied N-parts made in Europe, Southeast Asia."

[72]Director General, IAEA, *Implementation of the NPT Safeguards Agreement*, Feb. 20, 2004, pp. 6–7.

[73]Director General, IAEA, *Implementation of the NPT Safeguards Agreement*, May 28, 2004, pp. 4–5.

[74]Director General, IAEA, *Implementation of the NPT Safeguards Agreement*, Feb. 20, 2004, p. 4; Director General, IAEA, *Implementation of the NPT Safeguards Agreement*, May 28, 2004, Annex 1, p. 3.

[75]Dafna Linzer, "US Misled Allies About Nuclear Export North Korea Sent Material To Pakistan, Not to Libya," *Washington Post*, 20 March 2005, p. A1.

[76]Sanger and Broad, "Tests Said to Tie Deal on Uranium to North Korea," *New York Times*, February 2, 2005; see also Traynor, "North Korean nuclear trade exposed," *Guardian*, May 24, 2004, p. 12.

Weapons Designs

On the weaponization front, Libya received nuclear weapons design and fabrication documentation in late 2001 and early 2002. The documents were stored at the National Bureau of Scientific Research (NBSR) and included assembly drawings and manufacturing instructions for the detonator explosives and fissile material. The electronics and firing sets, however, were not covered in the documentation.[77] The documentation was reportedly related to a 10KT implosion device of a late 1960s Chinese origin for a warhead weighing around 453kg and which had previously been given to Pakistan.[78] However, the A. Q. Khan network reportedly did not supply one of the key part drawings. Moreover, the Libyans have since said that the utility of the information was not tested in any way.[79] Indeed, in 2004 IAEA inspectors did not identify any specific facilities in Libya involved in designing, producing, or testing weapons components.[80]

Despite all the procurement activity involving the A. Q. Khan network, Libya failed to make significant progress in its weapons program by late 2003, and there were several reasons for this. First, the program does not appear to have been managed in a coherent fashion and it also suffered from a lack of political direction and continuity. Although in 2004 the IAEA identified the NBSR as the organization in charge of the nuclear weapons effort in Libya,[81] very little else by way of a responsible bureaucracy for the program has since been identified. Moreover, IAEA investigations in Libya in 2004 did not uncover any specific facilities involved in activities related to designing, manufacturing, or testing nuclear weapons components.[82] This lack of evidence would seem to support the Qadhafi regime's claim that it had not even evaluated the weapons design information that it received via the A. Q. Khan network.[83]

Second, most countries were generally unwilling to export sensitive technology and assistance to Libya because of concerns about the regime's intentions. Third, and perhaps most important, the program suffered from a lack of local expertise in key areas such as centrifuges, and this stemmed from Libya's underdeveloped scientific, technological, and educational systems. According to the Libyan government, 800 nuclear specialists in all were involved in the program and 140 possessed advanced degrees, with some personnel educated in the United States and Europe.[84] However, as Mahley noted from his work in Libya during 2004, while he dealt with "knowledgeable, dedicated and innovative" individuals, he did so with the same people all the time as a result of what he termed the Libyans possessing "almost no bench."[85] Fourth, the A. Q. Khan network did not turn out to be an overly reliable partner in that it did not provide the 20 tonnes of UF6 that Libya requested, the two L-2

---

[77]Broad and Sanger, "Khan was selling a complete package," *International Herald Tribune*, March 22, 2005; Director General, IAEA, *Implementation of the NPT Safeguards Agreement*, May 28, 2004, p. 7.

[78]Andrew Koch, "The nuclear network: Khanfessions of a proliferator," *Jane's Defence Weekly*, 24 February 2004. "Chinese Warhead Drawings Among Libyan Documents," *Los Angeles Times*, February 16, 2004; Broad and Sanger, "Khan was selling a complete package"; "Libya Was Far From Building Nuclear Bomb," *Wall Street Journal*, February 23, 2004; "Libya nuke prints from China," *Associated Press*, February 15, 2004; Warrick and Slevin, "Libyan Arms Designs Traced Back to China," *Washington Post*, February 15, 2004, p. A1.

[79]Broad and Sanger, "Warhead Blueprints Link Libya Project to Pakistan Figure," *New York Times*, February 3, 2004; Frantz and Meyer, "For Sale."

[80]Director General, IAEA, *Implementation of the NPT Safeguards Agreement*, May 28, 2004, p. 3.

[81]Director General, IAEA, *Implementation of the NPT Safeguards Agreement*, May 28, 2004, p. 7.

[82]Director General, IAEA, *Implementation of the NPT Safeguards Agreement*, May 28, 2004, p. 3.

[83]Broad and Sanger, "Warhead Blueprints Link Libya Project to Pakistan Figure," *New York Times*, Feb. 3, 2004; Frantz and Meyer, "For Sale."

[84]Richard Stone, "Agencies Plan Exchange With Libya's Former Weaponeers," *Science*, 308/5719 (April 8, 2005), pp. 185–6.

[85]Mahley, "Dismantling Libyan Weapons," p. 7.

centrifuges were not in an operable condition, key parts were omitted from the 10,000 L-2 machines Libya ordered, some of the L-2 components delivered were unusable due to being damaged, and important information was not provided for the weapons design.

Giving Up the Pursuit of Nuclear Weapons

Libya's decision to give up its nuclear weapons program in late 2003 reflected the regime's calculation that doing so was in its best interests. In this respect the decision reflected the interaction of changes in the regime's domestic and international environments over the preceding decade or so.

Economic and Political Pressures

The U.S. embargo imposed in the mid-1980s for Libya's involvement in international terrorism, notably including attacks against U.S. citizens in Western Europe,[86] followed by the imposition of U.N. sanctions in 1992–93 over the Lockerbie incident, had targeted the Qadhafi regime's dependence on oil revenues—for example, by targeting the importation of oil industry equipment—to pay for his country's inflated public sector. The embargos also served to put off foreign companies from investing in the Libyan economy, notably the oil sector. Prior to the 1990s the regime had capitalized on oil revenues to secure and sustain popular support at home.[87] This was approached primarily through maintaining a significant welfare and educational system as well as employment in the public sector. All Libyan citizens were entitled to housing, healthcare, and utilities.[88]

Reduced oil prices in the 1980s and 1990s and the impact of the embargos significantly constrained the regime's oil sector revenues, resulting in reduced state revenues and spending. Sanctions severely curtailed Libya's exploration for oil and the expansion of the oil sector, leading to a situation in which oil production in the 1970s was twice that of 2003.[89] Moreover, whereas Libya had a gross domestic product roughly equal to that of the UAE in 1982, this decreased to around one seventh of the UAE GDP in 2003, in relative terms.[90] One outcome was that pay in the Libyan public sector remained effectively static between 1982 and 2003 and completely out of step with inflation.[91] Moreover, unemployment grew

---

[86]For example, in December 1985, five Americans and 15 others were killed in what the Reagan Administration later asserted were Libyan-backed terrorist attacks on Rome and Vienna airports. In April 1986, the administration later determined that evidence also implicated Libya in the bombing of LaBelle nightclub in West Berlin. This attack killed two Americans and one other person and injured 60 Americans plus 140 others.

[87]Oye Ogunbadejo, "Qaddafi's North African Design," *International Security*, 8/1 (Summer 1983), p. 155.

[88]Alison Pargeter, "Libya: All change for no change," *The World Today*, Aug.–Sept. 2000, pp. 29–31.

[89]See Stephen D. Collins, "Dissuading State Support of Terrorism: Strikes or Sanctions? An Analysis of Dissuasion Measures Employed Against Libya," *Studies in Conflict and Terrorism*, vol. 27, no. 1, Jan.–Feb. 2004, p. 11; Ray Takeyh, "Qadhafi and the Challenge of Militant Islam," *Washington Quarterly*, vol. 21, no. 3, Summer 1998, p. 163; Neil Ford, "Libya Springs a Surprise," *Middle East*, Feb. 2004, pp. 32–35; "Libya," *Country Analysis Briefs*, Energy Information Administration, U.S .Department of Energy, Jan. 2004, www.eia.doe.gov/emeu/cabs/libya.html; "Beating Swords into Oil Shares," *Economist*, Dec. 30, 2003.

[90]"Libya WMD Pledge: Result of Iraq War or 'Persistent Diplomacy'?" Issue Focus, U.S. Department of State International Information Programs, Office of Research, Dec. 24, 2003, via GlobalSecurity.Org; www.globalsecurity.org/wmd/library/news/libya/wwwh31225.htm.

[91]Binyon, "West beats path to forgive Libya its pariah status," *The Times*, Jan. 18, 2005; "Libya," *Country Analysis Briefs*, U.S. Department of Energy, Jan. 2004.

significantly because the state could no longer afford to take on all those looking for employment, with some 25% unemployed by 2003.[92] The upshot was that the Libyan state was no longer seen to be fulfilling its peoples' base requirements; this impression was typified by a declining standard of living. Domestic popular dissatisfaction with the current state of affairs contributed to the regime being confronted by a growing younger generation subject to high unemployment rates and disaffected from the political process. Opposition groups capitalized on the situation and attracted support because they offered an alternative to the status quo.[93]

Examples of such opposition groups included the Muslim Brotherhood, which supported political and economic reform based on Islamic ideals, and the National Salvation Front, which favored a political platform accommodating both secular and Islamic opponents.[94] Political opposition also encompassed violence perpetrated by militant Islamist groups against the regime in the mid-1990s, serving to demonstrate that a significant threat existed to the regime's security and survival. The Islamic Liberation Party (ILP) and the Islamic Martyrdom Movement (IMM) were two groups that supported armed resistance against the regime;[95] another group, the Libyan Islamic Fighting Force, had reported links with the al-Qaeda terrorist network.[96]

To bolster its own situation, therefore, the Qadhafi regime was confronted with the question of how to improve the country's economic and domestic political situation. This was evidently seen to require modernizing and expanding the economy, primarily the oil sector, and to do so by increasing investment from overseas. This could only be done if the embargoes on Libya were lifted, so an emphasis was placed on seeking their removal and reengaging with the international community. According to one report, the regime set up a committee in the spring of 1992, after multilateral sanctions were imposed, with the objective of reestablishing communications with the United States, but Libya's approach was rebuffed.[97] Over the next decade or so, however, Libya initiated a series of policy changes and took steps that ultimately ended both sanctions. This notably involved the regime's termination of its involvement in terrorism, handing over suspects in the investigation of the Lockerbie bombing and cooperation in resolving the case, refocusing Libya's external relations away from pan-Arabism and toward pan-Africanism, and deciding to give up its nuclear and other unconventional weapons programs. Notably, though it appears that Libya ceased its terrorist activities in the early to mid-1990s, the country was not taken off the U.S. State Department's list of state sponsors of terrorism until May 2006.

Negotiations

A key turning point for Libya came in August 1998, when the Qadhafi regime accepted a British-American proposal to put the Lockerbie suspects on trial in the Netherlands under Scottish law. A U.N. Security Council Resolution (1192) subsequently laid the groundwork for the suspension of multilateral sanctions if the suspects were handed over. In April 1999 the two suspects were finally given up and the sanctions were suspended.[98] However, it took until September 2003 for the U.N. sanctions to be fully removed, which required Libya's

---

[92]Pargeter, "Libya: All change for no change"; "Beating Swords into Oil Shares," *Economist*.

[93]Pargeter, "Libya: All change for no change"; Binyon, "West beats path to forgive Libya its pariah status."

[94]Takeyh, "Qadhafi and the Challenge of Militant Islam," pp. 167–168.

[95]*Ibid.*

[96]Ronald Bruce St John, "Libyan Foreign Policy: Newfound Flexibility," *Orbis*, vol. 47, no. 3, Summer 2003, p. 472.

[97]St John, "Libya is Not Iraq: Preemptive Strikes, WMD and Diplomacy," *Middle East Journal*, vol. 58, no. 3, Summer 2004, p. 390.

[98]St John, "Libyan Foreign Policy," p. 465.

complete fulfillment of Security Council demands, and an additional year before Washington removed its unilateral restrictions. This latter development was made conditional on Libya verifiably giving up the pursuit of nuclear weapons and other WMD, something that had been made clear to the Qadhafi regime by the Clinton and Bush Administrations in secret negotiations from 1999 to late 2003.

It is important to highlight several factors during this period that appeared to contribute to Libya's decision to disarm.

British Facilitation

The United Kingdom resumed official relations with Libya in 1999 and reestablished a diplomatic presence in Tripoli. This move had been made conditional on Libya taking responsibility for the shooting in London of policewoman Yvonne Fletcher in 1984 and agreeing to assist in investigating her murder.[99] British diplomats and intelligence officers subsequently played a pivotal role in facilitating the secret negotiations that resulted in the resolution of both the Lockerbie and WMD issues.[100] Most important, the United Kingdom served as a bridge between Tripoli and Washington since the Libyans were not in a position to directly approach the United State due to their antagonistic history. Interestingly, Col. Qadhafi and other senior Libyans cited the Blair government and Britain for doing the most to bring their country out of the cold following the public announcement on foregoing WMD in December 2003.[101] It is important to emphasize, however, that once the U.N. sanctions had been suspended, only the U.S. government could give the regime what it really wanted, and this entailed ending its unilateral embargo and reengaging with Libya.[102]

The Secrecy Imperative

Another important and defining feature of the negotiations was the emphasis all sides placed on maintaining the utmost secrecy, to prevent political opposition to the talks developing in the U.S., Libya, and elsewhere. For its part the Clinton Administration made it clear to the Libyans in 1999 that talks were conditional on them remaining secret, as well as Tripoli ending its efforts to have the sanctions removed, which it then did.[103] During 1999 and 2000, on the sidelines of the negotiations over Lockerbie, American officials informed the Qadhafi regime that it would also need to resolve WMD issues before Washington would completely accept Libya back into the international community; addressing Lockerbie would not be sufficient for achieving full reintegration.[104] An illustration of the importance attached to secrecy was the Clinton Administration's suspension of negotiations in 2000 due to its concern about potential leaks on the politically charged subject of talking with Qadhafi during a presidential election year.[105] This was evidently seen as a potential vote loser given that,

---

[99]St John, "Libyan Foreign Policy," p. 469.

[100]George Joffé, "Libya: Who Blinked and Why," *Current History*, 103/673 (May 2004), p. 221.

[101]St John, "Libya Is Not Iraq," p. 398.

[102]Michele Dunne, "Libya: Security Is Not Enough," *Policy Brief 32*, (Washington, D.C.: Carnegie Endowment for International Peace, Oct. 2004), pp. 1–7; www.carnegieendowment.org/publications/index.cfm?fa = view&id = 15921&prog = zgp&proj = zdrl.

[103]Martin S. Indyk, "The Iraq War Did Not Force Gadaffi's Hand," *Financial Times*, March 9, 2004; Flynt L. Leverett, "Why Libya Gave Up on the Bomb," *New York Times*, Jan. 23, 2004; St. John, "Libya Is Not Iraq," pp. 390–92; Paul Kerr, "IAEA praises Libya for disarmament efforts," *Arms Control Today*; Slavin, "Libya's rehabilitation in the works since early, '90s."

[104]Frantz and Meyer, "The Deal to Disarm Kadafi," *Los Angeles Times*, March 13, 2005.

[105]Indyk, "The Iraq War Did Not Force Gadaffi's Hand"; Leverett, "Why Libya Gave Up on the Bomb"; St John, "Libya Is Not Iraq," pp. 390–392; Kerr, "IAEA praises Libya for disarmament efforts"; Slavin, "Libya's rehabilitation in the works since early, '90s."

though the Lockerbie suspects had been given up, the issue still had a long way to go to resolution. There was particular concern about upsetting the American victims' families and the potential political capital that could be made out of this by the Republicans.

Following the 2000 Presidential election, the incoming Bush administration, though initially surprised to learn about the secret negotiations, subsequently resumed them but did so despite being similarly nervous about the political sensitivity of dealing with the Qadhafi regime.[106] The events of September 11, 2001, contributed to the resumption of negotiations as the three governments began to share information on the al-Qaeda network and affiliated organizations. The Bush Administration reportedly held at least six rounds of secret talks with Libya between October 2001 and December 2003.[107]

Flynt Leverett, who worked in the Bush Administration's National Security Council on Middle East policy "during two years of diplomatic negotiations beginning in 2002," [108]has noted that for Washington to take a "more constructive course with Libya [...] an informal coalition" of Secretary of State Colin Powell and National Security Advisor Condoleezza Rice approved the decision. Those in the administration who opposed offering positive incentives to countries like Libya to reform their behavior were sidelined in the process "when crucial decisions were made."[109]

One important milestone on the WMD issue came in August 2002, seven months before the Iraq war, when Mike O'Brien, then the U.K. Foreign Office Minister responsible for relations with North Africa, visited Libya.[110] The minister met with Qadhafi and discussed WMD, among other things, and was provided with "positive assurances of cooperation over the weapons issue."[111] The visit probably convinced the Libyan regime that London constituted a channel through which to negotiate a deal on WMD with the United States. The following month Prime Minister Tony Blair contacted Qadhafi, requesting a termination of Libya's WMD program.[112] It has been reported that President Bush knew that Blair was making the contact, and the Libyan leader is said to have responded with an indication that his foreign minister had been instructed to talk about "signing conventions" with London.[113]

Perhaps the most significant milestone, however, came in mid-March 2003, on the eve of the war to topple the Saddam Hussein regime in Iraq. This involved Libya's approach to British intelligence on the issue of Libya foregoing WMD in exchange for lifting all sanctions and full normalization of relations.[114] Qadhafi's son Saef al-Islam was directly involved in the approach, and this was taken as a sign that the Libyan leader was prepared to negotiate. Saef al-Islam was widely perceived to be a moderating influence on Qadhafi and had been working on bringing Libya back into the international fold, something reflected in an article he wrote in Spring 2003 on "Libya–American Relations," in which he wrote, "Libya is now ready to transform decades of mutual antagonism into an era of genuine friendship."[115]

---

[106]Slavin, "Libya's rehabilitation in the works since early '90s."

[107]Slavin, "Libya's rehabilitation in the works since early '90s"; St. John, "Libyan Foreign Policy," pp. 473–474.

[108]Kerr, "IAEA praises Libya for disarmament efforts."

[109]*Ibid.*

[110]Mary Dejevsky, "Libya decided 10 years ago against developing WMD without pressure, Foreign Minister says," *Independent*, Feb. 11, 2004, p. 6.

[111]Joffé, "Fear or Calculation?" *Middle East International*, Jan. 9, 2004, pp. 9–10; Joffé, "Libya: Who Blinked and Why," p. 223.

[112]St. John, "Libya Is Not Iraq," p. 398.

[113]Stephen Fidler, Mark Huband, and Roula Khalaf, "Return to the fold: how Gaddaffi was persuaded to give up his nuclear goals," *Financial Times*, July 27, 2004, p. 17.

[114]Michael Evans, "Libya knew game was up before Iraq war," *The Times*, March 13, 2004, p. 8; Frantz and Meyer, "The Deal to Disarm Kadafi"; Gertz, "Libyan sincerity on arms in doubt."

[115]Seif al-Islam Qadhafi, "Libyan–American Relations," *Middle East Policy*, 10/1 (Spring 2003), pp. 43–44.

The Iraq Effect

There has been great conjecture about whether or not the Iraq war played a role in pressuring Qadhafi to move forward on the WMD front, given that the initial approach took place just before hostilities were initiated. Moreover, British intelligence reportedly met Saef al-Islam as the first stage of coalition action unfolded.[116] Various Bush Administration officials argued that the "demonstration effect" of Iraq showed Qadhafi what would potentially happen to him if he did not give up his WMD programs. During his January 2004 State of the Union address, for example, the President himself stated in relation to Libya, "For diplomacy to be effective, words must be credible, and no one can now doubt the word of America."[117] It should be reiterated, however, that Libya had been negotiating since 1999 with the U.K. and U.S. and that WMD had been touched on during the talks. Indeed, the U.S. government had communicated to Libya since 1999 the requirement that WMD must be addressed if it was to fully reengage and lift unilateral sanctions. But various Libyan officials have highlighted that the pursuit and possession of WMD were no longer perceived to be in the regime's interests because this was seen to generate insecurity as opposed to security. The Libyan leader also reportedly spoke with Italian Prime Minister Silvio Berlusconi after the fall of the Saddam regime and said that he would do "whatever the Americans want."[118] Although the decision on WMD would probably have been forthcoming eventually because it was always part of the stated American requirement, the timing of the March 2003 approach does suggest that the Iraq war probably cemented the regime's view that its interests were best served by cooperating on this front. It might have given added momentum and accelerated the process, but the Iraq war is clearly not the sole or most salient reason for the outcome.

Intelligence

The U.K. and U.S. governments obviously wanted to secure a verifiable agreement from the Libyans on WMD.[119] Impetus in this direction increased from late September 2003 following the full removal of the U.N. embargo earlier in the month, after Libya had addressed the remaining issues of concern on Lockerbie, which included accepting responsibility for the actions of Libyan officials and agreeing to cover compensation for the families of the victims.[120] Intelligence played a key role in accelerating the process and securing the agreement, although its contribution might not have been critical.

Specifically, intelligence-derived information was provided to the Libyans to illustrate that London and Washington had detailed knowledge of the WMD and missile programs. Notably, the interception of gas centrifuge technology on the German-flagged *BBC China* between Dubai and Libya in early October 2003[121] demonstrated that they had in-depth knowledge about the regime's ongoing nuclear procurement activities; the ship was diverted to Taranto, Italy, where Italian and American officials removed this technology,[122] and it was

---

[116]Evans, "Libya knew game was up before Iraq war," p. 8.

[117]President George W. Bush, "State of the Union Address," Jan. 20, 2004; www.whitehouse.gov/news/releases/2004/01/20040120-7.html.

[118]Robin Gedye, "U.N. should fight for rights, says Berlusconi," *Daily Telegraph*, Sept. 4, 2003; James Kirchich, "Democratic critics are delusional about implications of the Bush Doctrine," *Yale Daily News*, Jan. 16, 2004.

[119]Evans, "Libya knew game was up before Iraq war," p. 8.

[120]"U.S. will not oppose ending of U.N. sanctions on Libya," Statement by the Press Secretary, Aug. 15, 2003, White House; www.globalsecurity.org/security/library/news/2003/08/sec-030815-usia04.htm.

[121]Ian Traynor, "Libya's black market deals shock nuclear inspectors," *Guardian*, Jan. 17, 2004.

[122]Frantz and Meyer, "The Deal to Disarm Kadafi."

then allowed to proceed to Libya. The presentation of such information to the Libyans has been described as a "vital lever" in influencing the regime to be forthcoming about its WMD capabilities in the October–December 2003 timeframe.[123] Indeed, the negotiations then moved forward more quickly, with British and American intelligence officials gaining access to some chemical and nuclear sites with Libyan cooperation.[124] In mid-December 2003, at a meeting in London, the Libyans eventually agreed to give up their chemical and nuclear weapons programs in exchange for an end to U.S. sanctions. Shortly afterward the Libyan foreign minister announced to the world that his country possessed chemical and nuclear weapons programs but was abandoning them—a move that was then publicly endorsed by Qadhafi.[125]

Dismantlement and Verification

A notable aspect of the subsequent dismantlement and verification phase was the cooperative approach adopted by British and American officials, which created an environment in Libya where the emphasis was on "verification" as opposed to "inspections," resulting in the provision of relatively straightforward access to all nuclear-related facilities, equipment, and staff as required. As Mahley has noted, the process was "not a punitive expedition" and the focus was not on "dragging things away from a protesting Libyan government."[126]

A related issue was the Qadhafi regime's insistence on involving the IAEA, not just the U.K. and U.S. governments, in verifying the dismantlement process. The Agency's role was complicated by the fact that it had been kept out of the loop during the secret talks, and the Bush Administration was initially reticent about having anyone other than the U.K. and the U.S. involved in the verification process when the Libya deal had been negotiated specifically by London and Washington. An agreement was subsequently reached in January 2004, however, under which the U.S. and the U.K. took on the task of dismantling, removing, and destroying Libya's nuclear capabilities, and the Agency verified that the program was dismantled correctly.[127] Libya's desire to involve the IAEA appeared to be based on a desire to avoid being seen to have given in to American pressure. It also appeared to reflect a view that working with organizations like the IAEA was an essential element of Libya's reengagement with the international community. The day after the decision was announced, a Libyan delegation held meetings at the IAEA and, among other things, it was agreed that Libya would implement measures to adopt an Additional Protocol to provide the Agency with broader inspection rights in the country.[128]

From an IAEA perspective, the ability to verify the dismantlement process meant that it was given an important opportunity to gain full knowledge of Libya's wholesale safeguards violations, including the contribution of the A. Q. Khan network.[129] On some occasions

---

[123]"Report to the President of the United States," Commission on the Intelligence Capabilities of the U.S. Regarding WMD, p. 252.

[124]Frantz and Meyer, "The Deal to Disarm Kadafi"; DeSutter, "U.S. Government's Assistance to Libya in the Elimination of its Weapons of Mass Destruction (WMD)," testimony before the Senate Foreign Relations Committee, Feb. 26, 2004; www.globalsecurity.org/wmd/library/congress/2004_h/DeSutterTestimony040226.pdf; Mahley, "Dismantling Libyan Weapons"; DeSutter, Testimony before the Senate Foreign Relations Committee, Feb. 26, 2004; Frantz and Meyer, "The Deal to Disarm Kadafi."

[125]*Ibid.*

[126]Mahley, "Dismantling Libyan Weapons," p. 4.

[127]DeSutter, Testimony before the Senate Foreign Relations Committee, Feb. 26, 2004.

[128]"IAEA Director General to Visit Libya," IAEA Press Release 2003/14, Dec. 22, 2003; www.iaea.org/NewsCenter/PressReleases/2003/prn200314.html.

[129]"ElBaradei: Libya helping with nuclear black market 'puzzle,' *USA Today*, Feb. 24, 2004.

IAEA teams did work in conjunction with joint U.S.-U.K. teams when material was being catalogued, packed, and removed from Libya.[130] The Agency's investigations focused primarily on nuclear material imports, uranium conversion activities, the centrifuge work, the irradiation of uranium targets at the TNRC, the separation of plutonium, nuclear weapon designs, and the contribution of A. Q. Khan.[131]

Phased Dismantlement and Engagement

A three-phased approach to the dismantlement process was agreed by the three governments and the IAEA in January 2004, influenced significantly by logistical challenges, including, for example, Washington's initial lack of a diplomatic presence in Libya. Although the dismantlement effort and diplomatic engagement were not linked at the start, they subsequently became aligned when the Bureau for Near East Affairs in the State Department began seeking milestones as a way to move the engagement process forward. The upshot was that the Bush Administration incrementally initiated actions to improve relations with Libya, including the termination of restrictions and sanctions.

Phase I focused on taking out of Libya the most proliferation-sensitive materials and equipment including UF6, P-2 centrifuges and the associated equipment and documentation, conversion modules, and the nuclear weapon designs. Beyond the nuclear realm, the United States also removed guidance systems from North Korean-supplied SCUD-C missiles. This merchandise was all removed for secure storage at the Oak Ridge National Laboratory in the United States by the end of January 2004.[132]

Prior to the removal of the weapon designs, IAEA officials had placed them under Agency seal. The IAEA also placed seals on some centrifuge equipment and documentation to be stored separately by the U.S.[133] The rewards for Phase I included lifting travel restrictions to Libya, allowing companies with pre-sanctions holdings in Libya to negotiate reentry, and permitting travel-related expenditures by U.S. government officials in the country.[134] A related development involved setting up an Interests Section in Tripoli and issuing an invitation for Libya to do the same in Washington.[135]

Congressional involvement also came to the fore in January when Representatives Tom Lantos and Curt Weldon travelled to Libya. Senator Joe Biden also visited in March.[136] These trips were significant because they signified a sea change in Congressional opinion on

---

[130] DeSutter, Testimony before the Senate Foreign Relations Committee, Feb. 26, 2004.

[131] See IAEA reports: Director General, IAEA, *Implementation of the NPT Safeguards Agreement*, May 28, 2004; Director General, IAEA, *Implementation of the NPT Safeguards Agreement*, Feb. 20, 2004. See also Jack Boureston and Yana Feldman, "Verifying Libya's Nuclear Disarmament," *Verification Yearbook 2004* (London: VERTIC 2003), pp. 90–92.

[132] DeSutter, Hearing before the Subcommittee on International Terrorism, Nonproliferation and Human Rights, Sept. 22, 2004; **DeSutter,** "Weapons of Mass Destruction, Terrorism, Human Rights and the Future of U.S.–Libyan Relations," Testimony before the House International Relations Committee, March 10, 2004; www.state.gov/t/vc/rls/rm/2004/30347.htm; Director General, IAEA, *Implementation of the NPT Safeguards Agreement*, May 28, 2004, pp. 4–5.

[133] DeSutter, Testimony before the Senate Foreign Relations Committee, Feb. 26, 2004.

[134] DeSutter, Hearing before the Subcommittee on International Terrorism, Nonproliferation and Human Rights, Sept. 22, 2004.

[135] Dunne, "Libya: Security Is Not Enough," pp. 1–7.

[136] Squassoni and Feickert, "CRS Report for Congress: Disarming Libya," p. 3.

Libya, since it was the Hill that had initiated the Iran Libya Sanctions Act during the mid-1990s to raise the economic and political pressure on Libya on the Lockerbie issue.[137]

The second phase of dismantlement focused on the remaining aspects of the nuclear program and started in the middle of February 2004. This phase involved the removal of approximately 1,000 tonnes of equipment from Libya,[138] and a deal was struck involving Russia, the IAEA, and Libya for the shipment of 16kg of HEU in fuel assemblies, delivered by Moscow for the IRT-1 research reactor, presumably in the early 1980s, for blending down into LEU. The U.S. Department of Energy paid for the shipment to Russia as part of the Tripartite Initiative under which the United States, Russia, and the IAEA are addressing security and safety issues by returning to the state of origin both spent and fresh nuclear fuel from Russian-designed reactors located abroad.

Phase II also began to address the redirection of former WMD staff in Libya; this has subsequently involved including significant work under the auspices of the U.S. Department of Energy's National Nuclear Security Administration (NNSA).[139] Moreover, the IAEA physically verified that Libya had imported 1,587 tonnes of yellowcake in 6,367 containers. These containers had been received from Niger between 1978 and 1981.[140] The two uranium producers also confirmed the total volume with the IAEA, as did documentation provided by Libya.[141]

U.S. engagement continued during the period including the removal, in April 2004, of most sanctions under the Iran Libya Sanctions Act.[142] On completion of the phase Washington also issued a general licence for trade and investment in Libya[143] and the establishment of a Liaison Office in Tripoli in June 2004 meant that direct diplomatic relations were formally restored.[144] At this time, however, the country remained on the U.S. State Department's list of designated state sponsors of terrorism; consequently some of the regime's assets continued to be frozen and defense exports prohibited.[145] Since the U.K. had already reestablished diplomatic relations in 1999, British Prime Minister Blair's meeting with Qadhafi in Tripoli during late March 2004 appeared to be a high-profile reward for the Libyan leader in return for implementing his decision on giving up the nuclear and other weapons programs.[146] A further reward from the United Kingdom followed some 25 months later when, in June 2006, the two countries signed a Joint Letter of Peace and Security. This pledged Britain to seeking U.N. Security Council action if Libya is attacked by another country

---

[137]Joffé, "Libya: Who Blinked and Why," p. 225.

[138]DeSutter, Hearing before the Subcommittee on International Terrorism, Nonproliferation and Human Rights, Sept. 22, 2004.

[139]"Removal of High-Enriched Uranium in Libyan Arab Jamahiriya," IAEA Staff Report, March 8, 2004, www.iaea.org/NewsCenter/News/2004/libya_uranium0803.html; DeSutter, Testimony before the Senate Foreign Relations Committee, Feb. 26, 2004; "Libya Sends Tajura HEU to Russia, Prepares to Convert Reactor to LEU," *NuclearFuel*, vol. 29, no. 6, March 15, 2004, p. 4.

[140]Director General, International Atomic Energy Agency, *Implementation of the NPT Safeguards Agreement of the Socialist People's Libyan Arab Jamahiriya*, May 28, 2004, Annex 1, p. 2.

[141]Director General, Implementation of the NPT Safeguards Agreement, May 28, 2004, p. 2.

[142]"U.S. Eases Economic Embargo Against Libya," White House, Office of the Press Secretary, April 23, 2004; www.whitehouse.gov/news/releases/2004/04/20040423-9.html.

[143]DeSutter, Hearing before the Subcommittee on International Terrorism, Nonproliferation and Human Rights, Sept. 22, 2004.

[144]Salah Sarrar, "U.S. Resumes Diplomatic Ties With Libya," Reuters, June 28, 2004.

[145]"U.S. Eases Economic Embargo Against Libya," White House, Office of the Press Secretary, April 23, 2004; www.whitehouse.gov/news/releases/2004/04/20040423-9.html.

[146]"Blair Visits Libya, Continuing a Thaw," Associated Press, March 25, 2004.

using chemical or biological weapons and to assisting Libya in "strengthening its defense capabilities." The two countries also "pledged to work jointly to combat the proliferation of weapons of mass destruction." From a U.K. perspective the move was designed to "encourage other countries to follow Libya's lead in abandoning its chemical and nuclear weapons programs."[147]

The final phase involved mostly verification work, and this encompassed interviewing Libyan personnel who had worked on the nuclear program, including procurement-related activities. This element of the dismantlement and verification process was mostly wound up by the end of September 2004.[148] The Bush Administration authorized the U.S. Export–Import Bank to support U.S. exports to Libya in September. This took place after Tripoli had announced in May that it would not work in the military field with states that it considered to be causes of proliferation concern.[149] Although the regime did not publicly list these states, it is believed that they included Iran, North Korea, and Syria.[150] The administration also terminated the National Emergency with Respect to Libya, imposed in 1986. Residual economic restrictions were removed on aviation services, flights, and some $1.3 billion in assets frozen under U.S. sanctions.[151] Nevertheless, Libya remained designated by the U.S. as a state sponsor of terrorism until May 2006, some 29 months after announcing the decision to give up WMD.

Finally, a Trilateral Steering and Cooperation Committee was set up by the three countries to serve as a mechanism for U.S., U.K., and Libyan experts to continue a confidential dialogue on nuclear and other WMD-related issues outside multilateral fora, including unresolved issues and the monitoring of Tripoli's compliance with its dismantlement pledge.[152] For example, one outstanding issue in the nuclear area was the conversion of the IRT-1 reactor to LEU fuel, and in December 2005 the Russian Atomic Energy Agency announced that it had returned 14 kg of LEU to the TNRC.[153]

Summary

The IAEA Board of Governors approved an Additional Protocol for Libya on March 9, 2004. It was signed the following day and subsequently took force on August 11, 2006.[154] The protocol provides the Agency with much greater inspection rights in Libya than in the period running up to December 2003. In this respect it is interesting to ponder whether Libya's safeguards violations in the past would have been picked up by the IAEA if such a

---

[147]Michael Nguyen, "U.K. offers Libya security assurances," *Arms Control Today*, Sept. 2006; www.armscontrol.org/act/2006_09/uklibya.asp.

[148]DeSutter, Hearing before the Subcommittee on International Terrorism, Nonproliferation and Human Rights, Sept. 22, 2004.

[149]Kerr, "Libya Pledges Military Trade Curbs, But Details Are Fuzzy," *Arms Control Today*, June 2004; www.armscontrol.org/act/2004_06/Libya.asp.

[150]Bolton, "Libya Ending Military Trade with States of Serious Weapons of Mass Destruction Proliferation Concern," Daily Press Briefing, May 13, 2004, www.state.gov/p/nea/rls/rm/32491.htm; Kerr, "Libya Pledges Military Trade Curbs, But Details Are Fuzzy."

[151]"State Department Highlights Positive Developments in Libya," Sept. 21, 2004, Office of the Spokesman, U.S. State Department; http://usinfo.state.gov/xarchives/display.html?p = washfile-english&y = 2004&m = September&x = 20040921190033ndyblehs0.3046381.

[152]DeSutter, Hearing before the Subcommittee on International Terrorism, Nonproliferation and Human Rights, Sept. 22, 2004.

[153]"Russia Delivers Low-Enriched Uranium to Libya," Radio Free Europe/Radio Liberty, Dec. 23, 2005.

[154]Strengthened Safeguards System: States with Additional Protocols, International Atomic Energy Agency; www.iaea.org/OurWork/SV/Safeguards/sg_protocol.html.

protocol had been in place that much earlier. Although it is impossible to answer this after the fact, it is possible to paint a fairly accurate picture of why successful and peaceful rollback took place in Libya.

Most important, the Qadhafi regime found itself, by the late 1990s and the first years of the 21st century, in a relatively weak political, strategic, and economic position, and this had a direct bearing on its willingness to trade in its nuclear program in return for reengagement with the outside world and an end to international isolation. Moreover, the position and role of the United States was undoubtedly of central importance to the final outcome, given Libya's perception of engagement with Washington as of paramount importance to realizing its core goals. The willingness of the Clinton and then Bush administrations to become involved in secret talks without insisting on major preconditions is also demonstrative of this centrality.

Another notable factor included the key role played by the United Kingdom as an intermediary in bringing the United States and Libya to the negotiating table, first on Lockerbie and subsequently on the weapons issue. Although the use of intelligence and interdiction, as well as the war in Iraq, certainly appeared to be important influences on Libyan decision making in 2003, they should be viewed as accelerators of a process that was already largely under way rather than as pivotal drivers.

An increasing volume of information on Libya's nuclear program began to be collected by U.S. and British intelligence agencies from 1999 to 2000 as a direct result of the regime's procurement of technology and materials via the A. Q. Khan network. The penetration of the network has been described by "intelligence community analysts" in the United States as being "critical to their understanding" of Libya's nuclear efforts.[155] Given the depth of Libya's involvement, it is evident that monitoring its procurement activities was also of pivotal importance to Britain and the U.S. in gaining as detailed and complete a picture as possible of how the network was structured and how it operated before seeking to close it down.

The Libyan leader has been rather damning of what his country has received since December 2003, and how quickly, from Washington in return for sacrificing its weapons projects. This has been despite Col. Qadhafi calling on other states with WMD capabilities and ambitions to give them up because Libya had "become an example to be followed," as he did during a visit to the European Union (EU) in April 2004.[156] In terms of an example, it is interesting to reflect on what the Libya experience can tell us about seeking to peacefully roll back the nuclear programs of other proliferators, notably Iran and North Korea. There are many lessons to learn, ranging from operational issues related to the dismantlement process all the way up to the invaluable role that third parties can play as diplomatic facilitators. Nevertheless, one key lesson stands out: For a negotiated rollback process to succeed, it cannot be a one-way street and it must take into account the broader strategic context. In this respect, Libya demonstrates how essential it is to apply a mix of positive (carrots) and negative (sticks) incentives to influence the decision-making calculus of the proliferating state in question. Moreover, Libya also illustrates that for success to be assured, specific negotiations on weapons programs must be one element of a broader strategic dialogue between the negotiating parties. Indeed, this fact has long been recognized by all the main parties in the context of North Korea's nuclear program, and hopefully it now appears to have been taken on board by all parties, including the United States, in the context of Iran.

---

[155]WMD Commission, p. 257.

[156]Qadhafi made his remark during a visit to the European Union in 2004. Michael Thurnston, "Kadhafi Urges World to Follow His Lead, Give Up Weapons of Mass Destrution," *Agence France-Presse*, April 27, 2004.

19 Elimination of Excess Fissile Material

Elena Sokova and Charles Streeper

Introduction

Fissile materials are essential for nuclear explosives and are often defined as radioactive materials that can sustain an explosive fission chain reaction. Fissile materials, although of different isotopic compositions, are also used in the generation of nuclear power. Concerns about a cascade of states acquiring fissile materials for weapons—highly enriched uranium (HEU) and plutonium (Pu)—and later over nonstate actors obtaining them were the primary motives for the introduction of a number of initiatives to control and eliminate nuclear weapons and their related materials. Over the many decades of the nuclear arms race and expansion of nuclear technology and industry, huge stocks of fissile materials usable for weapons have accumulated worldwide. Russia (the former Soviet Union) and the United States have produced the vast majority of these materials. Only with the end of the Cold War and the implementation of significant arms control reductions of their nuclear arsenals did the two states start a concerted effort to reduce their stockpiles of excess materials no longer needed for weapons programs.

The focus of this chapter is on the disposition of HEU and Pu—specifically on bilateral U.S.-Russian efforts to reduce fissile materials from their weapons programs. The chapter also provides a brief overview of other proposals and initiatives aimed at the reduction and elimination of fissile materials.

Weapons-Usable Materials and Their Availability

The materials best suitable for an explosive nuclear device are uranium (U) with a concentration of over 90% of the isotope ^{235}U and Pu with more than 90% of the isotope ^{239}Pu. Materials of this isotopic composition are often referred to as *weapons-grade* materials. However, U with a much lower concentration of ^{235}U and Pu of lower purity could also be used to make an explosive device. A 2006 report by the International Panel on Fissile Materials (IPFM) suggests that virtually all mixes of Pu isotopes are usable for making a bomb.[1] However, some isotopic mixes of Pu would be very difficult to use, and it is generally considered much easier for a state than for a terrorist group to create a nuclear device out of Pu. It is also important to note that any U or Pu that can be used for the manufacture of nuclear explosive devices without transmutation or further enrichment is considered

[1] *Global Fissile Material Report 2006: First Report of the International Panel on Fissile Materials*, Princeton University: Program on Science and Global Security, Sept. 25, 2006, p. 87.

weapons-usable. A few less common fissile materials—uranium-233, americium-242, and neptunium-237—can also be used for constructing a nuclear explosive device. However, the main focus of this chapter is on ^{235}U and Pu because they are the most common weapons-usable materials.

To denote the material's application or its origin, fissile materials are often divided into two broad categories: military and civilian. Most HEU and Pu were produced and used for nuclear warheads, i.e., for a military program. HEU is also used to fuel submarines, which is a military, though not a weapons, application. In addition to weapons and other military programs, weapons-usable fissile materials are used for a variety of civilian applications. For example, HEU is used as fuel for research and propulsion reactors and as targets for the production of medical isotopes. Some Pu derived from nuclear fuel irradiated in civilian power reactors is later fabricated into uranium-plutonium fuel or mixed-oxide (MOX) fuel and then recycled in commercial power reactors. Miniscule amounts of ^{239}Pu have also been used in sealed sources as a calibration device or as a neutron howitzer.

Global stockpiles of fissile materials include approximately 1730t[2] of HEU (military and civilian) and over 250t of separated Pu in weapons programs[3] and roughly 250t of Pu in civilian stocks. The vast majority of this material (over 90%) is in the possession of the nuclear weapons states (NWS). However, there are about 40 additional states that have some weapons-usable material on their territory.

Table 19.1 provides the most recent estimates of stocks of HEU and military Pu. These amounts take into account the already implemented reductions of HEU (downblended into low enriched uranium, or LEU) in Russia and the United States.

Even by the most conservative estimates, these stocks are enough for at least 100,000 nuclear bombs. According to the IAEA, as little as 8 kilograms (kg) of weapons-usable Pu and 25 kg of HEU are considered a "significant quantity," or the amount of material required to make a first-generation implosion bomb.[4] More sophisticated modern designs require even

Table 19.1 Estimates of Global Stocks of HEU and Military Pu,* 2005–2007 (metric tons).

Country	HEU (t) (93% Enriched Equivalent HEU)	Military Pu (t)
China	20	4
France	36.5	5
India	0.2	0.52
Israel	Not known	0.45
North Korea	Not known	0.035
Pakistan	1.3	0.064
United Kingdom	23.4	7.6
United States	654	92
Russia	985 ± 300	145 ± 25
Nonnuclear weapons states	10	—
Totals	1730 ± 300	254 ± 25

*Based on the *Global Fissile Material Report 2007*, by the International Panel on Fissile Materials, and on estimates of the 2005 *Global Stocks of Fissile Materials*, published by the Institute for Science and International Security.

[2]Tons (t) as used in this chapter are equal to metric tons.

[3]*Global Fissile Material Report 2007: Second report of the International Panel on Fissile Materials*, Program on Science and Global Security, Princeton University, 2007, pp. 10–14.

[4]"Safeguards Glossary: International Nuclear Verification Series 3," ed. 2001, Vienna: International Atomic Energy Agency, 2002.

smaller amounts of fissile materials. For example, according to a declassified report by the United States Department of Energy, "Hypothetically, a mass of 4 kilograms of Pu [...] is sufficient for one nuclear explosive device."[5]

Efforts to Control and Eliminate Fissile Materials

The idea to ban the production of fissile materials for weapons has been on the international security agenda since 1946, when the United States put forward the Baruch Plan. At that time the plan was rejected, but the very idea of controlling and stopping the production of materials for nuclear bombs continued to be discussed in the 1950s and 1960s, particularly during the negotiations of the Nuclear Nonproliferation Treaty (NPT).

Such a ban had little chance of success during the Cold War. The United States and the Soviet Union were competing against each other in the production of fissile material and converting it into additional warheads. Lack of transparency and hostility fueled practically unchecked growths of the stocks and arsenals of the two major rivals. In the mid-1960s the U.S. nuclear warhead arsenal peaked at around 30,000 and the Soviet Union's reached about 40,000 in the 1980s.[6] Other official NWS—the United Kingdom, China and France—were building up their nuclear arsenals as well. Only in the 1980s, after a significant advance in arms control negotiations between the United States and the Soviet Union, did a ban on future production of nuclear materials for weapons receive major backing by NWS. In 1982, the United Nations welcomed a proposal by the Soviet Union to cut off the production of materials for nuclear weapons.

The idea of developing a formal treaty banning the production of fissile materials also gained support. The United States, Russia, France, and the United Kingdom have all declared that they no longer produce fissile material for weapons. China has reportedly ceased its production as well. At the same time, de facto nuclear weapons states (India, Israel, and Pakistan) and later North Korea continued fissile material production. Iran is pursuing a uranium enrichment program, which is suspected of being intended for the production of HEU for weapons. A treaty that would ban the production of weapons-grade HEU, weapons-grade plutonium (WgPu), and ^{233}U for nuclear weapons or other nuclear explosive devices and would include all producers of nuclear weapon materials—inside and outside the NPT—was proposed by the United States in 1993 to the U.N. General Assembly. The proposal was approved by consensus and a negotiating mechanism was identified—the Conference on Disarmament in Geneva. The proposed treaty became known as the Fissile Material Cut-Off Treaty (FMCT).[7]

Despite initial support for the FMCT, disagreements about its scope, verification mechanisms, and proposed linkages with other disarmament and nonproliferation issues and initiatives prevented the negotiations from moving forward. To this date, little progress has been made toward the conclusion of a FMCT. At the same time, some experts believe that the original focus of the treaty on military fissile materials no longer reflects today's top security and terrorism risks and argue that a future treaty should also include restrictions on civilian HEU and Pu.[8]

---

[5]United States Dept. of Energy, Office of Declassification. *Restricted Data Declassification Decisions 1946 to the Present* (RDD-7), Jan. 1, 2001, Jan. 6, 2007; www.fas.org/sgp/othergov/doe/rdd-7.html.

[6]Robert S. Norris, and Hans M. Kristensen, "NRDC Nuclear Notebook: Global Nuclear Stockpiles, 1945, 2006," *The Bulletin of Atomic Scientists* 62.4 (2006): 64–66, Dec. 24, 2006.

[7]For more information on the FMCT, see the Fissile Material Cut-Off Treaty at www.fas.org/nuke/control/fmct/ and Jean Du Preez, "The Future of a Treaty Banning Fissile Material for Weapons Purposes: Is It Still Relevant?" *The Weapons of Mass Destruction Commission*, no. 9, 2006, Jan. 30, 2007; www.wmdcommission.org.

[8]Jean Du Preez, "The Future of a Treaty Banning Fissile Material for Weapons Purposes: Is It Still Relevant?" *The Weapons of Mass Destruction Commission*, no. 9, 2006, Jan. 30, 2007; www.wmdcommission.org.

Transforming Fissile Materials into Forms Not Usable for Weapons

Huge stocks of fissile materials require major efforts to continuously maintain and upgrade security systems for their use, storage, and transport. It is widely recognized that adequate control, accounting, and protection of fissile materials is not only an expensive but also a very challenging operation.[9] The arms reduction agreements between the United States and the Soviet Union in the 1980s and '90s prescribed significant reductions of each state's nuclear arsenals. Thousands of warheads were no longer necessary for the reduced number of nuclear weapons, and consequently the material in these weapons was no longer necessary for their respective military programs.

Security concerns apply to the material removed from nuclear weapons as much as to the weapons themselves. According to some proposals, material released from military programs should be placed in long-term secure storage, possibly under international control. However, the current approach to minimizing the risk of Pu or HEU diversion is that the amounts of these materials should be permanently reduced and excess quantities turned into a form unsuitable for weapons. This approach ensures the irreversibility of the elimination of the excess nuclear arsenals—the approach that both the United States and Russia agreed to apply in their bilateral agreements and efforts dealing with excess military fissile materials. This approach provides guarantees against the reincorporation of the material into the nuclear arsenals of the original possessor states, as well as deters theft or diversion. It also helps demonstrate the progress by the two major nuclear weapons powers in implementing their commitments to disarm under Article VI of the NPT.

As of February 2008, two large-scale programs were being undertaken by the United States and Russia to eliminate excess weapons-grade material: downblending of HEU into LEU and the disposition of Pu. Both efforts primarily involve material released from the dismantled warheads as the result of nuclear warhead reductions under several bilateral arms control agreements. These efforts aim at making surplus HEU and Pu unsuitable for weapons and disposing of it in a safe, secure, and internationally acceptable manner. Each program has its own technical characteristics, political and economic considerations, transparency measures, and other arrangements.

In addition, several relatively small-scale efforts exist to reduce the amount of HEU in the civilian nuclear sector and complement military HEU disposition. These efforts include international programs to convert research reactors to LEU and to repatriate and downblend HEU fuel previously supplied by the United States and Soviet Union to other states, as well as measures to reduce the number of facilities with HEU and the amount of HEU present at these sites by downblending surplus HEU to LEU.

HEU Disposition Efforts

HEU is one of the two primary fissile materials used in a nuclear weapon. Experts claim that HEU is also the choice material for building a crude nuclear explosive device.[10] Such a device could be constructed using a gun-type design—the least sophisticated type of nuclear weapon. A gun-type weapon uses explosives to propel one subcritical mass of HEU into another at high speed. When the two masses collide, they form a supercritical mass, which produces a nuclear explosion. The higher the enrichment level of uranium used in a gun-type bomb, the less HEU is required. Potentially, U enriched to 20% ^{235}U and even less could be

---

[9]For example, see the discussion of challenges of safeguarding fissile materials in Siegfried S. Hecker, "Toward a Comprehensive Safeguards System: Keeping Fissile Materials out of Terrorists' Hands," *Annals*, AAPSS, 607, Sept. 2006.

[10]Charles Ferguson, et al., *Four Faces of Nuclear Terrorism*, Routledge, 2005.

used to construct a crude nuclear bomb if a large enough amount of uranium were available.[11] HEU, which is uranium enriched to above 20% ^{235}U, is considered by the IAEA to be a "direct-use" weapons material.

Global HEU Stocks

The majority of HEU in the world was produced for military purposes. At the end of the Cold War, it is estimated that the Soviet Union had accumulated between 735t and 1365t of weapons-grade-equivalent HEU.[12] There are no publicly available official estimates of the amount of weapons-grade HEU produced by Russia. That is why such a wide range is used in describing Russian/Soviet HEU stocks. Prior to HEU downblending efforts, as of September 1996, the United States reported it had 740.7t of HEU.[13] The HEU stocks of Russia and the United States dwarf the stocks of other states.

Other official NWS—the United Kingdom, China, and France—are estimated to each have between 20t and 30t each of HEU. Most NWS have been reluctant to declare their military stocks; only the United States and the United Kingdom have disclosed their amounts. The de facto and one former state with nuclear weapons—Israel, India, Pakistan, and South Africa—are estimated to have stocks in the amounts of hundreds of kilograms. North Korea is suspected of clandestine efforts to enrich uranium, and Iran is working toward at least acquiring the capability to produce HEU.

HEU also became a choice material for several civilian applications, including research, test, space, and icebreaker reactors, as well as in the production of radioactive isotopes.

U.S. HEU Disposition

In 1994, the United States declared 174.3t of its HEU stock to be excess to military purposes and designated approximately 85% (more than 155t) of the surplus HEU to be converted into commercial reactor fuel. The remaining 15% (approximately 23t) of the surplus HEU is not usable for commercial-grade fuel and will be disposed of as waste. In 2001, the total amount of excess HEU was revised to 177.8t. In November 2005, the United States announced it was removing an additional 200t of HEU from its weapons stockpile. Of this amount, about 20t was to be downblended to LEU, another 20t would be reserved for space and research reactors, and 160t was set aside for use as naval reactor fuel.[14] Also in 2005, the United States announced an allocation of 17.4t of HEU to be downblended to LEU and contributed to a nuclear fuel reserve. These additional amounts increased the total of HEU

---

[11]For more information, see "HEU as weapons material: a technical background," prepared for the 2006 HEU Symposium in Oslo ,"Minimization of Highly Enriched Uranium (HEU) in the Civilian Nuclear Sector, www.nti.org/e_research/official_docs/norway/HEU_as_Weapons_Material.pdf, and Charles Ferguson, et. al., *Four Faces of Nuclear Terrorism*, p. 177, footnote 4. Also see *Global Fissile Material Report, 2006: First report of the International Panel on Fissile Materials*, www.fissilematerials.org/ipfm/site_down/ipfmreport06.pdf.

[12]David Albright, Frans Berkhout, and William Walker, *Plutonium and Highly Enriched Uranium 1996: World Inventories, Capabilities and Policies* (New York: Oxford University Press, 1997), p. 399.

[13]U.S. Dept. of Energy, National Nuclear Security Administration, Office of the Deputy Administrator for Defense Programs, *Highly Enriched Uranium: Striking a Balance, A Historical Report of the United States Highly Enriched Uranium Production, Acquisition, and Utilization Activities, 1945 Through September 30, 1996*, rev. 1, Jan. 2001; www.fas.org/sgp/othergov/doe/heu/index.html.

[14]U.S. Dept. of Energy, "DOE to Remove 200 Metric Tons of Highly Enriched Uranium from U.S. Nuclear Weapons Stockpile," Office of Public Affairs, Washington, D.C., Nov. 7, 2005, Feb. 10, 2007; www.energy.gov/print/2617.htm.

designated for downblending and disposition. By 2007, a total of 87t had been blended down.[15] An additional 123t is to be downblended to LEU and 23t disposed of as waste. A timetable for the disposition of different sets of HEU has been developed by the Department of Energy. The timetable envisions the complete disposition of all the material by 2050.[16]

Russian HEU Disposition

The United States and other nations were concerned over the fate of Russian HEU resulting from the dismantlement of nuclear warheads, particularly after the breakup of the Soviet Union and the deterioration of central control over nuclear facilities. The insecurity of the Russian material, along with concerns over the possible reintroduction of HEU into the weapons program, were powerful incentives to look for a permanent solution to reduce surplus HEU stocks in Russia and turn them into nonweapons-usable form.

In 1993, after more than a year of negotiations, Russia and the United States signed a governmental agreement concerning the disposition of HEU extracted from weapons. Russia agreed to downblend 500t of weapons grade HEU into LEU and to sell the resultant LEU to the United States for subsequent use in commercial power plants. The idea behind the agreement was to combine nonproliferation and disarmament goals with economic incentives. The revenue generated through this agreement was used to stabilize Russia's nuclear complex, fund safety and security upgrades, and provide employment to thousands of nuclear specialists at a time of political and economic turmoil in Russia.

The 1993 intergovernmental agreement was followed by a set of commercial agreements, including the so-called HEU Feed Deal, a contract to compensate Russia for the natural U component it would have used had LEU been produced from natural U rather than from HEU.[17] The program was projected to operate for 20 years and was originally set to generate about $12 billion for the Russian nuclear industry. It involved a fixed price for Russian LEU. Since the mid-1990s, sales under this agreement to the United States Enrichment Corporation (USEC), the U.S. executive agent for this program, amounted to $500 million annually. In 2002, a new pricing agreement was negotiated between USEC and Tenex (Techsnabexport), the Russian executive commercial agent for the agreement. The new pricing formula takes into account market fluctuations for enrichment services over a three-year period and provides a 10–15% discount to USEC. As of 2007, USEC had paid Russia over $5.1 billion.

The U.S.-Russian HEU downblending effort is often referred to as the HEU-LEU Purchase Agreement or the Megatons-to-Megawatts Program. The 1993 agreement on downblending envisioned that this material would be completely recycled into LEU by 2013. As of the beginning of 2008, the program had reached a significant milestone of downblending 325 tons of HEU to LEU. This amount is equivalent to the elimination of 13,000 nuclear warheads.[18] About 30t of Russian HEU are being downblended annually. Currently, the LEU originating from Russian nuclear warheads and shipped to the United States accounts for about 50% of the fuel for U.S. nuclear power plants.[19] Though not without its own controversies

---

[15]*Global Fissile Material Report 2007*, p. 19.

[16]*Global Fissile Material Report 2007*, p. 30.

[17]The distinction between the cost for enrichment services and the cost of uranium reflects the U.S. practice of paying separately for SWU units and the uranium itself.

[18]U.S.-Russian Megatons to Megawatts Program, April 2007; www.usec.com/v2001_02/HTML/Megatons-history.asp accessed on June 3, 2007.

[19]Linton Brooks, Ambassador, Under Secretary for Nuclear Security and Administrator, National Nuclear Security Administration, U.S. Dept. of Energy, "Countering Nuclear Terrorism," Remarks to Chatham House, London, Sept. 21, 2006, Feb. 10, 2007; www.nnsa.doe.gov/docs/speeches/2006/speech_Brooks_Chatham_House-21Sep06.pdf.

and challenges, this program is considered by both states as one of the most successful cooperative nonproliferation programs with Russia.

Technology of HEU Downblending

The process of downblending HEU into LEU involves a series of sophisticated technical operations. At the same time, it is a straightforward process that utilizes existing technologies and facilities. The relative simplicity of this process is particularly evident compared to the process of Pu disposition. There are two primary blending options: the mixing of liquids and the mixing of gases. The resultant material in both liquid and gas processes is not weapons usable and would require re-enrichment to regain weapons usability.

The United States primarily uses the first method. The existing commercial downblending facilities in the United States—BWX Technologies in Virginia and Nuclear Fuel Services in Tennessee—use the uranyl nitrate hexahydrate (UNH) method. Historically, the Y-12 Plant and Oak Ridge National Laboratory have also performed the molten U metal blending process in addition to the UNH method.[20] Blendstock material can be natural U, depleted U, or slightly enriched U (typically 1.5% ^{235}U).

Russia uses the gas (UF_6) mixing process. HEU from dismantled warheads is processed into metal shavings. It is then heated to convert the shavings into uranium oxide (U_3O_8) and to chemically remove all contaminants from the material (see Figure 19.1). The U_3O_8 is then converted into HEU hexafluoride (UF_6), which turns into gas when heated. UF_6 is then diluted to the desired enrichment level (usually from 2 to 4% ^{235}U but not exceeding 5%) by mixing it with 1.5% enriched ^{235}U. The 1.5% LEU blend stock is used to meet Western commercial fuel requirements.

Transparency and Verification

The Russian-US HEU-LEU program includes various transparency measures to verify its implementation. The transparency regime was established by the 1994 Protocol to the HEU-LEU Purchase Agreement. It is reciprocal in nature and includes specific monitoring, observation, and assessment activities permitted at both U.S. and Russian facilities. The regime, however, does not include access to the warhead dismantlement process or to the HEU metal

FIGURE 19.1 HEU monitor at Russian facility used to burn uranium metal and produce uranium oxide. (Photo credit: Lawrence Livermore National Laboratory.)

---

[20]"Uranium Downblending," World Information Service on Energy Uranium Project [WISE]: Arnsdorf, Germany. Aug. 20, 2005, Feb. 5, 2007; www.wise-uranium.org/eudb.html.

before it is transformed into shavings. Both states continue to view the composition and shape of HEU metal used in weapons as classified information.[21]

The United States made available part of its 177.8t of HEU stock designated for downblending to the International Atomic Energy Agency (IAEA) for monitoring. The IAEA monitored the downblending of about 50t of excess US HEU.[22] The 17.4t of HEU designated for fuel supply reserves is also expected to be converted into LEU under IAEA monitoring. The remaining U.S. HEU is reportedly not intended to be monitored by the IAEA.[23]

Possible Future Military HEU Reductions

Even when the United States and Russia complete downblending the material they have declared excess to military needs, both states will still have the largest stockpiles in the world and amounts far beyond the actual needs of their reduced arsenals and civilian nuclear industry.[24] Many nonproliferation experts believe that it is necessary and urgent that the United States and Russia declare additional large quantities of their HEU stocks excess and convert the material to LEU. One way to speed up the downblending process without having a negative impact on the uranium market is to blend HEU down to just below 20% ^{235}U, thus making the material unsuitable for weapons, but keep the downblended material off the commercial nuclear fuel market and downblend it to the required enrichment level at a later date.[25]

In 2006, the United States and Russia discussed the possibility of renewing or concluding a new HEU-LEU agreement, which would go beyond the 500t of HEU agreed in 1993. Russia was not enthusiastic about this proposal.[26] With the uranium price on the rise and anticipated increased demand for uranium and enrichment services, Russia prefers to sell its uranium and uranium products at market price. In February 2008, Russia and the United States signed a long-term deal to allow Russia to sell enriched uranium directly to U.S. utilities starting in 2011. This agreement became possible only after the U.S. government suspended antidamping sanctions imposed on Russian uranium in the early 1990s.[27]

---

[21]For more information on the HEU-LEU program transparency regime, see "Highly Enriched Uranium (HEU) Purchase Agreement Transparency Implementation," report at the National Nuclear Security Administration Website at www.nnsa.doe.gov/na-20/heu_trans.shtml.

[22]*Global Fissile Material Report, 2007*, p. 30.

[23]Michael Knapik, "DOE Has Limits on HEU Sales This Decade," *Nuclear Fuel*, Jan. 31, 2005, as cited in *Global Fissile Material Report 2006: First report of the International Panel on Fissile Materials*.

[24]See the discussion about U.S. and Russian excess military stocks in *Global Fissile Material Report 2006: First Report of the International Panel on Fissile Materials*, Sept. 25, 2006, pp. 14–15, and Robert L. Civiak, "Closing the Gaps: Securing High Enriched Uranium in the Former Soviet Union and Eastern Europe," May 2002, Federation of American Scientists, Feb. 12, 2007; www.fas.org/ssp/docs/020500-heu/full.pdf, p. 15.

[25]For proposals to accelerate the downblending of Russian HEU, see Arbman Gunnar, et al., "Eliminating Stockpiles of Highly Enriched Uranium: Options for an Action Agenda in Cooperation with the Russian Federation," April 2004; www.pugwash.org/reports/nw/heu-200415.pdf. The Nuclear Threat Initiative carried out a study in cooperation with Russian scientists on options for increasing the amount of HEU blended down each year. Amounts ranged from 30t to 60t per year. The study considered different possible enrichment levels for the final product—4.5%, 12%, and 19% uranium-235—taking into account technological, proliferation, and financial considerations.

[26]"Russia: No new HEU deal," *Fresh Fuel*, vol. 22, no. 865, June 12, 2006, p. 1.

[27]"United States and Russian Agreement Reached," U.S. Department of Commerce Press Release, Feb. 1, 2008.

In 2004–2005, the international community renewed its interest in multinational nuclear fuel arrangements, including the creation of a nuclear fuel reserve or bank. The idea of a nuclear fuel reserve is supported by the IAEA, the United States, Russia, and other major nuclear states. Several concepts of such banks and reserves exist. The United States, for example, in 2005 announced its intention to provide 17.4t of HEU to be used as future blend-down stock for assuring nuclear fuel supplies to states that forego enrichment and reprocessing.[28] At a special IAEA event in September 2006, the IAEA and the Nuclear Threat Initiative launched an initiative to create a nuclear fuel bank under the supervision of the IAEA. The creation of such banks or reserves might not only help contain the spread of enrichment and other sensitive technologies, it could also become a vehicle for downblending additional HEU quantities for use in these arrangements.

Civilian HEU and Efforts to Reduce and Eliminate Its Use

HEU is currently used in the civilian sphere to fuel research reactors, critical facilities, pulse reactors, and a few Russian icebreakers and in producing medical isotopes. During the 1950s and 1960s, the United States and the Soviet Union built hundreds of research reactors domestically and exported about 100 reactors to more than 40 other states. Over 120 of these facilities have 20 kg or more of HEU on site.[29] As of the end of 2003, there were about 50t of HEU in civilian power and research programs in over 40 states.

Civilian nuclear facilities usually have less protection than military facilities. This is particularly true for research and test reactors often located on university campuses in big cities. Some experts argue that for this reason civilian HEU is more vulnerable to diversion than military HEU. Moreover, the level of physical protection and security of these facilities varies from state to state. Although security concerns and protection measures have increased since 9/11, it is difficult to reconfigure a site that was not built with rigorous physical protection in mind. Most of these facilities store fresh and spent fuel on-site.

Several programs to reduce the use and exports of HEU for research reactor programs were put in place in the late 1970s. The work is far from being completed, and although there are no binding agreements to reduce civilian HEU stocks globally or to eliminate HEU use in other civilian applications, in the past three to four years a number of initiatives to address this issue have been launched.

The first efforts to reduce HEU use in civilian research reactors date back to 1978, when the United States initiated the Reduced Enrichment for Research and Test Reactors (RERTR) Program. This program was meant to reduce HEU supplies to reactors the United States had exported to other states. This effort included the development of new LEU fuels and conversion of research reactors to use the replacement LEU fuel. Fresh and spent fuel from converted reactors was to be shipped back to the United States, downblended, and used to fabricate replacement LEU fuels or disposed of in other ways. As of September 2006, about 3,300 kg of HEU had been transported back to the United States, which represents half of all the HEU provided to other states by the United States.[30]

That same year, Moscow also started a program to reduce enrichment at Soviet-built research reactors outside the Soviet Union and reduced the enrichment level of fuel for these reactors from 80–90% to 36%. Further efforts to reduce the enrichment level to below 20%

---

[28] "Fact Sheet on U.S. HEU for a Nuclear Fuel Reserve," U.S. Mission to International Organizations in Vienna, Sept. 27, 2005; http://vienna.usmission.gov.

[29] For more information on HEU uses in the civilian nuclear sector, see the Civilian HEU Reduction and Elimination database at the Nuclear Threat Initiative Website: www.nti.org/db/heu/index.html.

[30] U.S. Dept. of Energy, National Nuclear Security Administration, *NNSA Fact Sheet: NNSA Working to Prevent Nuclear Terrorism* (Washington, D.C.: Sept. 2006), Feb. 9, 2007, www.nnsa.doe.gov/docs/factsheets/2006/NA-06-FS07.pdf.

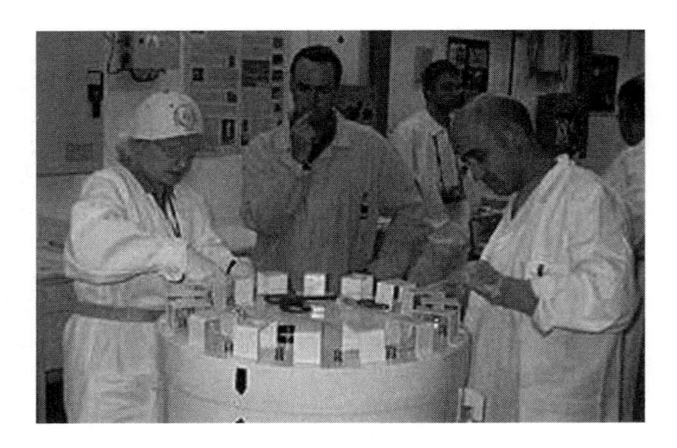

FIGURE 19.2 U.S. and IAEA officials monitor the sealing of HEU container prior to transportation to a blend-down facility. (Photo credit: International Atomic Energy Agency.)

were stalled by economic problems in the Soviet Union in the mid-1980s. After the collapse of the Soviet Union, Russia joined the RERTR program. Other states have initiated similar efforts and have also cooperated with the RERTR program. As of March 2007, 48 research reactors of the 106 originally targeted by RERTR had been either converted or shut down.[31] In 2007, additional reactors were added to the program, which now covers 129 out of the 207 reactors worldwide that operate on HEU fuel.[32] Some reactors cannot be converted, whereas others are not covered due to ownership issues. For example, dozens of research reactors, critical assemblies, pulsed reactors, and icebreakers in Russia contain significant amounts of HEU but are outside the RERTR efforts or similar programs.

In connection with the RERTR Program, the United States and Russia also launched an initiative to repatriate U.S. and Soviet-origin HEU from other states. The returned HEU is usually downblended to LEU. In Russia's case, HEU blend-down is required by agreements concluded for each repatriation mission. In May 2004, the RERTR program, fresh and spent HEU takeback and related programs were subsumed under the auspice of the Global Threat Reduction Initiative (GTRI), which was launched by the United States. Between 2002 and 2007, 446.3 kg of fresh and 143 kg of spent Soviet-origin HEU fuel were returned to Russia.

The United States and Russia have also been involved in the Material Consolidation and Conversion (MCC) project, which is a program complimentary to the U.S. Department of Energy's Material Control and Accounting (MPC&A) Program in Russia. Established in 1999, the program is focused on consolidating HEU to a smaller number of secure locations, reducing the number of areas that HEU is stored within each facility, and eventually reducing the total number of facilities housing HEU. Excess material identified in these operations is downblended to LEU. By April 2008, the MCC program had 10t downblended of HEU removed from Russian facilities.[33]

Many states recognize the risks posed by civilian use, storage, and commerce in HEU. At the same time, there are no global standards or arrangements that can effectively secure civilian HEU in use and storage and mandate the decommissioning of obsolete and redundant HEU-fueled research reactors and the expedited conversion of the operating research reactors to LEU. At the 2005 NPT Review Conference, Norway, on behalf of itself, and

---

[31] "U.S. and Russia Cooperate to Eliminate Dangerous Nuclear Material," U.S. National Nuclear Security Administration Press Release, April 24, 2008.

[32] Ibid.

[33] Matthew Bunn, *Securing the Bomb 2007*, p. 89.

Iceland, Lithuania, and Sweden submitted a working paper titled *Combating the risk of nuclear terrorism by reducing the civilian use of highly enriched uranium* in an effort to seek international consensus on this issue.[34] The June 2006 International Symposium on reduction of HEU in the civilian nuclear sector, held in Oslo, Norway, made significant progress in establishing a consensus among international experts on the technical feasibility of replacing HEU with LEU for most civilian uses. Achieving political support for this initiative, however, remains a challenge. Among the proposed measures to move forward with the phase-out of civilian HEU is the introduction of HEU guidelines, similar to Pu Transparency Guidelines, implemented within the IAEA framework. Other possible measures include the adoption of a voluntary HEU code of conduct, the establishment of regional centers of excellence and research reactor coalitions to limit the number of HEU-fueled reactors in operation and bring the security upgrades of the most viable of them to the highest standard. The key role and responsibility in this process lies on the United States, Russia, United Kingdom, France, China, Germany, and South Africa as the primary holders and users of civilian HEU.

Plutonium Disposition Efforts

In addition to HEU, the United States and Russia also had to address a large surplus of Pu from dismantled warheads. The two states' stocks account for the majority of global military Pu, which is currently estimated at around 254t.[35] The United States (in 1995) and Russia (in 1997) each declared 50t of Pu excess to military needs and agreed to dispose of it. The only other NWS that declared Pu excess to its defense program needs was the United Kingdom. In 1998, the U.K. had declared 3t of weapons-grade plutonium (WgPu) and an additional 4.1t of Pu as surplus. However, as of 2006, no official decision had been made on how to dispose of the material.

The disposition of Pu is a much more complex technical problem than the downblending of HEU. Unlike HEU, Pu cannot be diluted with another Pu isotope. This is a critical difference between these two elements. As all isotopes of the same element have the same chemical properties, the dilution of the fissionable isotope ^{235}U with a nonfissionable isotope of the same element, namely ^{238}U, precludes the easy path of chemical separation of the weapons usable ^{235}U. This approach is possible because of the natural abundance in nature of ^{238}U of an average of 99.3% in uranium. Therefore, ^{238}U is readily available and, in fact, is a byproduct of ^{235}U enrichment. Since Pu is not a naturally occurring element but is a product of transmutation in a nuclear reactor environment, the proportions of isotopes of Pu are dependent on the reactor environment. Therefore, dilution of Pu is only practical with other elements, but that would mean chemical separation would remain an easy path to reverse the process and obtain Pu. Initially, long-term storage was considered as a possible solution to deal with surplus Pu. However, the relatively easy reuse of the material in weapons, remaining security risks, and high costs associated with storage required a solution that would introduce intrinsic barriers to recovery of the material. Scientific studies and reports were commissioned before the United States and Russia were to officially commit to the task of choosing paths to eliminate surplus Pu.

---

[34]*Combating the risk of nuclear terrorism by reducing the civilian use of highly enriched uranium: Working paper submitted by Iceland, Lithuania, Norway and Sweden*, 2005 Review Conference of the Parties to the Treaty on the Non-Proliferation of Nuclear Weapons (New York: United Nations, Main Committee III, Working Paper 5, May 20, 2005), Feb. 13, 2007, http://daccessdds.un.org/doc/UNDOC/GEN/N05/352/69/PDF/N0535269.pdf?OpenElement.

[35]*Global Fissile Material Report 2006*, p. 20.

Studies on Pu Disposition

In 1992, a committee of the National Academy of Sciences (NAS) was asked to evaluate options for managing excess Pu. The NAS report was released in March 1994 and found that excess Pu posed a "clear and present danger to national and international security" and called for the transformation of excess Pu into a form resistant to proliferation.[36] The reports suggested the use of a spent fuel standard (SFS). Spent fuel—fuel discharged from a commercial reactor after irradiation—is highly radioactive and can only be manipulated remotely with extensive shielding and protection. This radioactive barrier increases the material's resistance to theft, proliferation, and reintroduction into a nuclear weapon program. The SFS, therefore, requires choosing a method that would make "plutonium roughly as inaccessible for weapons use as the much larger and growing quantity of plutonium that exists in spent fuel from commercial reactors."[37]

The NAS report examined several disposition options with one of the major focuses being on options that best met the SFS.[38] The NAS study determined that, despite the drawbacks of all the examined options, the best options were to burn the Pu in existing nuclear reactors as a plutonium-uranium mixed oxide (MOX) fuel or to immobilize it with high-level wastes.[39]

While the NAS study was in progress, the U.S. Department of Energy (DOE) tasked its Office of Nuclear Energy to begin conducting studies on disposition options.[40] In 1993, the DOE created the Office of Materials Disposition (OFMD), which took over the work started by the Office of Nuclear Energy and collaborated with the national laboratories to narrow the list of options to only the most realistic. In 1997, the DOE OFMD issued its "Record of Decision for the Storage and Disposition of Weapons-Usable Fissile Materials Final Programmatic Environmental Impact Statement." This decision, among other things, made a recommendation in support of the dual-track method of Pu disposition, i.e. utilizing both the MOX and immobilization options.

Russia also conducted its own studies. Russia's Ministry of Atomic Energy (MinAtom, later called the Russian Federal Nuclear Agency, and currently transformed into a state corporation called RosAtom) considered Pu a valuable energy resource and opposed immobilization and geological disposal. Although Russia had reservations about some aspects of the MOX option, it was considered much more acceptable than immobilization.[41] Russia believed that, because the WgPu isotopics are not changed in immobilization (unlike with MOX fuel), the WgPu in immobilized form, with no isotopic degradation, could be reconstituted into nuclear weapons. For this reason, Russia insisted that in any agreement on Pu disposition with the United States, the majority of U.S. Pu must not be immobilized.

The United States and Russia also conducted joint research into various disposition methodologies. In 1996, the U.S.-Russian Independent Scientific Commission on Disposition of Excess Weapons Plutonium was formed to further investigate disposition

---

[36]U.S. Congress, Committee on International Security and Arms Control, "Management and Disposition of Excess Plutonium" (Washington, D.C.: National Academy Press, 1994), p. 3.

[37]U.S. Congress, "Management and Disposition of Excess Plutonium," p. 12.

[38]The NAS reviewed a number of other Pu disposition options such as subseabed disposal, launching the material into the sun or out of the solar system, and even considered a Russian proposal to explode a nuclear device surrounded by Pu pits underground.

[39]U.S. Congress, "Management and Disposition of Excess Plutonium," p. 2.

[40]Some key participants of the early studies using nuclear reactors for surplus Pu disposition were Westinghouse, General Electric, Combustion Engineering, and General Atomics.

[41]A further explanation of the MOX fuel and immobilization options is provided in the next part of this section.

options.[42] The 1997 final report by this joint commission also recommended the MOX and immobilization approach to disposition.

Immobilization

Immobilization of Pu can be achieved by immobilizing it in a glass matrix (vitrification) or a ceramic matrix (ceramification). WgPu and high-level wastes (HLW) are combined before the mix is vitrified or ceramified. During the vitrification process, Pu is contaminated with HLW, which is then mixed with borosilicate glass to create a borosilicate matrix (glass logs). The logs are then sealed in stainless steel cylinders and stored for later geological disposal at a nuclear waste repository.

The ceramic immobilization process uses a titanate matrix instead of a borosilicate matrix. Both these options make the resultant material highly radioactive—in other words, create a barrier that meets the SFS. WgPu would account for no more than 5–10% of the mix. As a variant of immobilization, a can-in-canister option has also been explored. It involves taking Pu oxide and making ceramic (hockey puck-sized) Pu disks that are then put into cans. The cans are put into canisters and then suspended on a rack, at which point borosilicate and HLW is poured into the canister to make logs.

It should be noted that no Pu is destroyed by either one of the two immobilization methods and eventual disposal at a geological waste repository would be required.

MOX

The MOX option involves mixing Pu oxide with U oxide to make MOX fuel and then burning the fuel in commercial power reactors. The irradiated MOX fuel would have a radiation barrier that meets the SFS. However, the irradiated fuel would still contain Pu that could be separated later, after decades of cooling.

The processes for making MOX fuel from WgPu are similar to those used on a commercial basis in Europe for reactor-grade fuel. MOX fuel is fabricated from a mix of Pu oxide powder and U oxide powder. However, the production of MOX fuel from WgPu requires an extra step because WgPu comes in a metal form that is alloyed with gallium and trace amounts of other materials. Therefore, Pu metal must first be transformed into pure Pu oxide powder, removing any impurities, before it is suitable for the fabrication of MOX fuel. See Figures 19.3 and 19.4.

Once the Pu has been fabricated into MOX fuel assemblies, it can be burned in most commercial reactors, including light water and fast neutron reactors. However, some reactors require modifications to enable them to burn MOX fuel. Reactors also differ in the amount of MOX they are able to consume in one fuel cycle, but most commonly, MOX fuel accounts for only a third or a half of the fuel load. In addition, burning of MOX fuel does not destroy all the Pu in the fuel. Light water reactors destroy from 20% to 60% of their initial Pu load.[43] Fast reactors, if operated as net burners, can potentially consume almost all the Pu. See Figure 19.5.

Since the first investigations into Pu disposition, various types of reactors and fuel cycles have been discussed and continue to be advocated by some groups and experts as the best choice for Pu disposition. Among the most actively advocated reactors are the Gas-Turbine Modular Helium Reactor (GT-MHR), other types of reactors under development,

---

[42]Other joint U.S.-Russian studies were the "U.S.-Russian Plutonium Disposition Steering Committee" and the "Joint U.S.-Russian Working Group on Cost Analysis and Economics in Plutonium Disposition."

[43]U.S. Congress, "Management and Disposition of Excess Plutonium," p. 155.

FIGURE 19.3 Canister of mixed oxide. (Photo credit: Los Alamos National Laboratory and National Nuclear Security Administration.)

FIGURE 19.4 MOX fuel pellets. (Photo credit: Los Alamos National Laboratory and National Nuclear Security Administration.)

and the thorium-plutonium-uranium fuel cycle. However, none of these technologies is developed enough to be available for implementation any time soon. Government studies concluded that waiting for the development of future technologies "would result in substantial additional delays and higher costs, compared to beginning with the reactors that already exist and fuel cycle approaches that are already demonstrated."[44]

Although the civilian MOX fuel cycle has existed for some time in a number of states and has a history of domestic and/or international safeguarding, there have always been concerns about the ability to safeguard Pu in large bulk-handling plants such as those used for

[44]Bunn, *Securing the Bomb 2006*, "Reducing Excess Stockpiles: U.S. Plutonium Disposition."

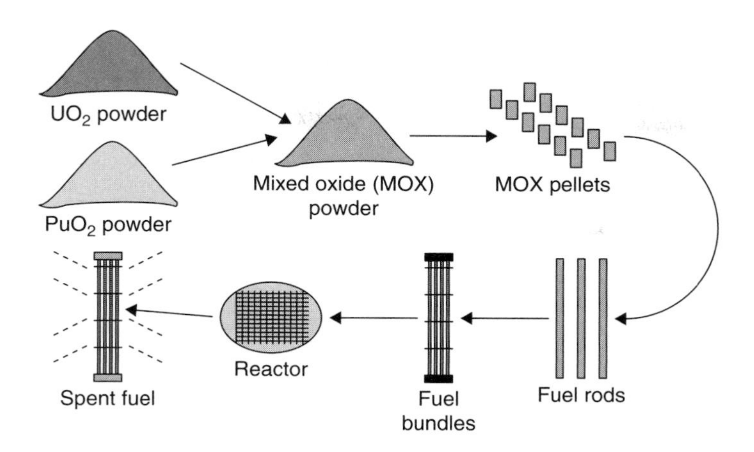

FIGURE 19.5 MOX flow diagram. (Diagram provided by the DOE in 1996 in *The Technical Summary Report for Surplus Weapons-Usable Plutonium Disposition*.)

the process of MOX fuel fabrication.[45] These concerns became even more pronounced with the proposal of using the MOX fuel cycle for recycling WgPu.

The 1994 NAS report, although promoting the MOX fuel cycle as one of the better disposition options, claims that the multistaged MOX fuel fabrication process is more vulnerable to "covert diversion and to theft" than the "fewer processing steps" entailed in vitrification. However, once the Pu is in its final vitrified or MOX fuel form, the vitrified Pu is considered more vulnerable to overt or covert diversion than MOX fuel because of a "modestly lower" isotopic barrier.[46] At the same time, a 1995 American Nuclear Society report states that the entire MOX fuel cycle can be sufficiently protected, managed, and safeguarded.[47]

The U.S.-Russian Plutonium Disposition Agreement

The United States and Russia's commitments to dispose of Pu have enjoyed significant support from the international community. In 1996, at the Summit on Nuclear Safety and Security, Russia and the G7 leaders confirmed the importance and urgency of the disposition of U.S. and Russian excess Pu. The G7 statement reiterated the idea of previous studies that when considering disposition options the emphasis should be on the threat of diversion and the safe, secure, irreversible, and expedited removal of excess Pu from weapons programs.[48]

In 1998, the U.S. and Russia signed an agreement, which laid the framework for scientific and technical cooperation and the parameters for approved methods of disposition

[45]American Nuclear Society, *Protection and Management of Plutonium: American Nuclear Society Special Panel Report: Panel on Protection and Management of Plutonium. Subpanel on Safeguards and Security*, La Grange Park, IL, Aug. 1995, p. 81.

[46]U.S. Congress, *Management and Disposition of Excess Plutonium*, p. 192.

[47]J. W. Tape, *Protection and Management of Plutonium: American Nuclear Society Special Panel Report: Panel on Protection and Management of Plutonium. Report of the Subpanel on Safeguards and Security*, p. 83.

[48]"Background Documents on Nuclear Safety and Security," G7+1 Moscow Nuclear Safety and Security Summit, April 20, 1996, University of Toronto, Feb. 13, 2007, www.g7.utoronto.ca/summit/1996moscow/background3.html.

of excess Pu.[49] That same year the Presidents of the U.S. and Russia both issued a statement that committed the two states to irreversibly remove 50t of Pu from their weapons programs and called for a binding intergovernmental agreement to dispose of the material. However, because of the anticipated high costs of this effort, Russia insisted on international funding for its portion of the program. To expedite the conclusion of the agreement, the United States and other G7 partners pledged money toward the initial costs of the program.[50] This gave the necessary impetus for the signing in September 2000 of the agreement on the management and disposition of plutonium designated as no longer required for defense purposes, commonly referred to as the Plutonium Management and Disposition Agreement or PMDA.[51]

The PMDA committed Russia and the United States to each dispose of 34t of declared excess Pu at a rate of 2t per year beginning no later than December 31, 2007. Only 34t out of the 50t of U.S. Pu was WgPu, whereas all of Russia's 50 t was WgPu. Consequently, not all the 50t originally declared surpluses was covered by the PMDA. The agreement does, however, leave open the possibility of future voluntary disposition of amounts above the 34t and declared excess amounts. For example, in September 2007 the U.S. Department of Energy announced that an additional 9t of Pu was being declared excess and would be suitable for conversion to MOX fuel.[52]

The PMDA does not give a timeline for completion of the disposition of the 34t. It was estimated that it would take between 15 to 20 years to fulfill the agreement. However, it mandates that the facilities used for disposition were to begin operating no later than December 31, 2007. The agreement has an "Annex on Schedules and Milestones," but the timeframe it provides for the construction of each state's necessary disposition facilities has to be revised and must take into consideration the changes from a November 2007 agreement (discussed later). The PMDA also commits each party to disposition of no less than 2t of excess Pu per year and to work, as soon as possible, with the international community in doubling that amount to 4t of Pu per year. The agreement does not allow the reprocessing of spent MOX fuel (a process that yields Pu) until all 34t are disposed in a once-through cycle and encourages refraining from reprocessing spent MOX fuel entirely, but does not prohibit the possibility of future reprocessing.

The PMDA also requires that the United States and Russia "cooperate in the management and disposition of plutonium, implementing their respective disposition programs in parallel to the extent practicable." The agreement also states that at every stage of disposition the

---

[49]The formal name of the agreement is The Agreement between the Government of the United States of America and the Government of the Russian Federation on Scientific and Technical Cooperation in the Management of Plutonium that Has Been Withdrawn from Nuclear Military Programs.

[50]The U.S. Congress adopted the FY 1999 Emergency Supplemental Appropriations Act presented by Senator Pete Domenici (R-NM) to appropriate $200 million to the Russian program on the condition of the conclusion of a U.S.-Russian agreement. Actual spending in 2000–2001 on plutonium disposition facilities in Russia was primarily limited to their design and planning and was funded separately from the emergency funds: $4.168 million in FY 2000 and $16.650 million in FY 2001.

[51]The formal name of the September 2000 agreement is the Agreement Between the Government of the United States of America and the Government of the Russian Federation Concerning the Management and Disposition of Plutonium Designated as No Longer Required for Defense Purposes and Related Cooperation.

[52]"U.S. to Add New Surplus Plutonium to Earlier Plans for Conversion into Reactor Fuel, Official Says," *Global Security Newswire*, Sept. 19, 2007, www.nti.org/d_newswire/issues/2007_9_19.html#72EB180F (Sept. 2007).

parties should ensure the "safety and ecological soundness of disposition plutonium activities" and apply effective control and accounting measures.[53]

Disagreement over the acceptability of immobilization as an option led Russia to choose MOX as its primary disposition method. Russia would not immobilize any of its 34t of WgPu slated for disposition and insisted that the U.S. use the MOX method for at least 25t of its material. The PMDA allows the remaining 9t of the U.S. material to be immobilized.

Implementation Challenges of the PMDA

Several open-ended issues were left unresolved by the PMDA, and their future resolution has become vital to the prospects for successful implementation of the agreement. Although some of these issues were either fully or partially resolved, their resolution did not occur without significant costs and delays. After the agreement was signed, several technical issues required investigation, including the design of MOX fuel fabrication facilities and reactor modifications. In addition, several key implementation issues had to be resolved, such as the development of a verification and monitoring/inspection regime; the negotiation of liability provisions; and securing funding for the operation of the program.

Infrastructure

Unlike the HEU downblending program, which to a large degree had a preexisting infrastructure, the Pu disposition program required the construction of costly new facilities and modifications to reactors not originally designed to use MOX fuel. Prior to negotiating the PMDA, the United States and Russia had investigated the use of MOX fuel with some limited use of Pu in reactors. However, these efforts were small-scale, had never used WgPu and were far from being capable of meeting the 2t of dispositioned material per year required by the PMDA. The implementation of the agreement requires the construction of a conversion facility to convert WgPu into Pu oxide. For full implementation of the agreement, both states needed to construct industrial size MOX fuel fabrication plants. The United States also required the construction of a pit disassembly and conversion facility, a facility for immobilization, and a site for waste solidification. Russia intended to use existing facilities for pit disassembly and conversion at Mayak and Seversk.[54]

The PMDA allowed light water reactors for MOX fuel use in the United States and Russia. The agreement also gave Russia permission to use its fast neutron reactors, the BN-600 reactor at Beloyarsk nuclear power plant and an experimental research reactor, the BOR-60 in Dimitrovgrad.[55] All the reactors approved by the PMDA would require modifications to burn WgPu. In 2005–2006, Russia renewed the construction of its second commercial fast neutron reactor, the BN-800, also located at the Beloyarsk plant. Russia prefers to use fast reactors rather than light water reactors to burn WgPu and hopes to bring additional fast reactors on line in the next decade. A Russian nuclear regularly agency, Gosatomnadzor (now Gostechnadzor), maintains that using MOX fuel in light water reactors would significantly

---

[53]Agreement Between the Government of the United States of America and the Government of the Russian Federation Concerning the Management and Disposition of Plutonium Designated as No Longer Required for Defense Purposes and Related Cooperation (Sept. 1, 2000).

[54]Bunn, *Securing the Bomb 2006*, "Reducing Excess Stockpiles: Russian Plutonium Disposition."

[55]PMDA, Sept. 1, 2000.

decrease their safety. Making the necessary modifications and licensing of modified reactors may cause significant delays.[56]

The use of other reactor types was also investigated. In 2000, through a joint agreement among the United States, Russia, and Canada called the Parallex Project, gram quantities of WgPu transformed into MOX fuel were shipped to the Atomic Energy of Canada Limited from Russia and the United States for the parallel experimental use of the material in a converted Canadian Deuterium Uranium (CANDU) pressurized heavy water reactor.[57] This test was the first of its kind and was a demonstration of the ability to disposition Pu in CANDU reactors and in parallel through a third party.[58] Although the initial shipments and tests were deemed a success, there have been no further shipments by the United States or Russia to Canada and current plans do not call for using CANDU reactors for Pu disposition.

Russia and the United States also had fairly well-developed programs for immobilizing radioactive wastes in glass, but once again, they lacked the experience of immobilization with WgPu and a large-scale infrastructure.

The construction and operation of the aforementioned facilities required significant financial commitments. Increasing costs became yet another unforeseen challenge for both sides. Early cost estimates were $4 billion for the United States and $1.7 billion for Russia. By 2000, estimates for the Russian program had reached $2 billion.[59] Similarly, by 2002 the cost estimate of the U.S. program had ballooned to $6.2 billion.[60] As of 2005, the cost estimates of the construction and design of just the U.S. MOX facility alone have risen to $3.5 billion, $2.5 billion more than the original estimates in 2002.[61] These estimates factor in the assumed value of energy captured by displacing low enriched or natural uranium with Pu in MOX fuel. A 1996 DOE report puts the energy recovered from using MOX at $1.4–2 billion, depending on the type of light water reactor used and the market price of uranium.[62] As of 2007, the price of uranium had risen considerably; however, the increasing costs of the Pu disposition program offset these changes in price.

Liability

The PMDA leaves liability provisions open to negotiation "at the earliest possible date."[63] In 2003, the tension over the scope of the liability coverage for U.S. contractors and participants

---

[56]German Solomatin, "*V Rossii politicheskiye resheniya po proizvodstvu MOKS-topliva operezhayut tekhnicheskiye vozmozhnosti, schitayut v Gosatomnadzore RF*," ITAR-TASS, Nov. 28, 2003; in Integrum-Techno, www.integrum.com.

[57]Elena Sokova, "Russia: MOX Fuel Overview," Center for Nonproliferation Studies: NIS Nuclear and Missile Database, May 4, 2001, Feb. 11, 2007, www.nti.org/db/nisprofs/russia/fissmat/mox/moxover.htm.

[58]U.S. Dept. of Energy, "Agreement Reached on Joint Non-Proliferation Experiment," News release R-99-229, Sept. 2, 1999.

[59]Elena Sokova, "Issue Brief: Plutonium Disposition," *Nuclear Threat Initiative*, July 2002, Feb. 5, 2007, www.nti.org/e_research/e3_11a.html.

[60]National Nuclear Security Administration, "Report to Congress: Disposition of Surplus Defense Plutonium at Savannah River Site" (Washington, D.C.: Office of Fissile Materials Disposition, Feb. 15, 2002), ch. 3, p. 3-3.

[61]U.S. Dept. of Energy, "Audit Report: Status of the Mixed Oxide Fuel Fabrication Facility" (Washington, D.C.: Office of Inspector General, Office of Audit Services, Dec. 21, 2005), p. 1.

[62]U.S. Dept. of Energy, *Technical Summary Report for Surplus Weapons-Usable Plutonium Disposition, rev. 1.* (Washington D.C.: Office of Fissile Materials Disposition, Oct. 31, 1996), ch. 4, pp. 4–6.

[63]Agreement Between the Government of the United States of America and the Government of the Russian Federation Concerning the Management and Disposition of Plutonium Designated as No Longer Required for Defense Purposes and Related Cooperation, Sept. 1, 2000.

in the Russian program led to a deadlock in negotiations for a liability agreement. The disagreement was based on the insistence of the United States that the U.S. workers and other representatives involved in PMDA projects in Russia would receive full indemnity from all wrongdoing, whether intentional or not. This stalemate was the reason behind the expiration of the 1998 agreement on technical cooperation in managing Pu. Although this agreement was essential for moving forward with the Pu disposition program, some efforts continued without liability coverage, albeit on a smaller scale, such as investigations into the design for a MOX fuel fabrication facility. Significant progress in the program was stalled because of the liability disagreements. A liability agreement was finally negotiated in the fall of 2005, but the actual signing of the agreement did not occur until September 2006.

Verification and Transparency

The PMDA underscores the importance of developing inspection/monitoring procedures and associated IAEA verification measures for Pu disposition. However, no details on the procedures or mechanisms for this regime are specified in the agreement. The creation of a verification and monitoring/inspections regime was complicated by the sensitivity of both states toward classified material and the need to sufficiently verify compliance of both parties to the agreement.

The PMDA calls for "consultations with the IAEA at an early date" to implement verification measures no later than the first delivery of Pu meant for disposition at a conversion or immobilization facility.[64] The PMDA also requires negotiation of a future agreement on monitoring and inspections to be concluded no later than December of 2002. This deadline has long since past.

One attempt to establish a verification regime to monitor fissile material excess to military needs was the 1996 launch of a cooperative effort among the IAEA, the United States, and Russia called the Trilateral Initiative. The parameters for IAEA verification were that the United States and Russia would voluntarily submit fissile material subject to IAEA inspections to the IAEA and that once the material was submitted, it would remain under IAEA management. Although significant technical, legal, and financial progress was made, especially in resolving the sensitive issue of verification of classified fissile material, several key unresolved issues prevented the further development of the verification regime. Among the unresolved issues were the organizational and funding challenges for the creation and operation of this regime as well as principal disagreements on key issues, such as permanent supervision by the IAEA of the submitted material and reluctance by Russia and the United States to commit any sites or material to IAEA monitoring.[65]

Political and Bureaucratic Impediments

When the Bush Administration came to office, it ordered a review of all nonproliferation programs with Russia, including the Pu disposition program. The administration doubted the feasibility and nonproliferation value of Pu disposition due to its high costs and operational uncertainties. Anticipation of the possible abolition of the Pu disposition program caused the Pu disposition effort to lose momentum at an early stage and raised doubts about committing additional financial support for the Russian portion of the program.

---

[64]Agreement Between the Government of the United States of America and the Government of the Russian Federation Concerning the Management and Disposition of Plutonium Designated as No Longer Required for Defense Purposes and Related Cooperation, Sept. 1, 2000.

[65]Bunn, *Securing the Bomb 2006*, "Monitoring Stockpiles: IAEA Monitoring of Excess Nuclear Material."

The National Security Council, charged with reviewing the program, completed its' review in December 2001. It recommended the continuation of the Pu disposition program but emphasized the need for it to be less costly and more efficient. The DOE reported that canceling the immobilization option would save the United States $2 billion in total program costs and accelerate closure of former nuclear weapons complex sites. As a result of the review, the United States dropped the immobilization method and the burning of MOX fuel in reactors became the only option for disposition of the mandated 34t of Pu.[66] Immobilization was later considered only for disposition of impure Pu and not WgPu.

In addition to the stalemate in securing funding for the Russian program, liability disagreements, and other issues, the requirement by the PMDA and the U.S. Congress that the U.S. and Russian programs run in parallel also impedes the progress of the U.S. program. In 2006, after many years of insisting on this requirement, the U.S. Congress indicated that it is inclined to decouple the U.S. program from Russia's. The DOE's Fiscal Year 2008 budget request includes funding for the U.S. disposition program and provides no new funding for the Russian program, relying entirely on past-year balances. However, it could take some time to finalize the decision on decoupling, and the implications and process of decoupling the U.S. program from Russia's remain unclear.

Progress to Date

In addition to resolving these issues, several other tasks have to be accomplished before full-scale implementation of the PMDA can proceed. These include the selection of sites and designs for the construction of the facilities, such as MOX fuel fabrication and other necessary facilities, as well as the completion of studies for the modification of existing reactors that would eventually use MOX fuel.

In 1999, the U.S. chose the Savannah River site (SRS) in South Carolina as the location for the immobilization, conversion, and MOX fuel fabrication facilities.[67] The SRS would receive all the excess Pu stored within the United States, thereby consolidating it all at one site and closing down all the others—a cost-saving measure that would save millions of dollars per year.[68] Also in 1999, the U.S. contracted with Duke, Cogema, and Stone & Webster and chose a French design that had been in use in France at the MELOX and La Hague MOX fuel fabrication plants since the 1960s. Although some money has been allocated to the project construction of the facility did not begin until August of 2007.

Russia selected the Siberian Chemical Combine in Seversk as the site for construction of its industrial-scale MOX fuel fabrication plant. Russia had two preexisting pilot MOX fuel production plants, both located at Mayak. Although the two Mayak plants had experimented with the fabrication of MOX fuel with WgPu, they have limited throughputs that are far below the industrial size needed for the PMDA mandated disposition rate of 2t per year.[69] Various options for the industrial scale MOX fuel fabrication plans were considered, including Russia's own research and design of a MOX fuel fabrication plant. The purchase of a prefabricated MOX plant from Hanau, Germany. France, Germany, and Japan

[66]National Nuclear Security Administration, *Report to Congress: Disposition of Surplus Defense Plutonium at Savannah River Site*, p. 2.

[67]U.S. Dept. of Energy, *Record of Decision for the Surplus Plutonium Disposition Final Environmental Impact Statement* (Washington D.C.: Office of Fissile Materials Disposition, Nov. 1999), ch. S-10-11.

[68]Bunn, *Securing the Bomb 2006*, "Reducing Excess Stockpiles: U.S. Plutonium Disposition."

[69]Anatoliy Diakov, "Status and Perspectives for MOX fuel production in Russia," *Energy and Security* 3 (1997). Institute for Energy and Environmental Research, Feb. 5, 2007, www.ieer.org/ensec/no-3/diakov.html.

funded and participated in several projects, particularly in the 1990s, devoted to the development of a MOX fabrication facility and other studies of Pu disposition. However, because of the time constraints and high costs associated with creating a new design and Germany's decision to sell its MOX plant to Japan, in 2002 the United States offered Russia the use of the same French design that was to be used by the U.S.[70] Russia's decision to use the French design was viewed as a positive step toward the implementation of the PMDA and its commitment to parallelism.

In November 2007, the U.S. and Russia reaffirmed their commitment to the PMDA and introduced several adjustments to the agreement. A joint U.S.-Russian statement allows Russia to use its fast breeder reactors (BN-600 and BN-800) to irradiate the Pu in MOX form. The original proposal to use light water reactors in Russia is no longer under consideration. According to the new plan, disposition will start in 2012, first using the BN-600 and followed by the BN-800 soon thereafter. The use of these reactors would provide disposal of an estimated 1.5t of Pu per year, i.e., at a slower rate than in the original agreement, and unless new reactors are brought on-line it may take 51 years to dispose of all 34t of Pu. The new joint statement also commits the U.S. to providing Russia $400 million toward this goal and calls for DOE/Rosatom efforts to secure funding from international donors to the timely implementation of the program in Russia. Finally, cooperation is encouraged between DOE/Rosatom on research and development of an advanced gas-cooled high-temperature reactor to determine if this may or may not speed up Pu disposition in the near future.[71]

Funding Commitments to Russia's Program

International financing of Pu disposition in Russia, compared to the estimated total cost of Russian Pu disposition of $2 billion ($1 billion in startup costs and $1 billion for operation of the program) and growing, has been significantly below the mark and continues to be one of the largest barriers to the initiation of the program. Russia has insisted on at least $1 billion being available before starting the groundbreaking of its MOX fabrication facility. By the beginning of 2008, international pledges, based on commitments from the G-8 summit in 2002, had met only $800 million of the $2 billion.[72]

In 2000, MinAtom offered to resolve the financing stalemate by leasing or selling the MOX fuel fabricated from WgPu to Europe for use in its reactors. The spent nuclear fuel would then be returned to Russia for permanent storage. The revenue from the lease of fuel was to be used to finance the operational costs of Russia's Pu disposition program. In addition, Russia would not have had to convert its own reactors to the specifications necessary for using MOX fuel. This proposal did not receive the support of European states, and thus a funding conundrum for the Russian portion of the program remains to be one of the key impediments to its implementation.

From the start, Russia insisted that the implementation of the PMDA in Russia was contingent on securing full international funding for its program. G-8 and other foreign partners insisted on Russia's demonstrated commitment to the PMDA and the resolution of outstanding issues, such as the liability dispute, prior to committing additional funds above the $800 million pledged in 2007. The successful resolution of the most pressing issues in 2006–2007 might help provide more confidence in the execution of the PMDA and thus might aid

---

[70]Christine Kucia, "Russia agrees to use U.S. MOX facility design," *Arms Control Today,* Jan./Feb. 2003, Jan. 20, 2007, www.armscontrol.org/act/2003_01-02/mox_janfeb03.asp.

[71]U.S. Dept. of Energy, *National Nuclear Security Administration. Joint Statement on Mutual Understanding Concerning Cooperation on the Program for the Disposition of Excess Weapon-Grade Plutonium* (Washington, D.C.: Nov. 19, 2007), Feb. 11, 2008, www.energy.gov/news/5742.htm.

[72]Bunn, *Securing the Bomb 2006*, "Reducing Excess Stockpiles: Russian Plutonium Disposition."

in securing additional funding for the program. At the same time, the increasing Russian oil and gas revenues and its relative economic prosperity make it harder for Western countries to justify funding a Pu disposition program that essentially helps Russia to develop a domestic civilian closed MOX fuel cycle.

In short, despite the recent advances, both the U.S. and Russian Pu disposition programs are significantly behind schedule and uncertainties continue to plague their implementation. The U.S. launch in 2006 of the Global Nuclear Energy Partnership and an ambitious Russian nuclear power expansion program have the potential to accelerate the creation of the infrastructure and technologies for a MOX fuel cycle. The Pu disposition program, as it is now interpreted by the new 2007 agreement, may serve to boost further development of a nuclear fuel cycle based on fast neutron reactors and MOX fuel. Some experts argue, however, that a closed fuel cycle, fast reactor technologies, and the increased amount of civil Pu associated with these technologies would increase proliferation risks.[73]

Civil Pu

Although there is still an ongoing debate on the use of civilian Pu in generating nuclear power, many experts argue that separated civilian Pu and its use in the fuel cycle are just as dangerous, if not more dangerous, than military Pu.[74] It is the possibility of building a bomb with civilian Pu, proven doable by India in 1974, that triggered the decision by the United States, through the Carter Administration's Presidential Decision Directive in 1977, not to reprocess spent fuel.

Most states with nuclear power reactors store spent fuel and do not reprocess it. Only France, India, Japan, Russia, and the United Kingdom reprocess and separate Pu from spent nuclear fuel. Approximately 2300t, or one third of all spent fuel generated, is reprocessed annually.[75] Only a fraction of the separated civilian Pu, however, is fabricated into MOX

Table 19.2 Civilian stocks of separated Pu (metric tons).*

Country	2003–2005 Data
Belgium	3.5 (+0.4 abroad)
France	79 (30 t foreign owned)
Germany	12.5 (+13.5 t in U.K. and France)
India	5.4
Japan	5 (+37 t in U.K. and France)
Russia	41
Switzerland	~3 (in U.K. and France)
United Kingdom	103 (26 foreign owned + 1 t abroad)
Total	~252

*Based on the *Global Fissile Material Report 2006*, by the International Panel on Fissile Materials.

[73]Thomas Cochran and Christopher Paine, "Peddling Plutonium: Nuclear Energy Plan Would Make the World More Dangerous," *National Resources Defense Council*, March 2006, Mar. 17, 2007, www.nrdc.org/nuclear/gnep/agnep.pdf.

[74]For an in-depth discussion on this issue, see Frank von Hippel, IPFM Research Report #3, *Managing Spent Fuel in the United States: The Illogic of Reprocessing*, Jan. 17, 2007; www.fissilematerials.org/ipfm/site_down/ipfmresearchreport03.pdf.

fuel and recycled in reactors. Most civil stocks are stored at reactor sites and their stocks continue to grow.[76] Table 19.2 gives details on civilian stocks of Pu.

Although there are calls from nonproliferation experts to stop the reprocessing of spent nuclear fuel in order to put an end to the growth of civilian Pu stocks and deter the further use of Pu in the civilian nuclear cycle, no government or international organization sponsored initiatives to date have addressed civilian Pu. There are also no concrete efforts to dispose of existing stocks of separated civilian Pu.

Summary

About 50 states have weapons-usable fissile material on their territory. The consensus is that nuclear materials in weapons programs are better protected than civilian stocks. However, security standards for both civil and military fissile materials vary from state to state. Therefore, the uneven and incomplete application of physical protection measures and safeguards worldwide is of great concern. The recently amended Convention on the Physical Protection of Nuclear Materials and Facilities, U.N. Security Council Resolution 1540, and other initiatives call for strengthened standards for controlling, accounting, and protecting fissile materials and make states responsible for their implementation. However, before these stringent security upgrades can be implemented, the new strengthened standards first need to be developed and approved.

Proper security of fissile materials requires a long-term commitment and resources. Common sense suggests that the less material and the fewer locations security measures need to be applied to, the lower the risk of their theft or diversion. In this context, the importance of the disposition of excess and vulnerable fissile materials can not be underestimated.

The elimination of surpluses and the clean-up of weapons usable materials worldwide is a sure way to reduce the threat of nuclear terrorism and proliferation. The U.S. and Russian programs to dispose of military HEU and Pu, though having the potential to make a significant contribution to eliminating excess fissile material, cover only a fraction of global stocks of weapons-usable fissile materials. Unilateral and bilateral U.S.-Russian efforts to downblend HEU from weapons programs have made good progress toward their established goals. However, the Pu disposition program has been plagued by delays and setbacks. Even after the completion of these two large-scale disposition efforts, the United States and Russia will still retain stocks of military fissile materials far beyond their defense program needs. Other states with weapons-usable fissile materials have not started their own efforts to eliminate or even declare their surpluses (except for the Pu surplus declared by the U.K.). The long overdue conclusion of an international agreement banning the production of new fissile materials that could be used in weapons might be helpful in setting a global norm to eventually reducing existing stocks of weapons usable fissile materials. As holders of the largest stocks, the United States and Russia will have a key role in this process. A voluntary promise to significantly reduce their stocks below current commitments would send a strong signal to other states and might expedite the conclusion of an FMCT and/or other states' decision to eventually declare and reduce their own stocks.

[75] *Global Fissile Material Report 2006*, p. 29.

[76] *Global Fissile Material Report 2006*, p. 30.

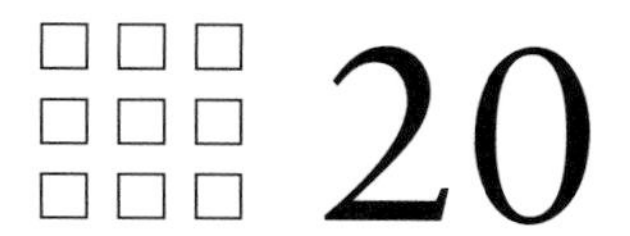

Case Study: Dismantlement and Radioactive Waste Management of DPRK Plutonium Facilities

George Baldwin and Jooho Whang

Introduction

The decommissioning and dismantlement problems for the Democratic People's Republic of Korea (DPRK's) two principal operational plutonium facilities at Yongbyun, the 5 MWe nuclear reactor and the Radiochemical Laboratory reprocessing facility, present a formidable challenge. Dismantling these facilities will create radioactive waste in addition to existing inventories of spent fuel and reprocessing wastes. The legacy wastes from these two facilities will result in at least 50 to 100 metric tons of uranium spent fuel, as much as 500,000 liters of liquid and other high-level waste, 600 metric tons of graphite from the reactor, an undetermined quantity of chemical decladding liquid waste from reprocessing, and hundreds of tons of contaminated concrete and metal from facility dismantlement. Various facilities for dismantlement, decontamination, waste treatment and packaging, and storage will eventually be needed. The shipment of spent fuel and liquid high-level waste out of the DPRK is also likely to be required.

Nuclear facility dismantlement and radioactive waste management in the DPRK are all the more difficult because of nuclear nonproliferation constraints, such as calls for *complete, verifiable, and irreversible dismantlement*, or CVID. It is desirable to accomplish dismantlement quickly, but many aspects of the radioactive waste management cannot be achieved without careful assessment, planning and preparation, sustained commitment, and long completion times. The radioactive waste management problem in fact offers a prospect for international participation to engage the DPRK constructively. DPRK nuclear dismantlement, when accompanied with a concerted effort for effective radioactive waste management, can be a mutually beneficial goal. This and other technical assessments of the dismantlement and radioactive waste management issues should prove useful for evaluating policy options during continuing negotiations with the DPRK regarding the elimination of its nuclear weapons program.

Background

The North Korean nuclear weapons program has drawn serious international attention since the early 1990s. In 1994, North Korea (DPRK) agreed to freeze its nuclear weapons program in exchange for energy and economic aid from the United States and South Korea (ROK), including the provision of two light water reactor (LWR) units and arrangement for interim energy alternatives. According to the so-called Agreed Framework between the U.S. and DPRK, the DPRK eventually agreed to dismantle its graphite-moderated reactors and related facilities. The U.S. and DPRK were supposed to cooperate in finding a method to store the spent nuclear fuel and to dispose of it in a safe manner that does not involve reprocessing in the DPRK. Although the technical and economic aspects of dismantlement were not treated in detail, the dismantlement was to be accomplished in coordination with the target date for the completion of the LWR project.

Unfortunately, the Agreed Framework broke down when International Atomic Energy Agency (IAEA) inspectors were expelled by the DPRK in 2002, and the DPRK denounced the Agreed Framework and resumed its push to develop nuclear weapons.

In 2003, negotiations on the DPRK's nuclear program resumed at the Six-Party Talks, which involve the DPRK, South Korea, the U.S., China, Russia, and Japan. The Six-Party Talks have gone through several rounds since then, with intervening lapses and diplomatic difficulties. A tenuous compromise agreement concluded in 2005 calls on the DPRK to "abandon" its nuclear weapons program.[1] The dismantlement of the DPRK's plutonium facilities will be an essential milestone in the fulfillment of the 2005 Joint Statement.

Elimination of the DPRK's nuclear weapons program and disposition of its nuclear materials and spent nuclear fuel are critical to improving regional stability. Many papers have dealt strategically[2,3,4] with how the North Korean nuclear weapons program would be frozen and eliminated in a complete, irreversible, and verifiable manner, if an agreement can be reached. Technical papers[5,6] have mostly been concerned with how to verify and reveal the status and history of its nuclear weapons program. This report highlights the technical challenges posed by the dismantlement and radioactive waste management of the DPRK's key plutonium facilities.

Objectives of Dismantlement

Denuclearization could involve several related issues or objectives. One objective is to discontinue current production of nuclear weapons-usable material, particularly plutonium. Another objective would be to prevent future production of the material. An additional objective would be to enable forensic analysis to determine the past production of plutonium. Further objectives include finding and accounting for all existing plutonium and then removing or otherwise disposing of that material.

A serious concern related to denuclearization is preventing the use of radioactive waste as source material for a radioactive dispersal device (RDD, or "dirty bomb"). Cleaning up

---

[1]See "Joint Statement of the Fourth Round of the Six-Party Talks Beijing, September 19, 2005," U.S. Department of State, www.state.gov/r/pa/prs/ps/2005/53490.htm (July 2007).

[2]Leon V. Sigal, *Disarming Strangers: Nuclear Diplomacy with North Korea,* Princeton University Press (1999).

[3]Duk-ho Moon, *North Korea's Nuclear Weapons Program: Verification Priorities and New Challenges,* Cooperative Monitoring Center Occasional Paper, SAND 2003-4558, Sandia National Laboratories (2003).

[4]David Albright and Corey Hinderstein, "Verifiable, Irreversible, Cooperative Dismantlement of the DPRK's Nuclear Weapons Program," *Proceedings of the Institute of Nuclear Materials Management,* 45th Annual Meeting (2004).

[5]Jared S. Dreicer, "How Much Plutonium Could Have Been Produced in the DPRK IRT Reactor?" *Science and Global Security,* vol. 8, pp. 273–286 (2000).

[6]David Albright, "North Korea's Current and Future Plutonium and Nuclear Weapon Stocks," ISIS Issue Brief, Institute for Science and International Security, Jan. 15, 2003.

the environmental mess associated with the nuclear program is still another important issue and would be of particular concern to advocates of Korean unification.

A political agreement would decide which of these objectives are encompassed. Whatever the eventual settlement, it is certain that radioactive waste management will be a critical issue.

Consequences of Dismantlement

Key steps in denuclearization include disposing of nuclear materials and dismantling the facilities. Each step consists of many procedural components for which technologies to be applied should be sought in economically viable and proliferation-resistant ways.

Dismantling the nuclear facilities will generate a large volume of radioactive waste containing various levels of radioactivity to be treated and disposed of safely. Considering that it might have inherited the poor radioactive waste management practices of the former Soviet Union, the DPRK might not be operating nuclear facilities in compliance with international safety standards.

Without help from other nations, the DPRK is unlikely to address the radioactive waste problem as part of denuclearization. This is unlike South Africa, which dismantled its weapons program and managed resulting radioactive waste without foreign assistance. Lack of knowledge of the nuclear weapons program infrastructure complicated IAEA verification of South Africa's dismantlement, which took approximately two years.

Need for Technical Work in Advance

Dismantling the infrastructure of a nuclear weapons program is both technically and economically complicated, requiring large resources of money, manpower, technology, and regulatory work. After reaching a strategic dismantlement agreement, substantial issues would still remain and require extensive discussion to resolve. Allocation of roles and responsibilities of the interested parties is one example.

To facilitate progress, much technical work can be considered in advance, including decisions on how to prepare and condition nuclear materials such as spent fuel for interim storage and shipping, reprocessing the spent fuel, solidifying high-level waste (HLW) stored in tanks, and storing and disposing of solid waste from dismantlement. It would help to assess the resources needed to dismantle the facilities and to treat the radioactive waste from the facilities. As a spin-off benefit, creating a good scheme for cooperation would be a starting point for, or component of, multilateral management of the nuclear fuel cycle in East Asia, potentially enhancing the nonproliferation regime.

Scope and Limitations

The scope of this discussion is limited to the plutonium-relevant nuclear facilities in the DPRK, specifically the 5 MWe graphite reactor and the associated Radiochemical Laboratory. We identify key issues and critical paths for dismantling the facilities and managing the resulting radioactive waste. We estimate the amount of waste resulting from dismantlement and draw on international experience with similar facility decommissioning to estimate timeframes and consider various alternatives from a technical point of view. The key plutonium facilities in the DPRK are relatively well known, which makes reasonable estimation possible.[7]

---

[7]Less is known about the suspected uranium enrichment program, which makes estimating the associated radioactive waste management from dismantlement difficult. It can be safely assumed, however, that such consequences only *add* to the consequences we deal with in this study. Also, from a technical point of view, the plutonium program dismantlement would probably be the most problematic.

Plutonium-Relevant Nuclear Facilities

Of the known nuclear reactor facilities in the DPRK, only two research reactors and a 5 MWe power reactor have been operational. The main facilities of concern include the 5 MWe graphite moderated reactor that had been operated since 1986 for the primary purpose of producing plutonium for nuclear weapons. Larger units, rated at 50 MWe, 200 MWe, and 1000 MWe, were under construction at the time of the Agreed Framework but were never operated. These reactors would be new threats if their construction resumed and operations began, but otherwise their decommissioning would not be a major issue. A large reprocessing facility (the Radiochemical Laboratory) has been operated to recover plutonium from the reactor spent fuel. All these facilities are located within several kilometers of each other, near Yongbyun, DPRK.

The 5 MWe Nuclear Reactor

Characteristics

The 5 MWe (20 MWth) graphite reactor at Yongbyun is a Magnox-type reactor. Magnox reactors have a graphite-moderated reactor core that is cooled by carbon dioxide gas. The uranium fuel is loaded and unloaded from the top of the reactor. Fresh fuel is supplied from a fuel fabrication plant located at the same site. Spent fuel withdrawn from the reactor is transferred to a storage pool.

The name "Magnox" comes from the cladding of the uranium fuel, which is an alloy of magnesium. The fuel itself is uranium metal, which corrodes much more easily than uranium oxide fuel found in light water reactors. In contact with water, metallic uranium converts to uranium oxide and uranium hydride. Both uranium metal and uranium hydride are pyrophoric, which presents a fire hazard.

Exact design features of the 5 MWe reactor have not been released by the DPRK, but it is reported to have adopted the design of the U.K. Calder Hall reactors. The Calder Hall reactors began operation in 1956, producing weapons-grade plutonium and electricity on a small scale. Estimates of the design characteristics of the 5 MWe reactor are presented in Table 20.1.[8]

Operational History and Status

The operational history of the 5 MWe reactor is a key factor with regard to nonproliferation objectives. This history is key to determining the amount of spent fuel discharged and, consequently, the amount of plutonium contained in the spent fuel.

The DPRK began construction of the 5 MWe reactor in 1980. It started operation in 1986 and had been in operation for eight years when the negotiated freeze began in 1994. The operation history of these eight years is not known except that the reactor experienced some abnormalities. Verifying the operational history of the 5 MWe reactor from 1986 to 1994 has been a concern to many investigators. Inconsistency between the DPRK's report to the IAEA and the IAEA's independent findings made investigators speculate on various operational options the DPRK might have taken.

In normal operation, the 5 MWe reactor could produce about 6 kg of plutonium per year.[9] Based on all available information, and making many assumptions to figure out the amount of spent fuel withdrawn and reprocessed, experts generally conclude that the DPRK has not drawn out more than two full cores (16,000 fuel elements). The DPRK restarted the

---

[8]David Albright and Kevin O'Neill, ed., *Solving the North Korean Nuclear Puzzle,* The Institute for Science and International Security, Washington D.C., 2000, 161–2.

[9]Albright and O'Neill, p. 144.

Table 20.1 Characteristics of the 5 MWe reactor (estimated).

Burnup,[1] maximum	1,370 MWD/tU
Burnup, average	635 MWD/tU
Initial fuel loading	50 tU
Graphite moderator	300 t
Graphite reflector	300 t
Reactor core:	
Effective height	590 cm
Effective diameter	643 cm
Number of fuel channels	812
Fuel elements/channel	10
Fuel elements:	
Core	Natural uranium, 0.5% aluminum
Cladding	Magnesium, 0.5% zirconium
Diameter	3.0 cm
Length	60 cm
Uranium mass	6.24 kg
Reactor vessel (steel)	
Inner diameter	880 cm
Height	1,680 cm
Thickness	4.0 cm
Shield:	
Thermal shield (steel) thickness	7.0 cm
Upper concrete thickness	450 cm
Radial concrete thickness	300 cm

[1]Specific burnup of the fuel is termed as the fission energy released per unit mass of the fuel, usually expressed in megawatt days per metric ton or per kg, i.e., MWD/tU or MWD/kgU.

5 MWe reactor in February 2003. IAEA inspectors announced the reactor was shut down as of mid-July 2007.[10]

Spent Fuel Arising

Following the 1994 Agreed Framework, the 5 MWe reactor was stopped and its full core of about 8,000 fuel elements was withdrawn. The 8,000 spent fuel elements were packaged into 400 stainless steel cans with inert gas and stored under water. IAEA inspectors sealed the cans to detect possible tampering. The spent-fuel canning operation began in April 1996 and finished in mid-1999. Canning took longer than expected due to acquisition of required tools, purification of pool water, and removal of sludge from the pool bottom. Since December 2002, when the IAEA inspectors were dismissed by DPRK, nothing has been confirmed about what happened to the spent fuel storage cans or their contents.

However, U.S. visitors to the Yongbyun site in late 2003 confirmed that there are no longer any cans in the spent-fuel storage pool.[11] The DPRK claims that it removed all 8,000

[10]"North Korea Closes All Yongbyon Nuclear Facilities, IAEA Says," *Global Security Newswire,* July 18, 2007, www.nti.org/d_newswire/issues/2007_7_18.html#4DED98AD (July 2007).

[11]Sigfried Hecker, testimony before the U.S. Senate Committee on Foreign Relations, Jan. 21, 2004, http://foreign.senate.gov/testimony/2004/HeckerTestimony040121.pdf.

spent-fuel elements from the storage pool and reprocessed them, starting in mid-January 2003 and finishing by the end of June 2003. The visitors were not able to determine that the Radiochemical Laboratory had been operated during the first half of 2003. Two possibilities are that all 8,000 spent-fuel elements were reprocessed as they claimed or that they simply moved the spent-fuel elements to another storage area. A third possibility is that the DPRK reprocessed only some of the spent-fuel elements while storing the remainder elsewhere.

The number of spent-fuel elements discharged from the 5 MWe reactor is a serious question from the viewpoints of both nuclear material nonproliferation and radioactive waste management. If the DPRK did not reprocess the spent fuel removed from the storage pool and simply stored it without special care in a dry pit or another water pool, the spent fuel may have undergone severe corrosion. Magnox spent fuel presents a spontaneous fire hazard due to uranium hydride and may leak fission products through corroded cladding, eventually contaminating the surroundings.

The spent fuel from the 5 MWe reactor has a relatively low average burnup (635 MWD/tU), which is about one tenth that typical of CANDU reactors. Although the fuel should be less radioactive than CANDU spent fuel, it still needs a thick layer of water or concrete for radiation shielding.

Radiochemical Laboratory

Characteristics

The Radiochemical Laboratory is the main reprocessing facility at Yongbyun. The building is 192m long, 27m wide, and six stories high. In the reprocessing building, six process cells are located on the first floor and three sampling cells are on the second floor. Spent fuel is processed in batch mode in two process lines, each with the following major components:[12]

- Cladding dissolver, 20 spent-fuel elements per batch
- Fuel dissolver, five spent-fuel elements per batch
- Thirty mixer-settlers, each 80 liter capacity
- Five glove boxes for further processing of plutonium

Construction of the radiochemistry lab began in 1984. In 1992, the IAEA experts saw that one process line was almost completed, lacking only the final step to reduce the volume of waste. In 1994, the construction of the second process line was observed to be complete. Since then it has not been known whether the DPRK added the final step for waste volume reduction. If the HLW passes through the volume reduction step, most of the aqueous part of the waste is removed, but the radioactivity carried by the solids remains the same. Thus, the HLW becomes more concentrated in radioactivity.

Different opinions exist regarding the Radiochemical Laboratory's capacity. One is that the facility is capable of reprocessing 220 to 250 metric tons of uranium fuel per year using two process lines;[13] the other is based on a statement the DPRK made to a visitor that the capacity of the Radiochemical Laboratory is 110 metric tons of uranium per line per year.[14]

An analysis laboratory is located to the north of the reprocessing building, and several waste-related tanks and buildings are within the facility boundary. A suspicious building was found by satellite images to connect to the reprocessing building. Its large containment structure, 67m by 24m by 9m, is probably used for storing waste from the reprocessing building. Half the volume may be for liquid waste storage and the other half for solid waste.

The DPRK adopted the Purex reprocessing method, a solvent extraction process, for separating plutonium. In a solvent extraction process, specific solutes dissolved in aqueous

[12]Albright and O'Neill, 154-6.

[13]Albright and O'Neill, 149.

[14]Hecker.

phase move to an organic solvent, due to their higher dissolution coefficients. The Purex method is commonly adopted by commercial reprocessing facilities for slightly enriched uranium fuel from power reactors. The basic method applies to fuel clad with stainless steel or zircaloy, but the DPRK Radiochemical Laboratory would have modified the process for Magnox fuel. Based on discussions so far, the DPRK-modified Purex process may include only the steps needed to produce plutonium and thus differs from the conventional Purex process:

1. Fuel with cladding is put into the fuel dissolver as a preparatory step in reprocessing. In the basic Purex process, only the fuel itself dissolves, leaving solid cladding hulls behind. But in the DPRK case, it is the cladding that is dissolved first, which is then removed as liquid waste. Solid fuel material moves on and is dissolved in the following step.
2. In the primary decontamination step in the Purex process, a large volume of nitric acid containing dissolved radioactive materials is removed as HLW (uranium and plutonium remain in the organic solvent). In commercial reprocessing plants, nitric acid recovery systems are added to reduce the volume of this HLW, but that might not have been done in the DPRK.
3. For light water reactors with enriched uranium fuel, impure uranyl nitrate from the reprocessing stream undergoes purification and conversion to recover enriched uranium. In the DPRK case, there is no incentive to recover the (natural) uranium in the uranyl nitrate, and so it becomes essentially another waste stream.

Operational History and Waste

The DPRK reported to IAEA in early 1989 that fresh fuel was used for a "cold test" of the reprocessing facility, followed by a hot test with 86 irradiated fuel elements. IAEA investigation found a discrepancy between the DPRK's report and sample analysis. Experts suspect, based on the results of sampling and analysis, that the DPRK might have reprocessed 25 to 50 metric tons of spent fuel, corresponding to 4,000 to 8,000 spent-fuel elements, during 1989 to 1991. In late 2003, the DPRK insisted that it had removed 8,000 canned spent-fuel elements from the storage pool and reprocessed all of them.

Reprocessing generates wastes of all kinds—gaseous, liquid, and solid—and of varying degrees of radioactivity. Wastes can be categorized by various criteria, such as the level of radioactivity, concentration in the waste, the nature of the radionuclides present in the waste, and the properties of the host medium. We use three classifications of high-, intermediate-, and low-level waste (HLW, ILW, and LLW, respectively).

The various waste streams involved in the Purex process include:

- The aqueous waste stream from primary decontamination, the first step in the solvent extraction, contains 99–99.9% of the fission products and is HLW.
- The ILW consists of the LLW concentrate, contaminated aqueous solutions from solvent washing, and other streams with appreciable solid content. Solutions from the Mg cladding dissolver ("chemical decladding waste") are ILW.[15]
- Liquid waste from plutonium purification and other decontamination steps is LLW. LLW streams are much less important and will not be considered here.

The DPRK is assumed to have operated the Radiochemical Laboratory for two main series of campaigns—the first from 1989 to 1991, as experts suspect, and the second during the first six months of 2003, as the DPRK insists. Further, it is assumed that each series reprocessed 50 metric tons of spent fuel and that there is no facility or process line to reduce waste volume (denitration of HLW, acid recovery, etc.).

---

[15] Jean-Marc Wolff, *Eurochemic 1956–1990: Thirty-five years of international co-operation in the field of nuclear engineering: The chemical processing of irradiated fuels and the management of radioactive wastes,* NEA/OECD, Paris, 1996.

Rough estimates of the waste volume associated with these two campaigns would suggest:

HLW: 500,000 liters (5,000 liter/tU × 50 tU/campaign × 2 campaigns)
ILW: 300,000 liters (3,000 liter/tU × 50 tU/campaign × 2 campaigns)
Uranyl nitrate: About 100 tU (50 tU/campaign × 2 campaigns)

Decommissioning the Facilities

International experience provides some technical basis for making estimates concerning DPRK facility decommissioning. To date, 90 commercial power reactors, more than 250 research reactors, and many other nuclear fuel-cycle facilities worldwide have been shut down. Some of them have been fully dismantled. Most of the Magnox-type reactors in the U.K., France, Italy, and Japan have also been shut down, some as far back as the 1980s. Shutting down Magnox reactors has sometimes been delayed to determine decommissioning strategies.

The IAEA, the OECD/NEA, and the World Nuclear Association have compiled the experience with decommissioning reprocessing plants from countries with advanced nuclear fuel cycles. More effort is required to compare cost data from many countries in a standard format. Nevertheless, proven techniques and equipment are available for decommissioning such nuclear facilities.

Many factors must be taken into account in developing a strategy from available options for decommissioning. These include safety and environmental issues, requirements for possible reuse of the plant and/or site, quantity and types of waste produced, availability of waste disposal sites, worker dose, cost and availability of funding, and consideration of sustainability and intergenerational equity arguments and stakeholder views. In the DPRK case especially, the decommissioning strategy must support nuclear nonproliferation and security issues.

Decommissioning Options

The IAEA has defined three options for decommissioning nuclear facilities: DECON, SAFSTOR, and ENTOMB:[16]

- DECON involves "immediate" dismantlement. Soon after the nuclear facility closes, equipment, structures, and portions of the facility containing radioactive contaminants are removed and decontaminated to a level that permits release of the property and termination of the license.
- SAFSTOR is a delayed DECON. A nuclear facility is maintained and monitored in a condition that allows the radioactivity to decay; afterwards it is dismantled.
- With ENTOMB, radioactive contaminants are encased in a structurally sound material, such as concrete, and appropriately maintained and monitored until the radioactivity decays to a level permitting release of the property.

The U.S. NRC requires that the decommissioning of a facility be completed within 60 years, whatever option or combination of options is taken. On the other hand, the strategy for most U.K. reactors involves deferrals of up to 100 years, until the radioactivity decays to a level that permits direct handling or removal work.

[16]IAEA, *Decommissioning of Nuclear Fuel Cycle Facilities*, Safety Standards Series, No. WS-G-2.2, 1991, Vienna, Austria.

Decommissioning the 5 MWe Reactor

Since the 5 MWe reactor is a graphite moderator type and is known to be modeled after the Calder Hall reactors in the U.K., it is beneficial to review the status of Magnox or graphite reactor decommissioning. There are more than 100 graphite reactors, experimental as well as plutonium production reactors, worldwide. Most of the older graphite reactors are already shut down and waiting to be decommissioned. In decommissioning nuclear reactors, about 99% of the radioactivity is safely contained in the fuel, which is withdrawn from the reactor following permanent shutdown. Besides surface contamination of the plant, the remaining radioactivity is from structural and piping materials, steel, and concrete that have been exposed to neutrons. In gas-cooled reactors (GCRs), a large amount of graphite, used as a neutron moderator and reflector, also becomes radioactive.

Most GCRs built in the era of nuclear introduction were shut down in the 1980s. Before decommissioning began or even before the decommissioning planning phase, owners or stakeholders of some of the reactors seemed to have iterated through many reviews to set up optimized strategies considering technical, safety, and, consequently, economic factors. The following are common findings from reviewing the decommissioning plans and experiences of several countries with GCRs:

- They spend long lead times, usually more than 10 years, before setting up a decommissioning plan.
- The decommissioning phase before reaching SAFSTOR takes 15 to 20 years.
- Decommissioning requires waste management facilities for packaging, storage, and disposal.

In general, graphite reactors are an order of magnitude more costly to decommission than light water reactors due to the larger volumes and types of waste involved.

Decommissioning of Key Elements

Nuclear Graphite Waste

A typical graphite-moderated reactor contains a few thousand tons of "nuclear-grade" graphite[17] to moderate and reflect neutrons during reactor operation. The 5 MWe reactor has 600 metric tons of graphite. Graphite has different characteristics than those of other radioactive waste due to its physical and chemical properties and radioactive content. Waste graphite contains tritium, ^{14}C, corrosion/activation products (^{36}Cl, ^{57}Co, ^{60}Co, ^{54}Mn, ^{59}Ni, ^{63}Ni, ^{22}Na), fission products (^{134}Cs, ^{137}Cs, ^{90}Sr, ^{152}Eu, ^{154}Eu, ^{155}Eu, ^{144}Ce), and small amounts of uranium and transuranium elements (^{238}Pu, ^{239}Pu, ^{241}Am, ^{243}Am).

Carbon-14 (half-life: 5,730 years) is usually the dominant contributor to the graphite activity. Fission products and transuranium elements exist in the graphite of reactors that have experienced fuel failure. Activation products arise from trace-level impurities. Chlorine-36 (half-life: 300,000 years), the activated residual chlorine from graphite purification, is another contaminant of importance due to both its long half-life and poor retardation by geologic medium. Radioactivity of graphite depends on the type of graphite reactor, extent of fuel failure, impurities or residues left at graphite manufacture, and time of operation.

As it exists, graphite would seem to meet most of the general disposal requirements for a solid radioactive waste form, since graphite keeps most of its good mechanical properties

---

[17]According to the IAEA Model Additional Protocol (InfCirc 540, corrected), "nuclear-grade" graphite has "a purity level better than 5 parts per million boron equivalent and a density greater than 1.50 g/cm^3."

after many years of irradiation. It is relatively insoluble and not chemically reactive. However, studies evaluating the radioactivity of graphite and other detailed characteristics show that the graphite from nuclear reactors cannot be accepted "as is" by existing disposal sites. Special treatment is required to increase its resistance to leaching and to remove specific radionuclide content. Thus, countries that have dismantled GCRs just store the graphite that is removed.[18]

Spent Fuel

Defueling and Canning Requirements

Once it is decided to shut down the 5 MWe reactor, the fuel would be removed (the reactor would be "defueled") according to a decommissioning plan. Areas near the reactor, especially the spent-fuel handling areas, are first decontaminated to facilitate defueling. The radiological situation of the reactor may be characterized, which would take several years, either before or at the same time as defueling.

The U.S. and the DPRK agreed in 1994 to dispose of the spent fuel from the 5 MWe reactor without reprocessing it in the DPRK. The Six-Party Talks may lead to a similar decision on spent fuel, i.e., to remove it from the DPRK. What is not clear is how long it would take before the fuel could be shipped and reprocessed. Magnox-type spent fuel is not easily stored for an extended time, even with carefully monitored, inert dry storage.[19]

If the eventual destination of the spent fuel is determined promptly, it could be stored in canisters for a relatively short period. Although this will not require long-lasting canisters, much effort will be given to decontaminate the fuel and the surroundings before starting the canning process. Since the preparation itself may take years, ensuring the safety of the stored spent fuel during preparation for shipping is essential.

If the situation requires a longer period of time for storing spent fuel before reprocessing, then the spent fuel will have to be placed in long-lasting canisters following a similar procedure that was taken after the Agreed Framework in 1994. Canning at that time took about five years, due to complications from contamination in the storage area and the lack of proper equipment.

International practices for the regular transport of Magnox spent fuel from one country to another for reprocessing (e.g., from Japan's Tokai-1 reactor to the U.K. or from Italy's Latina reactor to the U.K.) should be considered to determine the availability and capacity of packaging and transportation and their technical applicability to the DPRK case.

HLW, Storage, and Disposal

Countries with commercial reprocessing facilities receive spent fuel from their customer countries and return the vitrified HLW to the country of origin, i.e., the customer countries. Based on the experience at the Eurochemic reprocessing plant, reprocessing the DPRK's 16,000 spent-fuel elements from the 5 MWe reactor will yield about 27 cubic meters of high-level liquid waste. This waste must be processed and safely stored. One current practice is to vitrify the waste into glass blocks weighing 37 metric tons that occupy 260 stainless-steel containers. These stainless-steel containers could be returned to the DPRK and stored in a facility until a disposal repository is operational.

There are two generic interim storage options for the containers of vitrified high-level waste: air cooled and water cooled. The air storage option is typical of that in use at Marcoule, France, and at Rokkasho-mura, Japan. The storage facility requires a ventilation system to keep the glass in the waste container and concrete structures at safe temperatures.

---

[18]Countries with large volumes of radioactive graphite are the U.K. with 60,000 metric tons, the former Soviet Union with 50,000 metric tons, and France and the U.S. with a similar amount.

[19]*(U.K.) Radioactive Waste Management Advisory Committee's Advice to Ministers on: The Radioactive Waste Implications of Reprocessing* (Nov. 2000), "Annex 4: Dry Storage and Disposal of Magnox Spent Fuel," www.defra.gov.uk/rwmac/reports/reprocess/16.htm.

The time for the interim storage phase depends on availability of a repository and the heat and radioactivity loads that the repository can accept. Whether air cooled or water cooled, a facility is needed to store the 260 stainless steel containers for about 30 to 50 years before they are sent to a disposal site. The capacity of the storage facility should be about 400 containers to accommodate the containers from the vitrification of the liquid HLW that already exists in the site. This topic is covered in greater detail later in the chapter.

Although the HLW from the 5 MWe reactor is only about one tenth as radioactive as the HLW from a CANDU reactor, the time required for it to decay to a safe level is still several tens of thousands of years. Thus for HLW disposal, a deep and stable geological formation is required to ensure long-term safety.

Thorough characterization of a candidate site is expensive and time-consuming. Internationally, the U.S. is the most advanced country in developing a geological repository for HLW. Enabled by the Nuclear Waste Policy Act of 1980, the Yucca Mountain repository in Nevada has been explored for many years and recently obtained Congressional approval for HLW. Japan, Canada, France, and the U.K. have either passed legislation for the procedures to investigate and acquire sites for HLW repositories or established the organization(s) to undertake long-term procedures for HLW disposal.

Detailed Stages for Spent-Fuel Management

The details of spent-fuel treatment, from initial defueling to final disposal, can be grouped with respect to flow and activities, as shown in Table 20.2.

In reality, the detailed stages listed in Table 20.2 may be spread over a wide range of milestones. If properly scheduled and agreed, the defueling and reprocessing stages could take no more than 10 years. HLW storage lasts for several decades, and the HLW disposal stage is even longer.

Graphite Moderator and Reflector

The total amount of graphite that would need to be removed from the 5 MWe reactor after defueling is reported to be 600 metric tons, or about 400 cubic meters, in approximately equal amounts from the moderator (reactor core) and reflector (surrounding the reactor core).

Table 20.2 Spent-fuel treatment process.

Stage	Activities
Defueling	Defueling, including decontamination of necessary surroundings
	Decontaminating the spent-fuel storage pool
	Manufacturing canisters for interim storage
	Canning spent fuel for interim storage
Reprocessing	Manufacturing or leasing transportation casks/canisters
	Transferring spent fuel into the transportation casks
	Transporting the spent-fuel casks
	Reprocessing the spent fuel, including vitrifying liquid HLW
HLW storage	Transporting vitrified HLW back to the DPRK
	Constructing and operating a storage facility for vitrified HLW
HLW disposal	Planning the HLW disposal system
	Investigating sites for a HLW repository
	Characterizing sites and applying for license
	Constructing and operating the HLW repository

The graphite is in the form of blocks, usually stacked in several layers. The graphite blocks may be held together by steel restraint bands and interlaced by thermocouple wires and neutron flux measuring detectors. To remove graphite blocks from the reactor vessel, custom remote handling equipment must be used for cutting, clamping, and grabbing steel wires, plates, and graphite blocks. Designing and manufacturing the necessary remote operation equipment can take about two years, assuming a full knowledge of reactor core design specifications. This is normally done during the initial phase of decommissioning.

Since the graphite blocks may have various activity levels and different contaminations, sorting them based on their activity levels can help reduce the amount of graphite requiring the greatest decontamination effort.

Graphite blocks removed from the GCRs of the U.K., France, and other countries are encapsulated in concrete and placed in specially designed storage buildings. They must be stored in that way until proper repositories, conforming to regulations, can be constructed.

Besides the absence of proper repositories, waste-form stabilization is another issue. There have been many studies on how to prevent or reduce radionuclide leaching from graphite during storage. Metallization of the graphite surface using several methods has been investigated as one promising option for efficient leaching resistance. No single method, however, has been selected for mass application to graphite stabilization.

For the time being, graphite blocks removed from the 5 MWe reactor should follow the same procedure as in the U.K. and other countries: packaging and storing. Specially designed containers will be needed to hold the graphite. Both a packaging building and a storage building should be built near the reactor site. The storage building will have to be under regular surveillance and monitoring to detect radionuclide leaks.

Reactor Vessel, Steel Structures, and Concrete

Steel structures, such as the pressure vessel and thermal shield, are radioactive, having been activated by exposure to neutrons during normal reactor operation. Activation products include the radioisotopes ^{60}Co, ^{55}Fe, ^{59}Ni, ^{63}Ni, and ^{94}Nb. Concrete surrounding the reactor vessel also undergoes neutron activation and contains the activation products ^{3}H, ^{14}C, ^{36}Cl, ^{152}Eu, and ^{154}Eu. Because some of these activation products have long half-lives, the steel structures and concrete cannot generally be treated as LLW.

Because activation depends on the location of the material and its distance from the core, a detailed survey of the content of each radionuclide at different depths and locations should be performed. With the precise data of radioactivity distribution on and in the steel structures and concrete materials, one may set up a strategy to reduce the volume of ILW. An estimation study of decommissioning a GCR[20] shows that about 30% of the steel structure could be released through measurement and decontamination and another 30% packaged for ILW. Similarly, 97% of concrete could be released free.

The 5 MWe reactor will generate about 300 metric tons of steel (excluding pipes that were not subject to activation) and several thousand tons of concrete bioshield as waste. To reduce the volume of ILW, a sophisticated facility to segregate the waste according to its activity content should be provided. If this facility is equipped and operates as shown in the reference, one may reduce the intermediate-level steel waste to about 30% and concrete waste to about 3%.

At the least, a repository for ILW to accommodate 100 metric tons of steel and several hundred tons of concrete should be available. An interim storage facility for intermediate-level steel and concrete waste should be operating until a long-term repository is available. A waste packaging method for long-term storage should also be provided.

[20]R. J. Printz, U. Quade, and J. Wahl, *Packaging Requirements for Graphite and Carbon from the Decommissioning of the AVR in Consideration of the German Final Disposal Regulations: Technology for gas cooled reactor decommissioning, fuel storage, and waste disposal,* Proceedings of Technical Committee meeting held in Juelich, Germany, Sept. 8–10, 1997, pp. 275–285. IAEA.

Decommissioning the Radiochemical Laboratory

Liquid High-Level Waste

Characteristics

The volume of liquid waste from reprocessing depends on the mass of spent fuel, whether it is low- or high-burnup fuel. However, the amount of radioactive material (fission products and actinides, components of high-level waste, HLW) contained in the spent fuel, and therefore also in the liquid waste, depends mostly on burnup. To obtain weapons-grade plutonium, the DPRK operated the 5 MWe reactor for a very low burnup, averaging 635 MWD/tU, so the fuel contains a relatively small portion of fission products and actinides. Once vitrified, the liquid HLW from reprocessing the 5 MWe reactor fuel may result in much smaller solid volume than would result from waste vitrification associated with high burnup fuel.

Little is known about the status of the liquid HLW in the DPRK. To remove the traces of its past nuclear activities, the DPRK might have mixed the liquid HLW with other type of waste or with something else, which will not only cause difficulties in reducing volume as a preparatory step for vitrification but might make it totally impossible to segregate the liquid HLW from the mixed state. Thus, although the burnup of the spent fuel is quite low and the resulting volume of vitrified waste would be relatively small, it would be safe to assume that the liquid HLW in the DPRK contains almost the same amount of radioactive material as does the liquid HLW from high burnup fuel.

Reducing the Volume

The liquid HLW has to be immobilized for the ease and safety of long-term storage and final disposal in a repository. Immobilization is done by vitrifying concentrated liquid HLW. If a process line is provided to recover nitric acid and concentrate the HLW, the waste volume can be reduced to less than one tenth of the original volume.

Shipping, Vitrification, and Storage

Liquid HLW would need to be shipped in 200-liter drums with good shielding, leak tightness, and impact- and fire-proofing. If no volume concentration is done, the liquid HLW will require 2,500 drums. Usually reprocessing facilities are colocated with vitrification facilities, removing the chances of accidents occurring during transportation. Eurochemic, however, shipped about 50 cubic meters of HLW to PAMELA, a vitrification facility in Germany, in the 1980s.

As with the case of Eurochemic, the vitrification of the HLW—concentrated during the course of reprocessing 100 metric tons of natural uranium fuel—will result in approximately 260 stainless steel containers filled with 37 metric tons of glass. If the spent fuel is reprocessed outside the DPRK, the containers might be shipped back to the DPRK, where they must be stored until the final repository is available.

Liquid Waste from Chemical Decladding

Characteristics

The Radiochemical Laboratory in the DPRK is reported to have operated a process to dissolve fuel cladding. In the Eurochemic plant, magnesium cladding was dissolved in sulfuric acid, resulting in 3 cubic meters of waste solution per metric ton of uranium. It is categorized as ILW, for which storage is the only management solution until a specific repository is acquired. In the DPRK case, there would be 300 cubic meters of cladding waste solution after reprocessing 100 metric tons of fuel. To store the waste safely for a relatively long time, the solution has to be immobilized with a solidification agent.

Solidification Methods

Bitumen (asphalt) was used in the Eurochemic plant for immobilizing the chemical decladding waste. Bitumen is highly leach-resistant, has a low operating temperature, and possesses a good degree of plasticity. At its operating temperature, 99% of the water evaporates, resulting in a volume reduction of up to fivefold compared with conventional cementing techniques. Although some technical improvements have been added to the bituminization process, inherent drawbacks of the bitumen are its potential fire hazard and weak radiation resistance, resulting in the release of hydrogen gas. Typical bitumen products contain 40 to 60 weight percent waste solids.

Immobilization of radioactive waste by incorporating it into hydraulic cement has been practiced worldwide for many years. It does not present a fire hazard and has good radiation resistance. Its drawbacks are low leach resistance and low content of waste solids. The addition of sodium silicates or polymers has been studied and applied in other commercial fields to improve its characteristics.

Glass is a good medium to contain chemical decladding waste. Glass has high radiation resistance and leach resistance. It does not present a fire hazard. Vitrification, the process of embedding the waste in glass, is expensive and therefore is usually applied only to the solidification of HLW rather than ILW.

Solidification of a type of radioactive waste with a specific medium requires thorough study of possible interactions between the waste and the medium. Selecting the method to apply and developing the process takes a long time before an actual process is put into operation.

Storage and Disposal of Solidified Chemical Decladding Waste

It is difficult to estimate how much solidified volume of chemical decladding waste will result from each solidification method. Since Eurochemic had combined various kinds of chemical decladding waste for solidification, segregating the bituminized volume for the magnesium decladding waste seems to be impossible. From the estimated mass of the magnesium involved[21] and with proper chemical reactions assumed, one could calculate the volume of solidified chemical decladding waste.

For storage and disposal, the solidified waste from chemical decladding may share the same facilities with other waste streams, such as the graphite, steel, and concrete from the dismantling of the 5 MWe reactor.

Steel and Concrete

It is foreseeable that the decommissioning of the Radiochemical Laboratory and its related buildings will result in a large amount of steel and concrete waste. Most of the steel and concrete wastes from a reprocessing facility involve surface contamination rather than activated bulk material. The contamination is largely high-alpha radioactivity, which must be disposed of in a deep geological formation like the HLW.

The similarity of the Radiochemical Laboratory and the Eurochemic plant enables us to estimate the volume of materials to treat from the Radiochemical Laboratory. The Radiochemical Laboratory should produce roughly one half to two thirds as much solid waste: 500 to 1,000 metric tons of steel/metal and 16,000 to 24,000 metric tons of concrete (6,000 to 9,000 cubic meters). Assuming the same level of decontamination technology as was available for the Eurochemic plant, much of it can be released without any restriction: about 66% of metal (330 to 660 metric tons) and 93% of concrete (roughly 15,000 to 23,000 metric tons; 5,600 to 8,400 cubic meters). The rest will require special consideration for storage or disposal, and some portions of them will need to be disposed of as HLW.

[21]Each fuel element is clad with about 100 g magnesium. Considering a total of 16,000 spent-fuel elements, the mass of magnesium involved in chemical decladding is about 1.6 metric tons.

Basic Schemes for Decommissioning

The decommissioning schemes described in the following sections and depicted in the accompanying figures are based on the following assumptions:

- All the tasks required for decommissioning are performed in harmony with each other. The proper investment for those activities and infrastructure that require long lead times is done on time.
- The spent fuel withdrawn from the reactor is shipped to another country for reprocessing, while the vitrified HLW is then shipped back to the DPRK.
- Liquid HLW already generated by reprocessing in the DPRK will be shipped to another country for vitrification and then returned to the DPRK in vitrified form.
- Storage and final disposal of LLW, ILW, and vitrified HLW will be done in the DPRK.

Economic feasibility is not taken into consideration.

Spent Fuel and the 5 MWe Reactor

Figure 20.1 shows a decommissioning scheme for treating the spent fuel from the 5 MWe reactor. In this scheme, a key issue will be the timely shipping of the spent fuel out of the country, provided that a place to reprocess the spent fuel is secured. The vitrified HLW from both the reprocessing of the spent fuel and the vitrification of the liquid HLW will require interim storage on return to the DPRK. Eventually a small disposal repository for the HLW, with the capacity of about 400 disposal casks of HLW, will have to be provided as the final step of the scheme.

Figure 20.2 shows how the solid wastes are categorized and how much of each type of waste will be generated from dismantling the 5 MWe reactor. Some metal and concrete waste containing long-lived radionuclides should follow the same path as that for graphite. Special packaging methods have to be sought for long-term, leak-free waste forms for the graphite as well as for a relatively small amount of metal and concrete. A storage facility will serve until the disposal site for this waste is found.

Liquid HLW, Chemical Decladding Waste, and the Radiochemical Laboratory

Liquid HLW will follow the scheme in Figure 20.3. This assumes that the required preplanning has been done before deploying an HLW volume reduction facility. Analysis of this HLW is so critical in revealing the past history of reprocessing in the Radiochemical Laboratory that it could take longer than one or two years. The current impasse with regard to waste treatment will be whether we have enough experience in long-distance transport of HLW in liquid form. If the volume matters, a process line for volume reduction might have to be added to the existing facility, which could delay dismantling the Radiochemical Laboratory. Two years (shown in the figure) is an aggressively optimistic estimate for incorporating a volume reduction facility, but it's possible. Once the liquid HLW is vitrified, it will be stored in the interim HLW storage facility with the reprocessed spent fuel HLW until a disposal repository is available.

The chemical decladding waste should be solidified for interim storage and final disposal, as depicted in Figure 20.4. This will take a somewhat long time in characterizing the waste and selecting and constructing the solidification process.

Figure 20.5 shows the disposition of solid waste from decommissioning and dismantling the Radiochemical Laboratory. After surface decontamination, most of the concrete and metal may be released free or as LLW. Several hundred tons of concrete and metal remain as

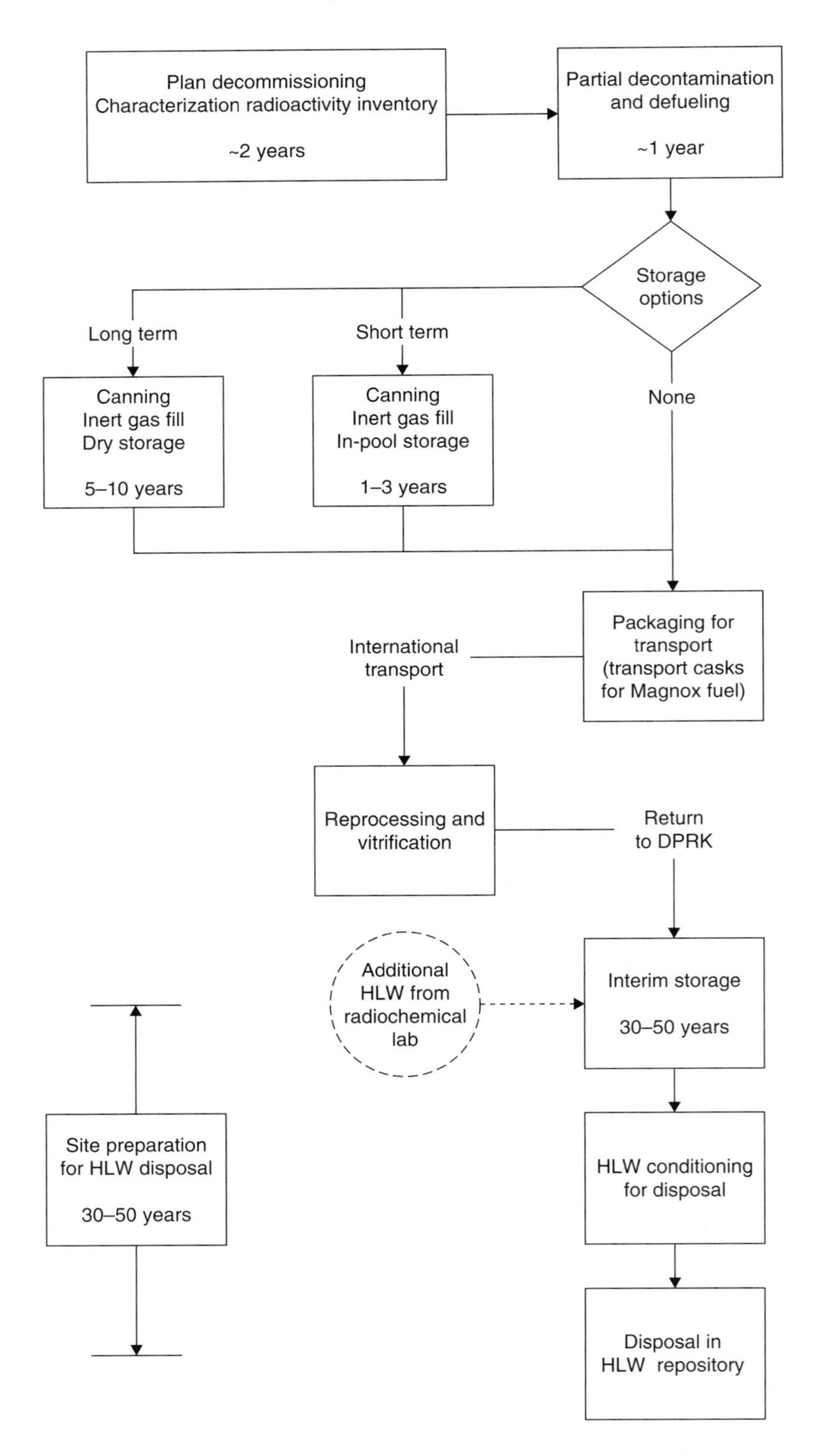

FIGURE 20.1 Schematic flow for spent fuel: from defueling the 5 MWe reactor to disposal of HLW.

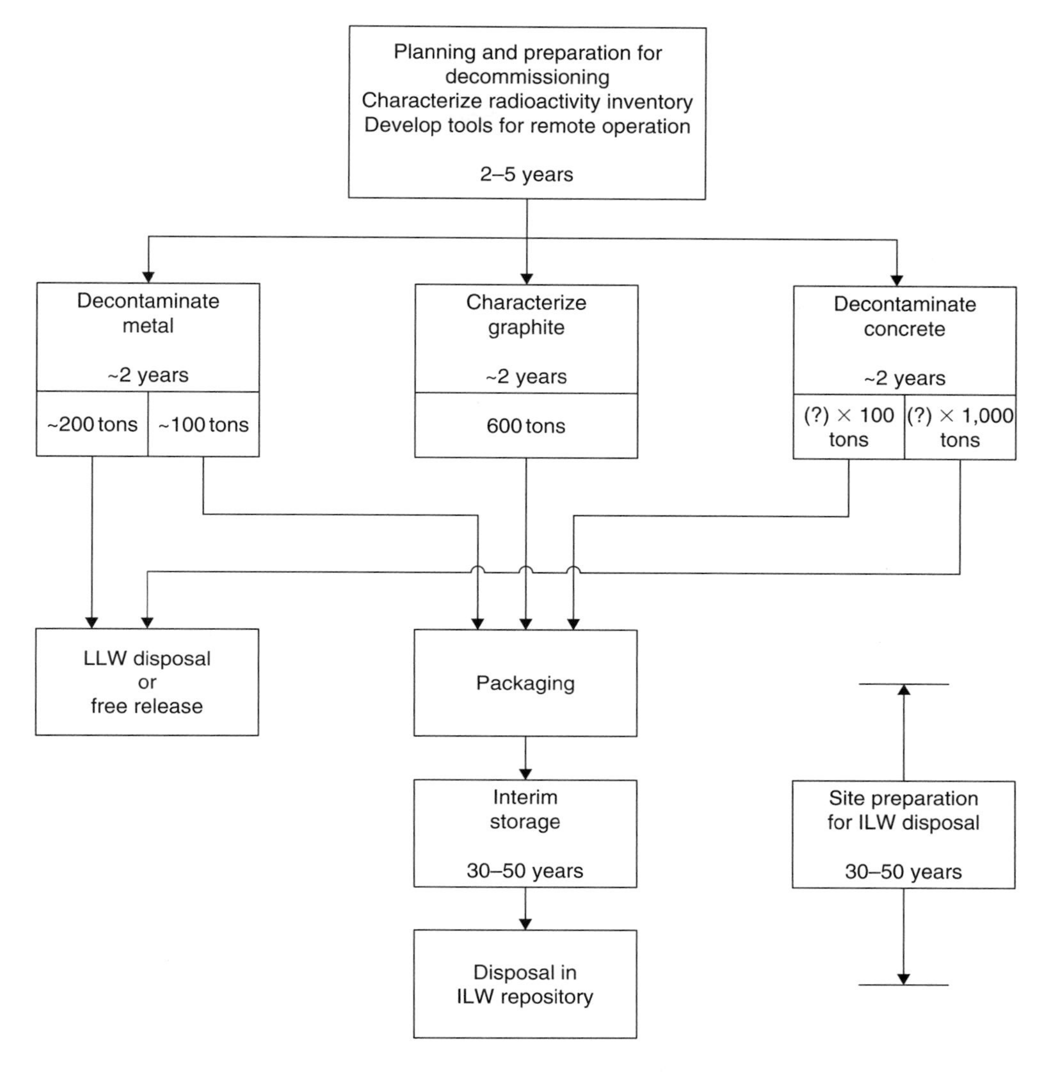

FIGURE 20.2 Schematic flow for solid wastes from dismantling the 5 MWe reactor.

intermediate waste due to high-alpha radioactivity after decontamination. A relatively small amount with very high activity of alpha-emitting radionuclides might have to be treated along with HLW.

Facilities Required for Waste Management in the DPRK

Reviewing the basic schemes shown in Figures 20.1–20.5, one might realize that operating and dismantling those facilities call for a wide range of activities, including treatment, packaging, storage, and disposal. Table 20.3 lists the facilities needed to manage the waste that results from decommissioning the 5 MWe reactor and the Radiochemical Laboratory.

In the initial phase of decommissioning, the processes of decontamination, canning, and packaging require dedicated facilities and the development of tools and package forms. Canning tools used during 1994–1996 might no longer exist. Facility operation should be supported by precise analytical capability.

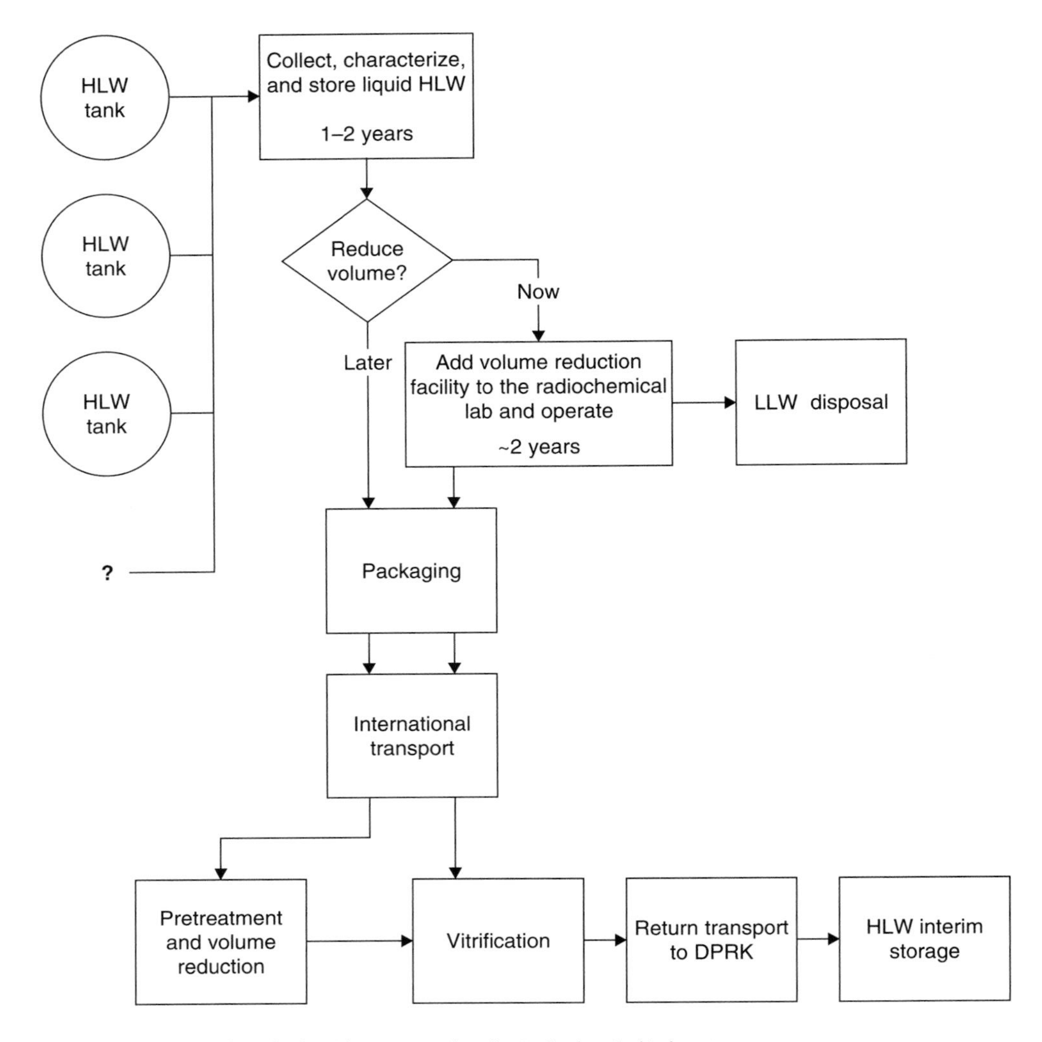

FIGURE 20.3 Schematic flow for liquid HLW stored at the Radiochemical Laboratory.

In the middle phase, interim storage facilities for HLW and ILW must be provided. The length of the storage period might depend on how soon the disposal sites can be acquired, which normally takes 30 to 50 years. A LLW disposal site should already be operating during this phase.

In the final phase, repositories for final disposal of both HLW and ILW are required.

Strategic Alternatives

Given the historical background and interests of the six negotiating states, plans to decommission and dismantle the 5 MWe reactor and the Radiochemical Laboratory are likely to be based on two strategic principles:

- *Decommissioning for nonproliferation.* Decommissioning the two nuclear facilities should be complete, verifiable, and irreversible so that these facilities cannot again be utilized for plutonium production. All possible technical means need to be considered for both effectiveness in accomplishing dismantlement and effectiveness in revealing past activities.

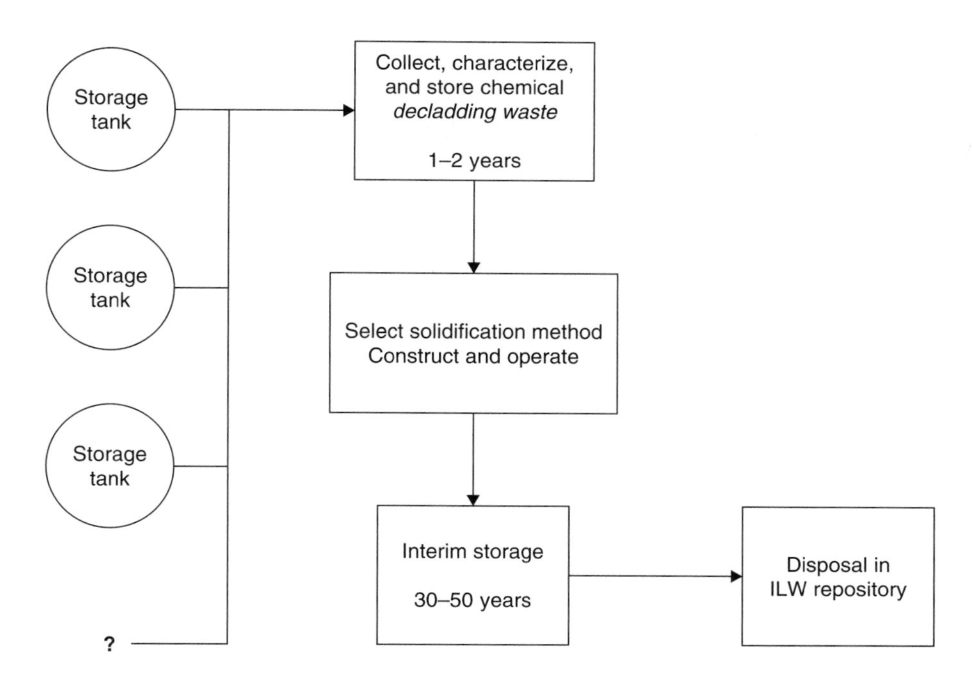

FIGURE 20.4 Schematic flow for chemical decladding waste stored at the Radiochemical Laboratory.

- *Decommissioning for the safety of the environment.* Planning to decommission the two nuclear facilities with consideration for international precedents and regard for safety will be critical in drawing cooperation not only from the DPRK but also from the other five parties, especially South Korea and the other states most likely to bear most of the costs.

Many additional factors can also play a role in developing the decommissioning strategy—cost and availability of funding, sustainability, and intergenerational equity arguments, stakeholder views, worker radiation health and safety, and others.

Fast-Track Decommissioning

If an outcome of the Six-Party Talks requires that the reactor and Radiochemical Laboratory be decommissioned as quickly as possible, we need to consider how soon the decommissioning project can be accomplished. For example, the 1994 Agreed Framework stipulated nominally 10 years. Such a "fast-track" approach might accept minimum achievement of decommissioning goals rather than allow for long lead times and adequate investment for infrastructure.

We assume that the spent fuel from the 5 MWe reactor will be shipped out of the country. It will either be reprocessed or simply stored for a long time in the country that receives the spent fuel. If the destination is selected and agreed on, the spent fuel may be shipped out of the country in 10 years. To finish this job in 10 years, canning the spent fuel for short-term storage before it is shipped should be done without any delay; shipping should also be prepared for well in advance with regard to packaging for overseas transport. To avoid difficulty in sustaining support, the parties have to agree that the HLW will not be shipped back to the DPRK after reprocessing.

Dismantling the 5 MWe reactor to a level of irreversibility may be done in 10 years or sooner, but it will generate various volumes and types of radioactive waste. The reactor and its cooling system could be the key components for irreversible dismantling. After dismantlement, the reactor must be periodically maintained and monitored to prevent radioactive leakage.

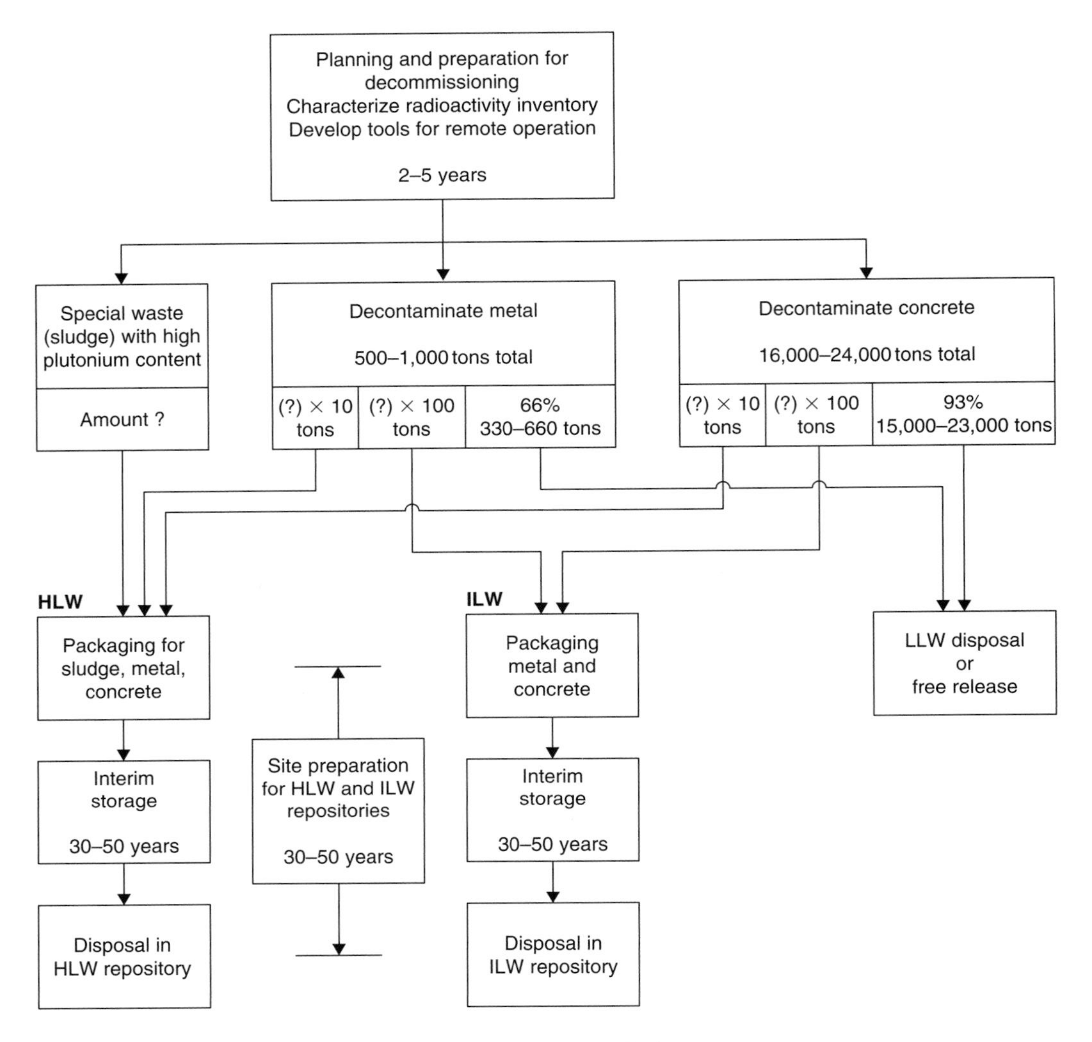

FIGURE 20.5 Schematic flow for solid wastes from the Radiochemical Laboratory.

Analyzing the liquid HLW for traces of past activities can be done in a relatively short time. Dismantling the key components of the Radiochemical Laboratory—such as dissolver, mixer-settler, and hot cells—to render it obsolete can be accomplished in less than 10 years; however, managing the resulting waste problems cannot be finished in a similarly short period of time.

Liquid HLW and ILW left at the site could be a serious problem. Liquid wastes might not only leak to the environment, but they could be attractive to those who would desire source material for a dirty bomb. The Radiochemical Laboratory must be well decontaminated to prevent radioactivity from leaking. It also requires periodic maintenance and monitoring.

Decommissioning activities that can be undertaken within ten years are limited. It should be possible to finish verification activities via analysis of the liquid HLW, spent fuel, and swipe samples. Although irreversible facility dismantlement also can be accomplished, much of the waste problem, which requires additional investment of time and resources, would remain unsolved. Thus the fast-track approach is not complete.

Table 20.3 Facilities needed to manage the wastes from decommissioning the 5 WMe reactor and the Radiochemical Laboratory.

Facility	Location	Application	Capacity
		Immediate Needs	
Canning station	At the 5 MWe reactor	Spent fuel is canned; cans filled with inert gas. *Requires decontamination before canning; special tools for broken fuel rods; shielding.*	50 or 100 tU (8,000 or 16,000 fuel elements)
Temporary dry storage pit	5 MWe reactor	Storage of canned spent fuel for 5–10 years	50 or 100 tU
Packaging station	5 MWe reactor	Canned spent-fuel prepared for overseas transport. *Requires transport casks for spent fuel.*	50 or 100 tU
Packaging station	Radiochemical Laboratory	Liquid HLW packaged for overseas transport. *Requires transport casks for liquid HLW.*	250,000 or 500,000 liters HLW
		Midterm Needs	
Decontamination and packaging station	Yongbyun Nuclear Research Center	Metals: Decontaminate	300 t from the reactor; 500–1,000 t Radiochemical Lab
		Package for ILW storage	~100 t from the reactor; (?) × 100 t Radiochemical Lab
		Concrete: Decontaminate	(?) × 1,000 t from the reactor; 16–24,000 t Radiochemical Lab
		Package for ILW storage	(?) × 100 t from the reactor; (?) × 100 t Radiochemical Lab
		Graphite: Characterize and package. *Requires waste package form.*	600 t from the reactor
Disposal site for LLW	DPRK	LLW from decommissioning both reactor and Radiochemical Lab	Depends on the free release limit
		Long-Term Needs	
Interim storage facility for solid HLW	Yongbyun Nuclear Research Center	Store HLW returned from overseas after vitrification, including HLW from both the reprocessed spent fuel and the existing liquid HLW inventory. *Dry storage requires heat removal; storage for 30–50 years.*	390 HLW canisters + special waste, metal, and concrete from the Radiochemical Lab decommissioning

(Continued)

Table 20.3 (Continued)

Facility	Location	Application	Capacity		
Interim storage facility for solid ILW	Yongbyun Nuclear Research Center		Metal	Graphite	Concrete
			~100 t	600 t	(?) × 100 t
		Reactor decommissioning	?	–	–
		Chemical decladding	(?) × 100 t	–	(?) × 100 t
		Radiochemical Lab decommissioning			
		Requires waste package form; storage for 30–50 years.			
Disposal site for HLW	DPRK	For vitrified HLW from reprocessed spent fuel and liquid HLW inventory	~390 HLW disposal casks		
		For special waste from the Radiochemical Lab	?		
		Requires a very long lead time.			
Disposal site for ILW	DPRK				
		Reactor decommissioning	Metal	Graphite	Concrete
		Chemical decladding	~100 t	600 t	(?) × 100 t
		Radiochemical Lab	?	–	–
		decommissioning	(?) × 100 t	–	(?) × 100 t
		Requires a long lead time.			

Decommissioning Supported by International Cooperation

The largest uncertainty in decommissioning is how to determine, fairly early, the responsible country (or countries) for long-term storage or reprocessing of spent fuel. That country should also be responsible for the vitrified HLW. Also, the verification of the decommissioning is a multilateral issue. One goal is for the DPRK to reestablish its membership to the NPT as a nonnuclear weapons state and conclude a new safeguards agreement with the IAEA. The IAEA would then become the primary monitoring and verification authority, expanding and formalizing the role it is now playing in confirming the shutdown of DPRK's plutonium facilities.

Decommissioning and dismantlement in the DPRK must be supported by international cooperation. The tasks for international cooperation should include:

- Finding technical dismantlement solutions consistent with nonproliferation goals
- Understanding accurately the technical challenges for waste treatment with regard to 5 MWe and the Radiochemical Laboratory
- Estimating the amount of work
- Defining a range of economically feasible paths for decommissioning

Regional sharing of responsibility for decommissioning and managing radioactive waste is the only practical approach, considering the need for and importance of cost sharing, technical experience, regional security and nonproliferation concerns.

Summary

Decommissioning and dismantling the DPRK's plutonium-relevant facilities will generate large volumes of solid and liquid wastes with a wide range of radioactivity. Prior international experience with decommissioning similar nuclear facilities suggests long timeframes and high costs for the states that perform these activities.

Although questions remain concerning inventories of previously-produced and separated plutonium, and even fabricated weapons, it is critically important to arrest the DPRK's existing capability to continue production. Partly dismantling the 5 MWe reactor and reprocessing facilities to accomplish nonproliferation goals, including removing key components and spent fuel quickly, would require the following tasks:

Task 1. Remove spent fuel and HLW

- Plan to remove and ship out of the DPRK the spent fuel and liquid HLW. Need to overcome technical hurdles as well as assign responsibility to a receiving party.
- Receiver of the spent fuel and the liquid HLW will either store or reprocess the spent fuel and store and dispose of vitrified HLW.
- Investigate associated technical, political, regulatory, and public acceptance issues.

Task 2. Support early decommissioning of facilities consistent with the IAEA's SAFSTOR guidelines. Include removing critical equipment and materials but also allow radioactivity levels to decline in abandon facilities prior to complete elimination.

- Need to investigate in advance radioactivity inventory estimation, removal tool development, canning and packaging of spent fuel and HLW, classification and segregation of waste, cost, etc.

Task 3. ILWs

- Investigate feasibility for providing interim storage and a disposal repository.

No matter how decommissioning might be limited to achieve minimum nonproliferation goals, large volumes of various types of waste will arise. To handle the wastes, the following facilities are needed:

- Immediate needs: Canning station, dry storage pit, packaging station for overseas transport, decontamination facilities, and LLW storage
- Midterm needs: Decontamination and packaging stations for metal, concrete and graphite, LLW disposal site
- Long-term needs: Interim storage facilities for HLW and ILW, disposal repository for HLW and ILW

Despite the formidable scope of the radioactive waste management problem resulting from DPRK nuclear dismantlement, it is clear that there are also encouraging opportunities. Effective radioactive waste management is intrinsically a mutually desirable goal. Concerted efforts by other countries—not only the other regional players, but also outside states with technical resources and capabilities and those with extensive experience in dealing with the similar dismantlement issues—can be brought to bear in a coordinated, cooperative engagement.

Such engagement is consistent with and provides opportunities for increased technical and economic interaction between the DPRK and other states. In this way the elimination of the DPRK's nuclear weapons program could foster long-term technical collaborations that establish a foundation for future industrial and commercial interactions in the areas of energy, waste management, chemical production, and construction.

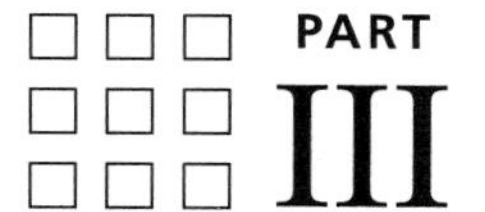

Preventing Nuclear Terrorism and Illicit Nuclear Trade

21

Why We Need a Comprehensive Safeguards System to Keep Fissile Materials Out of the Hands of Terrorists[1]

Siegfried S. Hecker

Introduction

The tragic events of September 11, 2001, underscored the urgency of preventing the nexus of terrorism and nuclear weapons. How difficult is it to keep nuclear weapons out of the hands of terrorists and what can be done about it? In the early and mid-1990s, the greatest concern was the possibility of "loose" nukes in Russia in the midst of the chaos that followed the dissolution of the Soviet Union. The publicity surrounding former Russian General Alexander Lebed's claim[2] that dozens of Russian suitcase-size nuclear bombs were missing focused the world's attention on the problem of theft or diversion of nuclear weapons from the stockpiles of nuclear weapons states. Although Lebed's claim was never proven, the United States and

[1]This article is based on one that appeared in the Sept. 2006 issue of the *Annals of the American Academy of Political and Social Science* on "Confronting the Specter of Nuclear Terrorism."

[2]Carey Sublette on Alexander Lebed. On Sept. 7, 1997, the CBS news magazine *60 Minutes* broadcast an alarming story in which former Russian National Security Adviser Alexander Lebed claimed that the Russian military had lost track of more than 100 suitcase-sized nuclear bombs, any one of which could kill up to 100,000 people. "I'm saying that more than a hundred weapons out of the supposed number of 250 are not under the control of the armed forces of Russia," Lebed said in the interview. "I don't know their location. I don't know whether they have been destroyed or whether they are stored or whether they've been sold or stolen, I don't know." Asked if it were possible that the authorities did know where all the weapons were and simply did not want to tell Lebed, he said, "No." During May 1997 Lebed said at a private briefing to a delegation of U.S. Congressmen that he believed 84 of the 1-kiloton bombs were unaccounted for. In the interview with *60 Minutes*, conducted in late August, Lebed said he now believed the figure to be more than 100. However, the Russian government denied Lebed's accusations.

Russia, both of which have thousands of nuclear weapons in their arsenals today, must keep their nuclear weapons secure, and they should commit to further, more drastic reductions to make it easier to protect the weapons that remain.

The number of nuclear weapons in all states except Russia and the United States is believed to be hundreds or less.[3] The small arsenals make it simpler to protect the weapons by strict physical protection measures. However, as more states acquire nuclear arsenals it becomes more likely that disgruntled or ideological insiders could conspire to circumvent even the strictest measures of protection and control. Should terrorists acquire an intact bomb, they must overcome additional barriers to use such weapons. For example, transportation of the weapons within or out of the state in question without detection presents formidable obstacles. Also, sophisticated weapons from nuclear weapons states are protected from unauthorized use by a variety of use-control measures. Simpler weapons from the newly declared states may be easier to detonate, but they also pose serious risk of detonating accidentally during handling or transportation.

In any case, we must assume that once in the hands of terrorists, nuclear weapons could be used to inflict catastrophic damage. Although no acts of nuclear terrorism have been committed to date, we cannot count on deterring all groups should they possess nuclear weapons. In fact, groups such as al-Qaeda have shown great interest in nuclear weapons.[4] The horrific destructive power of what today are considered rather rudimentary nuclear bombs was demonstrated in 1945 at Hiroshima and Nagasaki. With explosive yields of 15 to 20 kilotons, those bombs killed more than 100,000 people instantly and a similar number shortly thereafter. Bombs with only 10–20% of such destructive power could still kill tens to hundreds of thousands of people if detonated in one of the world's mega-cities. Nuclear destruction is overwhelming and almost instantaneous; therefore, the focus must be on prevention.

In 2002, a report by the U.S. National Academies[5] rated the probability of a state-owned nuclear weapon getting into the hands of terrorists over the subsequent five years as "moderate." The report assessed the threat level to be low from the U.S., the U.K., China, France, and Israel, whereas that from Russia, India, and Pakistan was rated as "medium." North Korea was not rated at that time; it is most likely low to moderate today. In spite of significant improvements in nuclear weapons security made by the Pakistani government since 9/11, Pakistan remains at the top of my list of concerns today due to its continuing political instability. Russia also continues to be vulnerable because of the substantial number of tactical nuclear weapons that remain in its stockpile. Protecting the security of nuclear weapons remains the single most important responsibility of every state that possesses nuclear weapons. This view is widely shared among security experts, as documented in the special September issue of the *Annals of the American Academy of Political and Social Science* on "Confronting the Specter of Nuclear Terrorism."[6]

Improvised Nuclear Devices from Inadequately Secured Fissile Materials

What is the likeliest route for terrorists to acquire a nuclear weapon? Theft or diversion of an intact weapon from a nuclear state is the most direct but not the most probable.

[3]David Albright, *Global Stocks of Nuclear Explosive Materials: Summary Tables and Charts*, Institute for Science and International Security, July 12, 2005, revised Sept. 7, 2005; Washington, D.C., www.isis-online.org/global_stocks/end2003/tableofcontents.html.

[4]National Commission on Terrorist Attacks Upon the United States, *The 9/11 Commission Report* (New York: W. W. Norton, 2004).

[5]National Research Council, *Making Our Nation Safer: The Role of Science and Technology in Countering Terrorism* (Washington, D.C.: The National Academies Press, 2002).

[6]G. Allison, ed., "Confronting the Specter of Nuclear Terrorism," *The Annals of the American Academy of Political and Social Science* 60 (Sept. 2006), pp. 1–202.

The National Academies report and other extensive studies[7,8] stress the importance of protecting nuclear weapons but conclude that improvised nuclear devices (INDs) built from stolen or diverted fissile materials, either plutonium or highly enriched uranium (HEU), pose a greater threat. Inadequate security of these materials around the globe represents the greatest danger. Unlike the weapons, which are limited in number and are located in relatively few, well-protected sites, weapons-usable materials are located in several dozen countries, often in inadequately secured locations and in forms that are difficult to safeguard.

Are terrorists capable of building, delivering, and detonating an IND? Much skepticism has been expressed because it is generally recognized that modern nuclear weapons are extremely complex devices, the creation of which requires the talent and facilities of some of the best scientific research laboratories in the world. The production of the nuclear arsenals in Russia and the United States, for example, has been performed in very large, closed facilities. Furthermore, the delivery of bombs or warheads with bombers or missiles requires specially trained armed forces. However, a terrorist's job of building a simple nuclear bomb that can be delivered by van, boat, or light airplane is much simpler. The basic design principles, manufacturing methods, and delivery were all demonstrated during the Manhattan Project 60 years ago.

The key ingredients for INDs are plutonium or HEU. Fortunately, the technologies and materials required to enrich uranium in the fissile isotope, U-235, or construct reactors to produce plutonium are considered beyond the reach of even the most sophisticated terrorist groups today. Moreover, procurement and construction activities are not easily carried out clandestinely, although recent revelations of the sophistication of A. Q. Khan's proliferation ring raise concerns.[9] So, the good news is that today terrorists are unlikely to make weapons-usable HEU or plutonium from scratch. The bad news is that they can potentially steal it or buy it because huge amounts are available worldwide, with some being inadequately secured. Keeping these materials out of the hands of terrorists is a much greater challenge than securing nuclear weapons.

The International Atomic Energy Agency (IAEA) requires the safeguarding of uranium with U-235 content of 20% or greater (HEU). Typical weapons-grade uranium is roughly 90% enriched in U-235. However, all concentrations down to 20% are weapons-usable, albeit significantly more uranium and high explosives will be required to produce a nuclear detonation. Weapons-grade plutonium is composed of greater than 93% isotope 239. Typical light water reactor spent fuel contains less than 60% Pu-239. However, as described by Mark et al.,[10] virtually any combination of plutonium isotopes may be used to build a bomb, although reactor-grade plutonium poses particularly challenging design and manufacturing hurdles.

HEU lends itself to the simplest bomb design: the so-called gun design, in which one subcritical mass of HEU is rapidly fired at another in a gun barrel. This design was used in the Hiroshima device without prior nuclear testing. Several tens of kilograms of weapons-grade HEU are sufficient to build a Hiroshima-like device that can yield more than 10 kilotons.

---

[7]R. A. Falkenrath, R. D. Newman, and B. A. Thayer, *America's Achilles' Heel: Nuclear, Biological, and Chemical Terrorism and Covert Attack* (Cambridge, Massachusetts: The MIT Press, 1998).

[8]C. Ferguson, W. C. Potter, A. Sands, L. S. Spector, and F. L. Wehling, *The Four Faces of Nuclear Terrorism* (Monterey, CA: Monterey Institute Center for National Security Studies, 2004).

[9]C. Braun and C. F. Chyba, "Proliferation Ring," *International Security* 29, pp. 5–49.

[10]J. Carson Mark, Theodore Taylor, Eugene Eyster, William Maraman, and Jacob Wechsler, "Can Terrorists Build Nuclear Weapons?" in P. Leventhal and Y. Alexander, eds., *Preventing Nuclear Terrorism: The Report and Papers of the International Task Force on Prevention of Nuclear Terrorism.* (Lexington Books, 1987), pp. 55–65.

The gun assembly does not work for weapons-grade plutonium because neutrons emitted by the spontaneous fission of Pu-240 will "preinitiate" the chain reaction and result in a fizzle. Hence, plutonium devices consist of a subcritical mass of plutonium surrounded by sufficient high explosives that rapidly implode plutonium to super-critical mass and nuclear detonation. The nuclear weapon dropped on Nagasaki used such a design, which can also be constructed of HEU if it is desirable to reduce the size of a device so that it is missile compatible.

As pointed out by the 9/11 Commission,[11] terrorists such as those involved in the 9/11 attacks were well educated and trained. Given the great reach of proliferation rings such as that run by A. Q. Khan, one cannot rule out that terrorists would be able to acquire the rest of the materials and components required to construct a simple bomb. Construction of an HEU bomb with a yield of 10 or more kilotons is possible. A plutonium implosion device is much more challenging because the design, fabrication, and detonation of the high explosive are difficult. However, one cannot rule out the possibility that a terrorist organization can succeed in detonating a bomb with 10–20% of the yield of the Nagasaki bomb. Moreover, if terrorists receive help from rogue nuclear technologists or detailed instructions that include manufacturing methods (such as those supplied by A. Q. Khan to Libya), the probability of success is increased substantially.

The consensus among nuclear weapons experts today is that terrorists would face significant but not insurmountable challenges to build a primitive but devastating nuclear device and that it would most likely be delivered to the intended target by truck, boat, or light airplane. Hence, highest priority for preventing nuclear terrorism must be assigned to safeguarding plutonium and HEU.[12]

That challenge is now widely recognized. Most recently, Presidents George W. Bush and Vladimir Putin addressed the matter in their 2005 Bratislava accord. The G-8 leaders pledged cooperation to this end in their Gleneagles Statement on Nonproliferation that same year, and it is also addressed by U.N. Security Council Resolution 1540 on Nonproliferation of Weapons of Mass Destruction. Moreover, several comprehensive treatises detail this challenge and offer potential solutions.[13] Allison[14] captures the essence of these studies with his challenge to governments around the world to keep fissile materials just as secure as treasures in the Kremlin Armory and gold in Fort Knox.

However, securing all fissile material around the world is considerably more challenging than locking it up to a "gold standard." This chapter first describes these challenges,

---

[11]National Commission on Terrorist Attacks Upon the United States, *The 9/11 Commission Report.*

[12]Allison, ed., pp. 1–202.

[13]Falkenrath et al., *America's Achilles' Heel: Nuclear, Biological, and Chemical Terrorism and Covert Attack;* Ferguson et al., *The Four Faces of Nuclear Terrorism;* D. Albright and K. O'Neill, *The Challenges of Fissile Materials Control* (Washington, D.C.: The Institute of Science and International Security, 1999); G. Allison, "How to Stop Nuclear Terror" *Foreign Affairs* 83, pp. 64-74; G. Allison, *Nuclear Terrorism: The Ultimate Preventable Catastrophe* (New York: Owl Books, Henry Holt and Co., 2005); M. Bunn and A. Weir, *Securing the Bomb 2005: A New Global Imperative* (Cambridge, MA, and Washington, D.C.: Project on Managing the Atom, Harvard University and Nuclear Threat Initiative, 2005); G. Perkovich, J. T. Mathews, J. Cirincione, R. Gottemoeller, and J. Wolfsthal, *Universal Compliance: A New Strategy for Nuclear Security* (Washington, D.C.: Carnegie Endowment for International Peace, 2005); National Research Council, *Protection, Control, and Accounting of Nuclear Materials: International Challenges and National Programs* (Washington, D.C.: The National Academies Press, 2006); National Research Council, *Strengthening Long-Term Nuclear Security: Protecting Weapon-Usable Material in Russia* (Washington, D.C.: The National Academies Press 2006).

[14]Allison, *Nuclear Terrorism: The Ultimate Preventable Catastrophe.*

which are both technical and political. Plutonium and highly enriched uranium are used in weapons, research, power reactors, and some industrial applications in forms that can be turned into weapons-usable materials with routine chemical processing. Such materials are processed, shaped, transported, stored, and used, and some inevitably wind up in waste streams. After exploring why securing fissile material is more difficult than is generally appreciated, the chapter then goes on to assess the components of a comprehensive safeguard system, addressing both the general and the specific challenges posed by the current threat environment. Many of the specific technical challenges that are dealt with here are covered in much greater detail in other articles in this book.

Five Characteristics of Fissile Materials

This section presents five reasons why securing fissile material is more difficult than generally appreciated. The characteristics of nuclear material must be understood to establish a comprehensive safeguards system.

Existing Inventories of Fissile Material Are Enormous, Whereas the Amount Required for a Nuclear Bomb Is Small

Most states with nuclear weapons have stopped producing weapons-grade plutonium and highly enriched uranium; in fact, the United States and Russia are reducing their inventories because they exceed current weapons requirements. The Institute for Science and International Security (ISIS) reports that approximately 1.9 million kilograms of highly enriched uranium and 1.83 million kilograms of plutonium exist worldwide.[15] Approximately 1.4 million kilograms of plutonium are found in highly radioactive spent fuel and would not be very attractive to terrorists. The remaining 2.3 million kilograms of weapons-usable fissile material, however, must be protected. But to truly prevent a nuclear terrorist attack, we must be able to account for a few tens of kilograms out of more than 2 million available worldwide.

The results of a recent study on plutonium in the United States underscore the problem of numbers.[16] The United States produced or acquired 111,400 kilograms of plutonium since 1943. In 1994, the total inventory was 99,500 kilograms. Although there are explanations for the "missing" 11,900 kilograms, the uncertainties between physical inventories and accounting are many times the amount required for a bomb.[17] This same study shows that

---

[15]Albright, *Global Stocks of Nuclear Explosive Materials: Summary Tables and Charts.*

[16]United States Department of Energy, *Plutonium: The First 50 Years* (Washington, D.C.: United States Department of Energy, 2004).

[17]The "missing" 11,900 kilograms were explained as follows: 3,400 kg expended in wartime and tests; 2,800 kg declared as inventory differences; 3,400 kg as waste (normal operating losses); 1,200 kg as fission and transmutation; 400 kg as decay and other removals; 100 kg in U.S. civilian industry; 700 kg exported to foreign countries; and a 100 kg rounding difference along with classified transactions. Inventory differences are defined as the difference between the quantities of material in accounting records compared to those determined in physical inventories. They were previously identified as "material unaccounted for," which included operating losses. Today, the operating losses are counted separately, and inventory differences result primarily from statistical measurement uncertainties; recording, reporting, and rounding errors; uncertainties of the amount of material held up in the processing plant; measurement uncertainties because of wide variations of material that contain fissile materials (material matrix) during processing; uncertainties associated with waste; and unmeasured materials associated with accidental spills or releases of materials. Waste (normal operating losses) is defined as intentional removals from the inventory as waste because they are technically or economically unrecoverable. Examples include discharges to cribs, tanks, settling ponds, or disposal facilities (burial sites).

even in the United States, where nuclear safeguard technologies and methodologies were first developed and applied, the accounting system alone cannot come close to ensuring that significant amounts of nuclear material are not missing. The investigation also illustrates that confidence in the security of nuclear materials should rest not on the numbers but instead must rely on the integrity of the nuclear safeguards system. It appears that a baseline study and inventory of plutonium and HEU has not been conducted in Russia.

Fissile Materials Exist in Every Imaginable Form

These materials are not like gold bricks at Fort Knox. Plutonium and uranium are highly reactive metals that oxidize rapidly, especially in humid conditions or in the presence of hydrogen. Furthermore, plutonium is constantly created and destroyed during reactor operation and, through radioactive decay, transmutes into other elements over time. For weapons applications, plutonium and uranium are used in metallic form—often alloyed with other chemical elements. For reactor applications, they are used in metallic or ceramic (principally oxide) forms. To make weapons or reactor fuel elements, they are processed using industrial processes such as dissolution in acids or salts; gasification; melting and casting; powder processing; electrochemical processing; shaping, machining, welding, or pressing; and waste processing and storage. For plutonium, all such operations are conducted in specially designed laboratories to prevent exposure to airborne plutonium. It is no surprise that operating losses and inventory differences are large when tons of plutonium or HEU are processed. Moreover, large-scale processing without adequate control and accounting leads to the potential of plant operators covertly diverting small but significant quantities of these materials.

Fissile Materials Exist in Many Locations, Not in Just a Few Storage Vaults

Plutonium and HEU exist in enrichment and fuel fabrication facilities, reactors, reprocessing plants, and storage facilities. The materials are typically well secured in weapons. Historically, however, security for nuclear research reactors and facilities has not been adequate. In states that reprocess spent fuel, plutonium also exists in reprocessing plants and mixed-oxide fuel fabrication plants. Of course, these materials are frequently transported, in ways that are not always secure.

Not only do these materials exist in many locations within one country, they exist in multiple countries. In addition to states with nuclear weapons programs, they exist in countries that have reprocessing plants or use mixed-oxide fuel. The greatest concern, however, is the use of highly enriched uranium in research reactors around the world. In the United States, the Atoms for Peace Program supported building such reactors in more than 40 countries. The Soviets had a similar export program. The security environment in many countries was inadequate to protect fresh HEU fuel. Today, roughly 120 research reactors in 40 countries still use HEU.

Fissile Materials Are Difficult to Measure and Handle

Safeguard systems must be able to measure fissile materials accurately. Monitoring and accounting of plutonium is hampered because plutonium must be handled in glove boxes or other ventilated enclosures and stored in airtight containers because of its radiotoxicity. Masses for inventories are measured by weighing, destructive assay methods employing wet chemistry, and nondestructive assay methods such as calorimetry (measuring heat content that is related to isotope concentrations) or neutron and gamma ray-based radiation measurements. The extraordinary scientific complexity of plutonium metal presents additional challenges. Plutonium exists in seven different crystal structures with varying densities. Adding a few atomic percent gallium or aluminum to pure plutonium will change its density by as much as 25%, complicating mass determinations.[18] Oxidizing plutonium metal to plutonium

---

[18] S. S. Hecker, "The Complex World of Plutonium Science," *MRS Bulletin* 26 (2001), pp. 672–678.

dioxide (typical for storage and reactor applications) drops the density of pure plutonium by nearly a factor of two.

Gamma-ray detectors are used to make nondestructive measurements of isotopic composition in plutonium and uranium. Chemical analysis is often required to ascertain the precise chemical composition, which is especially important for plutonium because it changes composition with time by transmutation. These measurements and analytical capabilities are not available in many locations that house plutonium or HEU.

Military Secrecy Hampers Safeguards and Transparency

In the early years, information regarding both bomb and reactor materials was classified. The Atoms for Peace Program declassified much of this information. But some details about plutonium chemistry and isotopic compositions were kept secret until the Energy Department released its plutonium study in 1994 (a similar study on HEU was never published). Russia and China still keep isotopic and chemical compositions of nuclear weapons materials secret, and most locations and amounts remain out of the public domain.

Although secrecy is necessary to protect a state's nuclear weapons program, excessive secrecy and, in particular, compartmentalization impede implementation of a rigorous safeguards system. There may be limited communication among sites that produce, use, and dispose of these materials. It can impede accounting, the establishment of systemwide inventories, and the sharing of best practices. Responsible government officials cannot assess systemic vulnerabilities due to a lack of transparency. Likewise, there is little information that allows other nations to judge the adequacy of each other's nuclear materials security. Excessive secrecy also precludes states from sharing crucial information about the chemical and isotopic composition of fissile materials stockpiles, which makes attribution in case of theft or a detonation more difficult.

For these five reasons, simply locking up all the materials is not a feasible course of action. Many states do not even know what "all" is. Implementation of a comprehensive safeguards system is imperative to protect weapons-usable materials worldwide.

Toward a Comprehensive Safeguards System

Each state that possesses weapons-usable fissile materials must provide for their physical protection, control, and accounting—the three pillars of a rigorous, comprehensive safeguards system. Such a system (nuclear materials protection, control, and accounting, or MPC&A) was first developed in the United States 40 years ago, with the Los Alamos National Laboratory playing the lead role. This system became the model for IAEA international safeguards. However, uneven and incomplete application of domestic and international safeguards contributes to inadequate fissile materials security worldwide today.

The international nuclear safeguards system is designed to assure the international community that states party to the Nuclear Nonproliferation Treaty (NPT) and similar agreements honor their commitments not to proliferate nuclear weapons. The traditional system attempts to verify nondiversion of declared nuclear materials; it focuses on correctness of a state's declaration. The strengthened safeguards system, which includes the Additional Protocol developed after the Gulf War, expands verification to provide credible assurance of the absence of undeclared nuclear materials; it focuses on completeness of declaration.[19]

Although international safeguards are necessary to prevent diversion of nuclear materials by a state, they are not sufficient to prevent theft of weapons-usable material by determined individuals or groups. Pellaud[20] points out that IAEA safeguards agreements with over

[19]P. Goldschmidt, "The IAEA Safeguard Systems Moves into the 21st Century," *Supplement to the IAEA Bulletin* (1991), pp. *S1:S20.*

[20]B. Pellaud, "IAEA Safeguards: Experiences and Challenges," presented at the IAEA Symposium on International Safeguards (Oct. 1997).

130 states cover some 900-plus facilities and locations but only 20,000 kilograms of HEU and 500,000 kilograms of plutonium (including 50 tons of separated plutonium) compared to the roughly 1.9 million kilograms that exist worldwide. Nuclear materials in military programs are not subject to international safeguards. The United States entered into voluntary IAEA safeguards agreements in 1977, but these exclude facilities with direct national security significance. India, Pakistan, and Israel never signed the NPT, and North Korea withdrew.

Adequate security, therefore, depends on rigorous application of domestic safeguards in addition to the international safeguards that may apply. The U.S. domestic safeguard system is designed to protect nuclear materials against external threats such as terrorists and against insider threats. The principal safeguard against external threats is physical protection. The more insidious insider threat also requires additional rigorous internal controls and accounting.

The Soviet Union focused on physical protection (guns, guards, and high fences) along with stringent personnel screening. Its nuclear materials security record was excellent because the Soviet police state, with its omnipresent KGB and a system of grave consequences, deterred the insider threat as well. However, with the social, political, and economic upheaval that followed the dissolution of the Soviet Union, its past practices become Russia's liability. Physical protection alone is no longer adequate.

Modern safeguard systems combine physical protection with MPC&A. Physical protection consists of measures to protect nuclear material or facilities (and their transportation) against sabotage and theft. Nuclear facilities that require physical protection include all research, development, production and storage sites, nuclear reactors, fuel cycle facilities, and spent fuel storage and disposal facilities. These measures include guards, fences, and exclusion areas around facilities, in addition to perimeter and interior intrusion detection systems. Measures also include limited access and egress to facilities, building, and rooms. Technologies employed include systems such as microwave, electric field, and infrared systems on the perimeter and ultrasound, infrared, and motion-detection closed-circuit television on the interior. Finally, neutron, gamma-ray, and metal detectors at points of egress add an important element of defense.

MPC&A systems are designed to offer accurate nuclear materials inventory information; control nuclear materials to deter and prevent loss or misuse; provide timely and localized detection of unauthorized removal of materials; and assure, in near real time, that all nuclear materials are accounted for and that theft or diversion has not occurred. Proper material control limits the handling of nuclear materials to only authorized and properly identified personnel and ensures that two persons are present during nuclear material transactions. It helps track nuclear material from one site to another, from facility to facility, and from room to room. It ensures that there are limited number of entries and exits and alarms alert authorities to potential theft or diversion. It identifies nuclear material for tracking purposes.

Modern material accounting also employs statistical and computer-based measures to maintain knowledge of quantities of nuclear material present in each area of a facility. It relies on inventories and material balances to verify the presence of material or to detect a loss. In the United States, the Nuclear Materials Management and Safeguards System (NMMSS) implemented in 1976 contains current and historical data on inventories and transactions involving source and special nuclear materials within the United States and on all exports and imports. It tracks all transactions, including domestic and foreign transfers, operating losses, inventory differences, and burnup (transmutation and fission). Reconciliation of facility books with NMMSS also ensures that control indicators are furnished to those who perform oversight responsibilities and that anomalies are identified.

I provide this level of detail to demonstrate the complexity of securing nuclear materials. Effective MPC&A systems must be integrated with operational and safety practices. In the United States, it remains a challenge to provide adequate protection against changing terrorist threats. In view of the DOE plutonium report, it is not possible to guarantee that kilogram quantities of plutonium are not missing. We must rely on the integrity of the MPC&A system and its application for our confidence that such materials are not outside of state control. In U.S. facilities, operators must account for every gram of these materials in virtual real time. To declare any of it as an "inventory difference" or "waste" requires rigorous justification and verification.

It is imperative that each state with nuclear facilities implement its own rigorous, comprehensive safeguards system to prevent theft or diversion of weapons-usable materials. Although both countries have made progress in recent years, Russia and China have much work to do to achieve a modern safeguards system. Little is known about Pakistan and India. States that currently employ such systems and the IAEA should significantly expand their efforts to provide technical assistance to these nations. The G-8 should reprioritize its nuclear security financial assistance to help states develop their own rigorous MPC&A systems. These efforts will also help states meet their counterterrorism obligations under UNSCR 1540. In addition, the international safeguards system should be strengthened by universal adoption of the Additional Protocol and greater access for IAEA inspectors, along with stricter enforcement by the U.N. Security Council.

Each state must also develop a complete registry of weapons-usable plutonium and HEU along the lines of the DOE plutonium study. The IAEA already has registry requirements for states that hold safeguarded materials, but as pointed out previously, that constitutes only a fraction of the total worldwide. Such registry studies (both public and classified) will help identify historical anomalies and potential vulnerabilities in nuclear material inventories.

Other Vulnerabilities

Because rigorous safeguard systems have not been in place since the advent of nuclear materials, and because many countries still fall short today, nuclear materials could already be in the wrong hands or at least outside state-controlled systems. Fortunately, there are few known incidents of theft to date. The IAEA illicit nuclear trafficking database shows 224 incidents involving nuclear materials from 1993 to 2005. Only 16 confirmed incidents involved fissile materials, three with kilogram quantities of HEU and three with gram quantities of plutonium (the rest amounted to less than 250 grams HEU total).[21]

Each state should enhance its internal detection and tracking capabilities for illicit trafficking of nuclear materials and enhance its border and port security. It is imperative that each state identify past weaknesses and anomalies in fissile material inventories. These efforts should be aided by international efforts such as the DOE Second Line of Defense Program, which has helped to train personnel and to install radiation detectors at airports, seaports, and border crossings in Russia and other states.

Efforts to interdict potential shipments of nuclear materials, such as the Proliferation Security Initiative, should be strengthened. Increased intelligence sharing is important. Cooperative sting operations may flush out material outside state-controlled systems. Enhanced emergency response capabilities will help manage the consequences of an attack and potentially help disable suspected terrorist devices. Finally, forensics and attribution will be important, both for response and for preventing repeat attacks.

Over the longer term, we must also guard against "mining" of low-grade materials such as nuclear waste, spent fuel, and lost or abandoned materials. We must also pay much greater attention to safeguarding alternate nuclear materials such as neptunium and americium, which have been produced in multiple-ton quantities and may eventually become a terrorist bomb threat.[22] We must safeguard any process or nuclear material that is easier to obtain and/or less costly than building an enrichment plant or a reactor. In addition, the

---

[21]International Atomic Energy Agency, "Illicit Trafficking and Other Unauthorized Activities involving Nuclear and Radioactive Materials," www.iaea.org/NewsCenter/News/2006/traffickingstats2005.html.

[22]Albright and O'Neill, *The Challenges of Fissile Materials Control.*

commercial nuclear industry must redouble its safeguards efforts, on both the front and back end of the fuel cycle, as nuclear power expands worldwide. It would also be beneficial to utilize fuel cycles that are inherently more proliferation resistant.

Why have terrorists not yet crossed the nuclear threshold? Perhaps it is the lack of access to weapons-usable fissile material. However, nuclear attacks may also present an unacceptable level of risk and uncertainty to terrorists—not only risk of injury or death in preparing the mission but potential failure of the mission. For example, even nuclear-capable states still experience criticality accidents that kill nuclear workers because of misjudgments in material handling. In addition, terrorists are much more certain of success using chemical explosives, with which they have much greater familiarity. Moreover, nuclear attacks might not fit the motives of the bulk of terrorist groups active today.

Terrorists have also not yet crossed the radiological dispersal bomb (dirty bomb) threshold. A dirty bomb will disperse radioactive materials but will not cause a nuclear detonation and mushroom cloud. Materials for dirty bombs include roughly a dozen radioisotopes that are ubiquitous in international use in millions of radiation sources for medicine, industry, and agriculture and readily available to determined terrorists. A dirty bomb will not kill many people, but it will cause enormous psychological trauma and economic disruption.[23] Regardless of whether or not terrorists are just about to cross the nuclear bomb threshold, we must assume that some of them eventually will. The best preventive measure is to keep the weapons-usable material out of their hands.

Today's Greatest Threats

To deal with today's urgent threats, it is important to consider specifically tailored solutions in addition to the generic recommendations made earlier in this chapter. To that end, this section briefly describes what I see as the greatest threats in the current security environment. These six threats represent the highest probability that several tens of kilograms of weapons-usable plutonium or HEU will get into the hands of terrorists by theft or diversion from existing state-controlled stockpiles. So, the terrorist danger arises not so much from state-controlled nuclear weapons or nuclear materials stockpiles, but instead from the loss of state control. We must assume that once terrorists are armed with such materials, they will eventually be able to build an improvised nuclear explosive device and detonate it somewhere in the world.

1. Pakistan heads the list. It has all technical prerequisites: HEU and plutonium; enrichment, reactor, and reprocessing facilities; a complete infrastructure for nuclear technologies and nuclear weapons; largely unknown but questionable nuclear materials security; and missiles and other delivery systems. It views itself as threatened by a nuclear India. It has a history of political instability; the presence of fundamental Islamic terrorists in the country and in the region; and uncertain loyalties of civilian (including scientific) and military officials, and it is home to A. Q. Khan, the world's most notorious nuclear black marketeer. Helping Pakistan secure its nuclear materials during these challenging times is made difficult by the precarious position of its leadership and the anti-U.S. sentiments of much of its populace. Yet, such cooperation is imperative.
2. North Korea is a threat because it has withdrawn from the NPT and has separated roughly 40 to 50 kilograms of plutonium.[24] Although it is unlikely that this material will be stolen, we cannot dismiss the possibility that plutonium (especially if more is accumulated) may be exported to terrorist groups or to Iran. This is most likely to occur when North Korea perceives the existence of its regime or its nation terminally

---

[23]Ferguson et al., *The Four Faces of Nuclear Terrorism.*

[24]S. S. Hecker and W. Liou, "Dangerous Dealings: North Korea's Nuclear Capabilities and the Threat of Export to Iran," *Arms Control Today* 37 (2007), pp. 6–11.

threatened. Recent agreements at the Six-Party negotiations have paved the way for the eventual abandonment of the North Korean nuclear program. However, in the meantime, preventing the export of plutonium must be the highest priority.

3. HEU-fueled research reactors around the world are still operating in about 40 countries, many with inadequate safeguards. Fresh fuel for these reactors takes little chemical processing to convert to weapons-usable HEU. These reactors have constituted a grave terrorist threat for three decades. Much has been done to close such reactors or retrofit them with low enriched uranium. The DOE Global Threat Reduction Initiative has increased the pace of these efforts during the past two years. Private organizations, such as the Nuclear Threat Initiative, have played a vital role in reducing this threat by helping to identify key problems and catalyzing the actions of governments. However, so long as any HEU exists in inadequately safeguarded facilities, it presents an unacceptable risk. The solution is an accelerated worldwide effort to better secure all HEU, support the U.S.-Russian-led effort to take back all HEU from research reactors and facilities, and eliminate the civilian use of HEU wherever possible.
4. The Russian nuclear complex was most vulnerable in the early and mid-1990s. We are fortunate that nothing really terrible happened in the Russian nuclear complex. Credit goes to the loyalty of Russian nuclear workers and to the Nunn-Lugar Cooperative Threat Reduction Program. Over the past six years, the Russian government has also greatly enhanced physical security at its sites and reduced economic hardship for its nuclear stewards. But the Russian complex remains excessively large, and the amount of weapons-usable materials is staggering. Whereas physical protection has been improved, cooperative efforts have yielded significant improvements in control and accounting in only a limited number of facilities. To my knowledge, Russia has neither a baseline inventory of fissile materials produced nor a reconciliation of what exists today with what has been produced and used, and there is apparently no incentive to pursue either. Enhanced physical protection and reemergence of strong security services provide only temporary protection. It is time for Russia to make a stronger commitment to and greater investment in a comprehensive, modern MPC&A system for all its facilities. The United States can help, but only if Russia takes the lead.
5. Kazakhstan returned Soviet nuclear weapons to Russia under the Nunn-Lugar program, but it did not return all weapons-usable material. Project Sapphire brought nearly 600 kilograms of HEU from Kazakhstan to the United States in 1994, but there are still HEU-fueled reactors and additional quantities of HEU in Kazakhstan.[25] Fortunately, NTI has also catalyzed actions of the U.S., Kazakh, and Russian governments to better secure and return some of these materials. However, Kazakhstan also inherited a Soviet BN-350 fast reactor along with several tons of lightly irradiated plutonium. It also now owns the huge former Soviet nuclear test site at Semipalatinsk. U.S.-Kazakh cooperation has enhanced security of reactor installations and the BN-350 fuel. The security of the test site, however, declined dramatically after the dissolution of the Soviet Union and the transfer of ownership to Kazakhstan, raising concerns about vulnerable materials that may have been left behind by the Soviets. In addition, the apparent decision to keep the spent BN-350 fuel in Kazakhstan creates significant risks should the country experience the political instability witnessed in some of its neighboring Central Asian states.
6. Iran is last on this short list because it is believed not to possess weapons-usable materials at this time. However, Iran is pursuing a course that will bring it very close. If it gets weapons-usable materials, it will move to second place because it will be difficult to keep such materials out of the hands of individuals or groups determined to use them. Hence, it is imperative that the current diplomatic efforts succeed in stopping Iran short of its own indigenous enrichment activities while not denying it

---

[25]Albright and O'Neill, *The Challenges of Fissile Materials Control.*

a longer-term path to civilian nuclear power. Likewise, it is important that Iran not put into operation the heavy water research reactor under construction in Arak. This reactor, if combined with reprocessing capabilities, would provide Iran with a source of plutonium for a potential nuclear weapons program.

This short list illustrates the extreme urgency of the threat posed by loose fissile material. But it also emphasizes the need for tailored nonproliferation strategies. Others may propose a different list with different priorities; indeed, my long list also includes China, India, and Israel as well as the additional incremental risk from anticipated large increases in commercial nuclear power worldwide, from nuclear wastes, and from the alternate nuclear materials mentioned previously.

I agree with the 2005 Gleneagles communiqué, which reads in part, "the proliferation of weapons of mass destruction and their delivery means, together with international terrorism, remain the preeminent threats to international peace and security." But the key element, keeping weapons-usable materials out of terrorists' hands, is much more difficult than is generally appreciated. A greater sense of urgency is required—not only on the part of the United States but on the part of states that have more benign views of the risks of nuclear terrorism and believe that nuclear proliferation is a U.S. problem. Quite the contrary, loose fissile material must be the top security priority of every nation.

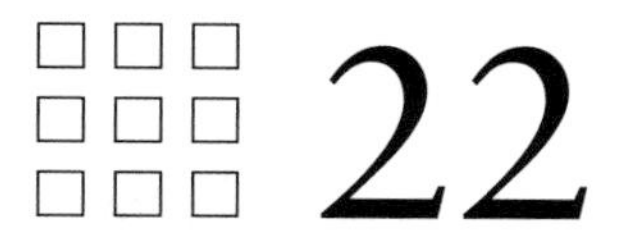

22 Illicit Trafficking of Nuclear and Radiological Materials

Galya I. Balatsky, Stacey Lee Eaton, and William R. Severe

Introduction

In recent years there has been a growing awareness of illicit trafficking in nuclear materials. During the early to mid-1990s, several cases of trafficking in nuclear material that could be directly used in nuclear weapons raised concerns that such materials could be obtained by terrorist or criminal organizations. These concerns were greatly increased by the events of September 11, 2001. Deterring access to and availability of nuclear and other radioactive materials is critical to combating terrorist capabilities to develop weapons of mass destruction. To address this issue, numerous programs have been implemented to secure nuclear weapons, nuclear materials, and other radioactive materials worldwide. Other programs have focused on detecting and preventing illicit trafficking of these materials as one means to mitigate the risk of nuclear terrorism. For purposes of this chapter, the term *illicit nuclear trafficking* comprises the unauthorized acquisition, provision, possession, use, transfer, or disposal of nuclear and other radioactive materials, whether intentional or unintentional and with or without crossing international borders.

More than 15 years of data have now been collected on nuclear trafficking. Despite well-founded concerns regarding the risk to our security posed by deliberate theft and misuse of weapons-usable material, the data demonstrate that most incidents of illicit nuclear trafficking don't indicate nefarious intent. Often, incidents indicate that a loss of control occurred during legitimate movements or transfers. In other cases, nuclear and other radioactive materials have been abandoned as a result of poor business practices, unauthorized disposal, or illegitimate attempts to avoid bureaucratic issues or costly fees. Despite the lack of a growing trend in the illegal movement of weapons-grade nuclear materials, the need to address this potential threat remains critical. There are thousands of tons of weapons-grade nuclear materials in the world, some inadequately secure. It only takes a few kilograms to make a nuclear weapon. Monitoring and analysis of all nuclear trafficking incidents are essential to gain insight into the vulnerability of materials, to understand the intent and motivation of nuclear traffickers, and to illustrate infrastructure weaknesses and potential pathways that may be exploited by an adversary.

Background: Post-9/11 Environment Highlights the Threat of Illicit Trafficking and Nuclear Terrorism

In his address to the General Assembly of the United Nations on December 8, 1953, President Dwight D. Eisenhower presented a new nuclear initiative to the world.[1] Talking

[1]Atoms for Peace speech. This address was given by President Dwight D. Eisenhower before the General Assembly of the United Nations on Peaceful Uses of Atomic Energy, New York City, Dec. 8, 1953.

about nuclear weapons, Eisenhower said, "The United States knows that if the fearful trend of atomic military buildup can be reversed, this greatest of destructive forces can be developed into a great boon, for the benefit of all mankind." He proposed to establish an international atomic energy agency and mentioned that the U.S. encouraged worldwide investigation into the most effective peacetime uses of fissionable materials.

Today the widespread of nuclear materials permeates various aspects of modern life, from power production to medical applications and process control. The International Atomic Energy Agency (IAEA) was established in 1957, and the system of safeguards to prevent diversion of nuclear materials for military uses was set up.[2]

Most states that received nuclear technology under the Atoms for Peace Program developed state systems of accounting and control, aimed chiefly at controlling nuclear materials and protecting public safety. The IAEA quickly developed a reputation for being both a global nonproliferation "watchdog" and a promoter of the safe use of nuclear and other radioactive materials.

Although illicit nuclear trafficking has likely taken place since the discovery of nuclear and radioactive materials, significant international concern about illicit nuclear trafficking was triggered by the end of the Cold War and the demise of the Soviet Union. The former Soviet states were perceived to have abundant stocks of unsecured nuclear materials, and the main concern was focused on rogue states acquiring special nuclear materials and establishing nuclear weapons programs. Several high-profile seizures caught the attention of the world, and numerous cooperative programs were initiated to address the security of the Soviet materials.

With events of September 11, 2001, however, nuclear terrorism came to be considered a serious and acute threat. In the past, there had been concerns over states acquiring special nuclear materials and establishing nuclear weapons programs but not over nonstate actors involved in clandestine efforts to build a nuclear device. Although international measures were in place to address proliferation by rogue states and a number of domestic measures were taken for public safety, no measures were specifically aimed at preventing terrorists from acquiring materials for an improvised nuclear device (IND) or a radiological dispersal device (RDD). Since the 9/11 events, the U.S. has stepped up its global efforts to improve the security of nuclear materials and to combat illicit nuclear trafficking. These efforts focus on securing international borders and supporting national programs to identify, secure, remove, and/or facilitate the disposition of vulnerable and excess nuclear and radiological materials.

The growing concern that terrorist nonstate actors may acquire, develop, traffic in, or use nuclear, chemical, and biological weapons and their means of delivery has been addressed in the U.N. Security Council Resolution 1540, which was adopted in 2004.[3] The 1540 resolution calls for the states to adopt the laws prohibiting "any nonstate actor to manufacture, acquire, possess, develop, transport, transfer, or use nuclear, chemical, or biological weapons and their means of delivery" or to assist or finance these activities; thus, the resolution seeks to criminalize trafficking of nuclear materials. (For more information on UN SCR 1540, see the corresponding chapter.)

Although instances of illicit trafficking predominantly involved materials in the former Soviet Union, other areas, such as Africa and South Asia, are of concern as well.

Materials: Today's Concern Is Not Only for Direct Weapons Material But for Radioactive Sources as Well

Before 9/11, the traditional materials of concern for illicit nuclear trafficking were those that could most directly be used by a rogue state to produce a nuclear weapon: highly enriched

[2] International Atomic Energy Agency, www.iaea.org.

[3] U.N. Security Council Resolution 1540 (2004), United Nations, New York City, New York, April 28, 2004.

uranium (HEU, typically defined as >20% U-235) and plutonium. These materials are still of concern, of course, since the estimates for their inventories worldwide are approximately 1800 MT plutonium and 1900 MT HEU.[4] At the end of 2004, only approximately 89 tons of fresh, separated plutonium and 32 tons of HEU were subject to IAEA safeguards.[5] With increased concern regarding terrorism, the threat is no longer limited to these materials; other nuclear materials (americium, neptunium, and thorium) and other radioactive materials (sources for various industrial and medical applications, byproduct materials, and naturally occurring materials) have become a concern as well because many of these materials can be used to make a "dirty bomb."

During the 1990s serious international concerns developed based on the growing awareness of accidents involving radioactive sources and their potential serious consequences. These concerns were heightened by the events of 9/11 as the potential for malicious use of radioactive material by terrorists became clear. Although there are some reports of malicious use of sealed sources, most information related to damage, serious injury and death that may have resulted from sealed sources is based on reports of misuse or accidents (some of the specific cases reported are discussed later in this chapter).

Dispersion of radioactive material by an explosively driven device (RDD) or other means has often been regarded as an event of less than catastrophic consequence. However, as demonstrated by the Goiania, Brazil, case of 1987,[6] the consequences of widespread radioactive contamination in an urban area would have a substantial social, environmental, and economic impact. A radiotherapy unit, containing a Cs-137 source, 1,375 Ci, in powder form, was abandoned by a Brazilian clinic in Goiania. The encapsulated source was subsequently opened by unsuspecting residents, resulting in an exposure that killed five people and made dozens sick. The authorities had to set up a special clinic to screen about 100,000 of the city's 1 million residents and performed extensive searches to locate all contaminated areas. Authorities identified seven main contaminated sites, including contamination of 46 residences and 50 public places. As a consequence of this incident, it took five years for the area to return to the level of economic output before the incident, and tourism plummeted to zero. During cleanup (see Figure 22.1), the authorities had to remove 6,000 MT of waste

FIGURE 22.1 Cleanup operation in Goiania, Brazil, 1987.

[4]www.isis-online.org.

[5]IAEA Annual Report for 2004.

[6]IAEA Report on Goiaia incident, Publication 815, 1988.

and to build temporary and long-term waste storage facilities. The last storage facilities were finished in 1997, 10 years after the incident.

International Monitoring of Nuclear Trafficking

There are a number of databases and organizations looking into illicit trafficking of nuclear materials.

The IAEA's Illicit Trafficking Database (ITDB) is the primary resource for the international community on information related to illicit nuclear trafficking. Established in 1995, the database's objective is "to facilitate exchange of authoritative information on incidents of illicit trafficking and other related unauthorized activities involving nuclear and other radioactive materials among the States."[7] The ITBD provides annual reports on state-confirmed incidents along with an assessment that provides information on common trends and patterns. IAEA member states participating in the ITDB are provided with regular updates and quarterly reports which contain more detailed information than is provided in the annual report. Additionally, the ITDB collects open-source and other information which the IAEA typically seeks to confirm with the states connected with such reports. The ITDB contained 827 confirmed incidents reported by the participating states as of December 31, 2005.

Additional Sources of Information on Illicit Nuclear Trafficking[8]

Open-Source Databases

- *Center for Nonproliferation Studies NIS Nuclear Trafficking Database, www.nti.org/db/nistraff/.* "The Newly Independent States (NIS) Nuclear Trafficking Abstracts Database includes cases and reported incidents of trafficking in nuclear and radioactive materials in and from the NIS. The database also features trafficking incident tables and summaries and analysis of relevant articles and reports."
- *Stanford's Center for International Security and Cooperation (CISAC) database of Nuclear Smuggling, Theft, and Orphan Radiation Sources (DSTO), http://cisac.stanford.edu.* Although once considered an authoritative source of trafficking information, the Stanford DTSO has not been updated since 2002. Its creators, Fritz Steinhausler and Ludmila Zaitseva, considered foremost experts in the field, moved to the University of Salzburg in Austria and now maintain a comparable database there with restricted access. In the past, the Stanford DTSO was divided into categories that included type of incident, type of material, suspected origin of material, perpetrators involved, reported destination, and intended use. Also included was analysis on historical trends in the evolution of exchanges and trafficking routes.

Regularly Updated Reports, Articles, and News Sources

- *NIS Export Control Observer, Center for Nonproliferation Studies at the Monterey Institute, http://cns.miis.edu/pubs/nisexcon/.* "The NIS Export Control Observer is devoted to the analysis of WMD export control issues in the NIS. It is published monthly in English and Russian for the NIS and Western export control community by the Center for Nonproliferation Studies, Monterey Institute of International Studies, with financial support from the U.S. Department of

[7]IAEA 2006 Illicit Trafficking Database Report

[8]Based on unpublished Los Alamos National Laboratory Report on smuggling databases, Paige Harper, Sept. 2005.

State." The *Observer* contains an illicit trafficking report in each issue; archives are available at the Website.

- *Jane's Information Group, www.janes.com.* Jane's Information Group is well known for its analysis of various threats and security issues in the world. Jane's publishes *Terrorism and Security Monitor.*
- *Transnational Threats Update, Transnational Threats Project at the Center for Strategic and International Studies (CSIS), www.csis.org/tnt/index.cfm.* "This update is produced by the Transnational Threats Project at the Center for Strategic and International Studies (CSIS) and provides monthly news on terrorism, drug trafficking, organized crime, money laundering, and other transnational threats. The *TNT Update* draws on several U.S. and international media sources, including Associated Press, Agence France-Presse, Reuters, Xinhua News Agency, World Tribune, Afghan News, and others." CSIS also has a task force dedicated to the nuclear black market that publishes reports on the subject.
- *Bellona Foundation of Norway, www.bellona.no/en/international/russia/nuke-weapons/nonproliferation.* Bellona's Website is devoted to news that pertains exclusively to nuclear issues in the spheres of energy and climate, Russia, the environment, and nonproliferation. Its extensive reporting, though not specifically devoted to a section on nuclear trafficking, does cover news items concerning loose or missing nuclear and radiological sources, especially in Russia.
- *Swedish Nuclear Power Inspectorate (SKI), www.ski.se/extra/tools/parser/index.cgi?url=/html/parse/index_en.html.* This site posts reports commissioned by SKI to study and combat illicit trafficking of nuclear and radioactive substances in the NIS.
- *Open Source Center (OSC), https://www.opensource.gov.* This news source run by the U.S. government requires subscription with OSC. It allows user to search through foreign media reporting and analytical products prepared by OSC. The Center also provides detailed topic- and region-specific searches, and subscribers can use automatic updates for news on specified topics.
- *World News Connection, http://wnc.fedworld.gov/subscription.html.* The World News Connection is the foreign news service of the U.S. government and requires a subscription to access its translations of unclassified foreign documents, scholarly works, research reports, serial publications, and other selected sources.
- *Yaderny Kontrol/Ядерный Контроль Center for Policy Studies in Russia (PIR Center), www.pircenter.org/russian/nrt/.* Yaderny Kontrol is an "electronic newsletter containing digest of the Russian press and other media (including regional media institutions and the closed cities), as well as various official documents." Archives are available at the Website.

Additional information of nuclear trafficking is available on many more sites, all of which would be impossible to list here, ranging from encyclopedias (*Britannica*, Wikipedia) to special-interest groups such as Green Peace. One good place to look is the database of radiological incidents and related events compiled by Wm. Robert Johnston (www.johnstonsarchive.net/nuclear/radevents/index.html). Though the database focuses on events that produced radiation casualties, it does contain information on trafficking cases.

The Los Alamos National Laboratory (LANL), center of the Manhattan Project, also collects data on illicit trafficking; the LANL database has incident reports from as early as the 1960s. The analysis of nuclear trafficking presented in the following discussion primarily draws on the LANL database. However, the trends and analysis presented are fully consistent with the information presented in the IAEA's *International Trafficking Database Report.*

Analysis of Illicit Nuclear Trafficking Data

With regard to special nuclear material traditionally of greatest concern, there have been several widely publicized seizures of both highly enriched uranium (HEU) and plutonium. Table 22.1 of the IAEA *Illicit Trafficking Database Report* from 2006 lists only 18 confirmed incidents involving these materials between 1993 and 2005; as of December 31, 2005, the total number of confirmed cases reported by the participating states in the IAEA's ITDB was 825. Although more cases were reported in open sources, they are not confirmed and not included in the statistics.

Of the 825 total confirmed cases depicted in Figure 22.2,[9] 516 cases, or 63%, involved radioactive materials, other than nuclear materials, mostly radioactive sources; 224 cases, or 27%, involved nuclear materials; 26 cases, both nuclear and other radioactive materials; and 50 cases, contaminated materials. An analysis of these cases reveals that most of the incidents occurred in the early 1990s (10 incidents by 1995), and the quantities of materials involved is relatively small. The largest single quantity reported is the ~3 kg of HEU seized in St. Petersburg, Russia, in 1994. The 2005 IAEA report, however, lists a case involving the possession and illegal transport of 170 g of HEU as recently as 2003.

Figure 22.3 shows the number of all plutonium and uranium cases per year since 1991 (included are all grades of these materials). Although there has been one recent notable seizure of HEU, the uranium cases mostly consist of low-grade materials such as low-enriched uranium (LEU) typically used as fuel in nuclear power plants. The overall trend in the number of uranium cases per year has been decreasing since the peak in 1993. The U.S. Department of Energy initiated the Material Protection, Control, and Accountability Program in 1994 to "prevent the theft and diversion of Russian nuclear weapons and nuclear weapons-usable material by consolidating, securing, and reducing the stocks of weapons grade fissile material."[10] It is encouraging to see indications that the success of this program is potentially responsible for the subsequent decline in the number of cases. The number of plutonium incidents per year has been fairly constant, but further analysis shows that the cases in recent years have mostly consisted of small sources of plutonium-238 or plutonium-239, typically in the form of Russian smoke detector sources.

Radioisotopes, on the other hand, account for a larger percentage (~40%) of the diverted/seized materials than do the nuclear materials. These are radioactive sources that have wide use in numerous industrial and medical applications. The top three radioisotopes involved in illicit trafficking incidents are cesium-137, cobalt-60 and strontium-90, respectively, comprising almost three-quarters of the total; these materials can be used to construct a radiological weapon. The location of these sources is not limited to the traditional nuclear facilities but rather is owned by thousands of private companies, universities, and hospitals worldwide. Figure 22.4 shows that their rate of involvement has been significantly increasing in recent years. It is uncertain at this point if this is actually due to their perceived heightened value on the black market or simply an artifact of better reporting and/or detection methods. In any case, their wide availability should motivate the continuation and possible enhancements of ongoing efforts to inventory, track, and protect these sources.

In Figure 22.4, the radioisotope cases are measured against other categories of materials typically seen in illicit trafficking incidents. Dual-use materials, which are being monitored by LANL and not included in the IAEA reporting, are those materials (such as tungsten and high-strength aluminum) that have legitimate industrial uses as well as weapons applications. "Other nuclear materials" refers to materials other than uranium and plutonium

---

[9]"Illicit Trafficking and Other Unauthorized Activities Involving Nuclear and Radioactive Materials," Illicit Trafficking Database (ITDB) Fact Sheet, IAEA, June 2006, www.iaea.org/NewsCenter/Features/RadSources/PDF/fact_figures2005.pdf.

[10]National Nuclear Security Administration, "MPC&A Program Strategic Plan," Department of Energy, July 2001.

Table 22.1 Incidents involving HEU and Pu confirmed to the ITDB, 1993–2006.

Date	Location	Material Involved	Incident Description
May 24, 1993	Vilnius, Lithuania	HEU/150 g	4.4t of beryllium including 140 kg contaminated with HEU were discovered in the storage area of a bank.
March 1994	St. Petersburg, Russian Federation	HEU/2.972 kg	An individual was arrested in possession of HEU, which he had previously stolen from a nuclear facility. The material was intended for illegal sale.
May 10, 1994	Tengen-Wiechs, Germany	Pu/6.2 g	Plutonium was detected in a building during a police search.
June 13, 1994	Landshut, Germany	HEU/0.795 g	A group of individuals were arrested in illegal possession of HEU.
July 25, 1994	Munich, Germany	Pu/0.24 g	A small sample of PuO_2-UO_2 mixture was confiscated in an incident related to a larger seizure at Munich Airport on August 10, 1994.
August 10, 1994	Munich Airport, Germany	Pu/363.4 g	PuO_2-UO_2 mixture was seized at Munich airport.
December 14, 1994	Prague, Czech Republic	HEU/2.73 kg	HEU was seized by police in Prague. The material was intended for illegal sale.
June 1995	Moscow, Russian Federation	HEU/1.7 kg	An individual was arrested in possession of HEU, which he had previously stolen from a nuclear facility. The material was intended for illegal sale.
June 6, 1995	Prague, Czech Republic	HEU/0.415 g	An HEU sample was seized by police in Prague.
June 8, 1995	Ceske Budejovice, Czech Republic	HEU/16.9 g	An HEU sample was seized by police in Ceske Budejovice.
May 29, 1999	Rousse, Bulgaria	HEU/10 g	Customs officials arrested a man trying to smuggle HEU at the Rousse customs border checkpoint.
December 2000	Karlsruhe, Germany	Pu/0.001 g	Mixed radioactive materials, including a minute quantity of plutonium, were stolen from the former pilot processing plant.
July 16, 2001	Paris, France	HEU/0.5 g	Three individuals trafficking in HEU were arrested in Paris. The perpetrators were seeking buyers for the material.
June 26, 2003	Sadahlo, Georgia	HEU/~170 g	An individual was arrested in possession of HEU upon attempting to illegally transport material across the border.
March to April 2005	New Jersey, United States	HEU/3.3 g	A package containing 3.3 g of HEU was reported lost.
June 25, 2005	Fukui, Japan	HEU/0.0017 g	A neutron flux detector was reported lost at NPP.
February 1, 2006	Tbilisi, Georgia	HEU/79.5 g	A group of individuals was arrested trying to illegally sell HEU.
June 30, 2006	Hennigsdorf, Germany	HEU/47.5 g	Authorities discovered trace amounts of HEU on a piece of tube found among scrap metal entering a steel mill.

Note: Incidents involving Pu in the form of radioactive sources are not included in this table.

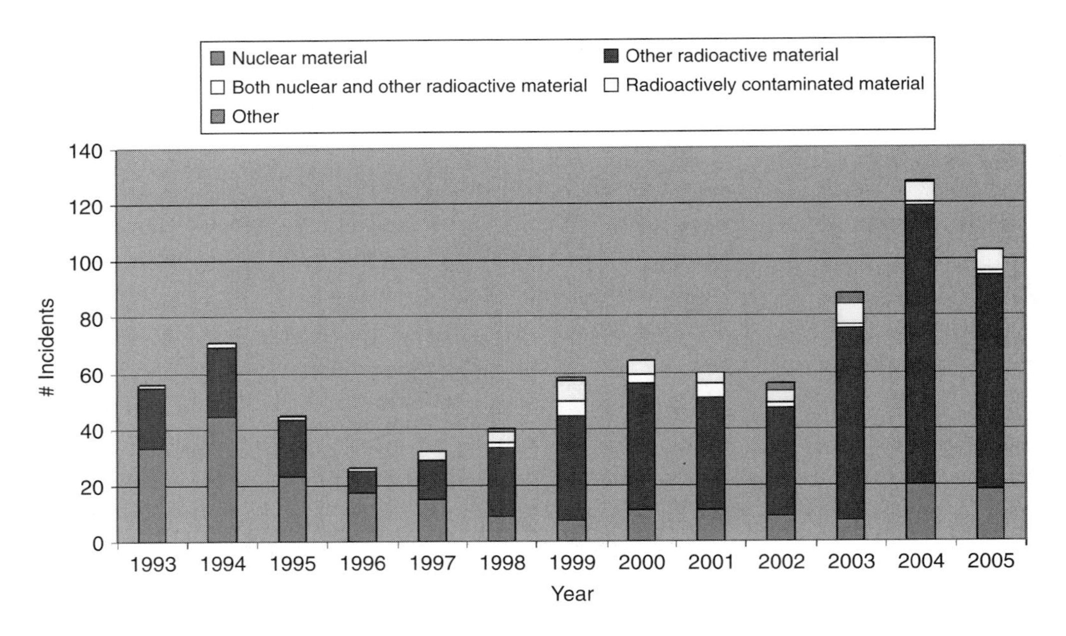

FIGURE 22.2 Number of confirmed incidents for 1993–2005 per IAEA's ITDB.

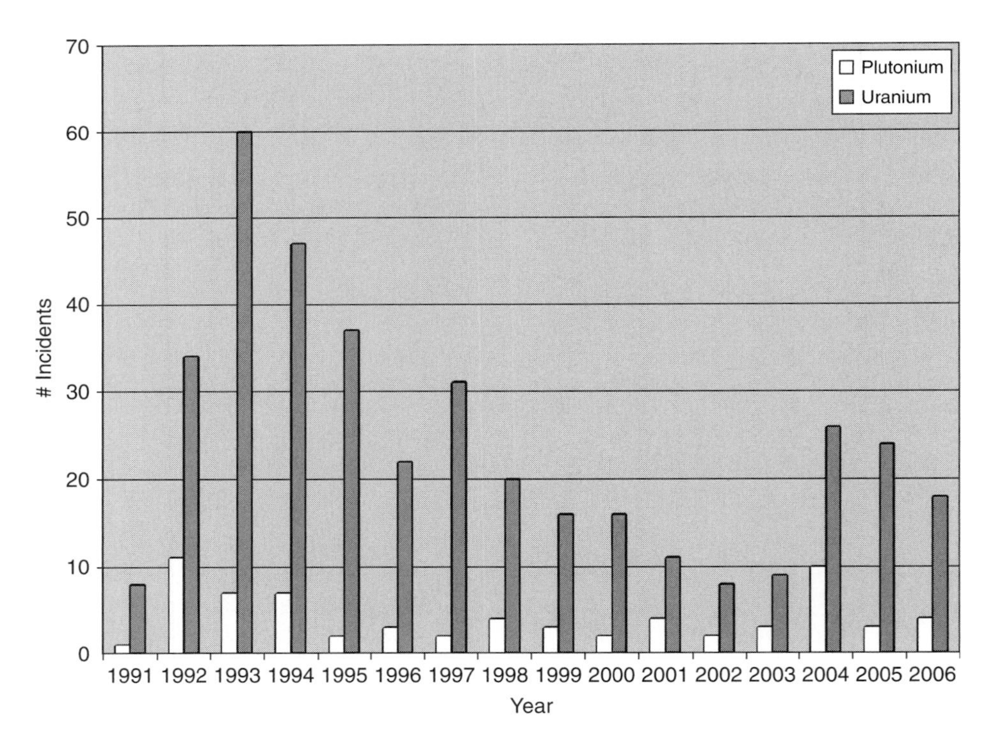

FIGURE 22.3 All grades uranium and plutonium cases per year, 1991–2006. (LANL data.)

(such as americium, thorium, and neptunium). The "other" category captures information on unknown materials or indirectly related items. Accurate analysis of the seized materials is often unavailable in press reporting, but the items are known to be radioactive in some fashion. The main components of this category, however, are the scam materials. A few materials,

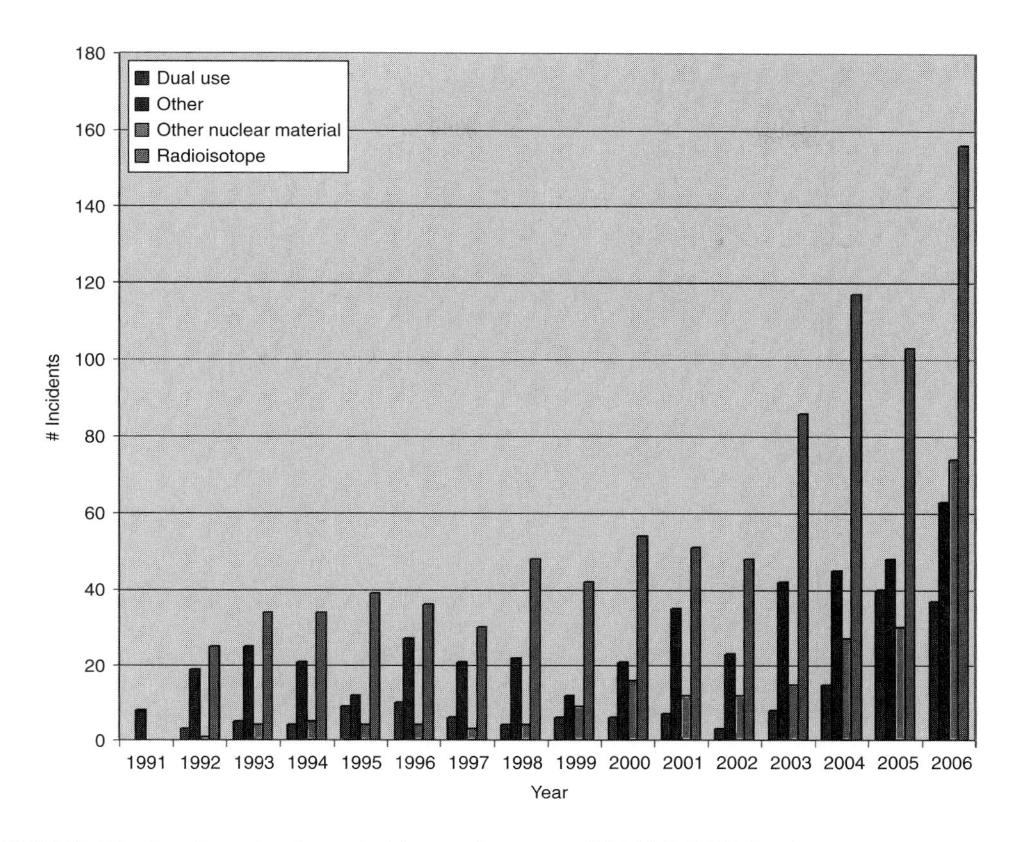

FIGURE 22.4 Number of cases in other material categories per year, 1991–2006. (LANL data.)

such as "red mercury" and osmium, in the illicit trafficking world have no known weapons applications, but their rumored nuclear value has become legendary, and there are still numerous reports of these materials being for sale.

Major Sources of Smuggled Materials

There are valuable insights to be gained about illicit nuclear trafficking through the evaluation of the associated locations. Determining where seized materials originated can indicate the security of nuclear facilities where they were stored. Preferred routes and movement methods can be determined by where materials traveled. Finally, where the materials were going (if known) can provide insight into the demand side of trafficking (perpetrators). Locations are most often referred to in geographical terms: countries, administrative regions, and specific cities. The roles these geographic locations have in smuggling include the origin of the material, the seizure location, the destination, or a transit location between the origin and destination/seizure point. The participation of specific locations, such as nuclear facilities or private companies, can be evaluated as well.

The dissolution of the Soviet Union, with its extensive nuclear infrastructure, in 1991 precipitated the rise in illicit trafficking events. Russia had the largest amount of materials and facilities and not surprisingly is by far the origin of the largest number of diverted/seized materials. Their number of cases per year has been declining overall, most likely due to the numerous cooperative programs designed to better protect their materials. Overall, the countries of the former Soviet Union, including Russia, account for almost 50% of both the reported origins and seizure locations.

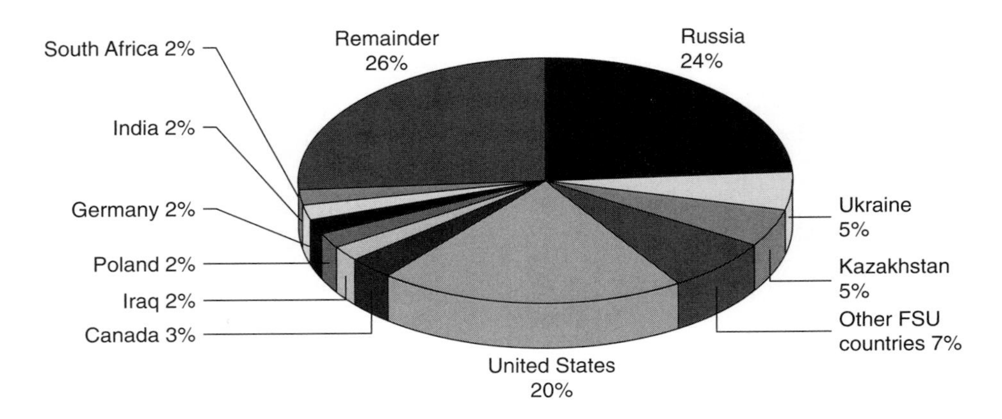

FIGURE 22.5 Distribution of illicit trafficking cases by the country of origin of the materials seized.

Nuclear smuggling, however, is by no means limited to Russia (see Figure 22.5). More than 100 countries are reported to have been involved in illicit trafficking in some role. Surprisingly, in a large number of cases, the United States was named as the origin of the material, with the majority of the cases occurring in the past two years. Most of these incidents involved the loss of industrial sources from private companies, usually inadvertently through the theft of company vehicles. A fair number of these sources are subsequently recovered. These examples, however, serve to illustrate the strong need for global efforts to better track and protect nuclear and radiological materials.

In illicit trafficking of nuclear materials, most often it is an individual that is arrested for possession of materials. Analysis shows that most individual perpetrators appear to be working mainly independently or within small groups of friends and relatives. They often do not have a prearranged buyer available but usually perceive a target of opportunity once they have possession of material. The main concern with individuals is the high risk associated with the insider threat. An "insider" is an employee of a facility with unescorted access to protected areas and/or nuclear materials. The U.S. Nuclear Regulatory Commission (NRC) recognizes this risk and regulates the security measures at nuclear power plants against a design basis threat that includes "an internal threat of an insider, including an employee (in any position)."[11]

Perpetrators' Intentions and Motives

To better combat illicit trafficking, it is important to learn about perpetrators' motives and the end use of trafficked materials. Weapons-grade materials always catch the attention of the press due to the high consequence of their potential use, but there are far more recorded cases and data available on incidents involving radioactive materials.

In an analytical publication on nuclear smuggling, L. Zaitseva and K. Hand analyzed 700 illicit trafficking incidents that occurred from 1991–2002.[12] They considered the supply and demand sides of the nuclear smuggling chain and categorized the participants as suppliers (outsiders and insiders), intermediaries, and end users.

---

[11]Nuclear Regulatory Commission, Code of Federal Regulations Website, 10 CFR 73.1, www.nrc.gov/reading-rm/doc-collections/cfr/part073/part073-0001.html.

[12]Lyudmila Zaitseva and Kevin Hand, "Nuclear smuggling chain: Supplies, intermediaries, and end-users," *The American Behavioral Scientist*, Feb. 2003, v. 46, iss. 6, pp. 822–844.

In a LANL study, an approach was made to classify all trafficking incidents according to the following categories:[13]

- Innocent and naïve accidents
- Economic opportunity
- Assaults, inflicting bodily harm (exposure, poisoning)
- Political considerations, nuclear terrorism

These are described in turn in the following sections.

Innocent and Naïve Accidents

There are numerous reports in which the loss of control of materials led to accidental but serious exposures and contaminations. For example, abandoned teletherapy sources containing cobalt caused an incident in Ciudad Juarez, Mexico, in 1983–1984. A scrap yard obtained used medical equipment with about 6,000 pellets of cobalt-60, totaling over 400 curies. The container with pellets was ruptured and the 1 mm pellets were spread throughout the scrap yard; later many of them ended up as scrap metal subsequently converted into steel products (rebars for buildings, table legs, and others). The radiation was detected when the truck carrying the products took the wrong turn and set off automatic radiation sensors installed at Los Alamos National Laboratory. It was estimated that 500–931 MT of contaminated steel had been shipped to the U.S. The products were collected in 40 U.S. states, and 21 contaminated areas in Mexico were found. Overall, 10 people were significantly exposed to radiation, several of them receiving 100–450 rem doses; one worker died and four were injured.

Economic Opportunity

The largest number of trafficking cases have been opportunistic incidents involving monetary reward. These vary from the traditional motive of selling nuclear material for a profit to companies simply trying to avoid costly disposal fees.

Zaitseva and Hand mention that collapse of the Soviet Union and the followup economic hardship made many nuclear workers commit a theft of the materials at their place of work. "A metalworker at the Electrostal plant, who diverted 115 kg of uranium dioxide pellets from his facility in 1993, said he did so out of desperation because he had three children to support and his salary, the only family income, had not been paid for months." They also concluded that the majority of theft cases were committed by impoverished insiders.

The low-level waste does represent a problem for many companies; it is expensive and time consuming to dispose of properly, and unscrupulous companies rely on dumping to save time, money, and effort. A foreign company dumped its waste at the Koko port in Nigeria in summer 1987, and some of the dumped chemicals were radioactive. Twenty-six workers who handled the waste were injured or exposed due to harmful chemicals.

Assaults and Inflicting Bodily Harm

In contrast to the innocent accidents just described, there are unfortunately numerous reports of radioactive materials being used to intentionally cause harm to individuals through exposures and/or poisoning.

As it is reported by Wm. Robert Johnston, sometime in the 1970s an individual in France placed radioactive graphite fuel element plugs under the driver's seat in a car, which caused a localized exposure to a person. The person received a 25–30 rad dose to his spinal

[13] All the cases quoted in the section were taken from Wm. Robert Johnston's database of radiological incidents and related events: www.johnstonsarchive.net/nuclear/radevents/index.html.

bone marrow and 400–500 rads to his testes. The perpetrator was eventually identified and convicted.

A Taiwanese graduate student had been repeatedly poisoned by a fellow student who placed acrylamide and in some cases radioactive phosphorus-32 in his drinks or on eating utensils at work on about 30 occasions from October 1992 to February 1996.

The most recent case is the polonium-210 poisoning and subsequent death of former Russian citizen and KGB agent Alexander Litvinenko, who had been granted political asylum in the U.K., in London in November 2006. U.K. authorities investigated Litvinenko's death as homicide but at the time of this writing no charges have been filed.

Political Consideration, Nuclear Terrorism

We can consider two potential scenarios for nuclear terrorism: an improvised nuclear device or a radiological dispersion device. Exploding a nuclear device will obviously result in total devastation of the area; the RDD will not be so damaging in terms of physical destruction but will result in significant indirect damages, such as economic losses, cleanup, and medical treatment costs in case of contamination as well as psychological trauma to the community.

The most infamous case of nuclear materials being used for political purpose occurred in Russia in 1995. Chechen separatist leader Shamil Basaev ordered placement of radioactive materials in a Moscow park and then informed Russian media. This was apparently intended as a demonstration that his group could perform a radiological attack in the future, although this attack was not undertaken.

The 2003 U.S. National Strategy for Combating Terrorism[14] states that the probability of a terrorist organization using a chemical, biological, radiological, or nuclear weapon or high-yield explosives has increased significantly during the past decade and that the new global environment makes it easier to do. This assessment is consistent with the view of many experts that terrorists are highly motivated to conduct a nuclear or radiological attack and are making efforts to acquire the capability to do so.

Illicit Trafficking Has Been Evolving

Overall, we can observe shifting trends in nuclear trafficking over the past 15 years. For example, according to LANL data, in the early 1990s numerous seizures were made of materials diverted by employees of Russian facilities. These workers were often underpaid and perceived a market for materials to which they had unrestricted access. Before upgrades were made to facility security measures, it was actually fairly easy to remove the materials without risk of detection. It was often an individual with no connections of his own who would choose to involve friends or relatives in the diversion or attempted sale of these materials.

The pathways for illicit nuclear trafficking were analyzed to determine geographic and regional trends; their shifting patterns also indicate the change. For example, it has been observed over time that trafficking out of Russia occurred in two general directions: west toward Europe/Eastern Europe and south toward Central Asia and the Caucasus region. In the early 1990s, the trend was movement of the material toward Europe, where the perceived market for such materials was located. There is speculation that this perception was aided by the active counter-smuggling operations in countries such as Germany. This trend declined in the late 1990s, while at the same time the movement appears to be increasing toward the Central Asia/Caucasus region. The number of cases seems to be evenly distributed between the two regions in the 2000s.

Since there is a concern over involvement of state-sponsored groups or even country-level illicit procurement of materials, it is important to watch for any indicators of organized

[14]National Strategy for Combating Terrorism, Feb. 2003.

involvement in trafficking. Analysis of past trafficking events currently shows a low level of involvement from larger, more sophisticated groups such as organized crime or terrorists. In the press, there has been a lot of speculation that the trafficked radioactive materials were headed to known terrorist regions, but very rarely is the ultimate destination of seized materials truly known. It is often the case that the possessors of the materials did not have an actual buyer in place, but it is worrisome that any true interest in procuring materials would go undetected. More detailed analysis and followup to seizures and incidents could help make more and better data available to aid counterproliferation programs. There is a definite risk that the more sophisticated attempts to smuggle nuclear materials will more likely go undetected, and the fear is that mainly unsophisticated individuals are currently being caught, with other traffickers remaining undetected.

Combating Nuclear Trafficking: Building an Integrated and International Response

Formulation of an effective state antitrafficking effort should encompass several elements on the state and international levels:

- A regulatory climate that supports the society, industry, and public intolerance toward trafficking in nuclear materials
- Effective detection measures and seizures (radiation monitoring equipment installed at border checkpoints and customs offices)
- Disposal pathways for unwanted, abandoned, and accidentally located nuclear materials
- More detailed analysis of illicit trafficking incidents, including forensics
- Cooperation between the agencies involved in combating nuclear trafficking and between the countries, especially across the border
- Assistance to countries that do not have nuclear capabilities or resources
- Preparedness and incident management and response capabilities

Some of these efforts have been ongoing; other initiatives have just started.

European Union Activities to Combat Illicit Trafficking

In its communication on nuclear trafficking of September 1994, the European Commission proposed a series of actions such as increased cooperation within the EU on customs and border checks and improved security and operations of nuclear facilities. The EU's Joint Research Center's Institute for Transuranium Elements (ITU) in Karlsruhe, Germany, is the primary resource for analysis and identification of materials involved in illicit nuclear trafficking. ITU collaborates closely with Europol and German police agencies to support EU efforts to counter nuclear trafficking.[15]

Nuclear Smuggling International Technical Working Group

The International Technical Working Group (ITWG) was established in 1995 at an International Conference on Nuclear Smuggling Forensics Analysis that was hosted by Lawrence Livermore National Laboratory. The ITWG meets at regular intervals to address

[15]European Commission, "Joint Research Center in Action, No. 1," April 2002, www.jrc.cec.eu.int/more_information/jrc-in-action/pdf/in-action2002-01_en.pdf (April 2002).

scientific and technical issues related to preventing illicit trafficking of nuclear materials, with an initial focus on nuclear forensics. The group includes technical experts from the Russian Federation, the EU, and the United States and is open to participation by scientists and law enforcement personnel from states and international organizations interested in preventing illicit nuclear trafficking.[16]

The ITWG is focused on nuclear forensic methods based on relations between material characteristics and process and production history as tools for analyzing intercepted materials and acquiring evidence to identify and attribute materials to prosecute the perpetrators. The ITWG has grown from its inception in 1996 and now includes about 30 member nations and organizations. The group also works closely with the IAEA to provide needed technical assistance. In 2004 the ITWG Nuclear Forensics Laboratories (INFL) was established to promote nuclear forensics sciences and technical development in the area.

Nuclear Forensics

The United States uses its network of National Laboratories to analyze material seized in nuclear trafficking incidents, not only to determine its type and quantity but also to establish its probable source. For example, the lead container and ampoule shown in Figure 22.6 were seized in May 1999 by Bulgarian customs officers. Documents found with the container described the material as 99.99% uranium-235. The U.S. Department of State arranged for the container to be sent to the Lawrence Livermore National Laboratory for forensic analysis. Complete analysis revealed that the ampoule contained uranium with 73% uranium-235 and 12% uranium-236. This is consistent with material that has been recycled from very highly enriched nuclear reactor fuel. Although weapons usable, the quantity was far short of that required for a nuclear device. Results indicated that the uranium originated from a facility in Eastern Europe.[17]

a.

b.

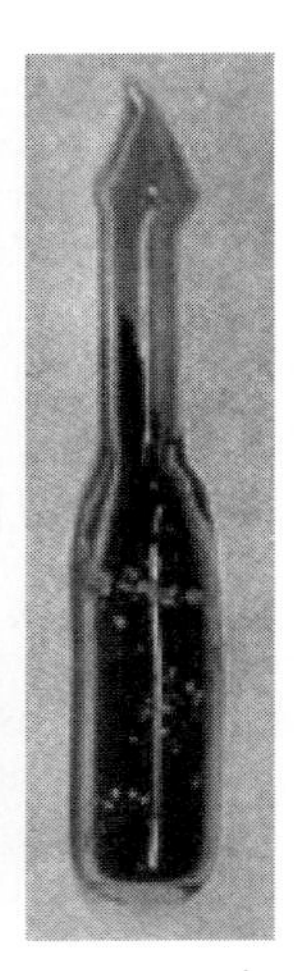

c.

FIGURE 22.6 a. A lead container shielding an ampoule filled with uranium was seized at the Bulgarian border in 1999 and sent to Livermore for analysis. b. The interior of the container was lined with yellow wax, which was used to cushion the nuclear material. c. The glass ampoule contained almost 4 grams of highly enriched uranium oxide.

---

[16]S. Niemeyer, L. Koch, and N.V. Nikiforov, "Synopsis of the International Workshop on Illicit Trafficking of Nuclear Material," Second ITWG Meeting, Obninsk, Russia, Dec. 2–4, 1996 (March 1997).

[17]"Tracing the Steps in Nuclear Material Trafficking," *Science and Technology Review*, March 2005.

Radiation Emergency Assistance

The World Health Organization (WHO) recognized the Oak Ridge Institute for Science and Education's Radiation Emergency Assistance Center/Training Site (REAC/TS) as a WHO Collaborating Center for Radiation Emergency Assistance in 1980.[18] The Center was established in 1976 and has a 24-hour emergency response program and a cadre of highly trained physicians, nurses, health physicists, radiobiologists, and emergency coordinators. The Center can provide information on medical care to radiation accident victims or consult on it globally. The Center, at the request of WHO and IAEA, provides services to foreign governments in the event of an actual radiation accident, provides information in cases of human radiation exposure, and assists in developing medical emergency plans to address large-scale radiation accidents. For example, the Center provided support for diagnostics evaluation of 10 people when an iridium-192 source was ruptured in Venezuela and developed followup procedures for people affected by the cesium-137 accident in Goiania, Brazil.

Collecting Abandoned Sources

IAEA has established a division of radiation and waste safety to assist countries in searching for, collecting, and disposing of abandoned radiation sources. According to IAEA data, many radiation sources were transported in the 1950s and 1960s before regulatory control was established, and no provisions for return or disposal were made.[19] Although the sources are quite old, fatal doses could be received if someone were to be exposed to some of the them.

The Offsite Source Recovery Project (OSRP) at LANL is tasked with removing and assisting with disposal of unwanted or disused U.S.-origin sources worldwide. The OSRP has a Website where any organization or user can register their unwanted sources and ask for assistance: http://osrp.lanl.gov. The project has recovered about 14,000 sources since 1999.

Second Line of Defense: "Megaports"

Effective export and border controls are of utmost importance to preventing the further spread of illicit nuclear trafficking. The Second Line of Defense Program is a U.S. cooperative program administered by the National Nuclear Security Administration (NNSA) designed to "prevent illicit trafficking in nuclear and radiological materials by securing international land borders, seaports, and airports that may be used as smuggling routes for materials needed for a nuclear device or a radiological dispersal device."[20] Such radiation monitoring equipment has already helped to interdict radioactive materials crossing international borders. The installation of improved detection technology and subsequent training will further assist these officials in stopping the movement of materials.

Implementation of 1540

Nevertheless, there are still challenges, the most common being lack of coordination and communications between the countries or agencies involved. For example, in May 2006, local press reporting indicated that authorities in U.S. and Mexico were not required to

[18]Oak Ridge Associated Universities, REAC/TS, www.orau.org.

[19]"Lost and Found Dangers: Orphan radiation sources raise global concern," *IAEA Bulletin*, 41/3/1999.

[20]Second Line of Defense Program Website: www.nnsa.doe.gov/na-20/sld.shtml.

notify each other about potential environmental hazards when workers working on a building that had previously been a cancer treatment facility found two pellets, presumably radium-226.[21]

Insufficient coordination among agencies that should be working together has been a common comment in the U.S. Government Accounting Office (GAO); the 2005 GAO report on combating nuclear smuggling listed in its finding "a common problem faced by U.S. programs to combat nuclear smuggling is the lack of effective planning and coordination among the responsible agencies."[22] It again illustrates the diverse nature of the trafficking issues and a need for an integrated, complex approach required to efficiently combat trafficking.

In addition, practical considerations should be taken into account. Too many incidents involving nuclear materials, however benign, consume resources and time. Reduction of the number of incidents, whether innocent alarms or simply finding some old materials, will improve the situation overall and allow authorities to concentrate on the more dangerous cases. Lessening the number of illicit trafficking cases of nonterrorist nature is crucial to properly addressing the threat of nuclear terrorism. If the states, users of radioactive materials, and international organizations take the measures outlined in this chapter, the number of such cases will decline.

Summary

For the foreseeable future, countries will continue to use nuclear and other radioactive materials. Nuclear power represents an attractive alternative to fossil fuels, and medical isotopes have become indispensable for the early detection and treatment of disease. Therefore, it becomes increasingly important to establish a comprehensive system to combat illicit trafficking in radioactive and nuclear materials at the local state and international levels. This objective has not yet been accomplished, but it must remain a goal if we want to be responsible in the use of nuclear sciences and technologies.

To effectively combat nuclear trafficking, it is important to clearly understand the issues associated with trafficking incidents and to establish coordination among various agencies and departments involved in countering nuclear trafficking. The cooperation among agencies should go beyond domestic offices and reach foreign, especially bordering neighboring countries. Another component is outreach to the public. Public awareness is key to establishing a culture that insists on consequent actions to investigate and follow up on all instances of trafficking in nuclear and radioactive materials.

Overall, the trend in illicit nuclear trafficking still appears to be on the rise. The number of cases per year rose sharply with the dissolution of the Soviet Union in the early 1990s. Cooperative programs between FSU countries and the U.S. appear to have slowed the illicit movement of true weapons-grade materials. The number of cases per year, however, continues to increase. Although most of the seized materials are of lower grades, they are still of great concern due to the known terrorist interest in them. The need has been demonstrated to continue and expand programs to inventory, track, and protect nuclear and radiological materials worldwide.

Many approaches can be taken to combat this continuing issue of nuclear trafficking. Nuclear facilities must be upgraded to adequately protect the materials of concern at the

---

[21]Diana Washington, "EP, Juarez don't tell each other of environmental hazards," *El Paso Times*, June 2006.

[22]"Combating Nuclear Smuggling, Efforts to Deploy Radiation Detection Equipment in the United States and Other Countries," testimony before the Subcommittee on the Prevention of Nuclear and Biological Attack and on Emergency Preparedness, Science, and Technology, Committee on Homeland Security, House of Representatives, June 21, 2005 (Washington, D.C.: Government Accounting Office, GAO-05-840T), p. 2.

source, perhaps to include better protection of radiological materials at private facilities. The capability to monitor passengers and cargo traveling by road, rail, air, and sea must continue to be enhanced and expanded. Homeland defense provides a final layer of protection against nuclear terrorism, with improved protection of U.S. borders and critical infrastructures. Finally, it is vital to continue to monitor the activities of perpetrators (of whatever form) through national technical means and intelligence gathering.

Through the end of 2006 and continuing to the present there have been more illicit nuclear trafficking incidents. In September 2007 IAEA reported[23] 1,080 confirmed cases in their database for the period 1993–December 2006 and broke them down into four categories: 275 illicit trafficking, 332 theft or loss, 398 other unauthorized activities, and 75 undetermined due to insufficient information. Despite current efforts, it is clear that illicit trafficking of nuclear and radiological materials is an ongoing problem and that to combat it effectively will require a concerted global effort from nuclear professionals, regulators, government officials, and the public.

Further Readings

Nuclear Terrorism: The Ultimate Preventable Catastrophe, Times Books, 2004.

Graham Allison, *Smuggling Armageddon: The Nuclear Black Market in the Former Soviet Union and Europe* (Rensselaer W. Lee, 1998).

Michael A. Levi, *On Nuclear Terrorism* (in press, release date: October 2007).

Charles D. Ferguson, William C. Potter, et al., *The Four Faces of Nuclear Terrorism* (Monterey, Calif.: Center for Nonproliferation Studies, 2004).

The Last Best Chance, film produced by the Nuclear Threat Initiative with additional funding from the Carnegie Corporation of New York and the Mac Arthur Foundation, 2005. It can be ordered at no charge from www.lastbestchance.org. The story is fiction but it is built on facts.

M. Bunn, *The next wave: Urgently needed new steps to control warheads and fissile material*, joint publication of Harvard University's Project on Managing the Atom and the Non-Proliferation Project of the Carnegie Endowment for International Peace, April 2000.

Robert F. Mozley, *The Politics and Technology of Nuclear Proliferation* (University of Washington Press, 1998).

[23]IAEA Illicit Trafficking Database Releases Latest Aggregate Statistics, Staff Report, 11 September 2007.

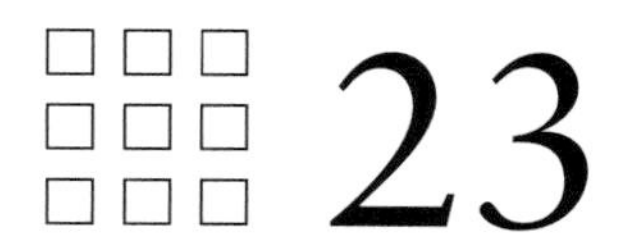

23 Nuclear Terrorism and Improvised Nuclear Devices

Charles D. Ferguson and William C. Potter

Introduction

Today the threat of nonstate actors detonating a nuclear weapon cannot be dismissed as bad fiction. Although few terrorist groups are likely to climb the escalation ladder to the highest levels of violence, we have a new breed of terrorist that covets high body counts and massive destruction. To achieve these goals, they could try to seize an intact nuclear weapon located in a nuclear-armed state's arsenal or, more likely, they might attempt to build an improvised nuclear device (IND)—a crude but devastating nuclear explosive.[1]

The Chain of Nuclear Terror

To detonate an IND at a high-value target such as a U.S. city, a terrorist group would have to carry out the following steps:[2]

1. The group must embrace extreme objectives and possess the necessary technical and financial resources.
2. The group must then choose to engage in an act of nuclear terrorism at the highest level of violence.
3. These terrorists must then acquire sufficient weapons-usable fissile material to make an IND, through gift, purchase, theft, or diversion.
4. They must next build the IND.
5. The group must transport the intact IND (or its components ready for final assembly) to a high-value target.
6. Finally, the terrorists must detonate the IND to complete their plan.

Although variants of this chain of causation can be imagined, this outline can help determine where to apply risk-reduction measures to reduce the probability of this act of

[1]A more detailed discussion of this issue is found in Charles D. Ferguson and William C. Potter with Amy Sands, Leonard S. Spector, and Fred L. Wehling, *The Four Faces of Nuclear Terrorism* (New York: Routledge, 2005).

[2]Matthew Bunn, Anthony Weir, and John P. Holdren, *Controlling Nuclear Warheads and Materials: A Report Card and Action Plan*, Project on Managing the Atom, Harvard University, March 2003. The report discusses in detail the components of this chain of necessary conditions in a section titled "Terrorist Pathway to the Bomb."

nuclear terror.[3] All these elements must combine for a terrorist IND attack to succeed, and intervention at any stage can be sufficient to avert catastrophe.

Terrorist Groups with Motivation to Build and Use an IND

Very few terrorist organizations are sufficiently motivated to attempt to detonate nuclear weapons. But the potential number of such groups is not known with precision and can vary over time. For example, new groups can arise or existing groups can change their motivations or form new alliances with groups already motivated to inflict high-consequence nuclear terrorism.

Traditional nationalist/separatist terrorist groups, such as the IRA in Ireland and the Tamil Tigers in Sri Lanka, are apt to be constrained from the pursuit of nuclear weapons use by the values of their base constituencies. In addition, they would be extremely vulnerable to retaliatory strikes and the fallout of a nuclear attack. Nationalist/separatist groups might, however, consider developing an IND (in contrast to using it) as a tool to gain international recognition or to blackmail adversary governments into making concessions. Single-issue terrorist organizations are also unlikely to want to cause massive destruction by using an IND, but extremist factions within such groups might consider doing so.

Currently, apocalyptic and political-religious groups stand out as the terrorist organizations with the strongest motivations to explode an IND. Aum Shinrikyo exemplifies the first type. Taking authorities by surprise, Aum in 1995 attacked the Tokyo subway system with sarin gas, killing 12 and injuring thousands. According to many analysts, this attack marked the advent of the new breed of terrorism. During the peak of its power, Aum enlisted the help of weapons scientists and bought a uranium mine in Australia to try to make an IND. Fortunately, Aum failed in this attempt. Although Shoko Asahara, the leader of Aum, has been apprehended and sentenced to death, apocalyptic groups seeking nuclear capabilities could arise again.

Political religious groups such as al-Qaeda appear the most likely to seek to inflict massive destruction with an IND. Al-Qaeda's attempts to obtain nuclear materials and weapons date back to the early 1990s. In 1998, Osama bin Laden said that it is "a religious duty" for al-Qaeda to acquire weapons of mass destruction. Documents found in Afghanistan during Operation Enduring Freedom show that members of al-Qaeda have studied primitive designs for nuclear weapons.

Al-Qaeda may also inspire other Islamist terrorist groups. For instance, although Chechen rebels are primarily driven to free Chechnya from Russian rule, exposure to al-Qaeda's operatives and ideas has been radicalizing many of the Chechen rebel factions. The massive killings resulting from the Dubrovka Theater siege in Moscow in October 2002 and the Beslan school siege in September 2004 have underscored the growing radicalization of these Chechen factions. In December 2004, the U.S. National Intelligence Council warned, "Russian authorities twice thwarted terrorist efforts to reconnoiter nuclear weapon storage sites in 2002."[4]

---

[3]One variant on the basic model would be to set off the device at a less than optimal site to reduce the risk of detection inherent in transporting the device across borders. Collaboration among terrorist organizations is another possibility. See, for example, Morten Bremer Maerli, *Crude Nukes on the Loose? Preventing Nuclear Terrorism by Means of Optimum Nuclear Husbandry, Transparency, and Non-Intrusive Fissile Material Verification*, Ph.D. Dissertation, Faculty of Mathematics and Natural Sciences, University of Oslo, 2004.

[4]National Intelligence Council, *Annual Report to Congress on the Safety and Security of Russian Nuclear Facilities and Military Forces*, Dec. 2004.

Types and Availability of Weapons-Usable Fissile Material

Conceivably, terrorists could make an IND from two types of weapons-usable fissile material: highly enriched uranium (HEU) or plutonium. According to the conservative figures used by the International Atomic Energy Agency (IAEA), only 25 kilograms of HEU or 8 kilograms of plutonium would be needed to manufacture a first-generation weapon.

HEU refers to uranium that has been processed to increase the portion of one isotope of uranium, uranium-235, from the naturally occurring level of 0.7% to the highly enriched level of 20% or more, at which use for weapons becomes practicable. Although all uranium enriched to more than 20% is termed "highly enriched," the ease of causing a nuclear detonation is greatly increased at higher enrichment levels. Specifically, terrorists would find it much easier to develop a workable IND with material enriched to 80% or more, and military programs prefer weapons-grade uranium, which is enriched to 90% or more.

Uranium enrichment requires expensive facilities and industrial resources only currently available to nation states. Although Aum Shinrikyo is the one known terrorist group that has tried to enrich uranium, it failed because enrichment is an extremely challenging process. Terrorist groups will have to seize HEU from existing stockpiles.

Based on the best unofficial estimates, these stockpiles are plentiful, containing about 1,850 metric tons globally—sufficient fissile material to build tens of thousands of nuclear weapons.[5] Most of the HEU is under military control. Russia and the United States possess an estimated 1,720 metric tons for weapons purposes and naval propulsion. Britain, China, and France have tens of metric tons. Pakistan and South Africa own stockpiles that contain upwards of several hundred kilograms—sufficient to power dozens of INDs. (South Africa formerly had a weapons program. South Africa's HEU stockpile is now used in its civilian isotope production program.) More than 40 other countries contain smaller amounts of HEU, which are dedicated to civilian scientific research, industrial, and medical isotope production programs. Many of the more than 120 research reactors and related facilities within these countries have enough HEU in each site to make a nuclear bomb.[6]

Plutonium production, like uranium enrichment, is apt to be beyond the technical capability of terrorists, who do not have state assistance. Reactors used to produce plutonium and reprocessing plants to extract plutonium from spent reactor fuel are expensive and currently available only to nation states. Consequently, to acquire plutonium, terrorists would have to target existing stockpiles.

Like HEU, plutonium exists in military and civilian sectors. Globally, military stockpiles exceed 250 metric tons of plutonium—sufficient to build tens of thousands of nuclear weapons. The United States and Russia own more than 90% of the military plutonium, and China, France, Great Britain, India, Israel, North Korea, and Pakistan possess the remainder.

Civilian plutonium also poses a risk for use in an IND. Although not ideal for weapons purposes, it is possible for reactor-grade plutonium to power a nuclear explosive. Some countries use reactor-grade plutonium in mixed-oxide (MOX) fuel. MOX fuel is a mixture of plutonium and depleted uranium oxides that can be used as a substitute for low-enriched uranium fuel in many nuclear power reactors. France, Germany, Great Britain, India, Japan,[7]

---

[5]David Albright and Kimberly Kramer, "Fissile Material: Stockpiles Still Growing," *Bulletin of the Atomic Scientists*, Nov./Dec. 2004, pp. 14–16. This reference estimates fissile material stockpiles as of the end of 2003. This chapter's estimates also account for the disposition of Russian weapons-grade uranium as of the end of 2005.

[6]U.S. General Accounting Office, *DOE Needs to Take Action to Further Reduce the Use of Weapons-Usable Uranium in Civilian Research Reactors*, GAO-04-807, July 2004, p. 28.

[7]Reprocessing of Japanese nuclear power plant fuel has been performed principally in France and Great Britain. The resulting plutonium is either stored in these countries or is being processed into MOX fuel for shipment back to Japan. In parallel, Japan is constructing its own commercial-scale reprocessing facility at Rokkasho-mura, which began testing in the spring of 2006.

and Russia have continued to separate plutonium from civilian nuclear power plant fuel. More than a dozen countries possess more than 230 metric tons of plutonium that has been separated from spent nuclear fuel. Because spent fuel tends to be highly radioactive, it provides a lethal barrier against acquisition by terrorists who do not have special protective handling equipment. Worldwide, spent fuel contains more than 1,300 metric tons of plutonium. Continued separation of this plutonium from spent fuel would increase the risk of terrorist acquisition unless consumption of the plutonium in new fuel kept pace with the rate of separation. However, in recent years, the rate of separation, or reprocessing, has exceeded the rate of use by several metric tons per year. Thus, every year this excess plutonium could power hundreds of terrorist- or state-built bombs.

HEU and plutonium that are contained outside of nuclear weapons are located at hundreds of sites worldwide. Although fissile material in any location is a potential target for terrorists, this chapter concentrates on three settings of particular concern:

- Russia, where hundreds of tons of these materials are used, processed, or stored at dozens of State Corporation Facilities
- Pakistan, where political instability and uncertain loyalties in the nuclear chain of command might result in fissile material coming into the hands of terrorists
- Research reactors using HEU fuel, including some Soviet-designed research reactor sites and research centers containing HEU outside of Russia and several U.S.-origin research reactors outside the United States

Acquisition of Fissile Material

In the nuclear terror chain, obtaining the necessary amount of fissile material probably presents the most difficult challenge for a terrorist organization. Terrorists could try to exploit many acquisition routes. In particular, a state might voluntarily share fissile material with a terrorist group or sell the material to it; a senior official or governmental element with authorized access to such materials might, for ideological or mercenary motives, provide it to terrorists, without the express approval of governmental leaders; the immediate custodians of the material, for money or ideology or under duress, might provide HEU or plutonium to the organization or assist it in seizing the material by force or stealth; or terrorists might obtain the material by force or stealth without insider help. Finally, nuclear weapons materials could come into the hands of terrorists during a period of political turmoil, including one brought on by a coup or revolution.

Deliberate Transfer by a National Government

Acquiring weapons-usable fissile materials directly from a sympathetic government would significantly simplify the requirements for terrorists, obviating the need to defeat security systems protecting such materials. Presumably, to further the purposes of the transfer, the state sponsor would also provide assistance in manufacturing an IND, perhaps by providing a design or the nonnuclear components or by machining the HEU or plutonium into appropriate shapes before handing it over. Such material might be provided to terrorist groups by a state that hoped to see an IND used against an opponent but wanted to be in a position to deny its involvement and reduce the threat of retaliation.

Authorized state transfers of nuclear materials to terrorist groups are unlikely because the state would risk suffering massive retaliation from the United States and its allies if the material were traced back to the state.[8] However, the greatest risk of such transactions would

---

[8]Jasen J. Castillo, "Nuclear Terrorism: Why Deterrence Still Matters," *Current History*, Dec. 2003.

likely involve states that are facing imminent regime change. These states might have little to lose by handing the ingredients for an IND to a terrorist group as a last means of striking against an opponent. For example, some expressed concern prior to the 2003 U.S.-led war against Iraq that regime change might provoke Saddam Hussein to transfer WMD-material to nonstate actors.[9] Thus, an unintended consequence of overthrowing the governments of states possessing HEU or plutonium could be to provoke them to aid or abet nuclear terrorists.

Unauthorized Assistance from a Senior Official

Although leaders of a state may have little or no interest in transferring the wherewithal for an IND to a terrorist group or, even if they are interested, may be deterred from carrying out such transactions, senior officials within that state may be inclined to provide access to nuclear assets. These officials might be motivated by greed or ideological alignment with the terrorists, and they may act without the knowledge or approval of the state's leadership.

Assistance from Fissile Material Production Workers and Custodians

Insiders at uranium enrichment or reprocessing plants have varying degrees of access to HEU or plutonium. Their motives for providing these materials to a terrorist group might include sympathy with the terrorists' goals, greed, or coercion through threats of violence or blackmail to friends, family members, or themselves. Identifying susceptible insiders and arranging for their assistance present substantial challenges. Terrorists might seek collaboration with organized crime to facilitate this method of acquisition. If, by taking advantage of the difficulty of accounting for fissile materials and/or weak security arrangements, the perpetrators were able to divert material without detection, they would gain the ability to mask their future actions—fabrication of an IND and transporting it (or its components) to the detonation site—without confronting intensive recovery efforts and heightened security at likely target locations. Poorly paid and demoralized nuclear workers and security guards in Russia might be vulnerable to subornation by terrorists or criminals. Moreover, the huge size and complexity of the Russian fissile material stockpile and production infrastructure greatly add to the difficulty of protecting HEU and plutonium.

Seizure Without Insider Help

A terrorist organization would need considerably greater effort and skill to seize fissile material without insider assistance because the organization would need to train and arm a force able to defeat all security measures protecting the materials. In addition, the terrorists would have to determine what security measures they would confront and would need to map out a secure means of escape, which could involve travel over long distances. Although assaults would be more problematic against fissile material storage or processing areas deep within large, secure complexes, fissile materials are also found at sites in city centers, at smaller suburban research parks, and at isolated, standalone plants, where armed assaults, perhaps accompanied by diversionary attacks, would be more practicable.

Coups d'état and Political Unrest

Political instability during a coup or a revolution could provide an opportunity for terrorists to gain control over fissile material. Insurgents allied to or cooperating with terrorists could trigger or be the main assault force behind a takeover of a state that has weapons-usable nuclear material. Even if such an insurrection were unsuccessful, however, nuclear sites could fall behind "enemy" lines, before fissile materials could be removed, permitting their transfer to terrorists or their allies. During a period of civil strife, response forces might join the conflict, leaving fissile material sites vulnerable to assault. Possibly during a period of political turmoil, nuclear custodians might abandon their posts or be swept aside in the tide of events.

---

[9]William C. Potter, "Invade and Unleash?" *Washington Post*, Sept. 22, 2002, p. B7.

Financial Resources and Technical Skills Needed to Make an IND

Financial and technical limitations could also impede a terrorist group that is otherwise motivated to make an IND. Nuclear weapons materials as well as the other components of an IND cost more than the parts of an improvised explosive device (IED)—a conventional bomb. Also, the technical skills needed for making an IND are more demanding than the skills required for building an IED. Nonetheless, as discussed in more detail later, some terrorist groups might be able to acquire the resources needed to make a gun-type IND, the easiest to make nuclear bomb.

A terrorist group would also likely require millions of dollars to bribe guards or corrupt officials or to obtain the firepower and other resources needed to attack nuclear weapons material storage sites. Moreover, the group would require considerable organizational skills, especially to coordinate an international operation in which the terrorists would acquire the material in one country, transport it to a safe haven in another country for construction into an IND, and then move the IND for detonation in the United States or some other country.

Finally, the group would need a considerable degree of technical competence. Most analysts have assumed that to accomplish this task, the terrorist group in question would have to assemble a small team of specialists with expertise in such varied areas as nuclear physics or engineering, metallurgy, machining, and conventional explosives.[10] However, as discussed in the next section, building the simplest type of IND, a gun-type device, might not require a large technical team.

Manufacture of an Improvised Nuclear Device

Assuming that terrorists would not have access to technologically sophisticated nuclear weapons design and fabrication infrastructures such as those possessed by a limited number of states, terrorists who seek to build an IND would favor nuclear weapons designs based on first-generation, well-proven technology. First-generation nuclear weapons draw on two designs: gun type and implosion type.

Gun-Type Devices

The simplest type of nuclear weapon is a gun-type device. Like a gun, it fires a projectile. The projectile in this type of weapon is a piece of highly enriched uranium. Moreover, like a gun, a gun-type device would use a gun barrel to direct the projectile. To ignite a nuclear explosion, the HEU projectile would travel down the barrel and collide with another piece of HEU. The HEU pieces would both be subcritical; that is, each one by itself could not sustain an explosive chain reaction. Following the collision, they would form a supercritical mass.

Weapons-grade HEU would be the most effective fissile material for a gun-type device because of its very high concentration of uranium-235.[11] Gun assembly is an inefficient

[10]See, for example, Carson Mark, Theodore Taylor, Eugene Eyster, William Maraman, and Jacob Wechsler, "Can Terrorists Build Nuclear Weapons?" in *Preventing Nuclear Terrorism: The Report and Papers of the International Task Force on Prevention of Nuclear Terrorism*, Leventhal and Alexander, eds., Lexington Books, 1987, pp. 55–65.

[11]The neutrons that cause fission (which is how energy is released in the bomb) would not have to travel as far before interacting with a uranium-235 nucleus in a mass of weapons-grade HEU compared to lower enrichments of HEU. Thus, more fissions can occur in a given period of time inside a mass of weapons-grade HEU.

means of exploding HEU, because it is a relatively slow way to form a supercritical mass (compared to implosion assembly, as described shortly) and it does not appreciably compress or change the density of the fissile material.[12] Therefore, a gun-type device requires relatively large amounts of HEU and would only fission a small fraction of the HEU during the explosive chain reaction. The IAEA's significant quantity for HEU is an amount of uranium containing 25 kg equivalent U-235. However, this amount is based on the assumption that a state could use this material to build the more technically challenging implosion weapon.

Most physicists and nuclear weapons experts have concluded that building a gun-type device would pose few technological barriers to technically competent terrorists as long as they have sufficient quantities of HEU.[13] Still, the question remains as to how technically competent the terrorists have to be and how large of a team would be needed. To make sure that the group could surmount any technical barriers, it would likely want to recruit team members who have knowledge of conventional explosives (needed to fire one piece of HEU into another), metalworking, draftsmanship, and chemical processing (for example, to extract HEU metal from other chemical forms, such as oxide or aluminum-based reactor fuel). A well-financed terrorist organization such as al-Qaeda would probably have little difficulty recruiting personnel with these skills. Concerning the size of the team and the preparation time required, Albert Narath estimated, "Once the HEU in metallic form is in hand it might require only a dozen individuals with the right set of skills to accomplish the design and construction over a period of perhaps a year."[14] Approximately one year's amount of preparation would allow for "rapid turnaround"—that is, "the device would be ready within a day or so after obtaining the material." As Carson Mark et al. also assessed, "Such a device could be constructed by a group not previously engaged in designing or building nuclear weapons."[15]

Because of its inherent simplicity, designing and constructing a gun-type device would be relatively straightforward. Testing the nonnuclear parts of the device would likely be

---

[12]The critical mass scales as the inverse of the density squared. Thus, if the density is increased by a factor of two, the required critical mass decreases by a factor of four. In contrast to the gun method, the implosion method significantly changes the density of the fissile material.

[13]See, for example, Carson Mark, Theodore Taylor, Eugene Eyster, William Maraman, and Jacob Wechsler, "Can Terrorists Build Nuclear Weapons?," 1987, op. cit.; Luis W. Alvarez, *Adventures of a Physicist*, Basic Books, 1988, p. 125; Frank Barnaby, "Issues Surrounding Crude Nuclear Explosives," in *Crude Nuclear Weapons: Proliferation and the Terrorist Threat*, IPPNW Global Health Watch Report No. 1, 1996; Morten Bremer Maerli, "Relearning the ABCs: Terrorists and 'Weapons of Mass Destruction,'" *The Nonproliferation Review*, Summer 2000; Matthew L. Wald, "Suicidal Nuclear Threat Is Seen at Weapons Plants," *New York Times*, Jan. 23, 2002, p. A9; Committee on Science and Technology for Countering Terrorism, National Research Council, *Making the Nation Safer: The Role of Science and Technology in Countering Terrorism*, National Academy Press, 2002; Richard L. Garwin and Georges Charpak, *Megawatts and Megatons: A Turning Point in the Nuclear Age*? Alfred A. Knopf, New York, 2001; Jeffrey Boutwell, Francesco Calegero, and Jack Harris, "Nuclear Terrorism: The Danger of Highly Enriched Uranium (HEU)," *Pugwash Issue Brief*, Sept. 2002; "Scientists' Letter on Exporting Nuclear Material," to W. J. "Billy" Tauzin, Sept. 25, 2003, Union of Concerned Scientists, available at www.ucsusa.org/global_security/nuclear_terrorism/scientists-letter-on-exporting-nuclear-material.html, accessed on May 21, 2006; and Gunnar Arbman, Francesco Calogero, Paolo Cotta-Ramusino, Lars van Dessen, Maurizio Martellini, Morten Bremer Maerli, Alexander Nikitin, Jan Prawitz, and Lars Wredberg, "Eliminating Stockpiles of Highly Enriched Uranium," report submitted to the Swedish Ministry for Foreign Affairs, SKI Report 2004:15, April 2004.

[14]Albert Narath, "The Technical Opportunities for a Sub-National Group to Acquire Nuclear Weapons," XIV Amaldi Conference on Problems of Global Security, April 27, 2002.

[15]Carson Mark et al., op. cit.

required, and an appropriate testing area would be needed (such as a terrorist training camp where other explosives were routinely used), to avoid arousing suspicions. Assuming such tests and a sufficient amount of HEU in the appropriate form, terrorists could have a moderate degree of confidence that their IND would result in a substantial nuclear yield. Notably, Manhattan Project scientists had such great confidence in the gun-type design that they believed it was unnecessary to test it through a nuclear detonation prior to its actual use over Hiroshima. Similarly, South African nuclear weapon designers had full confidence in the gun-type weapons they had built, even though that country is not known to have conducted a nuclear test. South Africa assembled these bombs in a warehouse—a relatively small building that escaped detection throughout its many years of operation.[16] Thus, the most formidable barrier to a gun-type weapon remains the acquisition of sufficient HEU.

A terrorist group might also attempt to extract HEU from fresh or spent HEU fuel used in research or propulsion reactors. Fresh fuel can contain up to 93% enriched uranium. Spent HEU fuel contains a reduced concentration of uranium-235 compared to fresh fuel, but if the original enrichment were high enough or if the fuel had been only partially used—or lightly irradiated—before it was seized, enrichment levels could easily remain close to the original concentration of uranium-235.[17] In fact, many research reactors are often used only intermittently, resulting in lightly irradiated fuel, which presents a reduced radiation safety hazard, greatly simplifying the HEU separation process.[18]

It is impossible to achieve a large nuclear explosion by employing plutonium in a gun-type device.[19] However, some authorities have concluded that a plutonium-fueled gun-type IND could produce a relatively small explosive yield. Both weapons-grade and reactor-grade plutonium would result in this fizzle yield.[20] This aspect of the weapon's impact would, in effect, be similar to a very large radiological dispersal device and would be especially dangerous inasmuch as small quantities of plutonium, if inhaled, are highly toxic. In sum, although HEU poses the greater threat by far because it could power a devastating gun-type device, terrorists could conceivably use plutonium to produce a significant but a lower order level of damage.

---

[16]David Albright, "South Africa and the Affordable Bomb," *Bulletin of the Atomic Scientists*, July/Aug. 1994.

[17]Alexander Glaser and Frank von Hippel, "On the Importance of Ending the Use of HEU in the Nuclear Fuel Cycle: An Updated Assessment," paper presented at the 2002 International Meeting on Reduced Enrichment for Research and Test Reactors, Nov. 3–8, 2002.

[18]Edwin Lyman and Alan Kuperman, "A Reevaluation of Physical Protection Standards for Irradiated HEU Fuel," 24th International Meeting on Reduced Enrichment for Research and Test Reactors (RERTR-2002), Nov. 2002, and Matthew Bunn and Anthony Wier, *Securing the Bomb: An Agenda for Action*, Project on Managing the Atom, Harvard University, Report Commissioned by the Nuclear Threat Initiative, May 2004, p. 37. Notably, during its crash program in 1991 to produce a nuclear bomb, Iraq planned to use both fresh and irradiated HEU fuel from its research reactors. David Albright, Frans Berkhout, and William Walker, *Plutonium and Highly Enriched Uranium 1996: World Inventories, Capabilities, and Policies* (SIPRI: Oxford University Press, 1997), pp. 344–349.

[19]Plutonium's spontaneous fission rate is much greater than uranium's. Before the gun-type device would be able to assemble plutonium into a supercritical mass, the neutrons emitted by the spontaneous fission would lead to a dud or a "fizzle" yield.

[20]Stanislav Rodionov, "Could Terrorists Produce Low-Yield Nuclear Weapons?" in National Research Council, National Academy of Sciences, in Cooperation with the Russian Academy of Sciences, *High-Impact Terrorism: Proceedings of a Russian-American Workshop* (Washington, D.C.: National Academies Press, 2002), pp. 156–159.

Implosion-Type Devices

To cause a nuclear explosion, an implosion-type device squeezes a sphere of fissile material from a relatively low-density subcritical state to a high-density supercritical state. If the implosion does not occur smoothly, the bomb will be a complete dud or result in a fizzle yield much lower than expected from a properly designed implosion weapon. Thus, in contrast to a gun-type device, an implosion-type device requires more technical sophistication and competence. A terrorist group, for example, would need access to and knowledge of high-speed electronics and high explosive lenses, a particularly complex technology. This equipment is necessary to result in a fast and smooth squeezing of the fissile material into a supercritical state. Unlike a gun-type device, an implosion-type device can employ HEU or plutonium because the speed of assembly is fast enough to allow the use of plutonium. The IAEA estimates that an improvised implosion-type weapon would probably require approximately 25 kg of weapons-grade HEU or roughly 8 kg of plutonium.[21]

Weapons-grade plutonium is the most desirable type of plutonium both from the perspective of a weapon scientist employed by a state and for a terrorist organization, since it is most readily detonated. Even reactor-grade plutonium could result in an explosive chain reaction, however, depending on the skill of the weapons designers and builders.[22] Because reactor-grade plutonium would have a much higher chance of pre-ignition, the bomb yield would likely be less than that of a weapon made from weapons-grade plutonium. Nonetheless, even if terrorists were only able to achieve a "fizzle" yield from the device, it would be far greater than the yield from a powerful conventional explosion, thus giving the terrorists a potent weapon.

Implosion-type weapons, using reactor-grade plutonium, weapons-grade plutonium, or HEU, would pose design and construction challenges much greater than those faced in building a gun-type HEU device. Even if terrorists obtained a workable design, manufacturing the components for the device and ensuring that they all worked together with the necessary precision would pose a daunting technical challenge, requiring considerable time and extensive testing of the non-nuclear "triggering package," both of which would increase the risk of detection. Given these challenges, terrorists would likely have far less confidence that their implosion-based device would work than they would have in the case of a far simpler gun-type assembly using HEU. Additionally, because terrorists presumably will have only limited quantities of plutonium available, a full-scale nuclear test undertaken simply to prove the design of the weapon the terrorists had built seems highly unlikely. The first detonation using plutonium would probably be at a target. Even if the device failed to produce a nuclear yield, its very existence would cause profound fear in the target state and permit blackmail based on the real or pretended existence of additional weapons.

In sum, given a choice between building a gun-type or an implosion-type device, terrorists almost certainly would choose to construct a gun-type device because it is more likely to result in a nuclear weapon producing a large explosive yield. If nuclear terrorists had access

---

[21]See Tariq Rauf, "Drawing Safeguards Conclusions," presentation to the 2004 NPT Preparatory Committee, April 29, 2004, www.iaea.org/NewsCenter/Focus/Npt/npt2004_ppt_2904.pdf (May 2007).

[22]In 1997, the U.S. government reemphasized earlier pronouncements that reactor-grade plutonium can fuel nuclear weapons. See U.S. Department of Energy, Office of Arms Control and Nonproliferation, *Final Nonproliferation and Arms Control Assessment of Weapons-Usable Fissile Material Storage and Excess Plutonium Disposition Alternatives*, DOE/NN-0007, Washington, D.C., DOE, 1997, pp. 37–39. In its report on nuclear terrorism, the U.S. National Research Council in 2002 stated, "Reactor-grade plutonium can be used to fabricate workable nuclear devices." See Committee on Science and Technology for Countering Terrorism, National Research Council, "Nuclear and Radiological Threats," Chapter 2 in *Making the Nation Safer: The Role of Science and Technology in Countering Terrorism*, National Academy Press, 2002, p. 40.

to only plutonium, they would be forced to build an implosion-type device to achieve high yields, or they could try to construct a low-yield gun-type device.[23]

Transporting the IND (or Its Components) to the Target Site

Assuming that terrorists acquired the necessary fissile material and manufactured an IND, they would then have to cross the next barrier to IND use. That is, they would have to deliver an IND to a target without being caught and stopped. The distance between the point of acquisition and the target could be quite substantial. If the loss of fissile material were detected, a massive hunt for the material would ensue, involving law enforcement and military personnel from many nations, assisted by nuclear specialists. In parallel, authorities would greatly intensify security over transportation links and points of entry. Unfortunately, for many scenarios, material might be diverted without detection for some time or the diversion might not be acknowledged, providing the opportunity for the terrorist organization to cover its tracks and move the material to a safe location to make an IND.

Transportation of an IND would not present insurmountable difficulties. Although an IND would likely be relatively heavy, perhaps weighing up to a ton, trucks and commercial vans could easily haul such a device. Moreover, container ships and commercial cargo airplanes could provide delivery means. Although the U.S. Department of Homeland Security in recent years has been checking shipping manifests, only a small fraction of the shipping containers entering the United States are physically checked. Nonetheless, terrorists would need extensive resources and networks of collaborators to move their IND over long distances, adding to the complexity of their plot. But detecting uranium or even plutonium in transit is difficult.

Every means of delivery, however, exposes terrorists to some risk of discovery. To reduce or eliminate this risk, a terrorist group might choose to detonate an IND on the spot where it was assembled. A devastating blast even in such a location would cause grave damage and many deaths—and provide terrorists the opportunity to threaten to destroy more impressive targets with INDs they could claim to possess. Terrorists might try to assemble and detonate a gun-type device, but probably not a more sophisticated implosion-type device, at a fissile material storage site, assuming that this site contained sufficient quantities of readily usable HEU metal and that the terrorists were suicidal and that the assault team included members versed in the relevant technical skills of gun devices.[24]

Detonation of the IND

Because terrorists who have constructed an IND would by definition thoroughly know its design, detonation of the device would pose little or no technical difficulties. However, as discussed previously, an implosion device presents a much greater chance of producing a dud or fizzle yield than a gun device.

Nuclear Materials Security at the State Level

If effective, national safeguards and security measures can block almost all paths to terrorist acquisition of weapons-usable nuclear materials, except when a sympathetic government might transfer such materials to a terrorist group. As previously outlined, Russia, Pakistan,

---

[23]Rodionov, 2002, op. cit., pp. 156–159.

[24]Matthew L. Wald, "Suicidal Nuclear Threat Is Seen at Weapons Plants," *New York Times*, Jan. 23, 2002, p. A9.

and certain research facilities containing HEU in a few dozen countries pose the greatest risks of terrorists acquiring these materials.

Russian HEU and Plutonium

The huge quantity of fissile material in Russia poses a uniquely dangerous risk of terrorist acquisition. The U.S. Department of Energy (DOE) has estimated that Russia possesses roughly 600 metric tons of weapons-usable plutonium and HEU outside of nuclear weapons, enough to make more than 20,000 nuclear warheads. These materials are stored at more than 50 military and civilian sites.[25] Numerous assessments, citing the general state of decay of Russia's nuclear infrastructure, decades of inadequate nuclear materials accounting, and the impoverishment of Russian nuclear workers and scientists, have concluded that large quantities of these materials are inadequately secured.[26] These findings are underscored by repeated reports of trafficking in Russian-origin weapons-usable nuclear materials. Analysts formerly at Stanford University estimate that about 40 kilograms of weapons-usable material have been stolen from the former Soviet Union.[27] Although most of the material involved in these reported incidents was recovered, the total attempts at theft or diversion remain unknown. Some evidence indicates that criminal organizations are becoming more interested in smuggling nuclear and radioactive material from the Soviet Union.[28]

The United States has several major programs, as summarized in Table 23.1, to help Russia secure, consolidate, and eliminate fissile materials. Many of these programs have made significant progress, but all are far from completion, and the acute dangers posed by Russian HEU and plutonium will continue beyond this decade. Concerning securing these materials, the Department of Energy's Material Protection, Control, and Accounting (MPC&A) Program has, as of the start of 2006,[29] only provided security upgrades to about half of Russia's fissile materials. More positively, the MPC&A program has helped improve the security at about 80% of the sites containing these materials.[30] Still, DOE continues to

---

[25]The Department of Energy has used the 600-ton figure for many years. It appears that new Russian production of several tons of separated plutonium annually (from military and civilian programs), together with fissile material removed from nuclear weapons and added to Russia's out-of-weapons fissile material stockpile, roughly balance the 30 tons of HEU per year that is removed from that stockpile through dilution of HEU into nonweapons-usable low-enriched uranium, pursuant to the U.S.-Russian HEU Purchase Agreement, discussed later in the chapter. Thus although the total quantity of Russian fissile material in and out of weapons is declining, the amount outside of weapons appears to be holding relatively constant at the 600-metric-ton level.

[26]For example, National Intelligence Council, "Annual Report to Congress on the Safety and Security of Russian Nuclear Facilities and Military Forces," Feb. 2002, and U.S. General Accounting Office, "Weapons of Mass Destruction: Additional Russian Cooperation Needed to Facilitate U.S. Efforts to Improve Security at Russian Sites," GAO-03-482, March 2003.

[27]Lisa Trei, "Database exposes threat from 'lost' nuclear material," *Stanford Report*, March 6, 2002.

[28]William C. Potter and Elena Sokova, "Illicit Nuclear Trafficking in the NIS: What's New? What's True?" *Nonproliferation Review*, Summer 2002, pp. 113–116.

[29]The status of many DOE programs described herein are as of 2005. These programs have continued to evolve, several in ways recommended by this chapter. Readers are encouraged to consult websites of the Nuclear Threat Initiative (www.nti.org), and the Department of Energy (www.energy.gov) for current status.

[30]Statement of Jerald S. Paul, Principal Deputy Administrator, National Nuclear Security Administration, Department of Energy, before the Senate Armed Services Committee, Subcommittee on Emerging Threats and Capabilities, March 29, 2006.

Table 23.1 U.S. programs to secure and reduce Russian fissile materials as of 2005.

Program	Goal	Status	Scheduled Completion Date
		Securing Fissile Materials	
Material protection, control, and accounting (MPC&A)	Secure fissile material outside weapons	Security upgrades provided for 49% of material and at 80% of the sites containing this material	2008
Mayak Fissile Material Storage Facility (FMSF)	Secure 50 tons of weapons-grade plutonium but could secure HEU as well	The FMSF remains empty despite the facility being turned over to Russia in December 2003; loading is waiting on completion of transparency agreement	2020?
		Eliminating Fissile Materials	
HEU purchase agreement	Downblend 500 metric tons of weapons-grade HEU for sale as commercial nuclear power plant fuel	About 260 tons of HEU rendered unusable for nuclear weapons as of start of 2006; additional conversion at the rate of 30 tons/year	2013
MPC&A HEU consolidation and conversion	Consolidate and downblend HEU from research centers and reactors in former Soviet Union and Eastern Europe	5.6 tons of HEU have been rendered unusable for nuclear weapons as of end of 2004; DOE plans to eliminate at least an additional 2 tons	2006
Plutonium disposition	Use 34 tons of weapons-origin plutonium as power reactor fuel, rendering it very difficult to use for weapons	Although progress was made in 2005 to complete a liability agreement, the agreement has yet to be finalized and further progress hinges on completion of the MOX fuel facility	2025
		Ending Production of Fissile Materials	
Elimination of weapons-grade plutonium production	End production of 1.2 tons/yr of weapons-grade plutonium by providing fossil fuel plants as alternative sources of heat and power for three Russian production reactors at Seversk and Zheleznogorsk	Revised agreement signed between the United States and Russia in March 2003; DOE began construction at Seversk in 2005 and expected to start construction at Zheleznogorsk in 2006	2011
Elimination of civilian plutonium separation; *no U.S. or international program*	End added accumulation of 1+ tons/yr of separated plutonium from Russian VVER nuclear power plants	No program	N/A

struggle to gain access to the remaining 20% of the sites, primarily Russia's weapons assembly and disassembly facilities, key locations where the largest quantities of the materials are housed.[31] Concerning eliminating Russian HEU and plutonium, considerably more work is urgently needed.

HEU is doubly dangerous compared to plutonium because of its relative ease of use in a gun-type IND, and it is used extensively in Russia in applications other than nuclear weapons and thus is more exposed to potential theft or diversion by terrorist groups.[32] Although the largest stores of HEU and plutonium in Russia outside of weapons are found at nuclear weapons assembly and dismantlement sites and at former fissile material production facilities, and although both are found in some research institutes, high-quality HEU is far more widely dispersed beyond these locations because of its additional uses:

- Some 40 operational research reactors, pulsed reactors, and critical assemblies in Russia use HEU. At least nine reactors rated above 1 MW power (a threshold above which a research reactor is considered of relatively high proliferation concern) employ 90% enriched HEU. Pulsed reactors, which are used to determine the effects of neutron bursts on materials, and critical assemblies, which are used to test reactor core designs, consume very little HEU. However, they typically contain large amounts of HEU, thus posing a proliferation and terrorism concern.[33] In addition, many of these reactor sites contain stores of fresh HEU fuel or lightly irradiated spent HEU fuel.[34]
- Russian submarine, cruiser, and icebreaker propulsion reactors also use HEU; some of the fuel for these vessels is reportedly enriched to 80% or more.[35] Most of the discharged submarine fuel would contain enrichment levels between 21% and 45%, far below the more easily weapons-usable 80% or greater enrichment levels; only two

[31]DOE and U.S. national laboratory officials respond that within the Russian system, these are the locations that are the most secure, even if they do not meet the level pursued in U.S. assistance programs; that for this reason, terrorists would be least likely to seek fissile materials at these sites; and that the Department has had considerable success in enhancing security at the initially more vulnerable facilities in other parts of the Russian nuclear complex.

[32]This situation would change if MOX fuel were to be widely used, thereby creating significant transportation and processing of plutonium and unirradiated plutonium-bearing MOX fuel. Under the U.S. Plutonium Disposition program, Russia is to convert 34 tons of weapons plutonium into MOX over a 17-year period and use the fuel in nuclear power reactors, thereby embedding the plutonium in highly radioactive spent fuel, rendering it far less accessible for potential use in nuclear weapons. The transportation and processing activities involved in this program, however, would create potential security risks that would need to be carefully addressed.

[33]Alexander Glaser and Frank N. von Hippel, "Global Cleanout: Reducing the Threat of HEU-Fueled Nuclear Terrorism," *Arms Control Today*, Jan./Feb. 2006.

[34]Both fresh and lightly irradiated HEU fuels are comparably dangerous because the radiation barrier in the lightly irradiated fuel would generally not be great enough to be lethal in the relatively short period of time required to process the fuel to extract HEU and fashion an IND.

[35]Don J. Bradley, *Behind the Nuclear Curtain: Radioactive Waste Management in the Former Soviet Union*, David R. Payson, ed., Battelle Press, Richland, Washington, 1997, p. 283; Oleg Bukharin and William Potter, "Potatoes Were Guarded Better," *Bulletin of the Atomic Scientists*, May 1995; Chunyan Ma and Frank Von Hippel, "Ending the Production of Highly Enriched Uranium for Naval Reactors," *The Nonproliferation Review*, Spring 2001, p. 91; and Mohini Rawool-Sullivan, Paul D. Moskowitz, and Ludmila N. Shelenkova, "Technical and Proliferation-Related Aspects of the Dismantlement of Russian Alfa-Class Nuclear Submarines," *Nonproliferation Review*, Spring 2002, p. 164.

classes of Russian submarine (November 645 and Alfa classes) were believed to have used weapons-grade HEU as fuel. Although tons of Russian naval spent fuel is stored under highly insecure conditions, in northwest Russia and in the Russian Far East, only a very small portion of this spent fuel would contain weapons-grade or near weapons-grade HEU. Reportedly, the *Kirov* battle cruiser (now called the *Admiral Ushakov*) uses weapons-grade HEU as fuel. Russian icebreakers use two types of reactor design: the OK-900A and the KLT-40. While the OK-900A uses fuel that is enriched between 45% and 75%, the KLT-40 employs fuel that is enriched up to 90%.[36] Thus, based on the fuel enrichment, the KLT-40 icebreakers pose more of a security concern than the OK-900A icebreakers.

- High-quality HEU may also be used in the floating reactors that Russia plans to employ in the Arctic region and potentially sell to other countries. Although the enrichment level of the floating reactors' fuel has not been openly published, some analysts believe that because the reactor design is based on the KLT-40 icebreaker design, weapons-grade HEU might be employed.[37]
- HEU of varying enrichment levels is also found in large quantities in fuel fabrication facilities, i.e., facilities where marine propulsion and research reactor fuels are manufactured from bulk HEU and at sites where these fuels are designed.
- In addition, weapons-grade HEU is processed in very large quantities under the U.S.-Russia HEU Purchase Agreement. The agreement provides that over the course of 20 years, Russia is to blend down 500 metric tons of HEU from or intended for nuclear weapons into low-enriched uranium. The latter material is suitable for use as nuclear power plant fuel but no longer usable for nuclear weapons. The blended-down material is to be purchased by the United States Enrichment Corp. for some $12 billion. As of early 2006, the HEU Purchase Agreement has resulted in the blending down of about 260 metric tons of Russian HEU, and each year 30 metric tons of the material must be taken from four weapons-disassembly sites, transported long distances by rail, and introduced into processing plants for blending. Significantly, for the first leg of this journey, the material transported is HEU metal, the form of HEU that could be most readily used by terrorists for an IND.
- At least one important research center near Moscow currently stores hundreds of kilograms of HEU metal, which, despite years of effort, have yet to be fully secured at a central storage site.

The U.S. DOE has focused on a number of these danger points. Its Material Consolidation and Conversion Program is gathering up smaller quantities of HEU from disparate sites in Russia and from Soviet-supplied research reactors abroad and downblending the material to nonweapons-usable low-enriched uranium. More than 5 metric tons of HEU have been rendered safe to date. The Department's MPC&A program has assisted the Russian Navy to secure virtually all fresh HEU submarine fuel, and it has given high priority to securing HEU fuel fabrication and development facilities. One initiative that could be considered is an expansion of Russian HEU blend-down activities to more rapidly eliminate greater quantities of fissile material.[38] Another possibility is to encourage Russia to use the Mayak Fissile Material Storage Facility to store excess weapons-grade HEU.

---

[36]Cristina Chuen, "Russia: Nuclear-Powered Icebreakers," NTI report, Dec. 2005, available at www.nti.org/db/nisprofs/russia/naval/civilian/icebrkrs.htm, accessed on May 23, 2006.

[37]V. M. Kuznetsov et al., "Floating Nuclear Power Plants in Russia: A Threat to the Arctic, World Oceans and the Non-Proliferation Treaty," report sponsored by Green Cross Switzerland and the Nuclear and Radiation Safety Programs of Green Cross Russia, Moscow, 2004, p. 47.

[38]For a matrix of options including costs for accelerating blend-down, see Laura S. H. Holgate, "Accelerating the Blend-Down of Russian Highly Enriched Uranium," paper presented at the Institute for Nuclear Materials Management Annual Meeting, July 2005; Michael Knapik, "NTI Study Presents Clearer Picture of Russian HEU Downblending," *NuclearFuel*, Aug. 29, 2005, p. 11.

Although high-quality HEU deserves the greatest attention in addressing the danger of terrorist construction of an IND, securing plutonium also remains highly important. It, too, could be used for an IND, although the device would be considerably more difficult to design and construct. In this regard, at a time when the United States is spending hundreds of millions of dollars to secure and eliminate fissile materials, Russia continues to increase its stocks of separated military and civil plutonium at a combined rate of roughly 3 metric tons (360 weapons, using IAEA standards) per year. The DOE has an active program to end production of Russian *military* plutonium, which DOE and Rosatom expect to complete by 2011, but there is no similar initiative to halt the separation of plutonium from spent fuel produced in certain Russian civil nuclear power plants.[39]

Although the United States has given the lion's share of assistance for nuclear material security to Russia, other countries have provided significant assistance with respect to particular physical protection issues, including the safeguarding of decommissioned Russian submarines and spent fuel. For an extended discussion of this assistance to Russia supported by the Global Partnership Against the Spread of Weapons and Materials of Mass Destruction, see the Global Partnership Update Website (www.sgpproject.org) compiled by the Strengthening the Global Partnership coalition, headed by the Center for Strategic and International Studies.[40]

Pakistani Fissile Material

Pakistan now produces both HEU and plutonium for weapons, although the bulk of its arsenal is thought to consist of HEU-based warheads. The relatively small quantity of fissile material in Pakistan (perhaps enough to make 30 to 50 weapons, including weapons already assembled, adding up to perhaps 1 metric ton) would make accounting and control of these materials significantly easier than is the case in Russia.[41] The danger that Pakistani fissile materials might fall into the hands of terrorists stems from the presence of extremist Islamic groups in that country and in the surrounding region, a history of political instability, and uncertain loyalties of senior officials in the civilian and military nuclear chain of command.[42]

Little information has been revealed concerning Pakistani security measures covering fissile materials. NBC Nightly News and press reports in January 2004, however, disclosed that the United States has been assisting Pakistan with improving the security of Pakistani nuclear material. It has been widely reported that during peacetime, Pakistan keeps the nuclear and nonnuclear components of its nuclear weapons separate. If true, this measure would greatly complicate efforts to seize an intact nuclear device and might also complicate the diversion of fissile material in the form of weapons components, since, presumably, these

---

[39]This activity takes place at the RT-1 reprocessing plant, at the Mayak Production Complex in Ozersk. The United States has proposed to assist Russia in the construction of spent-fuel storage capacity at the site where plutonium separation is now taking place (the RT-1 facility), contingent on Russia agreeing to end its nuclear cooperation with Iran, but Russia has not agreed to this arrangement. In its Dec. 2002 National Strategy to Combat Weapons of Mass Destruction, the Bush Administration declared that it "will continue to discourage the worldwide accumulation of separated plutonium." See National Strategy to Combat Weapons of Mass Destruction, National Security Presidential Directive 17 (unclassified version), Dec. 11, 2002, p. 4. To date, this prescription will lead to new U.S. initiatives aimed at discouraging Russia to end the separation of plutonium from nuclear power plant fuel.

[40]See also Chapter IV in Ferguson and Potter et al., *The Four Faces of Nuclear Terrorism.*

[41]Robert S. Norris et al., "Pakistan's Nuclear Forces, 2001," *Bulletin of the Atomic Scientists*, Jan./Feb. 2002.

[42]See Gaurav Kampani, "Nuclear Watch—Pakistan: The Sorry Affairs of the Islamic Republic," NTI Website, Jan. 2004, http://nti.org/e_research/e3_38a.html, accessed on Jan. 30, 2004.

receive the highest possible security within the Pakistani system.[43] Fissile materials that are in process, however, might be at greater risk. Through manipulation of material balances and other stratagems, insiders might be able to divert small quantities of fissile material from production or processing facilities over a period of months and avoid detection. The A. Q. Khan nuclear black market, which originated from Pakistan, and the meeting between two Pakistani nuclear scientists and Osama bin Laden in 2001 demonstrate that the threat of a conspiracy by insiders remains a significant concern.

Soviet-Origin HEU and U.S.-Origin HEU in Research Reactors

As noted earlier, some civilian nuclear programs use HEU in research reactors, as well as critical and subcritical assemblies. These programs include scientific research and production of radioisotopes for commercial applications. About half of the approximately 280 research reactors operating in more than 50 countries use HEU.[44]

The large number of research reactors using HEU fuel produced and supplied by the Soviet Union and, later, Russia is of particular concern. The U.S. government has identified more than 20 research facilities in 17 countries containing Soviet- or Russian-supplied HEU fuel.[45] These countries include Belarus, Bulgaria, China, Czech Republic, Egypt, Germany, Hungary, Kazakhstan, Latvia, Libya, North Korea, Poland, Romania, Ukraine, Uzbekistan, Vietnam, and Yugoslavia. Of these, there are 14 operational reactors in the 11 countries of the Czech Republic, Germany, Hungary, Kazakhstan, Libya, North Korea, Poland, Ukraine, Uzbekistan, Vietnam, and Yugoslavia.[46]

Recognizing the potential dangers of dispersing weapons-grade HEU fuels, the Soviet Union began in 1978 to produce and export 36% enriched fuel in lieu of more highly enriched material, when the new fuel was compatible with particular research reactor designs of its customers. Almost all the research and test reactors operating in former client states, such as Hungary, Poland, and Vietnam, have shifted to 36% enriched fuel, which they use today.[47] In 2005, Russia began converting some of these reactors to 20% enriched fuel.

Many reactor sites still house unused fresh, previously exported high-quality HEU fuel or spent high-quality HEU fuel, which retains its utility for an IND and is no longer so radioactive as to make handling the fuel difficult. The exact number of these facilities has not been openly reported.

For many years, the United States and Russia have worked together on several successful operations to bring fresh and spent HEU fuel back to Russia, where the material has been blended down into nonweapons-usable low-enriched uranium. To revitalize these efforts, on May 26, 2004, then Secretary of Energy Spencer Abraham launched the Global Threat Reduction Initiative (GTRI), which had the goal of repatriating all Soviet-origin fresh HEU

---

[43]David Albright, "Securing Pakistan's Nuclear Infrastructure," in *A New Equation: U.S. Policy Toward India and Pakistan After September 11*, Carnegie Endowment for International Peace, Working Papers, Number 27, May 2002.

[44]International Atomic Energy Agency, *Nuclear Research Reactors of the World*, Data Series 3, Vienna, 2000.

[45]T. Dedik, I. Bolshinsky, and A. Krass, "Russian Research Reactor Fuel Return Program Starts Shipping Fuel to Russia," paper for the 2003 International Meeting on Reduced Enrichment for Research and Test Reactors, Chicago, Illinois, Oct. 5–10, 2003, available at www.td.anl.gov/Programs/RERTR/RERTR25/PDF/Dedik.pdf, accessed on Feb. 18, 2004.

[46]"Research Reactors," World Nuclear Association, Dec. 2004, available at www.world-nuclear.org/info/printable_information_papers/inf61print.htm, accessed on May 31, 2006.

[47]Ibid, pp. 4–5. The bare critical mass for 36% enriched HEU is greater than 200 kg, indicating that it might be impracticable for terrorists to acquire such large amounts of this material and be able to fashion it into a workable weapon.

fuel to Russia. This work initially was planned to be complete in 2005, a date subsequently extended to the end of 2006. Moreover, DOE announced plans to repatriate all soviet-origin spent nuclear fuel by 2010. Although these targets subsequently have slipped, since 9/11 there have been numerous successful repatriation transfers of fresh Soviet-origin HEU including shipments from Bulgaria, the Czech Republic, the former East Germany, Latvia, Libya, Lithuania, Poland, Romania, Serbia, Uzbekistan, and Vietnam.[48]

The United States is still seeking to repatriate U.S.-origin HEU supplied to about two dozen countries. On May 26, 2004, Secretary of Energy Abraham stated that as part of the GTRI, DOE "will take all steps necessary to accelerate and complete the repatriation of all U.S.-origin research reactor spent fuel under our existing program from locations around the world within a decade."[49] The GTRI plans to complete the repatriation of fresh U.S.-origin HEU fuel by 2014 and spent HEU fuel by 2019.

Fissile Material Security in Other Settings

Fissile materials are found in hundreds of locations around the globe under varying levels of security. Although the risks posed by these materials are greatest in the three settings just described, their presence in many other contexts also creates potential targets for terrorists. Without offering a comprehensive analysis, here, it is worth briefly noting some of these other venues where the materials can be found, and where the need for high security is essential.

Nuclear Weapons Programs Outside Russia and Pakistan

All nuclear weapon programs produce, process, and machine fissile materials, steps that often also include their transportation among different sites. In many cases, nuclear testing also involved the transportation of fissile materials, where assembled into test devices at the test site. In addition, fissile materials are used in nuclear weapons research activities, which may employ still other locations and transportation links. For countries reducing their nuclear arsenals, comparable challenges can arise as materials are removed from weapons and stored, in some cases after additional processing. Each of these settings demands the highest levels of security against theft and diversion. In countries with smaller nuclear arsenals—France, Great Britain, China, France India, Israel, North Korea—this challenge is inherently more manageable than for the United States and Russia due to the smaller scale of activities involved. Nonetheless, in less developed countries, underlying weaknesses in national infrastructure, such as in rail and highway transportation systems, in communications, and in the level of guard force education and training, may erode security efforts.

Even in the United States, where security over fissile materials is generally deemed to be very stringent and where the issue has received added attention since September 11, 2001, evidence indicates that periodic deficiencies may exist at some facilities within the U.S. nuclear weapons complex.[50] In a very troubling episode in August 2007, six U.S. nuclear weapons were mistakenly transported by aircraft across the United States and were without adequate security protection for more than 24 hours.[51]

---

[48]NNSA Public Affairs, "Secret Mission to Remove Highly Enriched Uranium Spent Nuclear Fuel from Uzbekistan Successfully Completed," National Nuclear Security Administration, April 20, 2006.

[49]Spencer Abraham, Secretary of Energy, Speech at the International Atomic Energy Agency, Vienna, Austria, May 26, 2004.

[50]Richard W. Mies, Admiral, U.S.N. (retired), Memorandum to NNSA Administrator, Ambassador Linton Brooks, "Independent NNSA Security Review," May 2, 2005.

[51]"Second Review Set of Bomber Nuke Flight," *Global Security Newswire*, Sept. 21, 2007, www.nti.org/d_newswire/issues/2007_9_21.html#DD0366EB (Sept. 2007).

In sum, perfect security for nuclear weapons and fissile materials cannot be achieved by any country. Every nation faces challenges to providing the highest possible levels of security for these items. Russia and Pakistan face particularly difficult fissile material security issues, but all states have weaknesses in their security systems. Given the potential consequences of loss of nuclear weapons or fissile material to terrorists, it is extremely important that such weaknesses be addressed quickly and fully.

Naval Propulsion Systems

Several navies power ships with HEU. About 170 nuclear-powered vessels (including submarines, naval surface ships, and civilian vessels) are currently operational, all of which use pressurized-water reactors for propulsion. All U.S. and British nuclear ships, including submarines, use HEU fuel enriched to 93.5% U-235. French ballistic missile nuclear-powered submarines and France's single nuclear-powered aircraft carrier use HEU fuel enriched to 90%, whereas French attack nuclear-powered submarines use LEU fuel enriched to 7%. China, alone among the world's nuclear navies, uses only LEU fuel for its naval reactors, probably enriched between 3% and 5%. The nuclear submarine planned by India is likely to use nuclear fuel similar in enrichment to that of many Russian submarines, probably around 20%.[52]

Weapons-quality HEU used in these navies is present not only at naval fueling areas but also at sites where the HEU is produced, in fuel fabrication plants, and in transit to nuclear submarine bases. In addition, spent fuel, which may contain uranium enriched to 80% or more, is found at storage sites and in transit to those locations. No cases have been reported outside Russia involving thefts of, or illicit trafficking in, naval fuel. Nonetheless, Russia's experience—including the concerns of Russian Navy officers that led them to seek U.S. help in securing Russian nuclear submarine fuel—highlight the potential dangers in this sphere.

Plutonium in Civil Nuclear Power Programs and HEU in Nonmilitary Research Reactors in Industrially Advanced Countries

Until the late 1970s, it was widely assumed among nuclear energy planners that global uranium resources would be rapidly depleted and that it would be necessary to use plutonium, in the form of MOX fuel, as an alternative to low-enriched uranium fuel in most nuclear power programs. Because of slower-than-expected growth of nuclear power and the continuing discovery of new economically exploitable uranium reserves, however, uranium supplies have remained abundant while the costs of producing MOX fuel have increased significantly. These economic factors, together with concerns over the proliferation dangers posed by the widespread use of plutonium fuels, have led most nuclear power-using states to abandon such separation and "recycling" of plutonium in favor of the "once-through fuel cycle," in which spent nuclear power plant fuel is stored on an interim basis until emplaced in a permanent storage facility, usually planned for a stable geologic formation.[53]

---

[52]Chunyan Ma and Frank Von Hippel, "Ending the Production of Highly Enriched Uranium for Naval Reactors." Transfers to support nonexplosive military uses of nuclear materials are not prohibited under the nuclear Nonproliferation Treaty. James Clay Moltz, "Closing the NPT Loophole on Exports of Naval Propulsion Reactors," *The Nonproliferation Review*, Fall 1998, pp. 108–114.

[53]The Netherlands, Germany, Sweden, Spain, and the United States have cancelled domestic plutonium separation plans and/or have reduced or ended contracts for the separation of plutonium abroad and its return in the form of MOX. During the Communist period, Soviet satellite states were obliged to return spent fuel to Russia, where it was reprocessed; the resulting plutonium was not returned but stored in Russia.

For a variety of reasons, however, several states continue to pursue plutonium separation for civil nuclear energy purposes, most notably France, Great Britain, Russia, and Japan. Of these, only France has a successful recycle program that balances supply (newly separated plutonium) with demand (the fabrication and use of MOX fuel). Great Britain has no domestic program for using MOX fuel and its plutonium is stored after separation. Russia likewise has no domestic MOX program for civil plutonium. Although it stores spent fuel from its VVER-1000 reactors and RBMK units, it continues to reprocess spent fuel from VVER-440 reactors and store the resulting plutonium.[54]

Japan has contracted with France and Great Britain for the reprocessing of Japanese spent fuel; although Japan has a program for using the resulting plutonium as MOX in its nuclear power reactors, that program has been virtually frozen because of domestic opposition and other challenges. As a result, separated Japanese plutonium continues to accumulate in France and Great Britain. Notwithstanding this accumulation of tens of metric tons of separated plutonium in these countries, Japan has continued to work on a large-scale plutonium separation facility at Rokkasho-mura, which began test operations in early 2006. Once approved for full operation, the facility could process about 800 metric tons of spent fuel annually, separating up to seven tons of plutonium each year.

India also separates plutonium from spent nuclear power plant fuel. Its plan calls for the use of the plutonium in advanced, breeder reactors. Usually fueled with fuel containing about 20% plutonium, breeder reactors use excess power to irradiate additional uranium, thereby "breeding" new plutonium. All other countries, except Russia, have abandoned this technology as uneconomical. The sizable accumulation of separated plutonium stands in sharp contrast to extensive and costly Russian, G8, and U.S. efforts to eliminate fissile materials in other settings.

Regarding HEU use in research reactors in advanced countries, as discussed, the United States and Russia are working actively to reduce the use of HEU in research reactors they have previously exported (or to which they have provided fuel) and to repatriate and eliminate fresh and spent HEU fuels from these locations. In addition, both countries are gradually reducing the use of HEU fuels at home. Nonetheless, for years to come, more than a dozen major research reactors, located mostly in G8 countries (including the EU), will continue to use HEU fuels. The list includes several, such as the Petten High Flux Reactor in the Netherlands, that have formally agreed to switch to low-enriched fuels once they are available, as well as a number that are likely to use HEU fuels indefinitely, because of the unique research and/or isotope production these facilities support. Resisting the trend toward converting research reactors to low-enriched fuel, the German FRM-II reactor in Munich has been designed to use weapons-grade HEU fuel. The reactor owners have agreed to reduce the enrichment to 50% by December 2010, but meanwhile, the reactor will use bomb-grade HEU.

Summary

Given the vast quantities of fissile materials in all of the foregoing settings—Russia, Pakistan, Russian- or U.S.-supported research reactors around the globe, the nuclear weapons programs of the other seven nuclear-armed states, marine propulsion systems, and plutonium and HEU found in civilian nuclear programs—it appears that the threat of nuclear terrorism will be an unfortunate feature of the international security environment for decades to come.

---

[54]Under the U.S.-Russia Plutonium Disposition Program, Russia *is* planning to use MOX fuel in a number of its nuclear power reactors and hopes to build new, more advanced plutonium-fueled reactors. For the foreseeable future, however, all plutonium from these programs will come from stocks originating in the Russian nuclear-weapons sector, not material separated from civilian spent nuclear power reactor fuel.

24 Radiological Dispersal Devices

Greg Van Tuyle and James E. Doyle

Introduction

A radiological dispersal device, known as an RDD or dirty bomb, could be an extremely disruptive terrorist weapon, capable of spreading radioactive material over a wide area and possibly killing dozens or even hundreds of people if used in an urban area. An RDD causes no nuclear explosion but instead uses conventional or chemical explosives to disperse radioactivity. Materials and technology required to build radiation dispersal devices are much easier to obtain than those needed for a weapon producing a nuclear explosion.[1] In 2003, the International Atomic Energy Agency (IAEA) reported that more than 100 countries might have inadequate controls to prevent or even detect the theft of radioactive materials needed for an RDD.[2] However, because highly radioactive sources are used for the benign and essential purposes of cancer treatment, industrial radiography, electricity generation, and research, it becomes internationally impractical to secure and control every item.

This chapter explores a subset of high-risk isotopes that necessitate security and control. Although recognition of these materials' international vulnerability has only recently come to light, proper assessment and mitigation of risks to international security presented by radioactive sources is a high priority.

Radiological dispersal devices can take many forms: as containers of radioactive materials surrounded by conventional explosives, as aerosolized materials sprayed using conventional equipment, or even as manual dispersion of a fine powder into the environment.[3] Any of these RDD scenarios could cause long-term health effects and major financial loss. Potential health consequences include both immediate fatalities and long-term increases in cancer incidence. Environmental consequences range from long-term denial of property, disruption of services, and efforts to decontaminate, all of which disrupt social order well beyond the terror event. The initial public response to an RDD attack will nearly certainly cause public panic, irrespective of the amount and type of radioactive material actually dispersed.

---

[1] Charles D. Ferguson, Tahseen Kazi, and Judith Perera, "Commercial Radioactive Sources: Surveying the Security Risks," Center for Nonproliferation Studies, Occasional Paper #11, Jan. 2003.

[2] International Atomic Energy Agency, Division of Public Information, PR 2003/03 (March 13, 2003).

[3] Steven E. Koonin, Statement delivered before the Senate Foreign Relations Committee on Radiological Terrorism, March 6, 2002; http://units.aps.org/units/fps/newsletters/2002/april/cap02.cfm#a3 (May 2007).

The physical effects of an RDD are dependent on several factors; both the type and amount of radioactive material used in any device are especially critical. In most instances there would be a vast difference in lethality produced by different types of radioactive materials.

As this brief introduction implies, the key to reducing the threat of an RDD terrorist attack is to identify the most dangerous and vulnerable radioactive sources, eliminate them where possible, and protect those that must remain in use with the highest possible levels of security.

Overview of Nuclear and Radiological Materials

Nuclear and radiological materials have both military and civilian applications. Radioactive materials are generated as byproducts from the nuclear fuel cycle or are created by irradiating source materials. Extensive global production, use, and storage of commercial radioisotopes make these materials vulnerable to theft for use in an RDD. Yet, civilian reliance on radiological sources developed prior to recent concerns about terrorism, so many sources, whether in use or disuse, remain under conditions of minimal to no security.

At the most fundamental level, radiological sources are used for three purposes: to kill or otherwise alter organisms or tissue, to generate energy on a localized and/or remote basis, or to scan objects or provide other types of measurements. A hierarchy of radiological sources, grouped by purpose, is provided in Figure 24.1. Sources near the top of the chart can utilize thousands, sometimes millions, of curies of radioactive isotopes and are considered a primary concern for use in a large and damaging RDD.

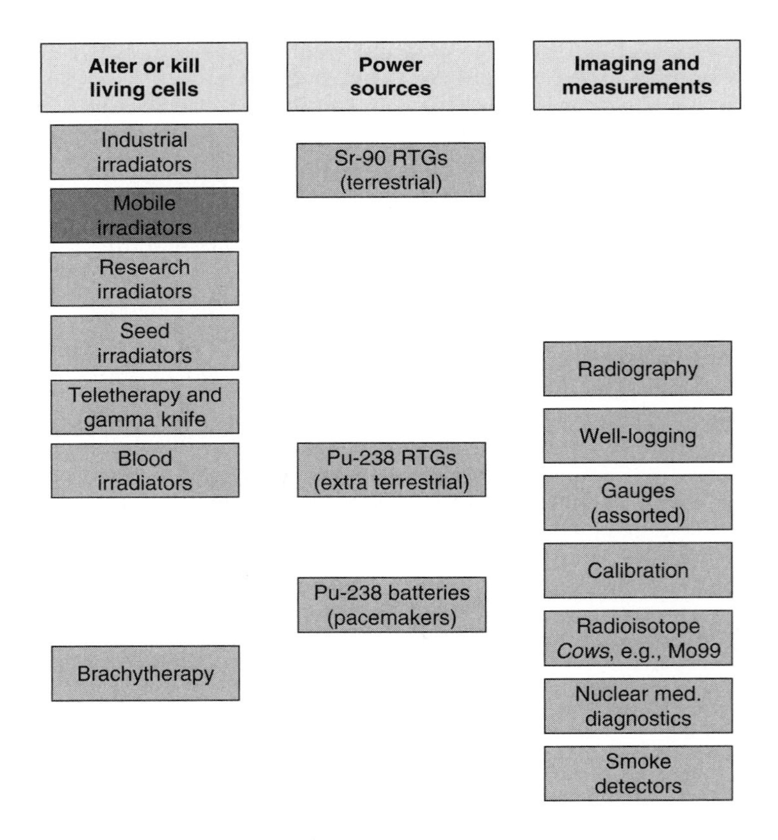

FIGURE 24.1 Hierarchy of radiological source applications.

Larger commercial applications, including industrial irradiators, research irradiators, seed irradiators, teletherapy units, blood irradiators, radioisotopic thermal generators (RTGs), radiography, and well-logging sources, use a limited number of radioisotopes: cobalt-60, cesium-137, strontium-90, iridium-192, plutonium-238, and americium-241. Other radioisotopes of concern include radium-226, californium-252, plutonium-239 but tend to be less widely available.

Radiological source materials that are used in large quantities are manufactured from waste products from nuclear fission events and fall into two categories. Fission products resulting from the splitting of uranium or plutonium fuel atoms include cesium-137 and strontium-90. Most spent nuclear fuel is not processed, but in cases where it has been processed, either to reuse the uranium and plutonium in fueling reactors or to use the plutonium in making weapons, fission products are waste byproducts. Although few facilities perform such spent-fuel separations, those that do generate an ample supply of fission products for use in radiological sources.

In contrast, the production of cobalt-60 is fairly costly and time consuming. Stable cobalt-59 is plated with nickel and placed in a target region of a reactor. During the process of neutron capture, lasting a few years, the cobalt-59 captures an additional fission-released neutron to become radioactive cobalt-60. Similarly, Pu-238 must be produced in a reactor.

Sources Used to Alter or Kill Living Cells

In high doses, radiation can kill living organisms, making it especially useful in sterilizing medical instruments, treating cancer, and killing bacteria and contaminants in food. Some of the more significant irradiators are described in the following sections.

Industrial Irradiators

Large industrial irradiators like the one shown in Figure 24.2 are typically highly shielded facilities used for medical supply sterilization and food irradiation. Shielding is necessary due to hundreds of Cobalt-60 pencils (small rods) that deliver gamma radiation during the irradiation process. Cesium-137 is alternatively used, though less commonly. Most industrial irradiators were produced by Canada and the U.K. and purchased through international assistance by other non-Western nations. Since 1980, the IAEA, through its technical cooperation

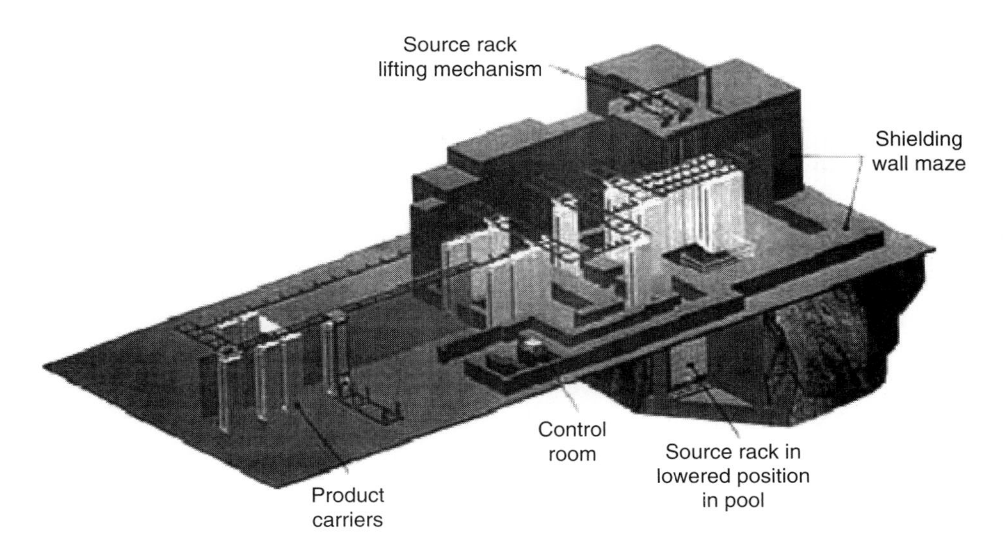

FIGURE 24.2 Large industrial irradiator. (Source: Tiffany Strub and Gregory J. Van Tuyle, "Large Radiological Source Production and Utilization and Implications Regarding RDDs," Los Alamos National Laboratory Report LA-UR-03-5432, July 2003.)

FIGURE 24.3 The Cobalt-60 Research Irradiator from Nordion.

program, has supplied 40 cobalt-60 irradiators to developing countries, along with the appropriate training and security.[4] Most of these units are equipped with significant safety and security features. Overall, approximately 191 industrial irradiation facilities reside in 63 countries.[5]

Mobile Irradiators

Mobile irradiators contain high-energy gamma emitters that are loaded into modified trucks driven to agricultural sites for food sterilization. Of particular concern are the irradiators manufactured by Argentina, which makes a 10-ton mobile 40,000-curie irradiator using cobalt-60, and China, which manufactures a 67-ton mobile 250,000-curie irradiator containing cesium-137. It is unclear to whom these mobile irradiators have been sold. Because produce irradiation can greatly extend food's shelf life, mobile irradiators may increase in commercial circulation.

Research Irradiators

Research irradiators like the one in Figure 24.3 vary from their industrial counterparts in both size and application. They are utilized in laboratory environments, often in connection with research institutes, whose more miniature machines use smaller cobalt and cesium sources and perform applications like dosimetry calibration, insect control, materials research, and smaller-scale food irradiation and medical sterilization.

Seed Irradiators

During the 1970s, Soviet scientists trucked mobile seed irradiators to parts of the Soviet Union as an agricultural research project to study the effects cesium chloride irradiation on

[4] "Private Sector Adopting Nuclear Techniques," *Inside Technical Co-operation*, vol. 3, no. 1, March 1997; www.iaea.or.at/worldatom/Periodicals/Bulletin/Bull391/nt.html, June 2003.

[5] G. J. Van Tuyle, et al., "Reducing RDD Concerns Related to Large Radiological Source Applications," Los Alamos National Laboratory Report LA-UR-6664, Sept. 2003.

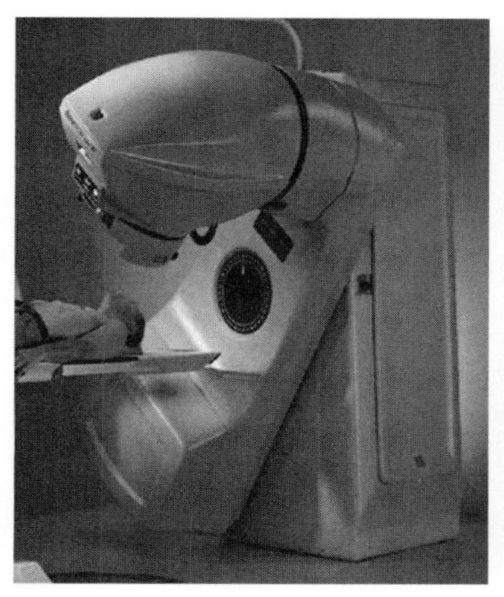

a.

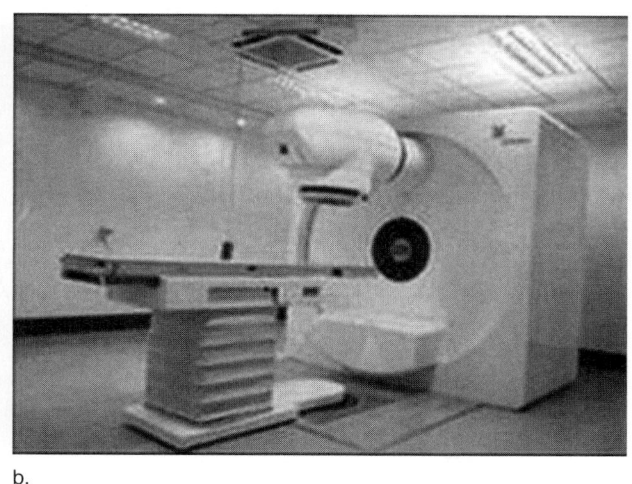

b.

FIGURE 24.4 Teletherapy units.

crops. Mobile seed irradiators disappeared after the 1970s due to their low capabilities and outdated conveyor systems. There is no known account of how many seed irradiators the U.S.S.R. produced or where they were sent. Estimates vary from 100 to 1,000 machines. Regardless, they are most likely poorly secured, and there has been very little success in recovering them. Since each unit was equipped with 3500 curies of cesium-137 chloride, mobile seed irradiators pose a significant concern.[6]

Teletherapy

Teletherapy, or external radiation treatment for cancer, places a radioactive source in the "head" of a device and focuses a beam of gamma radiation on the cancerous portion of the patient's body.[7] Whereas older teletherapy machines relied on cesium sources with a 30-year half-life, they now use cobalt material that is easily removed and replaced roughly every five to seven years to coincide with Co-60's five-year half-life. Most of the 5,300+ teletherapy units (see Figure 24.4) exist outside the United States because U.S. institutions replaced them with more sophisticated electron accelerators in the 1970s. Ironically, a major U.S. program to export excess teletherapy units—bringing lifesaving technologies to less developed nations—has succeeded in proliferating the units, each containing several thousand curies of cobalt-60 or cesium-137, around the world.

Identification of these widely distributed devices is much more difficult than for large irradiators. The U.S. company Neutron Products indicates that it would be difficult to identify the foreign institutions that received donated U.S. teletherapy equipment. Likewise, the IAEA has helped to establish teletherapy centers in many countries, including Mongolia, Ethiopia, Nigeria, and Ghana, but its Directory of International Radiotherapy Centers (DIRAC) cannot identify all recipient locations, since no nation was obligated to register its devices.[8] The DIRAC database reports that there are 5,347 registered radiotherapy centers in the

---

[6]*Ibid.*

[7]"C-188 Cobalt-60 Brochure," MDS Nordion, June 2001, www.nordion.com/master_dex.asp?page_id=122 (May 2003).

[8]"TC Projects: Radiotherapy," IAEA Department of Technical Co-operation Database, 2001, www-tc.iaea.org/tcweb/tcprogramme/projectsbyfacandapc/query/default.asp (May 2003).

world, housing roughly 2,350 cobalt-60 teletherapy devices and 45 cesium-137 units.[9] Realistically, a minimum of 10,000 sources are thought to exist worldwide.[10]

A less common variation of teletherapy is the gamma-knife. In this application, approximately 200 cobalt-60 sources are configured to expose a brain tumor from different angles. Around 10,000 curies of cobalt-60 are used in this application. Dozens of these devices exist, mostly in Western countries.[11]

Blood Irradiators

Blood irradiators like the one in Figure 24.5 sterilize blood using cesium-137 after the blood has been placed in bags and loaded into the ionizing chamber.[12] Blood irradiators are found primarily in Western hospitals; however, these hospitals often opt to purchase newer X-ray-based blood irradiators and export their older cesium units to poorer countries. In such cases, disposal of cesium sources becomes a major liability for hospitals, creating new problems of disused sources, orphan sources, or sources being resold outside the U.S.

Brachytherapy

Brachytherapy sources are typically needles or small pellets of radioactive materials that are inserted surgically into the tumors of cancer patients. Few radioisotopes are involved, with the selection driven by the organ that has been impacted by the tumor.

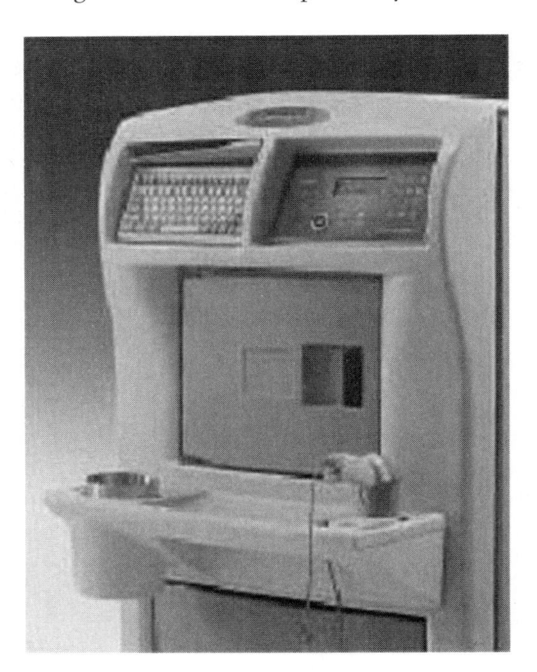

FIGURE 24.5 A blood irradiator.

---

[9]"Directory of Radiotherapy Centres," DIRAC, International Atomic Energy Agency and World Health Organization (V.2000.A) (Database).

[10]Abel J. Gonzalez, "Security of Radioactive Sources: The Evolving New International Dimensions," *IAEA Bulletin* 43/4/2001, April 2001, www.iaea.org/worldatom/Periodicals/Bulletin/Bull434/article8.pdf (May 2003).

[11]Carnegie Science Center, "Gamma Knife Module" of Website: www.carnegiesciencecenter.org/zapsurgery/00_gammaknife.asp (May 2003).

[12]The CISUS Inc. Website described the IBL-437 Blood Irradiator & Dose Writer: www.cisusinc.com/iblbody.htm (Aug. 2003).

Sources Used to Provide Power

Sr-90 RTGs (Terrestrial and Extra-Terrestrial)

Radioisotope thermal-electric generators (known as RTGs in the United States and RITEGs in Russia) use the heat emitted from source decay to produce power.[13] They are also referred to as *nuclear batteries* or *lighthouses*. An RTG is shown in Figure 24.6. Strontium-90 and its daughter yttrium-90 emit heat via beta particles; this energy is easily converted to an electric current. Plutonium-238 is also used in RTGs, but these units are much smaller and more expensive than the strontium units and reserved primarily for deep-space exploration missions. The half-lives of strontium-90 (29 years) and Pu-238 (87 years) contribute to their usefulness in RTGs. About 1,000 such units were deployed in the former Soviet Union to provide power for lighthouses along its north coast. Both the Soviets and the United States also deployed quite a few units for military purposes.

It is believed that there are 1,000 Soviet-produced RTGs, but exact number and locations remain unknown. The most detailed figures from the All-Russian Institute of Technical Physics and Automation indicate that 929 RTGs are currently operating in Russia, with an additional 169 in storage. Twenty-six RTGs are located in other former Soviet countries. With the radioactivity levels of the strontium RTGs ranging from 5,000 to 500,000 curies, the uncertainties are worrisome.[14]

The U.S. has manufactured 134 strontium-90 RTGs, according to the Off-Site Source Recovery Program at Los Alamos.[15] Of these, only 47 have been accounted for. It would be

FIGURE 24.6 A Soviet-made RTG.

---

[13]Fradkin et al., "Radioisotope Sources of Electric Power," Army Foreign Science and Technology Center Report AD/A-001 210, Sept. 1973.

[14]G. J. Van Tuyle, et al., "Reducing RDD Concerns Related to Large Radiological Source Applications," Los Alamos National Laboratory Report LA-UR-6664, Sept. 2003.

[15]Information compiled by the Off-Site Source Recovery Project at Los Alamos National Laboratory on behalf of the U.S. National Nuclear Security Administration, Office of Global Threat Reduction.

difficult to steal one of these RTGs due to its physical properties: It can weigh between 800 and 8,000 pounds and generates intense heat, and units are remotely located.

Extra-Terrestrial Pu-238 RTGs

In contrast to strontium-90, which is a waste product from nuclear fission, plutonium-238 must be specially produced in nuclear reactors and is quite expensive. The plutonium-238 RTG units are typically in the tens to hundreds of curies range and are closely controlled. Although 100 curies of plutonium-238 could make an especially potent RDD source, availability is so limited that such usage is unlikely.

Sources Used for Imaging and Measuring

Radiography

Industrial radiography is a method of checking welding errors in pipelines and buildings. Radiography sources are abundant mobile sources, typically mounted on carts and used to produce high-energy gamma scans of welds at construction sites. Radiography sources accompany sophisticated construction processes in nearly every part of the world. Although many of these sources are utilized by large multinational companies, it is equally common for smaller construction companies to use several sources apiece. Many radiography units use short-lived iridium-192 in quantities that are insufficient to qualify for urgent attention, but cesium-chloride radiography units should be considered candidates for replacement.

Well Logging

Well-logging sources like the one in Figure 24.7 are used in oil well drilling as well other drilling and mining operations to assess the geology surrounding exploratory bore-holes. Most well-logging sources are used by large international oil-exploration companies such as

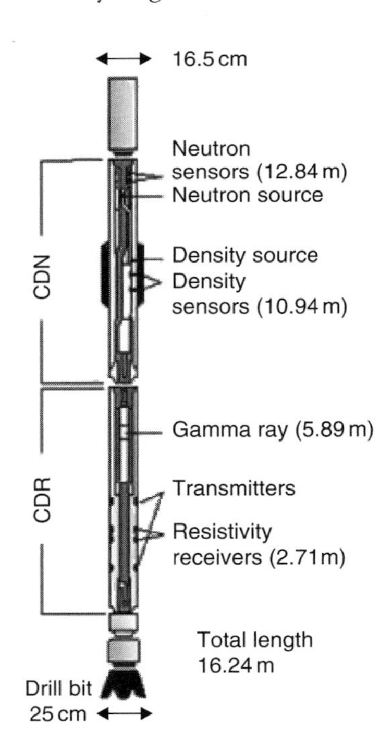

FIGURE 24.7 A well-logging instrument.

Schlumberger, Haliburton, and Baker-Hughes.[16] There may be 5,000 to 10,000 sources in use around the world, many of which contain a neutron source in the 15–20 Curie range to perform diagnostics. Well-logging sources typically also use a cesium source in the tens of curies to provide simultaneous density scans of the surrounding geology.

Radiological Source Producers and Suppliers Worldwide

Few, if any, countries do not have radioactive sources in use. In contrast, there are far fewer producers of radiological source materials, since the radioisotope production process requires nuclear reactors or particle accelerators, as well as sophisticated chemical separation processes. Though some isotope production can occur in power reactors, the vast majority of the radiological sources discussed in this report are manufactured in research reactors. Therefore, by identifying the research reactors in use (greater than 1 MW) and the companies/agencies that operate or have access to them, it becomes possible to identify the global pool of radioisotope source producers.

First-Tier Producers and Suppliers

Companies in seven different countries currently produce large amounts of cobalt-60, cesium-137, and/or strontium-90. In addition to being producers of radioactive sources, most of the companies in this category are involved in supplying the technology to accompany their sources. The top-tier isotope producers conduct a fair amount of business among themselves. Though it is often difficult to determine specific quantities of isotopes produced, a general understanding of the production size can usually be ascertained. Figure 24.8 identifies the largest isotope producers in the world.

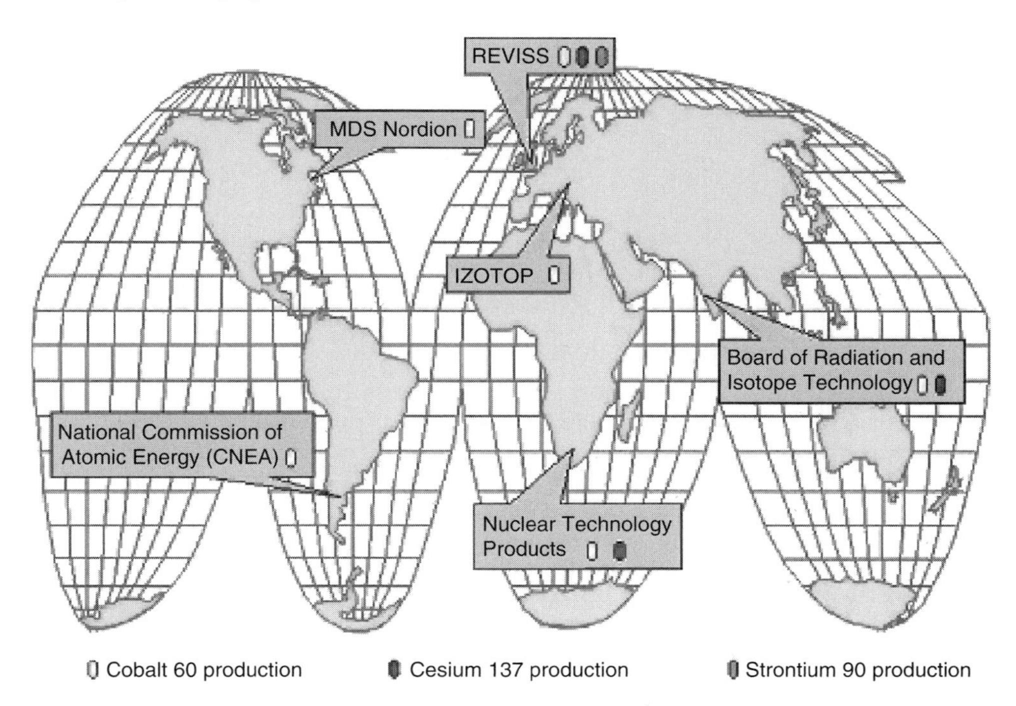

FIGURE 24.8 Large-scale manufacturers of radioisotopes: Canada, U.K., Russia, Argentina, South Africa, India, and Hungary.

[16]Stephen E. Prensky, "A Survey of Recent Developments and Emerging Technology in Well Logging and Rock Characterizations," published in *The Log Analyst*, vol. 35, no. 2, p. 15–45, no. 5, p. 78–84, 1994.

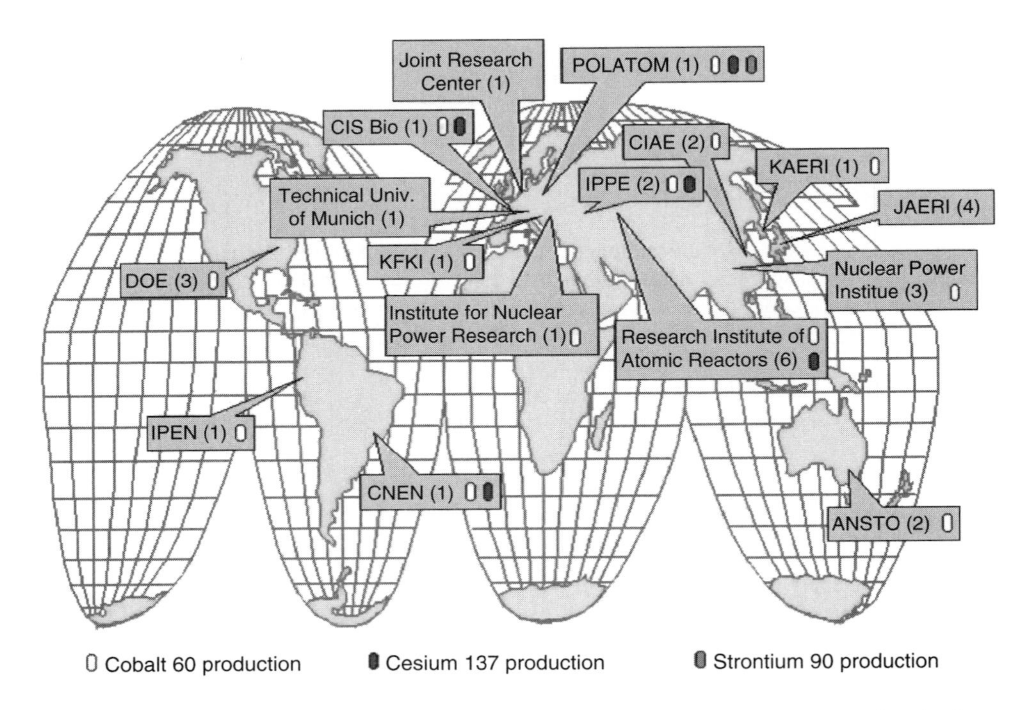

FIGURE 24.9 Secondary radioisotope suppliers and number of reactors: U.S., Peru, Brazil, Netherlands, Poland, France, Germany, Hungary, Romania, Russia, China, South Korea, Japan, and Australia.

Second-Tier Producers and Suppliers

The second tier of isotope producers generates much smaller quantities of radioactive sources in comparison to the previous group. Often these companies manufacture source material strictly for use only in their country or geographic region with smaller curie amounts, e.g., teletherapy sources as opposed to industrial irradiation sources. Additionally, these businesses and research institutes will often receive radioactive material from the first tier of producers. Shown in Figure 24.9 is a map of this second tier of source producers.

Third-Tier Producers and Suppliers

There is considerable uncertainty regarding information on the list of third-tier source producers. Although these national institutions possess nuclear reactors, have significant isotope-production capabilities, or are known to produce isotopes, information regarding their commercial production remains extremely limited. The map in Figure 24.10 includes details concerning this third tier, labeled as "possible producers and suppliers" because information on the source supply chains both to and from these organizations is limited. Several of these institutions could opt to enter the radioisotope market to produce cobalt-60, but uncertainty stems from the fact that several of these institutes are located in countries where controls on radioactive material are not well understood, such as Uzbekistan, Pakistan, Iran, and North Korea. Lack of transparent regulatory systems for radiological sources in these countries could present opportunities for terrorists to obtain source material for use in constructing an RDD. More information is needed to better assess these institutions and their radioisotope manufacturing operations.

Life Cycle of Radiological Sources

Radioactive source materials pose some level of concern from the moment they are created until the time of their disposal. Indeed, the vulnerabilities are apparent throughout the life cycle of the large radiological sources; see Figure 24.11.

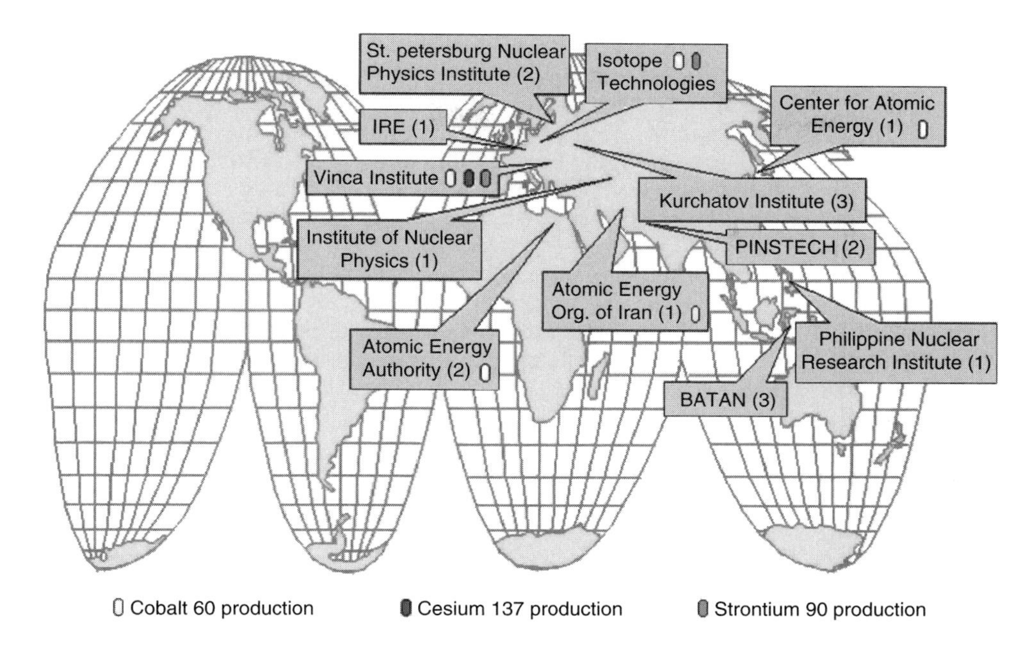

FIGURE 24.10 Possible third-tier radioisotope producers and suppliers: Vinca Institute, Yugoslavia; Isotope Technologies, Belarus; St. Petersburg Nuclear Physics Institute, Russia; Kurchatov Institute, Russia; Institute of Nuclear Physics, Uzbekistan; Center for Atomic Energy, North Korea; Atomic Energy Authority, Egypt; AEOI, Iran; PNRI, Philippines; BATAN, Indonesia; IRE, Belgium; PINSTECH, Pakistan.

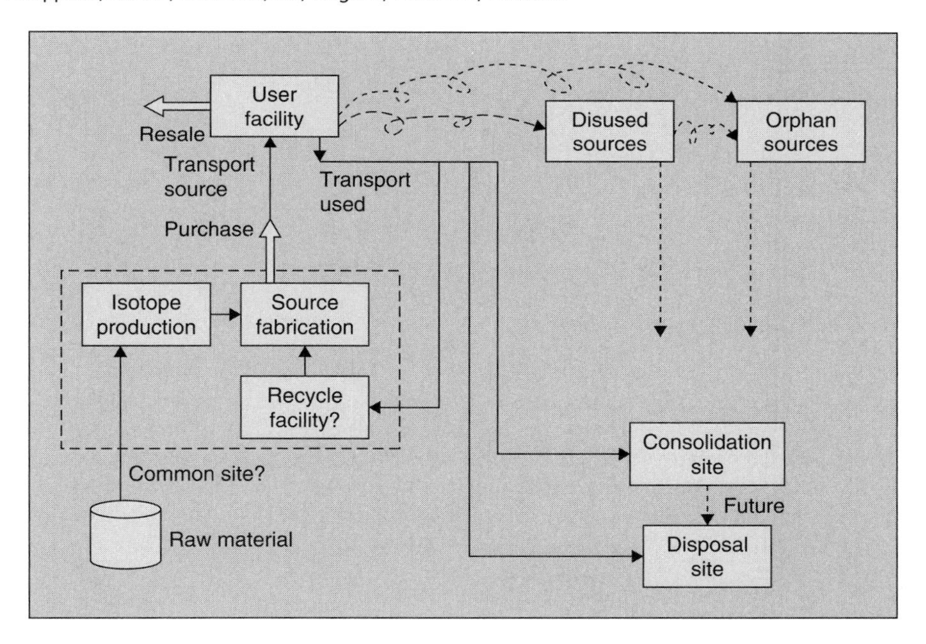

FIGURE 24.11 The life cycle for radiological source materials.

Defining the Life Cycle

Isotope production usually occurs in nuclear reactors, although some smaller radiological source materials are produced using particle accelerators. Some post-production processing is involved in isolating radionuclides of interest, usually associated with the isotope production facility (reactor). Source fabrication generally follows a mechanical/metallurgical processing step that could either be colocated with the radionuclide producers or a separate business. Radiological source users are nearly always located separately from the producers, sometimes introducing lengthy transportation routes. Ideally, when the end user has finished using the source, that user returns the source to the producer for recycling or ships it to a disposal site. Difficulties in doing so, however, lead to either a disused source being retained indefinitely at the user facility or, even worse, its loss, abandonment, or theft, rendering it an "orphan" source impossible to account for and track.

Different points in the life cycle render sources vulnerable to theft or diversion. The largest concentrations of materials are generally at production sites, but there security is greatest. User facilities contain large amounts of dangerous radiological sources, but security often depends on the purpose of the facility. "Orphans" present obvious concerns since they are very vulnerable if found. Disposal sites are uncommon in many parts of the world, and most of the sites that are open are used mostly for low-level wastes of little concern. Transportation between destinations can create significant vulnerabilities if effective security measures are not taken to deter and prevent successful hijacking or theft.

Locating commercial radioisotopes can be difficult. Manufacturers and suppliers are in the business of selling sources and are not responsible for tracking them throughout their life cycle. Large radiological source users are far more numerous and much more difficult to locate unless they are tracked through the source suppliers. Waste sites can be found through governmental/regulatory entities, particularly if they are large enough to house radiological sources of concern. Orphan sources are the most difficult to find, although the existence of orphans can be identified or confirmed through radiological source suppliers and users, provided they are still in business and have kept records. Therefore, an important means of locating large radiological sources is through the source producers and suppliers and the large users. States should consider regulations that require both parties to register and track sources throughout their lifetime.

Factors Contributing to Risk

Five factors for radiological materials contribute to their potential use in an RDD:

- The number of sources
- The radioactivity levels of the sources
- The hazard factor for a given source type
- The inaccessibility of the sources
- The level of security for a source

Number of Sources (Abundance)

The volume of sources available for large source applications and at any given stage of the life cycle is a function of its vulnerability. The more abundantly used teletherapy units increase the concern for them over the less used blood irradiators. Similarly, the number of disused and orphaned cobalt-60 sources is thought to be fairly limited, so cobalt-60 sources in use at industrial irradiators or teletherapy clinics pose greater numerical risks than limited disused or orphaned sources.

Radioactivity Level of Sources (Intensity)

The impact of an RDD increases in proportion to the radioactivity level, although very highly radioactive sources pose some handling challenges. Therefore, potential concern regarding a 10,000 curie radiological source is much greater than concerns about a 1 curie source.

Hazard Factor (Dose)

Although radioactivity level is a primary measure of concern, the impact per curie can vary significantly. Thus, the concern about 100 curies of plutonium is greater than the concern over 100 curies of strontium.

Inaccessibility of Sources

Some radiological sources are highly accessible, such as americium sources used in smoke detectors. In contrast, some RTGs are used in locations that are virtually inaccessible. Accessibility is thus defined as the measure of difficulty in locating and gaining access to a radiological source (irrespective of security features).

Security Level

Security features, provided for some radiological sources, are intended to restrict access to sources. Because many of the security features were designed prior to the recent increase in concern about RDDs, they most likely provide investment protection and safety features but might not be effective against a determined and knowledgeable adversary. Regardless, anything that deters source removal can provide a measure of protection and reduce concern.

Options for Reducing Radiological Source Concerns

Deny, Detect, Defend, Respond

The options for reducing the RDD risk, from a technical perspective, come down to four steps. First, deny access to the materials needed to create an RDD. Second, detect the radioactive materials while they are being transported from their point of acquisition to the target area. Third, defend potential targets. Fourth, prepare to respond to a successful attack, which involves both the emergency response and the post-event cleanup.

The remainder of this chapter focuses on potential actions to deny radiological materials to terrorists or unauthorized personnel. The following section considers the highest-risk radiological sources, their vulnerability, and stages of life-cycle usage and presents suggestions to more effectively deny and deter the unauthorized acquisition of RDD materials.

Recovery and Consolidation of Disused and Orphan Sources

In the current environment there is an almost unlimited number of disused and orphan sources, and the number of radiological source locations where security upgrades would be beneficial is daunting as well. The disused and orphan sources on the priority list include RTGs, seed irradiators, and well-logging sources.

The challenge is to consolidate these sources until they can be either recycled or disposed of permanently. Such source consolidation efforts have been ongoing in the United States for several years, and many disused and orphan sources have already reached holding points around the DOE complex. The IAEA has also been working to recover sources, sometimes in cooperation with the U.S. DOE. A recent and expanding effort involving the DOE NA-25 is addressing this problem globally as an outgrowth of nuclear materials controls efforts in Russia.

Physical Security Upgrades

Large shipments of cobalt traveling across continents to 190 industrial irradiators and 6,000 radiotherapy sites urgently call for bolstered security. Additionally, the 6,000 international radiotherapy centers present equally pressing security concerns.

The dilemma is how to proceed. One possible solution is to work with governmental and regulatory bodies to build a better framework, mandating improved transportation security and better security for hospital cancer treatment centers. Such a framework will directly impact countries that can help themselves and will strongly impact companies that transport sources. This could reduce the scope of the physical security upgrades efforts to poorer parts of the world. An enormous effort would remain, but working with allies at the IAEA, in Russia, and in Europe, such a mission might be feasible. Within the proposed framework, an effort should establish and secure waste consolidation sites in as many parts of the world as is feasible. Such action could help prevent the growing number of disused sources from making the current situation worse.

Improved Legal Framework and Infrastructure, Including Regulatory, Recycling, and Waste Disposal Provisions

There are four major areas in which action is needed: disposition of used sources, transportation security, user facility security, and regulation of the commerce in sources (sales):

- Disposition of used sources can include recycling, waste consolidation, and waste disposal. Laws that require used sources be returned to the source suppliers would be very helpful in the disposition of used sources because suppliers are more capable of recycling sources or managing the consolidation of used sources. International agreements can also help with the development of regional consolidation sites and possibly waste disposal facilities.
- Radiological sources are transported all over the world, but transportation security requirements vary between nations. The IAEA is working to try to improve and standardize the transportation security requirements, but agreements take time and implementation is likely to drag on and lack uniformity.
- User facility security improvements will come slowly on a global basis. Security upgrades at teletherapy facilities alone would require a massive undertaking. The best hope regarding source security is to forge international agreements regarding the necessary security at the various radiological source facilities and provide funding and technical assistance to foreign countries in need. The U.S. NNSA's Global Threat Reduction Initiative, other international donors, and the IAEA are taking this approach.
- Although suppliers do attempt to verify the legitimacy of the end user before completing sales and shipping of radioactive sources, those sources are often provided to countries where political stability is dubious and regulatory authorities nearly nonexistent. The IAEA can help by managing international registries of legitimate radiological source users, and it appears to be moving in that direction. The resale of sources is a worrisome gap in the system. It is not unusual to hear of source owners attempting to market a somewhat disused source into the international market as a convenient means of offloading a headache. International laws must be tightened to ensure that sources are not sold or resold to unknown parties.

Technology Options to Reduce Source Use

Although improvements in physical security, used source disposition, and the international regulatory environment can reduce the RDD risks, the root cause of the problem is the widespread use of large and dangerous radiological source materials. Each large application needs to be reevaluated to determine whether better options are available that do not put large radiological sources into circulation. Because the RDD threat is relatively new, there is likely to be significant room for improvement in many areas.

Four classes of options are of available, namely:

- Replace the application with something that presents fewer concerns regarding RDD materials.
- Replace the radioisotope utilized with something that presents a reduced concern.
- Alter the chemical and/or mechanical form to be more dispersion resistant.
- Modify the equipment to better resist theft of the device and/or the radiological source material.

Replacing Large and Dangerous Applications

As the largest radiological source applications, the large industrial sterilization units require consideration. Accelerator technology is a viable competitor within the U.S. and parts of the world where electricity is available and reliable. But it is not clear that the industrial irradiators present much of a risk in those parts of the world, since the facility already has security and it takes many hours for skilled personnel to unload/reload source materials. Perhaps the greater vulnerability is associated with the shipments of new Co-60 sources, which must occur frequently, given the 5.27-year half-life. But one huge disadvantage of attacking an industrial source or even a shipment of radioactive cobalt would be the involvement of law enforcement personnel, which would make theft of highly shielded Co-60 difficult. As a result, this very large application may be more a candidate for improved security features rather than an alternate technology as such.

The research irradiator user facilities present greater concerns regarding vulnerability due to their often poorly secured locations and the fact they use quantities of materials that are not so immediately life-threatening. Usually research facilities are most valued when they provide a range of research opportunities, so one should assume that these research irradiators are utilized to support a range of research programs. Thus, it is a little risky to offer blanket endorsements for alternative technologies. However, particle accelerators, including electron accelerators and cyclotrons, can produce a range of secondary particles. Accelerators are more costly and complex than research irradiators, but the added versatility and improved safety and security may justify the large investment.

Teletherapy units present an interesting dilemma. They have been largely replaced in the U.S. by electron accelerators, which are believed to deliver a more precise dose of radiation to the tumor site. But in less developed parts of the world, the low-tech teletherapy unit is more practical than the electron accelerators, which require both electricity and skilled technical staff. As a result, the alternate technology might not be viable in parts of the world. The gamma knife is a niche application because it is a special-purpose teletherapy device. It is not currently in widespread use, and it is not obvious that hospitals in less developed countries will pursue this special-purpose technology. It appears likely that normal teletherapy units could deliver an equivalent treatment, although the time and effort involved in bringing the beam in from 200 directions would likely force some compromises.

X-ray-based blood irradiators appear likely to replace cesium-based blood irradiators, if the hospitals currently using the cesium units could dispose of the unwanted cesium.[17] But the prospect of those hospitals trying to export the cesium sources or relegating them to disused status could increase their vulnerability to theft or misuse. It might be practical to discourage or possibly ban the sale of new cesium-based blood irradiators and begin a program to recover the partially utilized cesium sources currently in use or disuse.

Alternate technologies for RTGs are limited to devices that require special circumstances to succeed. Because of the often remote location, regular refueling or maintenance is impractical. Power devices based on solar or wind can work in some locations, but hostile climatic conditions could limit the viability of these alternatives. However, if one considers that the power requirement is often derived from the desire to run lighthouses along the north coast of Russia, then the range of options improves. Current satellite-aided navigational systems can pinpoint a ship or aircraft location within a few meters. The cost of equipping the ships that pass through such remote waters would not be insignificant, but it might be a feasible option for reducing reliance on RTGs used for maritime navigation.

An attractive alternative to the well-logging source was already being deployed when the industry changed its drilling practices and reinvigorated the use of the AmBe sources.[18] The alternative is deuterium-tritium (D-T) sources, which employ a small accelerator to drive the well-known fusion reaction to generate neutrons. The change in practice, called *logging while drilling*, involves attaching the neutron source to the drill bit and making measurements while drilling. Such a process is too stressful for the D-T sources, but the AmBe sources work well if they are big enough. It could be possible to ban the use of AmBe sources

---

[17]Brian S. Kirk, "Decommissioning and Disposal Options for Cesium-137 Blood Irradiators," *Rad Journal*, Sept. 28, 2001, www.radjournal.com/articles/Cesium/Cesiumdisposal.htm (Dec. 2002).

[18]Stephen E. Prensky, "A Survey of Recent Developments and Emerging Technology in Well Logging and Rock Characterizations," published in *The Log Analyst*, vol. 35, no. 2, p. 15–45; no. 5, p. 78–84, 1994.

and force the drilling companies to use D-T sources, but some resistance from the industry would be likely.

Using Alternate Radioisotopes

There are three types of substitutions for particular radioisotope uses that could help reduce risk. First, if the only chemical form associated with a radioisotope is bad and no substitute form is workable, it might be best to switch isotopes. Second, if an alpha emitter could be replaced by either a gamma or a beta emitter, the maximum potential inhalation or ingestion doses would decrease significantly. Third, a radioisotope with a long half-life could be a liability for centuries, long after the useful lifetime of the application. In some cases, an alternate radioisotope with a much shorter half-life could reduce the risk from sources that have fallen into disuse or disappeared.

In the case of the industrial irradiators, teletherapy applications, and blood irradiators, cobalt-60 offers three advantages over cesium-137. First, the higher-energy radiation from cobalt-60 requires about four times as much shielding mass, making it much harder to truck. Second, cobalt-60 has a market value that generates widespread recycling of the material, whereas cesium-137 is almost worthless and difficult to dispose of. Third, most large cesium sources are currently cesium-chloride, which is known to have dispersed very badly in an accident in Goiania, Brazil.[19]

Well-logging sources present a unique set of problems, and the use of several curies of any transuranic alpha emitter in a source that is transported and utilized around the world raises major concerns. Should it be impractical to substitute D-T sources for the large AmBe sources, an alternate to americium-241 should be considered. If the oil exploration industry could work with a 1 MeV monoenergetic neutron source, a couple of viable gamma emitters could be used. Were that substitution viable, the potential RDD dose impact would drop by around two orders of magnitude. If the higher-energy neutrons that result from the alpha *n* reaction are necessary, a couple of shorter-lived alpha emitters (isotopes of polonium and curium) could be substituted for the americium. The primary improvement would be a source that decays to insignificance in a decade or two, as opposed to many centuries.

Most of the newer radiography sources use irradium-192, which is not a particularly worrisome RDD source on a per-curie basis. When cesium-137 is utilized, it is usually in the form of a sealed ceramic source. There could be room for improvement in radiography sources, but this does not appear to be a high priority.

Deploying Alternate Chemical Forms

The discussion in this section is based on the experience with accidental dispersion of radiological source materials, since there have been cases where radiological sources have caused contamination problems.[20] The dispersion that could result from an RDD event would be highly scenario dependent given weather, physical environment, material form, and other factors, so it is not clear that the experience from accidental dispersions is a good indicator of what should be anticipated. It is therefore only *assumed* that sources that have behaved badly when accidentally dispersed would also behave badly for some fraction of the RDD attack scenarios and therefore constitute a concern.

The experiences regarding two of the radiological source materials are not encouraging. Cesium-chloride is a water-soluble powder that has been spread easily by accident and has caused significant cleanup problems. The AmBe sources are a fine mixture of americium-

---

[19]Alex Neifert, "Case Study: Accidental Leakage of Cesium-137 in Goiania, Brazil, in 1987," published online for the Camber Corporation: www.nbc-med.org/SiteContent/MedRef/OnlineRef/CaseStudies/csgoiania.html (Aug. 2003).

[20]Alex Neifert, "Case Study: Accidental Leakage of Cesium-137 in Goiania, Brazil, in 1987," published online for the Camber Corporation: www.nbc-med.org/SiteContent/MedRef/OnlineRef/CaseStudies/csgoiania.html (Aug. 2003).

oxide powder and beryllium powder that are blended together and compacted to optimize neutron production.

When cesium-137 is used for smaller sources, the most common form is a ceramic. Larger sources are not usually ceramic, perhaps because of poor heat conduction and other engineering factors. There exist some candidate alternate chemical forms, including cesium tetrafluoroborate, but more technical work is needed before it can be determined that these forms are good alternatives.[21]

The mixture of powdered americium-oxide and beryllium maximizes the probability that the alpha particle coming from the americium-241 would strike the beryllium and trigger the release of a neutron. The mixed powder is compacted and sealed within a capsule, but in the event the capsule should be ruptured, the potential for dispersion is evident. Because this source design was engineered before concerns about intentional dispersion developed, some reengineering might be appropriate. This could increase the cost and the amount of alpha-emitting material utilized, so some trade studies would be advisable.

Modifying Current Radiological Source Applications

For the large applications of radiological sources that involve sources and vulnerabilities that are worrisome, it would be better to eliminate the application or replace the problematic radioisotope or chemical form. However, in some cases this might not be practical and a combination of security gadgetry and materials tracking could provide the next best option.

For the large industrial sterilization units, an attack on the facility and an attempt to steal the source material would be very difficult. But with 190 such facilities in the world there is some chance of a poorly secured facility within a country where the law enforcement response would be minimal.[22] Because these facilities have such massive quantities of dangerous materials, some additional security gadgetry would be a wise investment. For example, radiation detection equipment could track the strength of the radioactive source and alert national authorities and possibly international responders if the source strength mysteriously drops by a significant fraction. The system could be designed to generate a periodic *all-is-well* signal, which then generates a red flag through either an alarm signal or a lack of any signal. Authorities could then contact the facility looking for an *all-is-well* password and an explanation and send a response team if the answers are unsatisfactory.

Whenever a large radiological source is being transported, including any mobile irradiator units, alert and track hardware should be built into the vehicle. If the vehicle departs from its planned itinerary, a timely response from law enforcement personnel could reduce the chances of theft.

Any new RTGs being deployed should also be designed to use part of the power supply to generate a couple of redundant *all-is-well* signals on a regular basis. If the power supply were removed, the signals would stop. If the entire RTG were to be moved with the power supply in place, the signals would register a changing global position report, alerting on the problem and providing a track beacon. An interrupted signal would also be treated as indication of possible theft and prompt a response from law-enforcement.

The large hospitals devices, particularly the teletherapy and blood irradiator units, should be provided better protection, regardless of the radioisotope in use. For teletherapy units, access to the source itself should require special tools and procedures. Attempts at unauthorized access should trigger alarms inside and outside the hospital. For blood irradiators, the fact the sources are welded in, combined with the bulk and mass of the units, will deter theft to some degree. It also provides an opportunity to encase some alert and track devices so that authorities can quickly find a stolen blood irradiator.

The mobility of well-logging sources and the dangers they pose are such that each unit should be rigged with alarm and tracking equipment. The process of removing the AmBe

---

[21]C. Mason, J. Conca, G. Van Tuyle, "An Alternative Matrix for Reducing the Threat of Radioactive Dispersal from 137Cs Sources," Los Alamos National Laboratory Report LA-UR-03-0048, Jan. 2003.

source from the *sonde* (an array of instruments attached to the drilling pipe in an exploratory shaft that can provide information on the geology and probability of finding oil) should be difficult so that the entire unit is more likely to be transported rather than just the source.

Most radiography sources are not large enough to require special gadgetry for tracking source materials and are not well suited for such an approach anyway. It is possible that a very large radiography source that is moved around on trucks might be suitable for alert and track devices, depending on the potential hazard posed by the source.

Materials tracking is a technology that could be transferred from NNSA's Materials, Control, Protection, and Accountability (MPC&A) programs used for special nuclear materials (nuclear weapons materials). Such technology should be applied selectively to high-priority items, such as large cesium and cobalt sources. Regarding the cesium sources in particular, MPC&A programs could help reduce the problems with disused and orphan sources.

Prioritizing Alternate Technology Options

Although the large industrial sterilization facilities utilize large amounts of radioactive cobalt-60, most of these facilities do not appear to be very vulnerable to theft. The primary concern is for facilities in countries of concern or cases where a facility is not properly secured, and these circumstances are not believed to be common. The alternate technology, based on particle accelerators, requires an infrastructure of expertise and electric power that could be unavailable in countries of concern. It is very possible that the accelerators will gain a competitive edge from the RDD concerns, so the number of cobalt-60 irradiators might be on a slow-growth pattern anyway. On the other hand, the step of wiring the sterilization facilities so that an ongoing attempt to steal a source becomes obvious to law enforcement would be prudent. A team of experts could conceivably steal the cobalt-60 source material, given enough time and some laxity of security, so provision of systems that would deny them the time they need could be a worthwhile investment.

Research irradiators are a concern because of the most common research environment, which is low security. Particle accelerators could provide the same capabilities but with much greater flexibility. The additional costs could be a concern, however, in most countries. If research irradiators are used, security features should be built in to compensate for the concerns.

The RTGs present some special problems, as there are few viable alternatives. The greatest weakness and the greatest advantage about these devices is their remote location. It would be hard to monitor these devices and hard to respond even if someone were tampering with an RTG. But it is also difficult for someone to travel to a remote location and transport these devices to a different location. The RTGs can and should be redesigned to make source removal very difficult, to make it obvious when the device is being tampered with, and to facilitate tracking and recovery of stolen units. A much more sweeping change could be the best approach. Ships equipped with GPS technology should not require lighthouses, and without lighthouses, the need for most RTGs would be eliminated.

The first priority on teletherapy units is to get rid of any remaining cesium units. With respect to the cobalt units, the need to frequently replenish the source strength raises a concern about potential source theft. When a source supplier visits the hospital to replace the cobalt source(s), special tools are required to access the chamber, providing a measure of theft resistance. Although this is a good start, the security features of this system would have been developed prior to the days of RDD threats and need to be reevaluated and upgraded.

The blood irradiators pose an interesting dilemma. The X-ray units appear to provide a viable and even attractive alternative, except for one big problem. If an X-ray blood irradiator is used to replace a cesium-based unit, the disposal of the cesium source becomes a major liability for the hospital. Thus, although the deployment of X-ray blood irradiators into facilities first acquiring the capability is helpful, the replacement of existing cesium-based units could create a new problem of disused sources, orphan sources, or sources being resold outside the U.S. The two options for dealing with this problem involve either providing disposal facilities for large cesium-chloride sources (easier said than done) or possibly recycling the cesium sources into a better chemical and/or mechanical source form.

The situation regarding well-logging sources is complex and requires interactions with representatives of the oil-exploration industry. Several options are available, and there's much room for improvement. It is apparent that the D-T sources can provide superior analysis of the geology around the bore-hole, if only they could withstand the hostile drilling conditions. The logging-while-drilling approach is relatively new and was developed to save time and money. If large AmBe sources were unavailable, the industry might well go back to using the D-T sources. If the industry insists on using so-called *chemical sources* (its jargon for AmBe sources and the equivalent), development of sources based on polonium or curium isotopes could greatly reduce the source lifetime, and there might even be the option of using a gamma-driven neutron source (if 1 MeV neutrons would suffice). It is clear that gamma-driven sources would reduce the RDD concerns, but the use of shorter-lived isotopes would reduce the liability, although much less significantly.

If these alternatives do not prove viable, the AmBe sources are candidates for reengineering. Many alternate design options will likely work, although most will be a bit less efficient and require somewhat more americium. Lastly, the well-logging source *sondes* could be fitted with alert and track hardware as something of a last resort. The gamma source in well-logging *sondes* is generally cesium-137, although a number of other radioisotopes could do the job as well. Most of these cesium sources are ceramic, so the value of deploying alternate source materials might not be very high in this case.

Many radiography units use short-lived iridium-192 in quantities that are insufficient to qualify for urgent attention to reduce RDD concerns. Where cesium sources are used, they are usually ceramic sources. If there are known instances where cesium-chloride is used, these should be considered candidates for replacement.

Summary

Radiological sources that are in common use around the world are vulnerable to theft, and some could be used to create very dangerous RDDs. Fortunately, the number of sources that could make a devastating weapon is a very small fraction of the total and is probably in the low 1,000s. The number of radiological source manufacturers is also very low compared to the number of end users; therefore, these manufacturers represent the point of origin for tracking large radiological sources throughout their life cycle. Unfortunately, the heightened concern over RDDs as a potential terror weapon is recent, and many dangerous sources have previously been manufactured and distributed throughout the world. The location and disposition of some of these sources are unknown, and many are known to be in use without adequate security systems to prevent their theft or misuse.

25 Responding to Radiological Threats

Leroy E. Leonard

Introduction

The terrorist attacks on the U.S. on September 11, 2001, fundamentally changed the way we view potentially hazardous materials in our environment. In particular, great concern has been expressed regarding the potential threat posed by radioactive material[1] in the environment. What if an individual or group acquired radioactive materials, not with the intention of making a weapon to cause a nuclear explosion but with the sole intention of creating a *radiological weapon*?[2] Such a device is commonly referred to as a *dirty bomb*. The potential impact of a radiological weapon, a radiological dispersion device (RDD), or a radiological emission device (RED) cannot be ignored.

We know there are individuals and groups bent on causing mass *destruction* or mass *disruption* within organized society, and we know they have considered the use of radioactive materials for attacks. We know also that there are large numbers of radiological sources, many containing very high levels of radioactivity, distributed in almost every country in the world. Together, the potential hazards to human health, the motivated potential nuclear terrorists, and the availability of suitable material create the radiological threat.

In this chapter we consider radiological materials in our environment, why we have them, and what kind of risk they present, both past and present. We also examine why we find ourselves in a position to be threatened by these materials. We discuss how we are addressing radiological threats today from the U.S. perspective and what will be needed over the next several years to reduce the threat. Finally, we examine how societal behavior must change to mitigate this threat in the longer term.

The Evolution of a Threat

Radioactivity was discovered by H. A. Becquerel in 1896; less than 20 years later, G. von Hevesy introduced the isotope polonium-210 (^{210}Po), separated from natural radium (^{226}Ra)

[1]The term *radioactive material* as used in this chapter means matter with unstable nuclei that spontaneously disintegrate, giving off radiant energy in the form of gamma rays and/or alpha and beta particles.

[2]The term *radiological* as used in this chapter is an adjective derived from the word *radiology*, which means the study of radiation. *Radiological material* is therefore radioactive material used in research, medical diagnosis or treatment, or industrial applications in which radiation emitted from the material performs a beneficial use.

FIGURE 25.1 Radium-226 cancer therapy kit containing a lead shield to hold radium needles, circa 1920 to 1940. (Photo from Oak Ridge Associated Universities Website Museum.)

as a radioactive tracer for biological processes in plants.[3] In the 1920s, long before scientists understood the potential of fissile material for nuclear weapons or nuclear reactors, scientists, engineers, and medical physicists were applying the ionizing effects of naturally occurring radium on biological tissue to treat various forms of cancer in humans (see Figure 25.1). It is the ionizing effect (the ability of radiation to alter the molecular structure of biological cells) that makes radiological sources both useful and dangerous.

The destructive effects of radiation on biological cells was understood early in the nuclear age, and the study of the effects of ionizing radiation on both plant and animal cellular tissue proceeded in parallel with the application of these new *radiological materials* to beneficial uses. In 1921 the British X-Ray and Radiation Protection Committee adopted dose standards for external radiation.[4] In this period the risk to public health and safety from these radiological materials was limited by the small quantities in existence. In 1939 the total amount of ^{226}Ra available worldwide was believed to be on the order of 100 curies.[5] But this was to change.

By the end of the Second World War the U.S. had accumulated millions of curies of new manmade isotopes as the residual byproducts of nuclear weapons production reactors. Most of this material was destined to be dealt with by the federal government as *high-level radioactive waste,* but some of these isotopes were separated and encapsulated into sealed radiological sources.[6] In the sealed form, their valuable radiological properties could be exploited for

---

[3] *Radioisotope Engineering*, Geoffrey G. Eichholz, ed., School of Nuclear Engineering, Georgia Institute of Technology, Marcel Dekker, Inc., New York, 1972 (ISBN: 0-8247-1156-4).

[4] *A Century of X-rays and Radioactivity in Medicine*, by Richard F. Mould, Institute of Physics Publishing, Techno House, Redcliffe Way, Bristol, BS16NX, U.K., 1993 (ISBN 0-7503-0224).

[5] Ibid.

[6] Radioactive material either fixed to a substrate or contained in some form of sealed encapsulation to prevent the escape or dispersion of the radioactive contents and permitting the encapsulated material to be safely used as a source of ionizing radiation for an appropriate beneficial application. Typically, sealed sources containing significant quantities of radioactive material are doubly encapsulated in welded stainless steel or similar durable metal.

beneficial uses. Simultaneously, scientists were learning that a myriad of additional manmade radioactive isotopes could be created in the early production reactors by neutron bombardment, yielding materials with specific radiological properties. By 1950, literally hundreds of new radiological applications for industry, medicine, and research were being found for these new materials; at the same time, the study of *health physics* and radiological health was making enormous advances in understanding the biological effects of radiation using the empirical results from the victims of the atomic bomb detonations over Nagasaki and Hiroshima.[7]

By 1954 the U.S. Congress had passed the Atomic Energy Act, which created a legal framework for the distribution, control, and regulation of radiological materials and sources to assure that public health was protected from the hazardous effects of these materials in the United States. By that time, however, radiological materials had been distributed worldwide as part of the promotion of peaceful uses of atomic energy and science. The U.S. and its industrialized allies were supporting the production, distribution, and use of radiological sources throughout the West, and the Soviet Union was doing the same within its sphere of influence.

The users of radiological sources throughout the world number in the tens of thousands and often have very limited economic and physical resources to safely store excess and unwanted radioactive material. By the 1990s, this problem of no access to disposal for radiological material was becoming a worldwide concern. In 1991, the IAEA published a technical document that highlighted the concern and concluded that "... the global situation as regards the management of spent radiation sources is unsatisfactory."[8] This situation further deteriorated with the collapse of the Soviet Union, which suddenly ended the central regulation and management of all radiological sources under its traditional sphere of control. The radiological threat has evolved over many decades as excess and unwanted radiological material and sources have accumulated in our environment, with no comprehensive plan to remove them. Today we understand that security of these materials is a critical factor that must be addressed to assure that they are not diverted to malicious use.

Evaluating the Threat

In a post-9/11 world, how can we effectively evaluate the radiological threat? Can we place it in its proper context of other threats that might be presented by other hazardous materials? If the threat from radiological materials is from their configuration into a credible weapon, the following conditions must exist for the threat to manifest:

- *Motivation.* There must be individuals or groups motivated to create a weapon. In this chapter we assume that such individuals or groups exist and that they have sufficient technical competency to design and configure a weapon if they can access the radiological materials.
- *Availability of material.* There must be sufficient radiological materials available. The materials must be reasonably accessible to the motivated individuals or groups via theft or other means, illicit or otherwise.
- *Technical feasibility.* There must be a sound technical basis to suggest that an RDD or RED can be configured from radiological material where the radiological material will enhance the effectiveness of the weapon. This topic is discussed later in greater detail.

Eliminating any one of these factors effectively eliminates the threat completely. Modifying any of the factors could have a profound effect on the potential threat. However,

[7]The study of the beneficial use of ionizing radiation while protecting workers and the public from potential hazards.

[8]*The Nature and Magnitude of the Problem of Spent Radiation Sources*, IAEA-TECDOC-620, IAEA Vienna, Austria, Sept. 1991 (ISSN 1011-4289).

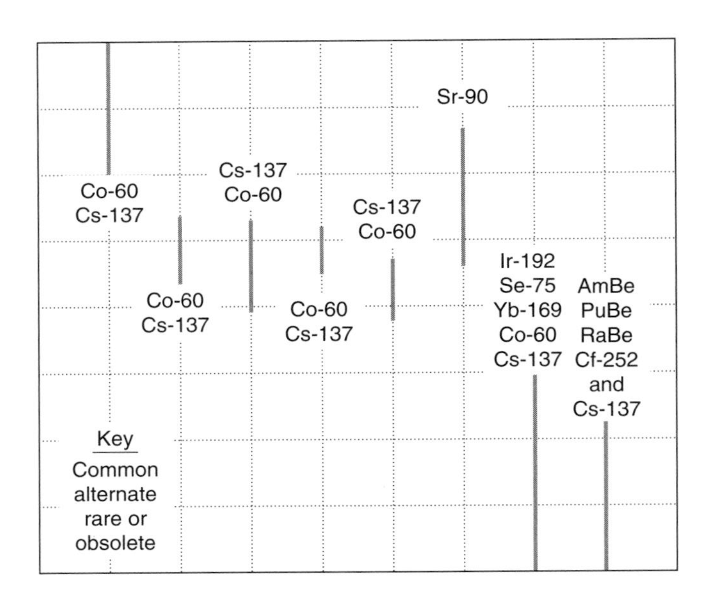

FIGURE 25.2 Large radiological source applications.

at this point we cannot expect to have much effect on the first factor. So, for the purposes of this chapter, let us consider the latter two factors in greater detail.

Availability of Material

We have discussed this factor qualitatively already; here we focus on the availability of material in a more quantitative way. Figure 25.2 is a graphic that shows the most common isotopes used in sealed-source radiological devices and their respective ranges of radioactivity for the various applications.[9]

Consider the size range of radioactivity described in the graphic for individual radiological sources and devices. In 1939 the total amount of radiological material available for use was around 100 curies. Around 1946, manmade isotopes began to be distributed, adding to the availability of radiological materials. These isotopes have continued to be distributed for 60 years. In Table 25.1, we estimate total quantities of the longer-lived radioactive material distributed under U.S. government programs for innovative and beneficial uses in medicine, industry, and research by the Atomic Energy Commission and later, the U.S. Department of Energy.

Manufacturers (some in government laboratories, some in private enterprises) purchased these isotopes as bulk material in various chemical forms and fabricated sealed radioactive sources in the physical sizes and activity levels necessary to meet the requirements of the many applications addressed. Figure 25.3 provides a graphic representation of the life cycle of radiological sources.

During the early phases of the life cycle, the security is good. Isotopes are usually produced in large government facilities and are transported under well-regulated processes to a relatively few manufacturers. These manufacturing facilities are also licensed and well regulated in the U.S. by state or federal agencies. These manufacturers are also limited in number. However, as sources are then distributed either directly to users or to other manufacturers

[9]Taken from Gregory J. Van Tuyle, Tiffany L. Strub, Harold A. O'Brien, Caroline F. V. Mason, and Steven J. Gitomer, *Reducing RDD Concerns Related to Large Radiological Source Applications*, LA-UR-03-6664, Los Alamos National Laboratory, Sept. 2003, used by permission of Gregory Van Tuyle.

Table 25.1 Primary long-lived isotopes distributed by U.S. government programs for the manufacture of radiological sources, 1946–present.

*Isotope (by isotopic mass, in kg)/ half-life	Pu-239/ 24,100 yrs.	Pu-238/ 87.8 yrs.	Am-241/ 433 yrs.	Cs-137/ 30 yrs.	Sr-90/ 28.6 yrs.
Total amount distributed	109.2	137	45	80	52
Total amount estimated to be recovered and secured, disposed of, or lost by isotopic decay	81.7	132	15	47	46
Amount estimated to be remaining in radiological sources and devices at unspecified locations throughout the world	27.5	5	30	33	6

*The quantities shown in this table represent best estimates based on a large number of historical references and ongoing operations. The data in these references are not always consistent and are limited in both accuracy and precision. These estimated quantities are intended to frame the issue of radiological material in the environment. (Source: Compiled by the Off-Site Source Recovery Project, Los Alamos National Laboratory, with valuable help and assistance of the Isotope Sales Program, Oak Ridge National Laboratory.)

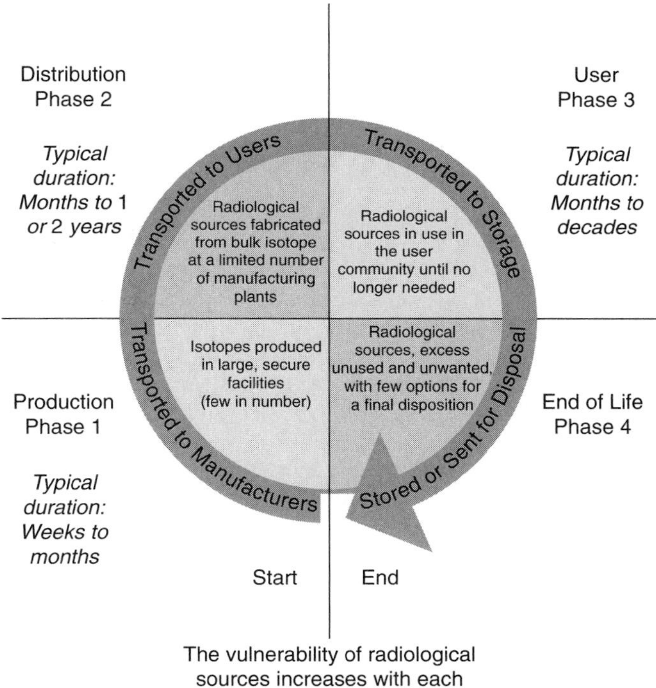

FIGURE 25.3 Life cycle of a radiological source.

who place the sources into a device or system that is then sold to user-customers, the security begins to erode. Although this third phase of the life cycle can be safe, secure, and well regulated, we have now tremendously increased the number of locations where radiological sources can be found.

Within the U.S., the regulatory authority for radiological materials is the U.S. Nuclear Regulatory Commission (NRC), along with the regulatory agencies of 33 states that provide their own intrastate regulation under agreement with the NRC. Together the states and the NRC administer over 20,000 individual licenses for the possession and use of radiological sources.[10] Depending on the specific nature of the use and the terms of the license, each licensee could possess one or perhaps hundreds of individual radiological sources of various activity levels. Additionally, sources can remain in use for long periods of time, making accurate recordkeeping more difficult. Although federal and state regulation of radiological material within the U.S. has been comparatively successful, it has its historical limitations when viewed from a post-9/11 perspective:

- With only a few exceptions, licensing requirements to possess and use radiological materials address safety of the material only.[11] Security from malicious theft or protection from acquisition by a determined adversary has not traditionally been part of the licensing regimen. This is beginning to change, but slowly.
- Radiological materials licenses typically permit licensees to possess a maximum activity of one or more isotopes. The specific number of sources that a licensee may possess or the specific nomenclature of individual sources such as model, serial number, and activity in the possession of a licensee has historically not been tracked on the license. Therefore, the total number of radiological sources under regulation in the U.S. today, their sizes, and their locations are not known.

Only a fraction of the 20,000 or so U.S. licensees (perhaps <10%) possess individually large sources that would fall into Category 1 or 2 (potentially most dangerous sources) in the IAEA categorization, but a significant fraction of licensees possess many smaller individual sources that, if aggregated, could fall within the Category 1 or 2 range.[12] Remember, unlike fissionable nuclear materials that can be found in quantities of concern at only a *few hundred* known locations worldwide, radiological materials (which are only rarely fissionable) in quantities of RDD concern are located at *many thousands* of locations in the U.S. alone and *tens of thousands* of locations worldwide. Furthermore, as discussed previously, the total number of radiological sources is not known.

Two additional important factors complicate the issue of the worldwide availability of radiological materials for malicious use:

- Over the several decades during which radiological materials have been in common use, regulatory infrastructures in many countries where materials have been distributed and routinely used have failed to evolve into effective institutions of control. Or, where viable regulation once prevailed, economic or political disruption has destroyed the ability of governments to maintain effective regulation of radiological sources. This was mentioned previously regarding the former Soviet Union.
- Governments and private institutions, both within the U.S. and internationally, have failed to provide for convenient and inexpensive disposal options for all but the lowest levels of radioactive waste. This means that when radiological sources and devices reach the end of their service life or for whatever reason become excess and unwanted by their owners, there is no place for most of them to go. The result has been that essentially all the larger and most dangerous radiological sources of RDD concern that were distributed from the 1950s through the 1990s have not been removed from the environment. Most remain excess, unwanted, and a burden to their owners.

---

[10] *2005–2006 Information Digest*, U.S. Nuclear Regulatory Commission, NUREG-1350 Vol. 17, Washington, D.C., 20555-0001, July 2005.

[11] In the relatively rare instances in which the NRC has licensed radiological sources containing special nuclear material (weapons-usable uranium or plutonium), the regulatory requirements ensure that the materials must conform to internationally accepted norms of safeguards and physical security for nuclear materials.

[12] Ibid.

These two factors, sometimes working together, sometimes independently, have created a large population of radiological orphan sources across the globe. In the U.S. on September 11, 2001, thousands of radiological sources were documented as excess and unwanted, leaving their owners burdened with a responsibility from which there was no legal exit. This situation has since been largely resolved by federal programs that will be discussed in the following sections, but no new disposal sites have opened since 9/11 and each week new radiological orphans are registered within the U.S.[13]

When potentially dangerous radiological materials become excess and unwanted with no simple disposal option, the material generally becomes at risk. Maintenance of unwanted radioactive material can be a costly and burdensome task, and experience in the field has shown that although safety of material in long-term storage is often maintained, security from theft is not.

Many governments around the world have become acutely aware of the radiological source problem and the risk of material availability. The IAEA is well aware of these problems and has published a Code of Conduct on the Safety and Security of Radioactive Sources.[14] The Code is a general description of the requirements needed for member states to properly control both the safety and security of radioactive sources. Over 80 member states of the IAEA have adopted the Code, but it is not binding. It does not define a uniform set of practices that could be considered a safeguards infrastructure, nor does it provide for any IAEA oversight or enforcement. The Code is a positive first step, but unless significantly upgraded and implemented, it will not significantly reduce the availability or improve the security of radiological materials.

In 2003 the IAEA issued a revised technical document categorizing the radioactive sources shown in Figure 25.1 and many other source types into five levels according to their potential danger to health and safety.[15] Category 1 represents the most dangerous and Category 5 the least dangerous. This document, however, considered "potential for radioactive sources to cause deterministic health effects ... comprised partly by the physical characteristics of the source and partly in the way the source is used." It did not specifically address the danger from intentional malicious acts involving such sources—in other words, their purposeful fabrication into radiological weapons. In short, there exist abundant quantities of potentially dangerous materials that are available to a determined adversary both in the U.S. and elsewhere in the world. Certainly the orphans and the unaccounted-for materials present the greatest risk, with the well-regulated and controlled radiological sources in the early phases of their life cycle presenting the lesser risk.

Technical Feasibility

If we concede that radiological material is available to the determined or motivated adversary, is there a technical basis to suggest that a radiological weapon could be effective? Research is being conducted to determine the levels of the most severe dispersal for various isotopes and the methods necessary to achieve such dispersals.[16] For example, Sandia and Los Alamos National Laboratories are using numerical models to quantitatively determine

---

[13]Information compiled by the Off-Site Source Recovery Project at Los Alamos National Laboratory on behalf of the U.S. National Nuclear Security Administration, Office of Global Threat Reduction.

[14]Code of Conduct on the Safety and Security of Radioactive Sources, IAEA/CODE/OC/2004, IAEA, Vienna 2004.

[15]Categorization of Radioactive Sources: Revision of IAEA-TECDOC-1191, Categorization of Radiation Sources, IAEA, Vienna, 2003, IAEA-TECDOC-1344.

[16]Frederick T. Harper and Stephen V. Musolino, "Emergency Response Guidance for the First 48 Hours After the Outdoor Detonation of an Explosive Radiological Dispersal Device," *Health Physics*, vol. 90(4), pp. 377–385, April 2006.

Table 25.2 Radioactive source incidents related to orphan sources.

Site, Source, Year (Deaths/Injuries)	Device
PRC, Sanlian orphaned source, 1963 (2/4)	Co-60 seed irradiator
Ciudad Juarez, Mexico orphaned source, 1983 (1/4)	Co-60 teletherapy
Goiania, Brazil orphaned source dispersal, 1987 (5/20)	Cs-137 teletherapy
Jilin, PRC orphaned source, 1992 (3/5)	Co-60 radiography
Tammiku, Estonia stolen source, 1994 (1/4)	Cs-137 teletherapy
Lilo, Georgia orphaned sources, 1996 (0/11)	Assorted
Istanbul, Turkey orphaned sources, 1998–1999 (0/10)	Co-60 teletherapy
Kingisepp, Russia orphaned source, 1999 (3/0)	Sr-90 RTG
Samut Prakarn, Thailand orphaned source, 2000 (3/7)	Co-60 teletherapy
Kandalaksha, Russia orphaned source, 2001 (0/4)	Sr-90 RTG
Liya, Georgia orphaned sources, 2001–2002 (0/3)	Sr-90 RTG
Kola Harbor, Russia orphaned sources, 2003 (0/1)	Sr-90 RTG

the total potential impact of an RDD using various weapon scenarios. The results of this research will continue to improve our understanding of the economic and societal disruption that a radiological dispersion device might produce. The general consensus is that although a severe radiological attack would produce low numbers of human fatalities (fewer than 100), it could potentially cause economic damage up to $100 billion.[17]

For those who want to look deeper into the specifics of a particular event, open-source data on radiological incidents and related events worldwide, including acts of nuclear and radiological terrorism with references and source material, are available. There have been several hundred incidents of criminal acts involving radiological material.[18] Table 25.2 provides a selected list of such incidents. As of August 2007 the accidental events have resulted in 112 deaths and 482 injuries. The criminal acts produced four deaths and one injury.[19] All these events involved radioactive sealed sources, sources that were no longer wanted or needed by their owners (commonly referred to as *orphan sources*) and ended up causing inadvertent injury or death.

From Table 25.2, perhaps the incident that most resembles the potential consequences of an RRD attack is the accident that occurred in Goiania, Brazil. This incident, which has

[17]B. A. Boughton and J. M. DeLaurentis, "Description and Validation of ERAD: An Atmospheric Dispersion Model for High Explosive Detonations," SAND92-2069. Also see H. Rosoff and D. von Winterfeldt, "A Risk and Economic Analysis of Dirty Bomb Attacks on the Ports of Los Angeles and Long Beach," *Risk Analysis*, vol. 27, no. 3, 2007, www.blackwell-synergy.com/doi/pdf/10.1111/j.1539-6924.2007.00908.x?cookieSet=1 (Aug. 2007).

[18]Charles Streeper, Marcie Lombardi, and Dr. Lee Cantrell, "Nefarious Uses of Radioactive Materials," Los Alamos National Security and California Poison Control System, San Diego Division, LA-UR 073686, presented at the Annual Meeting of the Institute of Nuclear Materials Management, July 2007.

[19]Database prepared by Wm. Robert Johnston and published on the Internet at www.johnstonsarchive.net/nuclear/radevents/. Also see "Illicit Trafficking and Unauthorized Activities Involving Nuclear and Radioactive Materials." *IAEA Fact Sheet*, 2005 (May 11, 2007), www.iaea.org/NewsCenter/News/2006/traffickingstats2005.html (Aug. 2007).

FIGURE 25.4 Cesium-137 contamination cleanup in Goiania, Brazil, 1987. (IAEA photo.)

been thoroughly evaluated by the IAEA,[20] was initiated when a group of scavengers looking for scrap metal stole and destroyed a teletherapy irradiator that held a Cs-137 source with approximately 1,200 curies of activity. During the dismantling of the device, the source was ruptured and the Cs dispersed. Over a short period of time this incident resulted in five deaths and 20 hospitalized individuals. The incident proved devastating for the economy of this city of 1 million and for the region. It resulted in the monitoring of 112,000 persons for radioactive contamination, the contamination of 101 houses, and the necessity to demolish six houses, as Figure 25.4 shows.

Goiania is an example of dispersing a relatively large radiological source. Small-scale radiological contamination can also have significant consequences. For example, small Cs-137 sources from radiological gauges have been accidentally melted with scrap metal. Without any injury to workers and within a closed industrial plant, the cost to the U.S. steel industry for a plant to recover from such an accidental melting of a Cs-137 sources of <1 curie has ranged from US$3 million to US$20 million.[21]

As shown previously, inadvertent dispersion of radioactive material from common radiological sources has shown it can have serious consequences on health, the environment, the economy, and the social and psychological health of a community. The intentional dispersion from an RDD is definitely technically feasible. At this time we can only speculate what impact and cost an intentional dispersion might have if it was carefully planned and executed in a location of maximum sensitivity—say, the National Mall in Washington, D.C., or the financial district in New York City. Perhaps in the future we can become more quantitative in our understanding.

Risk Assessment

Any discussion of threat would be incomplete without some discussion of the risk that relates to the threat. If there is a threat of rain and you leave the house without an umbrella, you run the risk of getting wet. This risk might be quite acceptable, so you have intuitively assessed the risk and it is acceptable. However, if there is a threat of a hurricane and you

[20]"The Radiological Accident in Goiania," International Atomic Energy Agency, Vienna, 1988.

[21]Ibid.

choose not to leave the area, the risk could turn out to be something that you would not willingly accept. Classically, risk is defined as:

$$R = PC$$

where:

R is the risk

P is the probability that an event will occur

C is the quantified consequence of the event

With respect to an exchange of nuclear weapons against cities, C, the consequences, have always been considered unacceptable when we assess the risk. The lives and property lost in even a small nuclear explosion are not acceptable under any circumstance. Therefore, C is made equal to 1, or unity. Risk, or R, for the nuclear weapons scenario, can only be reduced by lowering P, the probability of occurrence. This is what classic nuclear nonproliferation and international safeguards is focused on: lowering the probability of a nuclear event.

In considering radiological threat, in some instances, as with the threat of rain, a particular risk scenario could be acceptable. To use an extreme example, if someone attempted to disperse the Am-241 from the radioactive source found in a common household smoke detector in a public park (source activity 1.0E-6 curies), the resulting contamination would be very small. Such a scenario would pose no significant risk to human health or the environment. The consequence, the C, would be insignificant. In this case there would be no need to worry about the probability of occurrence.

On the other hand, if the dispersal was to involve six radiological sources such as those used in the oil industry as well-logging sources, which also contain Am-241 (16 curies each), 96 million times more radioactivity than is present in the smoke detector and a Category 2 Quantity under the IAEA Code of Conduct, and if the dispersal were to occur above the National Mall in Washington, D.C., on the Fourth of July during the annual fireworks display, the C, the consequence, would be quite different. In this case I believe we can agree that there would be ample concern to reduce wherever possible the P, or probability of such an occurrence, thus reducing the R, risk. With the large number of sites from which radiological materials could be acquired and the wide variety in the level of security placed on these materials, the P for the risk equation is extremely complex when we are trying to analyze the risk of radiological weapons.

Reflection on the Radiological Threat

What results might terrorists seek to achieve with an RRD attack? The answer is that we just don't know at this time. History has shown it is difficult to predict the exact nature of such attacks. However, U.S. Senator Richard Lugar of Indiana conducted a survey of national experts in issues of nonproliferation, including scholars, policy makers, diplomats, and technicians. Of 85 respondents who were asked to estimate the risk of radiological attack over the next five years, on average they believed the risk to be 27.1%, with the median at 25%. When asked to extend their horizon to 10 years, they increased their risk assessment to 40% for both average and median.[22] This probability estimate was twice as high as the median for a nuclear or biological attack over the same period.

With respect to the primary threat factors for radiological attack, it would appear that the three factors—motivated people, availability of material, and technical feasibility—all suggest that a radiological threat exists. Our national experts seem to think that there is a significant probability (40%) to expect an attack in the next 10 years. At this writing no

---

[22]Richard Lugar, *The Lugar Survey on Proliferation Threats and Responses*, 306 Hart Senate Office Building, Washington, D.C. 20510, June 2005, or http://ligar.senste.gov.

such attack has occurred. This could change, but if our experts are correct probabilistically speaking, there could be some time to reduce the threat.

Reducing the Threat

Of the factors that make up the radiological threat, the only one we currently have an opportunity to affect is availability of material. If adversaries have already acquired radiological material, we might have a chance to detect the material as it is moved from place to place before it is deployed as a weapon, but in general we know that if we can first keep such material out of the hands of terrorists, we have eliminated the threat. If radiological materials play an important and beneficial role in society and at the same time present a threat, we can in reality only reduce, not eliminate, the threat.

It is true that some sources present a greater risk than others, and those of highest potential risk should be addressed first. In general, options for reducing the availability of radiological materials include:

- *Convert.* Find alternative technologies that replace societal dependence on radiological material.
- *Protect.* Where radiological materials remain useful and in use, assure a strict degree of control and oversight through effective regulation, which equally stresses safety and security through a safeguards regimen. If materials are found in excess and unwanted, protect them as an interim measure.
- *Remove.* Where radiological materials are found as a legacy, excess, unwanted, or at risk because of inadequate control or lacking in security, the only assurance of achieving threat reduction is to recover and remove the material to secure storage or disposal.
- *Interdict.* Finally, if radiological materials are acquired for the purpose of creating a weapon, we can attempt to detect the movement of the material at transportation portals and interdict it before the weapon can be fabricated or deployed.

Since 2003 a U.S. federal program implemented by the National Nuclear Security Administration (NNSA) has been applying this strategy both within the U.S. and abroad. The Global Threat Reduction Initiative (GTRI) has, and continues to have, involvement in Convert, Protect, and Remove for both nuclear and radiological materials. Other NNSA programs deploy the technology and equipment to detect and interdict illicit movements of radiological as well as nuclear materials worldwide in cooperation with other nations and the IAEA. The U.S. Department of Homeland Security also deploys detection equipment on the U.S. borders and elsewhere within the country. In considering the life cycle of all long-lived radiological materials, it is important to note that both conversion and protection-in-place will sooner or later require a removal capability. It is only a question of when removal must be applied to achieve threat reduction for radiological materials.

There are two distinct parts to the world's stockpile of radiological materials:

- There is the excess and unwanted legacy fraction that has built up in the environment for over 60 years, with few options for end-of-life disposition.
- There is the in-use fraction that continues to be used in radiation technology industries for beneficial purposes, many of which have no nonradiological substitute.

Management scenarios for these two categories must be different to effect desired threat reduction.

Recovery and Removal

In the U.S., GTRI and its predecessors have focused the bulk of their radiological material threat-reduction resources on locating at-risk material and recovering it, removing it to a

FIGURE 25.5 Two excess 16-curie americium-241 well-logging neutron generators are placed in a shielded container for recovery and removal to Los Alamos National Laboratory under NNSA/GTRI Programs. (Photo supplied by Los Alamos National Laboratory Off-Site Source Recovery Project.)

FIGURE 25.6 A truck containing americium-241 and cesium-137 radiological sources, found abandoned by the licensee in Illinois, 1997. (Photo supplied by Los Alamos National Laboratory Off-Site Source Recovery Project.)

secure location, and, where possible, permanently isolating the material from the environment by geological disposal. This approach has proved effective for the excess and unwanted legacy fraction that has accumulated in the U.S. and around the world over decades.

From 1997 to the end of federal fiscal year 2006, nearly 14,000 excess and unwanted radiological sources (see Figures 25.5–25.8) have been recovered and removed from harm's way, primarily to Los Alamos National Laboratory in New Mexico under the GTRI and predecessor federal programs.[23] Additionally, pathways to permanent disposal in appropriate

[23]Recovery data provided by Shelby J. Leonard, Team Leader, Off-Site Source Recovery Program (OSRP), Los Alamos National Laboratory, Los Alamos, New Mexico, Sept. 2006. See also http://osrp.lanl.gov.

FIGURE 25.7 The rounded unit in the center of the photo is an RTG containing approximately 15,000 curies of strontium-90 after 30 years of decay. This unit is being buried at the DOE low-level waste disposal facility at the Nevada Test Site. (Photo supplied by Los Alamos National Laboratory Off-Site Source Recovery Project.)

FIGURE 25.8 As part of the NNSA's GTRI Program, an excessive and unwanted research irradiator containing approximately 200 curies of Cesium-137 is removed from a high school in San Antonio, Texas, 2005. (Photo supplied by Los Alamos National Laboratory Off-Site Source Recovery Project.)

repositories that did not exist on 9/11 are now available for the greatest number of these sources.

Removal of these at-risk materials has also served to reduce the burden on domestic regulatory agencies at the federal and state levels by continually reducing the amount of radiological material held under license that currently serves no useful purpose.

As this approach reduces the domestic threat over time, the U.S. government has established a basic removal infrastructure that can be transferred to the international community. GTRI recovery and removal efforts are beginning to impact the international legacy fraction as GTRI begins the process of repatriating radiological sources of U.S. origin from other countries, both bilaterally and in cooperation with IAEA.

The recovery and removal of sources internationally must be carefully planned and implemented. In many cases custodians of potentially dangerous radiological materials have gone years or decades with no viable options to safely, securely, or cost-effectively rid themselves of the material. They have done their best to maintain the safety of the sources with available resources. The relative anonymity of these custodians and their sources has offered a crude degree of security. This is especially true in developing countries. In such cases removal activities that draw attention to the existence of the sources should be conducted only when the removal infrastructure is effectively functioning and the disposal pathway clearly defined. When these conditions are in place and timely and safe removal can be assured, the international stockpile of excess radiological material will begin to shrink significantly because custodians of the material will have confidence that removal is in their best interests.

It is difficult to assess how much threat reduction has been achieved by source protection and removal activities to date. However, the total scope of the radiological material problem appears to be manageable. As shown in the lower row of radiological sources of concern presented in Table 25.1, it is clear that there is a finite quantity of each long-lived isotope distributed by the U.S. and currently existing somewhere in the world. If this material is removed from the environment and finally disposed of, a majority of the most dangerous material will have been adequately secured from theft or misuse. These are shown in Table 25.3.

In addition, current nuclear waste disposal practices followed within the United States can be utilized for disposal of the residual source quantities from Table 25.2. This waste disposal is robust. For example, if every curie of long-lived isotope distributed by the U.S. DOE and its predecessors that still exists somewhere in the world was removed from its owner and recovered to a DOE site for management, it would add only minor additional effort to current domestic waste disposal operations and would represent only a small fraction of the unprocessed waste inventory.

In summary recovery, removal, and repatriation to the U.S. of unused radiological sources is achievable. The challenge is to find the many individual sources, maintain an effective removal infrastructure, and apply secure, cost-effective removal strategies. This approach

Table 25.3 Estimated disposal volume for remaining long-lived radiological sources of U.S. origin (extrapolated from Table 25.2).

Radiological source type by isotope	Pu-239	Pu-238	Am-241	Cs-137	Sr-90
Approximation of the total number of units to be placed in a disposal site	275	2,550	3,000	2,870	96
Approximation of waste package disposal form	55 gal drums as transuranic waste for disposal at WIPP*	55 gal drums as transuranic waste for disposal at WIPP*	55 gal drums as transuranic waste for disposal at WIPP*	1 cubic meter shielded concrete waste container at 1,000 curies ea. LLW**	Existing package that contains RTG as LLW***

*These materials are currently being disposed of at the U.S. Department of Energy Waste Isolation Pilot Plant (WIPP) in Carlsbad, New Mexico.

**This is an estimated volume. Currently there are no large-source Cs-137 disposal activities at U.S. DOE disposal sites.

***Following disposal methods previously employed by U.S. Department of Energy Low Level Radioactive Waste Disposal Sites such as the Nevada Test Site in Las Vegas, Nevada.

would minimize the risk of recovering radiological material and maximize the benefit of removing it from vulnerable locations where it can add to the radiological threat.

Issues and Obstacles

At the time of publication, almost seven years after 9/11, there is still not a solid consensus on the scope and severity of the threat posed by radiological materials. For example, a full accounting of how much material was manufactured and distributed since the end of World War II is unavailable. Nor are current locations known for all sources for which manufacturing records do exist.

There is no sophisticated risk model that evaluates the most probable consequences of a radiological attack. Such a model could guide appropriation and deployment of resources to address radiological threat reduction. Nor is there a universally accepted safeguards system to assure the future security of existing and new radiological sources as they are manufactured and put into use. Neither is there universally available expeditious disposal for disused legacy sources or newer radiological sources as they need to be retired.

In areas where threat reduction infrastructure is in place and work is commencing, there remain a number of obstacles to efficient and effective recovery. These are listed here in no particular order of priority or measure of impact:

- There is no internationally accepted system of procedures or "safeguards" for radiological materials protection, control, and accountability. The IAEA Code of Conduct on the Safety and Security of Radioactive Sources and its supporting Technical Documents (TECDOCs) do not constitute such a system.[24] The Code is a useful but preliminary document that must evolve into a more detailed, prescriptive guidance for regulating the complete life cycle of radioactive sources. At this point, it is not clear how national authorities will implement radiological safeguards systems that meet or exceed the practices identified in the Code of Conduct. The isotope-producing nations bear the primary responsibility for developing and implementing such systems and assisting other states to create similar domestic regulations.
- The U.S. domestic regulatory structure that controls radiological material transportation has evolved over decades with the goal of ever-increasing assurance of safety and, to a lesser extent, security during the transportation process.[25] This approach needs to be improved with the results of more detailed risk/benefit analyses that acknowledge the increased risk of leaving dangerous sources in vulnerable locations. In a time when the options for radiological material transport are ever decreasing and the costs are increasing, it might be necessary to change or modify existing transportation regulations to facilitate removal of vulnerable material. Both domestic and international regulators would benefit from coordination with threat-reduction initiatives such as GTRI to increase the number of options and find rapid solutions for recovery and removal of radiological materials.
- More sophisticated models of total RDD risk impact are necessary to quantify the economic, social, and psychological impact of potential RDD attacks. Such models will permit statistically valid comparison with other quantified threats to better support resource allocation and operational prioritization.

---

[24]Ibid. (Footnote 15).

[25]Transportation of radiological material is regulated in the U.S. by the U.S. NRC under 10 CFR 71, by the U.S. Department of Transportation under 49 CFR Parts 100–185, and internationally by the IAEA.

Summary

Although the concept of a radiological weapon was known prior to the events of 9/11, those events increased the perception of the threat posed by radiological materials in the U.S. and internationally. Fortunately there has yet to be a serious attack using a radiological weapon. There are several possible explanations for this that do not refute the continued existence of a threat. Military and law enforcement organizations are always vigilant in an attempt to preempt clandestine or terrorist organizations that could have an interest in configuring and deploying a radiological weapon. In addition, efforts have increased worldwide to protect radiological materials and remove them from vulnerable locations. However, there is little that can be done directly to reduce the RDD threat other than to reduce the availability of radiological material to potential attackers.

In the longer term we must develop and implement a secure life cycle for radiological sources that effectively controls the risks while permitting the continued beneficial uses of these materials in medicine, research, and commerce. This requires a long-term commitment and the continued reevaluation of the critical factors that make up the evolving radiological threat. In the short term, the need for several actions is clear:

- Track and safeguard all radiological material through its entire life cycle.
- Recover and remove unused radiological material from the environment.
- Develop and implement sophisticated risk assessment methods to focus our threat-reduction resources on the greatest risks.
- Define and implement an effective safeguards regimen for radiological materials that is universally accepted, practiced, and monitored.
- Equip global transportation systems with equipment capable of detecting and identifying dangerous radiological materials.

26 Field Detection of Nuclear Materials

Mark Abhold and Christopher Lovejoy

Introduction

Reducing the risk from radiological and nuclear threats requires a layered defense with multiple elements; crucial layers include material protection, accountancy, and control at nuclear facilities; monitoring of potential threat materials during transportation, intelligence, law enforcement; search and emergency response; and consequence management. Each of these layers can rely, at least in part, on systems capable of detecting and/or characterizing special nuclear materials (SNM) and/or other radioactive threat materials such as intense industrial radionuclides that could be fashioned into a radiological dispersal device (RDDs). Applications that require radiation detection include the following:

- *Radiological screening at borders.* Preventing nuclear and radiological materials from being smuggled in cargo containers and vehicles crossing national borders requires quick and effective detection technology to be able to screen large amounts of traffic without undue delays or interference with commerce. Radiation portal monitors (RPMs) are deployed so that the traffic is forced to drive through the monitor at locations where customs inspectors can maintain positive control of the vehicle. A radiation alarm on the RPM triggers additional inspection activities that can be costly and time-consuming and thus potentially cause traffic delays. Therefore, low false alarm rates are needed and it is crucial to accurately distinguish threats from benign radiation sources in these environments.
- *Detection of threat materials in transit.* Monitoring all-mode traffic (railroads, highways, waterways) for threat materials during transit poses an especially difficult challenge. There is limited time to make radiation measurements since the time of encounter between the vehicle and the detector can be quite short, especially for vehicles at highway speeds. Separating the threat vehicle from other nearby vehicles in traffic requires the detection system to have high spatial and temporal resolution. Even in the case where detection is made, interdicting the vehicle before it can reach its intended target is difficult and expensive because a dedicated police response force on continuous alert is required. A low false alarm rate is necessary to reduce impact on traffic and to avoid unnecessary traffic stops and adverse impacts to individual freedom.
- *Search and emergency response.* Finding nuclear materials and other threat objects in unknown locations over wide areas requires detection technology and search strategies capable of rapid threat location and identification in all kinds of terrain and/or urban environments. Search personnel must be able to distinguish threats from other naturally occurring radioactive sources and distributed radiation

backgrounds. High sensitivity is required for efficient search strategies, as is the ability to enter buildings and access all areas in and behind structures that could shield the threat from detection. If a potential threat object is located, highly precise detection equipment is needed to characterize the potential threat to determine the safest approach to disable the threat.

- *Consequence management and recovery.* In the event that nuclear material is dispersed in a terrorist incident, radiation detectors are needed to determine the extent, quantity, and nature of the dispersed contamination, to provide information for incident management, and to monitor the decontamination and recovery process. Radiation detection technology in this area must be user friendly, robust, and accurate in extreme radiation environments.

All nuclear and radiological threat detection technology relies on measuring either emitted radiation signatures or signatures that can be induced by interrogating the threat material with other radiation. This chapter introduces the scientific basis of and technology used in the detection of nuclear and radiological materials.

Passive Gamma Detection

Passive gamma detection is possible for those nuclear and radiological materials that naturally emit penetrating gamma rays. Useful gamma signatures exist for special nuclear materials and many other radiological threat materials; however, the detection of these signatures can be complicated by similar signatures emitted from benign manmade and naturally occurring radioactive materials.

Gamma Signatures

Immediately after alpha or beta decay of a radioactive nucleus, when a radioactive nucleus spontaneously emits an alpha particle (the nucleus of a helium atom) or a beta particle (an energetic electron), the resulting daughter nucleus can energetically occupy the ground state, the lowest possible energy level, or it can be left in an excited state if it has an excess amount of energy. A nucleus in an excited state can release its excess energy by one of two mechanisms: internal conversion or gamma-ray emission. During *internal conversion* the daughter nucleus transfers the excess energy directly to one of the most tightly bound atomic electrons, resulting in the emission of an energetic electron, called an *Auger electron*, accompanied by X-ray emission. During *gamma emission*, one or more gamma rays, energetic photons with no mass or charge, are emitted by the nucleus. Gamma rays are similar to photons of visible light except they have much greater energy, but, like X-rays, gamma rays cannot be seen or felt.

Whereas X-rays come from energy transitions between energy states of atomic electrons, gamma rays come from energy transitions within the nucleus. Each emitted gamma ray has a discreet energy with a very specific frequency, in the same way that photons of visible light have specific frequencies that correspond to colors. These distinct gamma rays are often referred to as *gamma lines*, for the distinct lines they leave in a high-resolution plot of the gamma energy spectrum. The gamma lines are denoted by their energy as measured in kilo electron-volts (KeV) or million electron-volts (MeV). High-energy gamma rays in the range above 50 KeV can penetrate significant distances of air, perhaps more than 100 meters. They can also penetrate many other materials, and their energies are unique signatures, or fingerprints, of the radioactive nucleus that emitted them. Gamma rays are therefore useful to both detect radioactive isotopes and distinguish them as potential threat or benign materials.

Electrons and alpha particles emitted as a result of radioactive decay are not very penetrating, because they are charged particles and thus readily interact with materials with consequent rapid energy loss and absorption. Alpha particles can easily be stopped by a thin piece of paper; electrons can be stopped by a few millimeters of plastic or 10 to 20 meters of

Table 26.1 The major gamma-ray signatures and other characteristics of special and alternate nuclear materials.[12]

Isotope	Half-Life (yr)	Energy (KeV)	Activity (γ/g-s)	Mass Attenuation Coefficient[a] (cm^2/gm)	Comment
^{233}U	1.592×10^5	291.3	5.8×10^4	0.11	
		317.2	8.3×10^4	0.11	
		2614	$\sim 1 \times 10^5$	0.039	From ^{232}U Contamination
^{235}U	7.038×10^8	185.7	4.32×10^4	0.16	
^{238}U	4.468×10^9	766.4	2.57×10^1	0.069	From ^{234m}Pa
		1001.0	7.34×10^1	0.060	From ^{234m}Pa
^{237}Np	2.14×10^6	312.2	$\sim 1 \times 10^7$	0.11	
^{239}Pu	2.415×10^4	129.3	1.44×10^5	0.27	
		413.7	3.42×10^4	0.093	
^{240}Pu	6568	160.3	3.37×10^4	0.19	
^{241}Pu	14.35	148.6	7.15×10^6	0.20	
		208.0	2.04×10^7	0.14	From ^{237}U
^{241}Am	433.6	59.5	4.5×10^{10}	1.24	
		125.3	5.2×10^6	0.28	
^{241}Am	7380	117.7		0.31	
		142.0		0.22	

[a]The mass attenuation coefficient is a measure of the penetrability of a gamma ray at that energy as defined in the gamma ray shielding section below. The values listed are for attenuation by iron.

air. Consequently, charged particles are of little use in detecting radioactive materials at a distance and not useful to detect materials inside a container or vehicle, because they would not penetrate to the outside of the container.

Special and Alternate Nuclear Materials

Table 26.1 lists the gamma rays commonly used to detect SNM and alternate nuclear materials (ANM). SNM is defined by Title I of the Atomic Energy Act of 1954 as uranium enriched in the isotopes ^{233}U or ^{235}U, and any isotope of plutonium, Pu. ANM is defined to be the isotopes ^{241}Am, ^{243}Am, and ^{237}Np. Both SNM and ANM are proliferation concerns.

Highly Enriched Uranium

Highly enriched uranium (HEU), which is uranium artificially enriched in the isotope ^{235}U to a percentage greater than 20%, is of the greatest proliferation concern. The main passive signal detectable from unshielded HEU is the intense gamma line at 185.7 KeV from ^{235}U. Unfortunately, this line is very easily shielded owing to its low energy. A low-intensity line at 1001 KeV from the remaining ^{238}U content in HEU might be detectable, but in practice, detection of HEU threats using this line is difficult because this line also exists in natural uranium-bearing minerals that can be falsely identified as a threat. Additionally, a low-intensity line at 2614 KeV exists in some HEU contaminated with ^{232}U, a typical byproduct of reenriching uranium separated from spent reactor fuel. ^{232}U contamination does not exist in HEU from an enrichment facility that uses natural mined uranium if that facility has never processed uranium separated from spent fuel. In addition, this 2614 KeV line is shared with the decay products of natural thorium and is ubiquitous in the environment, making it very difficult to differentiate HEU from natural thorium-bearing materials based on this line. The

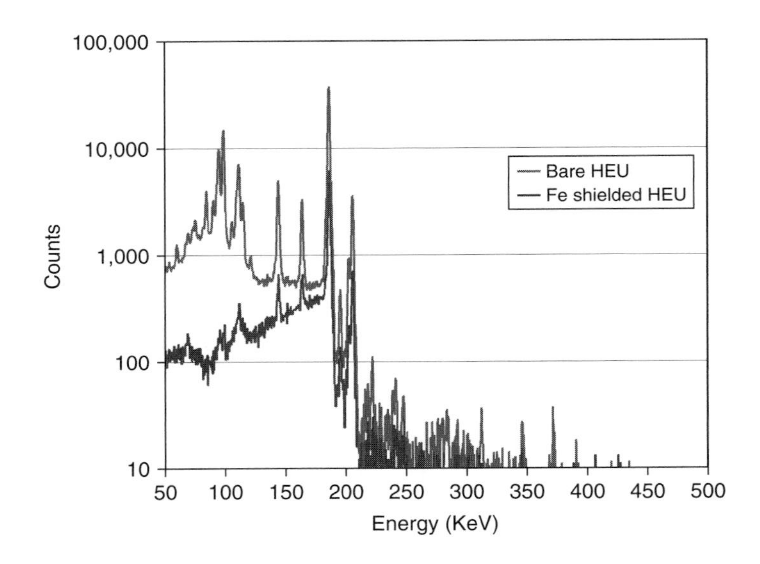

FIGURE 26.1 HEU gamma spectra. The top trace is bare HEU at 93% enrichment, the bottom trace is the same source shielded by 0.5 inches of steel. The most prominent line is at 185.7 KeV ^{233}Uranium.

detection capability for HEU is therefore determined primarily based on the 185.7 KeV line and is thus very sensitive to any shielding materials between the source and the detector.

Gamma energy spectra of HEU are shown in Figure 26.1 for bare 93% enriched ^{235}U and the same sample behind one half inch of steel. Comparison of these two spectra clearly shows an order of magnitude difference in the count rate of the 185.7 KeV line owing to the steel shielding and the depression of the lower-energy lines due to the difference in attenuation through the steel with the lower-energy lines attenuated far more than the higher-energy lines. This figure illustrates the severe attenuation of HEU signatures after passing through even small amounts of shielding.

^{233}U is formed by neutron capture on ^{232}Th and is the product of nuclear reactors designed to convert thorium. ^{233}U becomes a significant proliferation concern once it is chemically separated from the remaining thorium in the spent reactor fuel. Thorium-converting nuclear reactors are not as common as standard uranium/plutonium reactors, but some thorium-converting research reactors do exist, for example, in India. Other reactions in a thorium-converting reactor concurrently produce ^{232}U, which cannot be chemically separated from the other uranium isotopes. Separation of ^{232}U from ^{233}U using isotope enrichment techniques is also not practical owing to the single atomic mass difference between the two isotopes, and thus ^{232}U coexists with ^{233}U in concentrations well above those found in reenriched ^{235}U. The 2614 KeV gamma emission from the ^{232}U contamination in ^{233}U actually comes from a decay product of ^{232}U, ^{208}Tl. This gamma ray is intense and highly penetrating; in fact, it is so intense that radiologically safe handling becomes difficult in older separated ^{233}U because the amount of ^{208}Tl builds over several years with the decay of ^{232}U. ^{233}U is therefore relatively easy to detect.

Plutonium

The gamma signature of plutonium consists principally of a group of many gamma lines emitted from ^{239}Pu between 375 KeV and 425 KeV with a prominent line at 413.7 KeV. These plutonium gamma lines are intense and fairly penetrating; thus the gamma signature from plutonium is much more detectable through shielding than that from HEU. The gamma energy spectrum of Pu is complex; many gamma lines can be seen in the spectrum, as illustrated in Figure 26.2. Pu has a number of isotopes in addition to ^{239}Pu, principally ^{240}Pu and ^{241}Pu but also ^{242}Pu. The amount of these isotopes in separated Pu is determined by the plutonium's historical level of neutron irradiation in the production reactor, which is often

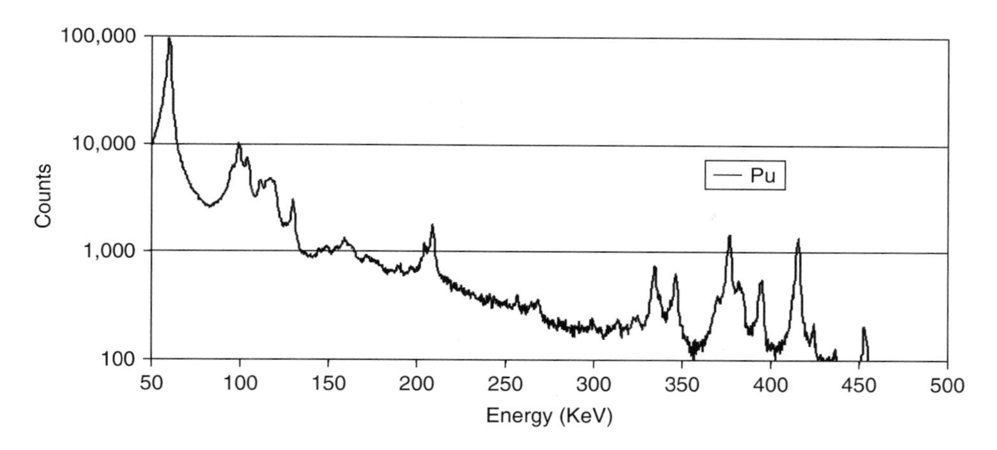

FIGURE 26.2 Plutonium gamma spectrum showing the characteristic gamma lines between 375 and 414 KeV.

referred to as the *burnup level* or simply *burnup*. Higher burnup results in proportionately more ^{240}Pu and higher atomic masses, and these additional isotopes increase the gamma radiation signature compared to low-burnup, "weapons-grade" Pu. High-burnup Pu is sometimes referred to as *reactor-grade* Pu.

Neptunium

Neptunium is produced as a byproduct of irradiating uranium with neutrons in a nuclear reactor. Typically, neptunium is not separated from the other fission products and actinides in spent fuel being treated as a waste product. However, if it is separated, as an ANM it is a proliferation concern. ^{237}Np emits a number of intense and fairly highly penetrating gamma rays, including a line at 312 KeV, and is therefore easier to detect than HEU.

Americium

^{241}Am is a beta decay product from ^{241}Pu produced in nuclear reactors. ^{241}Am coexists with plutonium except for freshly separated Pu. ^{241}Am emits an extremely intense gamma ray at 59.5 KeV, but this line is easily shielded. A less intense line at 125 KeV also exists, but this line is also fairly easily shielded. Other isotopes of Americium are not practical to separate from ^{241}Am and also contribute detectable signatures. Although Americium is a common radioactive element often used in smoke detectors, its detection is nevertheless treated seriously because it could indicate the presence of plutonium.

Commonly Encountered Radioactive Materials

Radioactive materials observed during radiation-monitoring operations vary with the location of monitoring and the type of cargo, vehicles, and people moving through the monitoring site. Naturally occurring radionuclides are most likely to be encountered where large quantities of materials are transported—for example, in shipments moving through seaports as well as trains and large truck traffic at land borders. Medical radionuclides taken up by patients are most likely to be encountered at vehicle border crossings and airports. Industrial radionuclides can be frequently seen near construction sites and research facilities and often are constrained to move along routes approved for the transportation of hazardous materials in and around cities.

Naturally Occurring Radioactive Materials

Naturally occurring radioactive materials (NORM) can be found virtually anywhere on Earth in detectable concentrations. The number of NORM isotopes is very large; however, many do not emit gamma rays or do not produce gamma rays at intensities high enough to interfere with the detection of threat materials. This section focuses on those NORM isotopes that are capable of creating innocent alarms in gamma detection hardware—actual alarms but innocent in that they result from radioactive materials but are otherwise benign.

NORM can be characterized as *primordial* (radioisotopes with half-lives so long they have survived to today from isotopes created before the creation of the Earth) or *cosmogenic* (radioisotopes formed as a result of cosmic-ray interactions with stable isotopes on Earth). NORM isotopes either occur singly or are a member of one of three distinct chains of radioactive isotopes formed by a series of alpha and beta decays from the parent nuclide. The three principle chains are:

- The uranium series starting with ^{238}U (Table 26.2)
- The thorium series starting with ^{232}Th (Table 26.3)
- The actinium series starting with ^{235}U

Table 26.2 Isotopes of the naturally occurring uranium series.[12]

Isotope	Half-Life
^{238}U	4.5×10^9 years
^{234}Th	24 days
^{234m}Pa	1.2 min
^{234}Pa	6.7 hours
^{234}U	2.5×10^5 years
^{230}Th	8.0×10^4 years
^{226}Ra	1622 years
^{222}Rn	3.8 days
^{218}Po	3.05 min
^{214}Pb	26.8 min
^{218}At	2 sec
^{214}Bi	19.7 min
^{214}Po	1.6×10^{-4} sec
^{210}Tl	1.3 min
^{210}Pb	22 years
^{210}Bi	5.0 days
^{210}Po	138 days
^{206}Pb	Stable

Table 26.3 Isotopes of the naturally occurring thorium series.

Isotope	Half-Life
^{232}Th	1.4×10^{10} years
^{228}Ra	5.75 years
^{228}Ac	6.13 hours
^{228}Th	1.91 hours
^{224}Ra	3.66 days
^{220}Rn	55.6 seconds
^{216}Po	0.15 seconds
^{212}Pb	10.64 hours
^{212}Bi	60.55 min
^{212}Po	0.3 μsec
^{208}Tl	3.07 min
^{208}Pb	Stable

Isotopes from the actinium series do not commonly cause innocent alarms owing to the low fraction of ^{235}U in natural uranium, so this series is not reproduced here.

^{40}K is by far the most ubiquitous of the singly occurring natural radionuclides of terrestrial origin. Nearly all gamma spectra taken in almost any environment will show substantial amounts of ^{40}K, which exists at an activity of about 855 pCi/g in all potassium (Ci stands for *curie*, the unit of radioactivity equivalent to 3.7×10^{10} disintegrations per second). The 1460 KeV line from this isotope is so commonly observed that it is often used to calibrate the energy response of gamma detectors.

The most common cosmogenic, singly occurring natural radionuclides are tritium ^{3}H, ^{14}C, ^{7}Be, ^{38}Cl, and ^{39}Cl. Of these, only ^{7}Be, ^{38}Cl, and ^{39}Cl have substantial gamma emissions, but typically only in accumulated, fresh rainwater are they seen by gamma detection equipment at levels high enough to cause interference. Other radionuclides from the uranium and thorium chains, principally radium and radon daughters, also wash out of the atmosphere after a rainfall, thus temporarily increasing the background at ground level to levels high enough to impact threat detection efficiency.

Of all of these NORM isotopes, the most frequently observed are ^{40}K, natural U plus daughters (^{238}U, ^{226}Ra), and ^{232}Th plus decay daughters. Table 26.4 lists the activity of these isotopes seen in some frequently transported materials. In addition to these materials, a wide variety of cargo has been observed during monitoring exercises to contain NORM, including ceramic tiles, liquefied petroleum gas, butane, welding rods, pumice, ceramics, clay, televisions, camera lenses, colored pencils, novelties, toys, candle supplies, dishes, computers, floor mats, crock pots, plastics, household supplies, polishing powder, tools, and many others. Although NORM is generally benign, there is a possibility that threat materials may be masked by embedding them in NORM-bearing radioactive cargo; thus any detection of radioactivity must then be followed by measurements to identify the source and categorize it as a potential threat or benign material.

Medical Radionuclides

Medical procedures involving the administration of radionuclides to a patient are common. These radiopharmeceutical treatments typically employ short-lived radionuclides with activities high enough to be easily detected. Some common medical radionuclides are ^{18}F, ^{67}Ga, ^{99m}Tc, ^{111}In, ^{124}I, ^{125}I, ^{131}I, ^{133}Xe, ^{192}Ir, and ^{201}Tl. These isotopes do cause innocent alarms in radiation monitors and therefore must be differentiated from other hazardous isotopes.

Industrial Radionuclides

Radionuclides are used in a multitude of industrial and research applications. Large sealed ^{137}Cs and ^{192}Ir sources are often used in industrial radiography and thickness-gauging applications.

Table 26.4 Frequently transported, normally occurring radioactive materials.[12]

Material	Approximate Activity Concentration in Bq/kg[a]		
	K–40	Ra–226	Th–232
Fertilizers	40–8000	20–1000	20–30
Granite	600–4000	30–500	40–70
Adobe	300–2000	20–90	32–200
Slate	500–1000	30–70	40–70
Sandstone	40–1000	20–70	20–70
Marble	40–200	20–30	20
Feldspar	2000–4000	40–100	70–200
Monazite sand	40–70	30–1000	50–3000
Concrete	150–500	40	40

[a]The Becquerel, Bq, is the fundamental unit of radioactivity equal to 1 disintegration per second.

^{210}Po sources are used for static elimination, and ^{60}Co is used in industrial and laboratory irradiators used to sterilize medical instruments and food. Other sources such as ^{63}Ni, ^{99}Tc, ^{3}H, and ^{46}Sc are sometimes used as tracers to understand industrial processes or to map well fractures in oil wells, to assist in oil recovery. When properly placarded and shielded, these sources are benign; however, the activity of some of these sources can be large enough to potentially be misused—for example, in an RDD (see Chapters 24 and 25 for more details on RDDs). It is therefore necessary to detect, identify, and determine the configuration and legal status of industrial radionuclides.

Gamma Detection Fundamentals

Gamma detectors used to detect SNM and other potential threat materials rely on detection materials capable of absorbing gamma energy and converting it to an electrical signal that can be amplified and counted. An understanding of how gamma rays interact with materials, how gammas can be shielded, and the statistical nature of radiation counting is necessary to understand the capabilities and limitations of various gamma detectors.

Gamma-Ray Interactions with Materials

Since gamma rays are massless and carry no charge, they interact mainly with the electromagnetic field of the atom or its electrons. There are a number of interaction processes, but the three main ones are:

- *Photoelectric effect.* The gamma ray transfers its entire energy to an atomic electron, resulting in the emission of an energetic electron and the complete disappearance of the gamma ray. This process dominates at low gamma energies out to a few hundred KeV, especially for high atomic number materials.
- *Compton effect.* The gamma ray transfers a portion of its energy to an atomic electron, resulting in the emission of an electron with enough energy to escape the atom and a lower energy photon that departs at an angle to the original gamma path. This process dominates at middle energies for all atomic numbers.
- *Pair production.* The gamma interacts in the field of the nucleus to create a electron-positron pair. This conversion requires a minimum of 1022 KeV, and any energy above this threshold is shared as kinetic energy of the electron and positron. This process dominates at energies above 3000 or 4000 KeV especially for high atomic number material.

Effect of Distance and Gamma-Ray Shielding

The flux of photons, or number of photons/cm^2-s that pass through any area, is reduced as the detector is moved away from the source by the geometric law:

$$\frac{\text{Flux}(R) = \text{source strength (photons/s)}}{4\pi R^2}$$

where R is the distance from the source to the detector in centimeters.

This reduction in flux with distance is often referred to as the "one over R squared" law. The flux is a useful concept in that the flux multiplied by the detector area gives the number of photons that are intercepted by a detector in 1 second of measurement time.

The flux, and consequently the number of photons intercepted by a detector, falls off very rapidly with distance. A detector that would count 10 counts per second at a distance of 1 meter from a radioactive source would only count 0.1 counts per second if moved to a 10 meter distance.

The intensity of a gamma-ray beam is also reduced as it passes though shielding materials as a result of the interaction mechanisms discussed in the previous section. This intensity loss, or *attenuation*, of a beam of gamma rays is estimated by the exponential attenuation law:

$$I = I_0 e^{-(\mu/\rho)\rho x}$$

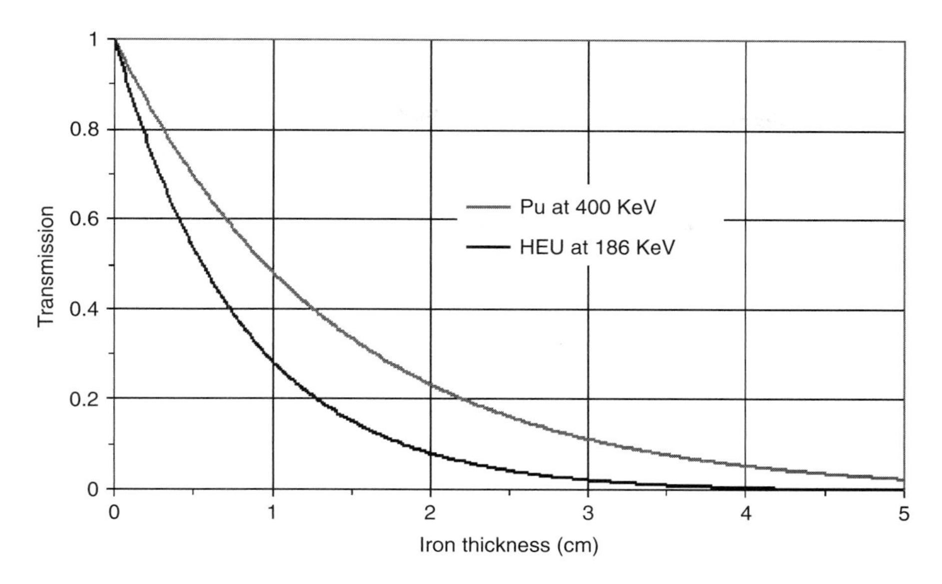

FIGURE 26.3 Attenuation of ^{235}U (186 KeV), and Pu (400 KeV) gamma rays in iron.

where I_0 is the incident gamma ray intensity (photons/cm^2-s) at the entrance face of the shield, I is the intensity at any depth x (cm) in the shield, and ρ (Greek symbol *rho*) is the material density (gm/cm^3). The mass attenuation coefficient, μ/ρ (Greek symbol *mu*), is a function of gamma energy and the material type.

A listing of mass attenuation coefficients as a function of gamma-ray energy can be found[1] for all elements and many compounds. A sample listing of mass attenuation coefficients in iron for gamma ray energies of interest can be found in Table 26.1.

Attenuation in iron of the 185.7 KeV gamma ray from ^{235}U and a 400 KeV gamma ray representative of the 375–425 KeV complex in Pu is shown in Figure 26.3. This figure shows the percentage of the initial gamma-ray beam transmitted though iron of varying thickness. Note that less than 2 cm of iron (at 7.8 gm/cm^3) is needed to attenuate the ^{235}U gamma ray to less than 10% of the incident intensity. This thickness could easily be encountered in items frequently transported in a large truck or cargo container; for example, a full load of steel in a cargo container (8 × 8 × 40 feet) can weigh as much as 65,000 lbs, an equivalent density of about 0.4 gm/cm^3. A gamma ray would encounter the equivalent of 6.2 cm of full-density iron on a path from the center of this cargo container to the closest outside wall, a distance of about 1.22 meters. This thickness of iron would reduce the 185.7 KeV gamma intensity to about 4.4×10^{-4} of the source strength, and when combined with the geometric reduction of 1/R^2, the total transmitted flux on the outside of the container is very small indeed and would be almost impossible to detect in the presence of background radiation.

Because uranium itself is a very good shielding material, the 185.7 KeV gamma rays can only escape from the outer surface of uranium metal. The gamma rays generated in the interior are absorbed before they can escape; therefore, the source strength of uranium metal scales as the surface area, which for compact metallic sources varies as the mass to the $^2/_3$ power.

Radiation Detection Processes

When a gamma ray interacts with a detector, all three of the main gamma interaction processes create energetic electrons in the detector material. As these electrons slow down they

[1]www.physics.nist.gov/PhysRefData/XrayMassCoef/intro.html.

create further interactions with the detector in ways that can be measured. In scintillator-type detectors such as sodium iodide (NaI) or plastic scintillators, the electron energy is transferred to excited electron states in the material that de-excite with the emission of light. The emitted light is typically detected with a photomultiplier tube, amplified, and counted. In gas-tube detectors such as Geiger counters and solid-state detectors such as high-purity germanium (HPGe), the energetic electrons cause further ionization in the detector materials, leading to the production of mobile positive and negative charges that are collected through the application of an electric field. The collection process yields an electric signal that is amplified and counted. For more details, an excellent overview of radiation detection can be found.[2]

Radiation detectors that count any photon independent of the photon's energy yield the *gross count*, the total number of gamma rays that interacted in the detector. An example of a gross-counting detector is the plastic scintillator often used in radiation portal monitors (RPMs). More sophisticated radiation detectors are capable of sorting counts into bins according to the deposited photon energy and thus record not only the gross counts but the spectrum of gamma energies. Spectroscopic detectors, those capable of recording the energy spectra, have varying energy resolution—the capability to resolve two gamma lines close in energy. Figure 26.4 shows an energy spectrum of background radiation taken by a HPGe detector cooled to liquid nitrogen temperatures compared with a low-resolution sodium iodide (NaI) detector, very low-resolution plastic scintillator, and medium-resolution but low-efficiency CdZnTe. High-resolution gamma detectors, such as HPGe, provide more capability to identify the radiation source by comparison of the spectra with known gamma lines but are typically quite expensive to purchase and operate. Low-resolution systems often cannot distinguish one isotope from another, but they can be afforded in much larger sizes and therefore much greater efficiency.

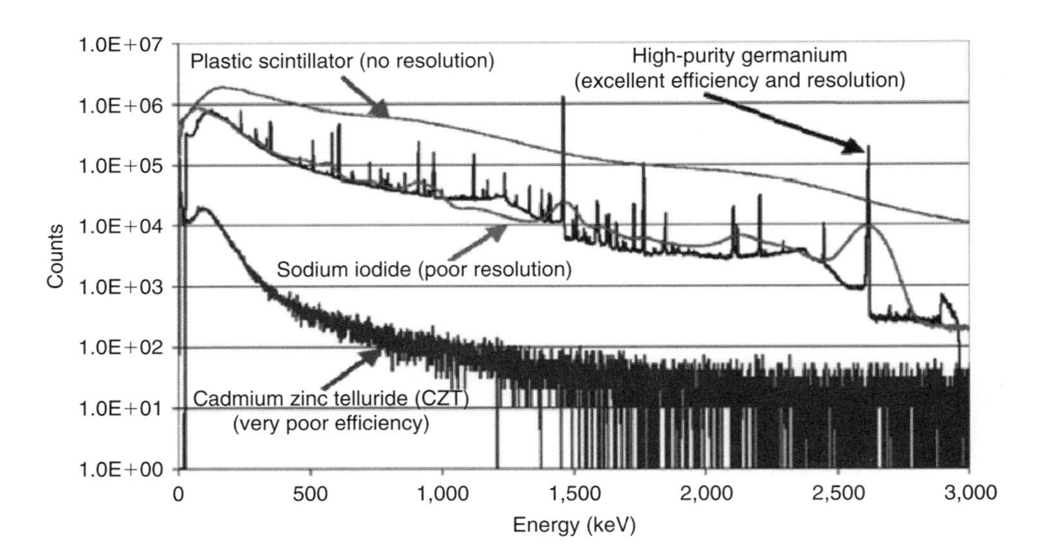

FIGURE 26.4 Gamma energy spectra as recorded by various detectors: a. very poor resolution but high efficiency plastic scintillator, b. high-resolution, high-purity germanium, c. low-resolution sodium iodide (NaI), and d. medium-resolution but low-efficiency CdZnTe.[12]

---

[2]G. Knoll, *Radiation Detection and Measurement*, third ed., New York: Wiley, 2000.

Counting Statistics

The counts recorded by a radiation detector are governed by the random fluctuations inherent in radioactive decay. A sample of radioactive nuclei will decay according to the following relationship:

$$N = N_0 e^{-\lambda t}$$

where N is the number of radioactive nuclei that remain after a time t, N_0 is the number of nuclei at time $t = 0$, and λ is the decay constant unique to each radionuclide. The half-life is the time necessary, on average, for half of the nuclei to decay. The half-life is related to the decay constant by:

$$T_{1/2} = ln(2)/\lambda$$

where $T_{1/2}$ is the half-life and $ln(2)$ is the natural logarithm of 2.

The average activity of a sample in decays per second is given by:

$$A\,(\text{decays/sec}) = \lambda N$$

In any sample of N radioactive nuclei, it can only be said that an *average* of λN will decay in one second, in any given second more or less may decay, and it is impossible to predict the exact number of decays. This is because radioactive decay is a truly random process determined by the rules of quantum mechanics. Because the number of decays is random, the number of counts in a radiation detector in a given time is also randomly distributed about the average. This distribution in counts is not caused by the detector; it is a consequence of the truly random nature of radioactive decay. For radioactive decay, the standard deviation of the count distribution is given by:

$$\sigma_C = \sqrt{C}$$

where σ_C (Greek symbol *sigma*) is the standard deviation of the count distribution and C is the total number of counts in a time interval. Therefore, a single count is always reported along with the uncertainty in the count, σ_C.

When measuring a source, the count rate is typically determined by measuring the background with the source removed and then bringing the source to the detector and repeating the measurement with the source present. Of course, the background is still present and must be subtracted to yield the source counts.

The count rate of just the source is then:

$$R = C/t_C - B/t_B$$

where R is the count rate (counts/second) due to the source alone, C is the total number of counts taken with the source and background present over a time t_C, and B is the total number of counts taken in a time t_B with just the background present. Both C and B are statistical distributions characterized by their standard deviations $\sqrt{C}$ and $\sqrt{B}$, respectively. Every time this measurement is repeated, the count rate R will vary within a distribution of possible answers characterized by the standard deviation of R, σ_R.

$$\sigma_R^2 = (1/t_C)^2(\sqrt{C})^2 + (1/t_B)^2(\sqrt{B})^2$$

Gamma Search

The statistical nature of radioactive counting provides a physical limit to the size of threat source that can be detected and the distance at which it can be detected in the presence of a radiation background.

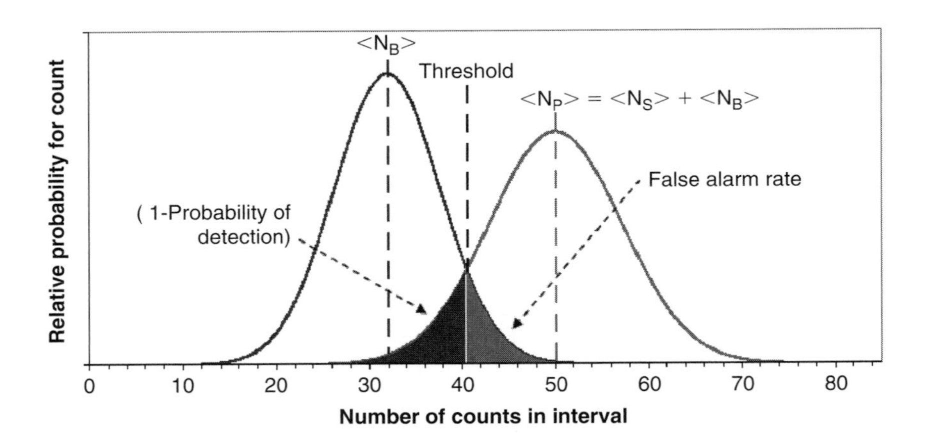

FIGURE 26.5 The statistical probability of source detection in a radiation background.

Figure 26.5 shows the count rate distribution that would be measured with only background sources present and also with a threat source added to the background. In this illustration, the average background count rate is given by $<N_B>$, about 32 cps. The source at the distance of detection adds an additional $<N_S>$ counts, or about 18 cps. When the source is measured in the background, the average count rate seen is $<N_S> + <N_B>$, or about 50 cps. However, the actual measured count rate will statistically fluctuate around the mean value. If an alarm threshold is set such that an alarm is issued whenever the measured count is above 41 cps, the false alarm rate, alarming when the source is not present, is shown by the area under the background radiation distribution above the threshold setting. When the source is present, there is a probability of not detecting the source given by the shaded area in the source plus background distribution below the threshold. At a fixed distance from the source, to increase the probability of detection without increasing count time or detector efficiency, one can only reduce the alarm threshold, but that comes at the expense of increasing the false alarm rate.

Larger, more efficient detectors and longer count times can improve the detection probability because with more counts in the time interval the source plus background distribution becomes more distinct from the background distribution. However, other practical considerations limit the usefulness of larger detectors and longer count times. For example, the background count rate often varies with location, which can dominate the errors. Efforts to improve detection probabilities include reducing the background count rate by collimating the detector with a gamma shield so that the detector can only see in the direction it's pointed, and by spectroscopic detectors capable of discriminating threat source gamma rays from background gamma rays by exploiting the difference in their energies.

The practical limit for detecting threat quantities of HEU in a 1 minute measurement is 10s of meters. Since plutonium gamma rays more easily escape plutonium metal, threat quantities of plutonium can be potentially be detected at distances of 100 meters or more.

Passive Gamma Detection Equipment

Passive detection equipment is commonly used to search for and to interdict radiological materials at borders and in commercial shipping by detection of gamma (and often, neutron) radiation. The equipment ranges in size and cost from small, handheld detectors to large, fixed radiation portal monitors (RPMs) cable of scanning pedestrians, vehicles, or shipping containers. Each of these categories of instruments ranges in cost and complexity from gross counting instruments to sophisticated spectroscopic detectors employing advanced computer algorithms to identify specific radioisotopes and special nuclear material.

The simplest and most commonly used gamma detection equipment utilizes gross counting of gamma radiation, with little discrimination of the gamma energy. Often plastic scintillators, available inexpensively in large sizes, are used. Pedestrians, vehicles, and cargo are scanned while they pass between paired detection portals, each portal containing one to

FIGURE 26.6 Radiation portal monitors being readied for simultaneous testing. Six vendors are represented.

four or more plastic scintillation detectors. These *portal monitors*, shown in Figure 26.6, were originally developed to screen radiological workers in the nuclear industries for facility security applications but now are widely applied to screening people and commerce for radiological sources. Measurements performed when the monitors are unoccupied are used to determine the radiation background and when occupied to determine radiation levels while the person or vehicle passes between the portals.

The monitors typically sound an audible alarm and flash a visible alarm when the detected radiation levels exceed the background radiation by some preset amount and require attendance by security personnel to stop traffic and conduct secondary searches of people or vehicles that cause an alarm. An analogous situation is encountered with metal detectors in airports and other areas with enhanced security concerns; primary inspection is performed with portal-type metal detectors and secondary inspection is performed with a handheld "wand" instrument. In both cases, the handheld detector is actually less sensitive in an absolute sense than the larger portal monitor, but it can be moved or scanned over a person or vehicle and thus placed in close proximity to a threat object.

With a metal detector conducting a secondary search of personnel, it is easy to locate a wristwatch or a belt buckle by scanning rapidly over a body, so secondary inspections are brief. Compared to metal detectors, the handheld radiation detectors have a fairly slow response so operators must be carefully trained to perform a secondary scan at a slow, controlled scan rate to achieve adequate sensitivity. The physical size of some vehicles and all shipping containers means that it may be difficult or impossible to achieve close proximity to a hidden threat object, even with carefully trained operators. This limits the sensitivity and usefulness of secondary inspection with handheld instruments; improvement of secondary inspection is a current area of research and development.

The measurement of gamma radiation yields a series of discrete values representing the number of "counts" in a given time interval and thus is subject to statistical constraints that require a tradeoff between sensitivity and the permissible duration of measurement. Longer measurements permit a finer discrimination of background versus above-background radiation levels. Quantitatively, the standard deviation of the measurements varies as the square root of the number of counts in a given interval (see 2.2, above). For example, the standard deviation of the background count rate B counts/second is $\sqrt{B}$; if averaged over a period of T seconds, the standard deviation of the average is $\sqrt{B/T}$. For a portal monitor operated with moving pedestrians or vehicles, the acceptable counting time is of order ~1–2 seconds, determined by the speed of passage of a threat object through the portal at typical speeds.

Alarm thresholds are set at a particular number of standard deviations above the background. The threshold is determined from consideration of the acceptable rate of innocent and

Table 26.5 Effect of alarm threshold on false alarm rate per trial and minimum detectable quantity of SNM (in arbitrary units).

Alarm Threshold (Units of $\sqrt{B}$)	False Alarm Probability	MDQ (Arbitrary Units)
3	1.35E-03	0.19
4	3.17E-05	0.44
5	2.87E-07	0.71
6	9.87E-10	1.00
7	1.28E-12	1.31

false alarms and the required detection sensitivity. False alarms arise from the statistical fluctuations in the measured count rate. The rate of false alarms can be estimated from the predicted frequency of measurement excursions greater than N standard deviations above background. For gamma detectors of practical size, the counting statistics can be closely approximated by a Gaussian or Normal distribution. This is shown schematically in Figure 26.5. Table 26.5 gives the predicted frequency of false alarms as a function of $N\sigma$ for a *single* 1 second measurement, along with the minimum detectable quantity of SNM in arbitrary units at a gamma background of 1,000 counts/sec. Note that a higher alarm threshold gives a less sensitive instrument, since more radiation is required to reach the alarm threshold, but with fewer false alarms. The background count rate is determined by geological factors such as the NORM content of the soil and of materials of construction and the altitude of the site, by the size and efficiency of the detectors, and by certain weather conditions. All other things being equal, a higher background means lower sensitivity for a given false alarm rate since the unit of measure is $\sqrt{B}$.

Acceptable false alarm rates are determined from practical considerations; for pedestrian traffic it might be ~0.001, but for vehicle traffic it might be higher or lower, depending on the availability of resources to perform secondary inspections and on acceptable restrictions on commerce. Most RPMs make multiple decisions per occupancy, often 5–10 decisions per second based on a running average of individual 0.1–0.2 second counts. The decisions thus show substantial correlation, since a high or low count in a given interval is used in multiple decisions.

In practical applications with commercial vehicles or shipping containers, innocent alarms far outnumber false alarms. Innocent alarms are caused by benign cargoes that contain NORM or by legitimate shipment of radioactive sources. For many such applications, an innocent alarm rate of ~0.01 might be considered acceptable in high-volume portals. False alarms thus make a negligible contribution to the total nuisance alarm rate.

We have described a situation in which the average count rate is at background, unless a source passes through the portal and the count rate rises, as shown in Figure 26.7a. Realistic cargos can also depress the count rate by absorbing ambient gamma radiation as the cargo passes through the portals, as in Figure 26.7b. The count rate still must exceed the alarm threshold to trigger an alarm, so a larger source is required in the presence of background suppression, Figure 26.7c. An average background suppression for fully loaded shipping containers is typically about 20% of the background count rate, for passenger vehicles 5%, and for pedestrians <1%. Background suppression can increase the minimum detectable quantity of SNM by a factor of two or more.

Commercially available radiation portal monitors are often built to a particular performance standard determined by federal purchasing requirements. Detection limits then vary with the radiation from each material. Table 26.6 shows typical detection limits for typical pedestrian and vehicle portal monitors when tested to a particular protocol such as those provided by the American Society for Testing and Materials (ASTM).[3]

[3]See ASTM Standard C 1169–97, 2003, "Standard Guide for Laboratory Evaluation of Automatic Pedestrian SNM Monitor Performance," ASTM International, West Conshohocken, PA, http://www.astm.org.

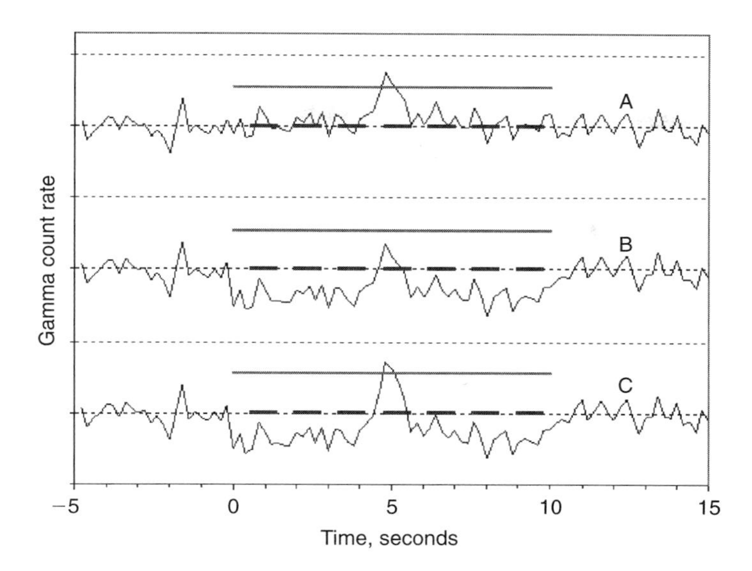

FIGURE 26.7 Schematic diagram showing the effect of background suppression on reducing monitor sensitivity: a. Simulated 10-second occupancy starting at $t = 0$, with no background suppression but with a compact gamma-emitting source in the center that exceeds the alarm threshold (solid line). b. The same simulated occupancy but with 20% background suppression during the occupancy; now the source does not exceed the alarm threshold (solid line). c. The same simulated occupancy as b. but with a 75% larger source that again exceeds the alarm threshold (solid line). The alarm threshold in each case is 4 standard deviations above the background prior to the occupancy (dashed line).

Table 26.6 Typical detection limits for Category II commercial radiation portal monitors tested to ASTM C1169, as stated by the manufacturers.

Type of Monitor	Gamma Detection of HEU and Pu (Grams)		Neutron Detection of Pu (Grams)
	HEU	**Pu**	
Pedestrian	10	1	120
Vehicle	1000	10	200
Rail	1000	10	200

Note: Detection limits depend on many factors, including the test protocol, radiation background, portal separation, and speed of passage. Pedestrian, vehicle, and rail monitors differ in the size and/or number of detectors and in the intended portal separation.

Because the sensitivity of a radiation detector varies with the square root of the background count rate, one can increase the sensitivity somewhat by limiting the range of gamma energies that are counted to a region of particular interest, thereby reducing the background. With inexpensive plastic scintillators the energy resolution is poor and a wide range of gamma energies is counted simultaneously, but the minimum detectable quantities of SNM may be improved by judicious choice of detection settings, at the cost of reduced sensitivity to species that emit higher-energy gammas. This is advantageous when searching for SNM but is not advisable in instruments used in nonthreat-reduction scenarios—for example, to prevent the accidental introduction of radioactive materials into steel scrap yards or other commercial processes.

With so-called "spectroscopic" instruments, the energy resolution is improved to a greater or lesser extent, in principle increasing the achievable sensitivity. However, the smaller

Table 26.7 Detector materials used in handheld radiation monitors.

Detector	Cost	Resolution	Size and Complexity	Sensitivity Per Dollar
High-purity germanium	High	High	High	Low
Sodium iodide	Medium	Low	Medium	Medium
Cadmium zinc telluride	Medium	Medium	Medium	Low
Polyvinyl toluene plastic	Low	Very low	Low	High

sizes of the commercially available spectroscopic detectors rapidly offset any gains in sensitivity from this effect, and costs are much higher for a given sensitivity. The usefulness of spectroscopic instruments is that often specific isotopes can be identified by their energy spectrum. This may be useful in identifying NORM cargoes with isolated gamma lines, such as K-40 in agricultural products; in identifying persons undergoing treatment with medicinal or diagnostic radioisotopes; and in confirming the composition of legal shipments of radioactive isotopes.

Handheld radioisotope identifiers are often used in secondary inspections to characterize the cargo that caused an alarm in primary inspection with sensitive, inexpensive plastic scintillators. It could take several minutes to obtain a spectrum of sufficient quality to identify the radioisotopes present, so it is impractical to use the handheld's "spectroscopic" qualities when conducting a physical search. However, one can search in "gross counting" mode and then switch to "spectroscopic" mode to characterize the emissions in a promising area. A number of materials have been investigated for handheld spectroscopic instruments; these are summarized in Table 26.7 along with typical sensitivity as stated by the manufacturers. Again, sensitivity is limited by the availability or cost of suitable detection materials.

A recent development is the availability of commercially manufactured "spectroscopic portals." By utilizing the largest-available samples of "spectroscopic" materials such as thallium-doped sodium iodide and high-purity germanium, it is possible to approach or equal the overall detection efficiency of commercially available plastic scintillator portal monitors, albeit at a substantial three- or fourfold increase in cost. Careful comparison of NaI spectroscopic portals against plastic scintillators with the same total detection efficiency and operated at the same false alarm rate shows a modest decrease in minimum detectable mass of SNM due to optimum use of spectroscopic information. However, one could easily equal this performance gain, at far lower cost, by increasing the volume or number of plastic scintillators used. Furthermore, the gamma emission from realistic "threat objects" ranges over several orders of magnitude, and both types of monitor will alarm over a large portion of this range. It therefore appears that in primary search mode, the plastic scintillator will likely continue as the instrument of choice.

In secondary inspection mode of commercial cargoes, however, the qualities of the spectroscopic portals are used to best advantage. Since actual threat objects are presumably rare, the overwhelming majority of "primary" alarms will always be innocent alarms caused by NORM cargoes. The ability of a spectroscopic portal to characterize and discriminate against the NORM material in such a cargo, while still detecting a hidden threat object, is worth the increase in cost. Note that by screening only 1/50 or 1/100 of the traffic that causes primary alarms, 50- to 100-fold longer count times become practical. Large-volume cargoes can be searched more rapidly and with higher sensitivity than can be achieved with handheld radioisotope identifiers.

Passive Neutron Detection

Neutron Signatures

Unlike gamma-ray sources for which there are a large number of naturally occurring terrestrial sources, there are no naturally occurring terrestrial sources that are capable of producing neutrons in appreciable quantities. The principal natural source of neutrons is production by cosmic-ray interactions in both the atmosphere and in materials on the ground. Therefore, the detection of neutrons at rates above the normal cosmic-ray background is fairly rare, and when they are detected it almost always indicates a manmade source. Since many manmade neutron sources are associated with the nuclear fuel cycle and SNM, the detection of excess neutrons is taken very seriously.

Cosmic-Ray Neutron Sources

Galactic cosmic rays are very high-energy ionizing particles that originate from outside the solar system primarily composed of hydrogen nuclei (~85%) but also helium nuclei (~12%), electrons (2%), and a small fraction of heavier ions (~1%).[4,5] These particles can have energies of up to 10^{14} MeV or more, and when they collide with nuclei of atoms in the atmosphere, they create cascades of secondary particles of every kind, including lighter nuclei, protons, neutrons, pions, muons, electrons, and gamma rays. These atmospheric-generated neutrons form a random background at ground level but also, along with the atmospheric pions and protons, can create bursts of secondary neutrons when they collide with high-z (high number of protons in the nucleus) materials such as cargo, building materials, or the detector itself.[6]

Manmade Neutron Sources

Manmade neutron sources create neutrons through two principal neutron production processes: alpha-n (α,n) reactions and spontaneous fission. Neutron sources are used for a variety of applications, including soil moisture gauges, well-logging sources (used to explore for oil by measuring the neutron energy reflected from geologic strata outside a bore-hole), and neutron medical therapy. The typical neutron source produces 10^4 to 10^5 neutrons per second and is either an (α,n) source such as Pu-Be or Am-Li or a ^{252}Cf spontaneous fission source. The possession of such sources is usually controlled by licensing arrangements. These types of sources emit neutrons continuously and cannot be turned off, so when they are encountered in commerce they may cause innocent neutron alarms.

Neutron emission can also be induced by bombarding certain materials with high-energy photons, protons, or other particles produced in a particle accelerator. These active neutron sources can be turned on and shut off because they emit neutrons only when the accelerator is operating. When turned off, they are cannot cause innocent neutron alarms.

Alpha-*n* Reactions

Alpha decay is a spontaneous process that occurs at appreciable rates only in the heaviest elements like thorium, uranium, neptunium, plutonium, americium, and others. The emitted alpha particles have energies from about 4 to 6 MeV and can produce neutrons through (α,n) reactions when the emitted alpha particles interact with light elements such as oxygen and fluorine, as in the following examples:

$$\alpha + {}^{18}\text{O} \rightarrow {}^{21}\text{Ne} + \text{n}$$
$$\alpha + {}^{19}\text{F} \rightarrow {}^{22}\text{Na} + \text{n}$$

Production of neutrons through the (α,n) reaction requires an intimate mixing of the alpha-producing isotope with the low-Z target material.

---

[4] G. Reitz, "Radiation Environment in the Stratosphere," *Radiation Protection Dosimetry*, 48 (1993), p. 5.

[5] T. K. Gaisser, *Cosmic Rays and Particle Physics*, Cambridge: Cambridge University Press, 1990.

[6] P. Goldhagen, "Cosmic-Ray Neutrons on the Ground and in the Atmosphere," *MRS Bulletin*, Feb. 2003, p. 133.

Table 26.8 Spontaneous fission neutron yields.[13]

Isotope	Number of Protons Z	Number of Neutrons N	Spontaneous Neutron Yield (n/s-gm)
^{232}Th	90	142	~6×10^{-8}
^{233}U	92	141	8.6×10^{-4}
^{235}U	92	143	3.0×10^{-4}
^{238}U	92	146	1.4×10^{-2}
^{237}Np	93	144	1.1×10^{-4}
^{239}Pu	94	145	2.2×10^{-2}
^{240}Pu	94	146	1.02×10^{3}
^{244}Cm	96	148	1.1×10^{7}
^{252}Cf	98	154	2.3×10^{12}

Spontaneous Fission

Neutrons can also be emitted spontaneously from a small number of isotopes having an excess number of neutrons in the nucleus. The isotopes capable of spontaneous fission with appreciable neutron emission are all heavy isotopes associated with the nuclear fuel cycle, including plutonium, americium, curium, and californium. Spontaneous fission rates are strongly dependent on the number of protons and neutrons in the nucleus with the spontaneous fission rate for so-called "odd-even" isotopes (odd number of protons, even number of neutrons) typically many orders of magnitude less than the rate for "even-even" isotopes. The neutron emission rates for isotopes of interest are shown in Table 26.8 Of particular importance is the low neutron yield of ^{235}U and relatively low yield of ^{238}U. These isotopes produce neutrons at such a low rate that it is not possible to detect their low neutron emissions in field conditions. However, the yield from ^{240}Pu is high enough that the ^{240}Pu component in threat quantities of plutonium can be readily detected by its neutron emission.

Neutron Detection Fundamentals

Neutron detectors used to detect SNM and other potential threats rely on detector materials that have nuclear reactions with neutrons and create reaction products that deposit energy in the detector in such a way that the deposited energy can be converted to a signal that can be amplified and counted. An understanding of how neutrons interact with materials and how neutrons can be shielded is necessary to understand the capabilities and limitations of neutron detectors. Neutron counting statistics are essentially the same as for gamma rays.

Neutron Interactions with Materials

A neutron, being without charge but having mass, interacts primarily with the nucleus of atoms. A neutron can undergo many different kinds of interactions, including:

- *Scattering.* The neutron's speed (energy) and direction change but the nucleus is left unchanged.
- *Absorption.* The neutron has a reaction with the nucleus, causing the nucleus to emit a wide range of other particles or to fission.

In absorption, the neutron is essentially incorporated into the nucleus, changing the configuration of the nucleus. For example, in an (n, γ) reaction the neutron is absorbed by the nucleus and excess energy is radiated away as gamma rays. In other reactions, such as the $(n, 2n)$ reaction, other nucleons are emitted; in this case two neutrons are emitted when the original neutron is absorbed. Finally, in a fission reaction, the nucleus splits into two smaller nuclei, typically with the release of a number of neutrons.

Threshold nuclear reactions require the neutron to possess at least a minimum amount of energy; an example is fission of ^{238}U, a fissionable nuclide that requires the neutron to have about 1 MeV of energy before the reaction can progress. Nonthreshold reactions can proceed even when the neutron has so little energy it has essentially come to thermal equilibrium with its environment. At room temperature, a fully thermalized neutron has about 0.025 eV of energy and is traveling, on average, about 2,200 meters per second. Thermal neutrons can cause fission to occur in fissile nuclides such as ^{235}U and ^{239}Pu.

Neutron Shielding

Effective neutron shielding must employ materials with a high probability of neutron absorption. However, the probability of absorption is low in all materials for high-energy neutrons such as those spontaneously emitted from plutonium which are typically emitted with an average near 2 MeV. The probability of absorption rises very steeply as the neutron loses energy, with the maximum probability occurring at thermal energy.

To shield neutrons effectively, they must first lose most of their energy through scattering reactions until they reach energies near thermal. The process of losing energy by scattering is referred to as *moderation*, and materials effective in moderation are called *neutron moderators*. The most effective neutron moderators contain hydrogen, because hydrogen, being nearly the same mass as a neutron, can take up on average one half of the neutron's energy in each collision. Heavier nuclides take up much less energy than hydrogen in a collision with a neutron. This effect is clearly demonstrated by observation of the cue ball in a game of pool. In a glancing blow, the cue ball, like the incident neutron, will lose a fraction of its energy to the target ball (like a hydrogen target with essentially the same mass as the neutron). However, if it collides head-on with the target ball, the cue ball can come to a complete stop, transferring all its energy to the target. This is only true for a target ball with the same mass as the cue ball. Experience tells us that a cue ball striking a bowling ball will never come to rest, even in a head-on collision. The lower-mass ball will retain most of its original speed and energy but will change direction (and might even completely reverse direction). In the same way, neutrons will change direction but retain most of their energy when they collide with any high-Z material such as lead. Therefore, lead or other high-Z materials are not good moderators.

Neutron shields must contain enough hydrogenous material, such as water or polyethylene, for moderation, along with enough absorbing material, such as boron or cadmium, to absorb the neutrons once they have thermalized. Neutron shields are therefore typically very thick compared with gamma shields. Pure water is both effective as a moderator and absorber and a very effective neutron shield in thicknesses of a meter or more. Water-containing cargoes such as fresh fruit or bottled wine have lower average density and require correspondingly larger thicknesses to be effective.

Neutron Detector Physics

The ^{3}He gas-filled tube is the most common type of neutron detector used to detect plutonium. ^{3}He occurs naturally at only about 1 ppm (parts per million) in natural helium, so it is usually prepared via the radioactive decay of tritium produced in nuclear reactors and is therefore quite expensive. A typical 1 meter long ^{3}He tube costs about $1,000, exclusive of the preamplifier and other counting electronics. The nuclear reaction that takes place in an ^{3}He detector is:

$$n + {}^{3}He \rightarrow {}^{3}H + {}^{1}H + 765\,KeV$$

This reaction, like most nonthreshold neutron reactions, is much more likely to take place with neutrons near thermal energy—in fact, about 5,000 times more likely for thermal neutrons than for unmoderated neutrons emitted from plutonium. Therefore, a moderator, typically polyethylene, is usually used to slow the neutrons before they enter the detector tube. A high voltage applied to the tube collects the ionization charge resulting from the proton (^{1}H) and triton (^{3}H) slowing down in the ^{3}He gas. The collected charge is then converted to an electronic signal that can be counted. Since the 765 KeV of reaction energy is shared between the two reaction products, the signal collected is very large and easily separated from the relatively small signals induced by gamma rays and other particles.

Neutron Search

The absolute efficiencies of neutron detectors are low; typically only a few percent of incident neutrons are detected even with large detectors. Handheld neutron detectors with small detectors have very low absolute efficiencies and are useful only for finding large neutron sources from relatively close distances. Many neutron detection applications require the suspect individual or vehicle to pass through a portal monitor that both affords large neutron detectors and ensures close detector-to-source distances.

Neutron Detection Equipment

Neutron portal monitors, like gamma-detecting portal monitors, were developed for facility security applications and later applied to screening vehicles and cargoes in threat-reduction scenarios.

Many detection materials have been investigated, but most commercial devices use polyethylene-moderated ^{3}He detectors for neutron detection. A polyethylene "box" is constructed containing one or more ^{3}He-tube detectors, shown in Figure 26.8. Neutrons are moderated to thermal energies by passage through the hydrogen-rich polyethylene and are absorbed efficiently in the ^{3}He, causing electrical pulses that are counted by the appropriate electronics.

Neutron backgrounds are substantially lower than gamma backgrounds. Measured with commercial neutron-detecting portal monitors, the backgrounds range ~20–40 counts/second at high-altitude sites down to ~2–5 count/second at many sea-level sites. The measurements are therefore governed by Poisson statistics, for which:

$$P(\mu, x) = \frac{\mu^x}{x!} e^{-\mu}$$

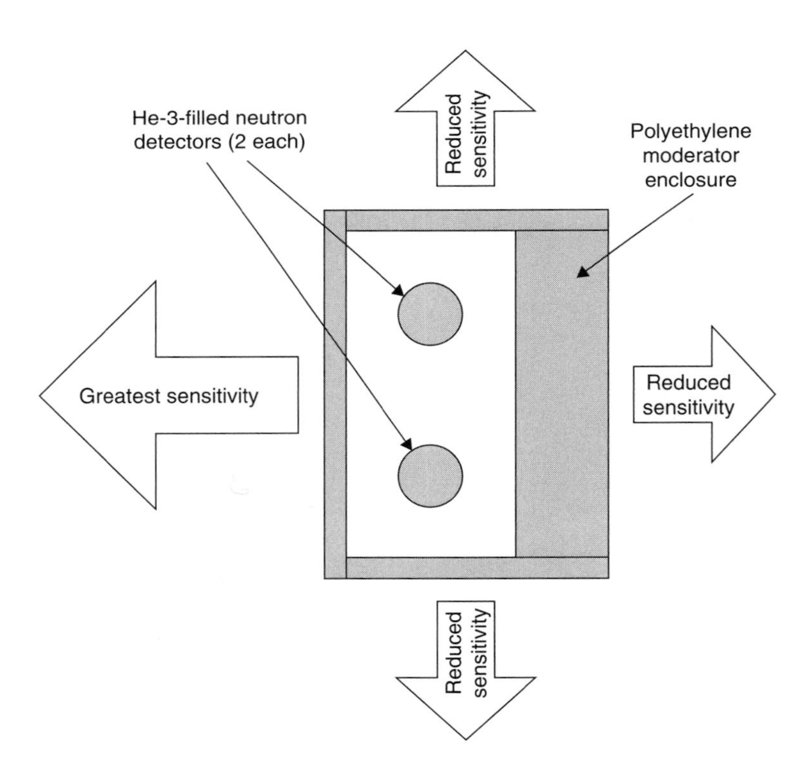

FIGURE 26.8 Schematic diagram of a cross-section through a moderated ^{3}He detector.

where $P(\mu,x)$ is the probability of recording a given integer count x for an average count rate of μ. Although an alarm threshold similar to that for gamma counting can be used, a more sensitive algorithm is the Sequential Probability Ratio Test. The background count rate is measured as with gamma detectors, as a T-second average when the monitor is unoccupied. When the monitor is occupied, a series of 1 second measurements is recorded and compared to the background as follows. The ratio of the probability that a given count comes from a hypothetical "alarm" distribution with average μ_A versus the probability that it comes from a "background" distribution with average μ_B is given by:

$$\frac{P(\mu_A, x)}{P(\mu_B, x)} = \left(\frac{\mu_A}{\mu_B}\right)^x e^{-(\mu_A - \mu_B)}$$

Repeated measurements that yield counts $x_1, x_2, \ldots x_N$ give the sequential ratio of probabilities as:

$$\frac{P(\mu_A, N)}{P(\mu_B, N)} = \left(\frac{\mu_A}{\mu_B}\right)^{x_1 + x_2 + \cdots + x_N} e^{-N(\mu_A - \mu_B)}$$

The Sequential Probability Ratio Test for a series of counts determines whether the observed counts are more likely to arise from the alarm distribution or the background distribution—in other words, whether the ratio exceeds 1. An alarm distribution is chosen to be that which gives a particular value of the rate of false alarms (occupancies that shouldn't alarm but do), with a given rate of false "negatives" (occupancies that don't alarm but should). The alarm distribution is determined as a function of background count rate B using Monte Carlo computer simulations and is often implemented in commercial radiation portal monitors as a lookup table. The sensitivity of this test, measured in grams of SNM required to alarm, is proportional to the inverse of the increase in count rate required to alarm $(\mu_A - \mu_B)^{-1}$. The minimum detectable quantity (MDQ) of SNM is smallest at low backgrounds and increases at high backgrounds and therefore must be specified at a particular neutron background. The sensitivity as a function of background for a particular portal monitor is shown in Figure 26.9. Commercial portal monitors typically alarm in the presence of ~150–200 grams of weapons-grade plutonium or other neutron source that emits 12,000–16,000 neutrons/second.

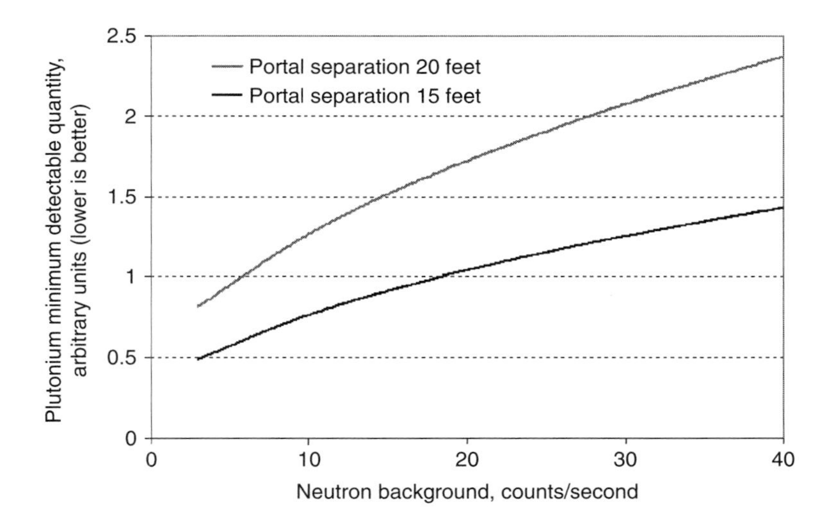

FIGURE 26.9 Effect of neutron background on minimum detectable quantity of plutonium.

Radiography

In some situations, it might be possible to detect threat objects by radiography. Radiographic machines typically use a source of ionizing radiation such as a high-energy X-ray machine or a gamma-ray source placed on one side of the object to be inspected and an array of detectors on the other side. The suspect object is then translated through the radiograph or the source and detector arrays are scanned across the object, which results in a two-dimensional transmission image similar to an X-ray image.

Recognizing all potential threat objects and differentiating threats from other objects that show up on a radiography image can be problematic. In the simplest case, with homogenous, low-density cargo, a radiographic image can easily identify dense objects that are out of place in that cargo. However, in complex, high-density cargo such as machine tools or automobile parts, the situation becomes more difficult. One difficulty stems from the sensitive or classified nature of many threats; it is not possible to share detailed information with radiograph operators that would allow them to reliably recognize the image of all potential threat objects and differentiate threats from other dense but benign objects.

Software designed to assist the operator by automatically identifying threat objects is also problematic. Automated threat recognition software requires the use of threat templates to describe the shape and characteristics of threats. Those templates, if they were to fall into the wrong hands, could be used to infer sensitive details about weapons design, and therefore the software must be protected against any terrorist attempt to extract the threat templates, realistically limiting its use to only secure environments. These limitations on training and threat templates means that threat recognition in the field can only be effective against those threats that resemble the simplest unclassified representation of real threat objects. One possibility to deal with these limitations is for the radiograph operator to send all complex radiograph images to a central secure, classified facility for image analysis by analysts with the proper security clearances.

Passive detection of SNM and other radiological threats cannot be assured under all conditions because radiation emissions from the threat can be shielded from the detector. Shielding is a special concern in monitoring bulk cargo, cargo containers, and large vehicles. Despite the limitations in recognizing classified threat objects, radiography systems capable of imaging the contents of containers and vehicles can not only provide the means to detect out-of-place purposeful shielding but configurations of cargo that might by itself preclude effective passive monitoring. With knowledge about the location and size of potential shields in the container obtained through radiography, the inspector in the field can make better decisions regarding which containers should be more intensively inspected or unpacked, if necessary.

Gamma and X-Ray Radiography

Gamma radiography equipment is already extensively used by customs inspectors to find contraband hidden in vehicles or shipping containers. These systems are also used to assist in inspections for SNM and other radiological threats. An example of a radiography system in widespread use is the SAIC Vehicle and Container Inspection System,[7] or VACIS. The VACIS operates by shining a collimated beam of gamma rays from ^{60}Co through the container to an array of NaI detectors on the other side. An example image of passenger vehicles on a truck transport is shown in Figure 26.10. Note that the engine blocks in this image are essentially opaque and a threat object would not be visible if placed along the same line-of-sight as the engine. An astute inspector would also note that HEU placed inside the engine block would not be detected by a passive radiation detector and could use the radiographic image to assist in deciding whether any additional inspection steps are needed.

---

[7] J. E. Gormley, personal communication, May 2, 2007. Figure courtesy SAIC Corporation, Rancho Bernardo, CA.

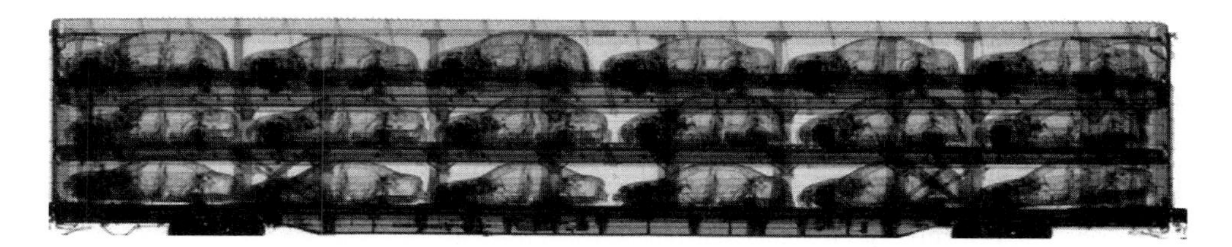

FIGURE 26.10 Example radiographic image of passenger vehicles on a vehicle carrier car.

Other currently deployed radiography systems use high-energy photon sources from X-ray generators as the source rather than an isotopic source. These systems can achieve higher photon energy, with commercial devices available up to about 9 MeV, which provides greater penetrability through dense, high-*z* cargo.

Neutron Radiography

A neutron shield found coincident with a gamma shield would strongly indicate an attempt to prevent detection of plutonium, because both kinds of shields together would be necessary to avoid plutonium detection by a combined neutron and gamma detector system. However, gamma and X-ray radiography is not effective in identifying neutron shielding since, being made mostly of low-*z* hydrogenous materials, neutron shields are not dense and might not be clearly visible on a gamma or X-ray radiograph. Neutron radiography has been suggested as a means to identify thick neutron shields where, instead of using an X-ray or gamma source, the container would be bombarded with high-energy neutrons and imaged with a pixellated neutron detector.

The capability to image neutron shielding is of limited practical value to augment combined gamma radiography and passive radiation detection systems, because *any* gamma shield, even a gamma shield not coincident with a neutron shield, would by itself cause concern that it could be concealing an HEU source. Since HEU cannot be passively detected by neutrons, essentially the same exact secondary inspection steps must be taken whether or not a neutron shield is discovered coincident with the gamma shield, and a neutron shield not coincident with a gamma shield would not by itself prevent threat detection by gamma emissions. The only situation where this might not be true is on a smuggling path where plutonium is the only known risk, with no possibility of HEU being smuggled.

Nevertheless, neutron radiography could provide useful information to augment active interrogation and is further discussed in the active interrogation section.

Other Radiography Concepts

Other concepts have been suggested to improve radiography capabilities. Examples include dual-axis radiography or tomography capable of generating 3-D images, dual- or multiple X-ray energy techniques capable of differentiating shield materials by atomic number, and Muon radiography,[8] an approach capable of generating a 3-D image using only cosmic-ray muons with no other external source. In general, all these approaches suffer from the inability to train the operators in classified threat recognition, but they could be used to assist the inspector in identifying shielding materials.

Active Interrogation

Shielded HEU can be difficult or impossible to detect using passive technology owing to its low radioactivity, low spontaneous neutron emission, and easily shielded gamma emissions.

---

[8]L. J. Schultz et al., "Image Reconstruction and Material Z Discrimination Via Cosmic Ray Muon Radiography," *Nucl. Inst. and Methods*, vol. 519, iss. 3, March 2004, pp. 687–694.

Reliable HEU detection in large containers will require development of safe, deployable, and operationally capable detection technologies capable of threat detection, even inside thick shielding materials. To meet this need, a number of active interrogation approaches are currently being developed that rely on interrogating the container with external radiation sources capable of inducing reactions in SNM.[9] In all approaches, the reactions induced in SNM result in the emission of relatively penetrating radiation that is observable outside the container. A brief outline of two active interrogation approaches follows.

Neutron Interrogation

Active neutron interrogation has been demonstrated to be highly effective for locating and assaying SNM in nuclear waste. A number of systems have already been successfully deployed that bombard the waste container with high-energy neutrons and measure the neutrons generated from induced fission in the SNM. One example is the crated waste assay monitor (CWAM) deployed at the Oak Ridge Y-12 Plant.[10] CWAM operates by interrogating 4 foot × 4 foot × 6 foot waste containers with 14 MeV neutrons generated by a deuterium-tritium accelerator source. The 14 MeV neutrons moderate in the CWAM cavity and are captured in the SNM, resulting in fission reactions causing fission neutrons to be emitted. These emitted fission neutrons are then detected in ^{3}He detectors, yielding a signal proportional to the amount of SNM in the waste container. This device can detect subgram quantities of HEU in waste. The active interrogation package monitor (AIPM) developed at Los Alamos National Laboratory is based on the same measurement technique used in CWAM but was developed and field tested to be able to detect SNM transported in small packages up to about 1 meter on a side.[11]

Existing deployed techniques show much promise for detecting SNM in small containers, but none of these techniques is expected to provide reliable SNM detection concealed in a fully loaded cargo container, thus prompting development of techniques better able to penetrate cargo of all kinds and densities. One example of neutron interrogation technology under development is the Lawrence Livermore National Laboratory "nuclear carwash."[12] This system interrogates cargo containers with a beam of neutrons in the energy range from 4 to 8 MeV that induce fission in concealed SNM. The system then measures the emission of gamma rays, with energy greater than about 3 MeV emitted following the beta decay of the neutron-induced fission products. These high-energy gamma rays are emitted sometime after the neutron interrogation with a delay characteristic of the short half-lives of radioactive-induced fission products. This beta-delayed gamma radiation following neutron interrogation is an abundant and robust signature for SNM detection, even in shielded cargo.

High-Energy Photon Interrogation

Another active interrogation approach is to use very high-energy photons to induce fission. The pulsed photonuclear assessment inspection system (PPA) under development by Idaho National Laboratory consists of a pulsed electron accelerator configured to produce high-energy

---

[9] C. E. Moss, C. L. Hollas, G. W. McKinney, and W. L. Myers, "Comparison of Active Interrogation Techniques," *IEEE Transactions on Nuclear Science*, vol. 53, no.4, Aug. 2006.

[10] S. G. Melton, R. J. Estep, and E. H. Peterson, "Calibration of the Crated Waste Assay Monitor (CWAM) for the Low-Level Waste Measurements for the Y-12 Plant," Los Alamos Report LA-UR-00-2468, presented at the 7th Nondestructive Examination Waste Characterization Conference, Salt Lake City, Utah, May 23–25, 2000.

[11] B. D. Rooney, R. L. York, D. A. Close, and H. E. Williams III, "Active Interrogation Package Monitor," in *Proc. IEEE Nuclear Science Symposium*, vol. 2, 1999, pp. 1,027–1,030

[12] D. Slaughter et al., "Detection of Special Nuclear Material in Cargo Containers Using Neutron Interrogation," Lawrence Livermore National Laboratory report UCRL-ID-155315, Aug. 2003.

photons and a neutron detection system capable of detecting the induced photo-fission emissions. These emitted particles are detected during the time between pulses of the accelerator.[13] The system detects beta-delayed neutrons from the photon-induced fission products; a clear indication of the presence of SNM. The current prototype has been used to assess performance using bremsstrahlung photons of up to 12 MeV, well above the approximately 6 MeV photo-fission threshold of SNM. One difficulty with this approach is that bremsstrahlung radiation, photons generated from the slowing of the accelerated electrons in a metal target, essentially the same mechanism used in the production of medical X-rays, has a wide energy range that extends well below the photo-fission threshold. The lower-energy photons do not contribute to the induced fission rate but add radiation dose to the container contents, potentially causing very high doses to be delivered—an unwanted effect. To overcome this problem, research is being conducted to develop high-energy monoenergetic gamma sources based on other reactions such as accelerated protons on ^{11}B targets.

---

[13] J. L. Jones et al., "Photonuclear-based Nuclear Material Detection System for Cargo Containers," *Nuclear Inst. and Methods in Phys. Research*, Sec. B: Beam Interactions with Materials and Atoms, vol. 241, iss. 1–4, Dec. 2005, pp. 770–776.

27 A Model for Attribution of Terrorist Nuclear Attacks

William S. Charlton, Mark Scott, David Burk, and Adrienne LaFleur

Introduction

During the Cold War, the U.S. had a small number of well-defined enemies. Nuclear weapons attacks would have almost certainly involved detectable military actions such as intercontinental missile or aircraft attacks whose point of origin could be quickly identified by early warning systems. Thus, in the event of a nuclear weapons attack on the U.S., identifying the country that attacked and what the appropriate response to this attack would be were relatively simple.

In today's security environment, the most likely threat of a nuclear or radiological attack is not from a state but from a subnational, or terrorist, group. This threat is not well defined and generally has no primary base of operations, and the local citizenry might be unaware of and not condone the group's actions. Thus, even determining who was responsible for such an attack will be difficult, as will be selecting an appropriate response. To help deter such attacks and respond appropriately, the United States would benefit from the ability to identify the origin of the nuclear materials involved in the attack. Determining origin does not prove culpability of the national owner of the material used in the attack, but it can help identify the attackers and alert the legitimate custodian of the nuclear material that it had been stolen and used in an attack.

To begin the process of identifying the perpetrators of a nuclear or radiological attack, many questions must be answered and useful attributes or signatures that can be linked to the attackers must be identified. These attributes could include many characteristics of the device, but one of the most useful characteristics for identifying the actors involved in the event would be the characteristics of the original nuclear material used in the device.

For determining the original material in the device, it is useful to divide the devices of interest here into five types:

- Radiological weapons using single isotope sources (for instance, medical isotopes, industrial irradiator isotopes, etc.)
- Radiological weapons using spent nuclear fuel
- Improvised nuclear device (IND) using highly enriched uranium
- IND using plutonium
- Thermonuclear weapons

For radiological weapons, there is no change of the materials in the weapon during detonation (meaning that the same nuclides are in the residue from the explosion that were in the original predetonated device). For nuclear weapons, the isotopes of the device will change during irradiation.

Of the types listed, only the spent fuel radiological weapon and HEU IND will be considered in detail here. The first type is not considered because its attribution is relatively simple (involving only the measurement of the principle isotopes used in the device as well as any impurities and then matching that to a database of manufactured sources). The two types chosen provide a general basis for the understanding of how post-nuclear event attribution can be performed.

Forward and Inverse Models for Material Attribution

Nuclear event attribution is based on inverse and forward models. Forward models are more commonly used and understood. Forward models use present-day observables and the physics of the system to simulate the way a system will evolve forward in time. This allows for a prediction of the state of the system at some future time based on knowledge of its present state and understanding of its physics.

Many scientists and engineers are not familiar with inverse models. Inverse models are used to predict the original state of a system, given knowledge of its present state. Thus, an inverse model may determine the source of material that is most likely to lead to a present-day observable. For example, given a present-day isotopic measurement, an inverse model can be used to predict the material attributes and the transpiring events that are most likely to have led to those recorded measurements. It must be noted that since the present-day observables contain measurement uncertainties, the inverse model is an inherently complicated problem. So, an exact solution can never be found, only the most likely solution. There might be other solutions that could also lead to the observable events. This must be kept in mind when the results of an inverse modeling exercise are used to inform a decision on responding to a nuclear or radiological attack.

The post-event attribution determination described in this chapter was reached using both forward and inverse models to characterize the original material used in the weapon. A general overview of the post-event attribution method is given in Figure 27.1. Following

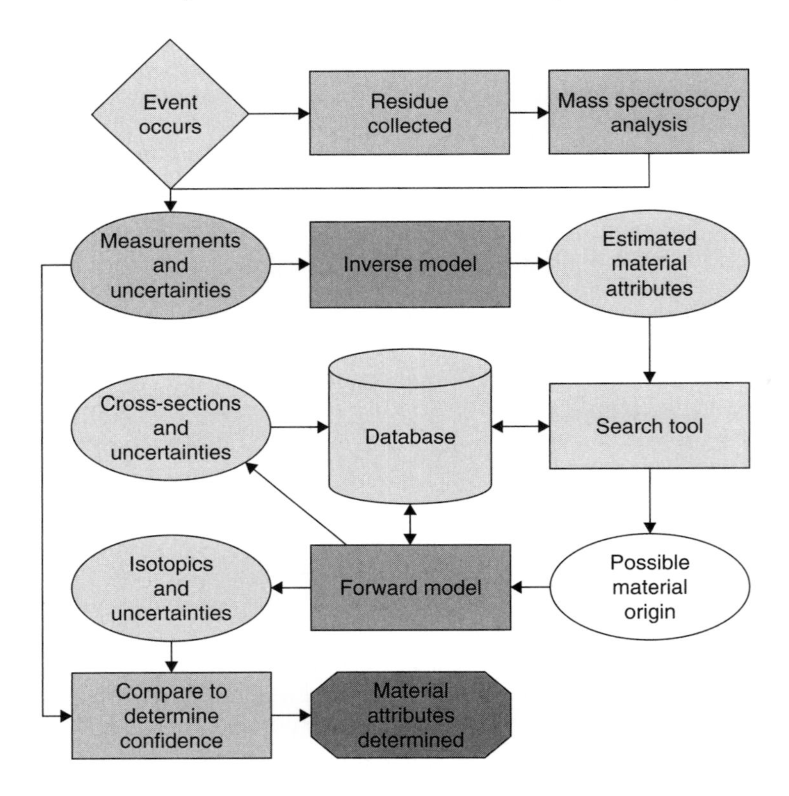

FIGURE 27.1 Overview of post-event attribution.

the event, residue will be collected from around the blast area. This residue will then be analyzed. This will involve a mass spectroscopic analysis to determine the isotopic content of the residue. The mass spectroscopy measurements are generally performed as isotopic ratios (usually relative to ^{238}U). The measured isotopic ratios and their uncertainties will then be used in an inverse model to predict the most likely attributes for the material used in the weapon. These attributes will then be compared to a database of known materials via a search tool to determine possible source materials that could have been in the weapon. We then use a forward model to simulate each of these possible materials forward in time to predict the present-day observables (i.e., residue isotopics) that would be expected following the detonation of the weapon. We then make a comparison of these predicted observables to the measured isotopics to determine the most likely source of material for the weapon.

In the next three sections, we discuss the particular models that can be used for a spent fuel RDD, HEU IND, or Pu IND nuclear event. All these models are based on analytical inversions of the buildup and depletion equations.

Spent Fuel RDD Attribution

If spent nuclear fuel is used as the radiological material for the weapon, it would prove to be a physical hazard and cause extensive sociological disruption. Described here is a methodology to attribute spent fuel to a source reactor assembly. The specific attributes determined via the inverse model are the spent fuel burnup, age from discharge, reactor type, and initial fuel enrichment. The characteristics found can then be used by a search tool. This search tool will look for facilities that match the reactor type and used fuel that matches the initial enrichment. It will then see whether the facilities produced spent fuel with the same age and burnup. If it finds any matches, it will check the suspect facilities with the database of known missing material. It is also possible that spent fuel would have been shipped from the reactor to another country for storage. Records of this activity should be available, and only one or two matches will likely be made.

To validate the inverse model results, a forward model of the suspect reactor will be created. The power history of the suspect fuel will then be retrieved and a forward model simulation performed for that specific fuel assembly. The results will be compared to the measured data, and if the data agree, the material will be considered attributed. Comparing the results of a forward model code to the measured values will generally allow for a more accurate match.

Before the inverse model can use the measurements, one-group cross-sections for each reactor type being considered must be stored in a database. These one-group cross-sections can be generated by a forward model lattice physics code (such as HELIOS, CASMO, TransLAT, APOLLO, or Monteburns). We chose to use Monteburns here.[1] Monteburns uses two other forward simulation codes, MCNP[2] and ORIGEN,[3] to make its calculations. This step can be performed in advance and the cross-sections can be stored in a database. Generating these cross-section sets is a well-known technique and will not be discussed further in this work.

The inverse model used for the spent fuel RDDs uses an iterative procedure to determine the fuel burnup, enrichment, reactor type, fuel age, and fuel power history. The iteration process is shown in Figure 27.2.

---

[1]D. L. Poston and H. R. Trellue, "User's Manual, Version 2.0 for Monteburns Version 1.0," LA-UR-99-4999, Los Alamos National Laboratory (1999).

[2]J. F. Briesmeister, "MCNP: A General Monte Carlo *N*-Particle Transport Code Version 4B," LA-12625-M, Los Alamos National Laboratory (1997).

[3]S. Ludwig, "ORIGEN: The ORNL Isotope Generation and Depletion Code," CCC-0371/17, Oak Ridge National Laboratory (2002).

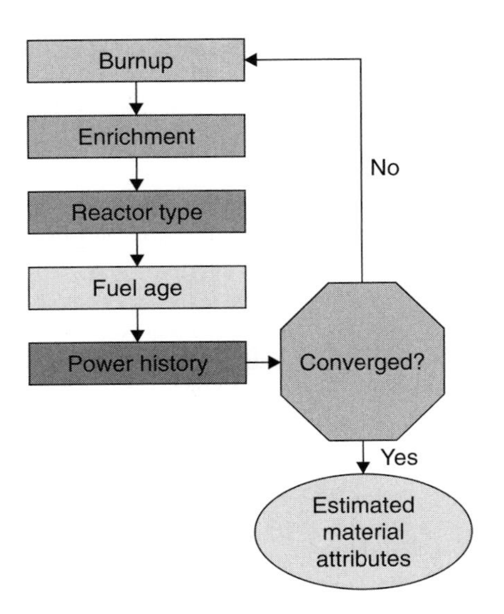

FIGURE 27.2 Iterative procedure for determining spent-fuel RDD attributes.

Monitor Isotopes and Methodology

Each fuel characteristic of interest to the forensics problem uses an isotopic monitor. The fuel characteristics of interest here are fuel burnup, fuel enrichment, reactor type, and fuel age. Each characteristic has unique requirements that the monitor must meet. A specific set of monitors must be used for determination of the enrichment. Burnup, reactor type, and age determination can be performed using any combination of isotopes that meet the requirements.

Burnup

The nuclear industry has developed accurate methods to measure fuel burnup.[4] In nuclear power plants, initially the fissions from ^{235}U (and, to a lesser extent, ^{238}U) provide the source of neutrons and power in the reactor. As the reactor burns the fuel, ^{236}U, ^{239}Pu, ^{240}Pu, and several other minor fissionable actinides are created and fissioned, which then contribute to the neutron flux. After the fuel is discharged from the reactor, the burnup can be measured indirectly using a burnup monitor. It has been shown that burnup, coupled with mass spectrometry, can be measured within 1% accuracy.[5]

A burnup monitor can be any fission product that is created directly proportional to the burnup. For this problem a more stringent restriction exists because the reactor type is unknown. Therefore, we need burnup monitors that are produced at the same rate, regardless of the reactor type. The neutron energy spectrum and fissioning isotopes are primarily what changes from one reactor type to another. So, for our purpose, a burnup monitor is any

---

[4]A.V. Bushuev, A. F. Kozhin, G. Li, V. N. Zubarev, A. A. Portnov, V. P. Alferov, and M. V. Shchurovskaya, "Determination of the Fuel-Assembly Burnup in a Research Reactor by Repeated Short-Time Irradiation Followed by G Spectrometric Measurements," *Atomic Energy*, **97**, 2 (2004).

[5]J. J. Giglio, D. G. Cummings, M. M. Michlik, P. S. Goodall, and S. G. Johnson, "Determination of Burnup in Spent Nuclear Fuel by Application of Fiber Optic High-Resolution Inductively Coupled Plasma Atomic Emission Spectroscopy," *Nuclear Instruments and Methods in Physics Research A*, **396**, 251 (1997).

nuclide whose fission yield is constant as a function of the neutron flux energy and fissioning isotopes. By identifying a burnup monitor that has a constant yield across reactor types, the burnup may be determined without knowing additional information about the fuel. Additionally, burnup monitors should have long half-lives (or be stable), have small absorption cross-sections, and be produced through simple fission product decay chains.

From the basic buildup and decay equations, we can derive equations for the burnup of an assembly (BU) at some irradiation time (T) as a function of the ratio of the burnup monitor's atom concentration (N^B) to the ^{238}U atom concentration (N^{U238}):

$$BU(T) = \left[\frac{N^B(T)}{N^{U238}(T)}\right]\left[\frac{N^{U238}(T)}{N_0^U}\right]\frac{N_A E_R}{Y_B} \tag{27.1}$$

where $N_0{}^U$ is the concentration of U atoms at time $T = 0$, N_A is Avagadro's number, E_R is the recoverable energy per fission, and Y_B is the cumulative fission product yield for burnup monitor B. The ratio of the ^{238}U atoms at time T to the total uranium atoms at time $T = 0$ is given by:

$$\left[\frac{N_0^U}{N^{U238}(T)}\right] = \frac{\left[\frac{N^{U235}(T)}{N^{U238}(T)}\right] + \left[\frac{N^{U238}(T)}{N^{U238}(T)}\right] + \left[\frac{N^{Pu239}(T)}{N^{U238}(T)}\right] + \cdots}{1 - \frac{M_0^U}{N_A E_R} BU(T)} \tag{27.2}$$

where $M_0{}^U$ is the initial uranium atomic mass at $T = 0$ and the series includes all fissioning isotopes. These two equations can be solved iteratively for the burnup by first setting $BU(T) = 0$ in Equation 27.2 and solving for $\left[N_0^U / N^{U238}(T)\right]$. The $BU(T)$ can then be solved for using Equation 27.1. This can then be repeated between Equations 27.1 and 27.2. The burnup equation is not very sensitive to changes in the initial uranium atom density, allowing it to converge after only a few iterations. Note that the initial uranium atomic mass is not known *a priori*, but it can be reasonably assumed to be equal to approximately the atomic mass of ^{238}U. Once the original fuel enrichment is known, this value can be updated with a more accurate value for $M_0{}^U$.

Enrichment

If no other fissionable isotopes are present while the fuel is in the reactor, the enrichment will change linearly with burnup. In reality, several other fissionable isotopes are produced and burned, thus complicating the determination of the initial enrichment. To perform this calculation we assume that:

- The only isotopes that fission are ^{235}U, ^{238}U, ^{239}Pu, ^{240}Pu, and ^{241}Pu.
- There is no production of ^{235}U and ^{238}U. All one-group cross-sections are constant with time.
- ^{239}Np and ^{239}U decay instantaneously to ^{239}Pu.
- The decay of all fissionable isotopes is neglected.

The only isotope that has a short enough half-life to significantly affect this calculation is ^{241}Pu. ^{241}Pu has a 14.1 year half-life, which will cause this calculation for initial enrichment to be slightly high. The higher the burnup and the longer the fuel has been discharged, the larger the effect will be on the results.

The initial enrichment is first predicted using an equation developed from a balance equation of the uranium atom density:

$$e_0 = \frac{N^{U238}(T)}{N_0^U}\left[\frac{N^{U235}(T)}{N^{U238}(T)} + \frac{N^{U236}(T)}{N^{U238}(T)}\right] + \frac{M_0^U}{N_A E_R} BU(T) - G^{238} - G^{239} - G^{240} - G^{241} \tag{27.3}$$

where

$$G^{238} \equiv \frac{\bar{\sigma}_f^{U238}}{\bar{\sigma}_a^{U238}} \left[1 - e_0 - \frac{N^{U238}(T)}{N_0^U} \right] \tag{27.4}$$

$$G^{239} \equiv \frac{\bar{\sigma}_f^{Pu239}}{\bar{\sigma}_a^{Pu239}} \left[-\frac{N^{Pu239}(T)}{N^{U238}(T)} \frac{N^{U238}(T)}{N_0^U} + G^{238} \right] \tag{27.5}$$

$$G^{240} \equiv \frac{\bar{\sigma}_f^{Pu240}}{\bar{\sigma}_a^{Pu240}} \left[-\frac{N^{Pu240}(T)}{N^{U238}(T)} \frac{N^{U238}(T)}{N_0^U} + G^{239} \right] \tag{27.6}$$

$$G^{241} \equiv \frac{\bar{\sigma}_f^{Pu241}}{\bar{\sigma}_a^{Pu241}} \left[-\frac{N^{Pu241}(T)}{N^{U238}(T)} \frac{N^{U238}(T)}{N_0^U} + G^{240} \right] \tag{27.7}$$

where e_0 is the initial enrichment, $\bar{\sigma}_f^X$ is the one-group fission cross section for isotope X, and $\bar{\sigma}_a^X$ is the one-group absorption cross-section for isotope X. Since e_0 is contained in both Equations 27.3 and 27.4, this solution requires iteration. In addition, benchmarking showed that the initial enrichment predicted from this solution can have errors as high as 15%, which is too high for accurate attribution. An additional forward model iteration step is instead implemented to improve the enrichment prediction using ORIGEN. This iteration scheme is shown in Figure 27.3. By adding this extra step, we reduce the initial enrichment error to less than 2%.

Reactor Type

The ability to predict the reactor type relies heavily on the accuracy of the one-group cross-sections and fission product yield values. To differentiate the reactor types from each other, isotopes with cross-sections and yields that change significantly from reactor type to reactor type are needed. To avoid the complication of decay, stable or long-lived isotopes are used. Using the known burnup and enrichment from above, these monitors are then used to determine the cross-section and yield sets (which essentially define the reactor type) that best fit the measured data.

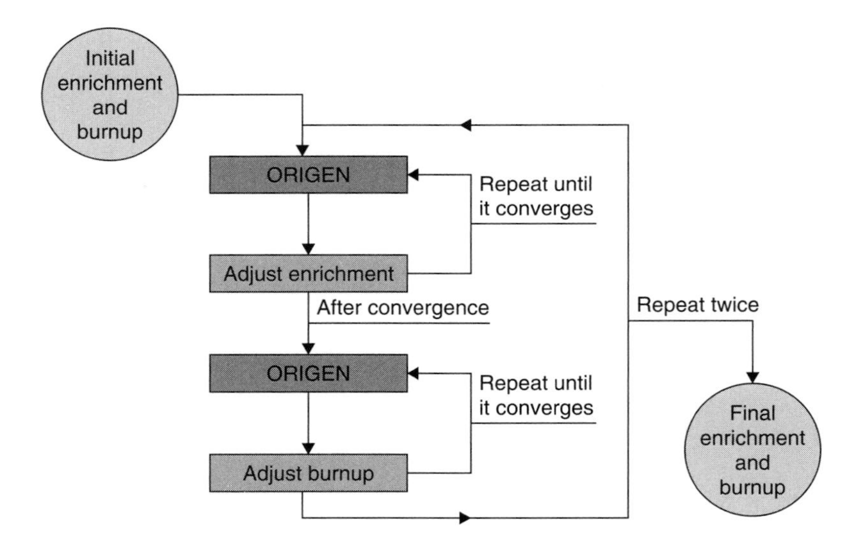

FIGURE 27.3 Enrichment iteration procedure.

It can be shown that the quantity of any long-lived fission product x in spent fuel which:

$$\frac{N^{x}(T)}{N^{238}(T)} = Y_x^{235}\sigma_f^{235}\frac{\int_0^T N^{235}(t)\phi(t)\,dt}{N^{238}(T)} + Y_x^{238}\sigma_f^{238}\frac{\int_0^T N^{238}(t)\phi(t)\,dt}{N^{238}(T)} + \cdots \tag{27.8}$$

where Y_x^{235} is the one-group cumulative fission product yield for isotope x from fissioning isotope ^{235}U and $\phi(t)$ is the one-group scalar flux as a function of irradiation time t. By substituting two simple definitions, we can acquire:

$$R^{x}(T) = Y_x^{235}\sigma_f^{235}F^{235}(T) + Y_x^{238}\sigma_f^{238}F^{238}(T) + \cdots \tag{27.9}$$

Note that the assumption of a stable-fission product is not a necessary assumption. It is only assumed here because it makes the equations simpler.

We can also show that the burnup of the fuel is given by

$$BU(T) = \frac{1}{\rho_0^U}\int_0^T \left[E_R^{235}\sigma_f^{235}N^{235}(t)\phi(t) + E_R^{238}\sigma_f^{238}N^{238}(t)\phi(t) + \cdots\right]dt \tag{27.10}$$

which can be expressed as:

$$BU(T) = \frac{N_A E_R}{M_0^U}\frac{N^{238}(T)}{N_0^U}\left[\sigma_f^{235}F^{235}(T) + \sigma_f^{238}F^{238}(T) + \cdots\right] \tag{27.11}$$

Via substitution, we can acquire:

$$\frac{BU(T)}{C} = \sigma_f^{235}F^{235}(T) + \sigma_f^{238}F^{238}(T) + \cdots \tag{27.12}$$

where C is a known constant. Equations 27.9 and 27.12 then produce a complete set of equations defining the production of any set of fission product isotopes. We have a value for the burnup from a previous step, and the values for $R^{x}(T)$ are measured in the post-detonation residue. We determine the number of fission products needed to uniquely identify the reactor type based mainly on the number of fissioning isotopes in the fuel. Typically, the larger the burnup of the fuel (and to a lesser extent, the lower the initial fuel enrichment), the more fissioning isotopes occur in the fuel.

In matrix form, the set of Equations 27.9 and 27.12 for four fissioning isotopes (^{235}U, ^{238}U, ^{239}Pu, and ^{240}Pu) and three independent fission products is given by:

$$\begin{bmatrix} \sigma_f^{235} & \sigma_f^{238} & \sigma_f^{239} & \sigma_f^{240} \\ Y_x^{235}\sigma_f^{235} & Y_x^{238}\sigma_f^{238} & Y_x^{239}\sigma_f^{239} & Y_x^{240}\sigma_f^{240} \\ Y_x^{235}\sigma_f^{235} & Y_x^{238}\sigma_f^{238} & Y_x^{239}\sigma_f^{239} & Y_x^{240}\sigma_f^{240} \\ Y_x^{235}\sigma_f^{235} & Y_x^{238}\sigma_f^{238} & Y_x^{239}\sigma_f^{239} & Y_x^{240}\sigma_f^{240} \end{bmatrix}\begin{bmatrix} F^{235}(T) \\ F^{238}(T) \\ F^{239}(T) \\ F^{240}(T)\end{bmatrix}^T = \begin{bmatrix} BU(T)/C \\ R^1(T) \\ R^2(T) \\ R^x(T)\end{bmatrix} \tag{27.13}$$

This can be expressed in shorthand notation as:

$$\boldsymbol{M}\cdot F = R. \tag{27.14}$$

To solve this system, we seek the matrix $\boldsymbol{M}$ such that we minimize $\{R - \boldsymbol{M}\cdot F\}$. This matrix $\boldsymbol{M}$ will then identify the most likely reactor type that could produce the fission product isotopes measured in the residue.

Fuel Age

Once the burnup, initial enrichment, and reactor type are known, finding the time since the discharge date is relatively straightforward. The quantity of the fuel age monitor at end-of-irradiation (EOI) can be predicted and the difference between its EOI activity and its measured activity can be used to determine the age since the fuel was discharged using the following:

$$t_d = -\frac{t_{1/2}}{\ln(2)} \ln \left\{ \frac{\left[\dfrac{N^C(T)}{N^{U238}(T)}\right]}{\left[\dfrac{N^C_{EOI}}{N^{U238}(T)}\right]} \right\} \tag{27.15}$$

where t_d is the decay time since discharge (or fuel age), $t_{1/2}$ is the half-life of the monitor isotope, $N^C(T)$ is concentration of the monitor isotope at the time of measurement, and $N^C{}_{EOI}$ is the concentration of the monitor isotope at the end-of-irradiation (or time of discharge from the reactor).

Power History

Several efforts have been made to develop monitors that yield power history information. Most of these efforts have proven unfruitful, but short-lived isotopes that saturate during irradiation can be used to determine the power level of the fuel, and the use of fission product isotopes that are produced mainly by absorption reactions in other isotopes can be used to yield information concerning the shutdown time lengths that occurred during irradiation of the fuel. A complete determination of power history from post-event isotopics has not been fully developed and is an area of continuing research.

Suggested Monitors

Numerous monitor nuclides could be used; however, Table 27.1 contains a list of suggested nuclides. These nuclides all have the proper characteristics to allow them to provide useful information for attributing the spent fuel. Note that no power history monitors are suggested in this table. Since this capability is still immature, it would be inappropriate to suggest any monitors at this time.

Attribution Testing

The basic methodology here can be tested using post-irradiation spent fuel examination data in lieu of any field data. Using spent fuel measurements from fuel discharged from the Mihama Unit 3 PWR, we can use the methodology above to predict the fuel burnup, enrichment, reactor type, and age and compare these results to the known attributes.[6]

Table 27.1 Suggested monitor nuclides.

Burnup monitors	^{140}Ce, ^{100}Mo, ^{98}Mo, ^{97}Mo, ^{138}Ba, ^{142}Ce, ^{148}Nd
Enrichment monitors	^{234}U, ^{235}U, ^{236}U, ^{238}U, ^{239}Pu, ^{240}Pu, ^{241}Pu
Reactor-type monitors	^{109}Ag, ^{153}Eu, ^{156}Gd, ^{143}Nd, ^{240}Pu, ^{108}Cd, ^{113}Cd, ^{149}Sm, ^{166}Er, ^{132}Ba, ^{98}Tc, ^{115}In, ^{72}Ge, ^{115}Sn
Age monitors	^{90}Sr, ^{93}Nb, ^{106}Ru, ^{101}Rh, ^{102}Rh, ^{125}Sb, ^{134}Cs, ^{137}Cs, ^{146}Pm, ^{147}Pm

[6] "List of Reactors in SFCOMPO on WWW Ver.2," www.nea.fr/html/science/wpncs/sfcompo/Ver.2/Eng, Nuclear Energy Agency, 2005.

The Mihama nuclear reactor is located in the Fukui prefecture in Japan. Nine samples were taken from three different spent-fuel assemblies from the third unit. Three samples were taken from the first assembly (JPNNM3SFA1), two samples from the second assembly (JPNNM3SFA2), and four samples from the third assembly (JPNNM3SFA3). The methodology described here predicted an LWR reactor type in all cases. The burnup, enrichment, and age results are shown in Tables 27.2, 27.3, and 27.4, respectively. The burnup was found to be within about 4.5% of the reported burnup and was systematically underpredicted. The enrichment calculation was within 1.25% or less (which is even better than expected) and did not appear to have a systematic bias. The age was reported to be five years, which is most likely an approximation and not exact. The age prediction had percentage differences

Table 27.2 Mihama-3 burnup results.

Sample No.	Reported Burnup (MWd/MT)	Predicted Burnup (MWd/MT)	Difference (%)
1	8,300	7,952	−4.19
2	6,900	6,678	−3.22
3	15,300	14,664	−4.16
4	21,200	20,399	−3.78
5	14,600	14,043	−3.82
6	29,400	28,394	−3.42
7	32,300	30,931	−4.24
8	33,700	32,371	−3.37
9	34,100	32,920	−3.46

Table 27.3 Mihama-3 enrichment results.

Sample No.	Actual Enrichment (a/o)	Predicted Enrichment (a/o)	Difference (%)
1	3.25	3.22	−1.08
2	3.25	3.27	0.62
3	3.24	3.20	−1.33
4	3.24	3.27	0.93
5	3.24	3.20	−1.23
6	3.25	3.21	−1.23
7	3.25	3.29	1.23
8	3.25	3.22	−0.92
9	3.25	3.21	−1.23

Table 27.4 Mihama-3 age results.

Sample No.	Actual Age (Years)	Predicted Age (Years)	Difference (%)
1	5	4.93	−1.36
2	5	4.89	−2.16
3	5	4.78	−4.36
4	5	4.86	−2.76
5	5	4.60	−7.96
6	5	5.05	1.04
7	5	4.93	−1.36
8	5	5.15	3.04
9	5	5.09	1.84

that ranged from 1% to 8%. The age is heavily dependent on the previous predictions and therefore is the most susceptible to error propagation.

Conclusions on Spent Fuel RDD Attribution

Spent-fuel RDD attribution capability is relatively complete and if an event occurred there is a reasonable confidence that the material used would be accurately attributed to a reactor post-detonation. This could lead to identification of the individual assembly from a specific reactor that was used in the device, which could help lead to the identification of the actors involved in the theft, construction, and detonation of the device.

HEU IND Attribution

If an HEU IND is detonated, the explosion of the weapon will leave behind fission product and actinide material as a residue. These fission products and actinides will not have the same isotopics as the predetonated material. Thus in this attribution step the inverse model must account for the change in material inventory during the detonation. Described here is an algorithm that uses measured isotopic ratios from fission product and actinide residue following the detonation of a nuclear weapon to compute the original attributes of the nuclear material used in the weapon. Although more accurate (and more computationally intensive) methods are being explored by others, the method described here could serve as a preprocessing step to a more detailed methodology (potentially saving on computational time). This would, in turn, expedite the process of determining where the device came from, eventually leading to identification of the terrorist group that perpetrated the event.

Given a measurement of the isotopics of residue post-detonation, the attribution interest here is to attempt to determine the following characteristics (in this order of importance): (1) predetonation ^{235}U enrichment, (2) predetonation $^{234}U/^{238}U$ isotopic ratio, (3) predetonation $^{236}U/^{238}U$ isotopic ratio, (4) enrichment method used to produce material, (5) preenrichment $^{234}U/^{238}U$ isotopic ratio, (6) preenrichment $^{236}U/^{238}U$ isotopic ratio, and (7) source (mine or otherwise) from which feed uranium was taken. It was acknowledged immediately that Steps 1–3 would have a likely chance of success and Steps 4–7 would be significantly more difficult (in fact, Step 7 is probably not possible, but is still an interesting problem to attempt to solve). For the purposes of this study, we have so far limited the analysis to gaseous centrifuge and gaseous diffusion enrichment methods, since these enrichment processes are very similar in physical process. It is expected that distinguishing most other methods (such as AVLIS or EMIS) would be much simpler.

^{234}U Isotopics in mines

Uranium mines throughout the world are characterized by different isotopic abundances of ^{234}U that can be used as a signature to indicate the geographic origin of the material. ^{234}U has a relatively short half-life and exists in secular equilibrium with ^{238}U. Thus, the ratio of ^{234}U to ^{238}U should equal to the ratio of the half-lives (53.8 ppm). Variations in the ratio of $^{234}U/^{238}U$ can result from processes that disrupt the decay chain of ^{238}U to ^{234}U.[7] Table 27.5 shows some of these variations in naturally occurring uranium.

Enrichment Methods

Weapons-grade HEU is typically enriched to about 90% ^{235}U. The method of enrichment is a signature that can indicate where the uranium was enriched. Methods used to enrich uranium

[7]S. Richter, A. Alonso, W. De Bolle, R. Wellum, and P. D. P. Taylor, "Isotopic 'Fingerprints' for Natural Uranium Ore Samples," *International Journal of Mass Spectrometry*, **193**, 19 (1999).

Table 27.5 Variations in natural uranium ^{234}U isotopic abundances from mines throughout the world.[7]

Country	Mine/Mill Facility	$^{234}U/^{238}U$ Atom Ratio
Gabon	Comuf Mounana	5.4344E-05
Canada	CAMECO Rabbit Lake Op.	5.4440E-05
Namibia	Roessing Uranium Mine	5.4604E-05

include gas centrifuge, gaseous diffusion, laser isotope separation, chemical/ion separation, and electromagnetic isotope separation. The two most common enrichment methods are gaseous centrifuge and gaseous diffusion, both of which separate the uranium isotopes in a gaseous state called uranium hexafluoride.

Both gaseous diffusion and gaseous centrifuge rely on the differences in mass between ^{235}U-containing molecules and ^{238}U-containing molecules, though they are based on different physical processes. Gas centrifuge is based on centrifugation, whereas gaseous diffusion is based on molecular effusion. Since ^{234}U is lighter than ^{235}U, it enriches even more in either the gas centrifuge or gaseous diffusion process than the ^{235}U. The following equation represents the ^{234}U enrichment for gaseous centrifuge:[8]

$$\left(\frac{N^{234}}{N^{238}}\right)_0 = 0.007731\left(\frac{N^{235}}{N^{238}}\right)_0^{1.0837}\left(\frac{M^{238}}{M^{235}}\right)^{1.0837} \tag{27.16}$$

where M^{238} is the atomic mass of ^{238}U and M^{235} is the atomic mass of ^{235}U. The following equation represents the ^{234}U enrichment for gaseous diffusion:[8]

$$\left(\frac{N^{234}}{N^{238}}\right)_0 = 0.008\left(\frac{N^{235}}{N^{238}}\right)_0\left(\frac{M^{238}}{M^{235}}\right) \tag{27.17}$$

Natural uranium contains essentially no ^{236}U (though small quantities are found in natural material due to the activation of ^{235}U from neutron background); however, enriched uranium of U.S. or Russian origin includes a significantly higher abundance of ^{236}U due to the reenrichment of naval fuel. The following equation represents the ^{236}U enrichment in U.S. origin fuel:

$$\left(\frac{N^{236}}{N^{238}}\right)_0 = 0.0046\left(\frac{N^{235}}{N^{238}}\right)_0\left(\frac{M^{238}}{M^{235}}\right) \tag{27.18}$$

Methodology

The forward model in this algorithm consisted of simulations to predict post-detonation (actually post-irradiation) isotopics given the original isotopics of the material and the number of fission (or yield) of the device. The data from the forward model was mainly used to test the viability of the inverse model. The inverse model predicted predetonation isotopics using analytical inversions of the buildup and decay equations and post-detonation isotopic measurements. The inverse model also included error propagations to allow for prediction of uncertainties in the attributes as well as to determine the sensitivity of the results to the input data.

Forward Model Development

The forward model simulations used the ORIGEN2 code.[3] ORIGEN2 calculates the buildup and depletion of isotopics from irradiation and decay. The code possesses a large set of libraries

[8] "Uranium Enrichment," www.urenco.de.

(each library corresponds to a specific type of reactor) with cross-section, decay, and fission product yield data. ORIGEN2 uses the matrix exponential method to solve a large system of coupled, linear, first-order ordinary differential equations. Although not a weapon burn code, ORIGEN contains sufficient capability to allow for analysis of the feasibility of the method developed here.

Four different uranium signatures from gaseous centrifuge and gaseous diffusion enriched uranium, both with and without ^{236}U present in the original material, were simulated. The uranium was enriched to 95 atom percent ^{235}U and the ^{234}U and ^{236}U concentrations were calculated for both methods of enrichment using Equations 27.16–27.18. Then ORIGEN was used to simulate the burnup of the material in the device given a 2 kT yield. The resultant isotopics from this burnup were then decayed for 1.0 day (assuming that it will take approximately 1 day or more to acquire measured resultants from residue post-detonation).

Inverse Model Development

The inverse model equations are all expressed in terms of atom ratios relative to ^{238}U (the ^{238}U concentration in the device is roughly constant during irradiation). The inverse model uses an iterative procedure where the predetonation $^{235}U/^{238}U$ ratio is set to an initial guess input by the user. The predetonation $^{234}U/^{238}U$ and $^{236}U/^{238}U$ (if applicable) ratios were calculated using Equations 27.16–27.18 and then combined with the initial guess for $^{235}U/^{238}U$ to calculate the ^{235}U enrichment of the original material using:

$$e_0^i = \frac{\left(\frac{N^{235}}{N^{238}}\right)^{i-1}}{\left(\frac{N^{234}}{N^{238}}\right)_0^{i-1} + \left(\frac{N^{235}}{N^{238}}\right)_0^{i-1} + \left(\frac{N^{236}}{N^{238}}\right)_0^{i-1} + 1} \qquad (27.19)$$

where e_0^i is the predetonation enrichment for step i and $\left(N^x/N^{238}\right)_0^i$ is the predetonation atom ratio of isotope x to ^{238}U from step $i-1$ (or from the initial guess for the first step).

The number of fissions in the device per unit mass was calculated using the measurement of two fission products: ^{95}Zr and ^{89}Sr. A single fission product could have been used, but when we use two fission products, iteration between the two yields a better prediction of the number of fissions. The following equation was used to calculate the number of fissions per unit mass in the device:

$$F_T^i = \left(\frac{N^{89}}{N^{238}}\right)_T \frac{N_A E_R}{M^{238} Y^{89}} e_0^i \left(\frac{N^{238}}{N^{235}}\right)_0^{i-1} \qquad (27.20)$$

where F_T^i is the number of fissions in the device following irradiation (i.e., at time T) per unit mass for step i, $\left(N^{89}/N^{238}\right)$ is the measured $^{89}Sr/^{238}U$ atom ratio post-detonation (i.e., at time T), N_A is Avagadro's number, E_R is the recoverable energy per fission, and Y^{89} is the cumulative fission product yield for ^{89}Sr. In using Equation 27.20 we assumed that the fission product yields and recoverable energy per fission from ^{235}U were adequate (i.e., this assumes that all fissions were from ^{235}U).

An updated $^{234}U/^{238}U$ value was then calculated using measurements of $^{232}U/^{238}U$ in the residue and the following equation:

$$\left(\frac{N^{234}}{N^{238}}\right)_0^i = \left(\frac{N^{232}}{N^{238}}\right)_T \frac{\sigma_f^{235}}{\sigma_{3n}^{234}} \frac{E_R}{F_T^i} \frac{N_A}{M^{235}} e_0^i \qquad (27.21)$$

where σ_f^{235} is the one-group microscopic fission cross-section for ^{235}U and σ_{3n}^{234} is the one-group microscopic $(n, 3n)$ cross-section for ^{234}U. This equation assumes that no ^{232}U existed in the original material and the measured ^{232}U concentration was produced only from the $^{234}U(n, 3n)^{232}U$ reaction.

An updated $^{235}U/^{238}U$ value was then calculated using measurements of $^{235}U/^{238}U$ in the residue and the following equation:

$$\left(\frac{N^{235}}{N^{238}}\right)_0^i = \left(\frac{N^{235}}{N^{238}}\right)_T + \left(\frac{N^{235}}{N^{238}}\right)_0^{i-1} \frac{\sigma_a^{235}}{\sigma_f^{235}} \frac{F_T^i}{E_R} \frac{M^{235}}{N_A} e_0^i \tag{27.22}$$

where σ_a^{235} is the one-group microscopic absorption cross-section for ^{235}U. This assumes that the change in ^{235}U is equal to its loss rate from absorption.

An updated $^{236}U/^{238}U$ value was then calculated using measurements of $^{236}U/^{238}U$ in the residue and the following equation:

$$\left(\frac{N^{236}}{N^{238}}\right)_0^i = \left(\frac{N^{236}}{N^{238}}\right)_T - \left(\frac{N^{235}}{N^{238}}\right)_0^i \frac{M^{235}}{N_A e_0^i} \frac{F_T^i}{E_R} \left\{\frac{\sigma_a^{235}}{\sigma_f^{235}} - 1 - \frac{\sigma_a^{235}}{2\sigma_f^{235}}\left[\left(\frac{N^{236}}{N^{235}}\right)_0^{i-1} + \left(\frac{N^{236}}{N^{235}}\right)_T\right]\right\} \tag{27.23}$$

This assumes the change in ^{236}U is equal to its production rate from radiative capture in ^{235}U minus the loss rate from the absorption of ^{236}U. Equation 27.23 was obtained by assuming that the ratio of ($^{236}U/^{235}U$) as a function of irradiation time was linear and therefore was easily integrated.

The new value for the enrichment can now be calculated using Equation 27.19. Equations 27.19–27.23 can then be repeated until the predetonation $^{235}U/^{238}U$ ratio converges to within some tolerance.

Inverse Model Results

The methodology described above was tested for a 2 kiloton detonation of a 95% enriched HEU device. The "measured values" were produced from ORIGEN simulations for four different uranium signatures from gaseous centrifuge and gaseous diffusion enriched uranium, both with and without ^{236}U present in the original material. The algorithm was insensitive to the initial guess for ^{235}U concentration. In all cases fewer than 10 iterations (less than 1 second computational time) were used to acquire a result. The results presented in Table 27.6 verify that for any positive initial guess of any order of magnitude input into the algorithm will be iterated to a reasonably correct answer.

The measured isotopic values generated from ORIGEN2 and the values computed in the inverse model for both centrifuge and diffusion enrichment processes (with and without ^{236}U) are presented in Tables 27.7 and 27.8. The results from the inverse model were consistently higher than the exact values for the original material attributes. The resulting error may be attributed to an assumption made when developing the algorithm that the atomic density of ^{238}U did not change with time.

Table 27.6 Comparison of the values calculated by the inverse model with various initial guesses for the ^{235}U concentration and the actual values for the original material attributes.

Enrichment Process	Initial Guess $(N^{235}/N^{238})_0$	Original Value $(N^{235}/N^{238})_0$	Inverse Model $(N^{235}/N^{238})_0$	Percent Error
Centrifuge (with ^{236}U)	1.00×10^{10}	42.4297	43.1132 ± 0.4309	1.6110%
Diffusion (no ^{236}U)	1.00×10^{-10}	22.4057	22.5538 ± 0.2254	0.6613%

Table 27.7 Comparison of the inverse model calculation and the exact value for the original material attributes for gaseous centrifuge and diffusion enrichment without ^{236}U.

Enrichment Process	Atomic Ratio	Measured Value (T = 1.1 days)	Original Value (T = 0 days)	Inverse Model (T = 0 days)	Percent Error
Centrifuge (no ^{236}U)	N^{235}/N^{238}	33.921569	35.500800	36.0751 ± 0.3605	1.6179%
	N^{234}/N^{238}	0.857089	0.868500	0.8831 ± 0.0127	1.6889%
	N^{236}/N^{238}	0.397059	0.000000	0.0047 ± 0.0075	–
Diffusion (no ^{236}U)	N^{235}/N^{238}	21.401713	22.405700	22.5538 ± 0.2254	0.6613%
	N^{234}/N^{238}	0.177368	0.179200	0.1816 ± 0.0047	1.3090%
	N^{236}/N^{238}	0.251309	0.000000	0.0273 ± 0.0026	–

Table 27.8 Comparison of the inverse model calculation and the exact value for the original material attributes for gaseous centrifuge and diffusion enrichment with ^{236}U.

Enrichment Process	Atomic Ratio	Measured Value (T = 1.1 days)	Original Value (T = 0 days)	Inverse Model (T = 0 days)	Percent Error
Centrifuge (with ^{236}U)	N^{235}/N^{238}	40.540784	42.429656	43.1132 ± 0.4106	1.6110%
	N^{234}/N^{238}	1.024335	1.037963	1.0552 ± 0.0151	1.6641%
	N^{236}/N^{238}	0.668319	0.195176	0.2041 ± 0.0110	4.5841%
Diffusion (with ^{236}U)	N^{235}/N^{238}	23.860440	24.980279	25.3754 ± 0.2536	1.5817%
	N^{234}/N^{238}	0.197745	0.199842	0.2042 ± 0.0029	2.2035%
	N^{236}/N^{238}	0.394534	0.114909	0.1213 ± 0.0065	5.5651%

Table 27.9 Comparison of the inverse model calculation and the exact value for the original material attributes for gaseous diffusion enrichment with and without ^{236}U.

Enrichment Process	Atomic Ratio	Original Values (T = 0 days)	Inverse Model (T = 0 days)	Percent Error
Diffusion (with ^{236}U)	N^{235}/N^{238}	24.980279	25.3754 ± 0.2536	1.5817%
	N^{234}/N^{238}	0.199842	0.2042 ± 0.0029	2.2035%
	N^{236}/N^{238}	0.114909	0.1213 ± 0.0065	5.5651%
Diffusion (no ^{236}U)	N^{235}/N^{238}	22.405700	22.5538 ± 0.2254	0.6613%
	N^{234}/N^{238}	0.179200	0.1816 ± 0.0047	1.3090%
	N^{236}/N^{238}	0.000000	0.0273 ± 0.0026	-

To determine valid signatures indicating the method of enrichment, the measured values for post-detonation ^{234}U concentrations were compared. For centrifuge enriched fuel, the ^{234}U concentration was approximately 5.0 times greater than the ^{234}U concentration for diffusion enriched fuel. These significant variations in ^{234}U were used as signatures indicating the enrichment process used.

After the enrichment process has been determined, whether ^{236}U existed in original weapons material must be established. The values computed in the inverse model for diffusion enriched fuel with and without ^{236}U are presented in Table 27.9. For diffusion enriched

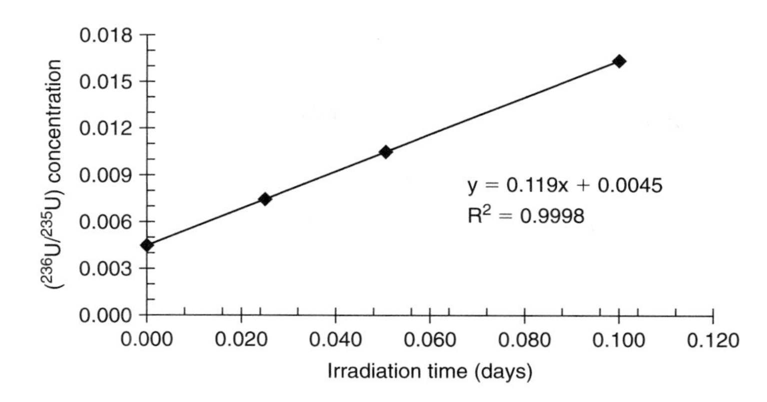

FIGURE 27.4 ORIGEN2 calculation of ($^{236}U/^{235}U$) isotopic ratio as a function of irradiation time for irradiation of 95% enriched uranium enriched.

fuel (with ^{236}U), the ^{236}U value from the inverse model was approximately 4.5 times greater than the ^{236}U value for diffusion enriched fuel (without ^{236}U).

In the derivation of Equation 27.23 it was assumed that the ratio of $^{236}U/^{235}U$ was a linear function with respect to time and could therefore be easily integrated. The assumption was verified in Figure 27.4, which depicts the ratio as a function of irradiation time. A linear trend line was used to fit to the data points.

Conclusions on HEU IND Attribution

The algorithm described above provides reasonable accuracy for determining predetonation attributes to a device from post-detonation isotopic measurements. From these isotopics it is likely that the type of enrichment process used to produce the material could be determined; then this information could be used to help identify the group responsible for the construction of the device. The following post-detonation isotopic ratios were used in the analysis: $^{89}Sr/^{238}U$, $^{95}Zr/^{238}U$, $^{232}U/^{238}U$, $^{234}U/^{238}U$, $^{235}U/^{238}U$, and $^{236}U/^{238}U$. The primary advantage gained from this methodology is that it provides reasonably accurate solutions with essentially no computational time required. A similar methodology could be developed for post-event attribution of a Pu-based IND.

Triage and ReachBack

In both attribution problems we've discussed, the algorithms developed were based on isotopic measurements from post-detonation residue. What was not discussed is all the detailed science and engineering that would need to be utilized in the case of an event, to even acquire the measured isotopic ratios. The national laboratory complex provides these capabilities in programs often referred to as Triage and ReachBack.

ReachBack consists of providing responders to nuclear, chemical, radiological, or natural disaster emergencies access to the full spectrum of technical and analytical expertise available at the laboratories. This could include analysis of samples beyond simply isotopic analysis.

Triage is a 24-hour-per-day, seven-day-per-week expert assistance program that provides first responders, using secure communications links, technical assistance in interpreting radiation spectra collected in the field. This allows the first responders to have some confidence in analyzing the event and knowing the appropriate steps to take in mitigating the effects of the event.

Summary

Two scenarios of post-nuclear event attribution were presented, providing the reader with some concept of the detailed science and technology needed to solve these problems. Both scenarios considered here were focused on event attribution using post-detonation isotopic residue. Numerous other technical skills are needed to fully characterize an event, including device modeling, radiochemistry, signature databases, and foreign weapons assessment needed for characterizing post-detonation debris to determine the origin of a nuclear device. Although these will not be discussed here, they are all important skills that must be continuously exercised and developed. Post-nuclear event attribution is an important field for providing additional response options following an event. Effective attribution of the nuclear materials involved in a clandestine attack can be a vital piece of information in determining who conducted the attack. If attackers believe that there is a good likelihood that they will be identified and face the prospect of a military response, they could be deterred from conducting such an attack.

28 Nonproliferation Export Controls

Carlton E. Thorne

Introduction

This chapter provides an overview of the evolution and implementation of nuclear export controls. Controls on nuclear exports are required by the domestic laws of most nations and often by their obligations under international treaties and agreements. Although most governments see nuclear export controls as an important instrument for implementing nonproliferation policies and national security objectives, private business interests have often viewed them as a hindrance to trade and an example of government overregulation. This chapter provides a history of nuclear export controls and illustrates how tensions between industry and government have affected their implementation since their inception with the Treaty on the Nonproliferation of Nuclear Weapons (NPT) that came into force in 1970.[1]

In addition to the NPT there have been other notable events in the history of nuclear export controls. These include:

- The creation of the NPT Exporters (Zangger) Committee in the early 1970s
- The detonation of a nuclear device by India in 1974
- Meetings of suppliers following the Indian test that led to the publication of the Nuclear Suppliers Guidelines and the creation of the Nuclear Suppliers Group (NSG)[2]
- Implementation by the NSG of nuclear-related dual-use export controls
- Adoption of a policy of requiring full-scope safeguards as a condition of supply in 1992
- Addition of technology controls to the NSG Guidelines in 1995

---

[1]As stated by the IAEA, the NPT is a landmark international treaty, the objectives of which are to prevent the spread of nuclear weapons and weapons technology, to foster the peaceful uses of nuclear energy, and to further the goal of achieving general and complete disarmament. The Treaty establishes a safeguards system under the responsibility of the IAEA, which also plays a central role under the Treaty in areas of technology transfer for peaceful purposes. The Treaty entered into force on March 5, 1970. Russia, the United Kingdom, and the United States are the three depositary governments. The full text of the Treaty can be found in IAEA Information Circular 140 at www.iaea.org/Publications/Documents/Treaties/npt.html.

[2]The official Website for the Nuclear Suppliers Group is found at www.nuclearsuppliersgroup.org. The site contains background information about the NSG and provides access to the latest versions of the Guidelines. There are two important characteristics of the Guidelines of particular relevance to this chapter. First, the Guidelines are informal and are not legally binding on NSG members. Second, all decisions by the NSG are by consensus.

- China becoming a member of the NSG in 2004
- The adoption of U.N. Security Council Resolution 1540
- U.S. initiative in 2006 to allow nuclear cooperation with India

Reasons for Control

The fundamental reason for controlling nuclear-related commodities and technologies is to ensure that transfers will be used solely for peaceful purposes and will not contribute to the proliferation of nuclear weapons. This objective is articulated in Article III, paragraph 2, of the NPT, which forbids states party to the treaty from providing (a) source or special fissionable material or (b) equipment or material especially designed or prepared for the processing, use, or production of special fissionable material to any nonnuclear weapons state that does not place such materials or equipment under international safeguards.[3]

It was recognized early on that some further clarification was needed of the term in Article III.2 of the NPT that referred to "equipment or material especially designed or prepared for the processing, use or production of special fissionable material." Thus, an informal group of nuclear suppliers party to the NPT began meeting soon after the NPT came into force, to create an illustrative list of equipment and material that met the criteria of especially designed or prepared (EDP) and thus should be controlled. That group, originally called the NPT Exporters Committee, is more commonly known today as the Zangger Committee, in honor of its first chairman, Professor Claude Zangger of Switzerland.

Other reasons for nuclear export controls are not related directly to curbing the proliferation of nuclear weapons. As a matter of national policy, a supplier state might want to control these items to some states for the purpose of industrial competitiveness or on the basis of their conduct in areas such as human rights, due to their association with terrorist activities, or for any other reasons where strategic trade controls or sanctions are chosen to make a point of principle or implement national policy. It might also be the policy of a supplier state to require licenses for certain sensitive nuclear technologies as a means of ensuring that any cooperation in those technologies is reviewed for consistency with its current policies.

States may also control nuclear exports for the purposes of safety and physical protection of certain materials or items. A supplier state may in some cases feel a responsibility to determine whether the intended recipient is likely to use the transfer in a safe manner or has the resources and expertise to safely use the transfer. A lesson learned from the Chernobyl experience is that a nuclear accident anywhere is potentially a problem everywhere. A supplier would not want to knowingly provide equipment or materials that would either pose a health and safety problem or be at risk of theft or sabotage.

History of International Nuclear Export Controls

Export Control Deficiencies of the NPT

From 1970, the year that the Treaty on the Nonproliferation of Nuclear Weapons (NPT) came into force until the nuclear explosion by India in 1974, multilateral nuclear export control policy was defined solely by Article III.2 of the NPT and clarified by the NPT Exporters Committee (known as the Zangger Committee). It was a time in which very little progress was made in strengthening nuclear export controls. The NPT does not require exporting

[3]Article III.2 of the NPT states: "Each State Party to the Treaty undertake not to provide: (a) source or special fissionable material, or (b) equipment or material especially designed or prepared for the processing, use or production of special fissionable material, to any non-nuclear-weapon State for peaceful purposes, unless the source or special fissionable material shall be subject to the safeguards required by this Article."

countries to insist that receiving countries accept international safeguards on its entire nuclear program, known as *full-scope safeguards*, as a condition of supply. Nor does the NPT control nuclear technology (knowledge and information) in addition to equipment and materials. For example, although the export of a nuclear reactor would fall within the requirements contained in NPT, Article III.2, these requirements would not apply to the export of design information explaining how to build such a reactor. In addition, the NPT does not obligate members to control so called "dual-use" items that have other legitimate nonnuclear uses but could be instrumental in a nuclear weapons development program as well. These deficiencies were recognized early by NPT member states and coordinated actions to address them were initiated shortly after the treaty entered into force.

The Origins of the Nuclear Suppliers Group

Following India's nuclear detonation in 1974, seven supplier states met in London for the purpose of strengthening nuclear export controls. These countries were Canada, France, the United Kingdom, the United States, the Soviet Union, West Germany, and Japan. This group offered a much better avenue for nuclear export control reform than the NPT Exporters (Zangger) Committee for at least two reasons: It would not be constrained by the limitations or vague language of Article III.2 of the NPT, and it would be aided by the presence of Japan and France, both major technology holders but non-NPT parties at the time and therefore not members of the NPT Exporters (Zangger) Committee.

The list of controlled items developed by the Zangger Committee was known as the Trigger List because export of those items triggers the requirement for the implementation of International Atomic Energy Agency (IAEA) safeguards on these items in the recipient state. The items are controlled under the Zangger Committee's understandings because if they were misused they could contribute to a nuclear explosive program. Examples of these items are plutonium, highly enriched uranium (HEU), reactors, reprocessing and enrichment plants, and equipment and components for such facilities.

After a series of meetings in London, the London Club (as the NSG was first called) had reached *ad ref* agreement on what the IAEA would publish in January 1978 as INFCIRC/254, the Nuclear Suppliers Guidelines. The NSG Guidelines are the first evolution of the Zangger Committee's list of items that would be controlled by NPT Article III.2. The nonproliferation objective of nuclear export controls is *implicit* in the introductory paragraph of Part 1 of the NSG Guidelines and is *explicit* in the objective of Part 2 of the NSG Guidelines.[4] One early accomplishment of the NSG Guidelines was the addition of heavy water plants and equipment to a trigger list that was otherwise identical to the original NPT Exporters (Zangger) Committee list. Another addition was a section on requirements for physical protection of nuclear materials and facilities. Finally, technology was mentioned in the context of exercising restraint in the transfer of sensitive technologies and in the replication of facilities using transferred equipment.

The negotiating record, however, shows that this small, elite group of countries, which met regularly over a period of two to three years, fell far short in many areas due to the strength of the commercial interests. For example, full-scope safeguards as a condition of supply were favored by some of the participants, but others would not support the proposal, so nothing in this area was achieved.

There were other compromises in the original Nuclear Suppliers Guidelines as well. In Paragraph 6, it says that suppliers should "encourage" rather than "require" recipients to accept, as an alternative to national reprocessing and enrichment plants, supplier involvement in appropriate multinational fuel cycle facilities. This meant that nonnuclear weapons

[4]Paragraph 1 of the Nuclear Suppliers Group Guidelines (Part 1) states: "The following fundamental principles for safeguards and export controls should apply to nuclear transfers for peaceful purposes to any non-nuclear-weapon State, and in the case of controls on retransfer, to transfers to any State."

states could decide to build their own reprocessing or enrichment plants and not be barred from buying needed equipment from other NPT members.

A final example of compromise is in Paragraph 8 of the Guidelines. It called for suppliers to endeavor to obtain agreement from the recipient for consent rights over any weapons materials derived from the transfer. This means that the recipient would agree in advance not to take any action with respect to manufacturing or utilizing weapons material derived from the transferred materials without the consent of the supplier. Here again the negotiations focused on whether to use "endeavor" or the word "require." Again there was no consensus for the stronger language, and so the weaker language was adopted.

Activity in the NPT Exporters (Zangger) Committee

The period following the publication of the Nuclear Suppliers Guidelines until the meeting of the adherents to the Guidelines in the Hague in 1991 was a very positive period for nuclear export controls. The number of states adhering to the Nuclear Suppliers Guidelines increased from 15 to 27. It was also a time of renewed vigor in the NPT Exporters (Zangger) Committee. Under the leadership of the United Kingdom, a major initiative to clarify export restrictions for gas centrifuge enrichment technology was completed in 1984. Following this pattern, a second upgrade exercise was led by the United States for the reprocessing guidelines and was completed in 1985. In 1990 the Trigger List entry for gaseous diffusion was clarified following an extended effort led by the Soviet Union. Although not finalized until 1992, the Canadian-led exercise to upgrade the heavy water production entry completed the informal work plan of the Committee that had begun with the United Kingdom in 1981. Together these upgrade exercises added large numbers of EDP equipment to the Zangger Committee Trigger List.

Revitalization of the Nuclear Suppliers Group

After 1991 the NSG began to meet regularly, although membership did not become a recognized concept until about 1993. Today, "adherent" is the status of a country that has informed the IAEA Director General of its intention to abide by either the NSG Guidelines or the NPT Exporters (Zangger) Committee Understanding and asks that he inform the Agency members of this decision. Adherence is an action that a country can take unilaterally without permission of any other state or group of states. Membership, on the other hand, in either arrangement is a status that can only be attained by consensus of the existing members in each arrangement.[5]

The Addition of Dual-Use Controls to the Nuclear Suppliers Group Guidelines

In mid-1990, the United States started serious talks with other major suppliers to examine the possibility of a multilateral agreement on dual-use items. On February 21, 1991, an informal meeting of the 26 adhering countries took place in the Hague. The purpose of this preliminary meeting was to solidify support for the creation of a working group to address dual-use controls at the upcoming meeting called by the Dutch and to introduce the participants to working drafts of the guidelines and dual-use list provided by the United States.[6] The Hague

[5]As of 2008 there are 45 members of the NSG. The European Union has been given permanent observer status. The members are Argentina, Australia, Austria, Belarus, Belgium, Brazil, Bulgaria, Canada, China, Croatia, Cyprus, Czech Republic, Denmark, Estonia, Finland, France, Germany, Greece, Hungary, Ireland, Italy, Japan, Kazakhstan, Korea (Republic of), Latvia, Lithuania, Luxembourg, Malta, Netherlands, New Zealand, Norway, Poland, Portugal, Romania, Russian Federation, Slovak Republic, Slovenia, South Africa, Spain, Sweden, Switzerland, Turkey, Ukraine, United Kingdom, and the United States.

[6]Carlton E. Thorne, the author of this chapter, was the U.S. Department of State official that led the dual-use consultations and chaired both the first informal meeting of the NSG adherents in 1991 and the working group that negotiated the dual-use guidelines and control list.

meeting resulted in the establishment of the Dual-Use Working Group to be chaired by the United States to examine the feasibility of a dual-use export control arrangement. Also agreed was that Finland would take the lead in harmonizing the NSG Trigger List with that of the NPT Exporters (Zangger) Committee, which as mentioned previously had undergone extensive updating and expansion.

The Dual-Use Working Group began a series of four weeklong meetings over the next nine months in the Hague, Brussels, and Annapolis, culminating with the final meeting at Interlocken, Switzerland, in January 1992. At the end of that final meeting there was agreement on the Dual-Use Guidelines and a Memorandum of Understanding, and an annex of equipment, materials, and related technology covering about 65 commodities. These documents were adopted at the NSG Plenary in Warsaw in April 1992.

Multilateral Agreement to Full-Scope Safeguards as a Condition of Supply

Parallel with the dual-use negotiations were efforts to bring closure on the issue of full-scope safeguards as a condition of supply. In 1991 more countries had announced the adoption of this policy. Most countries with a full-scope safeguards policy followed restrictions similar to those of the United States. The U.S. policy at the time consisted of a general statement requiring full-scope safeguards on any significant new supplier arrangement. Neither "full-scope safeguards" nor "new" nor "significant" were defined. As a matter of law the United States required de facto full-scope safeguards, i.e., all activities must be under IAEA safeguards at the time of export. At the Warsaw meeting, however, the states had to put into writing what full-scope safeguards actually meant.

In 1975 the seven states of the London Club were not able to agree on full-scope safeguards as a condition of supply for anything, not even the transfer of a complete enrichment plant. In 1992 there was agreement that the policy applied to every item, big or small, sensitive or nonsensitive, covered by the Trigger List. The full-scope declaration was adopted at the Warsaw meeting. Subsequently, with the addition of technology controls to the Trigger List in 1995, the full-scope safeguards condition of supply was extended to the technology for the development, production, or use of every item covered by the Trigger List. Two such momentous changes occurring in international export controls at one meeting was a major accomplishment.

Special Issues Associated with Technology Controls

International Technology Controls

Nuclear technology controls in the NSG are relatively new. The nuclear export control language of the NPT, found in Article III.2, speaks only of equipment and materials, not technology. When the Dual-Use Arrangement was agreed to in April 1992, it included comprehensive controls on the technology associated with the development, production, or use of the commodities on the control list. The resultant controls bore strong resemblance to the technology controls adopted by the Western allies' Coordinating Committee on Multilateral Export Controls (COCOM) from the Cold War era because many of the NSG members were familiar with those strategic technology controls designed to control exports to the Soviet bloc and China.[7]

[7]COCOM was the acronym for Coordinating Committee for Multilateral Export Controls. It was the mechanism for embargoing transfers during the Cold War from the West to East Bloc countries and China. The group of representatives from the 17 member states met in Paris on a regular basis to act on requests for exceptions to the embargo. COCOM ended in 1994, but its control lists were essentially adopted by the Wassenaar Arrangement, albeit under a different mandate.

With the adoption of the Dual-Use Guidelines and Annex (INFCIRC/254, Part 2) as the companion part to the existing Trigger List Guidelines (INFCIRC/254, Part 1), it soon became apparent to the members that technology was inadequately addressed in the Part 1 Trigger List. When the original Guidelines of Part 1 were negotiated in the 1975 to 1977 timeframe, it was not possible to get agreement on comprehensive technology controls, and any attention given to technology by the Guidelines dealt only with enrichment, reprocessing, and heavy water production.

The Addition of Technology Controls to the NSG Trigger List

At the 1993 NSG Plenary in Lucerne, Switzerland, tabled a discussion paper that pointed out the inadequacy of the treatment of technology in the NSG Part 1 Guidelines. The Swiss, in noting the differences in the treatment of technology between the two parts of the NSG Guidelines and further noting the need to harmonize the two parts of the NSG arrangement, called for the creation of a working group to resolve these differences.

After extensive consultations, the Technical Working Group reached *ad referendum* agreement on technology controls for Part 1, which was tailored after the technology controls of Part 2. The changes recommended by the Working Group were adopted without amendment at the 1995 NSG Plenary in Helsinki. The scope of the multilateral nuclear technology controls today is consistent between Parts 1 and 2 of the NSG Guidelines for the development, production, or use of the controlled commodities. Both parts exclude basic scientific research and information in the public domain.[8]

Intangible Technology

In response to growing concerns about the transfer of intangible nuclear technology (information, data, documents) via the Internet and by nuclear experts hired by foreign nuclear programs, a mandate was established to address the issue. Subsequent debate took place in the NSG on the question of whether intangible technology is controlled by the Guidelines. Intangible technology controls, though not explicitly referred to in Parts 1 and 2 of the NSG Guidelines, are nonetheless an integral part of the technology controls. The "Technology Controls" section in Parts 1 and 2 provides that "the transfer of 'technology' directly associated with any items in the Trigger List or the Annex will be subject to as great a degree of scrutiny and control as will the equipment itself, to the extent permitted by national legislation."

In the "Definitions" section of the Guidelines, "technology" is said to mean specific information required for the "development, production, or use of any controlled item." With respect to the intangible technology issue, technology is further defined to include technical assistance. The definition of "technical assistance" sheds further light on the issue of intangible technology controls. It says that technical assistance may take forms such as instruction, skills, training, working knowledge, and consulting services. In summary, the definition of technology in the NSG Guidelines clearly includes all forms of technology, both tangible and intangible. Moreover, all NSG members have an obligation to control both forms of technology under their NSG commitments.

Unless the membership of the NSG chooses to expand its mandate, it is important to keep in mind that the Guidelines of Parts 1 and 2 only address the *types* of technology controlled. The Guidelines do not address the *means* of technology transfer, nor enforcement measures to police the means of transfer. Much of the discussion of technology controls in the NSG has failed thus far to make this important distinction.

---

[8]In the U.S. regulation for controlling the transfer of nuclear technology (Title 10 of the Code of Federal Regulations, Part 810), the exemption for information in the public domain is explained. It says that public information that is given technical embellishment, enhancement, explanation, or interpretation, which in itself is not public, is no longer information in the public domain.

The Meaning of Transfer

Technology transfer in the context of this discussion is said to take place when controlled information moves from one country to another. The controlled information, as stated in the NSG definition of technology, can take the form of technical data (tangible) or technical assistance (intangible).

When the controlled information moves from one country to another, it can be transported by any of a number of means. The controlled information can be transferred electronically by fax machine, by telephone, or by computer via Internet. It can be transported by conventional means such as by ship, land vehicle, or air. In that regard, the information may be in such diverse forms as information in a package of blueprints or controlled information acquired by an individual or carried in the form of knowledge possessed by a passenger.

If the person transporting the acquired information in intangible form (i.e., as knowledge) is a foreign national, the transfer is said to take place at the time the knowledge is acquired, not when the foreign national leaves the United States. This is the so-called "deemed" export.[9] In the case of a U.S. citizen who leaves the country, his or her possession of controlled information as knowledge would not be considered a transfer unless there was an intention to pass that knowledge to persons in the country visited.

This NSG discussion of technology does not address transfers by illegal acts, such as espionage. Attempts by foreign agents to acquire restricted information are a special type of transfer and are dealt with by national security authorities and not a responsibility of the NSG to counter. Notwithstanding the fact that the NSG is not an enforcement agency, the NSG does have an obligation to seek to create a multilateral export control arrangement that reduces the risk of technology being illegally transferred by whatever means, ranging from exporters' inadvertent acts to espionage.

Reasons for Denial of a Nuclear Export

Now that some historical background has been provided on the evolution of nuclear export controls, it is useful to examine how a proposed export might be evaluated in the United States or other state following similar export control guidelines like those required by NSG members. The following are some possible reasons a nuclear-related transfer might be denied for nonproliferation purposes.

Nuclear Explosives Activities

The suspected use of a proposed transfer in a nuclear explosives activity is the most basic of all reasons for denial. It is prominent in the obligations of parties to the NPT and the commitments to the NSG Guidelines. For many supplier states it is an activity explicitly included in their so-called "catch-all" provisions.[10] To make a judgment on this criterion for denial, export control officials in supplier states have the advantage of access to information from intelligence sources and from confidential information sharing with other supplier states. However, one problem for the exporter is that the end users under this reason for denial are less likely to be publicly known. What then can an exporter do to evaluate this reason for denial?

A first step is to identify the few countries that are or might be engaged in nuclear explosive activities. This can be done by paying attention to governments' statements and to what is said in the open press. This would lead the exporter to construct a short list of countries for

[9]The "deemed export" rule does not apply to persons lawfully admitted for permanent residence or to persons who are "protected individuals" under the Immigration and Naturalization Act.

[10]*Catch-all* is a term used to describe export controls exercised by a supplier government over commodities not listed on any control list. It is sometimes referred to as *end-user controls*.

which a nuclear explosives program cannot be ruled out based on publicly available information. This list would include China, France, Russia, the United Kingdom, and the United States, the official nuclear weapons states,[11] plus others such as Pakistan, Israel, India, and North Korea, which have demonstrated or are widely believed to possess a weapons capability, and Iran, which has been reported to have nuclear weapons aspirations. Prior to the Iraqi War, Iraq would have been on this list of aspirants.

This list alerts the exporter that additional information on end use and safeguards within these countries will be essential before a decision to export was made or in some cases that that the likelihood of denial is high. With respect to proposed exports to countries not on such a list, the exporter can at least have the confidence that those countries have not been openly accused of having a nuclear weapons program.

Knowing whether or not the commodity can be used in a nuclear explosives program is also of value and is the second step in the evaluation of a proposed export for this reason for denial. It is in this area that technical support can be of value to both exporters and government export controllers. The list of controlled equipment, materials, and technologies contained in Part 1 of the NSG Guidelines are a guide to materials and items that have uses in the processing, use, or production of special nuclear materials and thus are applicable to fuel-cycle activities. For dual-use equipment, materials, software, and technologies applicable to a nuclear explosives program, it is useful to consult the Nuclear-Related Dual-Use Annex of Part 2 of the NSG Guidelines.

Unsafeguarded Nuclear Fuel-Cycle Activities

An export could be denied because the recipient state has unsafeguarded nuclear activities. Concern by supplier states over unsafeguarded nuclear fuel and fuel-cycle activities is reflected in the full-scope safeguards condition of supply found in Part 1 of the NSG Guidelines and the Basic Principle of Part 2 of the NSG Guidelines.

A difficulty export control officials and exporters face in evaluating proposed transfers against this criterion for denial is the broadness of the definition of unsafeguarded activities as contained in the NSG Guidelines. This definition "includes research on or development, design, manufacture, construction, operation or maintenance of any reactor, critical facility, conversion plant, fabrication plant, reprocessing plant, plant for the separation of isotopes of source or special fissionable material, or separate storage installation, where there is no obligation to accept International Atomic Energy Agency (IAEA) safeguards at the relevant facility or installation, existing or future, when it contains any source or special fissionable material; or of any heavy water production plant when there is no obligation to accept IAEA safeguards on any nuclear material produced by or used in connection with any heavy water produced therefrom; or where any such obligation is not met."

If it were simply a matter of identifying unsafeguarded nuclear fuel-cycle facilities, either under construction, operating, or shut down, the problem would be manageable. These facilities are generally known. However, the inclusion of research, development, design, and manufacture in the definition adds such diverse locations as laboratories, universities, and common industrial facilities.

As in the previous reason for denial, it is helpful to use a process of elimination. If a country is a party to the NPT, it has made a formal commitment to place its entire nuclear program, both present and future, under IAEA safeguards. If a country's conduct raises questions about its commitment, it is likely that considerable domestic or foreign media reporting on this issue will be available to exporters.

A list of facilities and activities under IAEA safeguards can also be helpful in the elimination process. The IAEA annually publishes a list of safeguarded facilities that is available

[11] Article IX.3 of the NPT defines a nuclear weapons state as one that had manufactured and exploded a nuclear weapon or other nuclear device prior to January 1, 1967.

online.[12] The user of this list should be aware that a facility might only be under IAEA safeguards when safeguarded material is located in the facility or being processed. Facilities where this is likely to occur are those engaged in conversion, fuel fabrication, reprocessing, or storage.

Other considerations for this reason for denial are the declared nuclear weapons states, all of which have nuclear activities that are not under IAEA safeguards. From a multilateral perspective this is not a problem. The "unsafeguarded" provisions in Parts 1 and 2 of the NSG Guidelines only apply to nonnuclear weapons states unless, with respect to Part 2, a supplier denies a transfer to an unsafeguarded activity in a nuclear weapons state because "it would be contrary to the objective of averting the proliferation of nuclear weapons."

A Lack of Full-Scope Safeguards

This reason for denial is related to the previous reason for denial, but with significant differences. A lack of a full-scope safeguards agreement in force with the IAEA is the principal basis for denial of commodities and technologies under Part 1 of the NSG Guidelines. A country, to meet this standard, must have a safeguards agreement in force with the IAEA covering all present and future nuclear activities. It is important to note that this reason for denial does not apply to the dual-use items on the NSG Part 2 list.

Contrary to Nonproliferation Principles

A proposed transfer should be denied if it is in conflict with the principles agreed to in the NSG. These include one of the following:

- The transfer would contribute to the proliferation of nuclear weapons or other nuclear devices (NSG, Part 1)
- The transfer is contrary to the objective of averting the proliferation of nuclear weapons (NSG Part 2)

The first of these subjective reasons for denial is found in the Proliferation Principle of Part 1 of the Nuclear Suppliers Guidelines; the second phrase is found in the Basic Principle of the NSG Dual-Use Guidelines in Part 2. What these reasons for denial really say is that regardless of whether a country is a member of the NPT or whether it has a full-scope safeguards agreement, if a supplier state has doubts about the country's actions and intentions, it can deny the export of any of the items on the NSG lists to end users in that country. They provide a type of "catch-all" justification for denial. This reason for denial has also been interpreted by the nonnuclear weapons states to mean that they can also use this justification to deny to the declared nuclear weapons states controlled items from Part 2 of the Guidelines for nuclear weapons purposes. This, of course, has always been their prerogative.

For exporters, this policy-related reason for a possible denial is unpredictable unless the country has been placed under an embargo. In addition to denials because of proliferation concerns about the country, a proposed transfer might be denied because the retransfer controls and enforcement mechanism of the country are judged to be too weak. In other cases a transfer might be denied because the risk of diversion (out of country) is unacceptable or because the government is unable to function due to war or internal strife.

Sensitive Technology Transfers

Transfers involving technologies deemed to be more sensitive than others are not necessarily prohibited by the NSG. This is best described as a possible reason for denial. What constitutes

[12]The list of safeguarded facilities can be accessed through www.iaea.or.at/OurWork/SV/Safeguards/sv.html.

sensitive nuclear activities can vary from country to country. However, the following nuclear fuel-cycle activities are widely considered to be sensitive:

- (Reprocessing) facilities for the chemical processing of irradiated special nuclear or source material
- Facilities for the production of heavy water
- (Enrichment) facilities for the separation of isotopes of source and special nuclear material
- (Plutonium or mixed-oxide fuel) facilities for the fabrication of nuclear reactor fuel containing plutonium

Additionally, nonpower reactors above 5 MW(thermal) in power should be considered sensitive because of their potential to produce weapons-grade plutonium. One of the most recent changes in the U.S. technology controls was the addition of accelerator-driven subcritical assembly systems to the list of sensitive technologies in the United States.

There are no special controls on commodities associated with these technologies in the multilateral commitments. In the case of reprocessing and enrichment, however, the NSG Guidelines for Part 1 say that suppliers should "exercise restraint" in their transfer. In Part 1 of the NSG Guidelines, the facilities and equipment are sufficiently identified and described for the exporter to determine what items are associated with these sensitive uses. In Part 2 of the NSG Dual-Use Guidelines, one of the relevant factors to be taken into account by supplier states is "whether the equipment, material, or related technology to be transferred is to be used in research on or development, design, manufacture, construction, operation, or maintenance of any reprocessing or enrichment facility."

It is not surprising that the U.S. government and others pay special attention to exports of reprocessing and enrichment equipment and technology, given the legal penalties that can be imposed on other countries that transfer or receive commodities and technology related to reprocessing and enrichment. Section 129 of the NNPA of 1978 (Conduct Resulting in Termination of Nuclear Exports) says that no nuclear materials and equipment or sensitive nuclear technology shall be exported to any nonnuclear weapons state that has "engaged in activities involving source or special nuclear material and having direct significance for the manufacture or acquisition of nuclear explosive devices" and, in another provision, has "entered into an agreement for the transfer of reprocessing equipment, materials, or technology to the sovereign control of a non-nuclear-weapon state."

Other elements of U.S. law also restrict cooperation between countries in enrichment and reprocessing. The so-called Glenn and Symington amendments bar U.S. economic or military assistance or military education and training or extending military credits or making guarantees to any country that has either transferred or received reprocessing or enrichment equipment, materials, or technology.[13]

Proposed exports related to heavy water production, to equipment associated with fuels containing plutonium and with the newly controlled accelerator-driven subcritical assembly systems, do not share the same legal and policy burdens as enrichment and reprocessing. Receptivity to license applications in these technologies will be heavily case dependent, involving the significance of the items and the credentials of the recipient. Fortunately for the exporter, the number of countries engaged in these activities is relatively small.

The U.S. Department of Energy's Part 810 regulations also give special attention to these end uses. Technical assistance or the transfer of technology associated with reprocessing, enrichment, heavy water production, and fuel fabrication containing plutonium to all

---

[13]In 1976 the Foreign Assistance Act of 1961 (22 U.S.C. 2429) was amended to require suspension of economic or military assistance to countries that either buy or sell enrichment or reprocessing technology and equipment to countries not party to the NPT or that have not accepted full-scope safeguards. The Glenn Amendment addressed reprocessing and the Symington Amendment applied to enrichment.

countries by U.S. entities requires a specific authorization from the Secretary of Energy. The DOE list of sensitive activities also includes technology related to production reactors, accelerator-driven subcritical assembly systems, and nonpower reactors with a power level greater than 5 MW (thermal).

Participation in Foreign Naval Nuclear Propulsion Plant Projects

This final reason for denial of an export application is one that is not a nuclear proliferation issue *per se* for commodities licensed by the Department of Commerce (DOC) and the NRC or for technology and technical assistance under the jurisdiction of the Department of Energy, but this reason is one that has come to be the responsibility of the nuclear export controllers by default.

Naval propulsion is not controlled by or prohibited by the NSG Guidelines. However, transfers for naval propulsion programs of items and technology covered by Part 1 of the NSG Guidelines, though not prohibited, would not be permissible under the Guidelines of Part 1 because nuclear submarines would not meet the peaceful purposes standard of Part 1. NSG Guidelines of Part 2 do not prohibit the transfer of items or technology for naval propulsion.

U.S. policy toward the participation of U.S. firms and individuals in foreign naval nuclear propulsion plant projects is contained in Section 744.5 of the Export Administration Regulations and Section 123.20 of the International Trafficking in Arms Regulations (ITAR). The essence of both policy statements is that such participation is prohibited unless the export comes under an Agreement for Cooperation in accordance with Section 123(d) of the Atomic Energy Act of 1954.[14] In the case of DOC controls, Section 744.5 says that it is the policy of the U.S. to encourage U.S. firms and individuals to participate in maritime (civil) nuclear propulsion plant projects in friendly foreign countries, provided that U.S. naval nuclear propulsion information is not disclosed. Fortunately there is no active "maritime" project to be encouraged since the restriction would appear to preclude any U.S. participation.

Nuclear-powered submarines have long been a part of the military forces of the five official nuclear weapons states (China, Russia, France, the United Kingdom, and the United States). In the future this situation could change. Media reports over the past decade indicate that Pakistan was reportedly negotiating with China in 1990 to purchase a nuclear submarine. This was apparently in response to India leasing a nuclear submarine from the U.S.S.R. There have been no recent reports of Pakistani interest in nuclear submarines, although as India's indigenous nuclear submarine program progresses, interest in Pakistan may increase. There were numerous reports in the past that both Argentina and Brazil were interested in the development of nuclear submarines. Neither country has made much, if any, progress in acquiring nuclear-powered submarines.

Most of the commodities and materials associated with fuel fabrication and nuclear reactors would presumably have some utility in a naval nuclear propulsion project. There is some uncertainty in terms of how broadly U.S. license reviewers will interpret what constitutes "participation" in a naval nuclear propulsion project. Will it tend to be closely associated with the propulsion system itself, or will "participation" include exports to support functions or to the organization in charge of the project?

---

[14]Section 123.20 of the ITAR precludes the export of items covered by Category VI unless the export comes under an Agreement for Cooperation for Mutual Defense concluded pursuant to the Atomic Energy Act of 1954, as amended. The exceptions to this rule are: (1) if the proposed export involves an article which is identical to that in use in an unclassified civilian nuclear power plant; (2) if the proposed export has no relationship to naval nuclear propulsion; and (3) if it is not for use in a naval propulsion plant. Readers should refer to Section 744.5 of the Department of Commerce's Export Administration Regulations for further guidance. Also, exporters should be aware that items exempted by the ITAR may be controlled by the Department of Commerce, the Nuclear Regulatory Commission, or the Department of Energy.

Additional Factors to Be Considered in Reviewing an Export Application

This part of the process to evaluate proposed exports examines and comments on factors that might be considered in the review of an export application. These factors tend to fall into two categories: those that relate to the nonproliferation credentials of the recipient country or end user and those that relate to the technical characteristics of the proposed transfer. Although these lists of factors to be considered were developed for the use of officials engaged in deciding whether to approve a proposed export, exporters can also gain a better appreciation and insight into the process by studying these factors as well.

Nonproliferation Credentials

Party to the NPT

Although this has been and continues to be a consideration for supply, many nuclear exporters have transferred equipment and technology to countries that are not party to the NPT. For example, the United States is trying to finalize an agreement to sell nuclear materials and technology to India. All countries of the world are parties to the NPT except India, Israel, and Pakistan, which never joined the Treaty, and North Korea, which withdrew from the Treaty.

An IAEA Full-Scope Safeguards Agreement in Force

All parties to the NPT are required to begin negotiations with the IAEA to conclude a full-scope safeguards agreement not later than the date of depositing the instruments of ratification or accession, and such agreements shall enter into force no later than 18 months after the negotiations begin. Several states party to the NPT are not in compliance with this requirement.[15] For commodities that are multilaterally controlled by Part 1 (the Trigger List) of the NSG Guidelines, having a full-scope safeguards agreement in force is a condition of supply. It is a favorable factor for proposed exports of commodities subject to NSG Part 2 controls.

An Additional Protocol in Force

The IAEA has undertaken a vigorous effort to negotiate an Additional Protocol to its safeguards agreement with all member states. Steady progress is being made on increasing the number of states that have implemented the Additional Protocol.[16] An Additional Protocol provides the IAEA with additional information on nuclear activities within states and expanded, complementary access to nuclear facilities.

---

[15]Nonnuclear weapons states party to the NPT without IAEA full-scope safeguards agreements in force as of late 2006 were Andorra, Angola, Bahrain, Benin, Burundi, Cape Verde, Central African Republic, Chad, Comoros, Congo, Djibouti, Equatorial Guinea, Eritrea, Gabon, Guinea, Guinea-Bissau, Kenya, Liberia, Mauritania, Micronesia, Moldova, Mozambique, Oman, Qatar, Rwanda, Sao Tome and Principe, Saudi Arabia, Sierra Leone, Somalia, Timor-Leste, Togo, and Vanuatu. The IAEA continues to urge states to negotiate and bring into force a safeguards agreement covering all nuclear activities. For an up-to-date status of safeguards agreements in force, refer to www.iaea.org/OurWork/SV/Safeguards/sv.html.

[16]The IAEA has made significant progress in getting states to bring into force an Additional Protocol. None of the nonnuclear weapon states party to the NPT that do not have a comprehensive safeguards agreement in force have significant nuclear programs. By contrast, several of the states that do not have the Additional Protocol in force do have significant nuclear programs. These include Argentina, Brazil, Egypt, Iran, Kazakhstan, and Mexico. For an up-to-date status of Additional Protocols in force, refer to www.iaea.org/OurWork/SV/Safeguards/sg_protocol.html.

Actions Consistent with Commitments

This factor can be addressed by identifying all countries that appear on official lists of U.S. nuclear export control documents that in some way have a negative connotation. It can be assumed that the country is on the list or lists for a valid reason. In some cases it might not be due to bad conduct in the field of nonproliferation but rather a failure to negotiate a full-scope safeguards agreement with the IAEA as required by the NPT, a failure to implement effective export controls, its location in a region of instability, or the country itself undergoing internal conflict. To an exporter, without further explanation, it is difficult to always discern the basis for the country being on a U.S. government list.

Cooperates in Nonproliferation Policy in General

A factor to consider is the extent to which a country has associated itself with treaties, agreements, and multilateral arrangements devoted to the nonproliferation of weapons of mass destruction and the means to deliver them. This would include being party to the Comprehensive Test Ban Treaty, the Chemical Weapons Convention, or the Biological Weapons Convention or membership in the Australia Group, the Missile Technology Control Regime, a nuclear weapons-free zone treaty, the NPT Exporters (Zangger) Committee, or the NSG.[17]

Derogatory Information About the End User

Some states and end-user entities are under sanctions by the United States or other nations for reasons related to other aspects of their foreign or domestic behavior. For example, from time to time the U.S. Department of State publishes a list of countries that it considers to be state sponsors of terrorism. Most nuclear exports are denied to these countries. Early in the license review process the export control official should ascertain whether an "essentially identical" transfer has been previously denied by another supplier. In addition, considerable weight should be given to whether other types of commodities have been denied to the end user in the past.

Technical Considerations

Appropriateness of Commodity to the Stated End Use and End User

In nearly all cases, an export license application requires the exporter to provide information on the end user of the proposed exported items and the specific end use or purpose for which the items are being exported. This factor forms the basis of the technical evaluation conducted in the license review process. In many cases knowledge of the nonnuclear uses of a commodity is also important. The exporter is often in a better position to evaluate the credibility of a proposed transfer than the licensing official who must make the decision on approval or denial, so it is in the interests of the exporter to make sure the information is complete and credible.

Significance for Proscribed or Controlled Purposes of the Proposed Transfer

To be able to properly evaluate this factor, the supplier state should have persons capable of determining whether a commodity meets the parameters of the control language, and second, determining whether the commodity is especially designed or prepared for the processing, use, or production of special fissionable material.

The U.S. System of Nuclear Export Controls

Overview of the U.S. Nuclear Export Control System

The U.S. system of nuclear export controls is a complex arrangement of laws and regulations that are administered by DOE, Department of Commerce (DOC), Department of State

[17]For a more complete analysis of the concept of "nonproliferation credentials," see Chapter 14, "Evaluating Nonproliferation Bona Fides," in this volume.

(DOS), and the NRC. Although not having direct licensing responsibilities, the Department of Defense (DOD) is also a key player in the process to ensure that national security concerns are addressed.

Nuclear Technology Controls of the Department of Energy

DOE has statutory responsibility for regulating the transfer of nuclear technology and technical assistance. In accordance with Section 57.b of the Atomic Energy Act, as amended, only the Secretary of Energy, with the concurrence of the DOS and after consulting with DOD, DOC, and the NRC, can authorize persons to engage, directly or indirectly, in the production of special nuclear material outside the United States. This provision applies to technology transfers and technical assistance related to all activities of the nuclear fuel cycle, including nonpower reactors. These transfers can take both tangible and intangible form.

Under the implementing regulation, Part 810 of Title 10, Code of Federal Regulations, the Secretary of Energy has granted a general authorization for transfers in nonsensitive nuclear activities to some countries. For other countries, a specific authorization from the Secretary is required for all activities that fall within the scope of the controls. For assistance involving sensitive technologies (production reactors, accelerator-driven subcritical assembly systems, enrichment, reprocessing, plutonium fuel, heavy water production, and nonpower reactors above 5 MW (thermal)), a specific authorization is required for transfers to all countries. Differentiation of countries involves many factors, including nonproliferation credentials, maturity of their export controls, location, and their nuclear trade relationship with the United States. Because of the complexity of the issues involved in these types of transfers, the time to process applications for nuclear technology transfers is considerably longer than for other export cases.

Inquiries from persons and companies seeking to transfer nuclear technology or to provide technical assistance are received and reviewed by DOE to determine whether the proposed activity falls outside the scope of the regulation, is generally authorized, or requires a specific authorization of the Secretary of Energy. If a specific authorization is required, DOE staff will prepare an analysis of the proposed assistance. If it is determined that the proposed activity is contrary to U.S. nuclear export control laws, regulations, or policy, the request is denied and the applicant is notified. If DOE staff intends to recommend approval to the Secretary of Energy, the analysis and the preliminary recommendation, along with any conditions on transfer, are sent to the other agencies for comment and, in the case of the DOS, for its approval. After the consultations are completed, the case is sent to the Secretary of Energy for signature.

For requests involving sensitive technologies, DOE will convene a panel of experts to determine whether the proposed activity meets the legal standard for Sensitive Nuclear Technology (SNT). [18] Assistance determined to be SNT requires much more stringent conditions of supply.

DOE must also maintain strict internal technology security controls. Due to its role in the development, production, testing, and disposition of nuclear weapons and thus its possession of vast amounts of sensitive information, DOE must be especially vigilant in protecting its own information. To meet this challenge, DOE has established programs to monitor DOE persons and contractors in their technical exchanges and travel; to track visits by foreign nationals to DOE facilities; and to review closely all transfers of publications, computer software, and technical data from DOE to other countries.

---

[18]Section 4 (a) (6) of the Nuclear Nonproliferation Act of 1978 defines sensitive nuclear technology this way: "'Sensitive nuclear technology' means any information (including information incorporated in a production or utilization facility or important component part thereof) which is not available to the public and which is important to the design, construction, fabrication, operation or maintenance of a uranium enrichment or nuclear fuel reprocessing facility or a facility for the production of heavy water, but shall not include Restricted Data controlled pursuant to chapter 12 of the 1954 Act."

Nuclear Material and EDP Equipment Controls of the Nuclear Regulatory Commission

In the U.S. system of nuclear export controls, essentially all commodities uniquely related to peaceful nuclear uses are under the jurisdiction of the NRC. Materials and equipment controlled by the NRC correspond to a great degree with commodities controlled by Part 1 of the NSG Guidelines, the so-called Trigger List.

NRC licensing authority comes from various sections of the Atomic Energy Act, as amended, and is implemented by Part 110 of Title 10, Code of Federal Regulations. Applications received by the NRC, an independent U.S. government agency, are transmitted to the DOS for a consensus Executive Branch recommendation for approval or denial. DOE is the Executive Branch agency that provides the technical evaluation of each NRC license application and for cases to be approved; the DOE obtains assurances from the recipient government on peaceful uses, retransfer, and physical security. An important aspect of the U.S. process for administering export controls on nuclear materials and equipment licensed by the NRC is the requirement that such cooperation be conducted in accordance with a bilateral agreement for cooperation in the peaceful uses of nuclear energy. These agreements are negotiated by the DOS with DOE participation and are signed by the Secretary of Energy on behalf of the U.S. government.

Dual-Use Equipment and Material Controls of the Department of Commerce

DOC has licensing authority for all commodities and technologies for use in peaceful nuclear programs that are not otherwise controlled by NRC and DOE. Working closely with DOC, experts from DOE and its laboratories created and maintain a Nuclear Referral List of about 90 entries controlled for nuclear nonproliferation purposes. With few exceptions, the nuclear-related commodities licensed by DOC are controlled multilaterally by the Dual-Use Annex found in Part 2 of the Nuclear Suppliers Group Guidelines.

DOE remains the key agency in the process for reviewing license applications for nuclear-related commodities submitted to DOC. Applications are forwarded to DOE from DOC for review. From DOE the applications are sent to the relevant technical experts at the DOE laboratories for technical and end-user analyses. In this high-volume and time-urgent process, DOE brings capabilities unique to the interagency process by being able to provide a comprehensive review encompassing both technical and policy aspects. This part of the system is by far the largest and the most labor intensive.

Military-Related Nuclear Equipment and Material Controlled by the Department of State

DOS has responsibility in the United States for all militarily related transfers, including equipment, materials, and technology for use in nuclear weapons programs and in naval nuclear propulsion programs. The legal authority of DOS in this area comes from the Arms Export Control Act of 1976 and is implemented by the International Traffic in Arms Regulations (ITAR) found in Title 22, Code of Federal Regulations, Parts 120–130. By virtue of U.S. legal and policy commitments, traffic in nuclear-related commodities licensed by DOS is very low. DOE closely monitors any license applications for commodities that could be of nuclear nonproliferation concern, including those applications related to nuclear submarine programs.

Summary

All three principal components of the nuclear nonproliferation regime (NPT, IAEA safeguards, and export controls) are dependent on the strength and viability of each other.

In many respects effective nuclear export controls face challenges identical to those faced by the other two elements of the regime. Some of these challenges are summarized here:

- *The U.S-India initiative.* On July 19, 2005, U.S. President George W. Bush and Prime Minister Manmohan Singh of India issued a joint statement declaring their resolve to transform the current relationship between their countries and establish a global partnership. Within the broad context of the proposed partnership were commitments by President Bush to conduct civil nuclear cooperation with India, a state not party to the NPT and possessing extensive unsafeguarded nuclear activities, including a mature nuclear weapons program.[19] President Bush said that he will work to achieve full civil nuclear energy cooperation with India; that he will seek agreement from Congress to adjust U.S. laws and policies; and that he will work with friends and allies to adjust international regimes to enable full civil nuclear energy cooperation and trade with India. Implementing these commitments will require far-reaching changes to U.S. and NSG export controls.
- *Russian and Chinese conduct.* Russian and Chinese cooperation with India and Pakistan, respectively, poses threats to the viability of the NSG. Both have exploited the two exemptions to the full-scope safeguards requirement of the NSG Guidelines.[20]
- When Russia announced in 1998 its intent to build two nuclear power reactors in India, it informed the NSG that it was doing so under the grandfather exemption of the NSG Guidelines. The basis for the Russian claim was an agreement made years earlier in 1988 by the Soviet Union and India for the construction of nuclear reactors in India. Notwithstanding the informal nature of the NSG Guidelines, this was the first time in the history of the NSG that a member had acted contrary to the provisions of the Guidelines. Soon thereafter, in 2001, Russia announced to the NSG that it was going to provide nuclear fuel to the Tarapur power reactor in India under the safety exemption. This was an even more incredulous declaration than the first exemption and showed that Russia was clearly in violation of the Guidelines. The NSG was powerless to act since the NSG Guidelines were not legally binding on the membership.
- Although not a clear violation of the Guidelines, Russia's cooperation with Iran in helping to build the Bushehr nuclear power plant can be interpreted as contrary to the Nonproliferation Principle of the NSG Guidelines. An example is the opinion of most other NSG members that Russia's cooperation with Iran on Bushehr will provide technology, experience, and knowledge that can assist with Iran's suspected development of nuclear weapons.
- China, on becoming a member of the NSG in May 2004, voluntarily announced a wide range of activities with Pakistan that it was declaring to be grandfathered. This too undermines the NSG because it essentially allows China a free rein in its nuclear commerce with Pakistan to the exclusion of all other suppliers.
- *The spread of nuclear technology.* Perhaps the greatest threat to the viability of export controls as an effective instrument of foreign policy has been the spread of nuclear technology to so many more countries today than when export controls

---

[19]The full text of the joint statement can be found at www.bilaterals.org/article.php3?id_article=2464> (Aug. 2007).

[20]There are two exceptions to the full-scope safeguards requirement under Part 1 of the Guidelines of the NSG. The first is a safety exception. In exceptional cases a transfer is permitted when an item is deemed essential for the safe operation of the facility, but that facility itself must be under safeguards. The second exception is the so-called "grandfather" clause. If an NSG member had entered into an agreement or contract prior to April 3, 1992, involving controlled items, the exports are permitted. Countries coming into the NSG after that date are "grandfathered" up to the date of membership.

> began. In the beginning there was a clear "fire break" between the few that possessed the technology to produce highly enriched uranium and plutonium and those that did not. Today all the countries of concern have some competency in the production of special fissionable material. In today's environment, export controls can at best delay or impede a country's nuclear program.

To keep pace with this diffusion of technology, the export control infrastructures of emerging nuclear suppliers must become as effective as possible. Several initiatives can help meet this challenge. First, all nations should take seriously their nonproliferation obligations under the NPT and the 2004 U.N. Security Council Resolution 1540. Resolution 1540 requires all states to establish and maintain appropriate effective national export and transshipment controls to prevent proliferation of weapons of mass destruction.[21] Second, additional steps need to be taken to strengthen the Nuclear Suppliers Group and incentives for its members to abide by the group's decisions. Some positive steps would include implementation of a decision to require the existence of effective export controls in the recipient state as a criterion of supply for nuclear materials, equipment, and technology and a factor for consideration for dual-use items and technologies. Other improvements would be a requirement that states have an Additional Protocol in force as a condition of supply and a further strengthening of NSG guidelines for enrichment and reprocessing technologies. Finally, as called for by Resolution 1540, states with large financial resources and highly developed export control infrastructures should provide assistance and training to help improve export control systems in developing states.[22]

---

[21]The text of the resolution can be found at www.un.org/News/Press/docs/2004/sc8076.doc.htm (Aug. 2007).

[22]For a description of such efforts by the United States and some of its allies, see Chapter 29, "The Growing Role of Customs Organizations in International Strategic Trade Controls," in this volume.

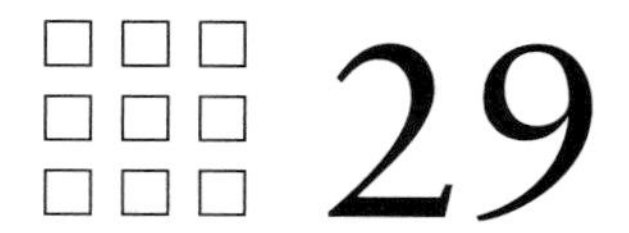

29 The Growing Role of Customs Organizations in International Strategic Trade Controls

Dr. Todd E. Perry

Introduction

Against the backdrop of export control developments described in the previous chapter, a new era of nonproliferation export control activity is emerging, one in which customs and other frontline enforcement organizations are making significant contributions to slow the spread of the materials, equipment, and technology required to manufacture weapons of mass destruction (WMD). Within a growing number of states, the knowledge of complex technology control lists developed by multilateral nonproliferation export control regimes is being distilled into information that customs and other enforcement officers need to identify controlled commodities during inspections. This process, in turn, enhances national export control systems by calling attention to the importance of strengthening national export control laws and to the training of technical specialists who support all aspects of national export control systems.

The addition of inspections of outbound and transiting cargoes to national export control capacities also underscores the imperative of information sharing between countries that are seeking to better detect and deter illicit WMD-related transfers without placing undue burdens on international trade. The resulting enhanced interdiction capabilities constitute an additional layer of defense against the inadvertent transfer by suppliers and transshippers of controlled commodities and are thus a high priority for states seeking to fulfill their legal obligations to prevent the transfer of WMD under the Nonproliferation Treaty (NPT) and under United Nations Security Council Resolution 1540 (UNSCR 1540).

This chapter first reviews the status of the proliferation threat and international developments that have promoted interdiction as a response to that threat. It then reviews the development of resources and training that have made it possible for customs and other enforcement organizations to interdict WMD-related commodities and outlines the prerequisites for supporting and sustaining these capacities. The chapter then focuses on the positive impact of export control enforcement-related customs training on national export control and border security arrangements and concludes with a review of how treaty-based international institutions are being used to catalyze export control enforcement capacities within the context of UNSCR 1540's overarching counter-proliferation mandate.

The Evolving Threat and the International Response

Several factors have worked together since the end of the Cold War to underscore both the threat of WMD proliferation and the urgency of enhancing national export control enforcement capacities. Of these, nothing better illustrates the threat more vividly than recent revelations about a wide range of illicit trading and manufacturing networks developed by the former Pakistani nuclear weapons program chief, A. Q. Khan.[1] The Khan network's supply of WMD-related commodities to Libya and other countries demonstrated conclusively that, like individual countries, nonstate networks can covertly supply proliferators.

Revelations about the Khan network emphasize two other aspects of the threat as well. The first involves the relatively new phenomenon of *onward proliferation*, where one proliferator supplies another. The role of onward proliferators like North Korea and Iran in this and related networks shows how these states participate in illicit WMD-related trade. In this sense, state and nonstate networks have combined forces, using the advantages of each to maximize illicit trading successes.

The second aspect of the threat involves the Khan network's capacity to "subcontract" the manufacture of controlled equipment in third countries. For example, the Khan network used Malaysia to manufacture thousands of centrifuge components that were subsequently sent to Libya. The absence of export control laws in Malaysia virtually ensured that these activities would go undetected. Based on this experience, it is now prudent to presume that any country with an industrial infrastructure must have a robust export control system if it is to have any hope of detecting the potential exploitation of its manufacturing base by proliferation networks.

These two aspects of the threat, when viewed through the lens of governments seeking to prevent the terrorist use of WMD, give rise to the alarming prospect that terrorists with access to stolen nuclear explosive fissile material could find the commodities needed to turn this material into a weapon without the knowledge of state suppliers. Although there is no conclusive evidence that illicit trading networks have supplied terrorist organizations with WMD-related commodities, the nonstate nature of these networks underscores the risk that terrorists might one day acquire the means to manufacture or assemble WMD from manufacturers that, as the Khan network has shown, can remain outside the control of state-based authorities.

Prior to the revelations about the Khan network and prior to the terrorist attacks of September 11, 2001, other proliferation events had already motivated governments to recognize the inadequacy of traditional "supply-side" export controls to the goal of preventing the transit and supply of sensitive items to state and possibly nonstate proliferators. Like past shocks to the Nuclear Nonproliferation Treaty (NPT) based nonproliferation regime, including the 1974 Indian nuclear test and the 1992 discovery of Iraqi WMD programs, 9/11 brought about a major transition in international efforts to curb the spread of weapons-usable commodities—only this time, instead of resulting in the creation or expansion of multilateral "regimes" of supplier countries devoted to developing lists of items needed to manufacture WMD and guidelines for export of these items, 9/11 resulted in an unprecedented, U.S.-led international effort to *implement and enforce* regime norms.

After 9/11, the WMD supplier regimes—the Nuclear Suppliers Group (NSG), the Australia Group (AG), and the Missile Technology Control Regime (MTCR) —adopted language directed at preventing the terrorist acquisition of WMD, and made significant adjustments and additions to their control lists. But the main focus of post-9/11 efforts to counter proliferation was specifically aimed at strengthening the operational capacity of countries to interdict the commodities these regimes aim to control. The first U.S. efforts, such as the Container Security Initiative (CSI), were established to preclude a repeat of 9/11 by preventing the use of containerized shipping to inflict damage on the United States. Other U.S.-led measures were broader in scope and designed to not only protect the U.S. homeland but to

---

[1]William Langewiesche, "How to Get a Nuclear Bomb," *Atlantic Monthly*, Dec. 2006.

also prevent illicit transfers of commodities needed to manufacture WMD. For example, the United States initiated the Proliferation Security Initiative (PSI) with the aim of interdicting shipments of WMD-related items worldwide and with the express aim of enhancing prospects for interoperability between countries' enforcement agencies.[2]

PSI and other interdiction and enforcement initiatives originally operated in the absence of an overarching international framework that fully captured the integration of the enforcement mission into the nonproliferation regime. The Khan network revelations changed this by catalyzing international consensus behind a universal mechanism that could capture and integrate "traditional" regime-based and newer antiproliferation-based approaches. This consensus resulted, with U.S. urging, in the unanimous passage of U.N. Security Council Resolution 1540 (UNSCR 1540) in 2004, which calls on states to criminalize WMD proliferation. UNSCR 1540 underscores the urgency of enacting *and* implementing controls to prevent the illicit transfers of WMD-related commodities.[3] Unlike the voluntary multilateral WMD regimes, UNSCR 1540 also makes it incumbent on all potential supplier *and* transit countries to implement effective export, import, transit, and transshipment controls on WMD-related commodities and to develop the air, land, and sea border controls needed to prevent WMD-related smuggling.[4]

UNSCR 1540 is thus the first treaty-based, international mechanism to explicitly connect border security to the export control mission.[5] Whereas past national export control efforts were rooted in national licensing systems working in conjunction with affected industries, the inclusion of export control *enforcement* in the new nonproliferation paradigm necessarily places a focus on the training of inspectors and analysts within national customs and other frontline enforcement organizations so that that they can recognize and interdict suspect WMD-related shipments pursuant to their national commitments. In this sense, UNSCR 1540 adds another "layer of defense" to existing treaty-based arrangements, multilateral regimes, and national systems of control.[6]

Implementation of the Export Control Enforcement Mission

Nonproliferation specialists have long been aware of the indispensable roles played by Customs authorities within countries possessing both modern licensing systems and fully

[2]U.S. Department of State, "The Proliferation Security Initiative: What Is the Proliferation Security Initiative?" June 2004, http://usinfo.state.gov/products/pubs/proliferation/.

[3]1540 Committee, http://disarmament2.un.org/Committee1540/report.html.

[4]United Nations Security Council Resolution 1540 (as adopted by the Security Council at its 4956th meeting, April 28, 2004), Paragraph 3(c): "Develop and maintain appropriate *effective border controls and law enforcement efforts to detect, deter, prevent and combat*, [italics added] including through international cooperation when necessary, illicit trafficking and brokering in such items in accordance with their national legal authorities and legislation and consistent with international law."

[5]UNSCR 1540 also calls for the protection of these commodities and related materials within national borders so that they are not vulnerable to theft by proliferators, again with the understanding that standards created by existing multilateral institutions constitute the best foundation for these controls. Thus, unlike voluntary standards developed by the International Atomic Energy Agency (IAEA), UNSCR 1540 creates a legal requirement to secure materials needed to manufacture WMD.

[6]United States Mission to the United Nations, "Statement by Andrew K. Semmel, Principal Deputy Assistant Secretary of State, Bureau of Nonproliferation, on Articles III and VII, Second Committee of the 2005 Review Conference of the Treaty on the Non-Proliferation of Nuclear Weapons," U.S.U.N. Press Release # 97 (05), May 19, 2005, www.un.int/usa/05_097.htm.

developed targeting and inspectional procedures. Customs organizations usually play a *de facto* role in a country's export licensing process, by validating that proper shipping manifests accompany licensed items, and in some export control systems, even exercising legal authority in the final clearance of licensed items as they leave a country's territory.

Traditionally, though, customs' responsibility for validating the legitimacy of exports or transshipments was almost exclusively related to the assessment of duties and tariffs on traded commodities and not to the confirmation of a correspondence between the physical appearance of an item and its accompanying paperwork. Detecting illegal activities in a shipper's declaration mostly meant detecting illegal drugs, weapons, or contraband regulated by states for internal security and law enforcement purposes. To be sure, customs organizations played important roles in the interdiction of WMD-related items when provided with intelligence tips or when pursuing or prosecuting companies known to be part of proliferation networks. But until recently, very few customs organizations had the knowledge or procedures to determine whether a so-called "dual-use" item might be controlled for proliferation purposes or even if an item was indeed dual-use in the first place, with potentially commercial *and* weapons applications.

Awareness among customs organizations of the need to ferret out the illicit shipment of items controlled for proliferation purposes emerged during the 1980s, in response to the knowledge about Iraq's attempts to acquire WMD-related commodities. The creation of the Nuclear Suppliers Group (NSG) dual-use control list and guidelines in 1992 was a reflection of growing knowledge among the supplier states that Iraq, like Pakistan before it, used legitimate shipping routes to hide illicit shipments of commodities needed to manufacture nuclear weapons.[7] In one of the first attempts to bring national customs authorities into the export control process, the NSG published in 1992 a *Guidebook for Customs Officers*, which was in effect an abbreviated version of the longer handbook created by NSG member state licensing officers for NSG member countries only. The abbreviated guidebook was made available to governmental officials in nonmember states, making it possible for the first time for transit state countries in particular to train frontline officers in the appearance of commodities on the NSG control lists.

The promulgation and publication of dual-use control lists by the NSG in 1992, followed by the emergence of parallel control lists and resources within the other supplier regimes, made it considerably more difficult for proliferators to acquire WMD-related commodities from traditional, state-based suppliers. Proliferators were thus forced to adopt the more sophisticated approach of shipping controlled items without proper licensing permissions. However, given the difficulty encountered by frontline enforcement agencies in identifying these items in the first place, and given the fact that certain supplier and transit states are known to have weak or nonexistent export control systems, proliferators can avoid detection by using false manifests and indirect shipping routes. This is why targeted inspections of suspect cargoes are *essential* to the goal of stemming illicit WMD commodity trade; smugglers do not apply for export licenses.

Challenges Associated with Identifying WMD-Related Commodities

Initial efforts to assist customs organizations in detecting illicit WMD-related shipments were first undertaken by the International Law Enforcement Academy (ILEA) beginning in the mid-1990s. These training sessions, attended by frontline enforcement officers from dozens of countries, provided information that would help officers identify potentially bad end users so that they could be called on to detain shipments based on intelligence tips.[8] The course also supplied trainees with the Nuclear Suppliers Group *Guidebook for Customs Officers*, thereby providing for the first time, on a systematic and multilateral basis, an opportunity

[7]Reference previous chapter.

[8]International Law Enforcement Academy, Roswell, New Mexico, 2004, www.ilearoswell.org/index.htm.

to expose senior customs officials to the existence of NSG control lists and to methodologies designed to defeat proliferators seeking to smuggle controlled commodities through the dramatically expanding system of global trade.

When ILEA initiated this course, customs officials and technical specialists alike believed that frontline customs inspectors could not be trained to distinguish between controlled and uncontrolled dual-use commodities. This belief resulted from two aspects of the dual-use control lists. The first is rooted in the term *dual-use*, which describes items that have legitimate commercial as well as potential WMD-related applications. Unlike guns or drugs, dual-use items are part and parcel of everyday commerce.[9] As noted previously, smugglers usually falsify the specifications of dual-use items or simply provide very generic information about a shipment. So long as these items are incorrectly or imprecisely manifested and offer, by their very nature, no sign of requiring a license, they can be shipped as normal cargo and are, in a very real sense, "hidden" in plain view.

The second aspect of WMD-related dual-use controls that postponed customs' active participation in the nonproliferation export control mission is the complex nature of the NSG control lists and, to a lesser extent, of the MTCR and AG control lists as well. These lists also often organize controlled commodities according to their potential use in a weapons program and include specifications and technology thresholds that are only fully understood by the trained industry and licensing and weapons specialists who wrote the lists in the first place.

The confusion that arises from attempting to use complex control-list specifications to determine whether or not an item is controlled is compounded by the redundancy of regime control lists. A wide range of controlled items such as manufacturing machinery and weapons testing and diagnostic equipment, not to mention basic metals and other materials, are referenced multiple times within individual control lists and are sometimes referenced within multiple regimes as well. To make matters worse, it is very difficult in a few cases, and virtually impossible in many others, to match the Harmonized Tariff Code (HTC) nomenclature, used internationally by customs to assess duties and tariffs on exported and imported goods, to the control-list specifications. *In short, unless an item is accompanied by an export license, there is no means of determining from the paperwork that accompanies it whether or not it is controlled.*

U.S. Efforts to Overcome the Challenge of Dual-Use Commodity Identification

The September 11, 2001, terrorist attacks put into motion a range of initiatives that laid the groundwork for beginning to overcome some of these challenges. As a result of the attacks, national governments and international institutions publicly took account of the overriding importance of assisting customs and other frontline enforcement organizations in the interdiction of potentially dangerous items. This did not immediately result in calls to educate customs officers on items on the WMD control lists, but it marked an important shift in how customs chiefs and their organizations around the world thought about themselves, with security becoming an important element of the customs mission. This rhetorical shift led in many cases to the realization among customs leaders that they would need several new capabilities: targeting systems capable of detecting chemical, radiological, and nuclear explosive devices; a better understanding of the role of dual-use items in international commerce; and

[9]Even many so-called "single-use" commodities on the NSG list that are especially designed or prepared for nuclear use can appear to the untrained eye as common industrial items. For example, a Malaysian police report on the illicit A. Q. Khan manufacturing operation in Malaysia notes that the single-use centrifuge components used for uranium enrichment were labeled as air-conditioning parts. "Press Release by the Inspector General of Police in Relation to Investigation on the Alleged Production of Components for Libya's Uranium Enrichment Programme," Feb. 20, 2004.

the technical expertise needed to identify assembled devices *or* the dual-use items needed to manufacture them, in the event that they were detected and detained.

As the target of the 9/11 attacks, the United States was the first country to take concrete action to address the importance of detecting chemical, radiological, or nuclear explosive devices, by creating the Container Security Initiative (CSI). Managed by the U.S. Department of Homeland Security (DHS), formed in 2002, CSI addresses the threats to border security and global trade posed by the potential terrorist use of maritime containers by enabling DHS to target and pre-screen shipments before their arrival in U.S. ports. Through CSI, U.S. officers work with host customs administrations worldwide to establish criteria for identifying containers deemed to be at high risk of being used to ship WMD or conventional explosive devices. The host customs organization then uses nonintrusive technology to quickly inspect these containers before they are shipped to U.S. ports.[10]

Similarly, in collaboration with CSI specialists, the U.S. Department of Energy's (DOE's) Megaports Initiative provides advice on the addition of radiological detection activities to port security operations, along with training on the use of detection equipment to scan outbound cargo.[11] Neither CSI nor the Megaports Initiative is tailored to help customs organizations search for WMD-related dual-use commodities. Their focus is primarily on detecting WMD or other weapons of mass effect being shipped to the United States, rather than on ensuring that the United States or partner countries do not themselves inadvertently proliferate to terrorists or countries of concern. Nevertheless, these initiatives have played an important awareness-raising role by elevating the importance of the WMD-related interdiction mission with a large number of customs organizations, thus providing an opportunity for other U.S. programs to directly address the commodity interdiction issue (see the following discussion).

By raising awareness of the proliferation threat and working with other countries to improve targeting, CSI also indirectly calls attention to the evaluation of potentially dangerous end users. Even prior to 9/11 senior CBP inspectors recognized that frontline officers responsible for in- *and* outbound shipments were capable of developing the means to identify suspicious dual-use shipments, based on a combination of manifest data such as the commodity shipper, consignee, and payment information. This knowledge emerged among ILEA-trained CBP inspectors and was reinforced by Iraq's record of U.S. commodity procurement during the late 1980s. After seeing how proliferators succeeded at exploiting their ports, officers started evaluating suspicious shipments, either through use of the Internet to check declared commodity specifications against manufacturers' databases or through specialists associated with the National Targeting Center and EXODUS Command Center (ECC). The ECC helps target suspect containers and connects CBP to DHS labs and to the various U.S. licensing agencies, including the Nuclear Regulatory Commission and the Departments of Commerce, Energy, and State.

The extensive training of CBP Officers and counterpart foreign government and border enforcement officials at the Volpentest HAMMER Training and Education Center in Washington State also helped prompt and further enhance an awareness of the dual-use smuggling threat.[12] Interdict/RADACAD courses offered at HAMMER focus on the detection, identification, and interdiction of materials, commodities, and components associated with the development or deployment of WMD, with a particular emphasis on special nuclear

---

[10]U.S. Customs and Border Protection, "Keeping Cargo Safe: Container Security Initiative," www.customs.gov/xp/cgov/borber_security/international_activities/csi/.

[11]National Nuclear Security Administration, "Megaports Initiative: Protecting the World's Shipping Network from Dangerous Cargo and Nuclear Materials," www.nnsa.doe.gov/megaports_ initiative.htm.

[12]Volpentest Hammer Training and Education Center, U.S. Department of Energy, www.hammertraining.com; and Interdict/RADACAD International and Domestic Border Security Training, Pacific Northwest National Laboratory, www.pnl.gov/interdict/training/.

and other radioactive materials. These training sessions help prepare frontline officers for the CSI mission as well as for U.S. domestic nuclear detection programs. In addition, HAMMER-based training provides brief sample overviews of dual-use commodities and in-depth field exercises simulating international seizures of radioactive materials *and* WMD-related dual-use commodities. DOE also now offers abbreviated training courses that focus on the importance of detecting WMD and related dual-use commodities at numerous U.S. ports for DHS inspectors and investigators.

The Role of Multilateral Regimes and of International Assistance Efforts

After 9/11, the various multilateral export control regimes also responded to customs organizations' needs by adopting consensus language urging regime members to make greater use of their customs authorities in the interdiction effort.[13] The NSG, MTCR, and AG created working groups alongside their annual plenary meetings, intended to facilitate greater participation of national customs authorities within the regimes and to promote the enforcement aspect of export controls among national governments. Supplementary guidebooks were created by the regimes for members' national customs organizations that explain, in general terms, the role of dual-use commodities in the manufacture of WMD. Perhaps most important for the longer term, the World Customs Organization (WCO) explicitly recognized, in its 2005 Framework of Standards language, the role played by its member organizations in the fight against proliferation and anticipated the importance of assisting member organizations in adhering to Framework objectives through the formation of a new capacity-building committee.

However, although multilateral and international organizations have had a great deal to say about interdiction, it has been primarily the United States, especially through export control and related border security assistance training programs, that has provided the tools and training needed for countries to understand and implement UNSCR 1540-based interdiction objectives. These programs have been managed since the mid-1990s by the Departments of Energy, Commerce, Defense, and Homeland Security and coordinated and partially funded by the U.S. Department of State's Export Control and Related Border Security (EXBS) Program.[14] Resources developed for international audiences by CBP include basic targeting and inspections training, whereas resources developed under contract by the Department of Commerce (DOC) include customized databases with images of controlled items and accompanying national laws and control-list specifications. These training sessions and resources do not provide frontline inspectors with actual knowledge about controlled commodities, but they do provide the basic skills and best practices for any frontline organization seeking to acquire this knowledge.

In contrast, the Commodity Identification Training (CIT) approach developed by DOE's International Nonproliferation Export Control Program (INECP) *does* endeavor to familiarize frontline inspectors with WMD-related, dual-use commodities and to simplify

---

[13]"Strengthening Measures to Prevent the Spread of Weapons of Mass Destruction," Australia Group, press release, June 2003, www.australiagroup.net/en/releases/press_2003_06.htm; press statement, Nuclear Suppliers Group Plenary Meeting, May 16–17, 2002, Prague, Czech Republic, www.nsg-online.org/PRESS/2002-03-press-prague.pdf; press release, Plenary Meeting of the Missile Technology Control Regime, Sept. 24–27, 2002, Warsaw, Poland, www.mtcr.info/english/press/warsaw.html.

[14]"The EXBS Program: Export Control and Border Security Assistance Program," U.S. Department of State, www.state.gov/t/np/export/ecc/20779.htm. In addition, other countries, including the United Kingdom, have provided basic customs training, and Japan's Trade Ministry, METI, has been a regional leader in providing licensing and basic export control awareness training. The European Union Commission also recently approved funding for export control outreach to non-EU countries but has not yet determined how to direct this outreach.

their identification. To accomplish this task, a team of DOE national laboratory specialists led by Argonne National Laboratory (ANL) took a different approach to the various regime control lists, grouping items together based on type (materials, industrial equipment, electronics) rather than the reasons for control and focusing primarily on appearance, notable features, and other readily identifiable criteria rather than detailed control specifications.

CIT has been offered and adapted by INECP in dozens of countries, with technical specialists in those countries receiving assistance from U.S. national laboratory specialists to customize and adapt commodity-specific modules to the trainings needs of their own national customs, border guard, border police and defense organizations. Training courses delivered by U.S. specialists and by their counterparts on a widespread basis have resulted in multiple interdictions of smuggled dual-use goods. CIT has become the internationally recognized means of systematically familiarizing frontline officers with commodities on the regime-based control lists.

CIT and related assistance is also provided by DOE on a regional basis to enhance information sharing between knowledgeable export control specialists from various national licensing and customs organizations. For example, Australian and Japanese technical specialists have joined their U.S. counterparts in providing CIT to a wide range of Asian audiences.[15] In Eurasia, Russian versions of regime-based nuclear commodity guides, developed by Russian specialists in collaboration with DOE's national laboratories, are being shared by their Russian developers with specialists in neighboring states. As a result, CITs are being delivered by national technical specialists in the Caucasus and Central Asia that are consistent with local and regional dual-use commodity manufacturing and transit trends.[16]

The Enforcement Mission as a Catalyst for Export Control System Reform

There is a broad range of operational challenges associated with the introduction of specialized export control enforcement trainings that have nothing to do with training content. For CIT to have the desired impact, customs organizations must have in place a targeting and risk management system *and* be prepared to physically inspect the very small percentage of cargoes that are identified as suspicious under such a system. DOE's WMD-related interdiction assistance, combined with CBP's assistance in the development of basic targeting and inspectional practices, has had a favorable impact in this regard, by demonstrating that the interdiction of WMD-related commodities requires not only the scanning of cargo for radiological sources but also the inspection of suspect cargo on the basis of a systematic review of manifest data.

DOE and CBP training teams join forces where possible to demonstrate, through combined CIT and enforcement training, the prerequisites for an enforcement approach that takes the WMD commodity interdiction mission into account. DOE-CBP partnerships include joint

---

[15]"NNSA Expands Training Efforts to Combat WMD Smuggling," Asian Export Control Observer, Issue 4, Oct./Nov. 2004, p. 20, http://cns.miis.edu/pubs/observer/asian/pdfs/aeco_0410.pdf.

[16]Central Asia: "Export Control Seminars Held in Kazakhstan," *NIS Export Control Observer*, June 2005, p. 21, http://cns.miis.edu/pubs/nisexcon/pdfs/ob_0506e.pdf; Sean Reid, "Commodity Identification Trainings Organized in Kyrgyz Republic and Kazakhstan," *International Export Control Observer*, May 2006, p. 18, http://cns.miis.edu/pubs/observer/pdfs/ieco_0605e.pdf; Caucasus: "Nuclear Dual-Use Commodity Identification Training in Georgia," *NIS Export Control Observer*, Aug. 2005, p. 7, http://cns.miis.edu/pubs/nisexcon/pdfs/ob_0508e.pdf; "Commodity Identification Training Workshop Held in Azerbaijan," *NIS Export Control Observer*, Dec. 2004/Jan. 2005, p. 25, http://cns.miis.edu/pubs/nisexcon/pdfs/ob_0412e.pdf; Richard Talley, "NNSA's Role in Preventing Weapons Proliferation: CIT Workshop Indigenization Moving Forward," *NIS Export Control Observer*, Sept. 2004, p. 3, http://cns.miis.edu/pubs/nisexcon/pdfs/ob_0409e.pdf.

training overseas and joint training for foreign customs officers who participate in CBP's International Seaport Interdiction Training (ISIT). EXBS has advanced the maritime interdiction effort too by supporting the integration of DOE's CIT by the U.S. Coast Guard in its inspectional and boarding training for foreign maritime partners. DOE also provides CIT for U.S. CSI officers posted to foreign ports, making CSI a *de facto* platform for the collaboration of U.S. and foreign customs officers to target and detect smuggled dual-use commodities.

Even when a country's frontline enforcement organizations have implemented robust targeting and inspectional practices, other obstacles remain. The development of CIT and other tools and training courses for frontline officers for the eventual purpose of interdicting WMD-related items presupposes the existence of licensing and other technical specialists capable of evaluating suspect shipments in a timely manner. This focus on technical "reachback" as part and parcel of DOE's CIT approach has in fact had the unexpected and positive effect of creating pressure on countries with limited or no legal basis for export controls to adopt the necessary laws and regulations mandating the creation of complete licensing organizations and accompanying civil and criminal penalties for ignoring their authorities.[17] These same pressures have prompted countries with understaffed systems and with inadequate information sharing about suspect manufacturers, shippers, and end users between licensing and enforcement agencies to undertake the legal and regulatory reforms needed to fill these gaps as well.

Future Challenges and Opportunities

Export controls have long been understood as an activity directed by national licensing organizations led by trade or defense ministries that regulate domestic industries through the licensing process. Licensing activities have traditionally involved answering questions concerning the potential WMD-related use of a commodity or associated information and the potential "end user" of an item proposed for export. These questions were usually answered by a combination of technical experts, intelligence specialists, and bureaucratic and sometimes political decision makers, resulting in the approval or denial of an export license request submitted by a manufacturer or shipping agent. These same experts were often responsible for providing technical support for their countries' participation in the multilateral nonproliferation regimes focused on maintaining up-to-date control lists and on providing guidance to members on how to best apply them.

The longstanding problem with these traditional export control practices is that short of an intelligence tip or a tip from industry, there is no means for even those countries with robust licensing systems to detect attempts to evade export control laws. The traditional approach has worked to the degree that manufacturers, motivated by penalties or even by the prospect of damage to their international reputation, have complied with licensing- and shipping-related export control requirements. Companies are also deterred from committing wide-scale export control violations by the fact that competitors are likely to notice unexpected sector-specific market successes and report them to national licensing authorities.

Small-scale diversions within countries with robust export control systems or even larger sensitive shipments from countries with underdeveloped or nonexistent systems of control are rarely discovered until well after the fact, if at all, since the only way to detect

---

[17]The installation of portal monitors to detect radiological materials by DOE's Second Line of Defense (SLD) has also resulted in the need to train technical experts who can provide reachback to assess detector alarms and intercepted nuclear materials. As with CIT, SLD reachback requirements have fostered national reforms by underscoring the importance of laws and regulations governing the transfer of controlled materials and of established procedures that enable enforcement officials to call on legal and technical experts to make determinations about potentially controlled items.

illicit shipments is to inspect and interdict outbound or transshipped cargoes. Inadvertent transfers even by states with very strong political and commercial commitments to nonproliferation represent a substantial and ongoing risk that traditional export control practices can only partially address. Export control enforcement is thus not simply an added layer in the fight against proliferation, it is also an *indispensable element* of any strategy that seeks to detect illicit shipments of WMD-related commodities.

UNSCR 1540 provides a framework for addressing these threats by connecting effective border controls to WMD proliferation. However, as with its other provisions, such as the protection of WMD-related materials, the establishment of licensing systems, and the initiation of outreach programs to manufacturers of WMD-related commodities, the Resolution's requirement for border controls to prevent the transit of WMD-related items is not self-enforcing. Indeed, as a cursory review of national implementation reports to the UNSCR 1540-mandated Experts Committee shows,[18] the universal treaty-based status of UNSCR 1540 offers no guarantee of immediate improvements in these areas, because it offers neither the mechanisms nor the resources to foster state-level reforms. Instead, interested states are left to seek and provide assistance, per the Resolution's mandate, from each other and from the various international arrangements and international organizations already in place to identify and remedy gaps in their existing export control systems.

It is therefore all the more remarkable that customs organizations worldwide have made progress in relation to the Resolution's objectives and that dozens of additional countries per year are adapting frontline enforcement tactics that are consistent with national regime-based export control strategies. The focus on interdicting illicit dual-use shipments relies on knowledge developed within the regimes on the scope of dual-use controls but provides only the most basic information to determine whether or not an item *might* be controlled so that it can be detained for further analysis by technical specialists. The necessary connection between frontline enforcement officers and technical specialists, in turn, helps call attention to shortcomings in national export control systems, especially on the imperative of further developing the human infrastructure needed to "staff" the main elements of these systems, including enforcement-related training and analysis. Related activities, such as regional exercises organized under the auspices of the PSI, are also, in part, designed to underscore the urgency of developing the laws and regulations required for the smooth functioning and interoperability of enforcement and licensing agencies.

Still other activities, such as national reporting requirements developed through treaty-based procedures such as the IAEA Additional Protocol (AP) and the Chemical Weapons Convention's Organization for the Prohibition of Chemical Weapons (OPCW) indirectly help member states strengthen national capacities to develop and enforce export control norms. These reporting activities can constitute small but important first steps in the development of internal procedures for national authorities seeking to better understand their countries' nuclear- and chemical-related trade patterns. CITs that cover these commodities have in fact fostered consideration by national governments of the need for more complete national systems of export control. Still other countries that have not yet developed complete export control systems have used data generated from AP and OPCW reporting requirements to make sure their customs organizations have detected shipments that correspond to transactions known to have taken place under the auspices of these international reporting mechanisms.[19]

---

[18]Richard T. Cupitt, "Export Controls and the Implementing UNSC Resolution 1540 (2004)," Carnegie Conference on Nonproliferation, Washington, D.C., Nov. 7–8, 2005, www.carnegieendowment.org/static/npp/2005conference/presentations/cupitt_export_controls.pdf.

[19]International Atomic Energy Agency, Model Protocol Additional to the Agreement(s) Between State(s) and the International Atomic Energy Agency for the Application of Safeguards, INFCIRC/540 (Corrected), Sept. 1997, www.iaea.org/Publications/Documents/Infcircs/1998/infcirc540corrected.pdf; Organization for the Prohibition of Chemical Weapons, 2006, www.opcw.org.

It is important to note as well the mutually supportive aspects of UNSCR 1540's export control and WMD-related materials accountancy and protection missions. While proliferators have so far chosen to manufacture WMD-related materials indigenously, it is widely believed that states and especially nonstate groups will attempt to shortcut that process by stealing fissile material. A wide range of ongoing DOE, DOD, and international initiatives thus seek to ensure the protection of nuclear, chemical, and biological materials needed to build WMD. Developing the capacity of frontline enforcement (and of internal police) organizations to recognize commodities on the supplier regime control lists, which include controlled materials, serves as an important complement to the material security and detection initiatives and is therefore of premier importance to both the export control *and* material protection elements of UNSCR 1540.

Achieving the levels of integration needed to implement the various provisions of UNSCR 1540 on a country-by-country basis and translating this action into the necessary regional and international sharing of information and best practices poses a tremendous and ongoing challenge to the nonproliferation effort. States and the international system itself are permeated as never before by organizations engaged in myriad unregulated and underregulated trading and manufacturing activities.[20] The international community's ability to capitalize on the positive developments described here, and on many others outside the scope of this chapter, will invariably remain largely with those governments most vested in strengthening the implementation of nonproliferation norms and in preserving the NPT-based nonproliferation regime.

[20]Moises Naim, *Illicit: How Smugglers, Traffickers, and Copycats Are Hijacking the Global Economy*, New York: Doubleday, 2005.

30
Case Study: The Khan Network

Sara Kutchesfahani

Introduction

Abdul Qadeer (A. Q.) Khan, widely regarded as the father of Pakistan's nuclear bomb, will be forever remembered as the leading black-market dealer in nuclear weapons technology and design. Revered in Pakistan as a national hero and loathed in the West for organizing illicit nuclear trade, he was responsible for supplying and receiving nuclear secrets through his infamous network: the A. Q. Khan network. Operating throughout the 1980s and 1990s, his global clandestine network sold uranium enrichment equipment and nuclear bomb designs to sworn enemies of the United States: Iran, North Korea, and Libya. His acquisition of gas centrifuge designs, through his network, allowed Pakistan to build a nuclear weapons arsenal in the 1990s. Suspicions over A. Q. Khan activities arose sporadically throughout the late 1990s, culminating in his downfall in December 2003. At the end of 2003, Col. Gaddafi of Libya publicly announced an end to Libya's hitherto secret active nuclear weapons program, most of which was procured through the Khan network. In February 2004, Khan went on Pakistani national television to admit trading his country's nuclear secrets. As a result of Khan's confession, the Pakistani authorities, headed by President Musharraf, agreed that Khan would not be handed over to anyone else, particularly to the United States, and would instead remain under house arrest in Pakistan.

The disruption of the A. Q. Khan network was an important achievement in preventing nuclear terrorism and illicit nuclear trade. However, before the network was exposed, nuclear weapons design and technology had already been shared and exchanged between several states. The Khan network also exposed the loopholes in international export controls. This chapter provides an insight into the Khan network, analyzes the network's exposure and Khan's subsequent admission, and highlights the many challenges that lie ahead in tackling illicit nuclear trade. Finally, it concludes that illicit trade in technology, equipment, and information related to nuclear weapons is a continuing threat to international security.

Insight into the Khan Network

Khan's Motivations

A. Q. Khan's motivations in launching his network cannot be tied down to one single reason alone but instead to a number of possible factors. The main reason was to obtain a nuclear weapon for Pakistan, but there are other factors that also need to be considered. Money, greed, fame, and politics all played a role in his motivation. Primarily, though, Khan's motivation was nationalism: It was to give Pakistan its own nuclear bomb, even if this meant not conforming to the nonproliferation system already in place. In Khan's mind, having a nuclear bomb would ensure the security and survival of Pakistan in the face of a nuclear-armed India. His highly successful illicit procurement network, launched in the 1970s, helped supply Pakistan's gas centrifuge program, used to produce weapons-grade uranium in Pakistan's nuclear weapons program, as well as designs for Pakistan's nuclear bomb.

It has been noted that in the late 1970s, Western nations tried to stop Khan from building a bomb, leading Khan to resent the nuclear weapon states telling others that they could not have the bomb, even if they needed it for their own security.[1] Khan hated this system and thought that by spreading details of getting the bomb, the system would be broken. With the benefit of hindsight, Khan's network has not resulted in the complete collapse of the nonproliferation regime, but it has resulted in North Korea, a former NPT signatory and a former A. Q. Khan consumer, becoming a nuclear weapons state. Khan's sense of nationalism and his hostility to Western controls on nuclear technology, coupled with the allure of generating millions and millions of dollars in buying and selling weapons designs, were what made his illegal network so successful and, for almost 30 years, free from disruption.

Details of the Transglobal Network

The international intelligence community has a relatively good idea of where Khan's network was operating, but questions still remain regarding which entities in the network were responsible for certain actions and whether all elements of the network have been shut down. The network spanned four decades and involved countries, companies, secret bank accounts, and individuals on four continents. The successful operation of Khan's network was possible because Khan was able to manipulate the globalization process and circumvent export controls. He created and used front companies all over the world in countries that have very flexible rules in internal and external trade, used the constant availability of communication and travel, and took advantage of the swiftness and anonymity of international finance.[2] Furthermore, it has been said that most of his network participants were more market-savvy than geopolitically inspired and so were willing and able to sell to people with the desire to buy.[3]

Khan's network really was transglobal. Factories, engineering and design services, and sales operatives were set up in Malaysia, South Africa, Turkey, Pakistan, Switzerland, the United Arab Emirates (UAE; Dubai in particular), and various other countries in Europe, the Middle East, and Africa.[4] In addition to the front companies set up in these places, the network also

---

[1]"Shopping for Bombs: Nuclear Proliferation, Global Insecurity, and the Rise and Fall of A. Q. Khan's Nuclear Network," discussion meeting between author Gordon Corera and Joanne J. Myers, Carnegie Council, Sept. 7, 2006, www.cceia.org/resources/transcripts/5391.html.

[2]Andrew Roch statement to the Hearing Before the Subcommittee on International Terrorism and Nonproliferation on "The A. Q. Khan Network: Case Closed?" May 25, 2006, p. 17, www.internationalrelations.house.gov/109/27811.pdf.

[3]*Ibid.*

[4]Most reports refer to these countries. Sources include Chaim Braun and Christopher F. Chyba, "Proliferation Rings: New Challenges to the Nuclear Nonproliferation Regime," *International Security*, vol. 29, no. 2 (Fall 2004); David Albright and Corey Hinderstein, "Unraveling the A. Q. Khan and Future Proliferation Networks," *The Washington Quarterly*, Spring 2005; Michael Laufer, "A. Q. Khan Nuclear Chronology," Carnegie Endowment for International Peace, Nonproliferation Issue Brief Volume VIII, Sept. 7, 2005, www.carnegieendowment.org/static/npp/Khan_Chronology.pdf; "Fact Sheet: Strengthening International Efforts Against WMD Proliferation," White House Fact Sheet, Feb. 11, 2004, www.whitehouse.gov/news/releases/2004/02/20040211-5.html; and "Shopping for Bombs: Nuclear Proliferation, Global Insecurity, and the Rise and Fall of A. Q. Khan's Nuclear Network," discussion meeting between author Gordon Corera and Joanne J. Myers, Carnegie Council, Sept. 7, 2006, www.cceia.org/resources/transcripts/5391.html.

depended on unaware manufacturing companies and suppliers in many countries.[5] The network used a factory in Malaysia, Scomi Precision Engineering (SCOPE), to manufacture key parts for centrifuges. The ultimate destination of these centrifuge components was hidden by transshipment through two different front companies set up in Dubai—one called Gulf Technical Industries and the other called SMB Computers. Dubai was the main hub to which parts from around the world were routed and then sent on to Khan's customers—in particular, Libya.

The network also reportedly used the Turkish electrical components firm Elektronik Kontrol Aletleri (EKA) to buy motors and frequency converters for the centrifuges.[6] It is known that Khan and his associates went to Syria, Saudi Arabia, Egypt, Chad, Mali, Algeria, Niger, and Sudan, among others, but what is not known is what exactly Khan did there.[7] According to Albright and Hindertsein at the Institute for Science and International Security, after the facilities produced the item (whether metals, equipment, or subcomponents), the network would send it to Dubai under a false end-user certificate, where it would be repackaged and sent to Libya.[8] Mohamed ElBaradei, Director General of the IAEA, put it very succinctly: "Nuclear components designed in one country could be manufactured in another, shipped through a third (which may have appeared to be a legitimate user), assembled in a fourth, and designated for eventual turnkey use in a fifth."[9] This is precisely the way in which the A. Q. Khan network was able to work on a transglobal basis.

As for personnel, several of the network's leaders, including Khan himself, were located in Pakistan, but other leaders were spread throughout the world, including in Switzerland, the United Kingdom, UAE, Turkey, South Africa, and Malaysia[10]—most of the time in cities were the network had set up bases.

Khan's Customers: Quid Pro Quo

From 1987 until about 2002, Pakistani nuclear technology was available for sale on the international black market. In what ElBaradei has called the "Wal-Mart [a large U.S. department store chain] of private sector proliferation,"[11] A. Q. Khan's network was responsible for providing blueprints and drawings of centrifuges used in a weapons program. As shown in Figure 30.1, Khan's customers were nations hostile to the United States. Iran was the network's first customer, and cooperation between the network and North Korea and Libya was the most recent. Information regarding what was given and exchanged is hard to come by in open-source literature, yet all the information available on Iran has been made public through IAEA reports pertaining to Iran's nuclear program.

---

[5]David Albright and Corey Hinderstein, "Unraveling the A. Q. Khan and Future Proliferation Networks," *The Washington Quarterly*, Spring 2005, p. 11400.

[6]Chaim Braun and Christopher F. Chyba, "Proliferation Rings: New Challenges to the Nuclear Nonproliferation Regime," *International Security*, vol. 29, no. 2 (Fall 2004), p. 15.

[7]Leonard Weiss statement to the Hearing Before the Subcommittee on International Terrorism and Nonproliferation on "The A. Q. Khan Network: Case Closed?" May 25, 2006, p.10, www.internationalrelations.house.gov/109/27811.pdf.

[8]Albright and Hinderstein, 2005 p. 115.

[9]Mohamed ElBaradei, "Nuclear Nonproliferation: Global Security in a Rapidly Changing World" (speech, Carnegie International Nonproliferation Conference, Washington, D.C., June 21, 2004), p. 3, www.ceip.org/files/projects/npp/resources/2004conference/speeches/ElBaradei.doc.

[10]Albright and Hinderstein, 2005 p. 114.

[11]Mark Landler, "U.N. Official Sees a 'Wal-Mart' in Nuclear Trafficking," *New York Times*, Jan. 23, 2004.

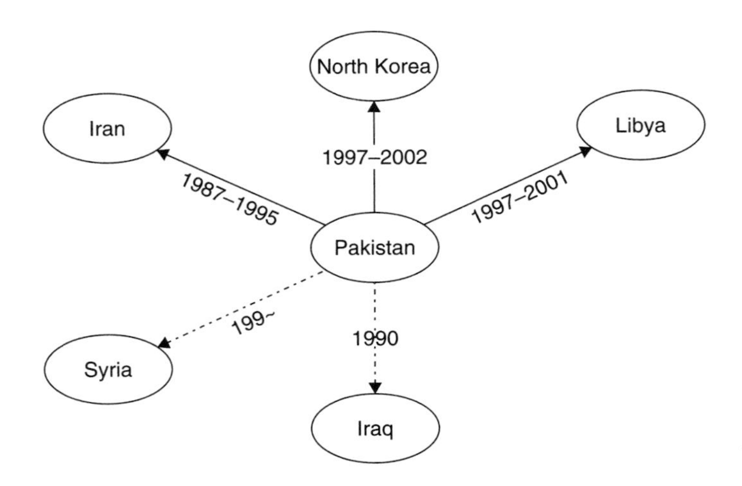

FIGURE 30.1 Interactions of the A. Q. Khan network. (Source: Diagram taken from Alexander H. Montgomery, "Ringing in Proliferation," *International Security*, vol. 30, no. 2 [Fall 2005], p. 173. Data for the diagram is found in Gaurav Kampani, "Proliferation Unbound: Nuclear Tales from Pakistan," CNS Research Story, Center for Nonproliferation Studies, Monterey Institute of International Studies, Feb. 23, 2004, http://cns.miis.edu/pubs/week/040223.htm [June 2007]. (*Note*: Declined offers of assistance are dotted; uncertain dates are marked – [mid-decade].)

It has been said that in late 1990, shortly after the U.N. Security Council imposed an embargo on Iraq as a result of Saddam's invasion of Kuwait, A. Q. Khan offered to help Baghdad produce gas centrifuges and design nuclear weapons. Khan's offer was discovered by the IAEA in the mid-1990s, and the Iraqis told the IAEA that they did not receive anything. Initially, the Iraqis were skeptical of Khan's offer, thinking it was a sting operation because Pakistan was a U.S. ally. Egypt, Syria, Saudi Arabia, and even Afghanistan were said to have been offered aid by Khan's network between 1997 and 2003.[12] Egypt is said to have turned down the offer, and Khan's network never provided assistance to Syria in the end.[13] Little is known about Saudi Arabia, even though the Kingdom may have been offered assistance. It is feared that during Khan's visit to Afghanistan in this period, he might have offered nuclear assistance to al-Qaeda, but these claims have yet to be substantiated.[14]

The main item for sale was centrifuges—what Pakistan have called the P1 and P2 centrifuges, the first two that Pakistan deployed in large numbers in its own nuclear program. As Albright and Hinderstein explain, "The P1 centrifuge uses an aluminum rotor, and the P2 centrifuge uses a maraging steel rotor, which is stronger, spins faster, and therefore enriches more uranium per machine than the P1 centrifuge's aluminum rotor."[15] The P1 centrifuges are similar to the early Dutch-designed aluminum CNOR/SNOR centrifuges, and the P2 design is based on the more advanced German-designed maraging steel G-2 centrifuge.[16] As Table 30.1 indicates, the P2 centrifuge is almost twice as efficient as the P1 design in separative work units per year. A separative work unit is a unit of measurement for the amount

[12]Albright and Hinderstein, 2005, p. 113.

[13]*Ibid.*

[14]*Ibid.*

[15]*Ibid.*, pp. 114–115.

[16]M. D. Zentner, G. L. Coles, and R. J. Talbert, "Nuclear Proliferation Technology Trends Analysis," Pacific Northwest National Laboratory, Sept. 2005, p. 35, www.pnl.gov/main/publications/external/technical_reports/PNNL-14480.pdf.

Table 30.1 Comparison of P1 and P2 gas centrifuges.

Type	P1	P2
Rotor material	Al (aluminum)	MS (maraging steel)
Speed (m/sec)	350	500
Length (m)	1–2	1
Kg SWU/yr	1–3	5

of uranium processed and the degree to which it is enriched. In other words, the P2 centrifuge can make more enriched uranium in less time.[17]

Iran

The IAEA Board of Governors reports on Iran's nuclear program released by the Director General and cited here are the most comprehensible and verifiable accounts of the nature of the Iranian program. The reports provide details of external assistance the Iranians received, yet the specifics are not mentioned. In other words, though A. Q. Khan's network is not mentioned by name, it is understood that his network was the main supplier to the Iranian nuclear program. U.S. government reports openly state that Iran (Libya and North Korea, too) were customers of the Khan network.[18] Policy papers, journal articles, and newspaper reports from all over the world also make the connection between A. Q. Khan and Iran.

Regarding what exactly the Iranians received from the Khan network, according to one of the most recent IAEA reports, Iran showed the IAEA a copy of a one-page document reflecting an offer it was said to have received in 1987 by a "foreign intermediary."[19] The document concerned the possible supply of a disassembled centrifuge (including drawings, descriptions, and specifications for the production of centrifuges); drawings, specifications, and calculations for a "complete plant"; and materials for 2000 centrifuge machines. The document also made reference to auxiliary vacuum and electric drive equipment; a complete set of workshop equipment for mechanical, electrical, and electronic support; and uranium reconversion and casting capabilities. To date, Iran has declined the Agency's request for a copy of the one-page document.[20] Furthermore, according to a U.S. Congressional report, in January 2006 Iran revealed that A. Q. Khan provided information on key processes related to weapons production, including uranium conversion into metal and casting uranium metal hemispheres.[21] It is not publicly known when this information was transferred.

---

[17]Victor Gilinsky, Marvin Miller, and Harmon Hubbard, "A Fresh Examination of the Proliferation Dangers of Light Water Reactors," Oct. 22, 2004, p.38, www.npec-web.org/Reports/Report041022%20LWR.pdf.

[18]See, for example, "Fact Sheet: Strengthening International Efforts Against WMD Proliferation," White House Fact Sheet, Feb. 11, 2004, www.whitehouse.gov/news/releases/2004/02/20040211-5.html; and Sharon Squassoni, CRS Report for Congress, "Iran's Nuclear Program: Recent Developments," July 20, 2006, http://fpc.state.gov/documents/organization/70030.pdf.

[19]"Implementation of the NPT Safeguards Agreement in the Islamic Republic of Iran," report by the Director General to the Board of Governors, Feb. 27, 2006, GOV/2006/15, p. 3, www.iaea.org/Publications/Documents/Board/2006/gov2006-15.pdf.

[20]*Ibid.*

[21]Sharon Squassoni, CRS Report for Congress, "Iran's Nuclear Program: Recent Developments," July 20, 2006, p. 3, http://fpc.state.gov/documents/organization/70030.pdf.

There are other reasons to believe that Pakistan was a primary source of nuclear technology for Iran. Christopher Clary, a research associate at the Center for Contemporary Conflict, U.S. Naval Postgraduate School, identifies three reasons:

- The IAEA suspects that sample swatches obtained in Iran containing traces of uranium enriched higher than 90% U-235 might have come from Pakistan.
- Centrifuge drawings acquired by Iran and given to IAEA inspectors resemble the design of the P-1 centrifuge.
- The IAEA discovered assembled centrifuges at the Doshan Tapeh military airbase near Tehran, which strongly resembled the P-2 centrifuge design.[22]

Added to these reasons, Iranian scientists were suspected of receiving nuclear training in Pakistan.[23] More details need to be released before the full picture can be painted, but with Iran repudiating IAEA requests to be transparent and the Pakistani government refusing to allow Khan to speak to anyone, it might take a few years before the international community knows for certain what exactly Khan sold to Iran.

North Korea

The Khan network's next customer was North Korea. The North Korean-Khan network collaboration enjoyed the benefits of a quid pro quo relationship. Pakistan's A. Q. Khan Research Laboratories (KRL; named after Khan in 1981 by President Zia ul-Haq in recognition of Khan's contributions to the operational enrichment facility at Kahuta, Pakistan) developed the Ghauri missile with North Korean assistance. It is said that in return, Khan could have transferred nuclear technology to North Korea, but very little is known about when any nuclear transfers began and what North Korea might have obtained from the Khan network. Most analysts point to 1992 as the date of the beginning of this cooperation, and many agree that it was not until 1997 that Pakistan transferred uranium enrichment technology to Pyongyang.[24] Transfers included old and discarded centrifuge and enrichment machines together with sets of drawings, sketches, technical data, and depleted uranium hexafluoride.[25] Centrifuge designs based on Pakistani versions of both early and second-generation centrifuges developed at the Urenco company enrichment plants in Almelo, Netherlands, and Gronau, Germany, were also sold to North Korea.[26] It should be noted that throughout the early 1970s, A. Q. Khan had been employed in the Almelo plant, and it has been said that he took design information and listings of component suppliers with him to Pakistan in 1975.[27] Armed with this information, the KRL Laboratories were able to develop and build the two centrifuge models that became an integral part of the Pakistani nuclear weapons program: the P1 and P2 centrifuges.

Libya

Cooperation with Libya appears to have begun in 1997 on an extensive scale and ended in 2003. Khan's support for Libya's nuclear weapons program was his most ambitious, but it also

---

[22]Christopher Clary, "Dr. Khan's Nuclear Wal-Mart," *Disarmament Diplomacy*, Issue No.76, March/April 2004, p. 3, www.acronym.org.uk/dd/dd76/76cc.htm.

[23]Leonard Weiss, "Turning a Blind Eye Again? The Khan Network's History and Lessons for U.S. Policy," *Arms Control Today*, March 2005, www.armscontrol.org/act/2005_03/Weiss.asp.

[24]See, for example, Clary, 2004, p. 3; Braun and Chyba, 2004; Laufer, 2005; Albright and Hinderstein, 2005.

[25]Laufer, 2005, p. 6.

[26]Braun and Chyba, 2004, p. 13.

[27]*Ibid.*

resulted in his eventual downfall. When Col. Gaddafi of Libya publicly announced an end to Libya's secret active nuclear weapons program in December 2003, A. Q. Khan and his network were exposed as having supplied Libya with uranium enrichment components and technology. Between 1997 and 2003, a vast network of companies and individuals in countries as far apart and as diverse as Switzerland, Malaysia, Pakistan, Spain, Turkey, South Africa, Germany, the United Kingdom, and the UAE supplied Libya with uranium enrichment components.[28] Table 30.2 shows what and when Libya received from the Khan network.

The Libyan purchases alone are estimated to have cost about $100 million,[29] with Tripoli almost receiving a turnkey nuclear weapons program with enough equipment to

Table 30.2 The Khan network's assistance to Libya.

Date	What Libya Received
No date provided, but probably at the early stage of the Khan-Libya relationship	Detailed nuclear weapons component designs, component fabrication information, and nuclear weapons assembly instruction.[1]
1997	Twenty assembled P-1 centrifuges and components for 200 additional units for a pilot enrichment facility.
September 2000	Two P2 centrifuges as demonstrator models. Libya places an order for components for 10,000 more models to build a cascade. Each centrifuge contains around 100 parts, implying approximately 1 million parts total for the entire P2 centrifuge cascade.
2001	1.87 tons of uranium hexafluoride. The amount is consistent with that required for a small pilot enrichment facility.
2001–2002	Blueprints for nuclear weapons plans.
October 2003	The German cargo ship *BBC China* is intercepted en route to Libya with components for 1,000 centrifuges.
March 2004	A container aboard the *BBC China* (the ship that was previously intercepted) arrives in Libya with one additional container of P2 centrifuge components. Gaddafi reports the arrival to U.S. intelligence and the IAEA. Furthermore, the Libyans warn U.S. officials that not all the components they had ordered had arrived and some might still show up in the future, which is why another container appeared on the *BBC China* five months after it was originally intercepted.
October 2005	All materials and components associated with Libya's nuclear weapons development program were removed and all associated activities have stopped.[2]

[1]David Albright and Corey Hinderstein, "Unraveling the A. Q. Khan and Future Proliferation Networks," *The Washington Quarterly*, Spring 2005, p. 114.

[2]Christopher M. Blanchard, "Libya: Background and U.S. Relations," *Congressional Research Service*, June 13, 2006, p. 29, http://italy.usembassy.gov/pdf/other/RL33142.pdf.

[28]Christopher M. Blanchard, "Libya: Background and U.S. Relations," *Congressional Research Service*, June 13, 2006, p. 23, http://italy.usembassy.gov/pdf/other/RL33142.pdf.

[29]Farhan Bokhari and Victoria Burnett, "Suspect quizzed over nuclear finance," *Financial Times*, March 26, 2004.

eventually construct a workable centrifuge enrichment plant.[30] If operated successfully, the centrifuge plant ordered would have been sufficient to produce enough highly enriched uranium to turn out approximately 10 nuclear weapons annually.[31] Ongoing technical assistance was offered for assembly and operation of the plant. As Albright and Hinderstein argue, had Libya continued to pursue its nuclear ambitions without exposing Khan's network, it could have assembled the centrifuge plant in about four or five years and produced significant amounts of highly enriched uranium.[32] For more details on Libya's nuclear weapon development efforts, see Chapter 18 by Wyn Bowen in this volume.

Pakistan's Imports

As discussed earlier, Khan's network was responsible for providing Pakistan with its initially covert nuclear weapons program. However, the international community is still at a loss as to what A. Q. Khan imported and where the imports came from. Pakistan refuses to tell investigators which items it imported from Khan's network. It has been said that China was the major supporter of the Pakistani program, having provided Pakistan with:

- A complete design of one of its early uranium nuclear warheads
- Sufficient quantities of highly enriched uranium for two such weapons[33]
- Short-range ballistic missiles and construction blueprints
- Assistance in developing a medium-range missile
- Support in developing second-generation uranium enrichment centrifuges, including the provision of 5,000 ring magnets in 1994–1995
- A 40 MW (th) heavy water plutonium and tritium production reactor located at Khushab[34]

Khan was instrumental in building his country's own nuclear arsenal and providing help to various other countries in building their own. His luck was bound to run its course after almost 30 years of operating underground since international intelligence agencies were soon latching on to him and his network. Pakistan's transfers were initially suspected by Western intelligence but were never prosecuted or successfully disrupted because of legal loopholes. However, by the end of 2003, with Gaddafi's declaration, the noose finally tightened around Khan, at last exposing his clandestine network operation.

The A. Q. Khan Network Exposed

In October 2003, the ship *BBC China* was intercepted by Italian authorities who seized sophisticated centrifuge components bound for Libya. When confronted with this evidence, Gaddafi accelerated his cooperation with investigators and renounced Libya's nuclear weapons program, exposing all the elements of the Khan network it had been dealing with.

---

[30]Clary, 2004, p. 4; Braun and Chyba, 2004, p. 16.

[31]Albright and Hinderstein, 2005, p. 113.

[32]*Ibid.*, p. 114.

[33]Chaim Braun and Christopher F. Chyba's paper ("Proliferation Rings: New Challenges to the Nuclear Nonproliferation Regime," *International Security*, vol. 29, no. 2 [Fall 2004], p. 21) makes reference to Albright and Hibbs, "Pakistan's Bomb," regarding this claim. Also see Leslie Gelb, "Pakistan Link Perils U.S.-China Nuclear Pact," *New York Times*, June 22, 1984, p. Al; and Leonard Spector et al., *Tracking Nuclear Proliferation*, Carnegie Endowment for International Peace, 1995, p. 49.

[34]Braun and Chyba, 2004, p. 21.

Pakistan subsequently faced intense pressure to deal with Khan and his cohorts. Pressure had already been mounting for the Pakistani government because in September 2003, the IAEA Board of Governors passed a resolution requesting all countries to help the IAEA resolve questions about Iran's nuclear program that was tied to Pakistan. Although Khan's network was not blamed directly, all fingers were pointing to him. With Iran's nuclear program called into question and being completely scrutinized, and then with the seizure of *BBC China*, it was time for Pakistan to give the international community some answers. Western intelligence had already garnered evidence against Khan and his network: the first stories of Pakistani proliferation broke in 2002.[35] Coupled with these two separate incidents unfolding, it was time for Khan and his network to account for their secret transfers.

Khan's Public Admission

Initially, Pakistani authorities were reluctant to arrest Khan, given that many Pakistanis considered him a national hero. But following requests from the United States and others, Khan was arrested. Khan received a conditional pardon from Pakistani President Musharraf and today he remains under house arrest at his home in Islamabad, with no access to outsiders. Many Pakistanis were detained when the scandal broke, but to date, none have been prosecuted. The Pakistani government has provided the IAEA and foreign governments with information about Khan's activities, serving as an intermediary, but no one outside the Pakistani government is allowed to interview Khan or the others who were detained. The IAEA has been allowed to submit written questions to Khan, but IAEA Director General ElBaradei has said that he would like to speak directly to Khan.[36] Khan's network spanned four continents, and many of his associates are still implicated. Prosecutions are taking place for individuals and companies involved in France, Germany, Japan, Malaysia, the Netherlands, South Africa, Spain, Switzerland, Turkey, and the United Kingdom. Because the network operated transglobally, information sharing among all states regarding the details and outcomes of these proceedings remains critical to the network's complete dismantlement.[37]

On February 4, 2004, Khan went on Pakistani national television, where he confessed to selling sensitive technology and equipment to Iran, North Korea, and Libya. In his statement, he took full and sole responsibility for the proliferation of nuclear weapons technology from Pakistan: "[...] It pains me to realize this in retrospect that my entire lifetime achievement of providing foolproof national security to my nation could have been placed in serious jeopardy on account of my activities, which were based in good faith, but on errors of judgment related to unauthorized proliferation activities."[38] He also claimed that his actions were not known by the Pakistani government: "I wish to place on record that those of my subordinates who accepted their role in their affair were acting in good faith, like me, on my instructions. I also wish to clarify that there was never ever any kind of authorization for these activities by the government."[39]

Questions arose regarding the supposed "no role" taken by the Pakistani government. The Pakistani government denies any involvement and knowledge of the Khan network and that of any senior official in the Pakistani army. Many experts view Khan's statement with

---

[35]Clary, 2004, p. 1.

[36]Q&A with Mohamed ElBaradei and Jonathan Mann, "Work in Progress," *IAEA Bulletin*, 47/2, 2006, www.iaea.org/Publications/Magazines/Bulletin/Bull472/htmls/nobel2005/work_in_progress2.html.

[37]Albright and Hinderstein, 2005, p. 117.

[38]A. Q. Khan, "I seek your pardon," *Guardian*, Feb. 5, 2004.

[39]*Ibid.*

skepticism and consider that it was intended not to embarrass the Pakistani government any further. How likely is it that Khan operated without the Pakistani government's knowledge or even approval? The fact that he has been placed under house arrest and is not allowed any interaction with the outside world suggests that there are a lot of things he could say that would embarrass the Pakistani government. Khan's claim that he acted alone is a hard pill to swallow, given that Musharraf's predecessor, the late Gen. Mohammed Zi ul-Haq, famously said, "It is our right to obtain [nuclear] technology. And when we acquire this technology, the entire Islamic world will possess it with us."[40] With the exception of North Korea, Khan's customers were all Muslim countries, suggesting on one hand his desire to spread nuclear weapons technology to his Islamic brethren and on the other his willingness to follow orders from his nation's president. Even though North Korea is not an Islamic country, Khan and Pyongyang developed a quid pro quo relationship in that Khan exchanged Pakistani nuclear weapons designs with North Korean help in missile technology.

It is hard to believe that the Pakistani authorities were unaware of Khan's activities. It is a known fact that Pakistan's nuclear arsenal is under control of the authorities and so nothing can be taken without those authorities' knowledge. How could it be possible for Khan to arrange the import and export of such huge quantities of nuclear-related material and technology without the knowledge of the Pakistani military or political leadership? Some of the transfers were even carried out on military planes. How could the Pakistani government not know about the network aggressively marketing its nuclear products, sometimes even with glossy brochures?[41] These glossy brochures were not hidden but were instead flaunted at arms shows; in November 2000, the former Washington, D.C., bureau chief for *Jane's Defense Weekly*, Andrew Koch, went to an arms show and saw the brochures, which he said stated that "Khan Research Laboratories is willing to offer a full range of nuclear products, including complete ultra centrifuge machines. On the back of this pamphlet and on an accompanying pamphlet it says more technical details. It clearly states that assistance is offered and provides contact numbers where you could go and get that assistance, and it says you can get a full range of help from assembling these machines to maintaining them and operating them."[42] These glossy brochures cannot be found on the Internet; it is uncertain whether they had been linked to the Dr A. Q. Khan Research Laboratory Website (www.krl.com.pk/) prior to his arrest, but what is certain is that they are not linked to the site today. Khan was not at this show, and as Koch asks, "One has to wonder was this really Khan and a few people, or is this an institutionalized program that was happening?"[43] Furthermore, how could the Pakistani government not be aware of the millions and millions of dollars Khan's network was amassing in Pakistan? President Musharraf speaks proudly of the Pakistan Army, saying it can account for "even a bolt of a rifle." But when pressed on this matter in an interview with CNN asking how nuclear technology could be transferred without his knowledge, his answer painted a very morbid picture: "Nuclear technology is on computers, on paper and in the minds of people."[44]

---

[40]Interview in *Akhbar al-Khalij*, March 13, 1986, p. F4, translated by Foreign Broadcast Information Service, FBIS-SAS-86-053, March 19, 1986.

[41]Edward R. Royce Prepared Statement to the Hearing Before the Subcommittee on International Terrorism and Nonproliferation on "The A. Q. Khan Network: Case Closed?" May 25, 2006, p. 2, www.internationalrelations.house.gov/109/27811.pdf.

[42]Andrew Koch prepared statement to the Hearing Before the Subcommittee on International Terrorism and Nonproliferation on "The A. Q. Khan Network: Case Closed?" May 25, 2006, p. 18, www.foreignaffairs.house.gov/archives/109/27811.pdf.

[43]Andrew Koch prepared statement to the Hearing Before the Subcommittee on International Terrorism and Nonproliferation on "The A. Q. Khan Network: Case Closed?" May 25, 2006, p. 18.

[44]Christiane Amanpour, "Full text of Musharraf interview," CNN, Jan. 23, 2004.

It is very easy to lay the blame entirely on the Pakistani government for its inability and/or unwillingness to stop Khan or to report his activities to the other nations in which he was operating, but the fault also lies with Western officials and the IAEA. This information was available to the IAEA because it was IAEA weapons inspectors who unraveled both Pakistan's assistance to the Iranian nuclear program and Khan's offer to Iraq. U.S. governments from the 1980s to date never made it their priority to expose Khan and put him out of business. Pakistan was an important U.S. ally in countering the Soviet invasion of Afghanistan in the 1980s. Even today, the Bush Administration's main priorities in Pakistan seem to be disrupting al-Qaeda terrorists living and/or operating from Pakistan. It is equally important to track down a full understanding of Khan's associates in Pakistan and where they are today.

The onus is on Pakistan's government more than on any other entity to ensure that it provides more assistance to investigators, including allowing the IAEA direct access to question Khan and his associates in person. Greater cooperation among the IAEA, Pakistan, the United States, and Western governments whose nations were involved in the network will allow for a more thorough investigation and for more unanswered questions to be resolved. Intelligence and information sharing can also help speed up the investigation and can serve as a tool to decrease the chances that such activities can occur again.

Challenges That Lie Ahead

Many unanswered questions and challenges still remain over Khan's network. Now that this particular network has been disrupted, how can the international community be assured that no other network is currently being formed? In what Stanford University academics Braun and Chyba have dubbed *proliferation rings*, the trend has already been set and the future looks bleak.[45] Braun and Chyba argue that the international community is faced with three interrelated challenges (or rings) to the nuclear nonproliferation regime. Ring One is *latent proliferation*, where nonweapons NPT members violate their NPT obligations and work on developing a weapons program. North Korea (as a former NPT state) and Iran are current examples of ring one; so was Libya. Ring Two is *first-tier nuclear proliferation*, of which A. Q. Khan's network is the prime example: Technology and/or material is sold to aid the development of nuclear weapons programs. Suppliers in this ring could be states that are not members of the Zangger Committee (the Nuclear Exporters Committee) or the Nuclear Suppliers Group or private entities within states that may be supplying nuclear weapons-relevant material on the international market. Finally, Ring Three is *second-tier nuclear proliferation*. This is where states in the developing world with different technical capabilities can trade among themselves to improve one another's nuclear weapons efforts. It is here that the next generation of A. Q. Khan's network can emerge. Thinking about such a possibility is exceptionally worrying, which is why these proliferation rings must be controlled; otherwise the international community will be faced with more nuclear weapons states. By addressing the challenges and remaining unanswered questions from the Khan example, it is hoped that the international community will reduce the risk of similar incidents in the future.

Some steps have already been taken to prevent some of the tactics that were used by Khan, including U.N. Security Council Resolution 1540, introduced in 2004, which seeks to stop WMD-related proliferation, as covered in more detail later in this chapter. Another measure that was introduced after the discovery of the Khan network was the Proliferation Security Initiative (PSI), launched in 2003 by President Bush as an initiative to coordinate efforts to impede sea, air, and land shipments of WMD, delivery systems, and related materials to and from states and nonstate actors. Furthermore, Chapter 29, "The Growing Role of Customs Organizations in International Strategic Trade Controls," by Todd Perry, in this

[45]Braun and Chyba, 2004.

volume describes how customs authorities are becoming central players in growing international efforts to detect illicit shipments of WMD-related commodities.

Remaining Unanswered Questions

A series of unanswered questions regarding Khan's legacy remains. Should these questions ever be answered, either by Khan himself or by the Pakistani government, more stringent policies can be implemented to ensure that there will not be another Khan-like network:

- Has the network been completely shut down? Or is it simply in a hibernation period waiting to strike again, but this time in a more damaging manner? As Joseph Cirincione, former director of nonproliferation at the Carnegie Endowment for International Peace, argues, "The network hasn't been shut down [...]. It's just gotten quieter. Perhaps it's gone a little deeper underground."[46]
- Does the international community know the full extent of the network? In other words, can we be certain that anyone who was part of the network has been identified and accounted for? Equally, have all the front companies been identified? Might there have been other countries from which the network was operating?
- What did Khan do in each country and city he visited? To whom did he speak?
- What kind of involvement did the Pakistani government and military have with the network?
- What did Pakistan import from the network? The network supplied Pakistan's covert nuclear weapons program, but Pakistan has refused to tell investigators which items they were.
- Have all of Khan's customers been exposed? Was it only North Korea, Iran, and Libya, or were there other nations that received equipment and have thus far remained quiet?
- Which nations received nuclear weapons designs from Khan? Are Pakistan's designs currently being used in Iran's program?
- How was China able to clandestinely provide Pakistan with the designs for its bomb and a supply of HEU in the 1980s?
- Does the international community know for certain that on his trips to Afghanistan, Khan did not meet with any al-Qaeda member? If he did, what information was exchanged?

Getting the answers to these questions can help in curbing the spread of nuclear weapons. Added to these answers, there are a series of preventative measures that the international community can seek to adopt, which are addressed here individually. First, there is a need to tighten export controls. The main reason Khan was so successful was that he very ably manipulated the international import and export control system. Second, there is a need to improve global intelligence and intelligence sharing. Western intelligence agencies failed to catch Khan even while he was operating on their soil. Third, there is a need for a long-term vision in dealing with proliferation. Finally, an effective response addressing the supply and demand sides of nonproliferation is needed.

Tightening Export Controls

Since the discovery of the Khan network, some key international measures have been introduced to strengthen international export controls and have been adopted by many nations. U.N. Security Council Resolution 1540 is one such measure. Introduced in April 2004, Resolution 1540 seeks to try to stop all WMD-related proliferation. Effectively, it seeks to improve all nations' import and export control systems by criminalizing WMD terrorism activities. Pakistan did not have a national export control system in place; hence

---

[46] Quoted in Louis Charbonneau, "Pakistan Reviving Nuclear Black Market, Experts Say," Reuters, March 15, 2005.

Khan's success. However, it has taken steps to create one, but the new system has yet to be implemented.

One of the key measures of Resolution 1540 calls for assisting countries in developing a national export control system because it is acknowledged that difficulties lie in the full implementation of the Resolution. One way forward would be for the United States and other Western allies, notably France and the United Kingdom (as permanent Security Council members), to make it their priority to help Pakistan, which would have to want that help, to work on an export control system as part of their commitment to the full implementation of 1540.

Improving Global Intelligence and Intelligence Sharing

Improving global proliferation intelligence should be a basic requirement if the international community seeks to stop proliferators and Khan-like networks. As the Khan network expanded over many different countries, an effective intelligence-sharing system should have been implemented. Today the exchange of intelligence data is improving, but this needs more formalization among cooperating states. Better global intelligence on international exports is needed to ensure that contraband items are not being smuggled. Without good intelligence, the Bush Administration's PSI cannot be successful.

Improved Multilateral Participation in Nonproliferation Initiatives

There needs to be greater multilateral cooperation in dealing with proliferation cases to prevent a reemergence of a Khan-like network. Multinational initiatives, including U.N. Security Council Resolution 1540, the PSI, the G8 Global Partnership against the spread of WMDs, and the Nuclear Suppliers Group are a good start. But for these initiatives to be completely effective, all measures must be implemented by all nations concerned. Nations that struggle to implement these measures, such as Pakistan, Malaysia, and other countries in the Khan network and their national export control system, should be aided by the permanent members of the U.N. Security Council. Efforts such as the U.S. Commodities Identification Training Program described by Todd Perry in Chapter 29 of this volume are critical in this regard and should be expanded.

Dealing with the Supply and Demand Sides of Proliferation

Finally, there is a need to deal with both the supply and demand sides of proliferation. Measures to deal with the supply side include limiting transfers of nuclear weapons-related technology and equipment. By improving the effectiveness of global nuclear export control systems with the participation of all states, the supply side can be improved. Dealing with the demand side is more difficult. Factors need to be addressed that fuel a country's desire to have a nuclear weapons program, such as the nation's security situation and sense of international status. (See James Doyle's chapter on rolling back nuclear weapons programs in this volume.) Addressing these factors can help reduce incentives and thus reduce internal demand for nuclear weapons.

Summary

A. Q. Khan will forever be synonymous with being the ringleader in trading nuclear weapons design and technology. His network famously exploited loopholes in international export controls and created a body of suppliers, manufacturers, and cargo fleets that provided secret nuclear technology to Iran, North Korea, and Libya—perhaps even to others. Khan helped build Iran and Libya's nuclear programs, and it was his help with Libya that ultimately led to his downfall. Had it not been for Khan, Iran's nuclear program would not be quite as advanced as it is today, nor would it be grabbing so much attention.

How Khan's transfers were never detected still remains a mystery. Western intelligence agencies had been suspicious, but it was not until Libya's announcement that it was dismantling its weapons program that Khan's luck ran out. For almost four decades, Khan was able to build his network on four continents. Millions and millions of dollars were exchanged, and people from all over the world were implicated in his network. Khan might be under house arrest now, but the international community did too little too late. Nuclear weapons designs and technology have traded hands all over the world, and not all blueprints have been found. Furthermore, many questions still remain unanswered today. Most of the answers can help to ensure that another Khan-like network will not emerge. But until these questions are answered, the international community is faced with the threat of terrorists getting their hands on controlled nuclear materials and technology. National governments must begin to take their proliferation commitments seriously, and they must engage in a global initiative. Otherwise, it will be only a matter of time before the threat of terrorists obtaining nuclear material becomes a reality.

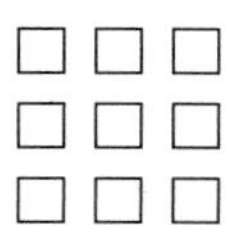

Subject Index

V

W